ELECTRON–MOLECULE COLLISIONS

PHYSICS OF ATOMS AND MOLECULES

ATOM–MOLECULE COLLISION THEORY: A Guide for the Experimentalist
Edited by Richard B. Bernstein

COHERENCE AND CORRELATION IN ATOMIC COLLISIONS
Edited by H. Kleinpoppen and J. F. Williams

VARIATIONAL METHODS IN ELECTRON–ATOM SCATTERING THEORY
R. K. Nesbet

DENSITY MATRIX THEORY AND APPLICATIONS
Karl Blum

INNER-SHELL AND X-RAY PHYSICS OF ATOMS AND SOLIDS
Edited by Derek J. Fabian, Hans Kleinpoppen, and Lewis M. Watson

INTRODUCTION TO THE THEORY OF LASER–ATOM INTERACTIONS
Marvin H. Mittleman

ATOMS IN ASTROPHYSICS
Edited by P. G. Burke, W. B. Eissner, D. G. Hummer, and I. C. Percival

ELECTRON–ATOM AND ELECTRON–MOLECULE COLLISIONS
Edited by Juergen Hinze

PROGRESS IN ATOMIC SPECTROSCOPY, Part C
Edited by H. J. Beyer and Hans Kleinpoppen

ELECTRON–MOLECULE COLLISIONS
Edited by Isao Shimamura and Kazuo Takayanagi

ISOTOPE SHIFTS IN ATOMIC SPECTRA
W. H. King

ELECTRON-MOLECULE COLLISIONS

EDITED BY

ISAO SHIMAMURA

RIKEN
The Institute of Physical and Chemical Research
Saitama, Japan

AND

KAZUO TAKAYANAGI

The Institute of Space and Astronautical Science
Tokyo, Japan

PLENUM PRESS • NEW YORK AND LONDON

Library of Congress Cataloging in Publication Data

Main entry under title:

Electron–molecule collisions.

(Physics of atoms and molecules)
Includes bibliographies and index.
1. Electron–molecule collisions. I. Shimamura, I. (Isao) II. Takayanagi, Kazuo.
III. Series.
QC794.6.C6E3735 1984 539 84-8364
ISBN 0-306-41531-3

A Division of Plenum Publishing Corporation
233 Spring Street, New York, N.Y. 10013

Printed in the United States of America

CONTRIBUTORS

J.N. BARDSLEY
Department of Physics, University of Pittsburgh, Pennsylvania, U.S.A.

B.D. BUCKLEY
Science and Engineering Research Council, Daresbury Laboratory, England

P.G. BURKE
Queen's University, Belfast, Northern Ireland

R.N. COMPTON
Oak Ridge National Laboratory, Oak Ridge, Tennessee, U.S.A.

R.I. HALL
Centre National de la Recherche Scientifique, and Laboratoire de Physique et Optique Corpusculaires, Université Pierre et Marie Curie, Paris, France

ARVID HERZENBERG
Department of Applied Physics, Yale University, New Haven, Connecticut, U.S.A.

C.J. NOBLE
Science and Engineering Research Council, Daresbury Laboratory, England

F.H. READ
Department of Physics, University of Manchester, England

DAVID F. REGISTER
Jet Propulsion Laboratory, California Institute of Technology, Pasadena, California, U.S.A.

ISAO SHIMAMURA
RIKEN, The Institute of Physical and Chemical Research, Saitama, Japan

KAZUO TAKAYANAGI
Institute of Space and Astronautical Science, Tokyo, Japan

SANDOR TRAJMAR
Jet Propulsion Laboratory, California Institute of Technology, Pasadena, California, U.S.A.

PREFACE

Scattering phenomena play an important role in modern physics. Many significant discoveries have been made through collision experiments. Amongst diverse kinds of collision systems, this book sheds light on the collision of an electron with a molecule.

The electron–molecule collision provides a basic scattering problem. It is scattering by a nonspherical, multicentered composite particle with its centers having degrees of freedom of motion. The molecule can even disintegrate, i.e., dissociate or ionize into fragments, some or all of which may also be molecules. Although it is a difficult problem, the recent theoretical, experimental, and computational progress has been so significant as to warrant publication of a book that specializes in this field. The progress owes partly to technical developments in measurements and computations. No less important has been the great and continuing stimulus from such fields of application as astrophysics, the physics of the earth's upper atmosphere, laser physics, radiation physics, the physics of gas discharges, magnetohydrodynamic power generation, and so on.

This book aims at introducing the reader to the problem of electron–molecule collisions, elucidating the physics behind the phenomena, and reviewing, to some extent, up-to-date important results. This book should be appropriate for graduate reading in physics and chemistry. We also believe that investigators in atomic and molecular physics will benefit much from this book.

The first chapter explains the basic formulation for scattering problems and provides an overview of the electron–molecule collisions as an introduction to the subject. The next three chapters give a detailed account of different kinds of dynamic processes, namely, rotational transitions (Chapter 2), vibrational transitions (Chapter 3), and dissociation (Chapter 4) in electron–molecule collisions. Chapter 2 emphasizes the basic physical concepts such as angular-momentum transfer, sudden-collision (or adiabatic-rotation) approximation, and division of the configuration space. Chapter 3 concentrates on resonance phenomena and threshold structures, but discusses also the sudden or adiabatic-nuclei approximation. Chapter 4 reviews the work on dissociative excitation, dissociative ionization, and dissociative electron capture, as comprehensively as possible within the space available. Resonance states are particularly important in dissociative capture processes. In contrast to these chapters discussing the collision mechanisms, Chapter 5 pays attention to the use of electron–molecule collisions as a spectroscopic tool for studying the molecular electronic structures. Negative-ion resonance spectroscopy and electron energy-loss spectroscopy on neutral molecules are dealt with here. The final two chapters review currently important experimental (Chapter 6) and computational (Chapter 7) techniques for the determination of cross sections.

Unfortunately, the time of completion of the manuscript varies from chapter to chapter. As a result some chapters omit very new references. Some overlaps remain; they appear necessary to some extent in the context of each chapter. When two or more chapters discuss closely related subjects, this is indicated in appropriate places. Every chapter is almost independent, and the reader with some basic knowledge of atomic collisions may begin with any chapter. Every chapter is followed by a list of references with titles of papers and books.

The reader will see that progress in the study of electron–molecule collisions is still in the early stage. Many important and challenging problems are left unsolved or untouched. We hope that this volume stimulates many readers to take up this field of research.

ISAO SHIMAMURA
KAZUO TAKAYANAGI

CONTENTS

Chapter 1. INTRODUCTION TO ELECTRON–MOLECULE COLLISIONS
Kazuo Takayanagi

Chapter 2. ROTATIONAL EXCITATION OF MOLECULES BY SLOW ELECTRONS
Isao Shimamura

Chapter 4. DISSOCIATION OF MOLECULES BY SLOW ELECTRONS
R.N. Compton and J.N. Bardsley

Chapter 5. MOLECULAR SPECTROSCOPY BY ELECTRON SCATTERING
R.I. Hall and F.H. Read

Chapter 6. EXPERIMENTAL TECHNIQUES FOR CROSS-SECTION MEASUREMENTS
Sandor Trajmar and David F. Register

Chapter 7. COMPUTATIONAL METHODS FOR LOW-ENERGY ELECTRON–MOLECULE COLLISIONS

B.D. Buckley, P.G. Burke, and C.J. Noble

INTRODUCTION TO ELECTRON-MOLECULE COLLISIONS

KAZUO TAKAYANAGI

1. Introduction

The purpose of this introductory chapter is twofold. First, readers of this book are not necessarily specialists in this particular field. Therefore, the editors thought that it would be appropriate to include an introductory chapter which is easily understood. Sophisticated techniques will not be used in this chapter. Only elementary quantum mechanics and basic mathematics will be required to understand this chapter. The theoretical side will be emphasized in this chapter since it is outside my ability to discuss experimental details.

Second, the theory of electron-molecule collisions, like the theory of other topics, was not developed in a day. There has been a gradual building up during the past half a century in this field of study. In the following chapters, readers will see that the research activity has been most active during the last decade for various reasons. People are often so busy catching up on the latest developments that they forget the efforts of earlier days. But it is instructive to look back on the history of theoretical developments before proceeding to more recent achievements.

Electrons play a central part in atomic and molecular physics. Because of their small mass, electrons are much more active than the nuclei in this microscopic world. Knowledge of the behavior of electrons is essential in understanding a large variety of problems, such as gaseous electronics, fusion plasmas, ionospheres, auroras, stellar atmospheres and interstellar gases. Fast electron scattering can be used as a probe to study the charge distribution within the target atom or molecule. Furthermore, electron scattering is closely related to the photoionization process. In the latter, the outgoing electron is in a scattered state. One of the big differences is that in photoionization there is no incoming electron wave. In this sense photoionization is a "half-collision".

Collision problems are sometimes regarded as contrasted with the bound-state problems in atomic and molecular physics. However, electron scattering

KAZUO TAKAYANAGI • Institute of Space and Astronautical Science, Komaba 4-chome, Meguro-ku, Tokyo 153, Japan.

from a neutral atom or a molecule is closely related to the negative ion states of the atom or the molecule. Similarly, electron scattering from a positive atomic or molecular ion is an analytical continuation of the bound-state problem in the neutral or ionized species. However, I shall not enter into such mathematical problems in this chapter.

Characteristics of the electron–molecule collisions as distinct from those of the electron–atom collisions are as follows:

1. Molecules have rotational and vibrational degrees of freedom. These degrees of freedom can be excited by a small amount of energy supply as compared with the electronic excitations. Thus the rotational and vibrational excitations are the dominant energy-loss processes for subexcitation electrons in molecular gases. (Subexcitation electrons are electrons with energy below the threshold of the lowest electronically excited state.) The slowing-down of electrons through rotational and vibrational excitations affects the rate of recombination and the rate of attachment because the rate of electron recombination with a positive ion and the rate of electron attachment to a neutral molecule usually increase rapidly as the mean electron energy decreases.

2. The electron–molecule interaction is essentially multicentered and nonspherical. First of all, molecules are characterized by the presence of more than one nucleus. Thus the interaction at short distances (or in the electron cloud of target molecule) is essentially multicentered. Even at large separations, the interaction is often strong and orientation dependent since molecules have various permanent electric multipole moments. Furthermore, the polarizability of the target molecule, which determines the polarization force (electron interaction with the induced dipole moment of the molecule), is also anisotropic. These features make the theoretical study of electron–molecule collisions more difficult than that of the electron–atom collisions.

3. Molecular targets can dissociate in collision with an electron. Simple dissociation and the dissociative ionization are fairly common:

$$e + \mathrm{AB} \longrightarrow \mathrm{A} + \mathrm{B} + e$$
$$\searrow \mathrm{A}^+ + \mathrm{B} + 2e.$$

The dissociative attachment

$$e + \mathrm{AB} \longrightarrow \mathrm{A} + \mathrm{B}^-$$

gives a unique collision process in electron–molecule interaction. Needless to say, this cannot take place if the fragment B does not have a positive electron affinity, although there is a possibility of the formation of a short-lived B^- which decays to emit an electron with a more or less definite value of kinetic energy.

The dissociative attachment usually occurs within a narrow range of collision energy (a resonance process).

2. Cross Sections

2.1. Concept of Cross Sections

Suppose a beam of electrons enters a collision chamber which contains a low-pressure gas. The beam intensity I decreases as a function of the distance x along the beam in the gas chamber. Since the fractional loss of the intensity should be proportional to the number density n of the gas molecules, we have

$$dI = -\sigma I n dx. \tag{1}$$

The constant σ has a dimension of the area. This quantity, called the collision cross section, gives us a convenient measure to represent the collision probability. If electron is a classical point mass and the target atom is a rigid sphere with a radius a, the collision cross section is simply given by the geometrical cross section πa^2. The real electron, however, follows the quantum mechanics. In other words, the wave nature of the electron motion comes in. Furthermore, the target atom or molecule has no well-defined boundary. For these reasons, the effective collision cross section is not a constant, but depends on the collision energy, i.e., on the kinetic energy of the relative motion.

From (1), the average number of collisions which each gas molecule on the way of the electron beam suffers in the unit time interval is

$$\nu = \sigma I = \sigma n_e v_e, \tag{2}$$

where n_e and v_e are the number density and the velocity of electrons in the beam. Strictly speaking, the target molecule also has a finite velocity, the distribution of which is determined by the gas temperature and the molecular mass. The collision cross section is a function of the relative velocity of the incident electron and the target molecule, rather than the electron beam velocity. The difference, however, is not very important unless the electrons are very low energy or the gas is at very high temperature, or both.

There is a large variety of collision processes. In elastic collision, the target atom or molecule does not change its internal state. Excitation and ionization are the typical inelastic processes. Excitation of a rotational, vibrational, or electronic state, or excitation of these degrees of freedom in combination can take place in a molecular target. If the target is initially in an excited state, the deexcitation is also possible. Ionization can be an ejection of one target electron, or, if the collision is a violent one, multiple ionization is also possible. Dissociation of the molecule takes place either by the direct excitation of a repulsive electronic state or by exciting a predissociating state. Predissociation is a nonradiative transition in an isolated molecule from a discrete energy level to a continuum state. A potential curve crossing is the usual cause of such a transition. Electron attachment is another source of the molecular dissociation. It is desirable to distinguish all these individual processes. Individual cross sections can be defined by extending (2) to each individual process. The effective cross section for a collision of type a is given by the probability per unit time, ν_a, for such a process per one target molecule divided by the flux of incident electron beam I. The sum of the individual cross sections over all the possible processes will give the total effective collision cross section.

Very often the angular distribution of the scattered electron is also required. When we look at a particular process a where the scattered electron goes into a small solid angle $d\omega$ in the direction (θ, ϕ), we can define the differential cross section $q_a(\theta, \phi)\, d\omega$. Evidently we have

$$\int q_a(\theta, \phi)\, d\omega = \sigma_a, \tag{3a}$$

$$\sum_a q_a(\theta, \phi) = q(\theta, \phi), \qquad \int q(\theta, \phi)\, d\omega = \sigma. \tag{3b}$$

For theoretical studies, it is most convenient to use the coordinate system where the center of mass of the electron–molecule system is at rest (the center-of-mass system, or c.m. system for brevity). Since the molecular mass is much larger than the electron mass, it is usually assumed that the center of mass of the molecule is at rest at the origin of the c.m. system. When we calculate the electron energy loss due to the momentum transfer to the target molecule as a whole, we have to take account of the finiteness of the electron mass.

In (3a) and (3b), scatterings into different directions are counted with equal weight. In some cases, however, we have to distinguish a large-angle scattering from a small-angle scattering because they have entirely different effects on some phenomena, such as the electron diffusion in a gas. Thus the integrated cross section with the weight function $(1 - \cos\theta)$, where θ is the scattering angle, i.e., the angle between the incident direction and scattered direction, is sometimes required. This cross section is called the diffusion cross section, or the momentum-transfer cross section:

$$\sigma_m = \int q(\theta, \phi)(1 - \cos\theta)\, d\omega. \tag{4}$$

For the accurate treatment of the behavior of electrons in gases, we need the following integrated cross sections[1]:

$$\sigma^{(l)} = \int q(\theta, \phi)[P_0(\cos\theta) - P_l(\cos\theta)]\, d\omega. \tag{5}$$

The momentum-transfer cross section σ_m is the special case of $l = 1$. In Eq. (5), P_l is the lth Legendre polynomial.

The momentum-transfer cross section is important also in determining the mean energy loss of electron per elastic collision with a gas molecule. Suppose that the initial velocity of the electron (mass m_e) is $\mathbf{v}$ and that of the molecule (mass M) is $\mathbf{V}$ and the relative velocity is $\mathbf{V}_r = \mathbf{v} - \mathbf{V}$. The velocities after the collision will be represented by the primed quantities: $\mathbf{v}'$, $\mathbf{V}'$, and $\mathbf{V}_r'$. By introducing the velocity of the center of mass of the system $\mathbf{V}_G$, we have

$$\frac{M\mathbf{V} + m_e\mathbf{v}}{M + m_e} = \mathbf{V}_G = \frac{M\mathbf{V}' + m_e\mathbf{v}'}{M + m_e},$$

$$\mathbf{v} = \mathbf{V}_G + \lambda\mathbf{V}_r, \qquad \mathbf{v}' = \mathbf{V}_G + \lambda\mathbf{V}_r', \qquad \lambda = \frac{M}{M + m_e}.$$

The electron energy loss (EEL) is given by

$$\text{EEL} = \frac{m_e}{2}(v^2 - v'^2) = \lambda m_e \mathbf{V}_G \cdot (\mathbf{V}_r - \mathbf{V}_r') + \frac{m_e}{2}\lambda^2(\mathbf{V}_r^2 - \mathbf{V}_r'^2). \tag{6}$$

In the elastic collisions, the last term vanishes. Furthermore, the average of the vector $\mathbf{V}_r'$ will be along the incident velocity $\mathbf{V}_r$. Therefore, $\langle \mathbf{V}_r - \mathbf{V}_r' \rangle = \mathbf{V}_r \langle 1 - \cos\theta \rangle$, where θ is the scattering angle in the center-of-mass system. Here the angle brackets represent the average over the scattering angle. Thus we have

$$\begin{aligned}
\langle \text{EEL} \rangle &= \lambda m_e \mathbf{V}_G \cdot \mathbf{V}_r \langle 1 - \cos\theta \rangle \\
&= \lambda m_e \langle 1 - \cos\theta \rangle \frac{M\mathbf{V} + m_e\mathbf{v}}{M + m_e} \cdot (\mathbf{v} - \mathbf{V}) \\
&= \frac{m_e M}{(M + m_e)^2} \langle 1 - \cos\theta \rangle \, [m_e v^2 - MV^2 + (M - m_e)\mathbf{v} \cdot \mathbf{V}].
\end{aligned}$$

If the electron energy is much higher than the kinetic energy of the gas molecules, $m_e v^2/2 \gg MV^2/2$. The last term containing the scalar product $\mathbf{v} \cdot \mathbf{V}$ will be small on the average. The collisions with $\mathbf{v} \cdot \mathbf{V} > 0$ and those with $\mathbf{v} \cdot \mathbf{V} < 0$ do not occur with the same frequency, because they correspond to different relative velocities $V_r = |\mathbf{v} - \mathbf{V}|$. However, since we have assumed that $v \gg V$, the two collision frequencies will be almost the same, so that on the average the scalar product $\mathbf{v} \cdot \mathbf{V}$ can be neglected. Therefore, one has

$$\langle \mathrm{EEL} \rangle = \frac{2m_e}{M} \langle 1 - \cos\theta \rangle \frac{1}{2} m_e v^2 = \frac{2m_e}{M} \frac{1}{2} m_e v^2 \frac{\sigma_m}{\sigma}. \tag{7}$$

Thus the $\langle \mathrm{EEL} \rangle \sigma$, which determines the average rate of energy loss of the electron due to the elastic scattering is given by

$$\langle \mathrm{EEL} \rangle \sigma \simeq \frac{2m_e}{M} \frac{1}{2} m_e v^2 \sigma_m. \tag{8}$$

When the inelastic collisions come in, the momentum-transfer cross section must be modified. If the internal degrees of freedom of the target molecule gains an energy ΔE, then the last term of (6) becomes $\lambda \Delta E$. Furthermore, we have

$$\langle \mathbf{V}_r' \rangle = \mathbf{V}_r \frac{V_r'}{V_r} \langle \cos\theta \rangle$$

since the equality $V_r = V_r'$ does not hold any more. Thus we have

$$\langle \mathrm{EEL} \rangle_a = \frac{2m_e}{M} \left\langle 1 - \frac{V_r'}{V_r} \cos\theta \right\rangle \frac{1}{2} m_e v^2 + \lambda \Delta E, \tag{9}$$

so that the momentum-transfer cross section should be modified as

$$\sigma_m^{(a)} = \int q_a(\theta, \phi) \left(1 - \frac{V_r'}{V_r} \cos\theta \right) d\omega \tag{10}$$

for an inelastic collision of type a. Using this notation, (9) becomes

$$\langle \mathrm{EEL} \rangle_a = \frac{2m_e}{M} \frac{1}{2} m_e v^2 \frac{\sigma_m^{(a)}}{\sigma_a} + (\Delta E)_a, \tag{11}$$

where λ has been replaced with unity because of the small mass ratio m_e/M.

The probability for a particular kind of collision a to take place when an electron proceeds in the gas by a small distance dx is $n\sigma_a dx$. The mean energy loss of the electron will be $\langle \mathrm{EEL} \rangle_a$. Therefore, the mean energy loss of an electron per unit distance is given by $n \Sigma_a \sigma_a \langle \mathrm{EEL} \rangle_a$, which is the stopping power of the gas. This quantity, when divided by the number density n, gives a quantity called the stopping cross section S:

$$S = \frac{2m_e}{M} E \sum_a \sigma_m^{(a)} + \sum_a (\Delta E)_a \sigma_a, \tag{12}$$

where $E = m_e v^2/2$ is the electron energy before collision. The stopping "cross section" has the dimension of area times energy. In order to make it a real "cross section", it must be divided by a unit of energy, such as the Rydberg unit of energy ($\simeq 13.6$ eV).

If the electrons are not in a beam form, but their velocities are distributed in direction, the mean number of collisions of type a in unit volume in unit time interval will be

$$nn_e \int v\sigma_a(v) f(\mathbf{v})\, d\mathbf{v}, \tag{13}$$

where $\sigma_a(v)$ is the effective cross section as a function of the electron velocity v and $f(\mathbf{v})$ is the electron velocity distribution function, which is not necessarily Maxwellian when an external electric field is applied or when there are rapid reactions going on in the gas. In (13), the molecular velocity has been neglected as compared with the electron velocity v. The integral in the expression is the effective collision rate constant. If $f(v)$ is the Maxwellian distribution, the rate constant becomes a function of the electron temperature T_e.

2.2. *Experimental Arrangements*

By integrating (1), we obtain

$$I(x) = I(0)\exp(-\sigma n x). \tag{14}$$

Thus, by measuring the attenuation of the electron beam, we can measure the total collision cross section σ. This is the way σ was measured by Ramsauer and Kollath and by some other pioneers in the late 1920s and early 1930s.

In order to distinguish different excitation processes, one can use the spectroscopic method where the intensity of photon emission from excited targets is measured at a definite wavelength corresponding to a definite molecular transition.

From these experiments, however, we can get no information about the angular distribution of the scattered electrons. In order to measure the differential cross section with respect to the scattering angle, we have to detect the scattered electrons. If we further measure the energy loss suffered by the electron, there is a possibility that we can identify the excitation process caused by that electron. For this purpose, the incident electron beam must have a reasonably uniform energy. This can be achieved by using an appropriate energy selector through which the electron beam passes before entering the collision chamber. After collision, the energy analyzer will be used to identify the electron energy loss. In this way, the energy losses can be detected down to such small values as several meV. Further information on the collision processes can be obtained if both the scattered electron and the emitted photon are detected (coincidence experiment).

In the ionization and dissociation processes, products (ion, or neutral fragment) are often detected and identified by mass analysis. The velocity distribution of these products gives us more insight into the collision processes.

In the early days, the target of electron collision was a gas in the collision chamber. Since the 1960s, technical developments have made it possible to carry out crossed-beam and merged-beam experiments, where the target particles are also in a beam. With this technique, it is now possible to study the electron collision with free atoms, free radicals, ions, and even with excited species provided that they have sufficiently long lifetimes. Unfortunately, it is rather difficult to determine the absolute cross section from this type of beam–beam collision experiments, since it is rather difficult to determine accurately the volume of space where the two beams collide and the effective solid angle with which the detector can be seen from the collision volume. For this reason, it is usual to normalize the observed cross section in relative scale into the absolute scale by comparing it with an already known cross section.

Another important type of experiment for electron collisions is the electron swarm experiment. This method is particularly important for low collision energies (much lower than 1 eV) where the beam experiment is increasingly more difficult. Here the behavior of an electron swarm in a gas is studied usually under an appropriate electric field. The drift velocity and the rate of diffusion of swarm electrons are measured. With the help of theoretical analysis (solution of the Boltzmann equation, for instance), various cross sections as functions of the collision energy can be derived from the observed data.[2]

For further details of the experimental analysis of the electron collision processes, readers are referred to Chapter 6.

2.3. Theoretical Studies of Electron–Molecule Collisions

In the early days, the goal of theoretical studies was to explain the experimental finding qualitatively, or to approximately reproduce the observed data. In recent years, however, with the help of rapid developments in computation techniques and computer facilities, it is sometimes possible to obtain cross sections theoretically which are claimed to be more accurate than observed data at least for some simple molecular targets. In the rotational transitions in electron–molecule collisions, for instance, the energy loss is so small (except for the hydrogen molecule) that it is difficult to distinguish experimentally individual rotational transitions at least in the standard energy-loss-spectrum technique. Theoretical calculation, though not an easy task, can provide the relevant cross sections with fairly high accuracy.

In low-energy electron–molecule collisions, possible processes include elastic collision, rotational or vibrational transitions, or both, and electron attachment. Except for the last-mentioned, the target molecule can be treated approximately as a vibrating rotor without other degrees of freedom. The electronic state of the molecule is taken into account only when an appropriate effective potential is chosen for the electron–molecule interaction. For elastic collisions and pure rotational transitions, the rigid rotor model is often adopted where the vibration of the molecule is entirely neglected. Once the interaction potential has been chosen, cross sections for various processes can be calculated in one way or another, depending on the nature of the process concerned and the accuracy required for the results. Even the simple Born approximation can be applied to rotational transitions in some cases. However, much more sophisticated approaches are usually required.

For higher energies, the electronic excitation and the ionization become dominant. Sometimes, the molecular dissociation is observed. In principle, the dissociation can take place through the rotational excitation or vibrational excitation without changing the electronic state of the molecule. However, the dissociation takes place more often when the molecule is left in a repulsive state as a result of electronic excitation or ionization process. In those cases, the first step is a simple excitation or ionization process. The nuclei in the molecule, which are much heavier than the electron, move apart slowly after the electron collision is over. For higher collision energies, the collision duration is so short that the energy and momentum transfers from the incident electron to the target molecule take place in one step and their magnitudes are expected to be small. That is, the first-order perturbation theory is appropriate to describe the collision processes. Thus the Born approximation, which is the first-order perturbation theory for collision processes, is a good starting point. The second Born, the Glauber, and some other approximation methods have been proposed to study the deviation from the simple Born approximation.

For intermediate energies, no convenient and reasonably accurate theory has been found. The first-order theory is no longer applicable to obtain reliable results. On the other hand, various calculation methods which have been successfully applied to the low-energy collision processes become more and more laborious when applied to collisions at higher energies.

Further descriptions of theoretical approaches to the problem will be found later in this chapter and in the other chapters of this book.

2.4. Examples of the Electron–Molecule Collision Cross Sections

In order to illustrate the magnitudes and their energy dependence of various collision cross sections, curves are shown in Figs. 1 through 5 for hydrogen and nitrogen as target molecules. These are the targets where there are more collision data available than for other molecules. It must be noted that the data

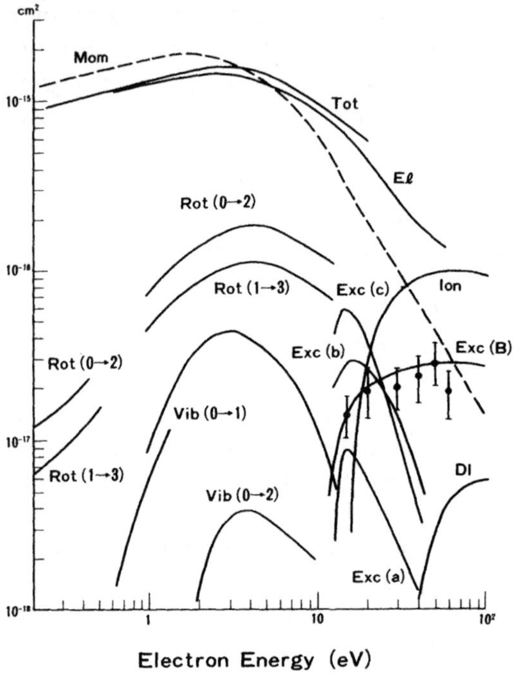

FIGURE 1
Cross sections of various processes in the electron–hydrogen molecule collision (see text for data sources)

adopted here are not necessarily the best available. The selection is arbitrary in some cases. In Fig. 1 are shown several cross sections for the electron–H_2 molecule collision. The curve "Tot" represents the total collision cross section measured by Golden *et al.*[3] It is of the order of 10^{-15} cm^2 and is slowly varying in the energy region below 20 eV where the measurement has been carried out. The "El" curve is for the elastic collision. This is based on the experimental data obtained by Linder and Schmidt[4] and those of Srivastava *et al.*[5] In order to make a smooth fit, the cross section reported in the Ref. 5 is arbitrarily increased by about 10%. It is seen that the elastic collision is dominant in the energy region below 20 eV. The momentum-transfer cross section "Mom" is taken from the recent compilation by Hayashi.[6] At very low collision energies, this is larger than the elastic cross section. This means that $\langle \cos\theta \rangle < 0$, that is, the backward scattering (the scattering angle θ is larger than 90°) is dominant. At higher collision energies, on the other hand, the forward scattering becomes dominant ($\sigma_m < \sigma_{el}$). This is understandable because as the incident energy increases, the collision duration becomes short and the momentum transfer between the electron and the target molecule becomes small. As to the inelastic processes, there is a drastic change at an energy around 10 eV. The electronic excitation and ionization processes are possible only above this energy. The vertical excitation energy ("vertical" means the "fixed internuclear distance") of the lowest excited state $b^3\Sigma_u^+$ is about 10 eV. The ionization energy of H_2 is 15.4 eV. Below 10 eV, only rotational and vibrational excitations are possible. The rotational cross sections are somewhat larger than the vibrational cross sections. In the region between 1 and 10 eV the curve "Rot($1 \to 3$)" in Fig. 1 for the rotational excitation $J = 1 \to 3$, J being the rotational quantum number, is taken from the beam experiment by Linder and Schmidt.[4] The curve "Rot($0 \to 2$)" for $J = 0 \to 2$ has been obtained by simply multiplying a "Rot($1 \to 3$)" by factor of 5/3. This approximate relation is an example of scaling laws (see Chapter 2 for details of the scaling laws). The "Rot" curves in the very low energy region are based on swarm experiments. Crompton *et al.*[7] have derived "Rot($0 \to 2$)," while Gibson[8] has reported "Rot($1 \to 3$)." If the molecule is in a rotationally excited state initially, the deexcitation to a lower state is possible. The cross section for a deexcitation process, generally, is related to that of the inverse process (that is, excitation) through the relation of the detailed balance

$$g_1 E\sigma(1, v \to 2, v') = g_2 E'\sigma(2, v' \to 1, v), \tag{15}$$

where "1" and "2" are two states of the target molecule, g_1, g_2 are their statistical weights, and $E = m_e v^2/2$, $E' = m_e v'^2/2$ are the electron energies at large distances from the target. The proof of this relation can be found in many textbooks of quantum mechanics.

The vibrational excitation cross section for $v = 0 \to 1$, v being the vibrational quantum number, and the one for $v = 0 \to 2$ are shown in Fig. 1. The "Vib($0 \to 1$)" curve below 1 eV is due to the swarm experiment by Crompton *et al.*,[9] while the curve above 1 eV is the result of the beam experiment.[4] The "Vib($0 \to 2$)" curve was obtained by Ehrhardt *et al.*[10] Experimental data are very scarce for the electronic excitation processes. The curves (a), (b), (c), corresponding, respectively, to the excitation of $a\ ^3\Sigma_g^+$, $b\ ^3\Sigma_u^+$, $c\ ^3\Pi_u$ states, are based on the two-state close-coupling calculation by Chung and Lin.[11] These states are spin triplet states. In other words, the spins of the two electrons in the molecule are parallel. In order to excite these triplet states from the singlet ground state, it is necessary to change the spin direction in one of the two molecular electrons. This is most efficiently done through the electron exchange, where the incident electron jumps into a bound orbital, while one of the molecular electrons is ejected. Such an electron exchange takes place with appreciable probability only when the incident and the molecular electrons have comparable velocities. As soon as the incident electron energy becomes higher, the probability of the exchange collision becomes smaller and smaller, so that the triplet excitation cross sections decrease rapidly. On the contrary, the excitation cross section for the $B\ ^1\Sigma_u^+$ state decreases rather slowly at higher energies. Here, the electron exchange is not necessary. Furthermore, this excitation is optically allowed. That is, the excitation of the B state is also possible by absorption of a photon of appropriate wavelength. As we see in Section 6.5, inelastic processes of this type, including ionization, become dominant at high collision energies. The dots with error bars in Fig. 1 are the results of experiment by Srivastava and Jensen.[12] The solid curve has been drawn rather arbitrarily passing through the error bars. Towards the high energy end, the curve deviates from the experimental data. This modification is based on the expectation that the cross-section curve tends smoothly to the asymptotic values provided by the Born approximation.[13] According to the Born approximation,[13] the excitation cross section for the $C\ ^1\Pi_u$ state is comparable with the one for the $B\ ^1\Sigma_u^+$ state. For most of the other excited states, there is very little information available on the relevant cross sections.

The total ionization cross section ("Ion" in Fig. 1) is given by

$$\sigma_{\text{Ion}} = \sum_n n\,\sigma_{\text{Ion}}^{(n)}. \tag{16}$$

$\sigma_{\text{Ion}}^{(n)}$ is the cross section for ionization process where n electrons are ejected from the target molecule. The curve "Ion" in Fig. 1 is the experimental result of Rapp and Englander-Golden.[14] It has a broad peak of about 1×10^{-16} cm^2 in the region of 65–70 eV.

So far, I have not mentioned dissociative processes. The excitation of the lowest excited state $b\ ^3\Sigma_u^+$ results in the dissociation because of the purely repulsive nature of the relevant interatomic potential. Experimentally, the effective cross sections have been measured for the production of excited atoms H($2p$) and H($2s$), and also for the emission of the major Balmer lines.[15] However, the nature of excited states producing these excited atoms is not definitely known. The cross sections for the H($2p$) and H($2s$) production (not shown in Fig. 1) are around 10^{-17} cm^2, while those for the Balmer line emission [or H(n=3) production] are around 10^{-18} cm^2 in the energy region $E \lesssim 100$ eV. The dissociative ionization, where H^+ is produced, has been studied by Rapp *et al.*[16] The "DI" curve in Fig. 1 is their result. It is one order of magnitude smaller than the total ionization cross section. Finally, the dissociative attachment

$$H_2 + e \rightarrow H + H^- \tag{17}$$

has a cross section less than 10^{-19} cm^2. It is too small to be shown in Fig. 1. The process has a strongly energy-dependent cross section as shown in Fig. 2. This is a resonance process as is discussed in Chapter 4, Section 5.1. A large isotope effect is also seen. The curves above 8 eV were obtained by Rapp *et al.*,[17] while the small peak slightly below 4 eV was found by Schulz and Asundi.[18]

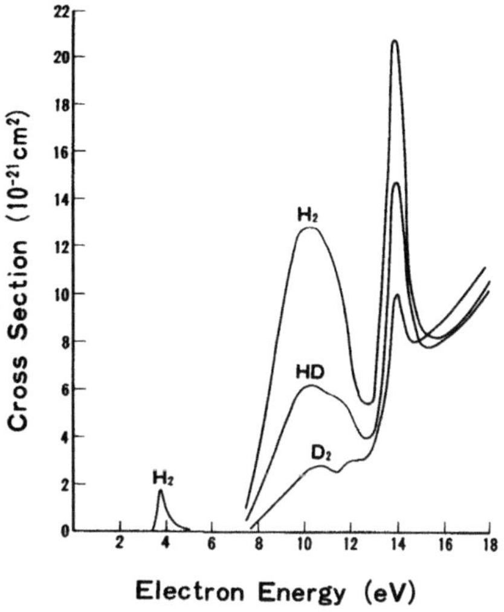

FIGURE 2
Cross section of dissociative attachment (production of either H^- or D^-) in electron collisions with H_2, HD, and D_2.[17,18]

The partial stopping cross sections [individual terms in Eq. (12)] derived from the cross sections in Fig. 1 are shown in Fig. 3. It is noted that "Vib$(0\to 1)$" contains the contribution of simultaneous rotational excitation. The rotational stopping power "Rot" is practically independent of the initial rotational state of the molecule, except in the very low energy region. I found this property empirically in the late 1960s when I was surveying the behavior of subexcitation electrons in various gases in collaboration with Mitio Inokuti (unpublished). Recently, Shimamura has given a general derivation of this property on the basis of the adiabatic theory. (See Chapter 2 for details.) In Fig. 3, it is also seen that the contribution of momentum-transfer is relatively small. For the electronic excitation of the *a*, *b*, *c*, and *B* states, the energy loss per collision $[(\Delta E)_a$ in Eq. (12)] is assumed to be 12, 10, 12, and 12.2 eV, respectively. In calculating the stopping cross section for ionization, it is necessary to estimate the energy loss ΔE. This consists of the three parts: (i) ionization energy (15.425 eV for H_2), (ii) excitation energy of the residual ion, and (iii) kinetic energy of the ejected electron ϵ. In obtaining "Ion" in Fig. 3, I have approximately assumed that all the ionization results in the H_2^+ ion in the electronically ground state $X\,^2\Sigma_g^+$ and vibrationally in the second excited state $(v = 2)$. Thus the excitation energy of the residual ion is about 0.523 eV. The mean value of the kinetic energy ϵ of the ejected electron has been calculated by using the empirical distribution function[19]

$$f(\epsilon) \propto \frac{1}{1 + (\epsilon/\zeta)^{2.1}}, \qquad \epsilon \leqslant \epsilon_{\max}. \tag{18}$$

The parameter ζ for H_2 is about 8.3 eV.[19] In the ionization process, two electrons, one scattered and one ejected, come out of the collision volume. The more

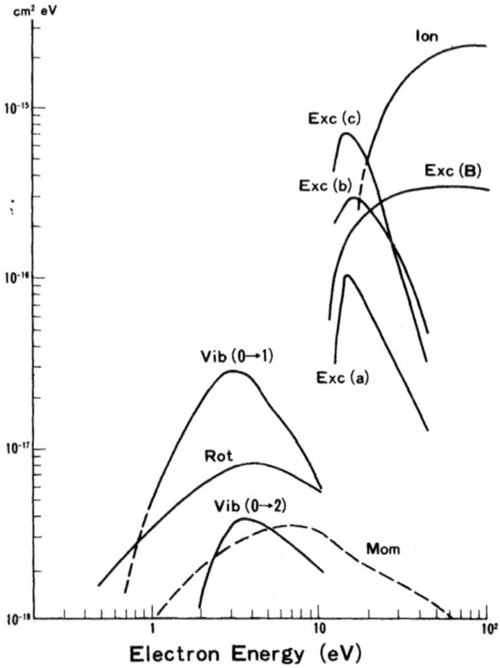

FIGURE 3
Stopping cross sections for the electron–hydrogen molecule collision, derived from the data in Fig. 1.

energetic one will be regarded as the scattered electron. With this definition, the maximum available energy $\epsilon_{\mathbf{max}}$ for the ejected electron is easily calculated from the incident electron energy. This formula was obtained at much higher incident energy (500 eV), so that the distribution (18) may not be very accurate in the energy region covered by Fig. 3.

The most important feature in this figure is the large jump of the stopping cross section at around 10 eV. Below this energy, it is of the order of 10^{-17} cm² eV. Above 10 eV, the stopping cross section increases rapidly and becomes of the order of 10^{-15} cm² eV. It should be remembered that there are many other electronic excitations which will further increase the stopping cross section in this region. As soon as the electron energy comes down below the excitation threshold of the first excited state, the rate of energy loss in the hydrogen gas becomes very small, so that the electron can remain in the superthermal energy region for a relatively long period of time. Electrons in this energy region are called the subexcitation electrons. Existence of subexcitation electrons strongly affects the rate of production of the negative ions (H^- in the case of H_2) and the rate of electron–positive ion recombination rate in ionized molecular gases.

Figure 4 shows various cross sections for the electron–nitrogen collision. The total cross section "Tot" in the lower energy region has conspicuous structures with a large peak in the 2–3-eV region. The data are from the measurement

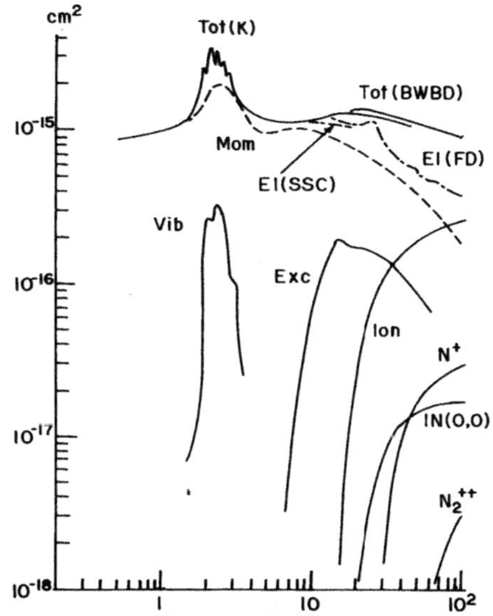

FIGURE 4
Cross sections of various processes in the electron–nitrogen molecule collision (see text for data sources).

by Kennerly.[20] The "Tot" in the higher energy region has been taken from the paper by Blaauw *et al.*[21] The elastic cross section data, "El", are by Shyn *et al.*[22] and by Finn and Doering.[23] The latter, marked as "El(FD)," smoothly join to the absolute cross-section measurement at 500 eV by Bromberg.[24] The momentum-transfer cross section "Mom" is from the compilation by Hayashi.[6] The rotational cross sections are not shown in Fig. 4. There are no experimental data available. There are some theoretical calculations.[25]–[27] The accuracy of these calculations cannot be estimated with certainty. However, these works indicate that the cross sections $\sigma_{\mathrm{Rot}}(J=0\rightarrow 2)$ and $\sigma_{\mathrm{Rot}}(0\rightarrow 4)$ are comparable in the energy region 2–10 eV and reach around 10^{-15} cm² in the resonance region (2–3 eV). Outside the resonance, these are of the order of 10^{-16} cm². The vibrational cross section "Vib" is the sum of the cross sections $\sigma_{\mathrm{Vib}}(v=0\rightarrow v')$. Schulz[28] measured the partial cross sections for $v'=1$–8 separately. Each partial cross section has conspicuous comblike structures. But the sum of them is much smoother than the individual components. The electronic excitation cross section "Exc" is the sum of 18 cross sections reported by Cartwright *et al.*[29] The curve "Ion" is the total ionization cross section obtained by Rapp and Englander-Golden.[14] The dissociative ionization (production of N^+) and the double ionization (production of N_2^{++}) cross sections are from the work of St. Halas and Adamczyk.[30] The emission of the first negative (0,0) band (wavelength $\lambda \sim 3914$ A) is due to the transition $B\,^2\Sigma_u^+$, $v=0\rightarrow X\,^2\Sigma_g^+$, $v=0$ of N_2^+. The excitation cross section for this emission[31] is also shown ["1N(0, 0)"].

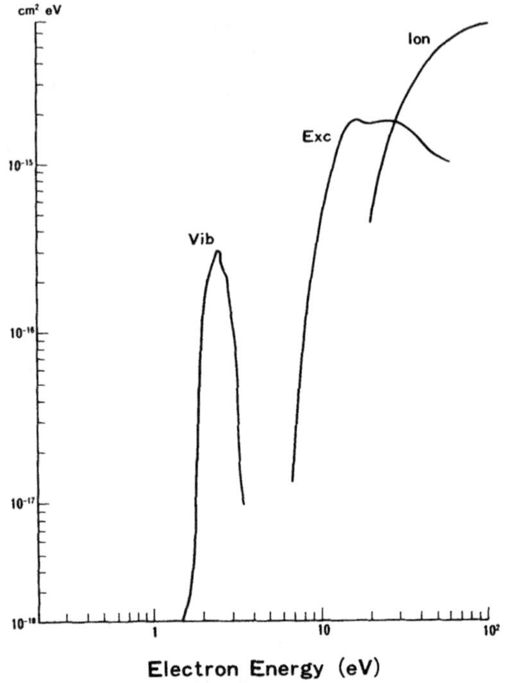

FIGURE 5
Stopping cross sections for the electron–nitrogen molecule collision, derived from the data in Fig. 4.

The stopping cross sections for the electron–nitrogen collisions are shown in Fig. 5. The momentum-transfer collisions contribute very little. It has a maximum of about 7.5×10^{-19} cm^2 eV in the energy region 60–70 eV. The rotational stopping cross section is around 10^{-18} cm^2 eV in the resonance region (2–3 eV) and much less outside this region. Therefore, below the excitation threshold of low-lying electronic states at several eV, the total stopping cross section is extremely small except in the resonance region where the vibrational stopping cross section is remarkably large. "Exc" is the sum of the stopping cross sections calculated from the 18 excitation cross sections.[29] Other excitation processes will further increase this value. The "Ion" curve has been obtained[32] in the same way as for the hydrogen target. The residual ion N_2^+ is assumed to be either in $X\ ^2\Sigma_g^+$ or in $A\ ^2\Pi_u$ states (branching ratio 1:1). The secondary-electron distribution function is again given by (18) with $\zeta = 12.7$ eV.

3. Electron–Molecule Interactions

In order to calculate cross sections accurately, we need accurate knowledge of the relevant interaction. Strictly speaking, the electron–molecule interaction is not unambiguously defined before we mathematically formulate the collision problem, since some parts of the interaction are of dynamical nature. However, it is often conceived that the interaction can be decomposed, to a good approximation, into three parts:

a. static part
b. exchange part
c. polarization part

The static part of interaction is the electrostatic interaction between the incident electron and the undeformed charge distribution of the target molecule. The exchange interaction is due to the possible exchange of the incident electron and a molecular electron. This gives rise to a nonlocal interaction because the exchange is controlled by the overall overlap between two single-electron wave functions. Finally, the polarization interaction comes from the deformation of the molecular charge cloud by the approach of the point charge of the incident electron.

3.1. Long-Range Interactions

The static field produced by the unperturbed molecule at a point sufficiently outside the molecular dimension is characterized by the electric multipole moments of the molecule. If the molecule is in an ionized state (net charge Ze), the Coulomb interaction has by far the largest effect on the motion of the incident electron at large separations, although this interaction does not directly induce any inelastic processes. If the molecule is neutral but has a permanent dipole moment D, the dipole interaction is important. The quadrupole interaction due to the molecular quadrupole moment Q is generally weaker, but not necessarily negligible. These interactions depend on the molecular orientation relative to the direction of the incident electron. This orientation dependence gives rise to a torque on the molecule, so that the rotational transitions can take place during the encounter. The moments D and Q depend on the internuclear separations within the molecule, so that the interactions under consideration can also give rise to vibrational transitions.

Generally, the static interaction is given in terms of the charge distribution $\rho(\mathbf{r})$ inside the molecule

$$V_{\text{static}}(\mathbf{r}) = -e\int \rho(\mathbf{r}')/|\mathbf{r}-\mathbf{r}'|\,d\mathbf{r}', \tag{19}$$

where $-e$ is the electron charge and $\mathbf{r}$ is the position vector for the incident electron. The charge distribution $\rho(\mathbf{r}')$ includes the point charges of the nuclei (represented by the delta functions) as well as the continuous distribution due to the molecular electron cloud. If the contribution to the integral from the distant region ($\mathbf{r}' > \mathbf{r}$) is neglected, we obtain by expanding $1/|\mathbf{r}-\mathbf{r}'|$

$$V_{\text{static}}(\mathbf{r}) = -e \sum_{n=0}^{\infty} r^{-n-1} \int \rho(\mathbf{r}') r'^n P_n(\hat{\mathbf{r}} \cdot \hat{\mathbf{r}}') d\mathbf{r}'$$

$$= -e \sum_{n=0}^{\infty} \sum_{m=-n}^{n} r^{-n-1} Y_{nm}^*(\hat{\mathbf{r}}) \frac{4\pi}{2n+1}$$

$$\times \int \rho(\mathbf{r}') r'^n Y_{nm}(\hat{\mathbf{r}}') d\mathbf{r}',$$

where $\hat{\mathbf{r}} \equiv \mathbf{r}/r$, $\hat{\mathbf{r}}' \equiv \mathbf{r}'/r'$ are the unit vectors. For linear molecules, the charge distribution is axially symmetric, so that if the z axis is taken along the molecular axis $\hat{\mathbf{R}}$, only the terms with $m = 0$ survive and we obtain

$$V_{\text{static}}(\mathbf{r}) = -e \sum_{n=0}^{\infty} r^{-n-1} P_n(\hat{\mathbf{R}} \cdot \hat{\mathbf{r}}) M_n, \tag{20}$$

where M_n is the 2^n-pole moment of the molecule

$$M_n = \int \rho(\mathbf{r}') r'^n P_n(\hat{\mathbf{R}} \cdot \hat{\mathbf{r}}') d\mathbf{r}'. \tag{21}$$

When the target molecule is an ion, the net charge $Ze = M_0$ is not zero. It is easy to show then that the dipole moment $D = M_1$ depends on the choice of the coordinate origin. Similarly, when the molecule has the dipole moment, the quadrupole moment $Q = M_2$ depends on the coordinate origin. Generally, higher multipole moments depend on the origin. However, since the kinetic energy terms in the Schrödinger equation are separable into the internal (rotational and vibrational) motion for fixed molecular center of mass and the incident electron motion relative to the molecular center of mass, it is convenient to adopt the molecular center of mass as the origin of the vectors $\mathbf{r}$ and $\mathbf{r}'$ in the above equations. When the incident electron is far outside the molecule, only the first few terms in the expansion (20) are important, so that

$$V_{\text{static}}(\mathbf{r}) \to -Ze^2/r - (De/r^2) P_1(\hat{\mathbf{R}} \cdot \hat{\mathbf{r}}) - (Qe/r^3) P_2(\hat{\mathbf{R}} \cdot \hat{\mathbf{r}}). \tag{22}$$

It is to be noted that at such a distance, the exchange part of the interaction can be entirely neglected.

The polarization interaction is given by the electron interaction with the induced dipole moment of the molecule. If the electron is far from the molecule, the electric field produced by the electron at the molecular site is practically uniform. Then, if the target system is spherical, the polarization interaction potential is given by

$$V_{\mathrm{pol}}(r) \to -\alpha e^2/2r^4, \tag{23}$$

where α is the electric dipole polarizability of the target and in this case it is a scalar. For most molecules, however, the polarizability is a tensor, depending on the orientation of the molecule relative to the applied electric field. For linear molecules, for instance, the polarizability is larger along the molecular axis than in the perpendicular direction, since the molecular electrons can move more easily along the axis than in the perpendicular directions. In this (linear molecule) case, the asymptotic potential (23) is replaced by

$$V_{\mathrm{pol}}(r) \to -\alpha e^2/2r^4 - (\alpha' e^2/2r^4)P_2(\hat{\mathbf{R}}\cdot\hat{\mathbf{r}}), \tag{24}$$

where α and α' are given by the polarizabilities $\alpha_{\parallel}$, $\alpha_{\perp}$ along the axis and along the direction perpendicular to the axis.

$$\alpha = (\alpha_{\parallel} + 2\alpha_{\perp})/3, \qquad \alpha' = (2/3)(\alpha_{\parallel} - \alpha_{\perp}). \tag{25}$$

When the electron approaches the molecule, the electric field it produces is no longer uniform over the molecular dimension. Therefore, the asymptotic expression (24) cannot be a good approximation. In Fig. 6 the adiabatic polarization potential $V_{\mathrm{pol}}^{\mathrm{ad}}$ calculated for $e + N_2$ by Morrison and Hay[33] is compared with

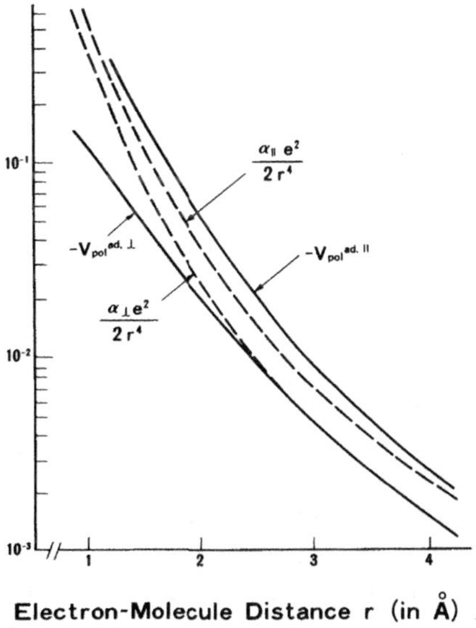

FIGURE 6
Adiabatic polarization potential[33] is compared with its asymptotic form for the system $e + N_2$ in the collinear and the equilateral-triangle configurations. r is the distance between the electron and the center of mass of N_2 and the energies are in atomic units (1 unit = 27.2 eV).

the asymptotic form (24). (A similar calculation has been done also by Eades *et al.*[24]). V_{pol}^{ad} is the difference, at each fixed value of r, between the total energy of the system for the case where the induced polarization of the molecular electron cloud is fully taken into account and that for the case where the molecule is assumed to be undeformed. It is seen that the asymptotic form (24) differs considerably from the adiabatic polarization potential over an important range of r. However, the calculation of the adiabatic polarization is not a simple task, so that very often a cutoff factor $f_c(r)$ is arbitrarily multiplied to the asymptotic form (24) and the resulting form is used down to small values of r. One of the frequently used cutoff factors is of the following form

$$f_c(r) = 1 - \exp\,[-(r/r_c)^6], \tag{26}$$

where r_c is an adjustable parameter. Unfortunately, there are the cases where the calculated cross sections depend sensitively on the adopted value of r_c.

The polarization of the molecule is represented by a superposition of the unperturbed molecular eigenfunctions. In the electron–molecule collisions, the molecular wave function is often expanded in terms of the electronic eigenfunctions of the isolated molecule. The polarization interaction is then automatically included, in principle at least. However, for practical reasons, the number of eigenfunctions taken into calculation is always limited, while the polarizability of the target is accurately given only when an infinite number of states are included. That this is so is easily demonstrated by the well-known formula for the polarizability in the simplest case of the spherical target atom

$$\alpha = (e^2\hbar^2/m_e) \sum_{n(\neq 0)} f_{on}/(E_n - E_0)^2, \tag{27}$$

where $h = 2\pi\hbar$ ($= 6.626 \times 10^{-27}$ erg s) is Planck's constant, f_{on} is the optical oscillator strength for absorption and E_n is the energy of the atom in the nth state. The summation extends not only over the discrete levels but also over continuum states. Very often, the contribution from the continuum states is appreciable. Thus the ordinary close-coupling approach, where only a limited number of target states are taken account of and a set of coupled differential (or integrodifferential) equations are solved, cannot give the full account of the polarization force. In order to improve the theory, some of the higher members of the set of basis functions are chosen in such a way that these are not necessarily eigenfunctions of the free molecule and these, combined with the other basis functions in the set, can give the correct polarization force at large separations at low collision velocities. These additional states other than the target eigenfunctions are called pseudostates.

When the incident electron velocity is low, the adiabatic polarization potential can be used for the polarization effect. When the incident electron velocity becomes higher, the collision duration becomes shorter, so that the molecular electron cloud cannot be adiabatically deformed. Therefore, the polarization interaction is expected to become less and less as the collision velocity becomes higher. Such a velocity dependence of the polarization interaction is also taken into account automatically in principle if the infinite set of coupled equations are solved.

In order to find out an approximate, but simple, expression for the velocity-dependent polarization potential, Onda[35] applied the semiclassical (impact parameter) method to the problem. A charged particle is assumed to move along a straight line with a constant velocity v at a relatively large distance from the target atom (or molecule). The electronic state of the target is calculated in the first-order perturbation theory. Using this perturbed wave function, the energy of the whole system is calculated at each instance. The difference of this energy from the corresponding energy obtained from the unperturbed target wave function gives the polarization potential. At the early stage of collision, where the electron–target atom distance is still large, Onda obtained

$$V_{\text{pol}} \sim -(e^4\hbar^2/m_e) \sum_{n(\neq 0)} \frac{f_{on}}{2(E_n - E_0)^2 r^4} \left\{1 - \frac{4\hbar^2 v^2}{(E_n - E_0)^2 r^2}\right\}. \tag{28}$$

If the excitation energies $E_n - E_0$ in the curly bracket are replaced by a mean excitation energy E_x, (28) becomes

$$V_{\text{pol}} \sim -\frac{\alpha e^2}{2r^4} \left\{1 - \frac{4v^2\hbar^2}{E_x^2 r^2}\right\}. \tag{29}$$

This is the first-order approximation which is valid only at large separations. For use at smaller separations, Onda suggested the following two simple expressions which approach (29) at large r:

$$V_{\text{pol}} = V_{\text{pol}}^{\text{ad}}(r) \cdot \frac{1}{1 + (4\hbar^2 v^2 / E_x^2 r^2)}, \tag{30a}$$

$$V_{\text{pol}} = V_{\text{pol}}^{\text{ad}}(r) \cdot \exp\left[-4\hbar^2 v^2 / E_x^2 r^2\right]. \tag{30b}$$

Onda and Truhlar[36] applied these approximate polarization potentials (30) to the elastic collision of electron with helium atom and found that these two forms gave practically the same result and that in the lower energy region studied (10–20 eV) these potentials gave phase shifts which were in good agreement with the more accurate theoretical calculations.

A somewhat different form has been suggested by Hayashi and Kuchitsu for the velocity-dependent polarization potential.(37–39) Here again, the semiclassical method is adopted. An incident charged particle is assumed to move along a straight line. The time-dependent perturbed state of the target is calculated and then the polarization force between the charge of the incident particle and the induced dipole moment of the target is calculated. Since the perturbed state is not the fully adiabatic state, the polarization force is not exactly along the position vector **r** of the incident particle. However, by looking at the radial component of the polarization force, equating it with $\partial U/\partial r$, and by integrating over r, the effective potential U can be obtained. This potential U is axially symmetric. The symmetry axis is along the incident direction. In order to derive a spherically symmetric effective potential, they used the eikonal approximation. Namely, they try to find out a spherical potential V_{pol} which has the same effect on the elastic collision as the original potential U. In this way they could obtain

$$V_{\text{pol}} = -\frac{e^2}{2r^4}\frac{e^2\hbar^2}{m_e}\sum_{n(\neq 0)}\frac{f_{on}}{(E_n - E_0)^2}F(\hbar v/r(E_n - E_0)). \tag{31}$$

Here, $F(x)$ is a function which is unity at $x = 0$. At large separations, therefore, (31) approaches the usual adiabatic potential [cf. (27)]. As x increases from zero, $F(x)$ first increases and soon reaches a maximum beyond which the function decreases monotonically. For large x, it has the asymptotic form $F(x) \to 0.84x^{-2}$. Hayashi and Kuchitsu applied this polarization potential to the elastic scattering of an electron by a H_2 molecule, and found close agreement with experimental values for collision energies above about 80 eV.

See Chapter 7, especially section 3.5, for further discussion on the polarization potentials and its applications to cross-section calculations.

3.2. Short-Range Interactions

The static interaction at short distances can be calculated from (19) if the accurate wave function is known for the target molecule.

The exchange effect is taken into account by antisymmetrization of the wave function of the whole system with respect to the electrons including the incident one. This leads to a set of coupled integrodifferential equations rather than a set of coupled differential equations. In other words, the exchange effect gives rise to a nonlocal interaction. In order to simplify the theory, the nonlocal exchange interaction is often replaced by an approximate local potential. The original idea was suggested by Slater[40] in the Hartree–Fock calculation of the bound-state wave functions of atoms and molecules. The essential point in the approximation is to evaluate the exchange interaction for the free-electron gas.

In the one-electron wave equation, the exchange term has a form

$$-\sum_n \phi_n(\mathbf{r}) \int \phi_n^*(\mathbf{r}') \frac{e^2}{|\mathbf{r}-\mathbf{r}'|} \psi(\mathbf{r}')d\mathbf{r}',$$

where $\psi(\mathbf{r})$ is the single-particle wave function for the incident electron under consideration and the summation is taken over the bound state orbitals (n), in each of which there is an electron with its spin parallel to that of the incident electron. If the above expression can be approximately transformed into the form

$$V_{\text{exch}}(\mathbf{r})\psi(\mathbf{r}),$$

where the function $V_{\text{exch}}(\mathbf{r})$ is the effective local exchange potential. Thus, the approximate local exchange potential is introduced by

$$V_{\text{exch}}(\mathbf{r}) = -\frac{\sum_n \int \phi_n^*(\mathbf{r}')\psi^*(\mathbf{r}) \dfrac{e^2}{|\mathbf{r}-\mathbf{r}'|} \psi(\mathbf{r}')\phi_n(\mathbf{r})d\mathbf{r}'}{|\psi(\mathbf{r})|^2}. \tag{32}$$

Following Slater, ϕ_n and ψ are replaced approximately by the plane waves

$$\phi_n(\mathbf{r}) = \frac{1}{\sqrt{V}} \exp(i\mathbf{k}_n \cdot \mathbf{r}),$$

$$\psi(\mathbf{r}) = \frac{1}{\sqrt{V}} \exp(i\mathbf{k} \cdot \mathbf{r}),$$

where V is the normalization volume. The real atoms and molecules have finite size, while the plane waves extend to infinity. In order to obtain realistic results, the integration over $\mathbf{r}'$ in (32) has to be cut down beyond a certain boundary. Slater[40,41] derived

$$V_{\text{exch}} \simeq -\frac{e^2}{2\pi^2} \int \frac{d\mathbf{k}'}{|\mathbf{k}'-\mathbf{k}|^2}. \tag{33}$$

The integral is taken over the region of $\mathbf{k}'$ occupied by the orbital electrons. By taking the average over the relative orientation of $\mathbf{k}$ and $\mathbf{k}'$, one obtains

$$\int \frac{d\hat{\mathbf{k}}'}{|\mathbf{k}'-\mathbf{k}|^2} = \frac{2\pi}{k'k} \ln \frac{k'+k}{k'-k}.$$

In the bound state problem, the electron under consideration is one of the bound electrons. Therefore, $\mathbf{k}$ is within the $\mathbf{k}'$ distribution. In other words, the integration over the magnitude of $\mathbf{k}'$ in (33) is divided into two regions corresponding to $k' > k$ and $k' < k$. In the collision problem, on the other hand, we always have $k > k'$. By integration over $\mathbf{k}'$, therefore, we obtain

$$\begin{aligned} V_{\text{exch}}(\mathbf{r}) &= -\frac{e^2}{\pi k} \int_0^{k_{\text{max}}} \ln\left(\frac{k+k'}{k-k'}\right) k' dk' \\ &= -\frac{e^2}{\pi} k_{\text{max}} \left(1 - \frac{k^2 - k_{\text{max}}^2}{2kk_{\text{max}}} \ln \frac{k+k_{\text{max}}}{k-k_{\text{max}}}\right). \end{aligned} \tag{34}$$

$k_{\text{max}}(\mathbf{r})$ is the maximum value of the wave number in the local bound-electron distribution and is given by

$$k_{\text{max}}(\mathbf{r}) = (3\pi^2 n_e)^{1/3}, \tag{35}$$

where n_e is the local number density of the bound electrons. For fast incident electrons ($k \gg k_{\text{max}}$), the effective exchange potential decreases as $(k_{\text{max}}/k)^2$.

In 1965, I suggested to S. Hara to use this local potential in a scattering problem. Hara used (34) for the first time in the electron–H_2 collision problem with encouraging results.[42] He adopted for k the following local wave number for the incident electron (see Fig. 7):

$$\hbar^2 k(\mathbf{r})^2/2m_e = \hbar^2 k_0^2/2m_e + \hbar^2 k_{\text{max}}^2/2m_e + I, \tag{36}$$

where k_0 is the incident wave number (outside the target) and I the ionization energy of the target molecule. For large values of r, this gives an incorrect value $(\hbar^2 k_0^2/2m_e + I)$ for the asymptotic energy. The correct value should be $\hbar^2 k_0^2/2m_e$. In such a region, however, the molecular electron density becomes small and $V_{\text{exch}}(\mathbf{r})$ goes to zero, so that the error introduced by the use of (36) should not be appreciable.

In fact, the same approximate exchange potential had been derived earlier by Mittleman and Watson.[43] We did not notice their work because their paper primarily aimed at high-energy collision problems.

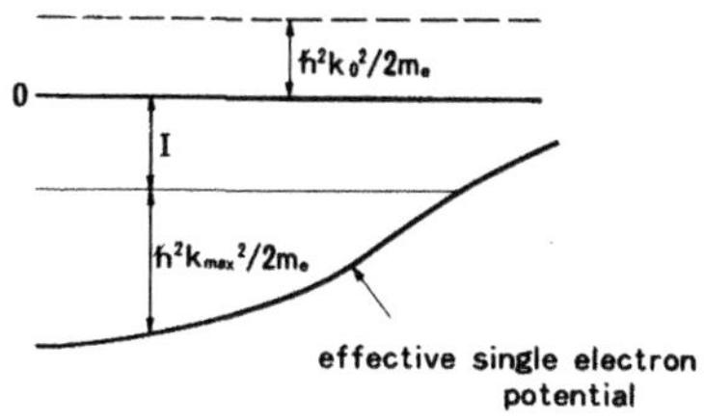

FIGURE 7
Illustration of the local wave number $k(r)$ [Eq. (36)].

Several years later, Furness and McCarthy[44] introduced still another local exchange potential. Their approach is sometimes called the semiclassical exchange approximation.

These and other local exchange approximations have been compared by Riley and Truhlar[45][46] and by Bransden *et al.*[47] More recently, Onda and Truhlar[48] applied the free-electron-gas approximation and the semiclassical exchange approximation to the 30-eV electron scattering from an N_2 molecule. They have found that the use of these local approximations in the intermediate energy region is encouraging.

Further discussion on the separable and model exchange potentials and their applications will be found in Chapter 7, Sections 3.3 and 3.4.

The concept of the "polarization" interaction becomes impractical as soon as the incident electron enters the molecular electron cloud. Under this circumstance the indistinguishability of electrons must be properly taken into account. Namely, the incident and the molecular electrons should be treated on the same footing, especially when the incident electron energy is low. Here, the polarization of the electron clouds by a particular electron is not a convenient concept. A more appropriate name for the effect is the correlation effect. The R-matrix method described in Chapters 2 and 7 is one of the promising ways to incorporate the correlation effect into the collision theory.

4. Resonance Effect

Resonances introduce conspicuous features in effective cross sections and have various important consequences. Briefly speaking, resonance is a temporary formation of a compound state between the incident electron and the target molecule with a lifetime appreciably longer than the electron passage time a/v (a is the linear dimension of the target molecule and v is the incident electron velocity). At some incident energies, the electron wave function has a large amplitude within the target. This happens only when the incident energy falls in one of the discrete (sometimes rather narrow) bands, where the incident electron finds a comfortable quasistationary orbit in the field of the target molecule. The quasistationary nature of the compound state is usually guaranteed by either of the following two mechanisms. The first possibility is the existence of an appreciable barrier in the effective potential. The attractive polarization force (Section 3.1) combined with the centrifugal force produces a barrier in the effective potential in the radial equation for the electron wave function. Once the electron has entered the region inside the barrier, it will take some time before the electron leaks out to the outer region again by the tunnel effect. Only at some appropriate energies can the incident electron find a comfortable quasistationary state in the inner region of the target, so it is only at these energies that the electron can penetrate the barrier with appreciable probability into the inner region. This type of resonance is called the shape resonance or the potential resonance since the resonance state is produced by an appropriate shape of the effective interaction potential between the electron and the molecule. One of the well-known examples of the shape resonance is found in the electron collision with N_2 molecule in the energy range 2–3 eV (see Fig. 4). A similar resonance is found also in electron–CO collision.

The second possibility arises when the inelastic channels are introduced. By exciting the target molecule, the electron loses its energy. Suppose that the incident electron energy is not so large that after the excitation the electron energy becomes negative and, furthermore, its value coincides with one of the bound-state energies allowed in the field produced by the excited target molecule. Then it will take some time before the electron gets its energy back from the target and escapes to the outside. Thus, one has a new type of resonance process. This description may not be rigorous, but it will give the readers a rough idea of this resonance process, which is called the core-excited type I resonance or the resonance of Feshbach type after the study of similar phenomena in nuclear collisions. There exist also shape resonances associated with the effective potential in an inelastic channel. They are called core-excited type II resonances or core-excited shape resonances.

These resonances affect the elastic collision and the rotational, vibrational, and electronic excitation processes. Sometimes the cross section is enhanced by the resonance effect by more than an order of magnitude. In many cases, these collision processes can also take place without passing through the resonance state. Then the scattered waves from the resonant and the nonresonant collisions interfere. This interference must be properly taken into account in order to interpret the observed energy and angular dependence of the relevant cross sections.

One of the collision processes closely associated with the resonance is the dissociative attachment. Examples are as follows:

$$e + H_2 \rightarrow H + H^-, \tag{37}$$

$$e + CH_4 \rightarrow CH_3 + H^-, \tag{38}$$

$$e + O_2 \to O + O^-, \tag{39}$$

$$e + HCl \to H + Cl^-. \tag{40}$$

The effective cross section of the dissociative attachment is of a peaked shape or a group of peaks, indicating a resonant nature of the process. In the case of H_2, (37), the cross section has a broad peak at the incident electron energy around 10 eV and a sharp peak at 14 eV. There is also a small peak at 3.75 eV, close to the threshold of process (Fig. 2). The different peaks correspond to different states of the negative ion H_2^-. Roughly speaking, the peak of the lowest energy corresponds to the compound state formed by the incident electron and H_2 in the ground state $X\,^1\Sigma_g^+$, while the broad peak corresponds to the incident electron in the field of the H_2 in the repulsive $b\,^3\Sigma_u^+$ state. See also Chapter 4, Section 5.1.

Resonance phenomena are discussed in great detail in Chapters 3 and 5. Other chapters also include accounts of the appearance of resonances in different kinds of processes.

5. Calculation of the Total Cross Section as a Potential Scattering Problem

The collision cross section is often calculated for a space-fixed molecular target, and then averaged later over the molecular orientation. This method gives a fairly accurate total (elastic + rotationally and/or vibrationally inelastic) cross section in the absence of the long-range dipole interaction; in its presence the cross section diverges for nonrotating molecules. The validity of this fixed-nuclei calculation will be discussed later briefly on the basis of the adiabatic theory.

5.1. Summary of the Theory for Scattering from a Central Field

Readers who know the topic well can skip this subsection. In nonrelativistic quantum mechanics the basic equation for the motion of a particle (mass m) in a central potential field $V(r)$ is the Schrödinger equation

$$\left[-\frac{\hbar^2}{2m}\nabla^2 + V(r)\right]\psi(\mathbf{r}) = E\psi(\mathbf{r}). \tag{41}$$

The incident particle is represented by a plane wave

$$\psi_0(\mathbf{r}) = e^{i\mathbf{k}_0\cdot\mathbf{r}}, \qquad E = k_0^2\hbar^2/2m. \tag{42}$$

This wave propagates to the direction of the wave vector $\mathbf{k}_0$. This is clearly seen when the time-dependent factor $e^{-iEt/\hbar}$ is multiplied by ψ_0. The time-dependent factor is always omitted in the stationary-flow treatment. If $V=0$, the plane wave (42) is a solution of (41). Because of the presence of V, the wave function ψ deviates from the unperturbed function ψ_0. This situation is explicitly shown by the integral equation

$$\psi(\mathbf{r}) = \psi_0(\mathbf{r}) - \frac{1}{4\pi}\frac{2m}{\hbar^2}\int \frac{e^{ik|\mathbf{r}-\mathbf{r}'|}}{|\mathbf{r}-\mathbf{r}'|} V(r')\psi(\mathbf{r}')d\mathbf{r}', \tag{43}$$

which is a formal solution of (41) as is seen by the direct substitution. There is another formal solution of (41) with $e^{-ik|\mathbf{r}-\mathbf{r}'|}$ instead of $e^{ik|\mathbf{r}-\mathbf{r}'|}$. However, we adopt (43) since the influence of $V(r)$ on $\psi(\mathbf{r})$ should propagate away from the origin, rather than converge into it. The converging (or incoming) wave solution, however, is sometimes useful as a mathematical tool in the theory of scattering.

If the potential V decreases sufficiently fast as $r\to\infty$ the equation (43) suggests that the asymptotic form of the wave function is

$$\psi(\mathbf{r}) \xrightarrow{r\to\infty} e^{i\mathbf{k}_0\cdot\mathbf{r}} + \frac{1}{r}e^{ikr}f(\theta,\phi). \tag{44}$$

The angles θ and ϕ specify the direction of scattering and $k=|\mathbf{k}_0|$. When we choose the z axis along the incident direction $\mathbf{k}_0$, we can drop out the dependence of the scattering amplitude f on ϕ because of the axial symmetry.

If $V(r)$ contains the Coulomb interaction which is proportional to r^{-1} at large separations, a careful consideration is required. I shall not discuss it in this chapter. For details of what I summarize in this section and for other related topics, readers are referred to the standard textbooks, such as *The Theory of Atomic Collisions* by Mott and Massey.[49]

The general plane wave $e^{i\mathbf{k}\cdot\mathbf{r}}$ can be expanded in terms of the Legendre functions $P_l(\cos\theta)$, where $\cos\theta = \hat{\mathbf{k}}\cdot\hat{\mathbf{r}}$, as

$$\begin{aligned} e^{i\mathbf{k}\cdot\mathbf{r}} &= \sum_l (2l+1)i^l j_l(kr)P_l(\cos\theta) \\ &= 4\pi\sum_{lm} i^l j_l(kr)Y^*_{lm}(\hat{\mathbf{r}})\,Y_{lm}(\hat{\mathbf{k}}). \end{aligned} \tag{45}$$

The functions Y_{lm} are the normalized spherical harmonics

$$Y_{lm}(\theta,\phi) = i^{|m|+m}\left(\frac{2l+1}{4\pi}\cdot\frac{l-|m|}{l+|m|}\right)^{1/2} P_l^{|m|}(\cos\theta)e^{im\phi}, \tag{46}$$

and $j_l(kr)$ is the spherical Bessel function

$$j_l(kr) = \left(\frac{\pi}{2kr}\right)^{1/2} J_{l+1/2}(kr) \xrightarrow{r\to\infty} (kr)^{-1}\sin\left(kr - \frac{1}{2}l\pi\right). \tag{47}$$

In quantum mechanics, the current density of the particle flow is given by

$$\mathbf{j} = \frac{\hbar}{2mi}(\psi^*\nabla\psi - \psi\nabla\psi^*). \tag{48}$$

As is seen from (45) and (47), a plane wave is a superposition of the incoming and the outgoing waves. Its total particles flux is zero, as is seen by substituting a plane wave into ψ in (48) and integrating the outward normal component of $\mathbf{j}$ over the surface S of a large sphere. On the other hand, the second term of (44) represents the outgoing wave only. One might think, therefore, that the particle number is not conserved in the asymptotic form (44). It should be kept in mind, however, that the wave function has to be interpreted as a whole. Then in the current density (48), which is of the second order in the wave function, there appears the interference term between the plane wave and the outgoing spherical wave. The interference is such that the particle conservation is assured when the whole wave function is taken into account. The second term of (44) thus has two roles. One is to represent the scattered wave and the other is to diminish the unscattered plane wave (particularly its outgoing part) to keep the number of particles constant throughout the collision process. The interference term is important only in the forward direction (small θ). Except in this direction, the particle flux scattered into a small solid angle $d\omega$ in the direction θ,ϕ is given by

$$j_r = v|f/r|^2 r^2 d\omega = v|f|^2 d\omega,$$

v being the incident velocity. The differential cross section q defined in Section 2.1 is now given by

$$q(\theta,\phi)d\omega = |f(\theta,\phi)|^2 d\omega. \tag{49}$$

In practice, the solution ψ of (41) is normally expanded in terms of $P_l(\cos\theta)$:

$$\psi(\mathbf{r}) = \frac{1}{r} \sum_l A_l \phi_l(r) P_l(\cos\theta). \tag{50}$$

Substitution into (44) gives the radial equation for ϕ's

$$\frac{d^2\phi_l}{dr^2} + \left[k^2 - U(r) - \frac{l(l+1)}{r^2}\right]\phi_l = 0, \tag{51}$$

where

$$U(r) = \frac{2m}{\hbar^2} V(r). \tag{52}$$

Since the wave function ψ should not be divergent at the origin, it follows from (50) that

$$\phi_l(0) = 0. \tag{53}$$

The radial equation (51) is now solved numerically with the condition (53). The resulting function will be normalized by

$$\phi(r) \xrightarrow{r\to\infty} k^{-1} \sin(kr - \tfrac{1}{2}l\pi + \eta_l). \tag{54}$$

This asymptotic form is correct, provided that the interaction V decreases faster than r^{-1}. The constant η_l is called the phase shift of the lth partial wave. The constant A_l in the expansion (50) is then determined in such a way as ψ has the required asymptotic form (44). When there is no long-range Coulomb type interaction, the result is

$$A_l = i^l \exp(i\eta_l)(2l+1) \tag{55}$$

and thus

$$f(\theta) = \sum_{i=0}^{\infty} \frac{1}{2ik} (2l+1)(e^{2i\eta_l} - 1) P_l(\cos\theta). \tag{56}$$

The integrated cross section is given by

$$\sigma = \int q(\theta) d\omega = 2\pi \int |f|^2 \sin\theta \, d\theta$$

$$= \frac{4\pi}{k^2} \sum_{l=0}^{\infty} (2l+1) \sin^2 \eta_l. \tag{57}$$

It can be shown that the series (57) is convergent if the potential $V(r)$ decreases faster than r^{-2} at large separations.

When the incident particle velocity v is large, the collision duration is short so that the scattering probability will be small. Then the standard perturbation method can be applied to solve eq. (41) [or (43)]. Here the potential $V(r)$ is regarded as a small quantity of the first order. Then the wave function ψ is expanded as

$$\psi = \psi_0 + \psi_1 + \psi_2 + \cdots, \tag{58}$$

where ψ_n is the quantity of the nth order. By substituting (58) into (41) and equating the terms of the same order in the both sides, we obtain the following set of equations:

$$[\nabla^2 + k^2] \, \psi_0 = 0 \tag{59a}$$

$$[\nabla^2 + k^2] \, \psi_1 = U(r)\psi_0 \tag{59b}$$

$$[\nabla^2 + k^2] \, \psi_2 = U(r)\psi_1 \tag{59c}$$

$$\cdots$$

With the asymptotic form (44) in mind, one can easily find that the solution of (59a) is simply the incident plane wave $e^{i\mathbf{k}_0 \cdot \mathbf{r}}$. It is also seen that all the other terms in the expansion (58) have simply the outgoing wave amplitudes asymptotically. The incident plane wave is now substituted into ψ_0 in the right-hand side of (59b). The outgoing wave solution of this equation is obtained as

$$\psi_1(\mathbf{r}) = -\frac{1}{4\pi} \int \frac{\exp(ik|\mathbf{r}-\mathbf{r}'|)}{|\mathbf{r}-\mathbf{r}'|} U(r') e^{i\mathbf{k}_0 \cdot \mathbf{r}'} d\mathbf{r}'. \tag{60}$$

Its asymptotic form gives

$$f(\theta) = -\frac{m}{2\pi\hbar^2} \int \exp[i(\mathbf{k}_0 - \mathbf{k}) \cdot \mathbf{r}] \, V(r) d\mathbf{r}, \tag{61}$$

where k is the vector with the magnitude of $|\mathbf{k}_0|$ and the direction (θ, ϕ). This is the scattering amplitude in the first-order Born approximation, usually called simply "the Born amplitude." By substituting (60) into (59c), we can derive the second-order Born approximation.

If we introduce a quantity

$$K = |\mathbf{k}_0 - \mathbf{k}| = 2k \sin(\theta/2), \tag{62}$$

(61) becomes

$$f(\theta) = \frac{2m}{\hbar^2} \int_0^\infty V(r) \frac{\sin(Kr)}{Kr} r^2 dr. \tag{63}$$

For electron scattering from the static field of an atom, the potential will have the form

$$V(r) = -\frac{Ze^2}{r} + e^2 \int \frac{\rho(\mathbf{r}')}{|\mathbf{r} - \mathbf{r}'|} d\mathbf{r}', \tag{64}$$

where Ze is the nuclear charge and $-e\rho(\mathbf{r})$ is the charge distribution of the electron cloud in the target atom. With this potential, (63) becomes

$$f(\theta) = \frac{me^2}{2k^2\hbar^2} [Z - F(\theta)] \csc^2(\theta/2), \tag{65}$$

where

$$F(\theta) = 4\pi \int_0^\infty \rho(r) \frac{\sin(Kr)}{Kr} r^2 dr \tag{66}$$

is known as the atomic form factor.

5.2. Independent-Atom Model

Sometimes, the electron–molecule interaction potential is assumed to be a sum of spherical atomic potentials

$$V(\mathbf{r}) = \sum_p v_p(r_p), \qquad r_p = |\mathbf{r} - \mathbf{R}_p|. \tag{67}$$

where $\mathbf{R}_p$ is the position of the pth center (or atom). This is an approximation even when the individual atoms have spherical charge distribution in the isolated state, since the chemical binding usually results in a considerable anisotropic distortion of the electron charge distribution. However, if the constituent atoms in the target are heavy atoms and if we are interested in the large-angle scattering of the fast electrons, the most important region of the interaction is the inner part of each atom, where the potential field is more or less spherical. Therefore, the assumption (67) is not as bad as one might think at the beginning.

Suppose that the wave function of the electron scattered from the pth atom has the asymptotic form

$$\psi_p(\mathbf{r}) \xrightarrow{r\to\infty} e^{i\mathbf{k}_0\cdot\mathbf{r}} + \frac{e^{ikr}}{r} f_p(\theta).$$

If the scattering center is shifted to $\mathbf{R}$, the solution of the wave equation will be $\psi_p(\mathbf{r}-\mathbf{R})$, which has the asymptotic form

$$\psi_p(\mathbf{r}-\mathbf{R}) \xrightarrow{r\to\infty} e^{i\mathbf{k}_0\cdot(\mathbf{r}-\mathbf{R})} + \frac{e^{i(kr-\mathbf{k}\cdot\mathbf{R})}}{r} f_p(\theta)$$

$$= e^{-i\mathbf{k}_0\cdot\mathbf{R}}\left[e^{i\mathbf{k}_0\cdot\mathbf{r}} + e^{i(\mathbf{k}_0-\mathbf{k})\cdot\mathbf{R}}\frac{e^{ikr}}{r} f_p(\theta)\right],$$

where use has been made of the relation $k|\mathbf{r}-\mathbf{R}| = k[r-(\mathbf{R}\cdot\hat{\mathbf{r}})]$ for large r and the definition of $\mathbf{k} = k\hat{\mathbf{r}}$ has been introduced. By dividing the above expression by the constant phase factor $e^{-i\mathbf{k}_0\cdot\mathbf{R}}$, we obtain the wave function for the standard incident wave $e^{i\mathbf{k}_0\cdot\mathbf{r}}$. The scattering amplitude is now given by $\exp[i(\mathbf{k}_0 - \mathbf{k})\cdot\mathbf{R}]\, f_p(\theta)$. If the multiple scattering within a molecule can be neglected, the amplitude for scattering from the molecule as a whole is simply given by

$$\sum_p \exp[i(\mathbf{k}_0-\mathbf{k})\cdot\mathbf{R}_p]\, f_p(\theta) \tag{68}$$

and the differential cross section by

$$\sum_{p,q} f_q^*(\theta) f_p(\theta) \exp[i(\mathbf{k}_0-\mathbf{k})\cdot\mathbf{R}_{pq}] \tag{69}$$

where $\mathbf{R}_{pq} = \mathbf{R}_p - \mathbf{R}_q$. By taking the average over the orientation of the molecule the differential cross section becomes

$$\sum_{p,q} f_q^*(\theta) f_p(\theta) \frac{\sin(KR_{pq})}{KR_{pq}} = \sum_p |f_p(\theta)|^2 + \sum_{p \neq q} f_q^*(\theta) f_p(\theta) \frac{\sin(KR_{pq})}{KR_{pq}}. \tag{70}$$

The first term is simply the sum of the scattered intensities from the individual atoms and the second term is the interference contribution. If the molecular vibration is taken into account, the factor $\sin(KR_{pq})/KR_{pq}$ must be averaged over the vibrational wave function.

For fast electron scattering from molecular targets, we can use this expression, where the atomic scattering amplitudes f_p's are calculated by (65) for each electron–atom system, although this does not necessarily give an accurate result, as mentioned already. By using more accurate atomic scattering amplitudes in (70), Massey and Bullard[(50)] obtained an angular distribution of scattered electrons which is in reasonably good agreement with experimental findings for N_2 target in the energy range 30–780 eV. For further details of the early investigations along this line, the readers are referred to Massey, *Electronic and Ionic Impact Phenomena*, Vol. 2[(51)]

More recent comparison between theory and experiment for N_2 in the independent-atom model was made by Herrmann *et al.*[(52)] A good agreement was obtained except in the small-angle region where the polarization interaction plays an important role. A better agreement with experiment in absolute scale can be obtained if the multiple scattering effect is taken into account.[(53)]

5.3. One-Center Expansion of Nonspherical Potential

The one-center treatment has often been used to study the electron–molecule collision problems. In this treatment, the Coulomb potential of the nuclei and the electron cloud is represented by an infinite series of functions, which must be truncated in the practical calculations. However, the formulation is simple since the familiar spherical harmonics are used as the basis functions. A brief outline of the formulation is given here. For simplicity, the molecule is assumed to be diatomic with the internuclear distance vector **R**.

The effective potential is expanded as

$$V = \sum_n v_n(r, R) P_n(\hat{\mathbf{r}} \cdot \hat{\mathbf{R}}), \tag{71}$$

and the wave function is expanded in terms of the spherical harmonics

$$\psi(\mathbf{r}) = \sum_{lm} r^{-1} f_{lm}(r) Y_{lm}(\hat{\mathbf{r}}). \tag{72}$$

If the molecular axis is chosen as the z axis, the magnetic quantum number m is made a good quantum number. By substituting (72) into the Schrödinger equation, the following set of equations is obtained for the radial functions f's:

$$\left[\frac{d^2}{dr^2} - \frac{l(l+1)}{r^2} - (lm|U|lm) + k^2\right] f_{lm}(r) = \sum_{l'(\neq l)} (lm|U|l'm) f_{l'm}(r). \quad (73)$$

The matrix element of the interaction (71) is given by

$$(lm|V|l'm) = \sum_n v_n(r, R) c^n(lm, l'm), \quad (74)$$

where

$$c^n(lm, l'm) = \left(\frac{4}{2n+1}\right)^{1/2} \int Y_{lm}^*(\hat{\mathbf{r}}) Y_{n,m-m'}(\hat{\mathbf{r}}) Y_{l'm'}(\hat{\mathbf{r}}) d\hat{\mathbf{r}} \quad (75)$$

can be evaluated by applying the general formula given in the book of Condon and Shortley, *The Theory of Atomic Spectra*[54] or using another expression given by Racah.[55] The formula is also found in Chapter 2, Section 3.3.

For each value of m, the coupled equations (73) are solved to obtain a set of radial functions of the following asymptotic form:

$$f_{lm}^{l_0}(r) \xrightarrow{r\to\infty} \delta_{ll_0} \exp\left[-i\left(kr - \frac{1}{2}l\pi\right)\right] - S_m(l|l_0) \exp\left[i\left(kr - \frac{1}{2}l\pi\right)\right]. \quad (76)$$

By taking a linear combination of such functions, a wave function with an appropriate asymptotic form is constructed:

$$\psi = \frac{2\pi i}{kr} \sum_{l_0} \sum_{lm} i^{l_0} f_{lm}^{l_0}(r) Y_{lm}(\hat{\mathbf{r}}) Y_{lm}^*(\hat{\mathbf{k}}) \quad (77)$$

$$\xrightarrow{r\to\infty} e^{i\mathbf{k}_0\cdot\mathbf{r}} + F_{\mathbf{k}_0}(\hat{\mathbf{r}}) e^{ikr}/r. \quad (78)$$

Recalling that the plane wave is decomposed as

$$e^{i\mathbf{k}_0 \cdot \mathbf{r}} = 4\pi \sum_{lm} i^l j_l(kr) Y_{lm}(\hat{\mathbf{r}}) Y^*_{lm}(\hat{\mathbf{k}}_0), \tag{79}$$

where $j_l(kr)$ is the spherical Bessel function, the scattered amplitude F is given by

$$F_{\mathbf{k}_0}(\hat{\mathbf{r}}) = \frac{2\pi}{ik} \sum_{lm} \sum_{l_0} i^{l_0 - l} [S_m(l|l_0) - \delta_{ll_0}] Y_{lm}(\hat{\mathbf{r}}) Y^*_{lm}(\mathbf{k}_0), \tag{80}$$

from which the differential cross section is calculated by

$$I(\hat{\mathbf{r}}) = |F_{\mathbf{k}_0}(\hat{\mathbf{r}})|^2. \tag{81}$$

At a large distance from the target, $\hat{\mathbf{r}}$ is nothing but the direction of the outgoing electron, so that $\hat{\mathbf{r}}$ can be replaced with $\hat{\mathbf{k}}$, the unit vector along the outgoing wave vector.

Practically, we need the mean differential cross section (averaged over the molecular orientation) as a function of the scattering angle relative to $\mathbf{k}_0$, not to $\mathbf{R}$. To derive this quantity, one has to take the average over the relative direction of $\mathbf{k}_0$ and $\mathbf{R}$, keeping the angle between $\mathbf{k}_0$ and the direction of observation [$\hat{\mathbf{r}}$ in (81), or $\hat{\mathbf{k}}$] fixed. After this procedure, one can get the mean differential cross section of the following form:

$$q = \sum_n I_n \frac{2n+1}{4\pi} P_n(\hat{\mathbf{r}} \cdot \hat{\mathbf{k}}_0). \tag{82}$$

The coefficient I_n can be calculated by multiplying (81) by $P_n(\hat{\mathbf{r}} \cdot \hat{\mathbf{k}}_0)$, integrating over $\hat{\mathbf{r}}$, and then taking the average of the resulting expression over the direction of $\hat{\mathbf{k}}_0$. That is, one has

$$I_n = \frac{1}{4\pi} \int d\hat{\mathbf{k}}_0 \int d\hat{\mathbf{r}}\, P_n(\hat{\mathbf{r}} \cdot \hat{\mathbf{k}}_0) I(\hat{\mathbf{r}}). \tag{83}$$

If the relation $P_n(\hat{\mathbf{r}} \cdot \hat{\mathbf{k}}_0) = (4\pi/2n+1) \sum_{\nu} Y_{n\nu}(\hat{\mathbf{r}}) Y^*_{n\nu}(\hat{\mathbf{k}}_0)$ is used, the integral (83) is expressed in terms of the quantity $c^n(lm, l'm')$ defined in (75). The total cross section and the momentum-transfer cross section (4) are given by

$$\sigma = I_0 \tag{84}$$

and

$$\sigma_m = I_0 - I_1. \tag{85}$$

Chapter 7, Section 3.1 discusses the single-center expansion of the potential including the exchange potential, and the techniques for solving the resultant coupled equations.

5.4. Separable Potentials

For electron scattering from a diatomic molecule, the mathematical formulation is often given in terms of the prolate spheroidal coordinates:

$$\xi = \frac{r_A + r_B}{R}, \qquad \eta = \frac{r_A - r_B}{R}, \qquad \text{and } \phi, \tag{86}$$

where r_A and r_B are the distances from the two nuclei A and B, respectively, and R is the internuclear distance. The Laplacian operator is written in terms of the spheroidal coordinates as

$$\nabla^2 = \frac{1}{d^2(\xi^2 - \eta^2)}\left[\frac{\partial}{\partial \xi}(\xi^2 - 1)\frac{\partial}{\partial \xi} + \frac{\partial}{\partial \eta}(1 - \eta^2)\frac{\partial}{\partial \eta}\right] + \frac{1}{d^2(\xi^2 - 1)(1 - \eta^2)}\frac{\partial^2}{\partial \phi^2}, \tag{87}$$

where $d = R/2$.

To begin with, a particular solution of the form $\Phi_m e^{im\phi}$ of the Schrödinger equation has to be found. It is easy to show that Φ_m satisfies the equation

$$[L_\xi^m - L_\eta^m + \kappa^2(\xi^2 - \eta^2) - W(\xi, \eta)(\xi^2 - \eta^2)]\,\Phi_m = 0, \tag{88}$$

where

$$W = \frac{2m_e}{\hbar^2} V d^2 = U d^2, \qquad \kappa = kd,$$

and

$$L_\zeta^m = \frac{\partial}{\partial \zeta}\left[(\zeta^2 - 1)\frac{\partial}{\partial \zeta}\right] - \frac{m^2}{\zeta^2 - 1}, \qquad \zeta = \xi, \eta. \tag{89}$$

The function Φ_m is expanded in terms of the spheroidal function $pe_n^m(\eta)$. These functions $pe_n^m(\eta)$, $m \geqq 0$, are the nonsingular solutions of the eigenvalue problem

$$(L_\eta^m + \kappa^2\eta^2)Y(\eta) = \lambda Y(\eta), \qquad |\eta| \leqq 1. \tag{90}$$

The corresponding eigenvalues are $\lambda = \lambda_{nm}$. When κ is zero, (90) becomes the well-known equation for the associated Legendre function $P_n^m(\eta)$ with the eigenvalue $\lambda = n(n+1)$. The functions pe_n^m are conveniently expanded in terms of P_n^m:

$$pe_n^m(\eta) = \sum_{l(\geqq m)}^{\infty} A_{n,l}^m P_l^m(\eta). \tag{91}$$

The coefficients A and the corresponding eigenvalues λ_{nm} can be determined by substituting (91) into (90) and solving the algebraic equations obtained.

It is important to notice that notation and normalization of the spheroidal functions vary from book to book. The above notation is that of Kotani.[56] In the book of Flammer,[57] the notation $S_{mn}(\kappa, \eta)$ is used for pe_n^m, while $S_{ml}^{(1)}(\kappa, \eta)$ used in the paper of Chu and Stratton[58] is proportional to the present $pe_{m+l}^m(\eta)$. The confusion arising from different normalization is avoided when one always writes $pe_n^m(\eta)/(N_{nm})^{1/2}$, where N_{nm} is a constant in the orthogonality relation

$$\int_{-1}^{1} pe_n^m(\eta) pe_{n'}^m(\eta) d\eta = \delta_{nn'} N_{nm}. \tag{92}$$

The function Φ_m is now expanded as

$$\Phi_m(\xi, \eta) = \sum_n f_{nm}(\xi) pe_n^m(\eta)/(N_{nm})^{1/2}, \tag{93}$$

which is then substituted into (88). In this way, the following set of equations for functions f is obtained:

$$\frac{d}{d\xi}\left[(\xi^2 - 1)\frac{df_{nm}}{d\xi}\right] + \left(\kappa^2\xi^2 - \frac{m^2}{\xi^2 - 1} - \lambda_{nm}\right) f_{nm} = \sum_{n'} W_{nn'}^m(\xi) f_{n'm}, \tag{94}$$

where

$$W_{nn'}^m = \frac{1}{N_{nm}} \int_{-1}^{1} \frac{pe_n^m(\eta)}{(N_{nm})^{1/2}} W(\xi, \eta)(\xi^2 - \eta^2) \frac{pe_{n'}^m(\eta)}{(N_{n'm})^{1/2}} d\eta. \tag{95}$$

The set of equations (94) is usually truncated and then solved. The wave function for our scattering problem will be obtained as a linear combination of these particular solutions:

$$\psi = \sum_{m=-\infty}^{\infty} \sum_{q} C_{q,m} e^{im\phi} \sum_{n} f_{nm}^{(q)}(\xi) pe_n^m(\eta)/(N_{nm})^{1/2}. \tag{96}$$

$f^{(q)}$ is the qth set of solutions of (94). The coefficients C should be determined by the usual requirement that the asymptotic form of the function (96) represents the sum of the incident plane wave and outgoing waves. For further details, see Takayanagi.[59]

A great simplification arises when the interaction potential has the following form:

$$W(\xi, \eta) = \frac{f(\xi) + g(\eta)}{\xi^2 - \eta^2}. \tag{97}$$

In this case, (88) has a solution of the form $F(\xi)\cdot G(\eta)$. If furthermore $g(\eta) = 0$, the ξ equation (94) becomes a set of uncoupled equations. In the latter case, the asymptotic form of the ξ functions have the form

$$f_{nm}(\xi) \xrightarrow{\xi \to \infty} \text{const} \times \frac{\sin(\kappa\xi - \frac{1}{2}n\pi + \eta_{nm})}{\kappa\xi}, \tag{98}$$

and one has for the scattering cross section

$$\sigma = \frac{4\pi}{k^2} \sum_{nm} \epsilon_m \sin^2 \eta_{nm}, \tag{99}$$

where $\epsilon_m = 2$ for $m \neq 0$ and $\epsilon_0 = 1$. The equation (99) is the formula for elastic scattering from homonuclear diatomic molecules used by Stier[60] and Fisk.[61,62] They both used an interaction of the form

$$V = V_c(\xi, \eta) u(\xi), \tag{100}$$

where

$$V_c(\xi, \eta) = \frac{4Ze^2}{R} \frac{\xi}{\xi^2 - \eta^2} = Ze^2 \left(\frac{1}{r_A} + \frac{1}{r_B} \right) \tag{101}$$

represents the Coulomb field of bare nuclei. Stier and Fisk used different, but simple forms for $u(\xi)$ and calculated cross sections for H_2, N_2, and O_2 with results comparable with experimental data. Later, Nagahara[63] calculated the scattering cross section of H_2 using a more realistic electrostatic potential.

However, he neglected both polarization force and exchange effect. More accurate calculation was done by Hara.[64] We shall return to his work later in the discussion of the rotational excitation cross sections. Other theoretical calculations in the spheroidal coordinates include that of Shimizu,[65] who studied the electron scattering from a finite dipole, that of Takayanagi and Itikawa,[66] who calculated scattering from a more realistic model potential for polar molecules, and a more recent work by Takagi and Nakamura,[67] who studied the electron scattering from H_2^+.

A great advantage of the two-center treatment is that the singular attractive potential produced by the nuclei can be represented by a simple function as in (101). A drawback of the method is that the complicated spheroidal functions are used as the basis set for the expansion of the wave function. A more serious limitation to the method is that it is no longer useful for polyatomic molecules.

5.5. Polyatomic Molecules

When the target molecule is nonlinear polyatomic, we can still use the one-center expansion approach by appropriately modifying the expansion (71) for linear molecules. However, a very large number of terms will be required to represent accurately the potential, particularly near the nuclei in the molecule. Thus, the one-center approach is not very convenient, except, perhaps, for nearly spherical molecules such as CH_4. Some other approaches must be found.

In the study of electronic structure of molecules, we often use the single-electron wave function (molecular orbital) which is a linear combination of atomic orbitals. The idea is that, when an electron is near one particular atom in the molecule, the dominant force the electron feels will be the one from that atom, so that the electron wave function in this region will be very much the same as the wave function (except for the normalization constant) obtained when all the other atoms in the system are ignored. This might suggest some new approach in the electron–molecule scattering problem. Onda[68] thus proposed a variational method. The form chosen for the trial function is $\psi(\mathbf{r}) = \psi^{\text{short}}(\mathbf{r}) + \psi^{\text{long}}(\mathbf{r})$, where $\psi^{\text{short}}(\mathbf{r})$ is a linear combination of quadratically integrable functions with their centers at constituent atoms of the molecule, and $\psi^{\text{long}}(\mathbf{r})$ is so chosen as to include distortion of the scattered wave caused by the long-range forces and to tend smoothly to the free particle wave function. Onda tested the applicability of his theory by a numerical example, where the elastic scattering of electrons by the static field of H_2 was calculated. His results are in reasonable agreement with Hara's calculations obtained by solving the coupled equations numerically. Unfortunately, Onda has not yet applied his approach to real polyatomic cases.

A somewhat different and more practical approach for polyatomic targets was proposed in the same year by Dill and Dehmer.[69] The method, called the continuum multiple scattering method (CMSM), is based on a multiple scattering approach, which had been applied successfully to nuclear scattering and electronic structure of solid states and molecules. The CMSM is another fixed-nuclei approach, where the electron scattering from the space-fixed target molecule is studied and then the cross section obtained is averaged over the orientation and over the vibrational coordinates of the molecule. In calculating the cross section, the whole space is divided into three kinds of regions. The region I consists of nonoverlapping spheres, each center of which is located at the nucleus of an atom in the molecule. If there are N atoms in the molecule, we thus have N such spherical regions, which are called $I_1, I_2, \ldots, I_N$. The radii of the spheres represent the atomic radii. At the same time, these radii are chosen so that the spheres are not isolated but come into contact with their neighbouring spheres. Then one defines a large sphere, inside which all the above-mentioned atomic spheres are contained. The region outside the large sphere is called region III, which extends to infinity. The remainder, namely, the interstitial space, is called region II.

In the spherical region I_i, the effective potential is assumed to be spherical. The static potential and the exchange effect are taken into consideration. For the exchange effect, one of the local approximations [cf. Section 3.2] may be used. Dehmer, Dill, and their colleagues have found that Hara's (or, one should say Mittleman, Watson and Hara's) free-electron-gas model, in combination with the CMSM approach, gives the best results so far.[70] In the region II, the effective potential is assumed constant. In the outer region III, the polarization potential is also taken into account. The potential there is often assumed spherical, but a nonspherical field due to the dipole and other multipole moments of the molecule can be taken into account when necessary. Further details of the approach will be given in Chapter 7, Section 3.6.

The CMSM approach has been successfully applied to $e + N_2$,[70–72] $e + SF_6$,[73] $e + CO_2$, OCS, CS_2.[74] These are the calculations of vibrationally elastic cross sections. More recently Dehmer *et al.*[75] have extended the theory to study the vibrational excitation in the $e + N_2$ collision. Not only the well-known resonant enhancement at 2.3 eV, but also a broad shape resonance contribution centered at about 26 eV has been demonstrated.

6. Theoretical Formulation of Inelastic Scattering Processes

6.1. Outline of the General Theory

The electron exchange effect is neglected for simplicity in this section, except in Section 6.6. The Schrödinger equation of the electron–molecule system is

$$[H_m + T + V]\,\Psi = E\Psi, \tag{102}$$

where H_m is the Hamiltonian for the free molecule, T is the kinetic energy of the relative motion of the electron–molecule system:

$$T = -\frac{\hbar^2}{2\mu}\nabla^2, \qquad \mu = \frac{m_e M}{m_e + M} \simeq m_e, \tag{103}$$

(M is the molecular mass), and V is the electron–molecule interaction:

$$V = \sum_j \frac{e^2}{|\mathbf{r} - \mathbf{r}_j|} - \sum_p \frac{Z_p e^2}{|\mathbf{r} - \mathbf{R}_p|} \equiv V(\mathbf{R}, \mathbf{r}_m, \mathbf{r}), \tag{104}$$

where $\mathbf{R}$ stands symbolically for the positions $\mathbf{R}_p$ of all the nuclei in the molecule, $\mathbf{r}_m$ for the whole set of coordinates $\mathbf{r}_j$ of the electrons in the molecule, and $\mathbf{r}$ is for the incident electron. The total energy of the system in the reference frame where the center-of-mass of the whole system is at rest is E.

We expand the wave function Ψ in terms of the target eigenfunctions ψ:

$$H_m \psi_\alpha = E_\alpha \psi_\alpha. \tag{105}$$

The suffix α indicates the rotational–vibrational state as well as the electronic state of the molecule.

$$\Psi = \sum_\alpha F_\alpha(\mathbf{r})\psi_\alpha(\mathbf{R}, \mathbf{r}_m). \tag{106}$$

By substituting the expansion (106) into (102) and using Eq. (105), we obtain a set of coupled equations for F's:

$$(\nabla^2 + k_\alpha^2)F_\alpha(\mathbf{r}) = \frac{2\mu}{\hbar^2}\iint \psi_\alpha^*(\mathbf{R}, \mathbf{r}_m)V(\mathbf{R}, \mathbf{r}_m, \mathbf{r})\Psi(\mathbf{R}, \mathbf{r}_m, \mathbf{r})\,d\mathbf{R}\,d\mathbf{r}_m \tag{107}$$

or

$$(\nabla^2 + k_\alpha^2)F_\alpha(\mathbf{r}) = \sum_\beta U_{\alpha\beta}(\mathbf{r})F_\beta(\mathbf{r}), \tag{108}$$

where

$$\frac{k_\alpha^2\hbar^2}{2\mu} = E - E_\alpha, \tag{109}$$

$$U_{\alpha\beta}(\mathbf{r}) = \frac{2\mu}{\hbar^2} \iint \psi_\alpha^*(\mathbf{R}, \mathbf{r}_m) V(\mathbf{R}, \mathbf{r}_m, \mathbf{r}) \psi_\beta(\mathbf{R}, \mathbf{r}_m) d\mathbf{R}\, d\mathbf{r}_m \tag{110}$$

By expanding F's further in terms of the spherical harmonics we obtain a set of coupled ordinary differential equations for the radial functions. In the standard close-coupling method, this set of coupled equations is truncated and then solved numerically.

Suppose the target molecule is initially in the state $\psi_0(\mathbf{R}, \mathbf{r}_m)$ and the incident electron approaches it with the kinetic energy of the relative motion $\hbar^2 k_0^2/2\mu$. If there is no electron–molecule interaction ($V = 0$), we have

$$\Psi = \exp(i\mathbf{k}_0 \cdot \mathbf{r}) \psi_0(\mathbf{R}, \mathbf{r}_m)\ [\equiv \Psi^0] \tag{111}$$

In reality, there is the interaction V, so that we have a deviation $\Psi - \Psi^0$ which asymptotically (namely, a long time after collision) becomes outgoing waves:

$$\Psi - \Psi^0 \longrightarrow \frac{1}{r} \sum_\alpha f(0 \to \alpha; \mathbf{r}) \exp(ik_\alpha r)\, \psi_\alpha(\mathbf{R}, \mathbf{r}_m). \tag{112}$$

It is noted that at an inelastic collision process the scattered electron loses or gains energy and the wave number changes accordingly. After the collision, the internal state ψ_α is associated with the outgoing electron wave number k_α so that the energy conservation is fulfilled. Thus the equation (44) in Section 5.1 or (78) in Section 5.3 for the potential scattering is now replaced by

$$F_\alpha(\mathbf{r}) \xrightarrow{r \to \infty} \delta_{\alpha 0} \exp(i\mathbf{k}_0 \cdot \mathbf{r}) + \frac{1}{r} f(0 \to \alpha; \hat{\mathbf{r}}) \exp(ik_\alpha r). \tag{113}$$

The formal solution which has this asymptotic form is obtained if the right-hand side of (107) is regarded as a known function. Thus we obtain

$$F_\alpha(\mathbf{r}) = \delta_{\alpha 0} \exp(i\mathbf{k}_0 \cdot \mathbf{r}) - \frac{1}{4\pi} \iiint \frac{\exp[ik_\alpha |\mathbf{r} - \mathbf{r}'|]}{|\mathbf{r} - \mathbf{r}'|} \psi_\alpha^*(\mathbf{R}, \mathbf{r}_m)$$
$$\times U(\mathbf{R}, \mathbf{r}_m, \mathbf{r}) \Psi(\mathbf{R}, \mathbf{r}_m, \mathbf{r}') d\mathbf{R}\, d\mathbf{r}_m\, d\mathbf{r}'. \tag{114}$$

From its asymptotic form, the scattering amplitude is obtained as

$$f(0 \to \alpha; \hat{\mathbf{k}}_\alpha) = -\frac{1}{4\pi} \iiint \exp(-i\mathbf{k}_\alpha \cdot \mathbf{r}) \psi_\alpha^*(\mathbf{R}, \mathbf{r}_m) U(\mathbf{R}, \mathbf{r}_m, \mathbf{r})$$
$$\times\ \Psi(\mathbf{R}, \mathbf{r}_m, \mathbf{r}) d\mathbf{R}\, d\mathbf{r}_m\, d\mathbf{r}. \tag{115}$$

Because of the unknown Ψ in the integral, this does not give the solution required. However, by substituting an approximate Ψ into the right-hand side of (114) we can obtain an approximate $F_\alpha(\mathbf{r})$.

The outgoing flux into the solid angle $d\hat{\mathbf{k}}_\alpha$ in the direction of $\mathbf{k}_\alpha$ is given by $v_\alpha|f(0\to\alpha;\hat{\mathbf{k}}_\alpha)|^2\,d\hat{\mathbf{k}}_\alpha$, where v_α is the velocity of the scattered electron when the target is left in the αth state. The incident flux per unit area is v_0. The ratio of these two quantities gives the differential cross section $q_\alpha\,d\hat{\mathbf{k}}_\alpha$ for the inelastic scattering into $d\hat{\mathbf{k}}_\alpha$. Thus the new factor $v_\alpha/v_0 = k_\alpha/k_0$ is brought into the formula [cf. (49) for elastic scattering]

$$q(0\to\alpha;\hat{\mathbf{k}}_\alpha) = \frac{k_\alpha}{k_0}|f(0\to\alpha;\hat{\mathbf{k}}_\alpha)|^2, \qquad (116)$$

from which the integrated cross section is also obtained

$$\sigma(0\to\alpha) = \int q(0\to\alpha;\hat{\mathbf{k}}_\alpha)d\hat{\mathbf{k}}_\alpha. \qquad (117)$$

6.2. Born Approximation

For fast collisions, the interaction between the electron and the molecule lasts for a very short time, so that the energy and momentum transferred between them are expected to be small. We can thus assume that Ψ is only a little different from Ψ^0. By substituting Ψ^0 into Ψ in the right-hand side of (114), we obtain the wave function in the Born approximation. By substituting this function again in the right-hand side of (114), we obtain the second Born approximation, and so on.

The same Born approximation can be also obtained by expanding F's in (108) just as we did for ψ in (58) and (59). The zeroth-order approximation is $F_\alpha^0(\mathbf{r}) = \delta_{\alpha 0}\exp(i\mathbf{k}_0\cdot\mathbf{r})$. By substituting this into the right-hand side of (108) and solving it for $F_\alpha(\mathbf{r})$, we obtain the first Born approximation, which is exactly the same as the one obtained above from the integral equations.

In this way, the (first) Born approximation for the scattering amplitude is derived as

$$f^{\mathrm{Born}}(0\to\alpha;\hat{\mathbf{k}}_\alpha) = -\frac{\mu}{2\pi\hbar^2}\iiint \exp(i\mathbf{K}\cdot\mathbf{r})\psi_\alpha^*(\mathbf{R},\mathbf{r}_m)V(\mathbf{R},\mathbf{r}_m,\mathbf{r})$$

$$\times\,\psi_0(\mathbf{R},\mathbf{r}_m)\,d\mathbf{R}\,d\mathbf{r}_m\,d\mathbf{r}, \qquad (118)$$

where

$$\mathbf{K} = \mathbf{k}_0 - \mathbf{k}_\alpha. \tag{119}$$

In the following subsections, the elastic and various inelastic processes are discussed based mainly on the Born approximation.

6.3. *Some Consequences of the Born Approximation, I: Elastic Collision and Rotational Transitions*

The molecular eigenfunction ψ_α is often approximated by a product of the rotational–vibrational and the electronic parts:

$$\psi_\alpha = \chi_A^a(\mathbf{R})\phi_a(\mathbf{R};\mathbf{r}_m), \qquad \alpha \equiv (A, a), \tag{120}$$

where ϕ_a is the normalized electronic eigenfunction which depends parametrically on the nuclear configuration $\mathbf{R}$, and χ_A^a is the normalized eigenfunction of the rotational and vibrational motion on the adiabatic potential curve (or surface) associated with the electronic state a. This is the Born–Oppenheimer approximation, which is usually a good approximation except for highly excited states.

For electronically elastic collisions, we obtain the (first) Born amplitude from (118) by integration over $\mathbf{r}_m$

$$f^{\text{Born}}(0 \to A; \hat{\mathbf{k}}_A) = -\frac{\mu}{2\pi\hbar^2}\iint \exp(i\mathbf{K}\cdot\mathbf{r})\chi_A^*(\mathbf{R})V_s(\mathbf{R},\mathbf{r})\chi_0(\mathbf{R})\,d\mathbf{R}\,d\mathbf{r}, \tag{121}$$

where

$$V_s(\mathbf{R},\mathbf{r}) = \int|\phi_0(\mathbf{R};\mathbf{r}_m)|^2\,V(\mathbf{R},\mathbf{r}_m,\mathbf{r})\,d\mathbf{r}_m \tag{122}$$

is the static interaction between the incident electron and the target molecule in the zeroth state. The polarization interaction is taken into account only through the higher-order terms in the Born approximation, so that this effect is not included in (122). However, we sometimes modify the static interaction V_s in the first-order expression (121) to include the polarization force without going to the second-order approximation. Similarly, a local approximation to the exchange interaction (Section 3.2) may be included. In this way, we obtain the first Born approximation for an appropriate effective potential $V_{\text{eff}}(\mathbf{R},\mathbf{r})$.

Since the Born approximation provides the simplest way to calculate the cross sections in quantum mechanics, this approach has been applied very often in early days to study the electron–molecule collisions. Here we shall first see briefly some early studies for homonuclear diatomic molecule and then those for polar molecules.

If the independent-atom model (Section 5.2) is applied to the homonuclear diatomic target, the scattering amplitude (68) becomes

$$f \simeq 2f_a \cos[(\mathbf{k}_0 - \mathbf{k}) \cdot \mathbf{R}/2], \tag{123}$$

where f_a is the scattering amplitude in the Born approximation due to an isolated single atom, and $\mathbf{R}$ is the internuclear distance vector. The dependence of f on $\mathbf{R}$ gives rise to both rotational and vibrational transitions. Morse(76) discussed rotational as well as vibrational excitations of homonuclear diatomic molecules on the basis of (123). Carson(77) made a more specific calculation for the case of H_2. They concluded that the fractional energy loss of the incident electron per collision due to rotational and vibrational excitations was comparable to the loss due to the momentum transfer to the target as a whole. The latter loss is of the order of $2m_e/M$ times the incident energy E [this should not be confused with E in (102) and (109)], M being the molecular mass. This conclusion, however, is unable to explain the large fractional energy loss derived from the observed behavior of electron swarms in molecular gases.

Gerjuoy and Stein(78,79) took an entirely different approach and put the emphasis on the long-range part of interaction. A homonuclear diatomic molecule is regarded as a point quadrupole in their treatment. They used the asymptotic form of the interaction [Q term in (22)] down to the origin and obtained an effective cross section of the following form for the rotational transition:

$$\sigma(J \to J' = J + 2) = \frac{8\pi}{15} a_0^2 \left(\frac{Q}{ea_0^2}\right)^2 \left(\frac{E'}{E}\right)^{1/2} \frac{(J+1)(J+2)}{(2J+1)(2J+3)}, \tag{124}$$

where E and E' are the initial and the final energy of the scattered electron, and $a_0 = 0.5292$ Å is the Bohr radius ($a_0^2 = 2.800 \times 10^{-17}\,\text{cm}^2$). This cross section could explain experimental data fairly well. This success is understood as follows.

First we shall rewrite the effective electron–molecule interaction (71) as

$$V = \sum_n v_n(r, R) \frac{4\pi}{2n+1} \sum_\nu Y_{n\nu}(\hat{\mathbf{r}}) Y_{n\nu}^*(\hat{\mathbf{R}}). \tag{125}$$

Here, only even values of n appear in homonuclear cases. If the nonspherical part (terms other than $n = 0$) is regarded as a small perturbation, the probability of transition $JMlm \to J'M'l'm'$ is proportional to $|(J'M'l'm'|V|JMlm)|^2$, where J, M (and J', M') are the rotational quantum numbers of the target molecule, while l, m (and l', m') are the orbital quantum numbers of scattered electron. The matrix element involves the integral

$$\int Y^*_{l'm'}(\hat{\mathbf{r}})\, Y_{n\nu}(\hat{\mathbf{r}})\, Y_{lm}(\hat{\mathbf{r}})\, d\hat{\mathbf{r}}.$$

This integral has a nonvanishing value only when

$$\Delta l \equiv l' - l = 0\ (l' = l = 0 \text{ forbidden}), \text{ or } \pm 2 \qquad \text{when } n = 2$$

$$\Delta l = 0\ (l' = l = 0 \text{ and } l' = l = 1 \text{ forbidden}),$$

$$\text{or } \pm 2\ (0 \nleftrightarrow 2 \text{ forbidden}), \text{ or } \pm 4 \qquad \text{when } n = 4.$$

...

The same selection rules apply for J [because of the factor $Y^*_{n\nu}(\hat{R})$ in (125)], so that transitions with $\Delta J =$ odd are forbidden. These selection rules remain valid even when we take account of the short-range interactions, provided that the electronic state of the molecule is of the $^1\Sigma$ type, i.e., the component Λ of the electronic angular momentum along the molecular axis is zero. In the case of $\Lambda \neq 0$, the rules do not apply, as demonstrated by Temkin and Faisal.[80] Similarly, in the molecule with a nonvanishing electron spin (such as O_2 in the ground $^3\Sigma$ state), ΔJ is not necessarily an even integer.[81]

Let us go back to the interpretation of the success of the Born approximation in the rotational transitions in H_2, N_2, etc. For a very slow electron ($\ll 1$ eV), only the s and p waves ($l = 0$ and 1, respectively) can penetrate sufficiently into the range of the molecular field. Within this limitation, the only possible combination of the initial and the final orbital quantum numbers compatible with the selection rules is the $p \to p$ transition arising from the "$n = 2$" interaction. Since the p wave will see only the outer part of the molecular field, the long-range quadrupole interaction is more important than the short-range interactions. It is also seen that the Born approximation is a fairly reasonable approach, since the interaction is not very strong at large distances. This explains[82] the success of Gerjuoy and Stein's theory.

When the incident energy increases, the p wave will see more and more of the inner part of the molecular field. First, the polarization interaction (24) becomes important. Dalgarno and Moffett[83] made a Born calculation using the Q term in (22) and α' term in (24) put together. Their result was

$$\sigma_{\mathrm{DM}}(J \to J+2) = \sigma_{\mathrm{GS}}(J \to J+2)\left[1 + \frac{\pi}{4}\left(\frac{\alpha'}{a_0^3}\right)\left(\frac{ea_0^2}{Q}\right)\left(\frac{\mathrm{Ry}}{E}\right)^{1/2} \frac{E - \frac{1}{4}\Delta E}{\mathrm{Ry}} + \frac{9\pi^2}{512}\left(\frac{\alpha'}{a_0^3}\right)^2\left(\frac{ea_0^2}{Q}\right)^2 \frac{E - \frac{1}{2}\Delta E}{\mathrm{Ry}}\right], \tag{126}$$

where σ_{GS} is the Gerjuoy–Stein cross section (124) and $Ry = e^2/2a_0 = 13.60\,eV$ is the Rydberg energy unit. As expected, their result agrees with the Gerjuoy–Stein formula when the incident energy is sufficiently low.

As the electron energy further increases, the p wave enters more and more deeply into the molecular field, where the plane-wave approximation for the scattered electron is no longer valid and the short-range interactions become important. The Born approximation becomes inapplicable above a certain energy (depending on the target), which is usually much below 1 eV. The distorted wave method is the next step of approach. However, as soon as the energy is further increased the excitation probability becomes so large that the close-coupling type calculation is required.

It is noted that the elastic cross section in the pure-quadrupole-interaction model is also obtained easily in the Born approximation. However, the elastic cross section in the real electron–molecule collision is considerably affected by the short-range interactions which have been neglected so far. Therefore, the resulting cross section is not so reliable as the rotational cross sections. Similarly, the momentum-transfer cross section calculated in this simple theory is not reliable.

Now we shall proceed to discuss the rotational transitions in electron–polar molecule collisions. First, we assume that the molecule is a linear one in the $^1\Sigma$ state. The relevant interaction is again expanded as in (71) and (125). Of the nonspherical terms, the dipole contribution ($n = 1$) has the longest range, and thus has the largest effect on electron collision, unless the dipole moment is very small. Massey[84] studied the rotational excitation of a polar symmetric-top molecule from the ground state ($J = 0$) to the first excited state ($J = 1$; note that the selection rule in the present case is $\Delta J = \pm 1$). The Born approximation was used and the point-dipole interaction [the D term in (22)] was adopted throughout the range of r. Since the dipole interaction does not excite the rotation around the molecular axis, his model is practically the same as for the linear molecule. Later, I[85] used the same approximation and obtained the cross section for the $J \to J + 1$ transition of a linear molecule:

$$\frac{\sigma(J \to J+1)}{a_0^2} = \frac{Ry}{\Delta E}\left(\frac{D}{ea_0}\right)^2 \frac{J+1}{2J+1} f\left(\frac{E}{\Delta E}\right), \tag{127}$$

where

$$f(x) = \frac{8\pi}{3x} \ln \frac{(x)^{1/2} + (x-1)^{1/2}}{(x)^{1/2} - (x-1)^{1/2}}. \tag{128}$$

Similarly, the deexcitation cross section is given by

$$\frac{\sigma(J \to J-1)}{a_0^2} = \frac{\text{Ry}}{\Delta E}\left(\frac{D}{ea_0}\right)^2 \frac{J}{2J+1} g\left(\frac{E}{\Delta E}\right), \tag{129}$$

where

$$g(x) = \frac{8\pi}{3x} \ln \frac{(x+1)^{1/2} + (x)^{1/2}}{(x+1)^{1/2} - (x)^{1/2}}. \tag{130}$$

By comparing (127) with (124), it is seen that at thermal energies the rotational excitation cross section due to the dipole interaction is many orders of magnitude larger than those of the nonpolar molecules unless $D/ea_0 \ll 1$. This is due to the long-range nature of the dipole interaction. Because of this nature of the interaction, the Born approximation is valid here in a wider energy range than in the electron–nonpolar molecule collision. The validity of the Born approximation is examined by decomposing the cross section and evaluating the relative importance of the partial cross sections:

$$\sigma(J \to J+1) = \sum_{l'} \sum_{m'} \sigma(l', m'). \tag{131}$$

Here l', m' are the orbital quantum numbers of the outgoing electron. The individual contributions $\sigma(l', m')$ can be calculated analytically. The calculation involves the evaluation of radial integrals of the form

$$\int_0^\infty r^{1-s} j_l(kr) j_{l'}(k'r)\, dr, \tag{132}$$

where s is the multipolarity of the molecular field. For the pure dipole interaction under consideration, $s = 1$. Inokuti[86] has presented an efficient and accurate method of evaluating (132) with arbitrary nonnegative integer s. For the incident electron energy less than a few times the threshold value ΔE, the contribution of the outgoing s wave ($l' = 0$) is the main part of the cross section (131). The s-wave contribution would be affected considerably by the short-range forces which have been neglected in the present discussions. However, the outgoing s wave is always associated with the incident p wave ($l = 1$). Since the p wave at very low collision energies does not penetrate deeply into the inner region of the target, the short-range interactions do not actually contribute much to the relevant element of the transition matrix, except through the phase shift introduced in the outgoing s wave. The latter effect will result in some modification of the Born result, but probably not drastically. Since the collision is a distant one, the perturbation approximation is not as poor as one might expect. As the collision energy E increases, the relative contribution of the first few partial waves decreases, so that the Born approximation becomes much more reliable. This was demonstrated by our model calculations,[87] where the Born approximation gave the rotational cross sections for e + CN, HCl fairly close to those obtained by the more elaborate close-coupling method.

The differential cross sections for the rotational transitions with the point dipole interaction are given in the Born approximation as

$$q(J \to J+1) = \frac{4}{3}\left(\frac{E'}{E}\right)^{1/2}\left(\frac{D}{ea_0}\right)^2 \frac{1}{K^2}\frac{J+1}{2J+1}, \tag{133}$$

and

$$q(J \to J-1) = \frac{4}{3}\left(\frac{E'}{E}\right)^{1/2}\left(\frac{D}{ea_0}\right)^2 \frac{1}{K^2}\frac{J}{2J+1}, \tag{134}$$

where $K^2 = |\mathbf{k} - \mathbf{k}'|^2 = [E + E' - 2(EE')^{1/2}\cos\theta]/(\mathrm{Ry}\cdot a_0^2)$ and θ is the scattering angle. Since the dipole interaction is of long range, the small-angle scattering is dominant. Unless the incident electron energy is very low, we may neglect the difference between E and E' to the first approximation. Then we have $K^2 \simeq 2E(1 - \cos\theta)/(\mathrm{Ry}\cdot a_0^2)$. Thus the momentum transfer cross section of the point dipole is

$$\sigma_m = \int (1 - \cos\theta) \sum_{J'=J\pm1} q(J \to J')\,d\omega = \frac{8\pi}{3}\left(\frac{\mathrm{Ry}}{E}\right)\left(\frac{D}{ea_0}\right)^2 a_0^2. \tag{135}$$

There is an additional contribution to the momentum-transfer cross section arising from the short-range interactions. However, unless the dipole moment is very small, the dipole contribution (135) becomes so large at low collision energy E that this dominates in the total momentum-transfer cross section. The formula (135) was first derived by Altshuler.[88]

When the momentum-transfer cross sections observed in various gases at room temperature are plotted against the dipole moment of the gas molecules, it is seen that the data points are somewhat above the curve representing the Born cross section (135). [See, for instance, Ref. 89.] This is probably due to the scattering by short-range forces which are neglected in the above discussions. In the region of the dipole moment near 2 esu. there are some molecules with σ_m much larger than the value given by (135). As the dipole moment increases further, the experimental points come below the Born curve, particularly when the dipole moment is larger than 3 esu. Here, probably the dipole interaction is so strong that the simple perturbation theory such as the Born approximation overestimates the cross section. Many articles have been published on this topic. Contribution of a resonance process was suggested and the critical dipole moment for negative ion formation was calculated for fixed as well as rotating dipole fields. Some trends of the experimental findings can be explained in terms of model calculations. However, quantitative explanation for the experimental values requires the detailed knowledge of the interaction including the short-range forces. For early works on the momentum-transfer cross section in the electron–strongly polar molecule collisions, the readers are referred to the review articles.[89,90]

So far, we have assumed that the molecule is linear. Crawford[91] extended (127) and (129) to the case of symmetric-top molecules. In this type of molecules, there are three constants of rotational motion: J, M, and K. $\hbar K$ is the angular momentum about the body-fixed symmetry axis and is a part of the total angular momentum $\hbar \mathbf{J}$ of the molecule. Thus we have $K \leqq J$. M is the z component (in space-fixed coordinates) of $\mathbf{J}$, and thus $M \leq J$. In quantum mechanics, both K and M may take values $-J, -J+1, \ldots, +J$. With the pure point-dipole interaction, the selection rules for transition in a single collision are

$$\begin{aligned} \Delta K &= 0 \\ \Delta J &= \pm 1, 0 \qquad (\text{if } K \neq 0) \\ &= \pm 1 \qquad (\text{if } K = 0). \end{aligned} \tag{136}$$

The effective cross section derived by Crawford in the Born approximation is

$$\frac{\sigma(J, K \to J+1, K)}{a_0^2} = \frac{8\pi}{3}\frac{\text{Ry}}{E}\left(\frac{D}{ea_0}\right)^2 \frac{(J_>^2 - K^2)}{(2J+1)J_>} \ln\left|\frac{\sqrt{E}+\sqrt{E'}}{\sqrt{E}-\sqrt{E'}}\right| \tag{137}$$

where $J_>$ is the larger of the initial and final J values.

Since the rotation around the symmetry axis of the molecule does not affect the dipole interaction with the incident electron, the quantum number K does not change as long as the pure dipole interaction is assumed. Itikawa[92] adopted a more general form of the interaction and calculated the rotational cross sections for $e + NH_3$ collisions, again in the Born approximation. Itikawa[93] has also shown how to incorporate the polarization effect into the Born formula for the symmetric top. He also applied the Born approximation to the rotational transitions in asymmetric-top molecules[94] and reported some numerical calculations for H_2O and H_2CO.

6.4. Some Consequences of the Born Approximation. II: Vibrational Transitions

We shall now discuss the vibrational transitions. Vibrational transitions take place through the dependence of the electron–molecule interaction on the vibrational coordinates of the molecule. In the Born approximation, we go back to the formula (121). The initial and the final molecular wave functions $\chi_0(\mathbf{R})$ and $\chi_A(\mathbf{R})$ now include not only the rotational wave functions but also the vibrational wave functions. Usually, each molecular wave function adopted is a simple product of the rotational wave function (a spherical harmonic function for the case of a linear molecule in the $^1\Sigma$ state) and the vibrational wave function (a function of the internuclear distance R only). The interaction V_s in (121) may be modified, as mentioned earlier, by including the polarization force and effective exchange force. The vibrational transition by electron impact then takes place through the R dependence of the interaction potential V. At room temperature, most simple molecules are in the ground state of vibration. Therefore, the amplitude of vibration is not so large. We may expand the interaction as

$$V(\mathbf{r}, \mathbf{R}) = V(\mathbf{r}, \hat{\mathbf{R}}, R = R_e) + (R - R_e) \left. \frac{\partial V(\mathbf{r}, \mathbf{R})}{\partial R} \right|_{R=R_e} + \cdots, \quad (138)$$

R_e being the equilibrium internuclear distance. The first term in (138) is the interaction for a rigid rotor, while the second and the higher-order terms cause vibrational transitions. By neglecting the higher-order terms and combining this interaction with the harmonic-oscillator assumption for the vibrational wave functions, we can easily derive a selection rule for the vibrational quantum number v such that

$$\Delta v = \pm 1. \quad (139)$$

Since the higher-order terms in (138) and the unharmonicity of molecular vibration are not necessarily negligible, the selection rule (139) is only an approximate one even within the framework of the Born approximation.

When the Born approximation is applied, it is tacitly assumed that the transition is a direct jump from the initial state to the final state and that the transition probability is small. The direct-excitation mechanism, however, cannot explain some observed vibrational excitations such as those in N_2 and CO, where transitions with Δv as large as several units take place with appreciable probability. Here a resonant process through an intermediate negative-ion state plays a dominant role. The idea of the resonance process was suggested first by Schulz(95) and then discussed theoretically by Chen and Magee(96) and in many later works. This topic is an important part of the theory of electron–molecule collisions, and will be discussed in detail in Chapter 3.

People are so much fascinated by the resonant processes that comparatively less attention has been paid to the direct processes. Though conspicuous, the resonance process is limited to certain ranges of collision energy, outside of which the background direct process is the sole excitation mechanism. Therefore, it is important to study also the direct excitation process in order to understand, for instance, the moderation of nonthermal electrons in molecular gases. The direct excitation of homonuclear diatomic molecules has been studied earlier by Massey,[97] Ta-You Wu,[98] Morse,[76] and Carson.[77] They all assumed the short-range electrostatic interaction to be responsible for the vibrational excitation and applied the Born approximation. In the 1960s many experiments were reported on the electron-induced vibrational excitations. For hydrogen, Engelhardt and Phelps[99] made an electron-swarm experiment. Schulz[28] made a beam experiment. Further beam experiments were reported by Ehrhardt *et al.*,[10] Ehrhardt and Linder,[100] and later by Linder and Schmidt.[4] The experimental cross section for the vibrational excitation $v = 0 \rightarrow v = 1$ of H_2 is found to be two orders of magnitude larger than the theoretical values. In 1965, I found that the R dependence of the polarization force [α-term in (24)] was just large enough to produce the magnitude of observed cross section. This was not a Born calculation, but a distorted-wave calculation where the distortion of the wave function of the scattered electron is approximately taken into account.[101] The asymptotic form of the polarization interaction is proportional to r^{-4}. If it is used down to $r = 0$, it is too singular. Therefore, some cutoff procedure had to be introduced. Unfortunately, the result depends considerably on the cutoff procedure, so that the conclusion was only semiquantitative. Breig and Lin[102] used the Born approximation to estimate the vibrational cross sections of H_2, N_2, and CO assuming the polarization interaction with cutoff. Bardsley *et al.*,[103] on the other hand, suggested that the vibrational excitation of H_2 was due to a resonance process. The observed peak of the cross section as a function of collision energy, however, is as broad as several eV in this particular case, so that if it is interpreted as due to a resonance process the lifetime of the negative ion state is very short, shorter than the period of molecular vibration. Therefore, the resonance effect is not so prominent as in the case of nitrogen. The excitation of the $v = 2$ state, for instance, has a cross section an order of magnitude smaller than that of $v = 1$. Later, using the adiabatic approximation and more realistic interactions, Henry and Chang[104] and also Faisal and Temkin[105] could obtain the theoretical cross sections for H_2, which are in reasonable agreement with experimental data.

For infrared-active molecules, a large contribution to the vibrational excitation or deexcitation comes from the dipole interaction, unless the relevant matrix element is too small. By assuming the pure dipole interaction and applying the Born approximation, the following excitation cross section formula is obtained:

$$\begin{aligned}\sigma(v, J \to v', J')/a_0^2 &= \frac{\text{Ry}}{E} \frac{|(v'|D(R)|v)|^2}{(ea_0)^2} \frac{J+1}{2J+1} f(E/\Delta E) \qquad \text{for } J' = J+1 \\ &= \frac{\text{Ry}}{E} \frac{|(v'|D(R)|v)|^2}{(ea_0)^2} \frac{J}{2J+1} f(E/\Delta E) \qquad \text{for } J' = J-1 \\ &= 0 \qquad \text{otherwise} \end{aligned} \tag{140}$$

where $f(x)$ is the function defined in (128) and $(v'|D(R)|v)$ is the vibrational matrix element of the dipole moment $D(R)$. Because of the different final states of rotation, the excitation energy ΔE for $J' = J+1$ and ΔE for $J' = J-1$ are slightly different. If this relatively small difference is neglected, the total cross section for the vibrational excitation $v \to v'$ will be

$$\begin{aligned}\sigma(v \to v') &= \sigma(v, J \to v', J+1) + \sigma(v, J \to v', J-1) \\ &\simeq \frac{\text{Ry}}{\Delta E} \frac{|(v'|D(R)|v)|^2}{(ea_0)^2} f(E/\Delta E), \end{aligned} \tag{141}$$

which is independent of J. This formula is essentially the result obtained by Ta-You Wu,[98] who studied the case of symmetric-top molecules for $v = 0$. [See also Ref. 85.] Itikawa has given a more general formulation of the vibrational excitation for polyatomic molecules in the Born approximation[106] and applied it to the excitation of H_2O.[107]

It is important to note that the nonpolar molecules such as CO_2 can be strongly infrared-active, since the molecular distortion in the course of vibration of some modes produces a large dipole moment. The molecule CO has a relatively small dipole moment at the equilibrium internuclear distance R_e. However, the derivative dD/dR is fairly large so that the matrix element $(v+1|D|v)$ is sufficiently large to make this molecule fairly strongly infrared-active.

The matrix element in (140) can be estimated from the infrared absorption intensity A (in cm^2 molecule^{-1} s^{-1}) as

$$|(v'|D(R)|v)|^2 = \frac{3\hbar c A}{2\pi\omega}, \tag{142}$$

where c is the volocity of light and ω the angular frequency of the radiation.

A partial-wave analysis such as was done for the rotational transitions [cf. Eq. (131)] is also applicable to the vibrational excitation. We are most interested in the energy region below 10 eV, where the vibrational excitation is one of the most important mechanisms of electron energy loss in molecular gases.

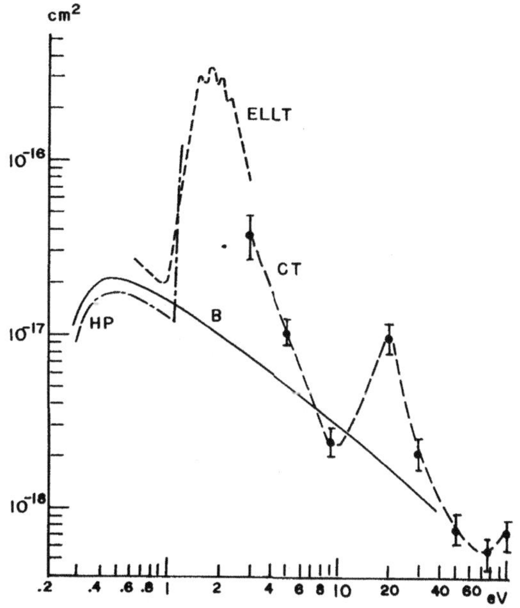

FIGURE 8
Vibrational excitation cross section for CO: $v = 0 \rightarrow v = 1$. B is the Born approximation with the pure dipole interaction. All the others are experimental data: CT, Chutjian and Tanaka[108]; ELLT, Ehrhardt, Langhans, Linder, and Taylor[10]; HP, Hake and Phelps.[109] The last one is the electron swarm experiment.

Since the vibrational quantum ΔE for simple diatomic molecules is usually 0.2–0.5 eV, we have to study the region where $E/\Delta E$ is less than 50. Under such a circumstance, the partial-wave analysis shows that the contribution of s, p, and d waves to the cross section amounts to 50% or more, and because of this we cannot expect a very accurate result from the Born approximation.

In Figs. 8 and 9, the Born cross section for the point–dipole interaction is compared with experimental cross sections for

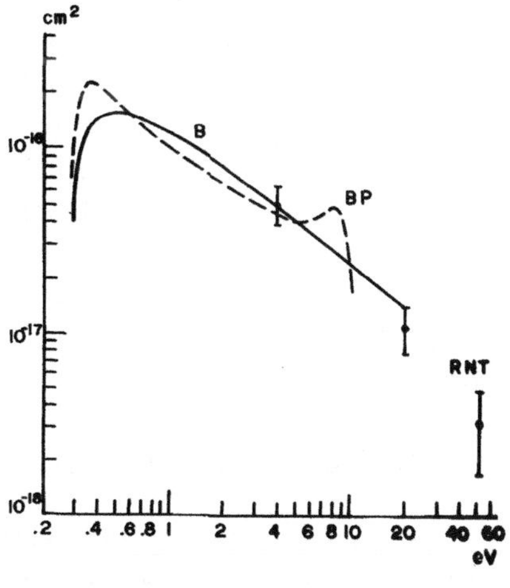

FIGURE 9
Excitation cross section of asymmetric vibration (001) of CO_2. B: Born approximation [Eq. (141)], dashed curve BP: Bulos and Phelps (swarm experiment)[111]; dots with error bars: Register *et al.* (beam experiment).[110]

$$e + CO(v = 0) \rightarrow CO(v = 1) + e$$

and

$$e + CO_2(000) \rightarrow CO_2(001) + e,$$

where (001) is the asymmetric vibration within the linear configuration. It is seen that the Born approximation is not too bad in these particular cases except in the regions of resonance peaks.

When the Born approximation is valid, the differential cross section is given by [cf. Eqs. (133), (134)]

$$\begin{aligned} q(v \rightarrow v') &= q(v, J \rightarrow v', J+1) + q(v, J \rightarrow v', J-1) \\ &\simeq \frac{4}{3}\left(\frac{E'}{E}\right)^{1/2} \frac{|(v'|D(R)|v)|^2}{(ea_0)^2} \frac{1}{K^2} \end{aligned} \tag{143}$$

where $K^2 = [E + E' - 2(EE')^{1/2} \cos\theta]/\mathrm{Ry} \cdot a_0^2)$ and θ is the scattering angle. Table 1 shows the value of $K^2 q(000 \rightarrow 001)$ for CO_2 calculated from the experimental differential cross section.[110] For the scattering angle below about 15°, the experimental $K^2 q$ is nearly constant in accordance with the prediction of the Born approximation (143). The absolute value is somewhat different from the theoretical value

$$K^2 q^{\mathrm{Born}}(000 \rightarrow 001) = 2.1 \times 10^{-2}\ \mathrm{sr}^{-1}$$

which is independent of the incident energy E and the scattering angle θ. For the matrix element, the value of $|(001|D(R)|000)|^2 = 1.588 \times 10^{-2}\ (ea_0)^2$ has been adopted on the basis of the known intensity of infrared absorption. The difference between the experimental values and the theoretical value is not unreasonable since there is probably some contribution of the resonance excitation in the experimental values and the estimated error in the experimental data is 20%–25%, while the Born approximation is not expected to be very accurate as mentioned before. More important is that the experimental value of $K^2 q$ decreases rapidly for larger scattering angles ($\theta > 15°$), corresponding to close collisions. The Born approximation is no longer valid here because of the strong interaction and the point–dipole interaction is not valid in these short distances.

6.5. Some Consequences of the Born Approximation. III: Electronic Transitions

We shall now turn to the electronic excitation of the target molecule. The ionization process is a special case where the final electronic state is in the

TABLE 1
Experimental Value of $K^2q(000 \rightarrow 001)$ for CO_2

	$K^2q(000 \rightarrow 001)$ (sr^{-1})		
θ (deg)	$E = 4$ eV	$E = 10$ eV	$E = 20$ eV
10	2.73 (−2)[a]	1.74 (−2)	1.78 (−2)
15	2.73 (−2)	1.48 (−2)	–
20	1.74 (−2)	1.05 (−2)	8.1 (−3)
30	1.28 (−2)	6.2 (−3)	4.3 (−3)
40	7.6 (−3)	4.7 (−3)	5.9 (−3)

[a]The number in parentheses indicates the power of ten by which the value has been multiplied.

ionization continuum. The direct ionization and the ionization through auto-ionizing states are distinguished. There are interferences between these different pathways leading to the same final state. If the final state of the excitation is a repulsive one, the dissociation of the molecule will follow.

Until about the mid-1970s, most of the theoretical cross sections for the electronic excitation of molecules had been obtained either (i) by the classical theory or (ii) by simple wave-mechanical approaches, such as the Born approximation, the Born–Oppenheimer approximation, or their modifications. In more recent years, people started to use the distorted-wave (DW) and the close-coupling (CC) methods.

The classical theory was proposed by Gryzinsky[(112)] for electron–atom and ion–atom collisions. It often gives results in fair agreement with experiments or with more elaborate calculations for ionization cross sections and for some excitation cross sections, particularly for low-lying states. The theory is very simple and easy to apply. However, the general range of its validity is not clear, so that no further details will be given here. (Readers who are interested in the classical approach to the electronic excitation of molecules may be referred to some calculations by Bauer and Bartky.[(113)])

The Born approximation can be used for quantitative understanding of high-energy collision processes. Even in lower-energy collisions, it gives us qualitative or semiquantitative information. For collision processes requiring the electron exchange, which becomes more important as the collision energy comes down, the Born–Oppenheimer approximation was the only way of approach in early days. In more recent years, the Ochkur approximation[(114)] or the Ochkur–Rudge approximation[(115)] is frequently applied. Another simple theory taking account of the electron exchange is the so-called first-order approximation.[(116)] However, this has not been applied to the electron–molecule collision yet. The Born approximation will be used in this section to learn some features of the electronic excitation in electron–molecule collisions. Some of the other approaches, including electron-exchange approximations, will be discussed briefly in later subsections.

6.5.1. The Born Cross Section

For simplicity, we write down the equations for the case where the target molecule is diatomic and the initial and final electronic states are $^1\Sigma$ states, so that the molecular eigenstates are of the following wave function:

$$\phi_a(\mathbf{r}_m, \mathbf{R}) \frac{1}{R} \chi_a^{vJ}(R)\, Y_{JM}(\hat{\mathbf{R}}). \tag{144}$$

Here the function ϕ_a represents the electronic state a for fixed $\mathbf{R}$, the internuclear distance vector, and $\mathbf{r}_m$ represents symbolically the totality of the electronic coordinates in the molecule. The corresponding energy of the system is a function of R and will be written as $V_a(R)$. $\chi_a^{vJ}(R)/R$ is the vibrational function, which, because of the centrifugal potential, depends on the rotational quantum number J. Finally, the rotational wave function Y_{JM} is the usual spherical harmonic function. The product form of the wave function as in (144) is called the adiabatic approximation or the Born–Oppenheimer approximation. (The latter should not be confused with the approximation of the same name for the treatment of the electron exchange scattering.) The adiabatic approximation is usually a good approximation for the ground and low-lying electronic states, since the orbital motion of electrons is much faster than the nuclear motion and the potential curves are well separated from each other. This is not the case for higher excited states since the electron motion in excited orbitals is slower and the potential curves often come close or cross each other.

From (118) and (144), we have the scattering amplitude

$$\begin{aligned} f^{\mathrm{Born}}(avJM, \mathbf{k} \to a'v'J'M', \mathbf{k}') &= -\frac{\mu}{2\pi\hbar^2} \iiiint \exp(i\mathbf{K}\cdot\mathbf{r})\chi_{a'}^{v'J'}(R)\, Y^*_{J'M'}(\hat{\mathbf{R}}) \\ &\quad \times \phi^*_{a'}(\mathbf{r}_m, \mathbf{R})\, V(\mathbf{R}, \mathbf{r}_m, \mathbf{r})\, \chi_a^{vJ}(R)\, Y_{JM}(\hat{\mathbf{R}}) \\ &\quad \times \phi_{a'}(\mathbf{r}_m, \mathbf{R}) dR\, d\hat{\mathbf{R}}\, d\mathbf{r}_m\, d\mathbf{r}, \end{aligned} \tag{145}$$

where

$$\mathbf{K} = \mathbf{k} - \mathbf{k}' \tag{146}$$

and

$$V = -\sum_s \frac{Z_s e^2}{|\mathbf{r} - \mathbf{R}_s|} + \sum_j \frac{e^2}{|\mathbf{r} - \mathbf{r}_j|}. \tag{147}$$

$Z_s e$ and $\mathbf{R}_s$ are the charge and the position vector of the sth nucleus in the molecule and $\mathbf{r}_j$ is the position vector of the jth electron. For electronically inelastic collisions, $\phi_{a'}$ and ϕ_a are orthogonal with respect to the integration over $d\mathbf{r}_m$, so that the first term in (147) does not contribute to the integral (145). The integration over $\mathbf{r}$ can be done by using the formula[117]

$$\int \frac{\exp(i\mathbf{K}\cdot\mathbf{r})}{|\mathbf{r}-\mathbf{r}_j|}\,d\mathbf{r} = \frac{4\pi}{K^2}\exp(i\mathbf{K}\cdot\mathbf{r}_j). \tag{148}$$

(NOTE: The straightforward manipulation leads to an integral of the form $\int_0^\infty \sin Kr\,dr$, which has no definite value. This difficulty is overcome by considering that the use of plane waves in the Born approximation is a mathematical idealization and wave packets should be used in more realistic description.) Thus we obtain

$$f^{\mathrm{Born}}(avJM, \mathbf{k}\to a'v'J'M', \mathbf{k}') = -\frac{2\mu e^2}{\hbar^2 K^2}\iiint \sum_j \exp(i\mathbf{K}\cdot\mathbf{r}_j)\chi_a^{v'J'}(R)\,Y_{J'M'}^*(\hat{\mathbf{R}})$$
$$\times\,\phi_{a'}^*(\mathbf{r}_m,\mathbf{R})\,\chi_a^{vJ}(R)\,Y_{JM}(\hat{\mathbf{R}})$$
$$\times\,\phi_a(\mathbf{r}_m,\mathbf{R})\,dR\,d\hat{\mathbf{R}}\,d\mathbf{r}_m. \tag{149}$$

The absolute square of this quantity, multiplied by the velocity ratio k'/k, gives the differential cross section of the process. The integrated cross section can be obtained by further integrating over the scattering direction $\hat{\mathbf{k}}'$.

Usually we are not particularly interested in specifying the initial and final rotational states. Thus the excitation cross section may be averaged over the initial rotational states and summed up over the final states. This procedure is approximately equivalent to calculating the excitation cross section for a fixed molecular orientation and averaging afterwards over the orientation. The outline of the argument is as follows: First, we suppose that the initial rotational energy is not so large. Then the final rotational energy is not large either, because a large change in the rotational quantum number in a single electron collision is improbable. If the rotational energy is small, the dependence of the vibrational wave functions χ on the rotational state is very little. Therefore, to a good approximation, we can neglect it. Now the rotational quantum numbers appear only in the spherical harmonics. When we sum up the absolute square of (149) over the final rotational state $J'M'$, we obtain

$$\sum_{J'M'} |f^{\mathrm{Born}}(avJM, \mathbf{k}\to a'v'J'M', \mathbf{k}')|^2 = \left(\frac{2\mu e^2}{\hbar^2 K^2}\right)^2 \left| \int d\hat{\mathbf{R}}\,|Y_{JM}(\hat{\mathbf{R}})|^2 \times\right.$$

$$\times \iint \sum_j \exp(i\mathbf{K}\cdot\mathbf{r}_j)\chi_{a'}^{v'}(R)\phi_{a'}^*(\mathbf{r}_m, \mathbf{R})\chi_a^v(R)\phi_a(\mathbf{r}_m, \mathbf{R})dR\, d\mathbf{r}_m\Big|^2. \tag{150}$$

Here the relation $\Sigma_{lm} Y_{lm}^*(\theta, \phi) Y_{lm}(\Theta, \Phi) = \delta(\Theta - \theta, \Phi - \phi)$ has been used. If we further average over the initial rotational state, the formula

$$\frac{1}{2J+1}\sum_M |Y_{JM}(\Theta, \Phi)|^2 = \frac{1}{4\pi}$$

gives for the differential cross section

$$q(av \to a'v') = \frac{k'}{k}\left(\frac{2\mu e^2}{\hbar^2 K^2}\right)^2 \Big|\int \frac{d\hat{\mathbf{R}}}{4\pi} \iint \sum_j \exp(i\mathbf{K}\cdot\mathbf{r}_j)\chi_{a'}^{v'}(R)\phi_{a'}^*(\mathbf{r}_m, \mathbf{R})$$

$$\times \chi_a^v(R)\phi_a(\mathbf{r}_m, \mathbf{R})dR\, d\mathbf{r}_m\Big|^2. \tag{151}$$

This quantity depends on the scattering angle θ through $K = (k^2 + k'^2 - 2kk' \cos\theta)^{1/2}$. However, as is expected from the orientation average, it is axially symmetric around the incident direction. Therefore, the integration over the direction of the scattered electron $\hat{\mathbf{k}}'$ can be replaced by $2\pi \sin\theta\, d\theta = 2\pi K\, dK/kk'$. In this way, the integrated cross section can be obtained:

$$\sigma(av \to a'v') = \frac{8\pi}{k^2}\int_{k-k'}^{k+k'} \frac{dK}{a_0^2 K^3} \Big\langle \Big| \int \chi_{a'}^{v'}(R)\phi_{a'}^*(\mathbf{r}_m, \mathbf{R})\chi_a^v(R)\phi_a(\mathbf{r}_m, \mathbf{R})$$

$$\times \sum_j \exp(i\mathbf{K}\cdot\mathbf{r}_j)dR\, d\mathbf{r}_m\Big|^2\Big\rangle_{av}, \tag{152}$$

where the symbol $\langle\,\rangle_{av}$ stands for the averaging over the molecular orientation and the reduced mass μ has been replaced, to a good approximation, by the mass m_e of the isolated electron $(m_e e^2/\hbar^2 = 1/a_0)$.

6.5.2. *The Generalized Oscillator Strength. Cross Section at High Energy*

We now introduce the quantity $F_{0\alpha}(K)$ defined by the expression

$$F_{0\alpha}(K) = \frac{E_\alpha - E_0}{\text{Ry}} \frac{\langle|\epsilon_{\alpha 0}(\mathbf{K})|^2\rangle_{av}}{a_0^2 K^2}, \tag{153}$$

where

$$\epsilon_{\alpha 0}(\mathbf{K}) = \iint \chi_{a'}^{v'}(R)\phi_{a'}^{*}(\mathbf{r}_m, \mathbf{R})\chi_a^v(R)\phi_a(\mathbf{r}_m, \mathbf{R}) \sum_j \exp(i\mathbf{K}\cdot\mathbf{r}_j)dR\, d\mathbf{r}_m \tag{154}$$

and E_0 and E_α are the internal energies of the molecule in the states "0" [$= (a, v)$] and "α" [$= (a', v')$], respectively. The dimensionless quantity $F_{0\alpha}(K)$ is called the generalized oscillator strength because in the limit of vanishing K it tends to the optical oscillator strength $f_{0\alpha}$:

$$\lim_{K\to 0} F_{0\alpha}(K) = f_{0\alpha}. \tag{155}$$

In the limit of large K, on the other hand, the exponential factor in (154) becomes a rapidly oscillating function so that the function $\epsilon_{\alpha 0}(K)$ and thus $F_{0\alpha}(K)$ quickly goes to zero for large K. One of the important properties of the generalized oscillator strength is that it satisfies the Bethe sum rule,

$$\sum_\alpha F_{0\alpha}(K) = N, \tag{156}$$

where the summation is taken over all the possible eigenstates including the ionized states and N is the total number of electrons in the molecule. This sum rule is a generalization of the well-known Thomas–Reiche–Kuhn sum rule for the optical oscillator strength

$$\sum_\alpha f_{0\alpha} = N \tag{157}$$

and these sum rules are exact as far as we can neglect the electron-to-nucleus mass ratio.[118]

In terms of the generalized oscillator strength, the differential cross section of an inelastic collision is given by

$$q(0\to\alpha;\theta) = \frac{4\mathrm{Ry}}{E_\alpha - E_0}\frac{k_\alpha}{k_0}\frac{F_{0\alpha}(K)}{K^2} \tag{158}$$

where K and the scattering angle θ are related through

$$K^2 = k_0^2 + k_\alpha^2 - 2k_0k_\alpha\cos\theta.$$

Conversely, the generalized oscillator strength can be defined by this relation, even beyond the range of validity of the Born approximation, if the accurate experimental or theoretical differential cross section is available. Lassettre and his colleagues[119] have shown that the generalized oscillator strength with this extended definition has the property that

$$\lim_{K \to 0} F_{0\alpha}(K) = f_{0\alpha}$$

Chapter V, Section 3.1 discusses the use of this formula for simulation of photon-impact experiments by electron-impact experiments. For further discussions on the generalized oscillator strength and on other topics related to the Born approximation for charged-particle collision with atoms and molecules, readers are referred to an excellent review article by Inokuti[120] (see also Inokuti *et al.*[121]).

In terms of the generalized oscillator strength, the integrated Born cross section (152) is transformed into

$$\sigma(0 \to \alpha) = \frac{8\pi a_0^2}{T_0/\text{Ry}} \frac{\text{Ry}}{E_\alpha - E_0} \int_{|k_0 - k_\alpha|}^{k_0 + k_\alpha} F_{0\alpha}(K) \frac{dK}{K} \tag{159}$$

where T_0 is the incident electron energy and the relation $k_0^2 a_0^2 = T_0/\text{Ry}$ has been used. For a sufficiently large incident energy, $k_0 + k_\alpha$ is very large, while the difference $|k_0 - k_\alpha|$ is very small. For an optically forbidden transition ($f_{0\alpha} = 0$), therefore, the lower and the upper limits of the integral in (159) can be replaced with 0 and ∞, respectively, without introducing an appreciable error. It is seen that the excitation cross section in such a case decreases as T_0^{-1} for fast electron collisions. [NOTE: Inelastic processes where the electron exchange is essential are not under consideration here.] For an optically allowed transition ($f_{0\alpha} \neq 0$), the lower limit of the integral cannot be replaced with zero since the integrand has a factor $1/K$. In this case, when the incident energy is sufficiently large, it can be shown that the cross section has the following asymptotic form:[120]

$$\sigma(0 \to \alpha) \simeq \frac{4\pi a_0^2}{\text{T}_0/\text{Ry}} f_{0\alpha} \frac{\text{Ry}}{E_\alpha - E_0} \ln \frac{4c_{0\alpha} T_0}{\text{Ry}} \tag{160}$$

where $c_{0\alpha}$ is a constant and its value can be calculated once the function $F_{0\alpha}(K)$ is known in the whole relevant range of K. If the experimental cross section is available at large electron energies, $c_{0\alpha}$ (and $f_{0\alpha}$) can be determined empirically. It must be noted also that when the electron velocity becomes comparable with the light velocity, the relativistic correction must be introduced.

6.5.3. Simplification of the Born Cross Section

We shall now go back to the original Born cross section (152) for inelastic electron–molecule collisions. For further simplification of the formula, a few different approximations may be introduced.

Approximation 1. Strictly speaking, k' depends on the final vibrational state v'. However the vibrational level spacing (typically of the order of 0.1 eV or less) is considerably less than the electronic excitation energy (typically 5–10 eV) and negligibly small as compared with the incident energy T_0 in the energy range where the Born approximation is reliable. Therefore, we shall neglect the dependence of k' on v'. As we see later, there are usually only a small number of final v' states to which the transition takes place with relatively large probability (Franck–Condon principle). Therefore, we can formally add up over all v' without changing the cross section in any serious way. The completeness of the set of vibrational eigenfunctions $\chi_{a'}^{v'}$ leads to

$$\sigma(a \to a') = \frac{8\pi}{k_0^2} \int_{|k-k'|}^{k+k'} \frac{dK}{K}$$

$$\times \left\langle \int dR\, |\chi_a^v(R)|^2 \left| \int d\mathbf{r}_m \phi_{a'}^*(\mathbf{r}_m, \mathbf{R}) \phi_a(\mathbf{r}_m, \mathbf{R}) \sum_j \exp(i\mathbf{K}\cdot\mathbf{r}_j) \right|^2 \right\rangle_{av}. \tag{161}$$

Approximation 2. The integral over the electronic coordinates

$$\int d\mathbf{r}_m \phi_{a'}^*(\mathbf{r}_m, \mathbf{R}) \phi_a(\mathbf{r}_m, \mathbf{R}) \sum_j \exp(i\mathbf{K}\cdot\mathbf{r}_j) \equiv \epsilon_{a'a}(\mathbf{R}, \mathbf{K}) \tag{162}$$

is assumed to depend little on R. If an appropriate constant $\bar{R}$ is substituted for R in $\epsilon_{a'a}(\mathbf{R}, \mathbf{K})$, the cross section (152) becomes

$$\sigma(a, v \to a', v') = \frac{8\pi}{k_0^2} \left| \int dR\, \chi_{a'}^{v'} \chi_a^v \right|^2 \int_{|k-k'|}^{k+k'} \frac{dK}{K^3 a_0^2} \langle |\epsilon_{a'a}(\bar{\mathbf{R}}, \mathbf{K})|^2 \rangle_{av} \tag{163}$$

where the vibrational factor

$$|\int dR \chi_{a'}^{v'} \chi_a^v|^2 \equiv q_{v'v}(a' \leftarrow a) \tag{164}$$

is called the Franck–Condon factor. For the representative separation $\bar{R}$, the so-called "r centroid"

$$\bar{r}_{v'v}(a' \leftarrow a) = \int dR \chi_{a'}^{v'} R \chi_a^v / \int dR \chi_{a'}^{v'} \chi_a^v \tag{165}$$

is often used.

Approximation 3. Combining the above two approximations, we obtain

$$\sigma(a \to a') = \frac{8\pi}{k^2} \int_{|k-k'|}^{k+k'} \frac{dK}{a_0^2 K^3} \langle |\epsilon_{a'a}(\bar{R}, \mathbf{K})|^2 \rangle_{av}. \tag{166}$$

Here, R_e, the equilibrium internuclear distance in the initial state, is usually substituted in $\bar{R}$ of $\epsilon_{a'a}(\bar{\mathbf{R}}, \mathbf{K})$. In this approximation, the internuclear separation is fixed at $\bar{R}$ and the molecular orientation is also fixed. The electronic excitation cross section is calculated and then averaged over the molecular orientation.

As an example of the electronic excitation of molecules, theoretical and experimental cross sections are shown in Fig. 10 for the process

$$e + H_2(X\,^1\Sigma_g^+) \to H_2(B\,^1\Sigma_u^+) + e. \tag{167}$$

The cross sections *B*1 and *B*2 are the Born calculations. *B*1 has been obtained by using the generalized oscillator strength $F(K)$ calculated by Miller and Krauss,[13] while *B*2 has been obtained by Fliflet and McKoy.[122] Both of them have been derived in the above-mentioned Approximation 2. The differences between these two cross sections are due to the different approximate molecular wave functions adopted. The Born calculation by Chung and Lin[11] has given the cross section close to that of Miller and Krauss. Miller and Krauss studied the excitation of *B*, *B*′, *C*, *D*, and *D*′ states of H_2 in the Born approximation. The *R* dependence of the electronic matrix element (162) has been studied. They found that the *R* dependence was strongest for the *B*-state excitation.

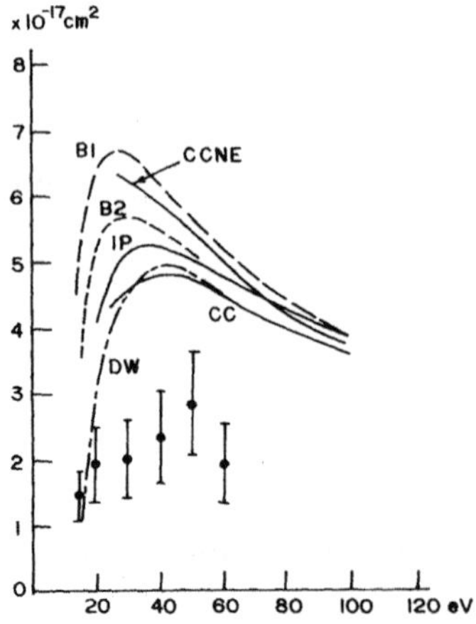

FIGURE 10
Electronic excitation for the $B\,^1\Sigma_u^+$ state of H_2. The dots with error bars are experimental data of Srivastava and Jensen.[12] See text for explanation of the theoretical curves.

However, even in this case, the approximation 2 with $\bar{R} = R_e$ agrees with results from the full cross section formula (152) within 20%. Therefore, the Franck–Condon factor gives fairly accurately the relative excitation probability within a band system. If the R dependence is larger, we have to apply the full formula (152). As is seen in Fig. 10, the Born cross sections are appreciably larger than the more elaborate close-coupling calculation (CC in Fig. 10)[11] and the distorted-wave calculation (DW).[122] An impact-parameter method (IP), where the incident electron is assumed to follow a classical trajectory, has been applied to the same problem by Hazi.[123] All these cross sections come closer at higher energies. The curve CCNE in the figure is the cross section obtained by the CC approach by neglecting the electron exchange effect.[11] The importance of the exchange effect at lower energies is evident. The small dots with error bars are the experimental values obtained by Srivastava and Jensen.[12] The experimental cross section is considerably smaller than the theoretical cross sections. The reason is not clear yet. See also Chapter 7, Section 5.2.

6.5.4. Franck–Condon Principle

The electronic jump takes place so rapidly that the position and velocity of the nuclei are expected to remain unchanged during the transition. (Franck–Condon principle) Suppose that A and B in Fig. 11 are the two potential curves of a target molecule and that the initial vibrational level is α. In the classical vibration, the velocity of the nuclear motion vanishes at the turning points, so that the probability of finding the internuclear distance within a unit range of R has maxima at both ends of the classically allowed range. When the molecule is near the inner turning point R_1, the transition to the electronic state B is likely to produce the molecule in the vibrational level β, since this is the vibrational state where the nuclear motion has nearly the zero velocity at the position R_1. Similarly, when $R \simeq R_2$, it is likely that the vibrational level γ is produced.

When we go to the wave mechanics, the Franck–Condon factor $q_{v'v}$

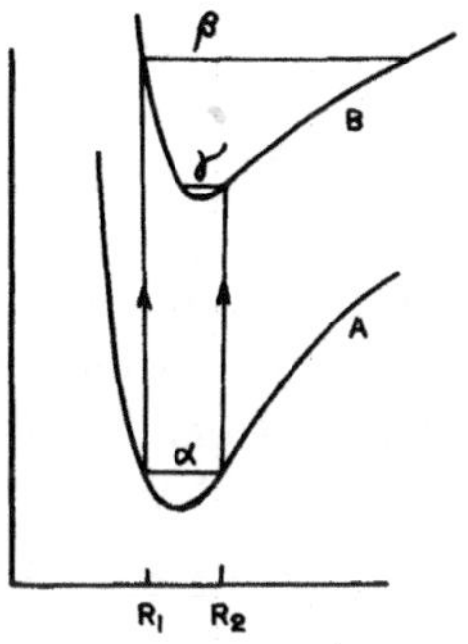

FIGURE 11
Franck–Condon principle.

contains the information of vibrational level distribution in the final state at least approximately. Except for the few lowest vibrational levels, the vibrational wave function oscillates many times within the classically allowed region, while it decays exponentially to zero in the classically forbidden region. The wave function has a large amplitude and a large local wavelength in the neighbourhood of the turning points of classical motion. In the intermediate region, on the other hand, both amplitude and wavelength are small. The product of vibrational wave function with another function, therefore, generally oscillates in the intermediate region and will only lead to a small contribution to the integral. It is expected from these considerations that the factor $q_{v'v}$ is small unless the two vibrational states have the classical turning points at nearly the same internuclear distance. For the lowest states, the situation is somewhat different. For the ground state, for instance, the vibrational wave function has the maximum amplitude not at the two classical turning points but at a middle point. For the transitions involving the low vibrational levels, therefore, we have to modify the Franck–Condon principle and state that the cross section is large only when the initial and the final vibrational wave functions have large amplitudes at about the same region of R. In the excitation process (167), for instance, the transition from $v = 0$ of the ground state $X^1\Sigma_g^+$ takes place preferably to the B state in $v' = 4$–6, corresponding to the vertically upward jump from the peak of the $v = 0$ vibrational wave function. Transition from $v = 3$ of $X^1\Sigma_g^+$, on the other hand, has four maxima of $q_{v',3}$ at about $v' = 0$, 4, 9, and 14, corresponding to the four loops of the $v = 3$ wave function. Among these four, only the maximum at $v' = 0$ has a large q value. The region where the $v' = 0$ wave function has the maximum nearly coincides with the outer turning point of the $v = 3$ level.

According to the Franck–Condon principle, the electronic excitation takes place vertically upward, roughly speaking. In Fig. 12a, the electronic state B is repulsive, so that excitation of this state leads to the dissociation. The states A, C, and D have bound levels, but the vertically upward jump from the ground vibrational level of the state X to the state C results in the repulsive part of the state, which is above the asymptotic energy value of this state for infinite internuclear separation. Thus a large fraction of the excitation of this state leads to the dissociation. The Schumann–Runge continuum of oxygen molecule excited either by electron impact or by photoabsorption,

$$O_2(X^3\Sigma_g^-) \to O_2(B\,^3\Sigma_u^-) \to O(^3P) + O(^1D) \tag{168}$$

is an example of this type of process.

The kinetic energy distribution of the dissociation products may be approximately estimated from the Franck–Condon factor q. Here the vibrational wave function of the upper level belongs to the continuum of the energy spectrum and is usually characterized by the asymptotic wave number κ. Thus we have

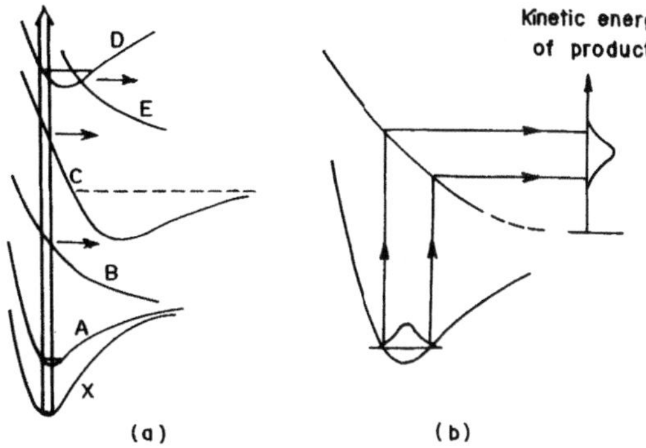

FIGURE 12
Dissociative excitation.

$$q_{\kappa,v} = |\int dR \chi_B^{\kappa}(R)\chi_X^{v}(R)|^2. \tag{169}$$

The probability of dissociation with the kinetic energy $\epsilon = \kappa^2\hbar^2/2M$, M being the reduced mass of the dissociating pair, is proportional to $q_{\kappa,v}$. The wave function χ_B^{κ} is usually normalized with respect to κ by

$$\int dR\, \chi_B^{\kappa}(R)\chi_B^{\kappa'}(R) = \delta(\kappa - \kappa'). \tag{170}$$

Very often, an approximation

$$\chi_B^{\kappa}(R) = \text{const}\, \delta(R - R_\kappa) \tag{171}$$

is adopted, where R_κ is the classical turning point associated with the potential curve B and the wave number κ. This substitution leads to

$$q_{\kappa,v} \sim |\chi_X^{v}(R_\kappa)|^2. \tag{172}$$

The kinetic energy distribution of the dissociating pair is now obtained by a method which is illustrated in Fig. 12b for $v=0$. This approximation is sometimes called the reflection approximation. The origin of the name is easily understood from Fig. 12b.

Molecular dissociation is possible also through the predissociation. This is a two-step process (strictly speaking, the two steps cannot be separable, though). The first step is an excitation of a bound state. The second step is a transition from this excited state to a repulsive state (from D to E in Fig. 12a) at a point of potential curve crossing.

Further discussions on the dissociative processes are given in Chapter 4.

6.6. Electron Exchange

6.6.1. Electron Exchange Effect

So far, we have neglected the exchange of the incident electron with electrons in the target molecule. The electron exchange has a considerable influence on the electron–molecule collisions, particularly in the low-energy encounters. For instance, the triplet state excitation from the singlet ground state in H_2 and other molecules becomes possible predominantly through the electron exchange. Even the elastic process is affected by the electron exchange. In this case, the effect is to introduce an additional potential, which is nonlocal.

Here, we shall see, as an example, how the electron–H_2 collision is treated when the electron exchange effect is included. In molecules consisting of atoms of low atomic numbers, the spin–orbit interaction is small. For this reason, the electronic eigenfunction of H_2 is simply given by a product of spatial wave function and a spin function. The hydrogen molecule, which has two electrons, is either in a singlet or a triplet state, corresponding to the total spin quantum numbers $S = 0$ and 1, respectively. For $S = 1$, there are three sublevels corresponding to $M_s = 1, 0, -1$, which represents the z component of the spin. The normalized spin functions for these states are:

Singlet:

$$\frac{1}{\sqrt{2}}[\alpha(\sigma_1)\beta(\sigma_2) - \alpha(\sigma_2)\beta(\sigma_1)], \qquad S = M_s = 0 \tag{173}$$

Triplet:

$$\left.\begin{array}{ll} \alpha(\sigma_1)\alpha(\sigma_2), & M_s = 1 \\ \frac{1}{\sqrt{2}}[\alpha(\sigma_1)\beta(\sigma_2) + \alpha(\sigma_2)\beta(\sigma_1)], & M_s = 0 \\ \beta(\sigma_1)\beta(\sigma_2), & M_s = -1 \end{array}\right\} S = 1. \tag{174}$$

Here $\alpha(\sigma)$ and $\beta(\sigma)$ are the usual normalized spin functions for a single electron representing, respectively, the spin states pointing in the $+z$ and $-z$ directions. σ's are the spin coordinates. In the following, $\alpha(\sigma_1)$, etc. will be written as α_1, etc. for brevity. Suppose that the molecule is initially in a singlet state ($S = M_s = 0$) and the incident electron is in the α state. Then S and M_s for the whole system are both 1/2, which do not change during the encounter when the spin–orbit interaction is negligible. If we neglect the symmetrization of the wave function, the initial spin state of the whole system will be

$$\frac{1}{\sqrt{2}}(\alpha_1\beta_2 - \beta_1\alpha_2)\alpha_3,$$

where 3 is the incident electron. When the molecule is excited to a triplet state, its spin state should be one of the functions in (174) or their linear combinations. The spin function for the whole system with $S = M_s = 1/2$ is then given by

$$-\frac{1}{\sqrt{3}}\frac{\alpha_1\beta_2 + \beta_1\alpha_2}{\sqrt{2}}\alpha_3 + \sqrt{\tfrac{2}{3}}\,\alpha_1\alpha_2\beta_3.$$

The coefficients $-1/\sqrt{3}$ and $\sqrt{\frac{2}{3}}$ are the Clebsch–Gordon coefficients (vector-coupling coefficients). The complete wave function of the system, properly symmetrized, is now written as

$$\begin{aligned}\Psi(\mathbf{r}_1, \sigma_1, \mathbf{r}_2, \sigma_2, \mathbf{r}_3, \sigma_3) = {} & \frac{1}{\sqrt{3}}\sum_{\lambda}\Bigg[\psi_\lambda^+(\mathbf{r}_1, \mathbf{r}_2)\frac{1}{\sqrt{2}} \\ & \times (\alpha_1\beta_2 - \beta_1\alpha_2)\alpha_3 F_\lambda^+(\mathbf{r}_3) + \text{term}\,(1, 2, 3 \to 2, 3, 1) \\ & + \text{term}\,(1, 2, 3 \to 3, 1, 2)\Bigg] - \frac{1}{\sqrt{3}}\sum_{\mu}\Bigg[\psi_\mu^-(\mathbf{r}_1, \mathbf{r}_2)\frac{1}{\sqrt{6}} \\ & \times \{(\alpha_1\beta_2 + \beta_1\alpha_2)\alpha_3 - 2\alpha_1\alpha_2\beta_3\}F_\mu^-(\mathbf{r}_3) \\ & + \text{term}\,(1, 2, 3 \to 2, 3, 1) + \text{term}\,(1, 2, 3 \to 3, 1, 2)\Bigg].\end{aligned} \tag{175}$$

Here the functions $\psi_\lambda^+(\mathbf{r}_1, \mathbf{r}_2)$ and $\psi_\mu^-(\mathbf{r}_1, \mathbf{r}_2)$ are the orbital parts of the normalized molecular eigenfunctions and they are, respectively, symmetric and antisymmetric for the interchange of the spatial coordinates: $\mathbf{r}_1 \leftrightarrow \mathbf{r}_2$. The term $(1, 2, 3 \to 2, 3, 1)$ stands for the term similar to the first term where the suffixes 1, 2, 3 are replaced with 2, 3, 1, respectively. In the following, $\psi(\mathbf{r}_1, \mathbf{r}_2)$, etc. will be abbreviated as $\psi(1, 2)$, etc. Substituting (175) into the Schrödinger equation for the e–H_2 system, in the fixed-nuclei approximation, and multiplying from the left by $\psi_\lambda^{+*}(1, 2)\sqrt{\frac{1}{2}}\,(\alpha_1\beta_2 - \beta_1\alpha_2)\alpha_3$ and integrating over $\mathbf{r}_1$, $\mathbf{r}_2$ and taking summation over $\sigma_1, \sigma_2, \sigma_3$, we have

$$[\nabla_3^2 + k_\lambda^2]F_\lambda^+(3) = \sum_{\lambda'}\iint \psi_\lambda^{+*}(1, 2)\psi_{\lambda'}^+(1, 2) \times$$

$$\times\, U(1,2;3)\, d\mathbf{r}_1\, d\mathbf{r}_2\, F^{+}_{\lambda'}(3) + \sum_{\lambda'} \iint \psi^{+*}_{\lambda}(1,2)\, \psi^{+}_{\lambda'}(2,3)$$

$$\times\, [\nabla_1^2 + k^2_{\lambda'} - U(2,3;1)]\, F^{+}_{\lambda'}(1)\, d\mathbf{r}_1\, d\mathbf{r}_2$$

$$+\sqrt{3} \sum_{\mu} \iint \psi^{+*}_{\lambda}(1,2)\, \psi^{-}_{\mu}(2,3)$$

$$\times\, [\nabla_1^2 + k^2_{\mu} - U(2,3;1)]\, F^{-}_{\mu}(1)\, d\mathbf{r}_1\, d\mathbf{r}_2, \qquad (176)$$

where

$$U(ij;k) = \frac{2m_e}{\hbar^2}\left(\frac{e^2}{r_{ik}} + \frac{e^2}{r_{jk}} - \frac{e^2}{r_{kA}} - \frac{e^2}{r_{kB}}\right), \qquad (177)$$

A and B being the two nuclei. Similarly, by multiplying the Schrödinger equation by $-(1/\sqrt{6})\psi^{-*}_{\mu}(1,2)]\,(\alpha_1\beta_2 + \beta_1\alpha_2)\alpha_3 - 2\alpha_1\alpha_2\beta_3]$, we obtain

$$[\nabla_3^2 + k^2_{\mu}]\, F^{-}_{\mu}(3) = \sum_{\mu'} \iint \psi^{-*}_{\mu}(1,2)\, \psi^{-}_{\mu'}(1,2)\, U(1,2;3)\, d\mathbf{r}_1\, d\mathbf{r}_2\, F^{-}_{\mu'}(3)$$

$$+ \sum_{\mu'} \iint \psi^{-*}_{\mu}(1,2)\, \psi^{-}_{\mu'}(2,3)$$

$$\times\, [\nabla_1^2 + k^2_{\mu'} - U(2,3;1)]\, F^{-}_{\mu'}(1)\, d\mathbf{r}_1\, d\mathbf{r}_2$$

$$-\sqrt{3} \sum_{\lambda} \iint \psi^{-*}_{\mu}(1,2)\, \psi^{+}_{\lambda}(2,3)$$

$$\times\, [\nabla_1^2 + k^2_{\lambda} - U(2,3;1)]\, F^{+}_{\lambda}(1)\, d\mathbf{r}_1\, d\mathbf{r}_2 \qquad (178)$$

The triplet state excitation from a singlet state becomes possible through the last line of (178). The situation is schematically shown in Fig. 13.

When the summations in (175) are truncated at a finite number of terms, the equations (176) and (178) provide us a set of coupled integrodifferential

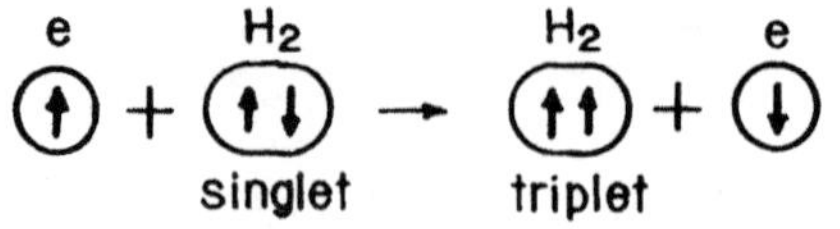

FIGURE 13
Singlet–triplet excitation.

equations. The numerical solution of these equations is the standard close-coupling (CC) method, which is discussed in detail in Chapter 7, Section 3.

6.6.2. Born–Oppenheimer and Bonham–Ochkur–Rudge Approximations

The first-order solution of these equations for the singlet–triplet transition is obtained by substituting the zeroth-order solution

$$F_\lambda^+ = \exp(i\mathbf{k}_0 \cdot \mathbf{r})\delta_{\lambda 0}, \qquad F_\mu^- = 0 \tag{179}$$

in the right-hand side of (178). The resulting scattering amplitude is

$$\begin{aligned} f_\mu^-(\hat{\mathbf{k}}_\mu) &= -\frac{\sqrt{3}}{4\pi} \iiint \psi_\mu^{-*}(1,2) \exp(-i\mathbf{k}_\mu \cdot \mathbf{r}_3) U(2,3;1) \psi_0^+(2,3) \exp(i\mathbf{k}_0 \cdot \mathbf{r}_1) \\ &\quad \times d\mathbf{r}_1\, d\mathbf{r}_2\, d\mathbf{r}_3 \\ &\equiv \sqrt{3}\, g_\mu(\hat{\mathbf{k}}_\mu). \end{aligned} \tag{180}$$

This scattering amplitude is a function of the internuclear distance R and the molecular orientation $\hat{\mathbf{R}}$. Through these dependences, the simultaneous rotational–vibrational transitions with the electronic excitation "$0 \to \mu$" can take place. We can calculate the scattering amplitude for such a joint transition by taking the matrix element of (180) with respect to the relevant rotational–vibrational eigenfunctions. If we are not interested in the rotational and vibrational levels, the differential cross section for the electronic excitation is calculated by

$$3\,\frac{k_\mu}{k_0}\,|g_\mu(\hat{\mathbf{k}}_\mu)|^2 \tag{181}$$

for fixed $\mathbf{R}$, and then averaged over the molecular orientation and vibration.

Similarly, a singlet–singlet excitation amplitude can be obtained by substituting the zeroth-order solution (179) into the right-hand side of (176). The resulting differential cross section (for fixed $\mathbf{R}$) is

$$\frac{k_\lambda}{k_0}\,|f_\lambda(\hat{\mathbf{k}}_\lambda) - g_\lambda(\hat{\mathbf{k}}_\lambda)|^2, \tag{182}$$

where the amplitude g_λ is calculated from (180), while f_λ is given by

$$f_\lambda(\hat{\mathbf{k}}_\lambda) = -\frac{1}{4\pi}\iiint \psi_\lambda^{-*}(1,2)\exp(-i\mathbf{k}_\lambda\cdot\mathbf{r}_3)U(1,2;3)\psi_0^+(1,2)\exp(i\mathbf{k}_0\cdot\mathbf{r}_3)$$

$$\times\, d\mathbf{r}_1\, d\mathbf{r}_2\, d\mathbf{r}_3$$

$$= -\frac{m_e e^2}{2\pi\hbar^2}\iiint \psi_\lambda^{-*}(1,2)\psi_0^+(1,2)\exp[i(\mathbf{k}_0-\mathbf{k}_\lambda)\cdot\mathbf{r}_3]$$

$$\times\left(\frac{1}{r_{13}}+\frac{1}{r_{23}}\right)d\mathbf{r}_1\, d\mathbf{r}_2\, d\mathbf{r}_3. \qquad (183)$$

These are the cross sections for the case where the initial target state is a singlet state. If the initial state is a triplet state, we have to take account also of the other possibility with $|M_s| = 3/2$.

This type of approximation was first introduced by Oppenheimer.[124] It is usually called the Born–Oppenheimer (BO) approximation. Unfortunately, this approximation does not agree with experimental findings at lower energies where the electron exchange effect is most important.[125]

Two modification procedures have been proposed to improve the BO approximation. In deriving the BO approximation, the relation $[\nabla^2 + k_0^2]\,F_0 = 0$ has been used since the zeroth-order solution (179) is a plane wave. According to (176) (with $\lambda = 0$), however, the quantity $[\nabla^2 + k_0^2]\,F_0$ is of the first order with respect to the interaction, so that it is comparable to the quantities retained in the BO approximation. Bell and Moiseiwitsch[116] have tried to improve this approximation by keeping all the first-order quantities in the theory. The first-order exchange approximation thus obtained gives a better agreement with experiment for electron–atom collisions than the original BO approximation. However, within the knowledge of the present author, this approach has not been used for electron–molecule collisions.

At about the same time, Bonham[126] and Ochkur[114] independently proposed another modification of the BO approximation. Bonham's work is not well known in this respect probably because the new treatment of the exchange effect is not indicated in the title and in the synopsis of the paper. Therefore, the new approach is often called the Ochkur approximation. The new formula for the exchange amplitude is essentially the first nonvanishing term of the corresponding BO amplitude when it is expanded in terms of the smaller of k_0^{-1} and k^{-1}. The higher-order terms are small at higher energies where the BO approximation is expected to be valid. For slow collisions, where the higher-order terms become appreciable, those effects which have been neglected in deriving the BO approximation will also be important. Therefore, it is meaningless to keep the higher-order terms from the original BO approximation. Of course, there is no strong reason for the validity of the first term in the low-energy region. However, the application of this approximation method to various electron-impact excitation processes has given promising results. For two-electron targets, such as He and H_2, the approximate exchange amplitude g_μ is given by[114]

$$g_\mu = \frac{2}{a_0 k_0^2} \iiint \psi_\mu^{-*}(2,1)\psi_0^{+}(1,2) \exp\,[i(\mathbf{k}_0 - \mathbf{k}_\mu)\cdot\mathbf{r}_1]\, d\mathbf{r}_1\, d\mathbf{r}_2. \quad (184)$$

Later, Rudge[115] introduced a correction factor which is important for low-energy collisions. He considered a variational expression, from which he derived an exchange amplitude

$$g_\mu = \frac{2}{a_0[k_\mu - (I/\mathrm{Ry})^{1/2}(i/a_0)]^2} \iint \psi_\mu^{-*}(2,1)\psi_0^{+}(1,2) \times \exp\,[i(\mathbf{k}_0 - \mathbf{k}_\mu)\cdot\mathbf{r}_1]\, d\mathbf{r}_1\, d\mathbf{r}_2, \quad (185)$$

where I is the ionization energy of the molecule in the initial state. Bonham–Ochkur and Bonham–Ochkur–Rudge approximations have been used repeatedly in the calculation of electron–atom and electron–molecule collisions. The earliest studies of electron–molecule collisions with these approximations include those of Khare[127][128][129] and that of Cartwright and Kuppermann.[130]

6.7. *Distorted-Wave Approximations*

The Born amplitude (145) involves the incident wave $\exp\,(i\mathbf{k}\cdot\mathbf{r})$ and the complex conjugate of the scattered wave $\exp\,(i\mathbf{k}'\cdot\mathbf{r})$, both of which are plane waves. The plane-wave approximation is appropriate only when the incident energy is so high that the distortion of the electron wave functions is negligible. As the collision energy decreases, the distortion of the electron wave functions becomes appreciable. Provided that the transition probability is still small, the major correction to be introduced in the theory is the effect of wave-function distortion. The incident and the outgoing plane waves are replaced, respectively, by appropriate distorted waves. The asymptotic form of the incident distorted wave should be the plane wave $\exp\,(i\mathbf{k}\cdot\mathbf{r})$ plus outgoing spherical waves due to the elastic scattering, while the scattered wave should be asymptotically a sum of incoming spherical waves plus the plane wave $\exp\,(i\mathbf{k}'\cdot\mathbf{r})$. In order to produce the plane wave directed to $\mathbf{k}'$ without violating the particle flux conservation under the influence of the distorting potential, there must be appropriate spherical waves coming in. In reality, however, the incident electron jumps into this final state while it is within the interaction region, so that the incoming part of the spherical waves is always vacant.

The effective potential producing the distortion effect is obtained from the electron–molecule interaction by taking the average over the internal degrees of freedom of the target molecule in one way or another. Depending on the choices of the effective potential, various distorted-wave (DW) approximations can be derived.

If the molecule is an ion, the long-range Coulomb interaction is of primary importance in distorting the wave functions of the scattered electron. In such a case, it is often assumed that the distortion is due only to the pure Coulomb interaction $-Ze^2/r$, where Ze is the net charge of the target molecule. Then the plane waves $\exp(i\mathbf{k}\cdot\mathbf{r})$ and $\exp(i\mathbf{k}'\cdot\mathbf{r})$ in the Born approximation are replaced by the Coulomb wave functions with the appropriate asymptotic forms. This approximation method is called the Coulomb–Born (CB) approximation. The earliest applications of this method to the electron–molecule collisions include the work of Stabler[131] and that of Boikova and Ob'edkov,[132] where the rotational and vibrational excitations are studied.

6.8. Adiabatic Approximations

The electron mass is so small that its velocity is very large even for a thermal electron at room temperature. (At 1/30 eV, the electron velocity is 100 km/s!) The collision duration for a thermal electron is usually somewhat shorter than the period of molecular vibration, and certainly much shorter than the rotational period. It is, therefore, a good first approximation to assume that the positions of the nuclei are fixed during an encounter.

For simplicity, only the rotational and vibrational degrees of freedom will be taken into account in the following discussions. Extension of the theory to the rotational and vibrational transitions associated with an electronic excitation is straightforward. As in the previous sections, the target molecule is assumed diatomic. A fairly accurate wave function of our scattering problem is thus of the form

$$\Psi \simeq \frac{1}{R}\chi^{vJ}(R)Y_{JM}(\hat{\mathbf{R}})\psi_{\mathbf{R}}(\mathbf{k}\,|\,\mathbf{r}), \tag{186}$$

where the three parts represent, respectively, the vibrational eigenfunction, the rotational eigenfunction, and the wave function of an electron scattered by the molecule for fixed $\mathbf{R}$. The incident wave vector $\mathbf{k}$ is explicitly indicated. As usual, the normalization of $\psi_{\mathbf{R}}$ is such that it represents asymptotically a plane wave with unit amplitude and an outgoing spherical wave.

An approximate scattering amplitude for a transition $v, J, M \to v', J', M'$ is obtained by substituting (186) into the right-hand side of (115):

$$f(v,J,M \to v',J',M'|\hat{\mathbf{k}}') = -\frac{1}{4\pi}\iiint \exp(-i\mathbf{k}'\cdot\mathbf{r})\chi^{v'J'}(R)Y^*_{J'M'}(\hat{\mathbf{R}})U(\mathbf{R},\mathbf{r})$$

$$\times\,\chi^{vJ}(R)Y_{JM}(\hat{\mathbf{R}})\psi_{\mathbf{R}}(\mathbf{k}|\mathbf{r})\,dR\,d\hat{\mathbf{R}}\,d\mathbf{r}. \qquad (187)$$

In many cases of practical importance, the electron energy is much larger than the rotational level spacings, so that it is a good approximation to assume that

$$|\mathbf{k}'| = |\mathbf{k}| \qquad (188)$$

at least for pure rotational transitions. Even for vibrational transitions, (188) becomes an increasingly good approximation as the collision energy exceeds the excitation threshold sufficiently. With this assumption (188), the integration over $d\mathbf{r}$

$$f_{\mathbf{R}}(\mathbf{k}\to\mathbf{k}') = -\frac{1}{4\pi}\int \exp(-i\mathbf{k}'\cdot\mathbf{r})U(\mathbf{R},\mathbf{r})\psi_{\mathbf{R}}(\mathbf{k}|\mathbf{r})\,d\mathbf{r} \qquad (189)$$

represents the exact amplitude for elastic scattering of an electron from the space-fixed target molecule. Thus we have

$$f(v,J,M \to v',J',M'|\hat{\mathbf{k}}') = \iint \chi^{v'J'}(R)\,Y^*_{J'M'}(\hat{\mathbf{R}})f_{\mathbf{R}}(\mathbf{k}\to\mathbf{k}')\chi^{vJ}(R)\,Y_{JM}(\hat{\mathbf{R}})\,dR\,d\hat{\mathbf{R}}. \qquad (190)$$

This approximation method was proposed originally by Chase[133] in a study of neutron scattering. In an article on the electron–polar molecule collision,[85] I suggested very briefly that the same approximation might be useful for studying the electron–molecule collision problems. In fact, Altshuler[88] had already applied this approximation to calculate the momentum-transfer cross section for the electron–polar molecule collisions. Oksyuk[134] made the first extensive calculation for the rotational excitations of H_2, N_2, and O_2 by making use of this approximation. However, he used the interaction potential suggested earlier by Fisk,[61,62] which was of the short-range type (as described in Section 5.4 of this chapter). Neither the long-range quadrupole nor the polarization interactions, important in the low-energy collisions, were taken into account. Later, Hara[135] made a more detailed calculation for the electron–H_2 collision, including the long-range interactions and also the electron exchange effect. The scattering amplitude $f_{\mathbf{R}}(\mathbf{k}\to\mathbf{k}')$ was calculated in the two-center coordinate system. The cross section for the rotational excitation $J=1\to 3$ is shown in Fig. 14 for four different interaction models. It is seen that the exchange effect

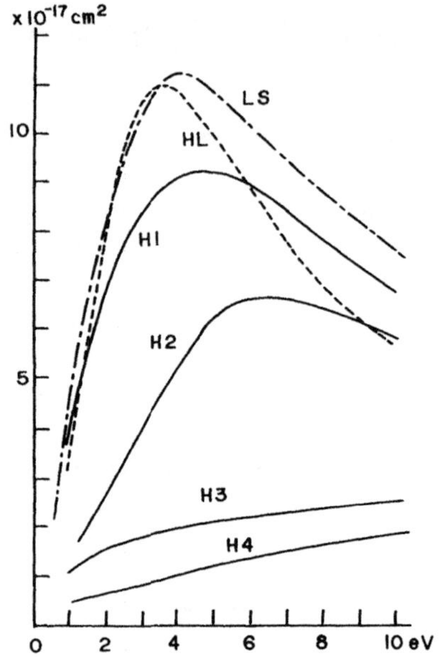

FIGURE 14
Cross section for rotational excitation $J=1\rightarrow 3$ of H_2. H1–H4: theoretical,[135] (H1: all interactions included, H2: polarization effect excluded, H3: exchange effect excluded, H4: exchange and polarization effects excluded), HL: theoretical,[136] LS: experimental.[4]

is very large in the energy range studied. The result of a close-coupling calculation by Henry and Lane[136] is also shown. They used a single-center potential. According to their calculation, not only the exchange effect but also the polarization effect is of considerable importance. In the region near the cross-section maximum, the Henry–Lane result is close to the experimental data obtained by Linder and Schmidt,[4] while at higher energies Hara's result is closer to the experimental values. The difference between the two calculations is partly due to the different approximation methods and partly due to the different interactions used. Oksyuk[134] made calculations for energies above 10 eV. In the energy range studied, his cross section is an increasing function of energy. At 10 eV, his rotational cross section for $J=1\rightarrow 3$ is about $10^{-17}\,\text{cm}^2$, which is too small to explain the experimental data.

We shall return to the adiabatic approximation. The differential cross section is given by

$$q(v,J,M\rightarrow v',J',M'|\hat{\mathbf{k}}') = \frac{k'}{k}\,|f(v,J,M\rightarrow v',J',M'|\hat{\mathbf{k}}')|^2. \qquad (191)$$

where the amplitude f is given by (190). In the energy region where the adiabatic approximation is well applicable, the factor k'/k can be omitted. However, with this factor retained, the cross section goes to zero at the excitation threshold as required, so that this factor gives some improvement of the theory near the threshold. The integral cross section $\sigma(v, J, M \to v', J', M')$ is obtained by integrating (191) over the scattering direction.

The adiabatic approximation is very useful because it provides us with various sum rules and other simple relations between the cross sections.[59,134,137] If we sum up (191) over the final states v', J', M', we can easily obtain, by making use of the completeness of the rotational and vibrational eigenfunctions,

$$\sum_{v'J'M'} q(v, J, M \to v', J', M' | \hat{\mathbf{k}}') = \iint q_{\mathbf{R}}(\mathbf{k} \to \mathbf{k}') |\chi^{vJ}(R) Y_{JM}(\hat{\mathbf{R}})|^2 \, dR d\hat{\mathbf{R}}, \tag{192}$$

where $q_{\mathbf{R}}(\mathbf{k} \to \mathbf{k}')$ is the differential cross section for the elastic scattering from the space-fixed target molecule. The summation over all the excited states does not conflict much with the assumption $k = k'$ since the contribution of the higher states to the sum is small. If we further take the average over M, the result is

$$\frac{1}{2J+1} \sum_{M} \sum_{v'J'M'} q(v, J, M \to v', J', M' | \hat{\mathbf{k}}') = \frac{1}{4\pi} \iint q_{\mathbf{R}}(\mathbf{k} \to \mathbf{k}') |\chi^{vJ}(R)|^2 \, dR d\hat{\mathbf{R}}. \tag{193}$$

When J and v are not too large, the vibrational state depends very little on the rotational state, so that the average differential cross section (193) does not depend much on the initial rotational state. Very often the energy loss or gain due to rotational transitions cannot be detected experimentally. The elastic cross section thus obtained is actually the sum of the elastic and the rotational cross sections. The above finding tells us that the result of such an experiment does not depend on the initial rotational state of the target molecule.

Derivation and validity of the adiabatic approximation and its consequences including various sum rules will be discussed in detail in Chapter 2, Section 3.

Once the scattering amplitude $f_{\mathbf{R}}(\mathbf{k} \to \mathbf{k}')$ has been calculated, all the rotational and vibrational cross sections can be calculated. Thus the adiabatic approximation is a powerful approach to these inelastic processes. Among others, the isotope effect can be studied easily in this approximation. Suppose we know the accurate scattering amplitude $f_{\mathbf{R}}$ for H_2 and we want to calculate the rotational cross sections for HD. Since the electron–HD interaction is nearly the same as for the electron–H_2 collision, we can use the same $f_{\mathbf{R}}$ except that the center of mass of the molecule is now shifted from the midpoint of the nuclei by an amount $\Delta\mathbf{R}$. A shift of the coordinate origin introduces a phase factor into the scattering amplitude. [See, for instance, Eq. (68).] We thus have[138]

$$f_{\mathbf{R}}^{\mathrm{HD}}(\mathbf{k}\rightarrow\mathbf{k}') = \exp(-i\mathbf{K}\cdot\Delta\mathbf{R})f_{\mathbf{R}}^{\mathrm{H_2}}(\mathbf{k}\rightarrow\mathbf{k}'), \qquad \mathbf{K} = \mathbf{k}-\mathbf{k}'. \quad (194)$$

This relation, combined with (190), has been used by Hara[139] to calculate the rotational cross sections for the system. Now that the inversion symmetry of the interaction potential with respect to the center of mass of the molecule has been lost, the rotational excitations of the type $J\rightarrow J\pm 1$ are possible. In the case of HD the shift of the origin is 0.12 Å. The numerical calculations show that the cross sections for $J=0\rightarrow 1$ and $J=1\rightarrow 2$ are an order of magnitude smaller than that of $J=0\rightarrow 2$. The latter is fairly close to that of the electron–H_2 collision.

Henry and Lane[140] made calculation of the rotational excitation cross sections of D_2 in the energy region below 1 eV. Their results are almost identical with those of electron–H_2 collision,[136] except in the very vicinity of the threshold. This finding can be easily understood in the adiabatic approximation.[141] Since Henry and Lane used the same interaction potential, the scattering amplitudes $f_{\mathbf{R}}^{\mathrm{H_2}}(\mathbf{k}\rightarrow\mathbf{k}')$ and $f_{\mathbf{R}}^{\mathrm{D_2}}(\mathbf{k}\rightarrow\mathbf{k}')$ are exactly the same. Thus the general formula (190) gives the same rotational cross sections for H_2 and D_2 (and also for T_2), except for the factor k'/k, which is practically unity when the collision energy is much higher than the excitation threshold.

The vibrational excitation cross section of H_2 was calculated by Henry and Chang[142] using the formulas (190) and (191). Their results were in better agreement with experiments[4][10] than a close-coupling calculation,[143] although the interactions adopted were not drastically different. Faisal and Temkin[144] independently studied the same problem in the adiabatic approximation and obtained comparable results. If the incident electron energy is not sufficiently high for the adiabatic approximation to be applicable to the vibrational excitation, one can use the approximation only to the rotational part of the transition and apply the close-coupling method to study the vibrational excitation (hybrid theory).[145] For this purpose, the molecular orientation $\hat{\mathbf{R}}$ is fixed in space and a set of coupled equations is solved to obtain the scattering amplitude $f_{\hat{\mathbf{R}}}(v\rightarrow v'|\hat{\mathbf{k}}')$ for the vibrational transition $v\rightarrow v'$. The rotational transition is then taken into account by

$$f(v,J,M\rightarrow v',J',M'|\hat{\mathbf{k}}') = \int Y_{J'M'}^{*}(\hat{\mathbf{R}})f_{\hat{\mathbf{R}}}(v\rightarrow v'|\hat{\mathbf{k}}')Y_{JM}(\hat{\mathbf{R}})\,d\hat{\mathbf{R}}. \quad (195)$$

See Chapter 7, Section 4.2.

When the interaction is of a long range, the effective collision duration is long, so that the adiabatic approximation is less adequate. For polar molecules, in particular, the differential cross section in the fixed-nuclei approximation diverges in the forward direction. This divergence, due to the distant collisions, is so strong that even the integrated cross section diverges. The validity of the adiabatic approximation may be discussed in terms of the frame-transformation theory of Chang and Fano,[146] which is presented in detail in Chapter 2, Sections 4.1 and 4.2. The body-frame (reference frame fixed to the target molecule) is adopted at shorter distances where the rotational Hamiltonian of the molecule can be disregarded as compared with the electron–molecule interaction, while the laboratory frame (coordinate system fixed to space) is more suitable at larger distances. At a certain distance, the wave functions in these two regions are smoothly joined through the frame transformation. If the electron energy is much larger than the rotational level spacings we can neglect the rotational Hamiltonian even in the outer region, and if furthermore the interaction decays faster than r^{-2} at large distances, the adiabatic approximation is derived. Otherwise, it is desirable to make a frame transformation at an appropriate distance. Then the correct cross section becomes finite even for the electron–polar molecule collision. This is easily seen in the total cross section from the Born approximation, which is given by the sum of the formulas (127) and (129). The Born approximation is valid at large distances.

Recently Nesbet[147] proposed a modification of the adiabatic approximation. His new theory, called the energy-modified adiabatic approximation, allows for the splitting and shifts of perturbed target-state energy levels. It gives the cross section which is qualitatively correct down to the excitation threshold. The theory is also applicable to resonance processes where the long collision duration makes the usual adiabatic approximation inapplicable. See Chapter 7, Section 4.1.

A few words must be said on the Glauber approximation.[148] The Glauber approximation is a kind of adiabatic approximation, where the scattering amplitude $f_{\mathbf{R}}(\mathbf{k} \to \mathbf{k}')$ for fixed nuclei is given by

$$f_{\mathbf{R}}(\mathbf{k} \to \mathbf{k}') = \frac{ik}{2\pi} \int [1 - \exp(i\chi)] \exp(i\mathbf{K} \cdot \mathbf{b}) d^2\mathbf{b}, \qquad (196)$$

where

$$\chi = -\frac{1}{\hbar v} \int_{-\infty}^{\infty} V(\mathbf{r}, \mathbf{R})\, dz, \qquad \mathbf{K} = \mathbf{k} - \mathbf{k}'. \qquad (197)$$

$v = \hbar k/m_e$ is the velocity of the incident electron, $\mathbf{b}$ is the projection of $\mathbf{r}$ onto a plane perpendicular to the z axis, while the latter is chosen along the bisector of the two wave vectors $\mathbf{k}$ and $\mathbf{k}'$. Among the earliest applications of the approximation to the electron–molecule collisions, calculated vibrational excitation cross section[149] and the differential cross section for electron scattering from strongly polar molecules[150] gave promising results. The Glauber approximation was originally a high-energy approximation. Nevertheless, it gives fairly good agreement with experiments in the intermediate energy range and sometimes produces quite acceptable results even in low-energy region. Further applications of the adiabatic, Glauber, and other approximation methods to electron–molecule collisions and their consequences will be found in Chapters 2, 3, and 7.

In this introductory chapter, I have discussed early developments (mainly those before 1970) in the theoretical studies of the electron–molecule collisions, supplemented with some new data. More advanced theoretical approaches, which have been excluded from this chapter, are discussed in other chapters of this book. Finally, some of the general references on the electron–molecule collisions published in recent years are included in the list of references.[151–156]

Acknowledgments

I would like to thank Drs. Yukikazu Itikawa and Kunizo Onda for helpful comments on the preliminary manuscript of this chapter.

References and Notes

1. See, e.g., S.L. Lin, R.E. Robson, E.A. Mason, Moment theory of electron drift and diffusion in neutral gases in an electrostatic field, *J. Chem. Phys.* **71**, 3483–3498 (1979).
2. L.G.H. Huxley and R.W. Crompton, *The Diffusion and Drift of Electrons in Gases,* John Wiley & Sons, New York (1974).
3. D.E. Golden, H.W. Bandel, and J.A. Salerno, Absolute total electron scattering cross sections in H_2 and D_2 for low electron energies, *Phys. Rev.* **146**, 40–42 (1966).
4. F. Linder and H. Schmidt, Rotational and vibrational excitation of H_2 by slow electron impact, *Z. Naturforsch.* **26a**, 1603–1617 (1971).
5. S.K. Srivastava, A. Chutjian, and S. Trajmar, Absolute elastic differential electron scattering cross sections in the intermediate energy region, I. H_2, *J. Chem. Phys.* **63**, 2659–2665 (1975).
6. M. Hayashi, Recommended values of transport cross sections for elastic collision and total collision cross sections for electrons in atomic and molecular gases, IPPJ-AM-19 (Institute of Plasma Physics, Nagoya Univeristy) (1981). The recommended momentum-transfer cross section for hydrogen has some error. In Fig. 1, revised data provided by the author have been used.
7. R.W. Crompton, D.K. Gibson, and A.I. McIntosh, The cross section for the $J = 0\text{–}2$ rotational excitation of hydrogen by slow electrons, *Aust. J. Phys.* **22**, 715–731 (1969).
8. D.K. Gibson, The cross section for rotational excitation of H_2 and D_2 by low energy electrons, *Aust. J. Phys.* **23**, 683–696 (1970).
9. R.W. Crompton, D.K. Gibson, and A.G. Robertson, Vibrational excitation of H_2 by low-energy electrons, *Phys. Rev. A* **2**, 1386–1395 (1970).

10. H. Ehrhardt, L. Langhans, F. Linder, and H.S. Taylor, Resonance scattering of slow electrons from H_2 and CO angular distributions, *Phys. Rev.* **173**, 222–230 (1968).
11. Sunggi Chung and Chun C. Lin, Application of the close-coupling method to excitation of electronic states and dissociation of H_2 by electron impact, *Phys. Rev. A* **17**, 1874–1891 (1978).
12. S.K. Srivastava and S. Jensen, Experimental differential and integral electron impact cross sections for the $B^1\Sigma_u^+$ state of H_2 in the intermediate energy region, *J. Phys. B* **10**, 3341–3346 (1977).
13. K.J. Miller and M. Krauss, Born inelastic differential cross sections in H_2, *J. Chem. Phys.* **47**, 3754–3762 (1967).
14. D. Rapp and P. Englander-Golden, Total cross sections for ionization and attachment in gases by electron impact. I. Positive ionization, *J. Chem. Phys.* **43**, 1464–1479 (1965).
15. D.A. Vroom and F.J. de Heer, Production of excited atoms by impact of fast electrons on molecular hydrogen and deuterium, *J. Chem. Phys.* **50**, 580–590 (1969).
16. D. Rapp, P. Englander-Golden, and D.D. Briglia, Cross sections for dissociative ionization of molecules by electron impact, *J. Chem. Phys.* **42**, 4081–4085 (1965).
17. D. Rapp, T.E. Sharp, and D.D. Briglia, Large isotope effect in the formation of H^- or D^- by electron impact on H_2, HD, and D_2, *Phys. Rev. Lett.* **14**, 533–535 (1965).
18. G.J. Schulz and R.K. Asundi, Isotope effect in the dissociative attachment in H_2 at low energy, *Phys. Rev.* **158**, 25–29 (1967).
19. C.B. Opal, W.K. Peterson, and E.C. Beaty, Measurements of secondary-electron spectra produced by electron impact ionization of a number of simple gases, *J. Chem. Phys.* **55**, 4100–4106 (1971).
20. R.E. Kennerly, Absolute total electron scattering cross sections for N_2 between 0.5 and 50 eV, *Phys. Rev. A* **21**, 1876–1883 (1980).
21. H.J. Blaauw, R.W. Wagenaar, D.H. Barends, and F.J. de Heer, Total cross sections for electron scattering from N_2 and He, *J. Phys. B* **13**, 359–376 (1980).
22. T.W. Shyn, R.S. Stolarski, and G.R. Carignan, Angular distribution of electrons elastically scattered from N_2, *Phys. Rev. A* **6**, 1002–1012 (1975).
23. T.G. Finn and J.P. Doering, Elastic scattering of 13 to 100 eV electrons from N_2, *J. Chem. Phys.* **63**, 4399–4404 (1975).
24. J.P. Bromberg, Absolute differential cross section of elastically scattered electrons. I. He, N_2, and CO at 500 eV, *J. Chem. Phys.* **50**, 3906–3921 (1969).
25. N. Chandra and A. Temkin, Hybrid theory calculation of simultaneous vibration-rotation in electron-N_2 scattering, *Phys. Rev. A* **14**, 507–511 (1976); Tabulation of hybrid theory calculated e-N_2 vibrational and rotational cross sections, *NASA Technical Note* TN D-8347 (1976).
26. B.D. Buckley and P.G. Burke, The scattering of low-energy electrons by diatomic molecules, *J. Phys. B* **10**, 725–739 (1977).
27. A.W. Fliflet, D.A. Levin, M. Ma, and V. McKoy, Discrete-basis-set calculation for e-N_2 scattering cross sections in the static-exchange approximation, *Phys. Rev. A* **17**, 160–169 (1978).
28. G.J. Schulz, Vibrational excitation of N_2, CO, and H_2 by electron impact, *Phys. Rev.* **135**, 988A–994A (1964).
29. D.C. Cartwright, S. Trajmar, A. Chutjian, and W. Williams, Electron impact excitation of the electronic states of N_2. II. Integral cross sections at incident energies from 10 to 50 eV, *Phys. Rev. A* **16**, 1041–1051 (1977).
30. S. Halas and B. Adamczyk, Cross sections for the production of N_2^+, N^+, and N_2^{2+} from

nitrogen by electrons in the energy range 16–600 eV, *Int. J. Mass Spectrom. Ion Phys.* **10**, 157–160 (1972/3).
31. W.L. Borst and E.C. Zipf, Cross section for electron-impact excitation of the (0, 0) first negative band of N_2^+ from threshold to 3 keV, *Phys. Rev. A* **1**, 834–840 (1970).
32. K. Takayanagi, Airglow excitation by precipitating low-energy electrons (in Japanese), *Bull. Inst. Space Aeronaut. Sci.* **10**, 235–246 (1974).
33. M.A. Morrison and P.J. Hay, Ab initio adiabatic polarization potentials for electron–molecule and positron–molecule collisions: The e-N_2 and e-CO_2 systems, *Phys. Rev. A* **20**, 740–748 (1979).
34. R.A. Eades, D.G. Truhlar, and D.A. Dixon, *Ab initio* self-consistent-field polarizabilities and electron–molecule adiabatic polarization potentials. III. N_2, *Phys. Rev. A* **20**, 867–878 (1979).
35. K. Onda, Electronic polarization of atoms in charged-particle impact, *Inst. Space Aeronaut. Sci. Univ. Tokyo. Rep.* **36**, 343–366 (1971).
36. K. Onda and D.G. Truhlar, New approaches to the quantum-mechanical treatment of charge polarization in intermediate-energy electron scattering, *Phys. Rev. A* **22**, 86–100 (1980).
37. S. Hayashi, Theoretical and experimental studies of elastic scattering of electrons by molecules, thesis, University of Tokyo (1975).
38. S. Hayashi and K. Kuchitsu, A quasi-adiabatic polarization potential for elastic electron–atom scattering. Helium and other atoms, *Chem. Phys. Lett.* **44**, 1–4 (1976).
39. S. Hayashi and K. Kuchitsu, A quasi-adiabatic polarization potential for elastic electron–atom scattering. I. General theory, *J. Phys. Soc. Jpn.* **42**, 621–627 (1977).
40. J.C. Slater, A simplification of the Hartree-Fock method, *Phys. Rev.* **81**, 385–390 (1951).
41. J.C. Slater, *Quantum Theory of Atomic Structure II,* Appendix 22, McGraw-Hill, New York (1960).
42. S. Hara, The scattering of slow electrons by hydrogen molecules, *J. Phys. Soc. Jpn.* **22**, 710–718 (1967).
43. M.H. Mittleman and K.M. Watson, Effects of the Pauli principle on the scattering of high-energy electrons by atoms, *Ann. Phys. (N.Y.)* **10**, 268–279 (1960).
44. J.B. Furness and I.E. McCarthy, Semiphenomenological optical model for electron scattering on atoms, *J. Phys. B* **6**, 2280–2291 (1973).
45. M.E. Riley and D.G. Truhlar, Approximations for the exchange potential in electron scattering, *J. Chem. Phys.* **63**, 2182–2191 (1975).
46. M.E. Riley and D.G. Truhlar, Effects of the Pauli principle on electron scattering by open-shell targets, *J. Chem. Phys.* **65**, 792–800 (1976).
47. B.H. Bransden, M.R.C. McDowell, C.J. Noble, and T. Scott, Equivalent exchange potentials in electron scattering, *J. Phys. B* **9**, 1301–1317 (1976).
48. K. Onda and D.G. Truhlar, Comparison of local-exchange approximations for intermediate-energy electron–molecule differential cross sections, *J. Chem. Phys.* **72**, 1415–1417 (1980).
49. N.F. Mott and H.S.W. Massey, *The Theory of Atomic Collisions,* Third Edition, Oxford University Press, New York (1965).
50. H.S.W. Massey and E.C. Bullard, The scattering of electrons by nitrogen molecules, *Proc. Cambridge Phil. Soc.* **29**, 511–521 (1933).
51. H.S.W. Massey, *Electronic and Ionic Impact Phenomena,* Vol. II, Chap. 10, Clarendon Press, Oxford (1969).
52. D. Herrmann, K. Jost, J. Kessler, and M. Fink, Differential cross sections for elastic electron scattering. II. Charge cloud polarization in N_2, *J. Chem. Phys.* **64**, 1–5 (1976).

53. S. Hayashi and K. Kuchitsu, Elastic scattering of electrons by molecules at intermediate energies. Calculation of double scattering effects in N_2 and P_4, *Chem. Phys. Lett.* **41**, 575–579 (1976).
54. E.U. Condon and G.H. Shortley, *The Theory of Atomic Spectra*, Cambridge University Press, Cambridge (1963).
55. G. Racah, Theory of complex spectra II, *Phys. Rev.* **62**, 438–462 (1942).
56. M. Kotani, An acoustical problem relating to the theory of Rayleigh disc, *Proc. Phys.-Math. Soc. Jpn.* **15**, 30–57 (1933).
57. C. Flammer, *Spheroidal Wave Functions*, Stanford University Press, Stanford (1957).
58. L.J. Chu and J.A. Stratton, Elliptic and spheroidal wave functions, *J. Math. Phys.* **20**, 259–309 (1941).
59. K. Takayanagi, Scattering of slow electrons by molecules, *Prog. Theor. Phys. Suppl.* **40**, 216–248 (1967).
60. H.C. Stier, Zur Deutung des Ramsauer-Effektes bei symmetrischen, zweiatomigen Molekülen, *Z. Phys.* **76**, 439–470 (1932).
61. J.B. Fisk, Theory of the scattering of slow electrons by diatomic molecules, *Phys. Rev.* **49**, 167–173 (1936).
62. J.B. Fisk, On the cross sections of Cl_2 and N_2 for slow electrons, *Phys. Rev.* **51**, 25–28 (1937).
63. S. Nagahara, The scattering of slow electrons by the diatomic molecules. II. Elastic scattering by the hydrogen molecules, *J. Phys. Soc. Jpn.* **9**, 52–55 (1954).
64. S. Hara, A two-center approach in the low energy electron-H_2 scattering, *J. Phys. Soc. Jpn.* **27**, 1009–1019 (1969).
65. M. Shimizu, Two-center Coulomb potential approximation, *J. Phys. Soc. Jpn.* **18**, 811–819 (1963).
66. K. Takayanagi and Y. Itikawa, Scattering of slow electrons by polar molecules, *J. Phys. Soc. Jpn.* **24**, 160–168 (1968).
67. H. Takagi and H. Nakamura, Elastic scattering of electrons from H_2^+: phaseshifts, quantum defects and two-electron excited states, *J. Phys. B* **13**, 2619–2632 (1980).
68. K. Onda, Theory of low-energy electron scattering by polyatomic molecules, I., *J. Phys. Soc. Jpn.* **36**, 826–838 (1974).
69. D. Dill and J.L. Dehmer, Electron–molecule scattering and molecular photoionization using the multiple-scattering method, *J. Chem. Phys.* **61**, 692–699 (1974).
70. J. Siegel, J.L. Dehmer, and D. Dill, Elastic-electron-scattering cross sections for N_2 from 0 to 1000 eV. Energy-dependent exchange potentials, *Phys. Rev. A* **21**, 85–94 (1980).
71. D. Dill and J.L. Dehmer, Total elastic electron scattering cross section for N_2 between 0 and 1000 eV, *Phys. Rev. A* **16**, 1423–1431 (1977).
72. J. Siegel, D. Dill, and J.L. Dehmer, Differential elastic electron scattering cross sections for N_2 from 0 to 30 eV, *Phys. Rev. A* **17**, 2106–2109 (1978).
73. J.L. Dehmer, J. Siegel, and D. Dill, Shape resonances in e-SF_6 scattering, *J. Chem. Phys.* **69**, 5205–5206 (1978).
74. M.G. Lynch, D. Dill, J. Siegel, and J.L. Dehmer, Elastic electron scattering by CO_2, COS, and CS_2 from 0 to 100 eV, *J. Chem. Phys.* **71**, 4249–4254 (1979).
75. J.L. Dehmer, J. Siegel, J. Welch, and D. Dill, Origin of enhanced vibrational excitation in N_2 by electron impact in the 15–35 eV region, *Phys. Rev. A* **21**, 101–104 (1980).
76. P.M. Morse, Excitation of molecular rotation-vibration by electron impact, *Phys. Rev.* **90**, 51–55 (1953).
77. T.R. Carson, The vibrational and rotational excitation of molecular hydrogen by electron impact, *Proc. Phys. Soc. A* **67**, 909–916 (1954).

78. E. Gerjuoy and S. Stein, Rotational excitation by slow electrons, *Phys. Rev.* **97**, 1671–1679 (1955).
79. E. Gerjuoy and S. Stein, Rotational excitation by slow electrons, II. *Phys. Rev.* **98**, 1848–1851 (1955).
80. A. Temkin and F.H.M. Faisal, Adiabatic theory of rotational excitation of non-Σ states, *Phys. Rev. A* **3**, 520–521 (1971).
81. S. Geltman and K. Takayanagi, Excitation of molecular rotation by slow electrons, II, *Phys. Rev.* **143**, 25–30 (1966).
82. K. Takayanagi, Excitation of molecular rotation and vibration by low energy electron impact, JILA Report No. 11 (1964); K. Takayanagi, Behavior of slow electrons in the atmospheric gases, Part I, *Rep. Ionos. Space Res. Jpn* **19**, 1–15 (1965).
83. A. Dalgarno and R.J. Moffett, The rotational excitation of molecular nitrogen by slow electrons, *Proc. Natl. Acad. Sci. India A* **33**, Part 4, 511–521 (1963).
84. H.S.W. Massey, The collision of electrons with rotating dipoles, *Proc. Cambridge Phil. Soc.* **28**, 99–105 (1931).
85. K. Takayanagi, Rotational and vibrational excitation of polar molecules by slow electrons, *J. Phys. Soc. Jpn.* **21**, 507–514 (1966).
86. M. Inokuti, On the evaluation of the Born partial-wave amplitude for electron scattering by any electrostatic multiple field, *J. Phys. B* **13**, 1221–1227 (1980).
87. Y. Itikawa and K. Takayanagi, Rotational transition in polar molecules by electron collisions: Applications to CN and HCl, *J. Phys. Soc. Jpn.* **26**, 1254–1264 (1969).
88. A. Altshuler, Theory of low-energy electron scattering by polar molecules, *Phys. Rev.* **107**, 114–117 (1957).
89. W.R. Garrett, Low-energy electron scattering by polar molecules, *Mol. Phys.* **24**, 465–487 (1972).
90. K. Takayanagi, Low-energy electron–molecule scattering, *Comments At. Mol. Phys.* **3**, 95–106 (1972).
91. O.H. Crawford, Scattering of low-energy electrons from polar molecules, *J. Chem. Phys.* **47**, 1100–1104 (1967).
92. Y. Itikawa, Electron-impact rotational excitation of a symmetric-top molecule: Application to NH_3, *J. Phys. Soc. Jpn.* **30**, 835–842 (1971).
93. Y. Itikawa, Effects of the polarization force on the rotational transition in polyatomic molecules by electron collision, *J. Phys. Soc. Jpn.* **31**, 1532–1535 (1971).
94. Y. Itikawa, Rotational transition in an asymmetric-top molecule by electron collision: Applications to H_2O and H_2CO, *J. Phys. Soc. Jpn.* **32**, 217–226 (1972).
95. G.J. Schulz, Measurement of excitation of N_2, CO, and He by electron impact, *Phys. Rev.* **116**, 1141–1147 (1959).
96. J.C.Y. Chen and J.L. Magee, Excitation of molecular vibration by slow electron impact, *J. Chem. Phys.* **36**, 1407–1411 (1962).
97. H.S.W. Massey, The excitation of molecular vibration by impact of slow electrons, *Trans. Faraday Soc.* **31**, 556–563 (1935).
98. Ta-You Wu, Excitation of molecular vibration by electrons, *Phys. Rev.* **71**, 111–118 (1947).
99. A.G. Engelhardt and A.V. Phelps, Elastic and inelastic collision cross sections in hydrogen and deuterium from transport coefficients, *Phys. Rev.* **131**, 2115–2128 (1963).
100. H. Ehrhardt and F. Linder, Rotational excitation of H_2 by slow electrons in a beam experiment, *Phys. Rev. Lett.* **21**, 419–421 (1968).
101. K. Takayanagi, Vibrational excitation of hydrogen molecule by slow electrons, *J. Phys.*

Soc. Jpn. **20**, 562–565 (1965); see also, K. Takayanagi, Excitation of molecular vibration by slow electrons, *J. Phys. Soc. Jpn* **20**, 2297 (1965).
102. E.L. Breig and C.C. Lin, Vibrational excitation of diatomic molecules by electron impact, *J. Chem. Phys.* **43**, 3839–3845 (1965).
103. J.N. Bardsley, A. Herzenberg, and F. Mandl, Vibrational excitation and dissociative attachment in the scattering of electrons by hydrogen molecules, *Proc. Phys. Soc.* **89**, 321–340 (1966).
104. R.J.W. Henry and E.S. Chang, Rotational-vibrational excitation of H_2 by slow electrons, *Phys. Rev. A* **5**, 276–284 (1972).
105. F.H.M. Faisal and A. Temkin, Application of the adiabatic-nuclei theory to vibrational excitation, *Phys. Rev. Lett.* **28**, 203–206 (1972).
106. Y. Itikawa, The Born cross section for vibrational excitation of a polyatomic molecule by electron collisions, *J. Phys. Soc. Jpn.* **36**, 1121–1126 (1974).
107. Y. Itikawa, Electron-impact vibrational excitation of H_2O, *J. Phys. Soc. Jpn.* **36**, 1127–1132 (1974).
108. A. Chutjian and H. Tanaka, Electron impact cross sections for $v = 0 \rightarrow 1$ vibrational excitation in CO at electron energies of 3 to 100 eV, *J. Phys. B.* **13**, 1901–1908 (1980).
109. R.D. Hake and A.V. Phelps, Momentum-transfer and inelastic-collision cross sections for electrons in O_2, CO, and CO_2, *Phys. Rev.* **158**, 70–84 (1967).
110. D.F. Register, H. Nishimura, and S. Trajmar, Elastic scattering and vibrational excitation of CO_2 by 4, 10, 20 and 50 eV electrons, *J. Phys. B* **13**, 1651–1662 (1980).
111. B.R. Bulos and A.V. Phelps, Excitation of the 4.3 μm bands of CO_2 by low-energy electrons, *Phys. Rev. A* **14**, 615–629 (1976).
112. M. Gryzinsky, Classical theory of electronic and ionic inelastic collisions, *Phys. Rev.* **115**, 374–383 (1959).
113. E. Bauer and C.D. Bartky, Calculation of inelastic electron–molecule collision cross sections by classical methods, *J. Chem. Phys.* **43**, 2466–2476 (1965).
114. V.I. Ochkur, The Born–Oppenheimer method in the theory of atomic collision, *Zh. Eksp. Teor. Fiz. USSR* **45**, 734–741 (1963); *Sov. Phys. JETP* **18**, 503–508 (1964).
115. M.R.H. Rudge, The calculation of exchange scattering amplitudes, *Proc. Phys. Soc.* **85**, 607–608 (1965).
116. K. Bell and B.L. Moiseiwitsch, First-order exchange approximation, *Proc. R. Soc. (London) A* **276**, 346–353 (1963).
117. H. Bethe, Zur Theorie des Durchgangs schneller Korpuskularstrahlen durch Materie, *Ann. Phys. (Leipzig)* **5**, 325–400 (1930).
118. J.O. Hirschfelder, W.B. Brown, and S.T. Epstein, Recent developments in perturbation theory, *Adv. Quantum Chem.* **1**, 255–374 (1964).
119. E.N. Lassettre, A. Skerbele, and M.A. Dillon, Generalized oscillator strength for $1^1S \rightarrow 2^1P$ transition of helium. Theory of limiting oscillator strengths, *J. Chem. Phys.* **50**, 1829–1839 (1969).
120. M. Inokuti, Inelastic collisions of fast charged particles with atoms and molecules – The Bethe theory revisited, *Rev. Mod. Phys.* **43**, 297–347 (1971).
121. M. Inokuti, Y. Itikawa, and J.E. Turner, Addenda: Inelastic collisions of fast charged particles with atoms and molecules–The Bethe theory revisited [*Rev. Mod. Phys.* **43**, 297 (1971)], *Rev. Mod. Phys.* **50**, 23–35 (1978).
122. A.W. Fliflet and V. McKoy, Distorted-wave-approximation cross-sections for excitation of the $b^3\Sigma_u^+$ and $B^1\Sigma_u^+$ states of H_2 by low-energy electron impact, *Phys. Rev. A* **21**, 1863–1875 (1980).

123. A.U. Hazi, Impact-parameter method for electronic excitation of molecules by electron impact, *Phys. Rev. A* **23,** 2232–2240 (1981).
124. J.R. Oppenheimer, On the quantum theory of electron impact, *Phys. Rev.* **32,** 361–376 (1928).
125. D.R. Bates, A. Fundaminsky, J.W. Leech, and H.S.W. Massey, Excitation and ionization of atoms by electron impact–The Born and Oppenheimer approximations, *Phil. Trans. R. Soc.* **A243,** 93–143 (1950).
126. R.A. Bonham, Inelastic scattering from atoms at medium energies, I. Bound states, *J. Chem. Phys.* **36,** 3260–3269 (1962).
127. S.P. Khare, Excitation of hydrogen molecules by electron impact, *Phys. Rev.* **149,** 33–37 (1966).
128. S.P. Khare, Excitation of hydrogen molecules by electron impact, II. Excitation to $D(3p\pi\,^1\Pi_u)$ state, *Phys. Rev.* **152,** 74–75 (1966).
129. S.P. Khare, Excitation of hydrogen molecules by electron impact, III. Singlet–triplet excitations, *Phys. Rev.* **157,** 107–112 (1967).
130. D.C. Cartwright and A. Kuppermann, Electron-impact excitation cross section for the two lowest triplet states of molecular hydrogen, *Phys. Rev.* **163,** 86–102 (1967).
131. R.C. Stabler, Rotational excitation of molecular ions by slow electrons, *Phys. Rev.* **131,** 679–683 (1963).
132. R.F. Boikova and V.D. Ob'edkov, Rotational and vibrational excitation of molecular ions by electrons, *Zh. Eksp. Teor. Fiz.* **54,** 1439–1444 (1968); *Sov. Phys. JETP* **27,** 772–774 (1968).
133. D.M. Chase, Adiabatic approximation for scattering processes, *Phys. Rev.* **104,** 838–842 (1956).
134. Yu.D. Oksyuk, Excitation of the rotational levels of diatomic molecules by electron impact in the adiabatic approximation, *Zh. Eksp. Teor. Fiz.* **49,** 1261–1273 (1965); *Sov. Phys. JETP* **22,** 873–881 (1966).
135. S. Hara, Rotational excitation of H_2 by slow electrons, *J. Phys. Soc. Jpn.* **27,** 1592–1597 (1969).
136. R.J.W. Henry and N.F. Lane, Polarization and exchange effects in low-energy electron-H_2 scattering, *Phys. Rev.* **183,** 221–231 (1969).
137. K. Takayanagi and Y. Itikawa, The rotational excitation of molecules by slow electrons, *Adv. At. Mol. Phys.* **6,** 105–153 (1970).
138. K. Takayanagi, Rotational excitation of the HD molecule by low-energy electron collision, *J. Phys. Soc. Jpn.* **28,** 1527–1535 (1970).
139. S. Hara, Rotational excitation of the HD molecule by slow electrons, *J. Phys. Soc. Jpn.* **30,** 819–823 (1971).
140. R.J.W. Henry and N.F. Lane, Polarization and exchange effects in low-energy electron–D_2 scattering, *Phys. Rev. A* **4,** 410–411 (1971).
141. K. Takayanagi, Low energy electron–molecule scattering, *Atomic Physics,* Vol. 4 (G. zu Putlitz, E.W. Weber, and A. Winnacker, eds.), Plenum Press, New York, (1975), pp. 435–447.
142. R.J.W. Henry and E.S. Chang, Rotational-vibrational excitation of H_2 by slow electrons, *Phys. Rev. A* **5,** 276–284 (1972).
143. R.J.W. Henry, Vibrational and rotational excitation of molecular hydrogen by electron impact, *Phys. Rev. A* **2,** 1349–1358 (1970).
144. F.H.M. Faisal and A. Temkin, Application of the adiabatic-nuclei theory to vibrational excitation, *Phys. Rev. Lett.* **28,** 203–206 (1972).
145. N. Chandra and A. Temkin, Hybrid theory and calculation of e-N_2 scattering, *Phys. Rev. A* **13,** 188–203 (1976).

146. E.S. Chang and U. Fano, Theory of electron–molecule collisions by frame transformations, *Phys. Rev. A* **6**, 173–185 (1972).
147. R.K. Nesbet, Energy-modified adiabatic approximation for scattering theory, *Phys. Rev. A* **19**, 551–556 (1979).
148. R.J. Glauber, High energy collision theory, in *Lectures in Theoretical Physics* (W.E. Brittin and L.G. Dunham, eds.), Interscience, New York (1959), Vol. 1, pp. 315–414.
149. T.N. Chang, R.T. Poe, and P. Ray, Glauber-theory approach for molecular vibrational excitations, *Phys. Rev. Lett.* **31**, 1097–1099 (1973).
150. K. Takayanagi, Low-energy electron scattering from strongly polar molecules, *Prog. Theor. Phys.* **52**, 337–338 (1974).
151. K. Takayanagi, The theory of low energy electron-molecule scattering, in *The Physics of Electronic and Atomic Collisions,* IX International Conference on the Physics of Electronic and Atomic Collisions, (J.S. Risley and R. Geballe, eds.) University of Washington Press, Seattle (1976), pp. 219–230.
152. Y. Itikawa, Electron scattering by polar molecules, *Phys. Rep.* **46**, 117–164 (1978).
153. *Symposium on Electron–Molecule Collisions,* (I. Shimamura and M. Matsuzawa, eds.), University of Tokyo, (1979).
154. Th. Rescigno, V. McKoy, and B. Schneider, eds., *Electron–Molecule and Photon–Molecule Collisions,* Plenum Press, New York (1979).
155. N.F. Lane, The theory of electron–molecule collisions, *Rev. Mod. Phys.* **52**, 29–119 (1980).
156. P.G. Burke, Theory of low energy electron–molecule collisions, *Adv. At. Mol. Phys.* **15**, 471–506 (1979).

[illegible]

2

ROTATIONAL EXCITATION OF MOLECULES BY SLOW ELECTRONS

ISAO SHIMAMURA

1. Introduction

This chapter covers the change in the rotational motion of molecules in collisions with electrons. Other degrees of freedom of the molecular motion may or may not change simultaneously with the rotational motion. The main emphasis will be on collisions at a few electron volts or at lower energies. The discussion, however, will sometimes include higher energies, or will sometimes apply to collisions at any energies. Some qualitative results are applicable even to rotational transitions caused by a wide class of mechanisms including collisions with heavy particles and perturbations by external fields.

The change in the rotational motion appears, in principle, as structures in the cross section plotted versus the energy loss by the scattered electron. Unfortunately, these structures are usually too fine to resolve in energy-loss measurements. Then, one obtains only an averaged, overall information on the rotational transitions. Techniques have recently been developed for distinguishing between molecules in different rotational states by use of lasers. There have been no reports, however, on applications of these techniques to electron–molecule collisions until the time of this writing.

Accurate *ab initio* computations taking full account of the rotational channels are possible for the smallest molecules, but are also difficult; usually the cross sections computed *ab initio* are either rotationally unresolved or for excitation from the ground rotational state only. In this sense reliable numerical information on the rotational transitions is far from sufficient even for the smallest molecules. Theoretical considerations together with the limited numerical information, however, have revealed various interesting aspects of the rotational response of the molecules to the incident electrons. Chapter 1 has briefly introduced many of these aspects. Chapter 7 will review computational techniques for determining the cross sections, and Chapter 6 will discuss experimental techniques. The present chapter mainly aims at elucidation of the

ISAO SHIMAMURA • RIKEN, The Institute of Physical and Chemical Research, Hirosawa 2–1, Wako-shi, Saitama 351, Japan.

physical concepts that are useful for understanding rotational-transition processes.

One may consider two classes of factors that determine the rotational-transition cross sections, namely, (a) factors that depend on the transition mechanism and on the molecular species, and (b) factors that are independent of the dynamics of the transition process. By clearly separating these two classes one can, on the one hand, derive general rules of systematics in the transition cross sections for any system and, on the other hand, study efficiently the collision dynamics of each particular system. Sections 2 and 3 discuss the dynamics-independent part of the collision physics, and Sections 4–6 deal with the theoretical basis for unraveling the collision dynamics. To help the beginner Sections 2 and 3 include an elementary account of the molecular rotational states and of the theory of angular-momentum coupling (or the Clebsch–Gordan coefficients), because they are essential for really understanding the physics of rotational transitions.

Review articles published recently (Refs. 1–8) and older reviews cited therein (in Ref. 4, in particular) complement the present chapter.

2. Classical Sudden Collisions and Angular-Momentum Transfer

2.1. Linear Rotators

Classical mechanics applies to the dynamics of electron–molecule collisions only in cases of a limited kind. Nevertheless, it is a good starting point of our discussion for the following reason. The concept of the transfer of angular momentum is quite general. It leads to important consequences that hold true irrespective of the dynamics of the rotational transitions. Many of these consequences are essentially common to both classical and quantum mechanics. The study of the change in the internal motion of a classical rotator due to a classical collision affords a clear physical insight into the quantum-mechanical rotational transitions to be studied later in this chapter.

2.1.1. Sudden Collision Involving a Linear Rotator

For defining the rotational-transition processes we need to define the rotational states of a molecule in the absence of an external perturber. We regard the molecule as a rigid rotator, and disregard any internal rotation.

A classical linear rigid rotator having the moment of inertia I and rotating with the angular velocity ω_r has the rotational angular momentum $\hbar\mathbf{J}$ with the magnitude $\hbar J = I\omega_r$. Here, $\hbar$ is the Planck constant divided by 2π, and J may take any positive value or zero. The rotational energy is

$$E_r = I\omega_r^2/2 = (\hbar^2/2I)J^2 = BJ^2. \tag{1}$$

Usually, the notation B is used for the frequency $\hbar/4\pi I$ in the quantum mechanical counterpart of Eq. (1). In this chapter, however, B denotes the energy $\hbar^2/2I$ for simplicity of notation. The angular momentum vector $\hbar\mathbf{J}$ may have any direction in space. The projection $J_z = M$ of $\mathbf{J}$ onto a z axis fixed in space may take any value that satisfies $-J \leqq M \leqq J$. All states with a common value of $\hbar J$ are degenerate in the rotational energy.

In a collision with the rotator a particle may transfer an angular momentum $\hbar\mathbf{J}_t$ to the rotator. Then the angular momentum of the final state of the rotator is

$$\hbar\mathbf{J}' = \hbar\mathbf{J} + \hbar\mathbf{J}_t, \tag{2}$$

as is shown in Fig. 1. For a specified value of J and for a specified value of J', the angular momentum transfer J_t (in units of $\hbar$) runs over the region

$$|J-J'| \leqq J_t \leqq J+J'. \tag{3}$$

In other words, all J_t that satisfy Eq. (3) contribute, in general, to the classical cross section $d\sigma(J\to J')/d\omega$ for the rotational transition $J\to J'$. $[d\sigma(J\to J')/d\omega]dJ'$ is the cross section for the rotator that initially has the angular momentum of magnitude $\hbar J$ to be found with its angular momentum of magnitude between $\hbar J'$ and $\hbar(J'+dJ')$ after the collision; ω is the solid angle of scattering.

Often the collision time is much shorter than the rotational period $2\pi/\omega_r$ of the rotator. Then, the sudden approximation is valid for the collision as compared with the rotational motion. In this approximation the rotator is (almost) stationary during the collision, i.e., the direction Ω of the axis of the rotator is (almost) fixed during the collision. The cross section $d\sigma_{\mathrm{AMT}}(J_t;\Omega)/d\omega$ for the angular-momentum transfer $\hbar J_t$ to the (almost) fixed rotator is independent of the initial angular momentum $\hbar J$, and hence, is equal to the cross section $d\sigma(0\to J_t;\Omega)/d\omega$ for excitation of the rotator initially at rest. Since both $\mathbf{J}$ and $\mathbf{J}'$ are perpendicular to the axis of the rotator, the three vectors in Fig. 1 are in a plane

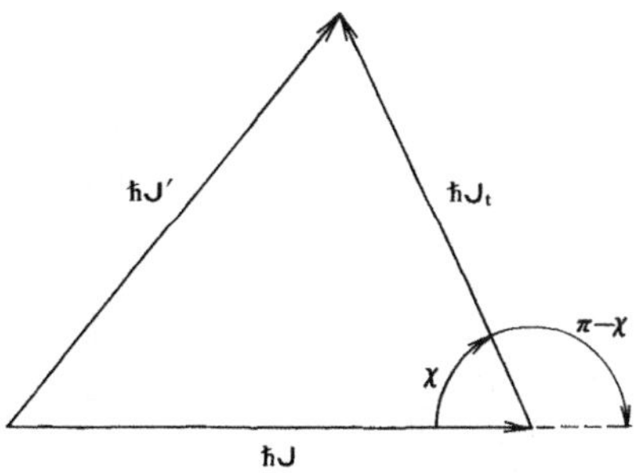

FIGURE 1
Transfer of an angular momentum $\hbar\mathbf{J}_t$ to a rotator with a rotational angular momentum $\hbar\mathbf{J}$.

fixed in space, and the angle $\pi - \chi$ between **J** and $\mathbf{J}_t$ uniquely defines the direction of **J** for a fixed $\mathbf{J}_t$ (Fig. 1). Naturally, all angles χ are equally probable.

Let $\{C(JJ_tJ'; 000)\}^2 dJ'$ be the probability of finding a resultant angular-momentum vector of length between $\hbar J'$ and $\hbar(J' + dJ')$, when adding two angular-momentum vectors (in a plane fixed in space) of lengths $\hbar J$ and $\hbar J_t$. (The meaning of the three zeros will become clear in Section 2.2.2.) The probability density $\{C(JJ_tJ';000)\}^2$ is zero outside the interval (3). The normalization condition satisfied by the probability reads

$$\int_{|J-J_t|}^{J+J_t} \{C(JJ_tJ'; 000)\}^2 \, dJ' = 1. \tag{4}$$

The cross section for a fixed Ω for the rotational transition $J \to J'$ is the superposition of $d\sigma_{\mathrm{AMT}}(J_t;\Omega)/d\omega$ $[= d\sigma(0 \to J_t;\Omega)/d\omega]$ weighted by the probability density $\{C(JJ_tJ';000)\}^2$. Thus we obtain, after taking an average over all Ω,

$$\frac{d\sigma(J \to J')}{d\omega} = \int_{|J-J'|}^{J+J'} dJ_t \{C(JJ_tJ';000)\}^2 \frac{d\sigma(0 \to J_t)}{d\omega}. \tag{5}$$

The cross section for a general rotational transition is a superposition of the cross sections for rotational transitions of the rotator at rest. The whole dynamics of the collision is contained in the angular-momentum-transfer cross section $d\sigma(0 \to J_t)/d\omega$, and not in the weight $\{C(JJ_tJ';000)\}^2$. The weight is determined solely by the property of the angular-momentum addition in general. This illustrates the separation of dynamics-dependent and dynamics-independent factors mentioned in Section 1.

2.1.2. *Sum Rules and High-J Formulas*

The normalization (4) of the probability together with the factorization relation (5) yields the sum rule

$$S(0;J) = \int_0^\infty \frac{d\sigma(J \to J')}{d\omega} dJ' = \int_0^\infty \frac{d\sigma(0 \to J_t)}{d\omega} dJ_t = S(0;0), \tag{6}$$

because the order of integration over J' and J_t is interchangeable.[9] The sum over the final rotational angular momentum $\hbar J'$ is independent of the initial rotational angular momentum $\hbar J$ in the sudden approximation. This result is quite natural; the summed differential cross section $S(0; J)$ is, indeed, the differential cross section (averaged over Ω) for all angular-momentum transfers $\hbar J_t$, and the effect of the initial rotation on it is negligible as long as the rotation is negligibly slow.

We call the square root $C(JJ_tJ';000)$ of the probability density the classical Clebsch–Gordan coefficient for the addition of two angular momenta in a plane fixed in space. From its definition, from the equal probability of χ, and from the second cosine formula

$$J'^2 = J^2 + J_t^2 + 2\mathbf{J}\cdot\mathbf{J}_t = J^2 + J_t^2 - 2JJ_t\cos\chi, \tag{7}$$

it follows that[10, 11]

$$\{C(JJ_tJ';000)\}^2 dJ' = d\chi/\pi = [(\partial\chi/\partial J')/\pi]\,dJ' = g(J')\begin{pmatrix} J & J_t & J' \\ 0 & 0 & 0 \end{pmatrix}^2 dJ', \tag{8a}$$

$$\begin{pmatrix} J_1 & J_2 & J_3 \\ 0 & 0 & 0 \end{pmatrix}^2 = \pi^{-1}[\beta(J_1J_2J_3)]^{-1/2}, \tag{8b}$$

$$\beta(J_1J_2J_3) = [(J_1+J_2)^2 - J_3^2][J_3^2 - (J_1-J_2)^2] = \beta(J_2J_3J_1) = \beta(J_3J_1J_2), \tag{8c}$$

where the factor

$$g(J) = 2J \tag{9}$$

is the density of the state J; recall that $-J \leqq M \leqq J$. We call $\left(\begin{smallmatrix} J_1 & J_2 & J_3 \\ 0 & 0 & 0 \end{smallmatrix}\right)$ the classical three-J (or 3-j) symbol. Its square is symmetric with respect to all the three angular momenta.

Bhattacharyya and Dickinson[9] have derived what is essentially the combination of Eqs. (5), (8), and (9), and have noted the sum rule (6).

Equation (4) follows directly from the integral $\int_0^\pi d\chi/\pi = 1$. Using Eqs. (1) and (7), we may calculate, in a similar way, the expectation value $\Delta E(J;J_t)$ of the energy gain by the rotator due to the angular-momentum transfer $\hbar J_t$:

$$\Delta E(J;J_t) = \int_{|J-J_t|}^{J+J_t} [E_r(J') - E_r(J)]\{C(JJ_tJ';000)\}^2 dJ'$$

$$= \int_0^\pi [E_r(J_t) - 2BJJ_t\cos\chi]\,d\chi/\pi = E_r(J_t). \tag{10}$$

Note that the term involving $\cos\chi$ [the inner product $\mathbf{J}\cdot\mathbf{J}_t$ in Eq. (7)] contributes nothing when averaged over χ. This is the reason why the second cosine formula (7) has led to a result that the average energy transferred to the rotator is the rotational energy corresponding to the angular-momentum transfer $\hbar J_t$.

Collection of the contributions from all possible angular-momentum transfers leads to a second sum rule[12]

$$S(1;J) = \int_0^\infty [E_r(J') - E_r(J)] \frac{d\sigma(J \to J')}{d\omega} dJ' = \int_0^\infty \Delta E(J;J_t) \frac{d\sigma(0 \to J_t)}{d\omega} dJ_t$$

$$= \int_0^\infty [E_r(J_t) - E_r(0)] \frac{d\sigma(0 \to J_t)}{d\omega} dJ_t = S(1;0). \tag{11}$$

In the sudden approximation the mean energy gain $S(1;J)$ by the rotator is independent of the rotational motion before the collision. Because a rotator at rest has no internal energy to lose, it follows that $S(1;0) \geqq 0$, and the mean energy gain $S(1;J)$ by a rotator in any state is positive or zero in the sudden approximation.

The set of cross sections for a fixed J and for all J' defines a continuous spectral distribution as a function of the energy change $E_r(J') - E_r(J)$. The global properties of this distribution are represented by the moments $S(\mu;J)$ defined by replacing $E_r(J') - E_r(J)$ in Eq. (11) by its μth power. We may compute them in a manner similar to the computation of $S(1;J)$ with a result that $S(2n;J)$ and $S(2n+1;J)$ are polynomials in $E_r(J)$ of degree n. The special case of $n = 0$ yields the J-independent sum rules (6) and (11). The moments $S(\mu;J)$ for any $\mu \geqq 0$ are calculable from a recurrence formula.[12]

For large J and for small $|J' - J|$ $(= |\Delta J|)$, simple expressions for the cross sections $d\sigma(J \to J \pm \Delta J)/d\omega$ follow. Expansion of Eqs. (8) in terms of the inverse powers of J gives an asymptotic formula for $J' = J \pm \Delta J$

$$\{C(JJ_tJ';000)\}^2 = \frac{1}{\pi}\{J_t^2 - (\Delta J)^2\}^{-1/2}\left[1 \pm \frac{\Delta J}{2J} + O(J^{-2})\right] \tag{12a}$$

$$= \frac{1}{\pi}\{J_t^2 - (\Delta J)^2\}^{-1/2}\{g(J')/g(J)\}^{1/2}[1 + O(J^{-2})]. \tag{12b}$$

Substitution of this formula into Eq. (5) leads to the form

$$\frac{d\sigma(J \to J \pm \Delta J)}{d\omega} = \frac{d\sigma(|\Delta J|)}{d\omega}\left[1 \pm c_1\frac{\Delta J}{J} + O(J^{-2})\right] \tag{13a}$$

$$= \frac{d\sigma(|\Delta J|)}{d\omega}\{g(J \pm \Delta J)/g(J)\}^{c_1}[1 + O(J^{-2})] \tag{13b}$$

with $c_1 = 1/2$ on the assumption that all J_t contributing appreciably to Eq. (5) satisfy $J_t < J + J'$, so that the upper limit of the integration is effectively independent of J and ΔJ. $d\sigma(|\Delta J|)/d\omega$ is independent of J.

Equation (13) expresses preference of excitation by $|\Delta J|$ to deexcitation

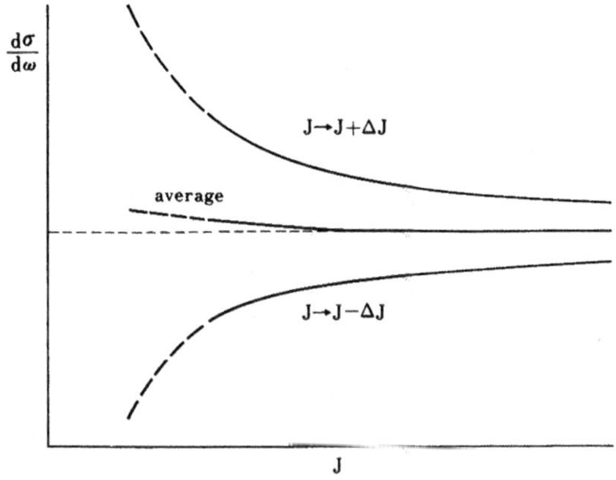

FIGURE 2
Dependence on J (at high J) of the rotational transition cross sections $d\sigma(J \to J \pm \Delta J)/d\omega$ and of their average for a fixed $\Delta J(>0)$.

by $|\Delta J|$ for high J. The two cross sections $d\sigma(J \to J \pm \Delta J)/d\omega$ plotted versus J approach a value $d\sigma(|\Delta J|)/d\omega$ common to both in the limit of high J, one from above and the other from below in a manner nearly symmetric with respect to the line of the constant cross section (Fig. 2). The ratio of excitation to de-excitation is

$$\frac{d\sigma(J \to J + \Delta J)}{d\omega} \Big/ \frac{d\sigma(J \to J - \Delta J)}{d\omega} = 1 + 2c_1 \frac{\Delta J}{J} + O(J^{-2})$$

$$= \left\{ \frac{g(J + \Delta J)}{g(J - \Delta J)} \right\}^{c_1} [1 + O(J^{-2})], \quad (14)$$

and the sum is

$$\frac{d\sigma(J \to J + \Delta J)}{d\omega} + \frac{d\sigma(J \to J - \Delta J)}{d\omega} = 2\frac{d\sigma(|\Delta J|)}{d\omega}[1 + O(J^{-2})]. \quad (15)$$

It follows from Eq. (13) that the elastic cross section $d\sigma(J \to J)/d\omega$ is constant within an error of order J^{-2}, which is consistent with Eq. (15).

2.2. Nonlinear Rotators

The angular momentum transfer to a stationary linear rotator is restricted in a plane fixed in space. This is not the case with nonlinear rotators. To see what modifications to the arguments in Section 2.1 are necessary, we first summarize the classical mechanics of nonlinear rigid rotators in Section 2.2.1,[13] and then generalize the theory of angular momentum transfer in Section 2.2.2.

2.2.1. Classical Mechanics of a Rigid Body

Let I_α be the moment of inertia of a rigid body about an axis α that passes through the center of mass. Plot a point on this axis at the distance $I_\alpha^{-1/2}$ from the center of mass. Such a plot for all possible directions of α defines a three-dimensional surface. We call this surface, which proves to be an ellipsoid, the (moment-of-)inertia ellipsoid.

The inertia ellipsoid has three principal axes a, b, and c, which are mutually perpendicular. The corresponding moments of inertia I_a, I_b, and I_c are the principal moments of inertia. They are usually labeled in increasing order as $I_a \leqq I_b \leqq I_c$. The principal moments of inertia are the three eigenvalues of the (moment-of-)inertia tensor. The orthogonal transformation that diagonalizes this tensor rotates a set of space-fixed coordinate axes to the principal axes of inertia a, b, and c.

Suppose that the rigid body rotates with the angular velocity $\boldsymbol{\omega}$, its projections onto the principal axes being ω_a, ω_b, and ω_c. The rotational angular momentum $\hbar\mathbf{J}$ is defined in terms of its components along the principal axes as $(\hbar J_a, \hbar J_b, \hbar J_c) = (I_a\omega_a, I_b\omega_b, I_c\omega_c)$. The rotational energy takes a form [cf. Eq. (1)]

$$\begin{aligned} E_r &= \frac{1}{2}I_a\omega_a^2 + \frac{1}{2}I_b\omega_b^2 + \frac{1}{2}I_c\omega_c^2 = \frac{\hbar^2}{2I_a}J_a^2 + \frac{\hbar^2}{2I_b}J_b^2 + \frac{\hbar^2}{2I_c}J_c^2 \\ &= AJ_a^2 + BJ_b^2 + CJ_c^2 \qquad (A \geqq B \geqq C). \end{aligned} \tag{16}$$

We call A, B, and C in Eq. (16) [and B in Eq. (1)] the rotational constants. If all of A, B, and C are unequal, the rigid body is an asymmetric top.

If two of them are equal, the rigid body is a symmetric top and has a body-fixed symmetry axis. In this chapter K denotes the projection of $\mathbf{J}$ onto the symmetry axis. A prolate symmetric top, shaped like an American football, has $B = C$ ($I_b = I_c$). The a axis (the axis of the smallest principal moment of inertia) is the symmetry axis. For an oblate symmetric top, shaped like a discus, we have $A = B$ ($I_a = I_b$). In this case the c axis (the axis of the largest principal moment of inertia) is the symmetry axis. The rotational energies of the symmetric tops are

$$E_r = B(J_b^2 + J_c^2) + AJ_a^2 = BJ^2 + (A - B)K^2 \qquad \text{(prolate symmetric top)}, \tag{17a}$$

$$E_r = B(J_a^2 + J_b^2) + CJ_c^2 = BJ^2 - (B - C)K^2 \qquad \text{(oblate symmetric top)}. \tag{17b}$$

A linear rotator is the limit of a prolate symmetric top as $I_a \to 0$. In this limit, $J_a \to 0$ $(K \to 0)$, $A \to \infty$, and $AK^2 \to 0$. Then, the term $(A - B)K^2$ in Eq. (17a) vanishes, and we recover Eq. (1).

A spherical top has three equal principal moments of inertia, viz., $A = B = C$. Thus the rotational energy reduces to the same form $E_r = BJ^2$ as Eq. (1). All states of a spherical top with a common J and with different values of K are degenerate, the degeneracy being $(2J)^2 = \{g(J)\}^2$.

2.2.2. *Sudden Collision Involving a Nonlinear Rotator*

For an almost stationary symmetric top we may consider the cross section $d\sigma_{\text{AMT}}(J_t K_t; \Omega)/d\omega$ for the angular-momentum transfer $\hbar\mathbf{J}_t$ with the magnitude $\hbar J_t$ and with the projection $\hbar K_t$ onto the symmetry axis. This cross section is independent of the initial rotational angular momentum $\hbar\mathbf{J}$, and hence, of J and K of the symmetric top. K_t is related to the symmetry-axis component $\hbar K'$ of the final rotational angular momentum by $K + K_t = K'$.

Let (θ, φ), (θ_t, φ_t), and (θ', φ') be the polar angles of $\mathbf{J}$, $\mathbf{J}_t$, and $\mathbf{J}'$ (Fig. 3).

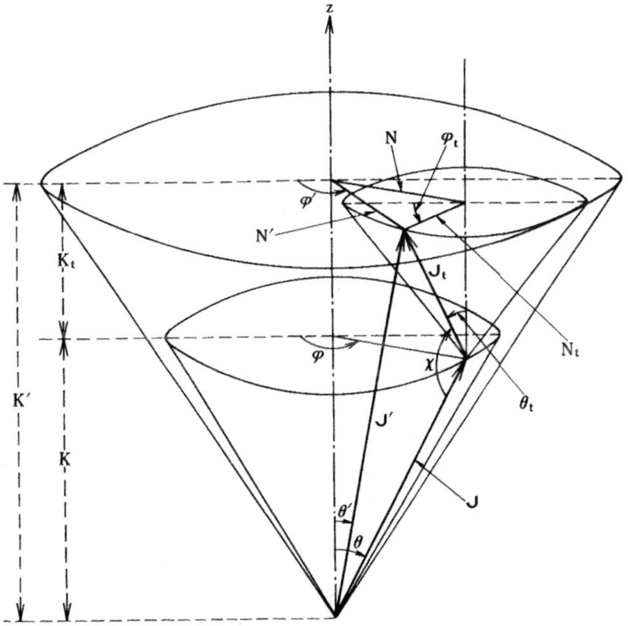

FIGURE 3
Transfer of the angular momentum $\hbar\mathbf{J}_t$ (with the magnitude $\hbar J_t$ and the z component $\hbar K_t$) to a symmetric top with the rotational angular momentum $\hbar\mathbf{J}$ (with the magnitude $\hbar J$ and the z component $\hbar K$). The symmetry axis of the top is chosen as the z axis.

We define projections of the three vectors onto a plane perpendicular to the symmetry (z) axis: $N = J\sin\theta$, $N_t = J_t \sin\theta_t$, and $N' = J'\sin\theta'$. The projection of the triangle ($\mathbf{J}$, $\mathbf{J}_t$, $\mathbf{J}'$) onto this plane naturally satisfies the triangular rules $|N - N'| \leqq N_t \leqq N + N'$, etc. These conditions are stronger than Eq. (3).

Equation (7) is still valid with $-\cos\chi = \cos\theta\cos\theta_t + \sin\theta\sin\theta_t\cos\phi$, ϕ being $\varphi - \varphi_t$. For fixed values of K, K_t, and K' the angles θ, θ_t, and θ' are fixed. The angle ϕ, however, may take any value between zero and 2π with equal probability. Because of the symmetry of $\cos\phi$ on both sides of $\phi = \pi$, consideration of the region $0 \leqq \phi \leqq \pi$ suffices.

Let $\{C(JJ_tJ'; KK_tK')\}^2 dJ'$ be the probability of finding a resultant vector of length between $\hbar J'$ and $\hbar(J' + dJ')$, when composing two vectors of lengths $\hbar J$ and $\hbar J_t$ and of projections $\hbar K$ and $\hbar K_t$. It follows from Eq. (7) that

$$\{C(JJ_tJ'; KK_tK')\}^2 dJ' = -d\phi/\pi = -[(\partial\phi/\partial\chi)(\partial\chi/\partial J')/\pi]\, dJ',$$

$$= g(J')\begin{pmatrix} J & J_t & J' \\ K & K_t & -K' \end{pmatrix}^2 dJ', \tag{18a}$$

$$\begin{pmatrix} J_1 & J_2 & J_3 \\ K_1 & K_2 & K_3 \end{pmatrix}^2 = \pi^{-1}[\beta(J_1J_2J_3) + 4(J_1^2K_2K_3 + J_2^2K_3K_1 + J_3^2K_1K_2)]^{-1/2}$$

$$\times\, \delta(K_1 + K_2 + K_3). \tag{18b}$$

Equation (18b) is symmetric with respect to the three vectors $\mathbf{J}_1$, $\mathbf{J}_2$, and $\mathbf{J}_3$.[10, 11] The Clebsch–Gordan coefficient $C(JJ_tJ'; KK_tK')$ and the 3-j symbol $\left(\begin{smallmatrix} J_1 & J_2 & J_3 \\ K_1 & K_2 & K_3 \end{smallmatrix}\right)$ reduce to those of Eqs. (8) for linear rotators, i.e., for $K = K_t = K' = 0$ (or $K_1 = K_2 = K_3 = 0$). Equations (18a) and (18b) are infinite on the boundaries $N_i = |N_j - N_k|$ and $N_i = N_j + N_k$ [$N_i^2 = J_i^2 - K_i^2$, etc., and $(i, j, k) = (1, 2, 3)$, etc.], and vary smoothly between them.

The 3-j symbol squared and the Clebsch–Gordan coefficient squared have many properties that their quantum-mechanical extensions have. For example, they are invariant under reflection in a plane perpendicular to the symmetry axis, as is obvious from their definition. In fact, Eq. (18b) is unchanged under simultaneous changes of signs of K_1, K_2, and K_3. Another set of examples follows from Eq. (18a) and from the symmetry property of the 3-j symbol:

$$\{C(JJ_tJ'; KK_tK')\}^2 = \{C(J_tJJ'; K_tKK')\}^2, \tag{19a}$$

$$g(J)\{C(JJ_tJ'; KK_tK')\}^2 = g(J')\{C(J'J_tJ; K' - K_tK)\}^2, \tag{19b}$$

$$g(J_t)\{C(JJ_tJ';KK_tK')\}^2 = g(J')\{C(JJ'J_t;-KK'K_t)\}^2. \qquad (19c)$$

The first of these indicates the equivalence of adding **J** to $\mathbf{J}_t$ to adding $\mathbf{J}_t$ to **J**. Equations (19b) and (19c) imply the principle of detailed balance.

The normalization condition for the probability follows immediately:

$$\iint \{C(JJ_tJ';KK_tK')\}^2 dJ'dK' = \int_0^\pi d\phi/\pi = 1, \qquad (20a)$$

$$\{C(0J_tJ';0K_tK')\}^2 = \delta(J_t - J')\delta(K_t - K'). \qquad (20b)$$

The Clebsch–Gordan coefficient squared $\{C(JJ_tJ';KK_tK')\}^2$ proves to coincide with the probability density of finding, for fixed values of J, J_t, J', and K', different values of K, and hence, $K_t(=K'-K)$.[10] Thus, we may show for a fixed K' another normalization condition

$$\iint \{C(JJ_tJ';KK_tK')\}^2 dK\, dK_t = 1. \qquad (20c)$$

The expectation value of the energy transfer to a prolate symmetric top due to the angular-momentum transfer (J_t, K_t) is

$$\Delta E(JK;J_tK_t) = \iint [E_r(J'K') - E_r(JK)]\{C(JJ_tJ';KK_tK')\}^2 dJ'dK'$$

$$= \int_0^\pi [E_r(J_tK_t) + 2AKK_t + 2BNN_t \cos\phi]\, d\phi/\pi$$

$$= E_r(J_tK_t) + 2AKK_t. \qquad (21)$$

For an oblate symmetric top, the rotational constant C replaces A. The second term in the right-hand side of Eq. (21) vanishes, if $\Delta E(JK;J_tK_t)$ for a given K_t is averaged over the degenerate initial states $\pm K$. Then, the average energy transfer is the rotational energy $E_r(J_tK_t)$ corresponding to the angular momentum (J_t, K_t) transferred to the top.

The cross section for the transition $JK \to J'K'$ (for fixed orientation Ω) of a symmetric top is the superposition of the cross sections for angular-momentum transfer weighted by the probability density $\{C(JJ_tJ';KK_tK')\}^2$. After averaging over all Ω we obtain an extension of Eq. (5)

$$\frac{d\sigma(JK \to J'K')}{d\omega} = \int dK_t \int dJ_t \{C(JJ_tJ';KK_tK')\}^2 \frac{d\sigma(00 \to J_tK_t)}{d\omega}. \qquad (22)$$

Due to the delta function in Eq. (18b) the integration over K_t allows only $K_t = K' - K$ to contribute to Eq. (22).

For a rotator having a plane of symmetry (chosen as the xz plane) the interaction is symmetric on reflection in this plane, leading to equality of the cross sections for the two transitions $(0,0 \to J_t, \pm K_t)$. Then from Eqs. (19b) and (22) follows the principle of detailed balance

$$g(J)\frac{d\sigma(JK \to J'K')}{d\omega} = g(J')\frac{d\sigma(J'K' \to JK)}{d\omega}. \tag{23}$$

For $K = K' = 0$ the detailed balance for a linear rigid rotator results.

The normalization (20a) used in Eq. (22) proves a sum rule

$$\begin{aligned} S(0;JK) &= \int_{-\infty}^{\infty} dK' \int_{|K'|}^{\infty} dJ' \frac{d\sigma(JK \to J'K')}{d\omega} \\ &= \int_{-\infty}^{\infty} dK_t \int_{|K_t|}^{\infty} dJ_t \frac{d\sigma(00 \to J_tK_t)}{d\omega} = S(0;00). \end{aligned} \tag{24}$$

The mean energy gain by the rotator having a plane of symmetry follows from Eq. (21) as

$$\begin{aligned} S(1;JK) &= \int_{-\infty}^{\infty} dK' \int_{|K'|}^{\infty} dJ' [E_r(J'K') - E_r(JK)] \frac{d\sigma(JK \to J'K')}{d\omega} \\ &= \int_{-\infty}^{\infty} dK_t \int_{|K_t|}^{\infty} dJ_t [E_r(J_tK_t) - E_r(00)] \frac{d\sigma(00 \to J_tK_t)}{d\omega} = S(1;00). \end{aligned} \tag{25}$$

[For a rotator having no plane of symmetry the average of $S(1;JK)$ over the degenerate initial states $\pm K$ is equal to $S(1;00)$; see the comment below Eq. (21).] Both $S(0;JK)$ and $S(1;JK)$ are independent of the initial rotational motion. A computational procedure for higher-order moments is a straightforward generalization of that for the first-order moment; Eq. (21) is easily generalizable for higher-order moments.

Formulas are simplified for spherical tops. Since the spherical-top states are degenerate in K, we may sum the cross section over K' and average it over K to define the cross section $d\sigma(J \to J')/d\omega$. Use the principle of detailed balance (19c) to change $C(JJ_tJ'; KK_tK')$ in Eq. (22) into $C(JJ'J_t; -KK'K_t)$. Then, carry out the integration over K and K' using the normalization (20c) with a result that

$$\frac{d\sigma(J \to J')}{d\omega} = \int_{|J-J'|}^{J+J'} dJ_t \left[\frac{g(J')}{g(J)g(J_t)}\right] \frac{d\sigma(0 \to J_t)}{d\omega}. \tag{26}$$

Sum rules similar to Eqs. (6) and (11) follow from Eq. (26); both the sum $S(0;J)$ and the mean energy transfer $S(1;J)$ are independent of J. Also, $S(2n;J)$ and $S(2n+1;J)$ $(n = 0, 1, \ldots)$ are polynomials in $E_r(J)$ of degree n.

Because the factor in the square brackets in Eq. (26) is $(1 \pm \Delta J/J)/2J_t$ for $J' = J \pm \Delta J$, transitions of a spherical top from different states J to states $J \pm \Delta J$ with a fixed $|\Delta J|$ are related by Eq. (13) in which $c_1 = 1$ and $O(J^{-2}) = 0$, provided that $2J - |\Delta J|$ is larger than any J_t that contributes appreciably to Eq. (26). Then, Eqs. (14) and (15) are valid with $c_1 = 1$ for a spherical top.

The detailed balance for a spherical top results from Eq. (23) as

$$\{g(J)\}^2 \frac{d\sigma(J \to J')}{d\omega} = \{g(J')\}^2 \frac{d\sigma(J' \to J)}{d\omega}; \tag{27}$$

note that the density of state J is $\{g(J)\}^2$.

3. Quantum-Mechanical Sudden Collisions and Angular-Momentum Transfer

3.1. Elementary Theory of Angular Momentum

In quantum mechanics differential operators define the components of the angular momentum. These operators do not commute with each other. The numerical values of the angular-momentum components are eigenvalues of these operators, and cannot be simultaneously well defined. Well-defined eigenvalues are discrete. The angular momentum transferred to a rigid rotator is also discrete. The Clebsch–Gordan coefficient and the 3-j symbol introduced in Section 2 need to be modified for description of the addition of the quantized angular momenta. The concept of the angular-momentum transfer in quantum mechanical sudden collisions leads to formulas very similar to those for classical sudden collisions.

3.1.1. Angular Momentum and Rotational States

We shall begin with the quantum mechanics of rigid rotators, because many molecules may be regarded as such, to a first and often good approximation, after being averaged over the vibrational motions. A correction due to the vibration-rotation interaction is usually small for diatomic molecules, but can be appreciable for polyatomic molecules. For electronic states with nonzero electronic spin angular momentum there is a correction due to the spin-rotation interaction. Polyatomic molecules with nonsinglet ground electronic states are rare.

The Hamiltonian operator for the rotation of the most general rigid body results from quantization of the classical rotational energy (16) as

$$H_r = A\hat{J}_a^2 + B\hat{J}_b^2 + C\hat{J}_c^2. \tag{28}$$

The operators $\hat{J}_a$, $\hat{J}_b$, and $\hat{J}_c$ are first-order differential operators involving the Eulerian angles $\Omega = (\varphi, \theta, \psi)$ defined in Fig. 4. The quantized angular momentum squared $\hbar^2\hat{\mathbf{J}}^2 = \hbar^2\hat{J}_a^2 + \hbar^2\hat{J}_b^2 + \hbar^2\hat{J}_c^2$ commutes with the rotational Hamiltonian H_r. The projection $\hat{J}_z$ of $\hat{\mathbf{J}}$ onto any space-fixed z axis also commutes with H_r. Then, we may choose any eigenstate $D_\Gamma(\Omega)$ (Γ denoting collectively the set of rotational quantum numbers) of H_r to be simultaneously eigenfunctions of $\hat{\mathbf{J}}^2$ and of $\hat{J}_z = -i\partial/\partial\varphi$; the eigenvalues are $J(J+1)$ ($J = 0, 1, 2, \ldots$) and M ($M = 0, \pm 1, \ldots, \pm J$). A function of φ, $(2\pi)^{-1/2}\exp(iM\varphi)$, factors out of the eigenfunctions $D_\Gamma(\Omega)$. All states are degenerate in M, this degeneracy being

$$g(J) = 2J + 1 \tag{29}$$

in contrast with the classical M degeneracy of $2J$ [cf. Eq. (9)].

For a symmetric top the symmetry-axis component $\hat{J}_c$ (or $\hat{J}_a$) commutes with H_r, $\hat{\mathbf{J}}^2$, and $\hat{J}_z$. We may choose $D_\Gamma(\Omega)$ to be also an eigenfunction of $\hat{J}_c$ (or $\hat{J}_a$) $= -i\partial/\partial\psi$. Its eigenvalue K may be any of $0, \pm 1, \ldots$, and $\pm J$. The (normalized) rotational wave functions have the form

$$D_\Gamma(\Omega) = D_{KM}^J(\varphi, \theta, \psi) = [(2\pi)^{-1/2}e^{iK\psi}][(J+\tfrac{1}{2})^{1/2}d_{KM}^{(J)}(\theta)][(2\pi)^{-1/2}e^{iM\varphi}]. \tag{30}$$

All eigenfunctions $D_{KM}^J(\varphi, \theta, \psi)$ form an orthonormal complete set in the three-dimensional space of Ω. The statements in this paragraph apply also to a spherical top, which is a kind of a symmetric top.

Suppose that the transformation of the frame of axes, described by the Eulerian angles $\Omega = (\varphi, \theta, \psi)$, brings the space-fixed coordinate axes (x, y, z) to the body-fixed axes (x''', y''', z''') as shown in Fig. 4. Define a unitary operator $R(\Omega)$ that transforms any function f of the coordinate $\mathbf{r} = (x, y, z)$ into a new function g of the coordinate $\mathbf{r}''' = (x''', y''', z''')$, so that $f(\mathbf{r}) = g(\mathbf{r}''') = [R(\Omega)f](\mathbf{r}''')$. The matrix element of the rotation operator $R(\Omega)$ with respect to the angular-momentum states $|JM\rangle$ and $|JM'\rangle$ proves to be[11, 14, 15]

$$\langle JM'|R(\varphi, \theta, \psi)|JM\rangle = \left(\frac{8\pi^2}{2J+1}\right)^{1/2} D_{M'M}^J(\varphi, \theta, \psi) \equiv \mathrm{D}_{M'M}^{(J)}(\varphi, \theta, \psi). \tag{31}$$

The function $|JM\rangle(\hat{\mathbf{r}})$ in the space-fixed frame is related to the functions $|JM'\rangle(\hat{\mathbf{r}}''')$ in the body-fixed frame by

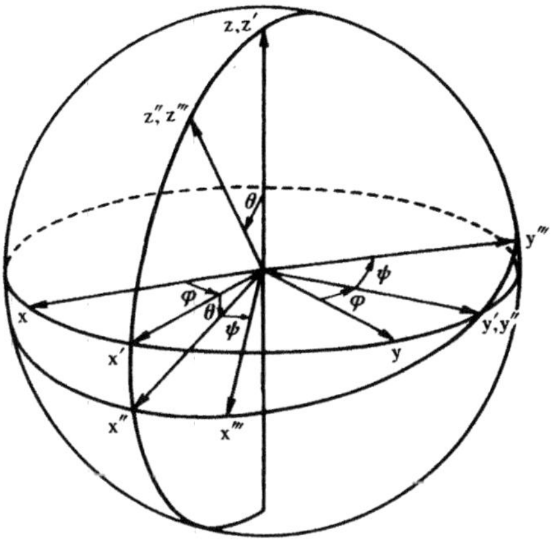

FIGURE 4
Definition of the Eulerian angles $\Omega = (\varphi, \theta, \psi)$ representing the successive rotations that bring the space-fixed coordinate axes (x, y, z) to the body-fixed principal axes of inertia (x''', y''', z'''). The rotations in the order to be performed are (1) a rotation φ $(0 \leqq \varphi < 2\pi)$ about the z axis, bringing (x, y, z) to (x', y', z'), (2) a rotation θ $(0 \leqq \theta < \pi)$ about the y' axis, bringing (x', y', z') to (x'', y'', z''), and (3) a rotation ψ $(0 \leqq \psi < 2\pi)$ about the z'' axis, bringing (x'', y'', z'') to (x''', y''', z''').[11,14,15] Note that θ and φ coincide with the polar coordinates of the z''' axis with respect to the space-fixed axes (x, y, z).

$$[R(\Omega)|JM\rangle](\hat{\mathbf{r}}''') = \sum_{M'} \mathsf{D}^{(J)}_{M'M}(\Omega)|JM'\rangle(\hat{\mathbf{r}}''') = |JM\rangle(\hat{\mathbf{r}}), \tag{32a}$$

$$\sum_{M} \mathsf{D}^{(J)*}_{M'M}(\Omega)|JM\rangle(\hat{\mathbf{r}}) = |JM'\rangle(\hat{\mathbf{r}}'''). \tag{32b}$$

The inverse transformation (32b) is due to the unitarity of the matrix D.

The rotational energies are obtained by changing J^2 in the classical energies (17a) and (17b) into an eigenvalue $J(J+1)$ of $\hat{\mathbf{J}}^2$. Thus we have

$$E_r(JK) = BJ(J+1) + (A-B)K^2 \qquad \text{(prolate symmetric top)}, \tag{33a}$$

$$E_r(JK) = BJ(J+1) - (B-C)K^2 \qquad \text{(oblate symmetric top)}, \tag{33b}$$

$$E_r(J) = BJ(J+1) \qquad \text{(spherical top)}. \tag{33c}$$

The degeneracy of the symmetric-top states is $(2-\delta_{0K})g(J)$. The spherical-top levels are $\{g(J)\}^2$-fold degenerate.

The linear rotator is the limiting case of the prolate symmetric top as $A \to \infty$, $K \to 0$, and $AK^2 \to 0$ (see Section 2.2.1). The rotational energy has the same form (33c) as the spherical top, but the linear-rotator levels are $g(J)$-fold degenerate. Because $K = 0$, the wave function (30) is independent of ψ. Then, we may define wave functions

$$Y_{JM}(\theta,\varphi) = (2\pi)^{1/2} D^{J}_{0M}(\varphi,\theta,\psi) = [(J+\tfrac{1}{2})^{1/2} d^{(J)}_{0M}(\theta)]\,[(2\pi)^{-1/2} e^{iM\varphi}] \tag{34}$$

normalized in the two-dimensional space of the solid angle $\Omega = (\theta, \varphi)$, the whole region V_Ω of the solid angle being 4π. All functions $Y_{JM}(\theta,\varphi)$, called spherical harmonics, form an orthonormal complete set in the space (θ,φ). The function

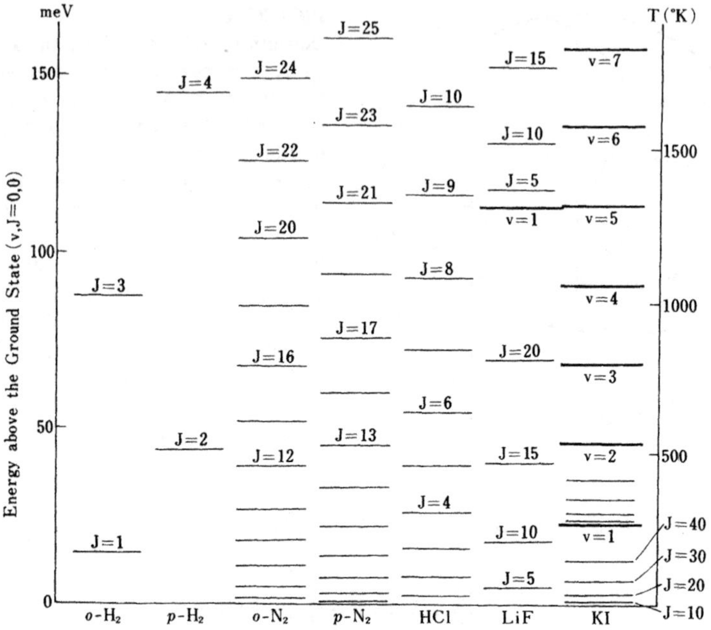

FIGURE 5
Vibrational-rotational levels (v, J) of some diatomic molecules. *o*-, ortho; *p*-, para.

$d_{0M}^{(J)}(\theta)$ is the associated Legendre function except for a multiplicative constant. (The definition of this multiplicative constant and the phase convention for the spherical harmonics and for the rotation matrix differ depending on the literature.)

The asymmetric top has well-defined quantum numbers J and M, but K is no longer well defined. The wave function may be expanded in terms of any complete set of basis functions that are eigenfunctions of $\hat{\mathbf{J}}^2$ and $\hat{J}_z$ with the eigenvalues $J(J+1)$ and M. Convenient basis functions are the $2J+1$ symmetric-top functions $D_{KM}^{J}(\Omega)$ for $K=0, \pm 1, \ldots, \pm J$. There are $2J+1$ different eigenstates of an asymmetric top for each set. We usually arrange these states in order of increasing energy, and label them as $|J_\tau\rangle$ by an index that ranges from $-J$ to J.

Figure 5 shows examples of rotational levels of some diatomic molecules that may be regarded as linear rotators.

3.1.2. Transfer of Angular Momentum

This subsection generalizes for quantum mechanics the theory of the transfer of the angular momentum $\hbar\mathbf{J}_t$ to a rotator with the angular momentum

$\hbar\mathbf{J}$. [See Eq. (2) and Figs. 1 and 3.] The angular momenta satisfy the same triangular relation (3) as in the classical theory, but J, J_t, and J' are now positive integers or zero. We consider a general case where the z axis in Fig. 3 may be any axis fixed in space; in the following M, M_t, and M' replace K, K_t, and K'.

To find a quantum mechanical analog of the classical probability density $\{C(JJ_tJ'; MM_tM')\}^2$ of Eq. (18), we consider the eigenstates $|JM\rangle$ of $\hat{\mathbf{J}}^2$ and $\hat{J}_z$, the eigenstates $|J_tM_t\rangle$ of $\hat{\mathbf{J}}_t^2$ and $\hat{J}_{tz}$, and the eigenstates $|J'M'; JJ_t\rangle$ of $\hat{\mathbf{J}}^2$, $\hat{\mathbf{J}}_t^2$, $\hat{\mathbf{J}}'^2$, and $\hat{J}'_z$, where $\hat{\mathbf{J}}' = \hat{\mathbf{J}} + \hat{\mathbf{J}}_t$; assume that all these states are normalized. The direct product state $|JM\rangle|J_tM_t\rangle$ may always be expanded in terms of the states $|J'M'; JJ_t\rangle$ with $M' = M + M_t$ and with different values of J' as

$$|JM\rangle|J_tM_t\rangle = \sum_{J'} C_{J'M}|J'M'; JJ_t\rangle = \sum_{J'} C(JJ_tJ'; MM_tM')|J'M'; JJ_t\rangle. \quad (35)$$

It is possible to choose all coefficients to be real. The square of $C(JJ_tJ'; MM_tM')$ represents the probability of finding the angular momentum $(\hbar J', \hbar M')$ when adding the angular momenta $(\hbar J, \hbar M)$ and $(\hbar J_t, \hbar M_t)$. We call $C(JJ_tJ'; MM_tM')$ the Clebsch–Gordan (or CG) coefficient. It is zero unless Eq. (3) is satisfied or unless the z components satisfy the relations $-J \leqq M \leqq J$, etc., and $M' = M + M_t$. For its phase convention see Refs. 11, 14, and 15.

The quantum mechanical 3-j symbol squared is the squared quantum mechanical Clebsch–Gordan coefficient divided by the quantum mechanical degeneracy $g(J') = 2J' + 1$ [cf. Eq. (18a)]. Both even and odd permutations of the columns in the 3-j symbol leave its square unchanged. Just as in the classical case, the Clebsch–Gordan coefficient squared and the 3-j symbol squared are invariant under reflection in a plane perpendicular to the z axis. In fact it follows that

$$C(JJ_tJ'; MM_tM') = (-1)^{J+J_t-J'} C(JJ_tJ'; -M, -M_t, -M'). \quad (36)$$

This implies the parity selection rule that $C(JJ_tJ'; 000)$ is zero unless $J + J_t - J'$ is even.

Orthogonality of the transformation (35) leads to the formulas[11,14]

$$|J'M'; JJ_t\rangle = \sum_{M} (C^{-1})_{MJ'}|JM\rangle|J_tM_t\rangle = \sum_{M} C_{J'M}|JM\rangle|J_tM_t\rangle$$

$$= \sum_{M} C(JJ_tJ'; MM_tM')|JM\rangle|J_tM_t\rangle, \quad (37)$$

$$\sum_{J'} C(JJ_tJ';M,M'-M,M')C(JJ_tJ';M'',M'-M'',M')$$

$$= \sum_{J'} (C^{-1})_{MJ'}C_{J'M''} = \delta_{MM''}, \quad (38a)$$

$$\sum_{M} C(JJ_tJ';M,M'-M,M')C(JJ_tJ'';M,M'-M,M')$$

$$= \sum_{M} C_{J'M}(C^{-1})_{MJ''} = \delta_{J'J''}, \quad (38b)$$

$$C(0J_tJ';0M_tM') = \delta_{J_tJ'}\delta_{M_tM'}. \quad (39)$$

The orthonormality (38) is similar to Eqs. (20a) and (20c). Equation (39), which corresponds to Eq. (20b), follows, e.g., from Eq. (38b).

Equation (19a) showing the equivalence of **J** and $\mathbf{J}_t$ follows also for the quantum mechanical Clebsch–Gordan coefficients. Equations (19b) and (19c) showing the principle of detailed balance are also valid with $g(J) = 2J + 1$, etc. With the correct phase factors the quantum mechanical relations are[11, 14]

$$C(JJ_tJ';MM_tM') = (-1)^{J+J_t-J'}C(J_tJJ';M_tMM'), \quad (40a)$$

$$\{g(J)\}^{1/2}C(JJ_tJ';MM_tM') = (-1)^{J_t+M_t}\{g(J')\}^{1/2}C(J'J_tJ;-M',M_t,-M), \quad (40b)$$

$$\{g(J_t)\}^{1/2}C(JJ_tJ';MM_tM') = (-1)^{J-M}\{g(J')\}^{1/2}C(JJ'J_t;M,-M',-M_t). \quad (40c)$$

The expectation value (21) of the energy transfer to the classical symmetric top is generalizable for the quantum mechanical top.[16] Because $\hat{J}'_a = \hat{J}_a + \hat{J}_{ta}$ and $\hat{\mathbf{J}} \cdot \hat{\mathbf{J}}_t = \hat{J}_a\hat{J}_{ta} + \hat{J}_+\hat{J}_{t-} + \hat{J}_-\hat{J}_{t+}$, where $\hat{J}_\pm = (\hat{J}_b \pm i\hat{J}_c)/\sqrt{2}$ and $\hat{J}_{t\pm} = (\hat{J}_{tb} \pm i\hat{J}_{tc})/\sqrt{2}$, quantization of the second cosine formula (7) shows that

$$\{B\hat{\mathbf{J}}'^2 + (A-B)\hat{J}'^2_a\} = \{B\hat{\mathbf{J}}^2 + (A-B)\hat{J}^2_a\} + \{B\hat{\mathbf{J}}^2_t + (A-B)\hat{J}^2_{ta}\}$$

$$+ 2A\hat{J}_a\hat{J}_{ta} + 2B[\hat{J}_+\hat{J}_{t-} + \hat{J}_-\hat{J}_{t+}]. \quad (41)$$

Compute the expectation value of the left-hand side with respect to the right-hand side of Eq. (35) (with M, M_t, and M' replaced by K, K_t, and K'). Compute also the expectation value of the right-hand side of Eq. (41) with respect to the left-hand side of Eq. (35). Equate the two results, and use the orthonormality (38a) of the Clebsch–Gordan coefficients and the orthonormality of the angular momentum functions. Then, we have

$$\Delta E(JK;J_tK_t) = \sum_{J'} [E_r(J'K') - E_r(JK)]\{C(JJ_tJ';KK_tK')\}^2$$

$$= E_r(J_tK_t) + 2AKK_t \tag{42}$$

except for an additional term, i.e., the expectation value of the last term in Eq. (41) with respect to the state $|JK\rangle|J_tK_t\rangle$. The operator $\hat{J}_+$ raises K by one and $\hat{J}_-$ lowers K by one, and similarly for $\hat{J}_{t\pm}$ and K_t. Therefore, the additional term is zero, just as the term $2BNN_t\cos\phi$ vanishes in the classical energy transfer. Equation (42) for the quantum mechanical top has the same form as Eq. (21) for the classical top. The comments made just below Eq. (21) apply also to Eq. (42).

3.2. *Sudden Approximation*

Regardless of the mechanics the collision system obeys, the assumption of the stationary rotator should be a good approximation, if the collision is sudden compared with the rotational period of the rotator. In quantum mechanics this expectation finds formal justification in the parametric Ω-dependence of the total wave function except for $D_\Gamma(\Omega)$ that factors out; the parametric dependence follows from the neglect of the rotational-channel dependence of the channel wave number. We shall prove it in Section 3.2.2. In the meantime, in Section 3.2.1, we shall assume it on physical grounds.

3.2.1. *Drozdov–Chase Method*

Consider a molecule regarded as a rigid rotator. In other words the rotational motion of the molecule is assumed to be much slower than the vibrational-electronic (abbreviated as vibronic) motion, and the two kinds of motion are separable from each other. The rotational energy $E_r(\Gamma)$ is much smaller than the vibronic energy $E_v(n)$. The molecular wave functions $\psi_{n\Gamma}(\mathbf{X}_T)$ are separable into the vibronic part $\Phi_n(\mathbf{x}_T) = \Phi_n(\mathbf{r}_T, R_T)$ and the rigid-rotator wave function $D_\Gamma(\Omega)$. Here, $\mathbf{r}_T$ denotes collectively the coordinates of the molecular electrons, R_T the nuclear coordinates representing the vibrational modes, $\mathbf{x}_T = (\mathbf{r}_T, R_T)$, and $\mathbf{X}_T = (\mathbf{x}_T, \Omega)$. The subscript T implies the target of the collision. We neglect the electron spins, because they are inessential in the following discussion to the extent that the spin-rotation coupling is negligible.

The collision system to be studied consists of the molecule in an initial state $\psi_{n\Gamma} = \Phi_n D_\Gamma$ and a particle without internal structure. The relative coordinate between them is $\mathbf{r}$. The wave vector $\mathbf{k}$ for the relative motion in the initial channel is related to the collision energy E by $E = \hbar^2 k^2/2\mu$ with the reduced mass μ of the collision system.

The wave function $\Psi_{n\Gamma}(\mathbf{r}, \mathbf{X}_T)$ for the total system is separable, according to our assumption on the physical grounds, as

$$\Psi_{n\Gamma}(\mathbf{r}, \mathbf{X}_T) = \Psi_n^{(s)}(\mathbf{r}, \mathbf{x}_T; \Omega) D_\Gamma(\Omega) \tag{43a}$$

$$\xrightarrow[r\to\infty]{} \left[\exp(i\mathbf{k}\cdot\mathbf{r})\Phi_n(\mathbf{x}_T) + \sum_{n'} \frac{\exp(ik_{n'}r)}{r} F(n\mathbf{k} \to n'\mathbf{k}_{n'}; \Omega)\Phi_{n'}(\mathbf{x}_T)\right] D_\Gamma(\Omega), \tag{43b}$$

where $\Psi_n^{(s)}$ depends on Ω only parametrically. It describes scattering by the molecule with its orientation Ω held fixed in space. F is the scattering amplitude for the vibronic transition $n \to n'$ in this fictitious scattering. Because the rotational motion is suppressed in $\Psi_n^{(s)}$, the final-channel wave number $k_{n'}$ is independent of the rotational channel. Expand the product $F(\Omega)D_\Gamma(\Omega)$ in terms of the complete set $\{D_{\Gamma'}(\Omega)\}$. Then we may rewrite the asymptotic form (43b) as

$$\Psi_{n\Gamma}(\mathbf{r}, \mathbf{X}_T) \xrightarrow[r\to\infty]{} \exp(i\mathbf{k}\cdot\mathbf{r})\Phi_n(\mathbf{x}_T)D_\Gamma(\Omega) + \sum_{n'\Gamma'} \frac{\exp(ik_{n'}r)}{r} f(n\Gamma \to n'\Gamma')\Phi_{n'}(\mathbf{x}_T)D_{\Gamma'}(\Omega) \tag{44}$$

with

$$f(n\Gamma \to n'\Gamma') = (D_{\Gamma'}|F(n\mathbf{k} \to n'\mathbf{k}_{n'})|D_\Gamma). \tag{45}$$

The asymptotic form (44) suggests that Eq. (45) is the scattering amplitude for the rotational-vibronic transition $n\Gamma \to n'\Gamma'$ in the approximation that the energy is no longer conserved with regard to the rotational motion.

We call this the (rotational) sudden approximation. We also often call it the adiabatic (-rotation) approximation because of the adiabatic separation between the rotational motion and the rest, or the (rotational) impulse approximation because of the impulsive interaction experienced. The name "adiabatic approximation" is sometimes confusing, because one may speak of the slow collision limit as adiabatic.

The approximation essentially in the form presented here has been introduced by Drozdov[17] and discussed later by Chase.[18] They have applied it to the collisions of neutrons with nuclei regarded as linear or symmetric-top rotators.[17-19] Appendix C to Chapter 3 discusses the relation between the Drozdov–Chase formulation and the formal theory of impulse approximation presented earlier by Chew and Goldberger.[20] Literature on the sudden approximation usually goes back only to the paper by Chase.[1-8] Applicability of the

Drozdov–Chase method to rotational transitions in electron–molecule collisions has been studied first by Altshuler.[21] He has also tested the method numerically within the first-order perturbation theory. Oksyuk has initiated numerical applications in a nonperturbative approach.[22]

Sometimes the vibrational motion is also separable from the total wave function. Then, we have the adiabatic-nuclei or the nuclear impulse approximation, in which the scattering amplitude is a matrix element of the fixed-nuclei amplitude (Chapter 3, Section 4 and Appendix C and Chapter 7, Sections 2.2 and 4.1). If the rotational motion only is separable, one may solve equations coupled in vibrational channels and adiabatic in rotational motion, which is referred to as the hybrid method (Chapter 7, Section 4.2).[23, 24]

3.2.2. Separability of the Rotational Motion

We shall now derive the separable form (43a) of the total wave function $\Psi_{n\Gamma}(\mathbf{r}, \mathbf{X}_T)$. (See, e.g., Ref. 25.) Convenient for this purpose is an integral equation for scattering, i.e., the Lippmann–Schwinger equation

$$\begin{aligned}\Psi_{n\Gamma}(\mathbf{r}, \mathbf{X}_T) &= \exp(i\mathbf{k}\cdot\mathbf{r})[\Phi_n(\mathbf{x}_T)D_\Gamma(\Omega)] - \frac{1}{2\pi}\int d\mathbf{r}'\int d\mathbf{X}'_T \\ &\quad\times \sum_{n'\Gamma'} \frac{\exp(ik_{n'\Gamma'}|\mathbf{r}-\mathbf{r}'|)}{|\mathbf{r}-\mathbf{r}'|}[\Phi_{n'}(\mathbf{x}_T)D_{\Gamma'}(\Omega)D^*_{\Gamma'}(\Omega')\Phi^*_{n'}(\mathbf{x}'_T)] \\ &\quad\times V(\mathbf{r}', \mathbf{X}'_T)\Psi_{n\Gamma}(\mathbf{r}', \mathbf{X}'_T).\end{aligned} \tag{46}$$

The incident particle and the molecule interact through the potential $(\hbar^2/\mu) \times V(\mathbf{r}, \mathbf{X}_T)$. The wave numbers $k_{n'\Gamma'}$ for the channels $n'\Gamma'$ satisfy the energy conservation $\hbar^2k^2/2\mu + E_v(n) + E_r(\Gamma) = \hbar^2k^2_{n'\Gamma'}/2\mu + E_v(n') + E_r(\Gamma')$. Obviously, we have $k = k_{n\Gamma}$.

The sum over n' and Γ' extends, in principle, over the complete set of functions $\psi_{n'\Gamma'}$. The separation of $D_{\Gamma'}$ from $\Phi_{n'}$, however, is invalid for high-lying vibronic states such as highly excited Rydberg and continuum states, and such as high vibrational states close to and above the dissociation limit. We assume the contribution from these high-lying states to be either negligible or represented by an effective potential that depends negligibly or only weakly on the scattering energy.

We also assume negligible contributions (or contributions described by an effective potential) from those vibronic states n' for which Ω is defined in a different way from that for the initial state n. Thus we exclude any state n' having an average nuclear geometry different from that of the state n.

An essential assumption valid for the sudden collision is as follows: For all intermediate states $n'\Gamma'$ that contribute appreciably to the processes to be studied, the dependence of the wave numbers $k_{n'\Gamma'}$ on the rotational channel Γ' is negligible in Eq. (46).

Replace all $k_{n'\Gamma'}$ for a common n' by a representative value $k_{n'}$. Then, the sum over Γ' in Eq. (46) reduces to the closure relation $\Sigma D_{\Gamma'}(\Omega)D^*_{\Gamma'}(\Omega') = \delta(\Omega - \Omega')$. Hence, it follows that

$$\Psi_{n\Gamma}(\mathbf{r}, \mathbf{x}_T, \Omega) = \exp(i\mathbf{k}\cdot\mathbf{r})[\Phi_n(\mathbf{x}_T)D_\Gamma(\Omega)] - \frac{1}{2\pi}\int d\mathbf{r}' \int d\mathbf{x}'_T$$

$$\times \sum_{n'} \frac{\exp(ik_{n'}|\mathbf{r}-\mathbf{r}'|)}{|\mathbf{r}-\mathbf{r}'|}[\Phi_{n'}(\mathbf{x}_T)\Phi^*_{n'}(\mathbf{x}'_T)]$$

$$\times V(\mathbf{r}', \mathbf{x}'_T, \Omega)\Psi_{n\Gamma}(\mathbf{r}', \mathbf{x}'_T, \Omega). \tag{47}$$

This equation involves the molecular orientation Ω only as a parameter and not as a variable. The dependence of $\Psi_{n\Gamma}$ on Ω enters only through the initial rotational state $D_\Gamma(\Omega)$ and the interaction potential $(\hbar^2/\mu)V$. Therefore, the wave function is separable as Eq. (43a).

Substitution of Eq. (43a) into Eq. (47) yields an equation

$$\Psi_n^{(s)}(\mathbf{r}, \mathbf{x}_T; \Omega) = \exp(i\mathbf{k}\cdot\mathbf{r})\Phi_n(\mathbf{x}_T) - \frac{1}{2\pi}\int d\mathbf{r}' \int d\mathbf{x}'_T$$

$$\times \sum_{n'} \frac{\exp(ik_{n'}|\mathbf{r}-\mathbf{r}'|)}{|\mathbf{r}-\mathbf{r}'|}\Phi_{n'}(\mathbf{x}_T)\Phi^*_{n'}(\mathbf{x}'_T)$$

$$\times V(\mathbf{r}', \mathbf{x}'_T; \Omega)\Psi_n^{(s)}(\mathbf{r}', \mathbf{x}'_T; \Omega). \tag{48}$$

This is the exact Lippmann–Schwinger equation for scattering by the orientation-fixed molecule. Therefore, the asymptotic form (43b) is appropriate. The vibronic degrees of freedom still remain unfrozen. The scattering amplitude $F(n\mathbf{k} \to n'\mathbf{k}_{n'}; \Omega)$ satisfies the relation

$$F(n\mathbf{k} \to n'\mathbf{k}_{n'}; \Omega) = -\frac{1}{2\pi}\int d\mathbf{r} \int d\mathbf{x}_T \exp(-i\mathbf{k}_{n'}\cdot\mathbf{r})\Phi^*_{n'}(\mathbf{x}_T)$$

$$\times V(\mathbf{r}, \mathbf{x}_T; \Omega)\Psi_n^{(s)}(\mathbf{r}, \mathbf{x}_T; \Omega). \tag{49}$$

The exact scattering amplitude for the vibronic-rotational transition $n\Gamma \to n'\Gamma'$ is calculable as

$$f(n\Gamma \to n'\Gamma') = -\frac{1}{2\pi}\int d\mathbf{r}\int d\mathbf{x}_T\int d\Omega \exp(-i\mathbf{k}_{n'\Gamma'}\cdot\mathbf{r})\Phi^*_{n'}(\mathbf{x}_T)D^*_{\Gamma'}(\Omega)$$

$$\times\, V(\mathbf{r},\mathbf{x}_T,\Omega)\Psi_{n\Gamma}(\mathbf{r},\mathbf{x}_T,\Omega) \tag{50}$$

in terms of the exact total wave function $\Psi_{n\Gamma}$. Use of the separable form (43a) for $\Psi_{n\Gamma}$ results in an approximation

$$f(n\Gamma \to n'\Gamma') = (D_{\Gamma'}|F(n\mathbf{k} \to n'\mathbf{k}_{n'\Gamma'})|D_\Gamma). \tag{51}$$

Note that, in the fixed-orientation amplitude F in Eq. (51), the wave vector $\mathbf{k}_{n'\Gamma'}$ is used, instead of $\mathbf{k}_{n'}$, for the final-channel plane wave. This is usually a slightly better approximation, and is a much better approximation for forward scattering of electrons by dipolar molecules (Section 5.3.2), than the sudden approximation (45). The sudden, adiabatic, or impulse approximation sometimes refers to Eq. (51). Superiority of Eq. (51) over Eq. (45) is sometimes discussed in the literature.[26] The Glauber theory (Section 5.3.2) is based on Eq. (51) rather than Eq. (45).[25,27] If F in Eq. (51) is calculated according to the first-order perturbation theory, the first Born approximation for $f(n\Gamma \to n'\Gamma')$ results (Section 5.2.2).

The differential cross section is given by

$$\frac{d\sigma(n\Gamma \to n'\Gamma')}{d\omega} = \left(\frac{k_{n'\Gamma'}}{k}\right)|f(n\Gamma \to n'\Gamma')|^2. \tag{52}$$

Usually the factor $k_{n'\Gamma'}/k$ is replaceable by $k_{n'}/k$ in the sudden collision. The correct factor, however, slightly extends the region of applicability of the approximation. Thus, the cross section approaches zero as it should, as the collision energy decreases down to the threshold. References 28 and 29 discuss and test the importance of the correct factor. See also Refs. 30–32 for the choice of the wave number in the sudden approximation in relation to the threshold behavior; there exists an ambiguity (of the order of the rotational transition energy) in the choice of the scattering energy parameter in F.

Physical reasoning suggests the condition for the separability of $\Psi_{n\Gamma}$: (A) The effective collision time should be much shorter than the characteristic rotational period of the molecule. On the other hand, the sudden approximation has followed from the condition (B) that the dependence of $\mathbf{k}_{n'\Gamma'}$ on Γ' be negligible in Eq. (46) and in $F(n\mathbf{k} \to n'\mathbf{k}_{n'\Gamma'})$. In this formalism the concept of collision time is obscure.

For each important n' let $\Delta k_{n'}$ be the spread of $k_{n'\Gamma'}$ for all Γ' that contribute significantly, and let $\Delta E_{n'}$ be the corresponding spread of the channel energies $\hbar^2 k_{n'\Gamma'}^2/2\mu$. Define an effective radius a_{eff} of the interaction region where the product $V\Psi_{n\Gamma}$ is significantly large. The condition (B) is[18]

$$1 \gg (\Delta k_{n'})a_{\text{eff}} \cong (a_{\text{eff}}/v_{n'})/(\hbar/\Delta E_{n'}) \tag{53}$$

for all important n', where $v_{n'} = \hbar k_{n'}/\mu$. This is roughly equivalent to the condition (A) for processes involving only a small number of low-lying intermediate states; $a_{\text{eff}}/v_{n'}$ is the collision time τ_c, and $\hbar/\Delta E_{n'}$ is the characteristic time of transitions between rotational states in an energy band of width $\Delta E_{n'}$. For scattering of electrons we estimate that

$$\tau_c \cong 10^{-16}\,\text{s}[a_{\text{eff}}/a_B]\,[E(\text{eV})]^{-1/2} \qquad (a_B = 5.29 \times 10^{-9}\,\text{cm}), \tag{54a}$$

$$\tau_{\text{rot}} \cong 2\pi\hbar/[(2\bar{J}+1)B] \cong 4 \times 10^{-12}\,\text{s}/[(2\bar{J}+1)B(\text{meV})]$$

$$\cong 10^{-11}\,\text{s}[T(\text{K})B(\text{meV})]^{-1/2} \tag{54b}$$

for linear rotators, where $\tau_{\text{rot}} = 2\pi/\omega_r \cong (2\pi\Delta J)(\hbar/\Delta E_{n'})$, and where $\bar{J}$ and ΔJ are a representative value and the spread of the rotational quantum numbers J of the states contributing to the process. The expression for τ_{rot} involving the gas temperature T is for $\bar{J} = \langle J \rangle$, the Boltzmann average of the initial state J.

Taking $\bar{J} = 2$ as an example (with excitation of molecules in the ground rotational state in mind), we estimate τ_{rot} to be $\sim 10^{-13}$ s for H_2 ($B \cong 7$ meV), $\sim 10^{-12}$ s for N_2 and CO ($B \cong 0.25$ meV), and $\sim 10^{-10}$ s for I_2 ($B \cong 5 \times 10^{-3}$ meV) and KI ($B \cong 8 \times 10^{-3}$ meV). For N_2 and CO at room temperature we have $\tau_{\text{rot}} \cong 10^{-12}$ s, and for I_2 at room temperature we have $\tau_{\text{rot}} \cong 10^{-11}$ s. The constant $2\pi\Delta J$ is normally of the order of 10^1. On the other hand, for $a_{\text{eff}} \cong 3a_B$, e.g., the collision time is $\sim 10^{-16}$ s for $E \cong 10$ eV, and $\sim 10^{-15}$ s for $E \cong 0.1$ eV. These values suggest the validity of the sudden approximation for electron–molecule collisions down to very low energies.

There are, however, exceptions. The approximation fails if a_{eff} is too large, such as for forward scattering by dipolar molecules (Section 5.3).

When $k_{n'\Gamma'}a_{\text{eff}} \ll 1$, not only the collision-time condition (53) but also a more stringent energy condition

$$\Delta k_{n'} \ll k_{n'\Gamma'} \quad \text{or} \quad \Delta E_{n'} \ll \hbar^2 k_{n'\Gamma'}^2/2\mu \tag{55}$$

is necessary for the condition (B) to be met.[18] The condition (55) is violated in the energy region close to the threshold of the final state or of any important intermediate state.

In resonance collisions the width Γ_{res} of the resonance determines τ_c. It is roughly the resonance lifetime $\tau_{\text{res}} \cong \hbar/\Gamma_{\text{res}}$. Therefore, the collision-time

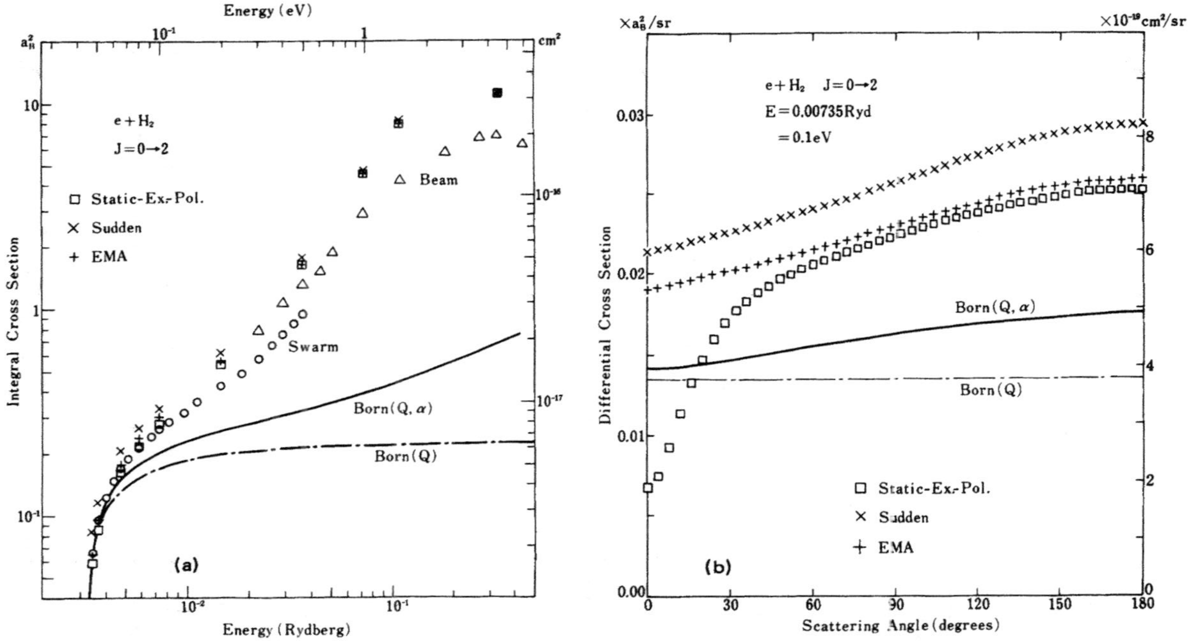

FIGURE 6

Breakdown of the rotational sudden approximation near the threshold of excitation $J = 0 \rightarrow J' = 2$ of H_2. □, converged close-coupling calculation in the laboratory frame in the static, model-exchange, cut-off-polarization approximation (Sections 5 and 6); ×, sudden approximation; +, energy-modified adiabatic approximation; ——, Born approximation with the quadrupole and polarization potentials (Section 5.5); —·—, Born approximation with the quadrupole potential only (Section 5.4); △, crossed electron-beam–molecular-beam experiment by Linder and Schmidt[34] for $J = 1 \rightarrow J' = 3$ modified into $J = 0 \rightarrow J' = 2$ by use of Eq. (62a) in Section 3.3.1, p. 122; ○, electron swarm experiment by Crompton *et al.*[35] (a) Integral cross sections. (b) Differential cross sections at $E = 0.00735$ Rydberg $= 2.28\,[E_r(J=2) - E_r(J=0)]$. (The data other than the experimental and the Born-quadrupole are taken from Feldt and Morrison.[33])

condition (A) takes a form

$$\hbar/\Delta E_{n'} \gg \hbar/\Gamma_{\text{res}} \cong 7\times 10^{-13}\,\text{s}/\Gamma_{\text{res}}(\text{meV}) \quad \text{or} \quad \Gamma_{\text{res}}(\text{meV}) \gg \alpha(\Delta J) \tag{56}$$

with a constant α. The τ_{rot} for $\bar{J}=2$ requires that $\alpha \cong 40$ for H_2, that $\alpha \cong 1$ for N_2 and CO, and that $\alpha \cong 10^{-2}$ for I_2 and KI. At room temperature, $\alpha \cong 4$ for N_2 and CO and $\alpha \cong 0.6$ for I_2. Many resonances in electron–molecule collisions satisfy this condition.

In resonance collisions of the Feshbach type channels with imaginary $k_{n'\Gamma'}$ ($\equiv i\kappa_{n'\Gamma'}$) play a decisive role. The condition (B) on imaginary wave numbers takes a form similar to (55), namely, $\Delta\kappa_{n'} \ll \kappa_{n'\Gamma'}$ or $\Delta E_{n'} \ll \hbar^2\kappa_{n'\Gamma'}^2/2\mu$. Feshbach resonances too close to any threshold violate this condition. It is satisfied, if inequalities (56) and $\Gamma_{\text{res}} < \hbar^2\kappa_{n'\Gamma'}^2/2\mu$ are simultaneously satisfied. The latter inequality requires that the resonance should not extend across any threshold.

In conclusion the rotational sudden approximation appears valid for most electron–molecule collisions except for excitation near threshold, except for narrow resonances, and except for forward scattering by dipolar molecules. Nesbet[30] proposes a modification of the sudden approximation, called the energy-modified adiabatic approximation, for extending the region of its validity and for remedying its shortcomings. (See Chapter 7, Section 4.1.) Figure 6 illustrates the breakdown of the sudden approximation near the threshold for rotational excitation $J=0\rightarrow 2$ of H_2 by electron impact.[33] Experimental data are also shown.[34,35]

3.3. General Relations for the Cross Sections

Classical cross sections in the sudden approximation satisfy various general rules (Sections 2.1.1, 2.1.2, 2.2.2). These rules are generalizable for quantum-mechanical sudden collisions. The crux of the theory hinges on the form of the cross section expressed by Eqs. (45) and (52).

3.3.1. Linear Dependence

Whatever dependence on Ω the fixed-orientation amplitude F has, it may be expanded in terms of the complete set of rotational wave functions $\{D_\Gamma(\Omega)\}$; Eq. (45) is expressible as

$$f(n\Gamma \rightarrow n'\Gamma') = \sum_{\Gamma_t} (D_{\Gamma_t}, F). \tag{57}$$

This form clearly separates the part (D_{Γ_t}, F) dependent on the dynamics from the matrix element independent of the dynamics.

If the initial rotational state is the ground state ($\Gamma = 0$), the wave function D_Γ is a constant $V_\Omega^{-1/2}$; $V_\Omega = 4\pi$ for a linear rotator and $V_\Omega = 8\pi^2$ for a symmetric top or an asymmetric top. Then the matrix element $(D_{\Gamma'}|D_{\Gamma_t}|D_0)$ is $V_\Omega^{-1/2}\delta_{\Gamma'\Gamma_t}$, and the sum in Eq. (57) reduces to a single term as $f(n0 \to n'\Gamma') = V_\Omega^{-1/2}(D_{\Gamma'}, F)$. Then it follows that

$$f(n\Gamma \to n'\Gamma') = \sum_{\Gamma_t} [V_\Omega^{1/2}(D_{\Gamma'}|D_{\Gamma_t}|D_\Gamma)]\, f(n0 \to n'\Gamma_t). \tag{58}$$

The scattering amplitude for any rotational transition is expressible as a linear combination of the amplitudes for transitions from the ground rotational state. Each of the latter amplitudes is that for the transfer of an angular momentum Γ_t. The coefficient in the square brackets is the probability amplitude of finding the resultant angular momentum Γ' when transferring the angular momentum Γ_t to Γ.

A relation between the cross sections rather than the scattering amplitudes would reveal a direct correspondence to the classical theory. Necessary to calculate is the quantity

$$\begin{aligned}|f(n\Gamma \to n'\Gamma')|^2 &= \sum_{\Gamma_t\Gamma_t'} [V_\Omega^{1/2}(D_{\Gamma'}|D_{\Gamma_t'}|D_\Gamma)]^*[V_\Omega^{1/2}(D_{\Gamma'}|D_{\Gamma_t}|D_\Gamma)]\\ &\quad \times f^*(n0 \to n'\Gamma_t')f(n0 \to n'\Gamma_t).\end{aligned} \tag{59}$$

In the following we consider the symmetric top and the linear rotator [so that $\Gamma = (JKM)$, $\Gamma_t = (J_tK_tM_t)$, and $\Gamma' = (J'K'M')$], and use a formula[14]

$$\begin{aligned}[V_\Omega^{1/2}(D_{\Gamma'}|D_{\Gamma_t}|D_\Gamma)] &= [g(J_t)g(J)/g(J')]^{1/2}C(JJ_tJ';MM_tM')C(JJ_tJ';KK_tK')\\ &= (-1)^{J-M}[g(J)]^{1/2}C(JJ'J_t;M,-M',-M_t)\\ &\quad \times C(JJ_tJ';KK_tK'),\end{aligned} \tag{60}$$

where the detailed balance (40c) has been used. Summation over the final degenerate states M' and averaging over the initial degenerate states M reduce Eq. (59) to an incoherent sum without cross terms

$$[g(J)]^{-1}\sum_{MM'}|f(n\Gamma \to n'\Gamma')|^2 = \sum_{J_t}\{C(JJ_tJ';KK_tK')\}^2\left[\sum_{M_t}|f(n0 \to n'\Gamma_t)|^2\right], \tag{61}$$

because $M_t = M_t' = M' - M$ and $K_t = K_t' = K' - K$, and because the summation over M gives $\delta_{J_t J_t'}$ due to the orthonormality (38b) of the Clebsch–Gordan coefficients.

We may rewrite Eq. (61) in terms of the cross sections as[19, 36–38]

$$\left(\frac{k}{k'}\right) \frac{d\sigma(nJK\mathbf{k} \to n'J'K'\mathbf{k}')}{d\omega} = \sum_{J_t=J_0}^{J+J'} \{C(JJ_tJ';KK_tK')\}^2 \left(\frac{k}{k_t}\right) \times \frac{d\sigma(n00\mathbf{k} \to n'J_tK_t\mathbf{k}_t)}{d\omega}, \tag{62}$$

where $J_0 = \max(|J-J'|, |K-K'|)$, and $\hbar^2 k'^2/2\mu = E'_{n'} - \{E_r(J'K') - E_r(JK)\}$ and $\hbar^2 k_t^2/2\mu = E'_{n'} - E_r(J_tK_t)$ with $E'_{n'} = \hbar^2 k^2/2\mu - \{E_v(n') - E_v(n)\}$.

Equation (62) relates transitions from different rotational states at a common collision energy. In view of the ambiguity in the energy parameter to be used in the fixed-orientation amplitude F [see below Eq. (52)], however, it is possible to derive relations between transitions at collision energies slightly different from each other.[39–41]

Equation (62) is essentially equivalent to the classical formulas (5) and (22), though the angular momentum is now quantized and the wave-number ratios are now introduced. [Remember that the integration over K_t in Eq. (22) merely imposes the condition that $K_t = K' - K$.] The following discussion until the end of Section 3 neglects the rotational-channel dependence of the wave numbers, so that $k/k_t = k/k' = k/k_{n'}$ for all J_t; the only exceptions are Eqs. (67), (62a), and (62b). (See also a comment in p.158.)

The form of the relation (62) remains unchanged, if the cross sections on both sides are integrated (with any weight function) over the scattering angle or the collision energy; similar relations hold for the integral and momentum-transfer cross sections and for the rate constant.

Standard textbooks contain formulas for computing the Clebsch–Gordan coefficients.[11,14,15,42] For linear rotators ($K = K' = K_t = 0$), in particular, we have

$$\{C(J, J_t, J \pm \Delta J; 000)\}^2 = T(J_+)T(J_-) \left(\frac{2J \pm 2\Delta J + 1}{2J + 2J_\pm + 1}\right) \left(\frac{T(J - J_\mp)}{T(J + J_\pm)}\right)$$

$$[J_\pm = (J_t \pm \Delta J)/2], \tag{63}$$

where $T(x) = (2x)!/(x!)^2$, for $J_t = |\Delta J|, |\Delta J| + 2, |\Delta J| + 4, \ldots, 2J \pm \Delta J$. The Clebsch–Gordan coefficients for other values of J_t vanish because of the triangular rule (3) and of the parity selection rule mentioned below Eq. (36). Note that $J_\pm$ are zeros or positive integers. Table 1 lists the numerical values of some Clebsch–Gordan coefficients.

Most molecules in a gas at room temperature or higher occupy high rotational levels due to the small rotational constants. Useful in view of this fact is an asymptotic expression for large J and for small ΔJ, namely,[43,44]

$$\{C(J, J_t, J \pm \Delta J; 000)\}^2 = 4^{-J_t} T(J_+) T(J_-) \left[1 \pm \frac{\Delta J}{2J} + O(J^{-2})\right] \tag{64a}$$

$$= 4^{-J_t} T(J_+) T(J_-) \{g(J \pm \Delta J)/g(J)\}^{1/2} [1 + O(J^{-2})]. \tag{64b}$$

[Compare this with the classical formula (12).] For $J_t \leqq 4$, for example, neglect of the term of order J^{-2} in Eq. (64a) introduces an error of less than 10% for $J \geqq 5$ and less than 3% for $J \geqq 9$.[43,44]

Equation (64) used in Eq. (62) shows that the asymptotic form (13) with $c_1 \doteq 1/2$ of the classical cross sections applies also to the quantum-mechanical cross sections $d\sigma(nJ \to n'J \pm \Delta J)/d\omega$,[29,44] for which

$$\frac{d\sigma(|\Delta J|)}{d\omega} = \sum_{J_t \geqq |\Delta J|} 4^{-J_t} T(J_+) T(J_-) \frac{d\sigma(n0 \to n'J_t)}{d\omega}. \tag{65}$$

Equation (13) satisfies the principle of detailed balance (23) within an error of order J^{-2}. The behavior of the cross sections as J increases is as shown in Fig. 2 schematically. The quantum-mechanical cross sections also satisfy Eq. (14) for the ratio of excitation by ΔJ to deexcitation by ΔJ, and Eq. (15) for the sum of them.

Figure 7 illustrates the J dependence of the differential cross sections for the $e + N_2$ collisions. The data are obtained by analysis[43] (to be explained in Section 3.3.3) of energy-loss spectra of electrons.[45] The asymptotic formula (13) is in fact accurate even for small J.

The properties of the Clebsch–Gordan coefficients (or the properties of the vector addition itself) as we have seen in Section 3.1.2 lead to some general properties of the cross sections.[29] First, the triangular rule (3) holds for J, J_t, and J'; the cross sections for the angular-momentum transfers J_t less than max ($|\Delta J|$, $|\Delta K|$) and greater than $J + J'$ contribute nothing to the cross section $d\sigma(nJK \to n'J'K')/d\omega$. Second, for the linear rotator, only odd (even) J_t contribute to the cross sections for odd (even) ΔJ. [See below Eq. (36).] In other words, the transitions with odd ΔJ are completely independent of those with even ΔJ, and vice versa. Third, the principle of detailed balance for the rotational degrees of freedom, i.e., Eq. (23) with $g(J) = 2J + 1$ holds for a rotator having a plane of symmetry; see Eq. (19b) or Eq. (40b). Note that, if $n \neq n'$, we should understand the cross sections on both sides of Eq. (23) to be for the same

TABLE 1

Clebsch–Gordan Coefficient Squared $[C(JJ_tJ';000)]^2$ for Use in the Cross-Section Formula (62) for the Rotational Transition $J \to J' = J + \Delta J$

$[\Delta J = 0]$ $J \to J$	$J_t = 0$	$J_t = 2$	$J_t = 4$	$J_t = 6$	$J_t = 8$	$J_t = 10$	...		
$0 \to 0$	1								
$1 \to 1$	1	2/5							
$2 \to 2$	1	2/7	2/7						
$3 \to 3$	1	4/15	2/11	100/429					
$4 \to 4$	1	20/77	162/1001	20/143	490/2431				
$5 \to 5$	1	10/39	2/13	80/663	490/4199	756/4199			
$\infty \to \infty$	1	1/4	9/64	25/256	$35^2/4^7$	$63^2/4^8$	...		
$[\Delta J = 1]$ $J \to J+1$	$J_t = 1$	$J_t = 3$	$J_t = 5$	$J_t = 7$	$[\Delta J = -1]$ $J \to J-1$	$J_t = 1$	$J_t = 3$	$J_t = 5$	$J_t = 7$
$0 \to 1$	1				$1 \to 0$	1/3			
$1 \to 2$	2/3	3/7			$2 \to 1$	2/5	9/35		
$2 \to 3$	3/5	4/15	10/33		$3 \to 2$	3/7	4/21	50/231	
$3 \to 4$	4/7	18/77	180/1001	35/143	$4 \to 3$	4/9	2/11	20/143	245/1287
$4 \to 5$	5/9	20/91	2/13	280/1989	$5 \to 4$	5/11	180/1001	18/143	280/2431
$5 \to 6$	6/11	7/33	80/561	420/3553	$6 \to 5$	6/13	7/39	80/663	420/4199
$\infty \to \infty + 1$	1/2	3/16	15/128	350/4096	$\infty \to \infty - 1$	1/2	3/16	15/128	350/4096
$[\Delta J = 2]$ $J \to J+2$	$J_t = 0$	$J_t = 2$	$J_t = 4$	$J_t = 6$	$[\Delta J = -2]$ $J \to J-2$	$J_t = 0$	$J_t = 2$	$J_t = 4$	$J_t = 6$
$0 \to 2$		1			$2 \to 0$		1/5		
$1 \to 3$		3/5	4/9		$3 \to 1$		9/35	4/21	
$2 \to 4$		18/35	20/77	45/143	$4 \to 2$		2/7	100/693	25/143
$3 \to 5$		10/21	20/91	7/39	$5 \to 3$		10/33	20/143	49/429
$4 \to 6$		5/11	20/99	28/187	$6 \to 4$		45/143	20/143	252/2431
$5 \to 7$		63/143	1400/7293	6300/46189	$7 \to 5$		21/65	280/1989	420/4199
$\infty \to \infty + 2$		3/8	5/32	105/1024	$\infty \to \infty - 2$		3/8	5/32	105/1024

$[\Delta J = 3]$ $J \to J + 3$	$J_t = 1$	$J_t = 3$	$J_t = 5$	$J_t = 7$	$[\Delta J = -3]$ $J \to J - 3$	$J_t = 1$	$J_t = 3$	$J_t = 5$	$J_t = 7$
$0 \to 3$		1			$3 \to 0$		1/7		
$1 \to 4$		4/7	5/11		$4 \to 1$		4/21	5/33	
$2 \to 5$		10/21	10/39	21/65	$5 \to 2$		50/231	50/429	21/143
$3 \to 6$		100/231	7/33	168/935	$6 \to 3$		100/429	49/429	1176/12155
$4 \to 7$		175/429	1400/7293	6804/46189	$7 \to 4$		35/143	280/2431	20412/230945
$5 \to 8$		56/143	490/2717	360/2717	$8 \to 5$		56/221	490/4199	360/4199
$\infty \to \infty + 3$		5/16	35/256	189/2048	$\infty \to \infty - 3$		5/16	35/256	189/2048

$[\Delta J = 4]$ $J \to J + 4$	$J_t = 0$	$J_t = 2$	$J_t = 4$	$J_t = 6$	$[\Delta J = -4]$ $J \to J - 4$	$J_t = 0$	$J_t = 2$	$J_t = 4$	$J_t = 6$
$0 \to 4$			1		$4 \to 0$			1/9	
$1 \to 5$			5/9	6/13	$5 \to 1$			5/33	18/143
$2 \to 6$			5/11	14/55	$6 \to 2$			25/143	14/143
$3 \to 7$			175/429	504/2431	$7 \to 3$			245/1287	1176/12155
$4 \to 8$			490/1287	504/2717	$8 \to 4$			490/2431	4536/46189
$5 \to 9$			882/2431	420/2431	$9 \to 5$			882/4199	420/4199
$\infty \to \infty + 4$			35/128	63/512	$\infty \to \infty - 4$			35/128	63/512

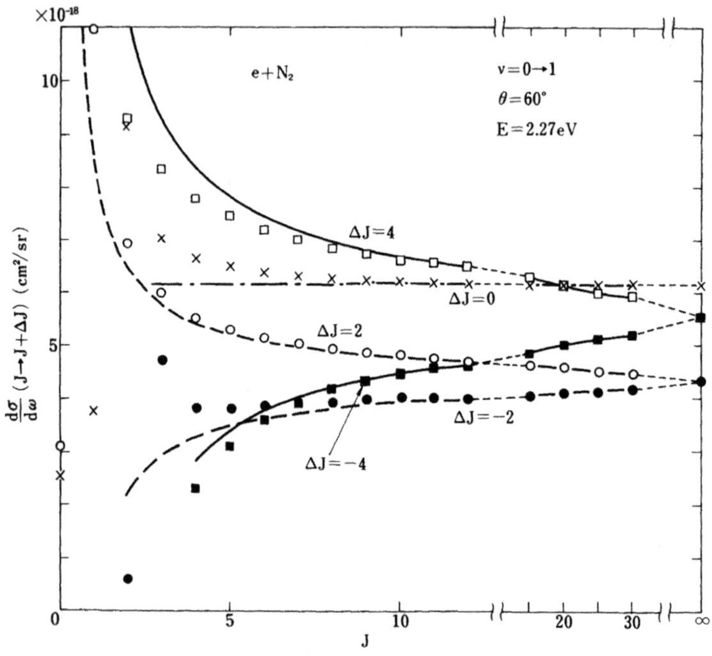

FIGURE 7
Example of the J dependence of the differential cross sections for the vibrational-rotational transitions $(v, J) = (0, J) \to (1, J + \Delta J)$. Scattering of electrons of energy $E = 2.27$ eV by N_2 in the ground electronic state $(X\,^1\Sigma_g^+)$ into the scattering angle $\theta = 60°$. Analysis by Shimamura[43] of energy-loss spectrum measured by Wong and Dubé:[45] ×, $\Delta J = 0$; ○, ●, $\Delta J = \pm 2$; □, ■, $\Delta J = \pm 4$. Asymptotic formula (13a) with the term of order J^{-2} neglected: —·—, $\Delta J = 0$; – – –, $\Delta J = \pm 2$; —, $\Delta J = \pm 4$ (taken from Shimamura[44]).

vibronic transition $n \to n'$, and not $n' \to n$ on the right-hand side. Fourth, inspection of the numerical values of the Clebsch–Gordan coefficients reveals many general inequalities between the cross sections for different rotational transitions. For example, the cross section for the transition $n00 \to n'00$ is never greater than that for the transition $nJK \to n'JK$, because $\{C(J0J; K0K)\}^2 = 1$ according to Eqs. (39) and (40a). Similarly, the cross section for $n10 \to n'00$ is the smallest among all the cross sections for $nJ0 \to n', J-1, 0$ for all J, the cross section for $n20 \to n'00$ is the smallest among all $nJ0 \to n', J-2, 0$ for all J, and so on.

An electron and a molecule usually exchange only a small number of angular-momentum quanta. Therefore, the sum over J_t in Eq. (62) is practically restricted to a few low-lying rotational states $|J_tK_t\rangle$, other angular-momentum-transfer cross sections being negligibly small. Then, only the cross sections for a few transitions from the ground rotational state are linearly independent. Study of the transitions from excited rotational states adds no new information (except that for the check of consistency), if the sudden approximation is valid. If it is invalid, but not very much so, a convenient set of physical quantities to be studied consists of the angular-momentum-transfer cross sections $d\sigma(n00 \to n'J_tK_t)/d\omega$ and the nonsudden indices η_0 defined by

$$\eta_0(nJK \to n'J'K') = \left[\frac{d\sigma(nJK \to n'J'K')}{d\omega} \bigg/ \frac{d\sigma_0^{(s)}(nJK \to n'J'K')}{d\omega}\right] - 1, \quad (66)$$

where $d\sigma_0^{(s)}/d\omega$ is the cross section calculated from $d\sigma(n00 \to n'J_tK_t)/d\omega$ by use of Eq. (62). In this way independent and physically transparent information emerges from numerical data containing otherwise mixed-up information.

For the spherical top we may sum the cross section over K' and average it over K to define the degeneracy-averaged value $d\sigma(nJ \to n'J')/d\omega$. The use of Eq. (19c) in Eq. (62) and summation over K using Eq. (38b), in parallel to the classical theory in Section 2.2.2, lead to a result that [16,43]

$$\left(\frac{k}{k'}\right)\frac{d\sigma(nJ\mathbf{k} \to n'J'\mathbf{k}')}{d\omega} = \sum_{J_t=|J-J'|}^{J+J'} \left[\frac{g(J')}{g(J_t)g(J)}\right]\left(\frac{k}{k_t}\right)\frac{d\sigma(n0\mathbf{k} \to n'J_t\mathbf{k}_t)}{d\omega} \quad (67)$$

similar to Eq. (26). Neglecting the rotational-channel dependence of the wave numbers as before, we may prove Eqs. (13)–(15) with $c_1 = 1$, Eq. (65) in which the coefficient should be replaced by $1/g(J_t)$, and the principle of detailed balance (27) with $g(J) = 2J + 1$ with the understanding that the two cross sections are for the same vibronic transition $n \to n'$.

For the asymmetric top the relation (58) between the (complex) scattering amplitudes does not seem to reduce to a relation involving only the usual (real) cross sections, because the cross terms in Eq. (59) survive.[46,47]

Equation (62) shall now be applied, as an example, to scattering of low-energy electrons by the ground-state hydrogen molecules. Because the small nuclear charges make the potential singularities at the nuclear positions weak, and because the nuclei are close to each other ($R_e \cong 1.4$ a.u.), the charge distribution of the ground-state H_2 is nearly spherically symmetric except deep in the electron cloud. Therefore, the electron and H_2 exchange the angular momenta inefficiently except in head-on collisions. The integral and small-angle differential cross sections for the rotational transitions with $|\Delta J| = 2$ are much smaller than the rotationally elastic cross sections $\Delta J = 0$;[34,48,49] see Figure 11a in Section 5.1. The $|\Delta J| \geqq 4$ cross sections are negligibly small.[34,48] The transitions with odd ΔJ are forbidden (see Section 4.2.2), so that ortho-para conversion never occurs in collisions with electrons.

Neglect of the terms with $J_t \geqq 4$ in Eq. (62) results in the relations

$$\frac{d\sigma}{d\omega}(nJ\mathbf{k}\to n'J\mathbf{k}'_{nJ\to n'J}) = \frac{d\sigma}{d\omega}(n0\mathbf{k}\to n'0\mathbf{k}'_{n0\to n'0}) + \left(\frac{k'_{n0\to n'0}}{k'_{n0\to n'2}}\right)\frac{J(J+1)}{(2J-1)(2J+3)} \times \frac{d\sigma}{d\omega}(n0\mathbf{k}\to n'2\mathbf{k}'_{n0\to n'2}),$$

$$\frac{d\sigma}{d\omega}(nJ\mathbf{k}\to n',J\pm 2,\mathbf{k}'_{nJ\to n'J\pm 2}) = \left(\frac{k'_{nJ\to n'J\pm 2}}{k'_{n0\to n'2}}\right)\frac{3J_>(J_>-1)}{2(2J+1)(2J_>-1)} \times \frac{d\sigma}{d\omega}(n0\mathbf{k}\to n'2\mathbf{k}'_{n0\to n'2}), \tag{62a}$$

where $J_>$ is the larger of J and J' $(=J\pm 2)$. The second relation is just what the Born approximation with the quadrupole interaction predicts (Section 5.4 of this chapter and Chapter 1, Section 6.3), because only the $J_t = 2$ term survives in Eq. (62) in this approximation. Equations (62a) are valid even when the Born approximation fails.[37] The cross section $d\sigma(nJ\to n', J+2)/d\omega$ of Eq. (62a) is a monotonically decreasing function of J, and $d\sigma(nJ\to n', J-2)/d\omega$ is a monotonically increasing function of J. The cross section $d\sigma(nJ\to n'J)/d\omega$ decreases monotonically from $J=1$ onwards, but $d\sigma(n0\to n'0)/d\omega$ is the smallest. When the ratio $R^{02}_{nn'}$ of $d\sigma(n0\to n'0)/d\omega$ to $d\sigma(n0\to n'2)/d\omega$ is much larger than unity, as is the case with the vibronically elastic scattering by H_2, the cross section $d\sigma(nJ\to n'J)/d\omega$ is almost independent of J according to Eq. (62a). Lane and Geltman,[48] for example, observe this numerically on H_2.

Electron energy-loss experiments (Section 3.3.3) on H_2 yield direct information on the ratio $R_{vv'}$ of the rotationally elastic cross section for $vJ\to v'J$ averaged over the rotational distribution to the cross section for the vibrational-rotational transition $v1\to v'3$.[34,49,50] This ratio is related to the ratio $R^{02}_{vv'} = [d\sigma(v0\to v'0)/d\omega]/[d\sigma(v0\to v'2)/d\omega]$ by

$$R_{vv'} = \frac{\langle d\sigma(vJ\mathbf{k}\to v'J\mathbf{k}'_{vJ\to v'J})/d\omega\rangle}{d\sigma(v1\mathbf{k}\to v'3\mathbf{k}'_{v1\to v'3})/d\omega} = \frac{k'_{v0\to v'2}R^{02}_{vv'} + k'_{v0\to v'0}\langle C^2\rangle}{(3/5)k'_{v1\to v'3}}, \tag{62b}$$

where $\langle C^2\rangle$ is the average of $J(J+1)/\{(2J-1)(2J+3)\}$ over the rotational distribution. This relation allows one to extract from experimental data the purely dynamical quantity $R^{02}_{vv'}$ suitable for theoretical studies. Note that Eq. (62b) is valid provided that the *rotational* sudden approximation is good and that the transitions $v0\to v'J'$ with $J'\geqq 4$ are negligible. The dependence of the experimentally observed ratio $R_{0v'}$ on v' has attracted much theoretical attention; see Ref. 4 for summary.

3.3.2. Sum Rules

Application of the unitarity relation (38a) to Eq. (62) results in a generalization of the classical sum rules (6) and (24), namely,

$$S(0;JK) = \sum_{K'J'} \frac{d\sigma(nJK \to n'J'K')}{d\omega} = \sum_{K_tJ_t} \frac{d\sigma(n00 \to n'J_tK_t)}{d\omega} = S(0;00). \tag{68}$$

Similarly, application of Eq. (42) leads to a generalization of Eqs. (11) and (25), namely,[16,47,51]

$$\begin{aligned} S(1;JK) &= \sum_{K'J'} [E_r(J'K') - E_r(JK)] \frac{d\sigma(nJK \to n'J'K')}{d\omega} \\ &= \sum_{K_tJ_t} [E_r(J_tK_t) - E_r(00)] \frac{d\sigma(n00 \to n'J_tK_t)}{d\omega} = S(1;00) \geqq 0 \end{aligned} \tag{69}$$

for a rotator having a plane of symmetry. For a rotator without a plane of symmetry the average of $S(1;JK)$ over $\pm K$ satisfies the sum rule; see the remark just below Eq. (25). [In Eqs. (68) and (69) the summations over K' and over K_t should be dropped for the linear rotator.] Further average over K of Eqs. (68) and (69), which are independent of K, shows that $S(0;J)$ and $S(1;J)$ for the spherical top are independent of J. Higher-order moments exhibit other regularities. In particular, the moments $S(2s;J)$ and $S(2s+1;J)$ for the linear and spherical-top rotators are polynomials in $E_r(J)$ of degree s.[12] The principle of detailed balance transforms the sum rules into those for the sums over the initial rotational state, giving useful information on transitions to a specified rotational state.[52]

A more general proof of the sum rules warrants an explanation. Drozdov,[17,19] and later, Altshuler[21] have noted that, due to the completeness of the rotational wave functions,

$$\begin{aligned} \sum_{\Gamma'} \frac{d\sigma(n\Gamma \to n'\Gamma')}{d\omega} &= (k_{n'}/k) \sum_{\Gamma'} (D_\Gamma | F^* | D_{\Gamma'})(D_{\Gamma'} | F | D_\Gamma) \\ &= (k_{n'}/k)(D_\Gamma | F^* F | D_\Gamma). \end{aligned} \tag{70}$$

The average of $|D_\Gamma|^2$ over all M is an invariant V_Ω^{-1}.[14,51] This proves that

$S(0;\, JK)$[19,22,36] and $S(0;\, J\tau)$ defined similarly for the asymmetric top[51] are indeed independent of (J, K) or of (J, τ).

This argument may be carried over, with modifications, to the mean-energy-transfer cross section with a result that

$$\{S(\mu; JK) \text{ or } S(\mu; J\tau)\} = V_{\Omega}^{-1}(k_{n'}/k)(F|H_r^{\mu}|F) \geqq 0 \qquad (\mu = 0, 1) \quad (71)$$

for the linear, symmetric-top, and asymmetric-top rotators.[51] The result for the simple sum ($\mu = 0$) has been included here.

Table 2 illustrates the sum rules for $\mu = 0$ and 1 for a linear rotator on the basis of the rotational-sudden calculations by Chandra and Temkin.[53]

Molecules in a gas usually occupy many different rotational states. The summed cross section and the mean-energy-transfer cross section for the gas are averages of Eqs. (68) and (69) or Eq. (71) over the rotational distribution. Owing to the sum rules these averages are independent of the rotational distribution or of the temperature T of the gas. The mean energy transfer from electrons to molecules is always positive or zero according to Eq. (69) or Eq. (71). As the electron temperature T_e approaches T from above, Eqs. (68)–(71) begin to fail; k/k' and k/k_t begin to depend appreciably on the rotational channel, and at even lower T_e the sudden approximation to the scattering amplitude begins to fail. Assuming the sudden approximation to be still valid, we may correct Eqs. (68) and (69) for the rotational-channel dependence of the wave-number ratios to the first order. Then, $S(0; JK)$ is still independent of the initial state, but $S(1; JK)$ depends weakly on $J(J + 1)$ and K^2. After averaging over the rotational distribution we obtain[12]

$$\langle S(1; JK)\rangle_{\text{av}} \cong S(1; 00)(1 - T/T_e). \qquad (72)$$

In the discussions on the sudden approximation we have assumed the incident particle to have no internal structure. In fact modification necessary for a particle with structure is easy. Essentially, we have only to regard n as designating not only the vibronic state of the molecule but also any internal state of the incident particle. For the regularities of the sudden-collision cross sections and amplitudes for atom–molecule and molecule–molecule collisions, see Refs. 39–41, 46, and 54, and other references cited therein. Ostrovsky and Ustimov[55] extend the discussion to various angular-momentum coupling schemes for the molecular electronic states.

The relations in this subsection and in the preceding subsection have resulted from the form of Eq. (52) with Eq. (45). Except for the wave-number ratio this is just the form of the transition probability. In this sense the relations are quite general. They depend only on the property of the angular-momentum transfer, and not on the dynamics of the transitions, as has been emphasized in Sections 1 and 2. Thus, for example, the photoionization cross sections satisfy these relations in the sudden approximation for the final continuum state.[38,56,57]

TABLE 2

Illustration of the Sum Rules $S(\mu; J) = \Sigma_{J'} [E_r(J') - E_r(J)]^{\mu} \sigma(J \to J')$ ($\mu = 0, 1$) for Purely Rotational Transitions $J \to J'$ of N_2[a]

E (eV)	$\sigma(0 \to 0)$ (a_B^2)	$\sigma(0 \to 2)$ (a_B^2)	$\sigma(0 \to 4)$ (a_B^2)	$S(0;0)$ (a_B^2)	$S(1;0)$ (meV $\cdot a_B^2$)
0.1	8.698	0.240	3.3×10^{-4}	8.938	0.358
0.3	11.92	0.287	5.4×10^{-4}	12.21	0.430
1.0	14.29	0.513	0.022	14.82	0.870
2.3	22.06	3.861	8.697	34.62	48.88
5.0	11.56	3.086	1.204	15.85	10.56
E (eV)	$\sigma(1 \to 1)$ (a_B^2)	$\sigma(1 \to 3)$ (a_B^2)	$\sigma(1 \to 5)$ (a_B^2)	$S(0;1)$ (a_B^2)	$S(1;1)$ (meV $\cdot a_B^2$)
0.1	8.793	0.144	2.2×10^{-4}	8.938	0.359
0.3	12.04	0.173	3.4×10^{-4}	12.21	0.430
1.0	14.49	0.317	0.012	14.82	0.870
2.3	23.61	6.182	4.832	34.62	48.89
5.0	12.79	2.386	0.669	15.85	10.56

[a]The integral cross sections $\sigma(J \to J')$ for electrons of energy E are taken from Chandra and Temkin.(53) They calculate the cross sections using their hybrid method in which the vibrational channels are closely coupled and rotational-sudden approximation is used. a_B is the Bohr radius.

3.3.3. *Application: Analysis of Energy-Loss Spectra*

The cross sections for excitation and deexcitation of molecules in collisions with electrons can be measured, in principle, as follows (Chapter 6, Section 4). A beam of monoenergetic electrons (with energy E) coming out of an electron monochromator is crossed with a beam of molecules in a collision chamber. Electrons scattered within the acceptance angle of an analyzer is energy analyzed and detected. A peak in the detected intensity $I(E')$ (or the energy-loss spectrum) as a function of the energy E' of the scattered electrons (or of the energy loss $\Delta E = E - E'$) is an evidence of a molecular transition having the transition energy $E - E'$. In this way we may determine from $I(E')$ relative values of the cross sections for different transitions.

In practice the spectrometer has a finite ($\gtrsim 10^1$ meV) energy resolution, and closely spaced peaks cannot be resolved. Rotational transition peaks in the energy-loss spectra have been resolved so far only for H_2, for which the peak spacings ($\sim 4B$) are ~ 30 meV,(34,49,50,58,59) and for D_2, for which the spacings are $4B \sim 15$ meV.(60) Because of the small peak spacings of ~ 1 meV, the rotational transition peaks for N_2 overlap each other and form a broad feature.(45,61–63) This makes it difficult to determine the cross sections; too many rotational transitions contribute to the shape of the broad feature. We have seen, however, that not all the cross sections are linearly independent. Equation (62) greatly reduces the number of independent cross sections, and makes line-shape analysis practicable. We note that for molecules heavier than the hydrogen the sudden approximation is valid at the collision energies $\gtrsim 10^{-1}$ eV at which crossed-beam experiments are feasible; the only exceptions are very narrow resonances, very close to a vibronic or vibrational excitation threshold, and forward scattering by dipolar molecules (Section 3.2.2).

Suppose that the spectrometer produces an elastic scattering peak $I_0(E')$. This function, called the apparatus function, may be measured by replacing the molecular beam by a beam of rare gas atoms, because the elastic peak is well separated from inelastic peaks for scattering by atoms without rotational or vibrational motion. The peak for a transition $nJ \to n'J'$ of a linear rotator is shifted from the elastic peak by the transition energy $\Delta E(nJ \to n'J')$. The peak height is proportional to the differential cross section and to the number $N_n(J)$ of molecules in the state $|nJ\rangle$. Therefore, the energy spectrum of scattered electrons corresponding to the vibronic or vibrational transition $n \to n'$ is the sum of the peaks for all rotational transitions, i.e.,

$$I(E') \propto \sum_{JJ'} N_n(J) \frac{d\sigma(nJ \to n'J')}{d\omega} I_0(E' + \Delta E). \tag{73}$$

We have assumed the cross section to be almost constant within each peak.

Substitution of Eq. (62) into Eq. (73) and neglect of the kinematic factors lead to a linear combination[43]

$$I(E') \propto \sum_{J_t} \bar{I}_{J_t}(E') \frac{d\sigma(n0 \to n'J_t)}{d\omega}, \tag{74a}$$

where

$$\bar{I}_{J_t}(E') = \sum_{JJ'} N_n(J) I_0(E' + \Delta E) \{C(JJ_tJ'; 000)\}^2. \tag{74b}$$

Because the Boltzmann distribution is usually a good approximation to $N_n(J)$, the basic spectra (74b) may be assumed known.

For electron scattering usually the cross sections $d\sigma(n0 \to n'J_t)/d\omega$ for only a few J_t values are appreciable. Odd J_t are absent from the summation in the case of $^1\Sigma$ homonuclear diatomic molecules and of linear polyatomic molecules with a center of symmetry (point group $D_{\infty h}$). For low-energy $e + N_2$ scattering, for example, only $J_t = 0, 2$, and 4 need to be retained in Eq. (74a).

The smallness of the number of terms in Eq. (74a) allows determination of relative values of $d\sigma(n0 \to n'J_t)/d\omega$ by a least-squares fit of Eq. (74a) to an observed energy-loss spectrum. The absolute values of these cross sections may then be obtained from the absolute value of the sum of them. This sum is equal to the rotationally unresolved cross section measured at any temperature T according to the sum rule. The absolute cross sections $d\sigma(n0 \rightarrow n'J_t)/d\omega$ being determined, the cross sections for any rotational transitions $nJ \rightarrow n'J'$ are calculable from Eq. (62). The cross sections in Fig. 7 have been extracted from energy-loss spectra in this way.[43,44] Tanaka *et al.* also apply this line-shape analysis.[62]

The spectrum $\bar{I}_0(E')$ has contributions only from rotationally elastic scattering $nJ \rightarrow n'J$, and has the same shape as the apparatus function $I_0(E')$. For a least-squares fit to be accurate the contributions to Eq. (74a) from terms other than $J_t = 0$ must be appreciable. In other words the spectrum must be broadened by the basic spectra $\bar{I}_{J_t}(E')$ for $J_t > 0$, which are broader than $I_0(E')$. This broadening has been discussed by Read[64] and observed for N_2 first by Comer and Harrison.[61] The broadening is better observable for lighter molecules and at higher T (because ΔE is larger for larger B and J), for higher energy resolution, and for larger ratios of $d\sigma(n0 \rightarrow n'J_t)/d\omega$ for $J_t > 0$ to $d\sigma(n0 \rightarrow n'0)/d\omega$. With the resolution of the contemporary spectrometers the broadening is observable only for molecules lighter than or comparable in weight to N_2. In the absolute measurements of the rotationally unresolved cross sections one may lower the energy resolution and get higher signal intensities.

Reference 43 discusses the use of the asymptotic form (64) for the Clebsch–Gordan coefficient in Eq. (74b). Read[64] suggests an infinite-J approximation, which amounts to retaining only the leading term in Eq. (64a). In this approximation an energy-loss spectrum may be divided into branches labeled by ΔJ. Wong and Dubé use this approximation together with the principle of detailed balance to analyze the energy-loss spectra measured by themselves.[45]

Neglect of the term of order J^{-2} in Eq. (64b) is equivalent to the use in Eq. (73) of the asymptotic cross sections (13b) with Eq. (65). Jung *et al.* apply this high-J approximation to scattering by N_2 and by CO, and extract high-J cross sections.[63]

Finally, we note that the above line-shape analysis is applicable with a minor change to spherical tops and symmetric tops.[43] It is applicable even to asymmetric tops, in principle, although the number of adjustable parameters is larger.[47] It is possible also to analyze energy spectra of photoelectrons using the same principle.

4. Division of the Configuration Space

4.1. Dynamics in Different Regions

Consider an electron coming from far away and slowly approaching a molecule. The electron first finds the net point charge, if any, of the molecule by feeling the Coulomb potential that behaves as $\sim r^{-1}$. Because this potential is independent of the orientation of the molecule, the rotational motion is unaffected at this stage.

On further approach to the molecule the electron sees the charge distribution. First, the electron notices only the charge bias, or the dipole moment, through the point dipole potential behaving as $\sim r^{-2}$. Then, the electron sees a finer structure, such as the quadrupole moment through a potential proportional to r^{-3}, octupole moment through a potential proportional to r^{-4}, and so on. The electron also perturbs the molecular charge distribution, and recognizes the charge deformation through the polarization potential behaving as $\sim r^{-4}$ at large r.

These long-range potentials depend on the molecular orientation. If the electron is moving slowly enough, it learns that the molecule is rotating. The rotational motion is in turn perturbed, but not very much, because the interaction in this region is weak. Then, the rotational angular momentum $\hbar \mathbf{J}$ of the molecule is only weakly coupled with the angular momentum $\hbar \mathbf{l}$ of the electron. We may conveniently use the rotational wave functions $|\hbar \mathbf{J}\rangle$ and the spherical harmonics Y_{lm} for the electron to expand the total wave function for the collision system. An appropriate coordinate system has an orientation fixed in space. In other words, the laboratory frame of reference is appropriate in this region (Fig. 8).

As the electron–molecule distance becomes shorter, the interaction between them becomes stronger. The electron acquires a high velocity in this region, even if its initial velocity is low. Its motion is strongly coupled with the

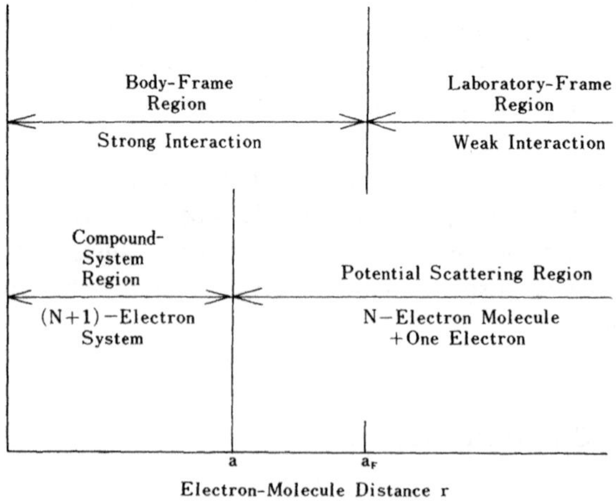

FIGURE 8
Division of the configuration space into regions with qualitatively different features of the electron–molecule interaction (after Chang and Fano[67]). The abscissa, not drawn to scale, represents an effective distance between the electron and the molecule, and is not necessarily the radial distance of the electron from the center of mass of the molecule.

molecular axes, and the symmetry-axis component $\hbar\lambda$ of the angular momentum of the electron is a fairly good quantum number. Thus the coordinate system fixed on the molecule, i.e., the molecule-fixed or body-fixed frame of reference is appropriate here (Fig. 8). A convenient basis set for the expansion of the total wave function consists of the spherical harmonics $Y_{l\lambda}$ referred to the body-fixed frame.

Whichever frame of reference we may choose and whichever basis set we may use, it makes no difference to the resultant wave function and to the resultant scattering parameters, if we fully use the complete set without approximation. The two approaches are mathematically equivalent; one is related to the other simply by a unitary transformation.[65–67] The use of the laboratory frame at large electron–molecule distances and the body-fixed frame at small distances, however, allows truncation of the complete sets to small basis sets without any serious error, according to the above discussion on physical grounds. For connecting the wave functions in the two regions we transform the frame of reference and the basis set on the boundary between the regions. This is the essence of the frame-transformation theory of Chang and Fano.[67]

That the electron velocity is high in the body-frame region suggests the approximate separability of the rotational part from the total wave function, even if the collision time is long. Then, the rotational channels are uncoupled. This greatly simplifies the treatment in this region regardless of the collision energy or the collision time.

When the electron is outside the molecular electron cloud, the electron–molecule interaction is expressible as a potential matrix. Description as a two-body multichannel scattering is appropriate here (Fig. 8). The potential is a smoothly varying function of the position of the electron relative to the molecule. The single-center expansion of the potential converges rapidly.

Once the electron enters the region of the molecular electron cloud, it is indistinguishable from the molecular electrons; the Pauli principle is important. All electrons in the system are strongly correlated and must be treated on a par as forming a compound system (Fig. 8). The electronic channels of two-body scattering are ill defined. Furthermore, the field is essentially multicentered and has a singularity at each center.

The wave function in the compound-system region is to be connected smoothly on the surface of the molecular electron cloud with the wave function of the potential-scattering type outside the molecule. This is the basic idea of the R-matrix theory of electron–molecule collisions as discussed by Burke *et al.*[68]

Almost every conventional approach to the atomic collision relies on a single representation of the wave function in the whole region of the configuration space. In view of the qualitatively different physical situations depending on the electron–molecule distance, a desirable approach is to divide the configuration space into some regions in most general cases; different descriptions of the system in different regions will clarify the physical picture and probably simplify the theoretical approach. The division into regions in the light of the rotational motion invokes the frame-transformation theory. That in the light of the electronic motion calls for the R-matrix theory briefly mentioned above, though the idea of the R-matrix theory is more general. The frame-transformation theory is generalizable for the vibrational degrees of freedom.[67] The scope of this chapter, however, excludes discussion on this generalization.

Fano stresses the importance of expanding the scattering wave function into partial-wave components.[69] In the real scattering problem a particle comes from infinity, is scattered by another particle, and goes to infinity in any direction. The partial-wave expansion converts this problem into another problem of following a fictitious particle emerging from the origin and going towards infinity always in one direction on a single coordinate axis q, which may be but need not be the single-particle radial distance r. This way of looking at the scattering problem shows clearly the importance of the angular momenta, and simplifies the definition of the separate regions and the subsequent mathematical manipulations.

4.2. *Frame Transformation*

4.2.1. *Basis Sets of Expansion*

Section 4.2 deals only with the symmetric-top rotator without spin for simplicity. The spin of the incident electron is also neglected. The angular momentum $\hbar\mathbf{J} = (\hbar J, \hbar M)$ of the rotator added to the angular momentum $\hbar\mathbf{l} = (\hbar l, \hbar m)$ of the electron makes the total angular momentum $\hbar\mathbf{J} + \hbar\mathbf{l} = \hbar\bar{\mathbf{J}} = (\hbar\bar{J}, \hbar\bar{M} = \hbar M + \hbar m)$, which is a constant of motion in the absence of an external perturbation.

A convenient complete set of functions for describing the rotational motion during the collision consists of eigenfunctions of $\bar{\mathbf{J}}^2$ and $\bar{M}$. In the laboratory-frame region where $\hbar\mathbf{J}$ and $\hbar\mathbf{l}$ are weakly coupled, the most convenient eigenfunctions are formed by coupling these two as in Eq. (37):

$$X_J^L(\Omega, \hat{\mathbf{r}}; \bar{J}\bar{M}lK) = \sum_M C(Jl\bar{J}; Mm\bar{M}) D_{KM}^J(\Omega) Y_{lm}(\hat{\mathbf{r}}). \tag{75}$$

For the linear rotator for which $K = 0$ this basis function has the parity of $(-1)^{J+l}$ under inversion $\hat{I}$ of all coordinates.

In the body-frame region the electron coordinate $\mathbf{r}'$ refers to the body-fixed frame. According to the arguments in Section 4.1 eigenfunctions of $\bar{\mathbf{J}}^2$ and $\bar{M}$ appropriate as basis functions are[4,67]

$$X_\lambda^B(\Omega, \hat{\mathbf{r}}'; \bar{J}\bar{M}lK) = D_{\bar{K}\bar{M}}^{\bar{J}}(\Omega) Y_{l\lambda}(\hat{\mathbf{r}}') \qquad (\bar{K} = K + \lambda). \tag{76}$$

A unitary matrix $\mathbf{U}$ transforms one basis set to the other as

$$X_J^L = \sum_\lambda X_\lambda^B U_{\lambda J} \quad \text{or} \quad \mathbf{X}^{L\dagger} = \mathbf{X}^{B\dagger}\mathbf{U}. \tag{77a}$$

Here, $\mathbf{X}^L$ and $\mathbf{X}^B$ are column vectors, and the daggers indicate transposed row vectors. We may compute the matrix elements $U_{\lambda J}$ using Eqs. (75) and (76); we transform $Y_{lm}(\hat{\mathbf{r}})$ from the laboratory frame to the body-fixed frame according to Eq. (32a), and use the integral formula (60) and the orthogonality (38b) of the Clebsch–Gordan coefficients. Then we have

$$U_{\lambda J} = (X_\lambda^B, X_J^L) = \{g(J)/g(\bar{J})\}^{1/2} C(Jl\bar{J}; K\lambda\bar{K}). \tag{77b}$$

Chang and Fano[67] construct basis functions that are eigenfunctions of the inversion operator $\hat{I}$. Harter *et al.*[70] review extension of the frame-transformation theory to more general polyatomic molecules in a different context.

4.2.2. Laboratory Frame of Reference

The Hamiltonian H for the total collision system consists of the kinetic energy $-(\hbar^2/2\mu)\Delta$ of the electron, the vibronic Hamiltonian H_v and the rotational Hamiltonian H_r of the molecule, and the interaction $(\hbar^2/\mu)V$ between the electron and the molecule. The total wave function $\Psi(\mathbf{r}, \mathbf{X}_T)$ satisfies the Schrödinger equation

$$(H-E)\Psi(\mathbf{r}, \mathbf{X}_T) = [H_v + H_r + (\hbar^2/2\mu)(-\Delta + 2V) - E]\,\Psi(\mathbf{r}, \mathbf{X}_T) = 0. \tag{78}$$

The formulation in the laboratory-frame region, to be outlined in the following, involves an expansion of the wave function[71]

$$\Psi^L(\mathbf{r}, \mathbf{X}_T) = \sum_p r^{-1} f_p^L(r)\Phi_n(\mathbf{x}_T) X_J^L(\Omega, \hat{\mathbf{r}}; \bar{J}\bar{M}lK), \tag{79}$$

where $p = (nJKl\bar{J})$; omission of $\bar{M}$ from this index will shortly prove to be appropriate. The expansion (79) is over a finite number of channels p in any practical treatments. Especially, the vibronic states Φ_n are usually assumed to be separable into electronic and vibrational parts, and to belong to only one electronic state, or a few at most.

Assume that the functions Φ_n are known. The only unknown functions are $f_p^L(r)$. We determine these functions so as to optimize variationally the scattering parameters extracted from $\Psi^L(\mathbf{r}, \mathbf{X}_T)$. This is accomplished by using Eq. (79) in the Schrödinger equation (78) and by projecting this equation onto each basis function $\Phi_{n'}(\mathbf{x}_T)X^L_{J'}(\Omega, \hat{\mathbf{r}}; \bar{J}\bar{M}l'K')$. This projection involves integration over all coordinates but r. Because $H_v\Phi_n = E_v(n)\Phi_n$ and $H_rX^L_J = E_r(JK)X^L_J$, we obtain, after interchange of p and p', coupled ordinary differential equations for $f^L_p(r)$[71]

$$\left[\frac{d^2}{dr^2} + k_p^2 - \frac{l(l+1)}{r^2}\right] f^L_p(r) - 2\sum_{p'} V^L_{pp'}(r) f^L_{p'}(r) = 0, \tag{80}$$

$$V^L_{pp'}(r) = (\Phi_n(\mathbf{x}_T)X^L_J(\Omega, \hat{\mathbf{r}}; \bar{J}\bar{M}lK)|V|\Phi_{n'}(\mathbf{x}_T)X^L_{J'}(\Omega, \hat{\mathbf{r}}; \bar{J}\bar{M}l'K')), \tag{80a}$$

$$E = \hbar^2 k_p^2/2\mu + E_v(n) + E_r(JK); \tag{80b}$$

see also Chapter 1 for the derivation of similar coupled equations.

Because $\bar{J}$ and $\bar{M}$ are conserved, channels with different $\bar{J}$ or $\bar{M}$ uncouple. Also, $V^L_{pp'}$, and hence f^L_p, are independent of $\bar{M}$ owing to the invariance of V under the rotation of all coordinates, which leaves the relative positions of all particles unchanged. This justifies the exclusion of $\bar{M}$ from the index p. Furthermore, among the vibronic states regarded as linear rotators, channels with different parities uncouple because of the even parity of V. In other words channels with even $J + l$ are uncoupled from channels with odd $J + l$. For the homonuclear diatomic molecule in a $^1\Sigma$ state the parity of J is fixed by the nuclear permutation symmetry (s or a), and no coupling exists between even-l and odd-l channels. This is also the case with the polyatomic linear molecule in a $^1\Sigma$ state, such as the ground-state CO_2, that has the $D_{\infty h}$ symmetry, i.e., that has a center of symmetry through which reflection of the identical nuclei may occur. Chang and Fano[67] discuss parity uncoupling for the symmetric top with $K \neq 0$. See also Temkin and Faisal.[72]

The physical solutions to the coupled equations (80) are matched to the wave function in the body-frame region on the boundary $r = a_F$ between this region and the laboratory-frame region; see Eq. (91) in Section 4.2.3. If there are n_p coupled channels, there are n_p linearly independent solution vectors, each having n_p elements $f^L_p(r)$ and satisfying the boundary condition at $r = a_F$. With a second index p' labeling these vectors the functions $f^L_p(r)$ satisfy the asymptotic boundary condition

$$f^L_{pp'}(r) \underset{r\to\infty}{\sim} k_p^{-1/2}\left[\sin\left(k_p r - \frac{l\pi}{2}\right)\delta_{pp'} + \cos\left(k_p r - \frac{l\pi}{2}\right)K^L_{pp'}\right] \qquad (k_p^2 > 0), \tag{81a}$$

$$f^L_{pp'}(r) \underset{r\to\infty}{\sim} 0 \qquad (k_p^2 < 0). \tag{81b}$$

This defines the reactance matrix $\mathbf{K}^L$. If the molecule is an ion, the arguments in the sine and cosine functions are different from the above form because of the asymptotic Coulomb scattering (see Chapter 7, Section 2.1).

$\mathbf{K}^L$ is related to the scattering and transition matrices $\mathbf{S}^L$ and $\mathbf{T}^L$ by

$$\mathbf{T}^L = 1 - \mathbf{S}^L = -2i\mathbf{K}^L(1 - i\mathbf{K}^L)^{-1}. \tag{82}$$

To relate the transition matrix to the scattering amplitude $f(n\Gamma \to n'\Gamma')$ we require that Eq. (79) should have the asymptotic form (44) without the approximation (45). Then we may prove that[71]

$$f(n\Gamma \to n'\Gamma') = (\pi/k_p k_{p'})^{1/2} \sum_{\bar{J}ll'} i^{l-l'} \{g(l)\}^{1/2} \times C(Jl\bar{J}; M0\bar{M}) C(J'l'\bar{J}; M'm'\bar{M}) T^L_{pp'} Y_{l'm'}(\theta, \varphi). \tag{83}$$

The two Clebsch–Gordan coefficients stem from Eq. (79) for the initial and final channels. Because of the choice of the z axis in the direction of the incident electron, the z component m of $\mathbf{l}$ vanishes.

The differential cross section (52) involves summation over $\bar{J}_i$, l_i, and l'_i ($i =$ 1, 2), the subscripts $i = 1$ corresponding to $f(n\Gamma \to n'\Gamma')$ and $i = 2$ to $f^*(n\Gamma \to n'\Gamma')$. The Clebsch–Gordan series[11, 14]

$$4\pi Y_{l'_1 m'}(\theta, \varphi) Y^*_{l'_2 m'}(\theta, \varphi) = \delta_{l'_1 l'_2} P_0(\cos\theta) + \sum_{s=1}^{\infty} g(s) c_s P_s(\cos\theta) \tag{84}$$

(with constants c_s expressed in terms of the Clebsch–Gordan coefficients) casts the degeneracy-averaged cross section into the form of a series

$$\frac{d\sigma(nJK \to n'J'K')}{d\omega} = \sum_{s=0}^{\infty} \{g(s)/4\pi\} A_s P_s(\cos\theta). \tag{85}$$

The integral cross section is simply A_0, because the integral of $P_s(\cos\theta)$ vanishes for $s \neq 0$. The momentum-transfer cross section is

$$\sigma_m(nJK \to n'J'K') = A_0 - (k_{p'}/k_p) A_1. \tag{86}$$

Calculation of A_0 is easy. Because $l'_1 = l'_2$ for $s = 0$ due to Eq. (84), the summation of $C(J'l'_1\bar{J}_1; M'm'\bar{M}) C(J'l'_2\bar{J}_2; M'm'\bar{M})$ over M' shows that $\bar{J}_1 = \bar{J}_2$ according to the orthonormality relation (38b). Then, the summation of $C(Jl_1\bar{J}_1; M0\bar{M}) C(Jl_2\bar{J}_2; M0\bar{M})$ over M after the interchange of l_i and $\bar{J}_i$ according to the detailed balance formula (40c) shows that $l_1 = l_2$ due to Eq. (38b). The coherent sixfold summation reduces to an incoherent triple summation as

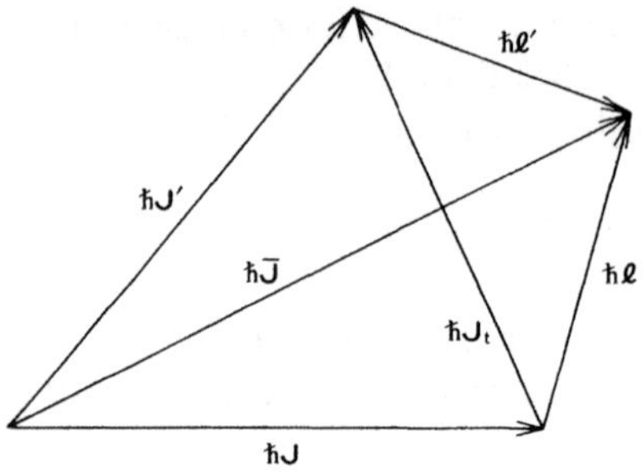

FIGURE 9
Relation between the total-angular-momentum ($\hbar\bar{\mathbf{J}}$) representation and the angular momentum-transfer ($\hbar\mathbf{J}_t$) representation. The angular momentum transferred from the electron to the molecule is $\hbar\mathbf{J}_t = \hbar\mathbf{J}' - \hbar\mathbf{J} = \hbar\mathbf{l} - \hbar\mathbf{l}'$.

$$\sigma(nJK \to n'J'K') = A_0 = (\pi/k_p^2) \sum_{\bar{J}=0}^{\infty} \{g(\bar{J})/g(J)\} \sum_{l=|\bar{J}-J|}^{\bar{J}+J} \sum_{l'=|\bar{J}-J'|}^{\bar{J}+J'} |T_{pp'}^{L}|^2. \tag{87}$$

Expressions for other coefficients A_s ($s \neq 0$) are complicated in general, because of the full sixfold summation; the differential cross section is more difficult to calculate than the integral cross section. Especially, the *coherent* double infinite series over $\bar{J}_1$ and $\bar{J}_2$ necessitates, even if truncated at a finite value $\bar{J}_{\max}$, calculation of the fourfold sum over (l_1, l_2, l_1', l_2') $\bar{J}_{\max}^2$ times. This complexity arises from the coupling between **J** and **l** to form $\bar{\mathbf{J}}$ first (Fig. 9), and from the $\bar{J}$ representation of the transition (or **T**) matrix. In fact, transformation of the **T** matrix into the angular-momentum-transfer representation simplifies the expressions. In this representation **J**′ and −**J** are coupled first as $\mathbf{J}' - \mathbf{J} = \mathbf{J}_t$ (Fig. 9), and the **T** matrix elements are labeled by J_t rather than by $\bar{J}$.[38,67] The transformation into the J_t representation involves a single (infinite) sum over $\bar{J}$. Then in terms of the new **T** matrix, A_s involves only an *incoherent* single summation over J_t in addition to the summation over (l_1, l_2, l_1', l_2').[67,73] In low-energy electron–molecule collisions only a small number of angular-momentum quanta are transferred, and the summation over J_t is reduced to only a few terms. This should be compared with a large $\bar{J}_{\max}$ that is usually necessary. The J_t representation is more appropriate than the $\bar{J}$ representation for the description of the physical situation.

4.2.3. Body-Fixed Frame of Reference

In the body-frame region we expand the total wave function $\Psi(\mathbf{r}, \mathbf{X}_T)$ as

$$\Psi^B(\mathbf{r}, \mathbf{X}_T) = \sum_q r^{-1} g_q^B(r) \Phi_n(\mathbf{x}_T) X_\lambda^B(\Omega, \hat{\mathbf{r}}'; \bar{J}\bar{M}lK), \tag{88}$$

where $q = (nKl\lambda\bar{J})$. The Schrödinger equation (78) projected onto the basis functions $\Phi_{n'}(\mathbf{x}_T) X_{\lambda'}^B(\Omega, \hat{\mathbf{r}}'; \bar{J}\bar{M}l'K')$ is[74]

$$\left[\frac{d^2}{dr^2} + k_q^2 - \frac{l(l+1)}{r^2}\right] g_q^B(r) - 2\sum_{q'} V_{qq'}^B(r) g_{q'}^B(r) = 0 \tag{89}$$

with

$$V_{qq'}^B(r) = (\Phi_n(\mathbf{x}_T) X_\lambda^B(\Omega, \hat{\mathbf{r}}'; \bar{J}\bar{M}lK) | V | \Phi_{n'}(\mathbf{x}_T) X_{\lambda'}^B(\Omega, \hat{\mathbf{r}}'; \bar{J}\bar{M}l'K')) \tag{89a}$$

and

$$E = \hbar^2 k_q^2 / 2\mu + E_v(n), \tag{89b}$$

provided that H_r is negligible in Eq. (78) as compared with V. Indeed, the arguments in Section 4.2.1 suggest the choice of the body-frame region in such a way that V is much stronger there than $E_r(JK)$ of the relevant rotational states. This choice of the body-frame region is assumed in the following.

The interaction potential V depends only on the relative position $\mathbf{r}' = (r, \hat{\mathbf{r}}')$ of the electron referred to the molecule, and not on the molecular orientation Ω. Therefore, the matrix $\mathbf{V}^B$ is diagonal in and independent of $\bar{J}$, $\bar{K}$, and $\bar{M}$ [see Eq. (76)]. Furthermore, within any submatrix for which $K = K'$, the potential is diagonal in $\lambda (= \bar{K} - K)$. For the homonuclear diatomic molecule no coupling exists between even-l (gerade) channels and odd-l (ungerade) channels. Chang and Fano[67] discuss further uncoupling due to the parity under inversion $\hat{I}$. With all these kinds of uncoupling the number of coupled equations in (89) required for attaining convergence is much less than the number of those coupled laboratory-frame equations (80) which would be required if used in the body-frame region.

The solutions to Eqs. (89) relevant to the scattering problem satisfy a boundary condition, to be discussed in Section 4.3, on the surface of the compound-system region. Starting from there, we solve Eqs. (89) outward, and determine the body-frame wave function Ψ^B. We match it to the laboratory-frame wave function Ψ^L at the boundary $r = a_F$ as $\Psi^L = \Psi^B$ or, in terms of the vectors $\mathbf{X}^B$ and $\mathbf{X}^L$ in Eq. (77a) generalized to include the vibronic index n, as

$$\mathbf{X}^{L\dagger} \cdot \mathbf{f}^L = \mathbf{X}^{B\dagger} \cdot \mathbf{g}^B = \mathbf{X}^{L\dagger} \mathbf{U}^{-1} \mathbf{g}^B. \tag{90}$$

Hence, a matching condition on the channel functions follows from the orthogonality of $\mathbf{U}$ as

$$f_{nJKl\bar{J}}^L(a_F) = \sum_\lambda U_{\lambda J} g_{nKl\lambda\bar{J}}^B(a_F). \tag{91}$$

In practical calculations of a limited basis-set size the resultant scattering parameters can be sensitive to the choice of the matching radius a_F. Indeed,

Chandra[75] has experienced this sensitivity in applications to low-energy scattering of electrons by the CO molecule. Le Dourneuf *et al.*[76] test three matching procedures on this collision system using a model potential, and demonstrate great reduction of the sensitivity to a_F by a proper choice of the procedure. They also suggest possible improvements made by retaining the diagonal elements of the H_r matrix in the body-frame region.[24,77] Chandra proposes a similar technique of retaining only the diagonal elements of a transformed H_r matrix in the laboratory-frame region.[78] His technique simplifies solution of the laboratory-frame equations for dipolar molecules (Section 5.3.3).

Physical arguments readily suggest that the proper value of a_F depends on the channel.[79] For channels with large channel wave numbers, for example, a_F may be large, because H_r is negligible up to large r. The use of different values of a_F for different channels, however, complicates the actual calculations.

Collins and Norcross[7] study a simple limiting case of the frame-transformation method. They choose $a_F = \infty$ for small angular momenta l and $a_F = 0$ for large l. In other words, they employ the body-fixed frame for $l \leqq l_F$ in the entire region of r, and switch into the laboratory frame for $l > l_F$ in the entire region of r. This frame-switching procedure, referred to by the authors[1] as the angular-frame-transformation method, is based on the following expectation. For high angular momenta the use of a small a_F will introduce only a small error, because the electron hardly penetrates the centrifugal potential barrier to enter the region of strong interaction. For low angular momenta for which the interaction in the body-frame region dominates the collision, the use of a large a_F will introduce only a small error. See Section 5.3.3 for more details.

Further simplification arises if the body-frame region (in which H_r is negligible) extends to infinity for all l. This amounts to the neglect of H_r altogether in the whole region of the configuration space, H_r being certainly negligible in the compound-system region. In other words we make the sudden approximation (Section 3.2.2). The asymptotic form of the wave function defines the body-frame reactance, scattering, and transition matrices, namely,

$$g^B_{qq'}(r) \underset{r\to\infty}{\sim} k_q^{-1/2} \left[\sin\left(k_q r - \frac{l\pi}{2}\right)\delta_{qq'} + \cos\left(k_q r - \frac{l\pi}{2}\right)K^B_{qq'}\right] \qquad (k_q^2 > 0), \tag{92a}$$

$$g^B_{qq'}(r) \underset{r\to\infty}{\sim} 0 \qquad (k_q^2 < 0), \tag{92b}$$

and

$$\mathbf{T}^B = 1 - \mathbf{S}^B = -2i\mathbf{K}^B(1 - i\mathbf{K}^B)^{-1}. \tag{93}$$

Equation (91) relates $\mathbf{T}^B$ to $\mathbf{T}^L$ by the orthogonal transformation

$$\mathbf{T}^L = \mathbf{U}^\dagger \mathbf{T}^B \mathbf{U}, \tag{94a}$$

or

$$T^L_{pp'} = \sum_{\lambda\lambda'} [\{g(J)g(J')\}^{1/2}/g(\bar{J})]\, C(Jl\bar{J};K\lambda\bar{K})C(J'l'\bar{J};K'\lambda'\bar{K})T^B_{qq'}. \tag{94b}$$

For $K = K'$ the double summation reduces to a single summation, because $\lambda = \lambda'$ then.

The two Clebsch–Gordan coefficients in Eq. (94b) complicate the apparent form of the scattering amplitude (83). In the sudden approximation, however, Eq. (62) relates a general transition $nJK \to n'J'K'$ to the transitions from the ground rotational state. Therefore, formulas for only the cross sections $d\sigma(n00 \to n'J'K')/d\omega$ suffice. For $J = K = M = 0$ we have $\bar{J} = l$, $\bar{K} = \lambda$, and $\bar{M} = 0$ due to Eq. (39). It follows from Eqs. (83) and (94b) that

$$f(n0 \to n'\Gamma') = (\pi/k_p k_{p'})^{1/2} \sum_{ll'} i^{l-l'} \{g(J')/g(l)\}^{1/2}$$
$$\times\, C(J'l'l;M'm'0)Y_{l'm'}(\theta,\varphi) \sum_{\lambda} C(J'l'l;K'\lambda'\lambda)T^B_{qq'}. \tag{95}$$

The use of Eq. (84) yields the coefficients A_s in Eq. (85) in terms of $\mathbf{T}^B$. In particular, the integral cross section is expressible, after summation over M' according to Eq. (38b) followed by the use of Eq. (40b), as

$$\sigma(n00 \to n'J'K') = A_0 = (\pi/k_p^2) \sum_{ll'\lambda_1\lambda_2} (-1)^{\lambda_1+\lambda_2}$$
$$\times\, C(ll'J';-\lambda_1,\lambda_1',-K')C(ll'J';-\lambda_2,\lambda_2',-K')T^B_{q_1q_1'}T^{B*}_{q_2q_2'}. \tag{96}$$

Here, q_1 and q_2 (q_1' and q_2') differ only in $\lambda(\lambda')$. If we sum Eq. (96) over J' using Eq. (38a), only the terms with $\lambda_1 = \lambda_2$ survive, and we have

$$\sum_{J'} \sigma(n00 \to n'J'K') = (\pi/k_p^2) \sum_{ll'\lambda} |T^B_{qq'}|^2. \tag{97}$$

4.3. R-Matrix Theory

Wigner's R-matrix theory of nuclear reactions[80-82] has proved to be powerful also for calculations in atomic physics.[83] Chapter 7, Section 3.8 discusses applications to electron–molecule collisions. The R-matrix theory provides a boundary condition on the body-frame coupled equations (89) on the surface of the compound-system region. This subsection presents the essence of the theory by starting with a simple problem, namely, scattering of a spinless electron by a structureless particle A through a spherically symmetric potential $(\hbar^2/\mu)V(r)$ $(0 \leqq r < \infty)$.[80-84]

4.3.1. Internal and External Regions

The partial-wave expansion decouples channels with different angular momenta. We discuss only the s-wave channel for simplicity. There is no essential difference for higher partial waves.

We consider first the finite region $I_0 = [0, a]$, where the compound system A^- is formed. This region is called the internal or inner region in the R-matrix theory. The Schrödinger equation is written as

$$(H - E)\psi^{(i)}(r) = \frac{\hbar^2}{2\mu}\left[-\frac{d^2}{dr^2} + 2V(r) - k^2\right]\psi^{(i)}(r) = 0 \qquad (0 \leqq r \leqq a), \tag{98}$$

the superscript on the wave function implying "internal."

The electron emerges from the origin with the boundary condition $\psi^{(i)}(0) = 0$. For any energy $E = \hbar^2 k^2/2\mu$ we may follow the outward motion of the electron by integrating Eq. (98), and may calculate the logarithmic derivative (multiplied by r and made dimensionless) $R^{-1}(a) = r(d\psi^{(i)}/dr)/\psi^{(i)}|_{r=a}$ on the boundary. Figure 10 shows a schematic plot of the surface quantity $R^{-1}(a)$ vs E. Such a plot contains information on the dynamics of the compound system A^-, and is determined completely by the Schrödinger equation in the internal region and by the boundary condition at the origin.

There is no essential qualitative difference between positive-energy and negative-energy solutions, because the electron is still ignorant of the threshold of the continuum. The distinction between bound and continuum states is made only through the asymptotic value of the potential and the asymptotic boundary condition.

A boundary condition $R^{-1}(a) = \beta$ at $r = a$ defines an infinite and enumerable number of discrete eigenvalues E_α and eigenfunctions $\psi_\alpha^{(i)}$, called the R-matrix states, corresponding to the points in Fig. 10. These eigenfunctions form

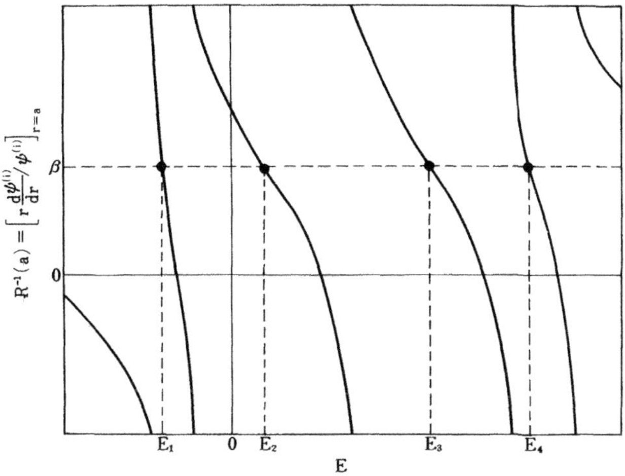

FIGURE 10
Schematic plot of the logarithmic derivative of the internal wave function $\psi^{(i)}(r)$ on the boundary $r = a$.

a complete set; any wave function *in the internal region* may be expanded in terms of the functions $\psi_\alpha^{(i)}$.

The scalar quantity $R(a)$ is the R matrix for the single-channel problem. A slightly more general R matrix is defined by

$$\psi^{(i)}(a) = R(a, b)[r d\psi^{(i)}/dr - b\psi^{(i)}]_{r=a}, \tag{99}$$

where the parameter b shifts the R-matrix poles. If we choose b to equal β, $R(a, \beta)$ has a pole at each E_α. Indeed, the expansion of the wave function for E in terms of $\{\psi_\alpha^{(i)}\}$ leads to a formula (Section 4.3.2)[80]

$$R(a, \beta) = (\hbar^2/2\mu a) \sum_\alpha \{\psi_\alpha^{(i)}(a)\}^2/(E_\alpha - E). \tag{100}$$

We shall deal mostly with the case of $b = 0$ and write $R(a) = R(a, 0)$.

Because the R matrix is essentially the logarithmic derivative at $r = a$, it may be used to connect the wave function into the external or outer region $r \geqq a$. For $E > 0$ we introduce two linearly independent solutions $S(r)$ and $C(r)$ for $r \geqq a$ having the asymptotic forms

$$\begin{aligned} S(r) &\underset{r\to\infty}{\simeq} k^{-1/2} \sin kr, \\ C(r) &\underset{r\to\infty}{\simeq} k^{-1/2} \cos kr. \end{aligned} \tag{101}$$

The external wave function is a linear combination of $S(r)$ and $C(r)$

$$\psi^{(e)}(r) = S(r) + C(r)K. \tag{102}$$

The coefficient K is determined by the matching condition

$$S(a) + C(a)K = R(a)[aS'(a) + aC'(a)K] \tag{103}$$

as

$$K = -[C(a) - R(a)aC'(a)]^{-1}[S(a) - R(a)aS'(a)]. \tag{104}$$

The asymptotic forms (101) used in Eq. (102) show that K is the one-dimensional reactance matrix, and is the tangent of the phase shift δ.

The negative-energy solution in the external region is a linear combination of exponentially growing and exponentially decaying functions. At particular values of E the coefficient, determined by a relation similar to Eq. (103), of the exponentially growing function vanishes. These values of E are bound-state energies.

4.3.2. Basis-Set Expansion in the Internal Region

We derive Eq. (100) by the expansion of the internal wave function[84]

$$\psi^{(i)}(r) = \sum_i c_i v_i(r) = \mathbf{c}^\dagger \cdot \mathbf{v}(r) \qquad (0 \leq r \leq a) \tag{105}$$

in terms of orthonormal functions $v_i(r)$ satisfying $v_i(0) = 0$. With this basis set the internal Schrödinger equation in matrix notation is

$$(\mathbf{H} - E)\mathbf{c} = 0. \tag{106}$$

The elements of the Hamiltonian matrix $\mathbf{H}$ follow from partial integration of the kinetic energy as

$$H_{ij} = \frac{\hbar^2}{2\mu}\int_0^a v_i(r)\left[-\frac{d^2}{dr^2} + 2V(r)\right] v_j(r)dr = -\frac{\hbar^2}{2\mu} v_i(a)v_j'(a) + (H_s)_{ij}, \tag{107}$$

where $\mathbf{H}_s$ is a symmetric matrix with elements

$$(H_s)_{ij} = \frac{\hbar^2}{2\mu}\int_0^a [v_i'(r)v_j'(r) + 2v_i(r)V(r)v_j(r)]\,dr. \tag{108}$$

Being defined over a finite region, the matrix **H** is non-Hermitian. The asymmetric surface term in Eq. (107) may be rewritten as

$$-\frac{\hbar^2}{2\mu}v_i(a)v_j'(a) = -\int v_i(r)\left[\frac{\hbar^2}{2\mu}\delta(r-a)\frac{d}{dr}\right]v_j(r)dr \equiv -(v_i|L_b|v_j). \quad (109)$$

The operator L_b has been introduced by Bloch,[85] and is called the Bloch operator. It allows the division of the non-Hermitian Hamiltonian operator H into the Hermitian part $H+L_b$ and the non-Hermitian part $-L_b$.

Rewriting the internal Schrödinger equation (106) as

$$(\mathbf{H}_s - E)\mathbf{c} = (\hbar^2/2\mu)\mathbf{v}(a)\mathbf{v}'^{\dagger}(a)\mathbf{c}, \quad (110)$$

formally solving this for **c** on the left, and multiplying it by $\mathbf{v}^{\dagger}(a)$ from the left, we obtain a relation

$$\mathbf{v}^{\dagger}(a)\cdot\mathbf{c} = (\hbar^2/2\mu a)[\mathbf{v}^{\dagger}(a)(\mathbf{H}_s - E)^{-1}\mathbf{v}(a)]\,[a\mathbf{v}'^{\dagger}(a)\cdot\mathbf{c}]. \quad (111)$$

The left-hand side is $\psi^{(i)}(a)$ according to Eq. (105). Also, we have $a\mathbf{v}'^{\dagger}(a)\cdot\mathbf{c} = r d\psi^{(i)}/dr|_{r=a}$. Therefore, comparison of Eq. (111) with Eq. (99) yields a formula[84,86]

$$R(a) = (\hbar^2/2\mu a)[\mathbf{v}^{\dagger}(a)(\mathbf{H}_s - E)^{-1}\mathbf{v}(a)]. \quad (112)$$

The pole-shifted R matrix $R(a,b)$ follows from replacement of $\mathbf{H}_s$ by $\mathbf{H}_s(b) = \mathbf{H}_s - (\hbar^2 b/2\mu a)\,\mathbf{v}(a)\mathbf{v}^{\dagger}(a)$.

Diagonalization of $\mathbf{H}_s$ gives eigenvalues E_α and eigenvectors $\mathbf{c}_\alpha$. Because $(\mathbf{H}_s - E)^{-1} = \Sigma\mathbf{c}_\alpha(E_\alpha - E)^{-1}\mathbf{c}_\alpha^{\dagger}$, Eq. (112) reduces to Eq. (100) with $\psi_\alpha^{(i)}(a) = \mathbf{c}_\alpha^{\dagger}\cdot\mathbf{v}(a)$.

The solution of the linear equations

$$(\mathbf{H}_s - E)\mathbf{x} = \mathbf{v}(a) \quad (113)$$

at each E also gives the R matrix

$$R(a) = (\hbar^2/2\mu a)\,[\mathbf{x}^{\dagger}\cdot\mathbf{v}(a)]. \quad (114)$$

The R matrix together with the asymptotic boundary condition defines the external wave function $\psi^{(e)}$. The matching condition $\mathbf{v}'^{\dagger}(r)\cdot\mathbf{c} = d\psi^{(e)}/dr$ at $r=a$ used in Eq. (110) gives the internal wave function

$$\psi^{(i)}(r) = (\hbar^2/2\mu)d\psi^{(e)}/dr|_{r=a}\,[\mathbf{x}^{\dagger}\cdot\mathbf{v}(r)] \qquad (0 \leq r < a). \quad (115)$$

If the basis set $\{v_i(r)\}$ is complete in the internal region, the R matrix of Eq. (112) or Eq. (114) is exact, and so is the wave function (115) for $0 \leqq r < a$. If a finite basis set is used, the results are approximate, and Eq. (112) is variationally optimum in the finite function space.[87]

For a finite basis set the differentiation of Eq. (115) reveals that $d\psi^{(i)}/dr$ coincides with $d\psi^{(e)}/dr$ at $r = a$ only when $\mathbf{x}^\dagger \cdot \mathbf{v}'(a)$ happens to be $2\mu/\hbar^2$. In general the internal wave function determined by the R-matrix method *does not* connect smoothly to the external wave function.

The inverse matrix in Eq. (112) represents the effect of the interaction over the internal region as a whole. This effect, described in the space of the basis functions, is projected onto the boundary due to the vectors $\mathbf{v}(a)$ and $\mathbf{v}^\dagger(a)$, leading to the surface quantity $R(a)$. In contrast with the calculation of $R(a)$ by the outward integration of the Schrödinger equation, the form of Eq. (112) allows us to avoid tracing the electron position by position. We may study the dynamics of the compound system as a whole. This aspect makes the R-matrix theory particularly suitable for applications to electron–molecule collisions, in which all electrons are indistinguishable in the compound-system region, and in which the position of the scattered electron is difficult to single out. In this respect the R-matrix method has the potentialities of better describing the physics in the compound-system region than the coupled-equations approach of Sections 4.2.2 and 4.2.3. The possible success all depends on the choice of the basis functions, which contain the whole dynamics that is described by this method. Indeed, Burke and collaborators choose particularly appropriate basis functions and extensively demonstrate the great power of the method in electron–atom collisions and atomic photoionization.[83]

Just like the division of the potential scattering region in Fig. 8 into the laboratory-frame and body-frame subregions, the external region in the R-matrix theory sometimes needs to be subdivided. A subregion has two boundaries in general. The Schrödinger equation (110) generalized for a subregion $[a_1, a_2]$ involves two surface terms as

$$\begin{aligned}(\mathbf{H}_s - E)\mathbf{c} &= (\hbar^2/2\mu)\,[\mathbf{v}(a_2)\mathbf{v}'^\dagger(a_2) - \mathbf{v}(a_1)\mathbf{v}'^\dagger(a_1)]\,\mathbf{c} \\ &= (\hbar^2/2\mu)\,[\mathbf{v}(a_2)\psi'(a_2) - \mathbf{v}(a_1)\psi'(a_1)]\,. \end{aligned} \tag{116}$$

The symmetric matrix $\mathbf{H}_s$ here is defined by Eq. (108) with the integration range changed into $[a_1, a_2]$. After solving Eq. (116) for $\mathbf{c}$ we may multiply either $\mathbf{v}^\dagger(a_1)$ or $\mathbf{v}^\dagger(a_2)$ from the left to obtain $\psi(a_1)$ or $\psi(a_2)$. We have

$$\begin{pmatrix} -R(a_1) & 0 \\ 0 & R(a_2) \end{pmatrix} \begin{pmatrix} -a_1\psi'(a_1) \\ a_2\psi'(a_2) \end{pmatrix} = \begin{pmatrix} \psi(a_1) \\ \psi(a_2) \end{pmatrix} = \begin{pmatrix} \mathbf{v}^\dagger(a_1)\cdot\mathbf{c} \\ \mathbf{v}^\dagger(a_2)\cdot\mathbf{c} \end{pmatrix}$$

$$= \begin{pmatrix} \mathbf{R}_{11} & \mathbf{R}_{12} \\ \mathbf{R}_{21} & \mathbf{R}_{22} \end{pmatrix} \begin{pmatrix} -a_1 \psi'(a_1) \\ a_2 \psi'(a_2) \end{pmatrix} \tag{117a}$$

with the generalized R matrix

$$\mathbf{R}_{nm} = (\hbar^2/2\mu a_m)\mathbf{v}^\dagger(a_n)(\mathbf{H}_s - E)^{-1}\mathbf{v}(a_m). \tag{117b}$$

For Eq. (117a) to have a nontrivial solution it is required that

$$\det \begin{vmatrix} \mathbf{R}_{11} + R(a_1) & \mathbf{R}_{12} \\ \mathbf{R}_{21} & \mathbf{R}_{22} - R(a_2) \end{vmatrix} = 0, \tag{118a}$$

or

$$R(a_2) = \mathbf{R}_{22} - \mathbf{R}_{21}[\mathbf{R}_{11} + R(a_1)]^{-1}\mathbf{R}_{12}. \tag{118b}$$

This allows propagation of the R matrix from $r = a_1$ to $r = a_2$ by use of the generalized R matrix. The ordinary R matrix (112) is a special case of Eq. (118b) for $a_1 = 0$, for which $\mathbf{v}(a_1) = 0$ and $\mathbf{R}_{12} = \mathbf{R}_{21} = 0$.

4.3.3. Multichannel Scattering

In electron–molecule collisions the wave function in the external region takes a multichannel form (88), i.e., $\Psi^{(e)} = \Sigma r^{-1} g_q^{(e)}(r)\psi_q(\hat{\mathbf{r}},\mathbf{X}_T)$. Let all linearly independent solution vectors $\mathbf{g}_{q'}^{(e)}(r)$ be represented by a matrix $\mathbf{G}^{(e)}(r)$, of which the elements are $G_{qq'}^{(e)}(r) = g_{qq'}^{(e)}(r)$. If there are n_q channels, $\mathbf{G}^{(e)}(r)$ is an n_q-by-n_q matrix.

For connecting $\Psi^{(e)}$ to the internal wave function $\Psi^{(i)}$, we write the latter in the same channel representation $\Psi^{(i)} = \Sigma r^{-1} g_q^{(i)}(r)\psi_q(\hat{\mathbf{r}}, \mathbf{X}_T)$ as that for $\Psi^{(e)}$ near the boundary between the internal and external regions. If $\Psi^{(i)}$ has a different form, the channel functions $g_q^{(i)}(r)$ are calculable by projecting $\Psi^{(i)}$ onto ψ_q on and close to the boundary. We define a matrix $\mathbf{G}^{(i)}(r)$ by $G_{qq'}^{(i)}(r) = g_{qq'}^{(i)}(r)$.

A generalization of the R matrix of Eq. (99) for the n_q-channel problem is an n_q-by-n_q matrix $\mathbf{R}(a, \mathbf{b})$ (with any constant matrix $\mathbf{b}$) defined by

$$\mathbf{G}^{(i)}(a) = \mathbf{R}(a,\mathbf{b})\,[r d\mathbf{G}^{(i)}/dr - \mathbf{b}\mathbf{G}^{(i)}]_{r=a}. \tag{119}$$

The following discusses the case of $\mathbf{b} = 0$, and $\mathbf{R}(a)$ shall denote $\mathbf{R}(a,0)$.

Molecules are nonspherical in general. Nevertheless, the definition (119) of the R matrix assumes a spherical internal region, i.e., a spherical compound system. This is merely for mathematical simplicity. We choose the sphere that just encloses the compound system. An internal region of a more realistic shape can be defined in a natural way for the diatomic molecule by use of the prolate spheroidal coordinates.[88,89]

Suppose that the n_q-channel coupled equations in the external region are solvable. Let $S_{qq'}(r)$ and $C_{qq'}(r)$ be the linearly independent solutions that satisfy the asymptotic boundary conditions [cf. Eq. (101)]

$$\begin{aligned} S_{qq'}(r) &\underset{r\to\infty}{\sim} k_q^{-1/2} \sin\left(k_q r - \frac{l\pi}{2}\right) \delta_{qq'} \\ C_{qq'}(r) &\underset{r\to\infty}{\sim} k_q^{-1/2} \cos\left(k_q r - \frac{l\pi}{2}\right) \delta_{qq'} \end{aligned} \tag{120}$$

if $k_q^2 > 0$, and that behave as $S_{qq'}(r) \to 0$ if $k_q^2 < 0$. The wave function $\mathbf{G}^{(e)}(r) = \mathbf{S}(r) + \mathbf{C}(r)\mathbf{K}$ [cf. Eq. (102)] has the asymptotic form (92). Following the formulation for the single-channel problem and matching $\mathbf{G}^{(e)}(r)$ to $\mathbf{G}^{(i)}(r)$ at $r = a$ by use of Eq. (119), i.e., solving Eq. (103) changed into a matrix form, we obtain the reactance matrix

$$\mathbf{K} = -[\mathbf{C}(a) - \mathbf{R}(a)a\mathbf{C}'(a)]^{-1}[\mathbf{S}(a) - \mathbf{R}(a)a\mathbf{S}'(a)]. \tag{121}$$

This has the same form as Eq. (104), though matrices now replace the scalars.

Expansion of an approximate internal wave function in the form

$$\Psi^{(i)} = \sum_{q=1}^{n_q} r^{-1}\{\mathbf{c}_q^\dagger \cdot \mathbf{v}_q(r)\}\psi_q(\hat{\mathbf{r}},\mathbf{X}_T) + \mathbf{c}_0^\dagger \cdot \mathbf{\Phi}(\mathbf{r},\mathbf{X}_T) \tag{122}$$

allows compact presentation of the result, the first term being convenient for determining $\mathbf{G}^{(i)}$. The elements $\phi_i(\mathbf{r},\mathbf{X}_T)$ of the vector $\mathbf{\Phi}$ are functions that die off before reaching the boundary.

The symmetrized Hamiltonian matrix $\mathbf{H}_s$ consists of submatrices

$$\mathbf{H}_{qq'} = (r^{-1}\mathbf{v}_q\psi_q|\mathsf{H}^s|r^{-1}\mathbf{v}_{q'}^\dagger\psi_{q'}) \qquad (q \neq 0, q' \neq 0) \tag{123a}$$

as

$$\mathbf{H}_s = \begin{pmatrix} \mathbf{H}_{00} & \mathbf{H}_{01} & \cdots & \mathbf{H}_{0n_q} \\ \mathbf{H}_{10} & \mathbf{H}_{11} & \cdots & \mathbf{H}_{1n_q} \\ \cdot & \cdot & \cdot & \cdot \\ \cdot & \cdot & \cdot & \cdot \\ \cdot & \cdot & \cdot & \cdot \\ \mathbf{H}_{n_q0} & \mathbf{H}_{n_q1} & \cdots & \mathbf{H}_{n_qn_q} \end{pmatrix} \tag{123b}$$

For $q=0$ (or $q'=0$) $\boldsymbol{\Phi}$ (or $\boldsymbol{\Phi}^\dagger$) replaces the vector $r^{-1}\mathbf{v}_q\psi_q$ (or $r^{-1}\mathbf{v}_{q'}^\dagger\psi_{q'}$) in Eq. (123a). The operator $\mathbf{H}^s$ is a generalization of $H+L_b$ discussed in Section 4.3.2. Its matrix elements (123a) are the same as those of the total Hamiltonian H except that the integral of $-\mathbf{v}_q d^2\mathbf{v}_{q'}^\dagger/dr^2$ over r is replaced by the integral of $(d\mathbf{v}_q/dr)(d\mathbf{v}_{q'}^\dagger/dr)$ if $q\neq 0$ and $q'\neq 0$. For $q=0$ or $q'=0$, $\mathbf{H}^s$ is identical with H, because the surface terms vanish due to $\boldsymbol{\Phi}$.

The multichannel extension of the surface vector $\mathbf{v}(a)$ is a matrix

$$\mathbf{V}(a) = \begin{pmatrix} \mathbf{0} & \mathbf{0} & \cdots & \mathbf{0} \\ \mathbf{v}_1(a) & \mathbf{0} & \cdots & \mathbf{0} \\ \mathbf{0} & \mathbf{v}_2(a) & \cdots & \mathbf{0} \\ \cdot & \cdot & \cdot & \cdot \\ \cdot & \cdot & \cdot & \cdot \\ \cdot & \cdot & \cdot & \cdot \\ \mathbf{0} & \mathbf{0} & \cdots & \mathbf{v}_{n_q}(a) \end{pmatrix} \tag{124}$$

The variational optimized R matrix resembles Eq. (112), namely,

$$\mathbf{R}(a) = (\hbar^2/2\mu a)[\mathbf{V}^\dagger(a)(\mathbf{H}_s-E)^{-1}\mathbf{V}(a)]. \tag{125}$$

For the case of subdivided external region generalization of $\mathbf{R}_{nm}$ in Eq. (117b) for multichannel scattering is straightforward in terms of $\mathbf{V}^\dagger(a_n)$ and $\mathbf{V}(a_m)$. With the scalar quantities changed into matrices Eq. (118b) may be used to propagate the multichannel R matrix outward. During the R-matrix propagation we may orthogonally transform the channel basis set $\{\psi_q\}$ on any boundary; the R matrix transforms similarly to the T matrix in Eq. (94a). We may combine the R-matrix theory with the frame-transformation theory in a natural way. Burke *et al.*[(68)] take advantage of this fact to study electron–molecule collisions. Figure 8 suggests that we propagate $\mathbf{R}(a)$ to $\mathbf{R}(a_F)$ either by the variational estimate (117b) together with Eq. (118b)[(90–92)] or by solving the body-frame coupled equations (89) numerically, transform $\mathbf{R}(a_F)$ into the laboratory-frame channel space, solve the laboratory-frame coupled equations (80) from a_F to infinity, and finally use Eq. (121) to obtain the asymptotic K matrix. The details of implementation of the R-matrix method for electron–molecule collisions is saved for Chapter 7, Section 3.8.

In concluding Section 4 we note that there are other approaches (see Ref. 84 for an elementary account and Ref. 82 for a review in detail) similar to the R-matrix theory, e.g., the eigenchannel theory[(82, 93–95)] and the quantum defect theory.[(96–100)] Differing in technical details, they are based on the same philosophy of separating the internal region dynamics and the external region dynamics. Fano and collaborators combine the frame-transformation theory with the multichannel quantum defect theory to analyze electron–ion collisions and photoabsorption of neutral atoms and molecules.[66,99]

5. Physics in the Potential Scattering Region

5.1. General Trend of the Cross Sections

Figure 11 summarizes integral cross sections for scattering of electrons by several molecules measured by crossed electron-beam–molecular-beam technique.[101–113] Figure 12 gives an example of the differential cross section $d\sigma/d\omega$ for a dipolar molecule. We observe the following general trend.

The integral cross section σ differs greatly depending on the energy E and the dipole moment D. The σ is larger for larger D and smaller E, and is nearly proportional to D^2/E when σ is large (Fig. 11b). For large D and small E the cross section σ is much larger than the geometrical cross section of order 10^{-16} cm^2. The angular distribution peaks sharply toward small scattering angles θ, the peak dominating σ when σ is large.

The large cross section implies a large effective interaction radius even at a crude lower-bound estimate, i.e., even on the assumption of the scattering probability of unity for all impact parameters (or partial waves) contributing to the cross section. In other words the interaction in the potential scattering region in Fig. 8 dominates the collision. Furthermore, the trend of the cross section with the change of D suggests the importance of the charge–dipole long-range interaction, proportional to D/r^2, between the electron and the molecule. The importance of the long-range interaction is also clear from the dominance of the forward scattering or the distant collisions. For distant collisions the interaction must be so weak as to act as a small perturbation. The crude D^2 dependence of σ supports this view in a zeroth approximation, because the scattering amplitude in the first Born approximation is of first order in the interaction potential.

Further systematics in the small-angle cross sections for highly dipolar molecules become manifest if $d\sigma/d\omega$ is put on a scale of reduced quantities. Figure 13 shows the reduced differential cross section[29]

$$d\bar{\sigma}/d\omega = (E/D^4)d\sigma/d\omega \tag{126a}$$

as a function of the reduced momentum transfer

$$\bar{q} = 2D(\hbar q/\hbar k) \cong 4D \sin(\theta/2), \tag{126b}$$

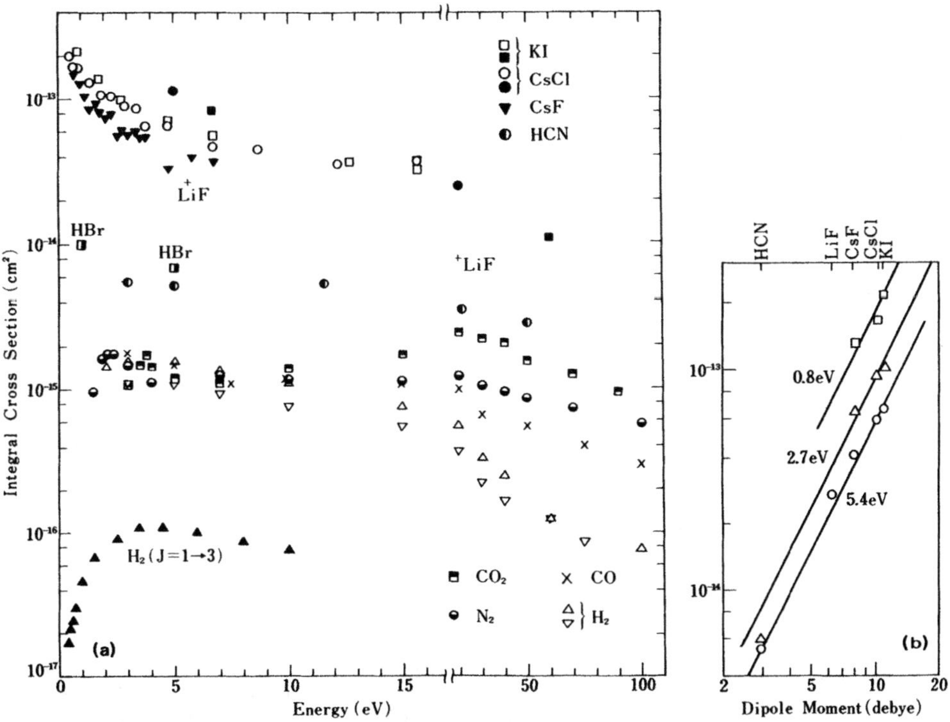

FIGURE 11

(a) Integral cross sections for scattering of electrons by several molecules measured by the crossed electron-beam–molecular-beam technique. The method of extrapolation of the measured differential cross sections to small-angle and large-angle regions that are not really covered in the experiments differs depending on the case. The method of normalization of the measured data differs also depending on the case. The cross sections given here are taken directly from each original paper. Molecular-beam recoil technique (Chapter 6, Section 3.1.2): ▼, CsF (all transitions);[101] ○, CsCl (all transitions);[102] □, KI (all transitions).[103] Energy-loss measurements: ■, KI (electronically elastic);[104] +, LiF (electronically elastic);[105] ●, CsCl (electronically elastic);[106] ◐, HCN (vibrationally elastic plus $0 \rightarrow 1\ \nu_2$ vibrational excitation);[107] ◨, HBr (vibrationally elastic);[108] X, CO (vibrationally elastic);[109] ⬒, CO_2 (vibrationally elastic);[111] ◒, N_2 (vibrationally elastic);[112] ▽, H_2 (vibrationally elastic);[110] △, H_2 (vibrationally and rotationally elastic);[113] ▲, H_2 (rotational transition $J = 1 \rightarrow 3$).[34] (b) Some measured integral cross sections from Fig. 11a (for large dipole moment D and at small energy E) plotted versus D. In the case where the cross section at a given energy is unavailable, it has been inferred by interpolation. The straight lines have a slope showing the D^2 dependence, and are not the best fit to the data.

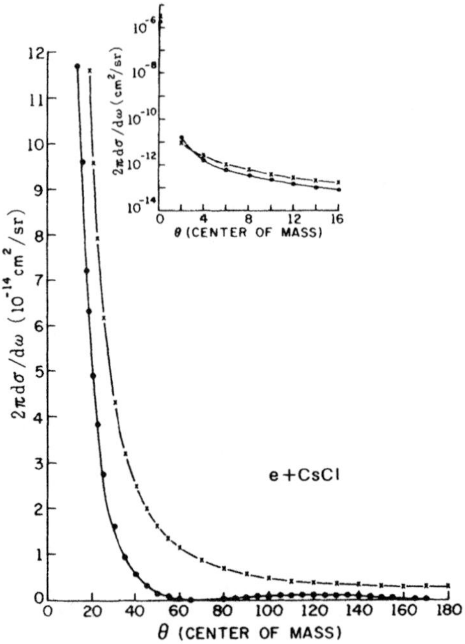

FIGURE 12
Differential cross section ($2\pi d\sigma/d\omega$) for e + CsCl scattering measured at 4.77 eV by use of molecular-beam recoil technique (—●—●—). The dipole–Born cross section (Section 5.3.2) is also shown for comparison (—×—×—). (Reproduced from Becker *et al.*[102])

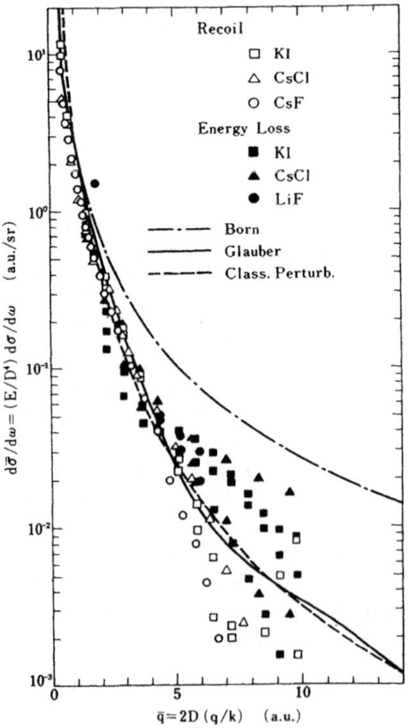

FIGURE 13
Reduced differential cross section $d\bar{\sigma}/d\omega = (E/D^4)d\sigma/d\omega$ (at scattering angles $\theta \lesssim 70°$) plotted versus reduced momentum transfer $\bar{q} = 2D(q/k) \cong 4D \sin(\theta/2)$, D being the dipole moment, q the momentum transfer, E the collision energy, and k the corresponding wave number of the electron. Molecular-beam recoil measurements [see caption of Fig. 11a]: ○, CsF (1.80 eV);[101] △, CsCl (4.77 eV);[102] □, KI (6.74 eV,[103] 15.7 eV[104]). Energy-loss measurements [see caption of Fig. 11a]: ●, LiF (5.4 eV, 20.0 eV);[105] ▲, CsCl (5 eV, 20 eV);[106] ■, KI (6.74 eV, 15.7 eV, 60 eV).[104] The procedure for normalizing the measured cross sections is nonunique, especially for the energy-loss measurements having large error bars at small θ where simple theories are accurate. —·—, dipole–Born approximation ($8/3\bar{q}^2$). ——, dipole–Glauber approximation. – – – –, classical perturbation theory with the dipole potential ($\pi/\bar{q}^3$). (Taken from Shimamura.[29])

where $\hbar\mathbf{q}$ $(=\hbar\mathbf{k}-\hbar\mathbf{k}')$ is the linear momentum transferred to the molecule from the electron. Both the energy and the dipole moment are measured in atomic units in Eqs. (126); the atomic unit of energy is $e^2/a_B = 27.21$ eV and the atomic unit of dipole moment is $ea_B = 2.542$ D. An approximate scaling relation appears to hold for the differential cross section over many orders of magnitude. See Section 5.3.2 for further discussion on this scaling.

5.2. Single-Center Expansion of the Interaction Potential

Multipole moments of the molecular charge distribution give rise to long-range potentials (Section 4.1). Chapter 1, Section 3.1 introduces the multipole expansion of the interaction potential for linear molecules. The present subsection first generalizes this expansion for nonlinear molecules in Section 5.2.1. Then, Section 5.2.2 studies the first Born approximation under the multipole potentials in view of the importance (see Section 5.1) of the weak interaction through the long-range potentials. This is a generalization of the discussion in Chapter 1, Section 6.3, where only the linear rotators are treated in detail.

5.2.1. Multipole-Moment Tensors and Multipole Potentials

Assume that the vibronic wave functions $\Phi_n(\mathbf{x}_T)$ are separable into the electronic and vibrational parts as $\phi_e(\mathbf{r}_T;\, R_T)\chi_v(R_T)$ (Section 3.2.1), and that only the ground electronic state $\phi_0(\mathbf{r}_T; R_T)$ contributes to the scattering process of interest. This is the electrostatic or static potential approximation. Let $\rho_0(\mathbf{r}''; R_T)$ denote the charge density, including that of nuclei, of this state at the position $\mathbf{r}''$. The coordinate axes are fixed to the frame of the molecule at the nuclear configuration R_T.

In the potential scattering region the position $\mathbf{r}' = (r, \hat{\mathbf{r}}')$ of the incident or scattered electron is outside of the molecular charge distribution. Therefore, the electrostatic potential $(\hbar^2/\mu)V_{00}^B$ is expressible, due to the converging expansion

$$|\mathbf{r}'-\mathbf{r}''|^{-1} = \sum_{L=0}^{\infty} \frac{r''^L}{r^{L+1}}\left[\frac{4\pi}{g(L)}\right] \sum_{\lambda_L} Y_{L\lambda_L}(\hat{\mathbf{r}}')Y_{L\lambda_L}^*(\hat{\mathbf{r}}'') \qquad (r > r''), \tag{127}$$

as

$$\left(\frac{\hbar^2}{\mu}\right) V_{00}^B(\mathbf{r}';R_T) = -\int \frac{e\rho_0(\mathbf{r}'';R_T)}{|\mathbf{r}'-\mathbf{r}''|}\,d\mathbf{r}'' = \left(\frac{\hbar^2}{\mu}\right) \sum_{L\lambda_L} r^{-(L+1)}\bar{V}_{L\lambda_L}(R_T)Y_{L\lambda_L}(\hat{\mathbf{r}}'), \tag{128a}$$

$$\begin{aligned}\bar{V}_{L\lambda_L}(R_T) &= -\left(\frac{\mu e}{\hbar^2}\right)\left[\frac{4\pi}{g(L)}\right]\int r''^L Y_{L\lambda_L}^*(\hat{\mathbf{r}}'')\rho_0(\mathbf{r}'';\, R_T)\,d\mathbf{r}'' \\ &= (-1)^{\lambda_L}\bar{V}_{L,-\lambda_L}^*(R_T).\end{aligned} \tag{128b}$$

The quantities $\bar{V}_{L\lambda_L}(\lambda_L = -L, \ldots, L)$ constitute an electric multipole moment tensor of rank L to within a multiplicative constant. (Note that there are several different definitions for multipole moments in the literature.[114]) For a molecule with an n-fold symmetry axis, which is chosen as the body-fixed z axis, the components of the multipole moment tensors are nonzero only for $\lambda_L = ns(s = 0, \pm 1, \pm 2, \ldots)$. Therefore, only one independent scalar quantity is necessary to determine the multipole tensor of rank L less than n.[114] For a linear molecule we have $n = \infty$, and any multipole moment tensor has only one component with $\lambda_L = 0$. When there are symmetry elements besides that provided by an n-fold axis, it may also be possible to describe multipole tensors of rank $L > n$ by scalars. For example, a single scalar moment is sufficient to describe tensors of rank up to at least six, if a molecule has tetrahedral T_d symmetry.[114]

When only the component $\bar{V}_{L0}$ is nonzero we refer to the quantity

$$\mathscr{M}_L(R_T) = \int r''^L P_L(\cos\theta'')\rho_0(\mathbf{r}''; R_T)d\mathbf{r}'' = -\frac{\hbar^2}{e\mu}\left[\frac{g(L)}{4\pi}\right]^{1/2} \bar{V}_{L0}(R_T) \quad (129)$$

as the multipole moment. Thus, $\mathscr{M}_1(=D)$ is the dipole moment, $\mathscr{M}_2(=Q)$ is the quadrupole moment, $\mathscr{M}_3(=O)$ is the octupole moment, etc., and $\mathscr{M}_0$ is the net charge. Combination of Eqs. (128a) and (129) recovers Eq. (22) of Chapter 1, Section 3.1.

The first nonzero multipole moment tensor of a given molecule is independent of the origin of the coordinate system, but all tensors of higher ranks depend on the coordinate origin. Multipole moments are often defined relative to the center of mass of the molecule. For heteronuclear diatomic molecules, the direction of the positive z axis is from the center of mass towards the lighter atom. If the center of mass coincides with the center of symmetry, the direction of the positive z axis is immaterial.

The potential matrix $\mathbf{V}^B$ of Eq. (89a) in the body-frame coupled equations is calculable from the wave function X_λ^B of Eq. (76) and the electrostatic potential of Eq. (128a). Before integration over R_T we have

$$\begin{aligned} V_{qq'}^B(r; R_T) &= [X_\lambda^B(\Omega, \hat{\mathbf{r}}'; \bar{J}\bar{M}lK)|V_{00}^B(\mathbf{r}'; R_T)|X_{\lambda'}^B(\Omega, \hat{\mathbf{r}}'; \bar{J}\bar{M}l'K')] \\ &= -\sum_L (-r)^{-(L+1)}\bar{V}_{L\lambda_L}(R_T)\left[\frac{g(L)}{4\pi}\right]^{1/2} \\ &\quad \times C(lLl'; 000)C(l'Ll; \lambda'\lambda_L\lambda)\delta_{\lambda-\lambda', K'-K}, \end{aligned} \quad (130)$$

where we have used the formulas (60) and (40b). For a linear rotator, in particular, Eq. (130) reduces to the form

$$V^B_{qq'}(r;R_T) = (\mu e/\hbar^2) \sum_{L=0}^{\infty} (-r)^{-(L+1)} \mathscr{M}_L(R_T) C(lLl';000) C(l'Ll;\lambda'0\lambda)\delta_{\lambda\lambda'}. \tag{130a}$$

The potential matrix $\mathbf{V}^L$ of Eq. (80a) in the laboratory-frame coupled equations is the unitary transformation $\mathbf{U}^\dagger \mathbf{V}^B \mathbf{U}$ in terms of the $\mathbf{U}$ of Eq. (77b).

5.2.2. The First-Order Born Approximation

The scattering amplitude in the Born approximation follows from Eq. (51), if the fixed-orientation amplitude $F(n\mathbf{k} \to n'\mathbf{k}_{n'\Gamma'})$ is calculated only to the first order in the potential (Section 3.2.2). The first-order amplitude $F^{B1}(n\mathbf{k} \to n'\mathbf{k}_{n'\Gamma'})$ results from Eq. (49) by replacement of $\mathbf{k}_{n'}$ by $\mathbf{k}_{n'\Gamma'}$ and of $\Psi_n^{(s)}$ by the unperturbed function $\exp(i\mathbf{k}\cdot\mathbf{r})\Phi_n(\mathbf{x}_T) = \exp(i\mathbf{k}\cdot\mathbf{r})\phi_0(\mathbf{r}_T;\, R_T)\chi_v(R_T)$. Then, F^{B1} involves the initial and final wave numbers only in the form of the linear momentum $\hbar\mathbf{q} = \hbar\mathbf{k} - \hbar\mathbf{k}_{n'\Gamma'}$ transferred to the molecule; F^{B1} depends on the rotational channels through and only through $\hbar\mathbf{q}$. Therefore, the linear-dependence formula (62) connects the Born cross sections for different rotational transitions of a symmetric top for a common value of the momentum transfer rather than for a common scattering angle. The following discusses only symmetric tops and not asymmetric tops; only cross sections for transitions from the ground rotational state $\Gamma = 0$ need to be considered, they being the only linearly independent cross sections.

In the static-potential approximation over the whole region of $\mathbf{r}$ the scattering amplitude in the Born approximation is

$$f^{B1}(v0 \to v'\Gamma') = -2^{-1/2}(2\pi)^{-2} \int d\Omega\, D^*_{\Gamma'}(\Omega) \int d\mathbf{r} \exp(i\mathbf{q}\cdot\mathbf{r}) \times \int dR_T \chi^*_{v'}(R_T) V^B_{00}(\mathbf{r}';R_T)\chi_v(R_T), \tag{131}$$

since $D_0(\Omega) = (8\pi^2)^{-1/2}$. Substitute V^B_{00} with its multipole expansion (128a), use the relation $Y_{L\lambda_L} = (-1)^{\lambda_L} Y^*_{L,-\lambda_L}$, and expand the body-frame spherical harmonics $Y^*_{L,-\lambda_L}(\hat{\mathbf{r}}')$ in terms of the laboratory-frame ones $Y^*_{Lm}(\hat{\mathbf{r}})$ according to Eq. (32b). Also use the Rayleigh expansion[14]

$$\exp(i\mathbf{q}\cdot\mathbf{r}) = 4\pi \sum_{\lambda\mu} i^\lambda j_\lambda(qr) Y^*_{\lambda\mu}(\hat{\mathbf{q}}) Y_{\lambda\mu} \tag{132}$$

of the plane wave, where $j_\lambda(qr)$ is a spherical Bessel function. The integration over the angle $\hat{\mathbf{r}}$ gives the overlap between Y_{Lm}^* and $Y_{\lambda\mu}$, i.e., $\delta_{L\lambda}\delta_{m\mu}$. The integration over Ω gives the overlap between $D_{\Gamma'}^*$ and $D_{\Gamma''}$ with $\Gamma'' = (L, -\lambda_L, \mu)$, i.e., $\delta_{LJ'}\delta_{-\lambda_L K'}\delta_{\mu M'}$. These are due to the angular-momentum transfer through the first-order perturbation interaction. As a result only a single term in Eq. (128a) contributes to any particular transition from the ground rotational state. Because the sum of $|Y_{J'M'}(\hat{\mathbf{q}})|^2$ over M' is $(2J'+1)/4\pi$,[14] and because the integral of $x^{1-J'}j_{J'}(x)$ over $[0, \infty)$ is $[(2J'-1)!!]^{-1} = [(2J'-1)(2J'-3)\cdots 3\cdot 1]^{-1}$ for $J' > 0$, it follows that[115,116]

$$\frac{d\sigma^{B1}(v00\mathbf{k} \to v'J'K'\mathbf{k}'; q)}{d\omega} = \pi^{-1}(k'/k)[q^{J'-2}/(2J'-1)!!]^2 |(\chi_{v'}|\bar{V}_{J',-K'}|\chi_v)|^2 \tag{133a}$$

for $J' > 0$ in the Born approximation. Note that Eq. (133a) is a function of only q except for the factor k'/k, a property of the Born approximation in general. The Born cross sections for other transitions are calculable from

$$\left(\frac{k}{k'}\right)\frac{d\sigma^{B1}(vJK\mathbf{k} \to v'J'K'\mathbf{k}'; q)}{d\omega} = \sum_{J_t} \{C(JJ_tJ'; KK_tK')\}^2 \left(\frac{k}{k_t}\right) \times \frac{d\sigma^{B1}(v00\mathbf{k} \to v'J_tK_t\mathbf{k}_t; q)}{d\omega}. \tag{133b}$$

Naturally, the Born cross sections satisfy the sum rules discussed in Section 3.3.2, if the rotational-channel dependence of $\mathbf{k}'$ is negligible.

For scattering by a molecular ion, for which the net charge $\mathscr{M}_0 \neq 0$, the Coulomb–Born approximation must be used instead of the plane-wave Born approximation discussed so far.[117–119] Chu derives formulas for symmetric-top ions.[120] See also Dickinson and Muñoz for first-order time-dependent perturbation theory using hyperbolic classical paths.[121]

5.3. Charge–Dipole Interaction

In the multipole expansion (128) of the electrostatic potential the dipole term ($L = 1$) proportional to r^{-2} distinguishes scattering by dipolar molecules from that by nondipolar molecules. This subsection studies the conspicuous effect of this long-range potential on rotational transitions. First, Section 5.3.1 discusses the exactly solvable problem of scattering by a central field c/r^2 as a preliminary. Then, Sections 5.3.2 and 5.3.3 deal with scattering of electrons by dipolar linear and symmetric-top rotators. Molecules having an n-fold symmetry axis with $n \geqq 3$ are linear or symmetric tops. Sections 5.3 and 5.4 consider only such molecules, although there exist symmetric tops having no n-fold axis with $n \geqq 3$. For $n \geqq 3$ both the dipole and quadrupole moments are scalars, the $\lambda_L \neq 0$ components vanishing.

5.3.1. *Preliminary: Scattering by a Spherical Potential c/r^2*

Consider scattering by a spherically symmetric potential having the asymptotic form c/r^p. If $p \leqq 2$, the potential is of so long a range that the integral cross section is infinite, the differential cross section diverging towards the zero angle due to distant collisions (Ref. 122, Section 124; Ref. 123, Chapter 12). Because the tail of a long-range potential is weak, the Born approximation is applicable to distant or high-partial-wave collisions. On the other hand the classical picture is valid for a particle with even a long de Broglie wavelength, if the potential varies very slowly with the distance r. In fact, both the quantum-mechanical Born approximation and the classical theory reproduce the exact Rutherford cross section for $p = 1$. For $p = 2$ the Born and the classical approximations are complementary, the former being valid for small-angle scattering and the latter for larger-angle scattering; see below.

We shall study scattering by the potential $(\hbar^2/\mu)V(r) = (\hbar^2/\mu)g/r^2$ $(0 \leqq r < \infty)$.[124, 125] The Schrödinger equation for each partial wave reads

$$k^{-2}\left[\frac{d^2}{dr^2} + k^2 - \frac{l(l+1)}{r^2} - \frac{2g}{r^2}\right]\psi_l(kr) = \left[\frac{d^2}{dx^2} + 1 - \frac{L_g(L_g+1)}{x^2}\right]\psi_l(x) = 0, \tag{134}$$

where $x = kr$, and L_g is either of the two roots of $L_g(L_g + 1) = l(l + 1) + 2g$, namely, $L_g = -\frac{1}{2} \pm \{(l + \frac{1}{2})^2 + 2g\}^{1/2}$. Because the energy and the distance appear only in the combination kr, the differential cross section having the dimension of r^2 is proportional to k^{-2}, i.e., to E^{-1}. The angular distribution retains the same shape at all E.

Equation (134) is satisfied by the two functions $x^{1/2}J_\nu(x)$, where $J_\nu(x)$ are the Bessel functions of orders $\nu = L_g + \frac{1}{2}$. The function $x^{1/2}J_\nu(x)$ for $\nu \geqq 0$ satisfies the boundary condition at the origin. Then, if $g \geqq g_c = -(l + \frac{1}{2})^2/2$, we have $\psi_l(x) = x^{1/2}J_{L_g+1/2}(x)$ with $L_g + \frac{1}{2} = \{(l + \frac{1}{2})^2 + 2g\}^{1/2} \geqq 0$, and the asymptotic form of $\psi_l(x)$ determines the phase shift to be (Ref. 124, Chapter 2)

$$\delta_l = (l - L_g)\pi/2. \tag{134a}$$

If $g < g_c$, then $L_g + \frac{1}{2}$ is imaginary, and the phase shift is ill defined due to the ill behavior $x^{1/2}J_{\pm i\nu}(x) \cong \text{const} \times x^{1/2} \exp\{\pm i\nu \ln(x/2)\}$ of the Bessel functions close to the origin. In other words these functions oscillate an infinite number of times as $r \to 0$. This is because the particle with a low angular momentum falls into the infinitely deep potential well at the origin (Ref. 122, Section 35; Ref. 123, Chapter 12; Ref. 125). This phenomenon is physically irrelevant, because any interaction potential in atomic physics is free from such a high singularity.

For $g \gtrsim g_c$ the use of the exact phase shift (134a) in the partial-wave expansion of the scattering amplitude

$$f(\theta) = (2ik)^{-1} \sum_{l=0}^{\infty} (2l+1)\{\exp(2i\delta_l) - 1\} P_l(\cos\theta) \tag{135}$$

yields the exact differential cross section. The phase shift behaves as $\delta_l \cong -\pi g/(2l+1)$ ($\equiv \delta_l^{B1}$) for large l, as is given by the first-order perturbation theory. The slow convergence as $l \to \infty$ makes the forward scattering cross section infinite; note that $P_l(\cos\theta) = 1$ for $\theta = 0$. The use of the first-order phase shift δ_l^{B1} for all l down to the s wave and the approximation that $e^{2i\delta_l} - 1 \cong 2i\delta_l$ in Eq. (135) give the scattering amplitude to the first order in the potential, i.e., in the first-Born approximation. This approximation is valid for small θ, for which the low-l waves contribute little to the cross section, and gives the reduced cross section

$$d\bar{\sigma}^{B1}/d\omega = (E/g^4) d\sigma^{B1}/d\omega = 2\pi^2 \bar{q}^{-2} \tag{136}$$

in atomic units, where $\bar{q}$ is the reduced momentum transfer

$$\bar{q} = 2g(\hbar q/\hbar k) = 4g \sin(\theta/2). \tag{137}$$

The cross section shows a strong divergence in the forward direction.

Classical scattering by the potential $(\hbar^2/\mu)g/r^2$ is well defined only when $g > 0$. The exact reduced cross section is $d\bar{\sigma}^C/d\omega = (E/g^4) d\sigma^C/d\omega = g^{-3}[\pi^2(\pi - \theta)/\{\theta^2(2\pi-\theta)^2 \sin\theta\}]$ in atomic units (Ref. 123, Chapter 5). When $g < 0$, the classical deflection angle is infinite for any low angular momentum $J\hbar$ that satisfies $-J^2/2 = g_c > g$, and the particle spirals an infinite number of times towards the origin. This phenomenon is similar to the falling into center in quantum mechanical scattering for $g < g_c$ (Ref. 123, Chapter 5; Ref. 125).

In the linear-trajectory classical perturbation theory the momentum change $\hbar k_y$ perpendicular to the trajectory is equal to the impulse $\int F_y dt$ from the perpendicular (or y-) component F_y of the force. The classical deflection angle for an impact parameter b is $\theta(b) \cong k_y/k = (\pi/2)(g/b^2)(\hbar^2/\mu)/E$.[126] In the approximation that $\theta \cong \sin\theta \cong 2\sin(\theta/2)$, the reduced cross section becomes[29, 126]

$$d\bar{\sigma}^{CP}/d\omega = (E/g^4) d\sigma^{CP}/d\omega = 2\pi\bar{q}^{-3}. \tag{138}$$

This agrees with the exact classical reduced cross section $d\bar{\sigma}^C/d\omega$ within the relative error of 3% *at all* θ, all E, and all g.[29] Incidentally, the same classical perturbation theory applied to Coulomb scattering yields $\theta = c/Eb$ and reproduces the exact Rutherford cross section.[126]

For the classical theory to be applicable the uncertainty Δy in the position y must be smaller than b, and the uncertainty Δp_y in the momentum component p_y must be smaller than $\hbar k_y \cong \hbar k\theta$. On the other hand the uncertainty principle requires that $(\Delta y)(\Delta p_y) \gtrsim \hbar$, or $kb\theta \gtrsim 1$. Indeed, the ratio of the quantum-mechanical to classical cross sections for the potential c/r^p with $p \geqq 2$ is $1 + (2kb\theta)^{-2} + O((kb\theta)^{-4})$ in the limit of large $kb\theta$.[126] For Coulomb scattering $kb\theta$ is independent of θ, and the uncertainty principle places no restriction on θ. For scattering by $(\hbar^2/\mu)g/r^2$, however, the inequality $kb\theta \gtrsim 1$ tells that the classical theory is valid only when $\theta \gtrsim 1/\pi g$ or $\bar{q} \gtrsim 2/\pi$. Equation (138) is really a good approximation in this range of $\bar{q}$.[29] For smaller scattering angles ($\bar{q} \lesssim 1/\pi$) the Born cross section (136) is accurate. The cross sections (136) and (138) intersect at $\bar{q} = 1/\pi$.

Explanation of one of the semiclassical eikonal approximations, i.e., the one due to Glauber, is now in order (Refs. 25 and 27; Ref. 123, Chapter 18). In this approximation as applied to the present problem the first-order phase shift δ_l^{B1} replaces δ_l in Eq. (135), and the terms of all orders in δ_l^{B1} are retained in contrast with the Born approximation. Then, the large-l small-θ approximation $P_l(\cos\theta) \cong J_0((2l+1)\sin(\theta/2))$ and the semiclassical approximation $kb \cong l + \frac{1}{2}$ are made, and the summation over l is replaced by integration over b. The resultant scattering amplitude is

$$f^E(\theta) = ik\int_0^\infty b\,db\,(1 - e^{i\chi})J_0(qb) \tag{139a}$$

$$= (ik/2\pi)\int d^2\mathbf{b}(1 - e^{i\chi})e^{i\mathbf{q}\cdot\mathbf{b}}, \tag{139b}$$

where $\chi = 2\delta^{B1}_{kb-1/2} = -\pi g/kb$ for our problem, and where $\mathbf{b}$ is the two-dimensional impact-parameter vector perpendicular to $\mathbf{k}$. The reduced cross section in Glauber's eikonal approximation is a function only of $\bar{q}$. Its leading term for small $\bar{q}$ coincides with the Born cross section (136), as is expected from the derivation of Eq. (139). For $\bar{q} \gtrsim 2/\pi$ the eikonal and the classical-perturbation theories are in good agreement with each other (the difference being less than 14%).[29] The eikonal cross section smoothly connects near $\bar{q} = 1/\pi$ the Born cross section for smaller $\bar{q}$ with the classical-perturbation cross section for larger $\bar{q}$. In spite of the small-angle approximation the eikonal amplitude (139) accurately reproduces the exact cross section *at all* θ. For $g \lesssim 1$, for example, the relative error is always a few percent or less.[29] Incidentally, the eikonal amplitude for Coulomb scattering is exact at all θ and all E except for a phase factor independent of θ.[25,27]

Another derivation of Glauber's eikonal approximation (139b) elucidates its physical meaning.[25] Write the solution $\Psi(\mathbf{r})$ of the time-independent Schrödinger equation in terms of its amplitude and phase as $\Psi(\mathbf{r}) = A(\mathbf{r}) \exp [iS(\mathbf{r})]$. In the short-wavelength and weak-potential approximations $\nabla^2 A/A \ll k^2$ and $V \ll k^2$ the Schrödinger equation becomes

$$(\nabla S)^2 \cong k^2 - 2V \cong (k - V/k)^2. \tag{140}$$

Then, the shift in the phase of the wave $\Psi(\mathbf{r})$ or the eikonal $S(\mathbf{r})$ is the integral of the local wave number $k - V/k$ along the path of a geometric ray. The rays are perpendicular to the wave fronts, i.e., to the surfaces of constant $S(\mathbf{r})$. Assume that all ray paths are almost straight lines (parallel to the z axis along $\mathbf{k}$) implying small-angle scattering. Then an approximate solution satisfying the conservation of flux ($A^2 \cong$ const) is

$$\Psi(\mathbf{r}) \cong \exp(i\mathbf{k}\cdot\mathbf{r}) \exp\left\{-i \int_{-\infty}^{z} dz' \,(V/k)\right\}. \tag{141}$$

Use this form in the expression

$$f(\theta) = -(2\pi)^{-1} \int d\mathbf{r} \exp(-i\mathbf{k}'\cdot\mathbf{r}) V(\mathbf{r}) \Psi(\mathbf{r}) \tag{142}$$

for the exact scattering amplitude. Assume, in the light of the small-θ approximation, that the momentum transfer $\mathbf{q} = \mathbf{k} - \mathbf{k}'$ is perpendicular to the ray paths. Then follows, after some algebra, Eq. (139b) with

$$\chi = -\int_{-\infty}^{\infty} dz\,(V/k). \tag{143}$$

Equation (143) is the change in the phase of the wave (141) when it propagates from $z = -\infty$ to $z = +\infty$. Although linear in V, Eq. (143) is, in general, unequal to twice the Born phase shift δ_l^{B1} even for a spherical potential. As the energy E increases for a given l, the impact parameter $b \cong (l + \frac{1}{2})/k$ decreases. We may use r in place of $z = (r^2 - b^2)^{1/2}$ for high E, which reduces Eq. (143) to $2\delta_l^{B1}$ for $l = kb - \frac{1}{2}$ (Ref. 123, Chapter 18). For $V = g/r^2$ Eq. (143) happens to coincide with $2\delta_{kb-1/2}^{B1}$ at all k and all l.

5.3.2. Scaling of the Differential Cross Section

In accordance with the remark in the beginning of Section 5.3 we study rotational-vibrational transitions in the collision of an electron with a symmetric-top molecule interacting only through the dipole potential

$$(\hbar^2/\mu)V = (\hbar^2/\mu)r^{-2}\bar{V}_{10}(R_T)Y_{10}(\hat{\mathbf{r}}') = -eD\cos\theta'/r^2. \tag{144}$$

The main differences of this problem from scattering by a potential c/r^2 are the anisotropy of the interaction and the existence of the internal state of the target. We note that the rotational sudden approximation is valid except at extremely small scattering angles (see below), except very close to a threshold of excitation, and except in a very narrow resonance (Section 3.2.2). Consider a case where not only the rotational but also the vibrational sudden approximation is valid. The internal states are inessential, as far as the cross section summed over the final state is concerned. Even the cross section for any specified transition is determined by the amplitude for potential scattering. Because of the r^{-2} dependence of the potential (144) the argument just below Eq. (134) holds equally true for the present problem; the differential cross section is proportional to E^{-1} to the extent that the sudden approximation with $k \cong k'$ is valid.

Due to the slowly varying nature of the potential (144) the classical perturbation theory deserves consideration.[127–130] According to this theory the deflection angle $\omega = (\theta, \varphi)$ as a function of the impact-parameter vector $\mathbf{b} = (b, \varphi_b)$ and of the orientation $\Omega = (\theta_T, \varphi_T)$ of the dipole is ($\theta = eD\sin\theta_T/Eb^2$, $\varphi = 2\varphi_b - \varphi_T + \pi$). (Dickinson[129] discusses the exact classical solution.) In contrast with the situation in scattering by the spherical potential c/r^2 two orbits having the same b but with φ_b differing by π contribute to the same deflection angle. Averaging the differential cross section over all dipole orientation Ω and using the small-angle approximation that $\theta \cong \sin\theta \cong 2\sin(\theta/2)$ we obtain the reduced cross section [as defined by Eq. (126a)] [29]

$$d\bar{\sigma}^{CP}/d\omega = \pi\bar{q}^{-3} \tag{145}$$

in atomic units, where $\bar{q}$ is the reduced momentum transfer of Eq. (126b).

The classical picture applies to our problem only when $kb\theta \gtrsim 1$ according to the uncertainty principle. This inequality gives the classical condition that $\bar{q} \gtrsim 1$ a.u., if the heavily weighted orientation $\theta_T \cong \pi/2$ is adopted in the evaluation of $\theta(b)$. Figure 13 indicates the approximate validity of Eq. (145) for $\bar{q} \gtrsim 1$ a.u.

For use at smaller scattering angles we study the quantum-mechanical first–Born approximation. Equations (133a) and (133b) as applied to the dipole potential (144) lead to the differential and integral cross sections[115]

$$d\sigma^{B1}(vJK\mathbf{k} \to v'J'K'\mathbf{k}'; q)/d\omega = (k'/k)Aq^{-2} \tag{146a}$$

and

$$\sigma^{B1}(vJKk \to v'J'K'k') = (2\pi/k^2)A \ln |(k+k')/(k-k')|, \tag{146b}$$

where

$$A = (4/3)(\mu e/\hbar^2)^2 D^2_{v'v}\{C(J1J';K0K')\}^2 \tag{147a}$$

$$= \begin{cases} (4/3)(\mu e/\hbar^2)^2 D^2_{v'v}\delta_{KK'}K^2/\{J(J+1)\} & (J=J'\neq 0) \\ (4/3)(\mu e/\hbar^2)^2 D^2_{v'v}\delta_{KK'}(J^2_> - K^2)/\{J_>(2J+1)\} & \\ \qquad (|J-J'| = 1) & \\ 0 \qquad (|J-J'| \geqq 2; J = J' = 0). & \end{cases} \tag{147b}$$

Here, $J_> = \max(J,J')$ and $D_{v'v} = (\chi_{v'}|D|\chi_v)$. The selection rules that $|\Delta J| = |J'-J| \leqq 1$, that $K=K'$, and that $K \neq 0$ if $J=J'$ have followed from the Clebsch–Gordan coefficient in Eq. (147a).

Due to large-l distant collisions the differential cross section (146a) is strongly peaked in the forward direction. It diverges at $\theta = 0$ for elastic scattering $vJK \to vJK$ ($K \neq 0$), making the integral cross section (146b) infinite. For inelastic scattering the momentum transfer $\hbar q$ takes a small but positive value $\hbar|k-k'|$ at $\theta = 0$, making the differential cross section large (especially so for a purely rotational transition) but finite. In that sudden approximation in which k' is assumed to equal k, the inelastic cross sections also diverge at $\theta = 0$, the integral inelastic cross sections being infinite. This stems from the inapplicability of this sudden approximation to distant collisions because of the large effective potential range or of the long effective collision time. The breakdown of the sudden approximation at small θ *at any* E is due entirely to the long-range tail ($\propto r^{-2}$) of the potential, and not to the Born approximation or to the inaccurate description of the interaction at shorter distances.

At small θ where the linear-dependence relation (62) based on the rotational sudden approximation fails, the Born approximation appears to be usually valid,[7] and the relation (133b) may be used. In other words Eq. (62) is applicable at all scattering angles (except close to a threshold of excitation and except in a narrow resonance), if the cross sections on both sides are understood to be for a common value of q rather than of θ. Using q instead of θ is crucially important for *forward* scattering by *dipolar* molecules, but otherwise it makes only negligible difference. Note that the cross sections for use in Eq. (62) sometimes need analytic continuation beyond the physical region $[|k-k'|, k+k']$ of q, because this region depends on the channel.

Since $q^2 = k^2 + k'^2 - 2kk' \cos\theta = (k-k')^2 + 4kk'\sin^2(\theta/2)$, the sudden approximation with $k \cong k'$ or $q \cong 2k\sin(\theta/2)$ may be made only for θ satisfying[129]

$$2k\sin(\theta/2) \gg |k - k'| \quad \text{or} \quad \theta \gg |\Delta E|/2E. \tag{148}$$

The ratio $|\Delta E|/2E$ is often extremely small for purely rotational transitions. In the angular region (148) the dipole–Born cross section (146a) summed over the final rotational state gives the reduced cross section [see Eq. (38a)]

$$d\bar{\sigma}^{B1}/d\omega = (E/D^4)\sum_{J'} d\bar{\sigma}^{B1}(vJK\mathbf{k} \to v'J'K\mathbf{k}')/d\omega = (8/3)\bar{q}^{-2} \tag{149}$$

in atomic units with D in Eqs. (149) and (126) denoting $D_{v'v}$. The classical-perturbation and first-Born cross sections intersect at $\bar{q} = 3\pi/8 = 1.18$ a.u.

The true integral and small-angle differential cross sections for the transitions allowed by the simple dipole–Born formulas (146a) and (146b) are much larger than those for the forbidden transitions, and are usually reproduced fairly well by Eqs. (146a) and (146b). Although useful in this sense, the dipole–Born approximation has its limitations. (a) Transitions with $\Delta K \neq 0$, with $\Delta J = 0$ for $K = K' = 0$, and with $|\Delta J| \geqq 2$ are completely forbidden. (b) Unitarity is violated for a molecule with a large dipole moment. (c) Potentials of shorter range than the dipole are neglected.

In Fig. 13 the dipole–Born cross section (146a) deviates from the observed cross sections for $\bar{q} \gtrsim 1$ a.u. The upward instead of downward deviation cannot be due to limitation (a), because any contribution from transitions forbidden by the dipole–Born approximation would increase the summed cross section. That the deviation is due mainly to limitation (b) rather than to (c) shall be made clear by studies in the Glauber approximation with the dipole potential (144).[29,131,132]

In Glauber's theory for scattering by a particle with internal structure the scattering amplitude takes a form (51) of the sudden-approximation amplitude.[25,27] The fixed-orientation (fixed-Ω) amplitude F there is calculated according to Glauber's eikonal formulas (139b) and (143). The momentum transfer $\mathbf{q}$ in Eq. (139b) allows for the difference between k and k' for inelastic collisions, keeping the forward differential cross section finite. Because the Born amplitude also takes the form (51), and because the fixed-Ω eikonal amplitude (139b) approaches the fixed-Ω Born amplitude for small θ, the Glauber amplitudes for the transitions allowed in the Born approximation also approach the Born amplitudes for small θ.[25,27,132]

The transitions forbidden in the dipole–Born approximation are allowed in the dipole–Glauber approximation, because the latter involves, although approximately, terms of all orders in the potential. The form (51) of the scattering amplitude and the dependence of F on the rotational channels only through q guarantee that the linear-dependence relation (133b) holds also for the Glauber cross sections for a common value of q; the cross sections for any rotational transitions are expressible in terms of those for the transitions from the ground rotational state.

The reduced cross section, similar to Eq. (126a), for any rotational transition $J \to J'$ is a function of only $\bar{q}$ in the Glauber approximation except for the factor k'/k.[132] Figure 14 compares the dipole–Glauber reduced cross sections with the results of numerical close-coupling and hybrid S-matrix (Section 5.3.3) calculations[7,133,134] on purely rotational transitions of some highly dipolar molecules from the ground rotational state. The latter calculations take account not only of the dipole interaction but also of the interaction of shorter range approximately. A scaling law is seen to relate approximately the cross sections for different molecules and at different energies. The only exception that fails to obey the scaling law is the set of hybrid S-matrix differential cross sections for CsCl at 20 eV. In other cases the reduced cross sections are close to the Glauber universal curves at small and medium scattering angles θ, but depart from the Glauber theory beyond critical θ_c. The θ_c is larger for molecules with larger D, because the interaction of shorter range than the dipole begins to dominate over the dipole interaction at smaller r for larger D. This makes the scaling law useful only for highly dipolar molecules.

Figure 14 shows that the dipole–Born-allowed transition $J = 0 \to 1$ has a cross section much larger at small θ than the dipole–Born-forbidden transitions. The dipole–Born cross section proves to be a good approximation to the true cross section for $J = 0 \to 1$ at small θ. Note also that, according to Fig. 14, the dipole interaction is dominant at small and medium θ even in the dipole–Born-*forbidden* transitions $J = 0 \to 0,2$ of highly dipolar molecules.

In the sudden-collision region (148) of θ the sum of the dipole–Glauber reduced cross section over the final rotational state is a function of only $\bar{q}$. Owing to the sum rule of Section 3.3.2 this summed reduced cross section is independent of the initial rotational state, and is appropriate for comparison with experimentally observed cross sections that are rotationally unresolved. This comparison is made in Fig. 13. The dipole–Born cross section, which is also shown in Fig. 13, starts to deviate from the dipole–Glauber cross section at smaller $\bar{q}$ ($\bar{q} \cong 1$ a.u.) than that at which the Glauber curve starts to deviate from experiments on highly dipolar molecules. This indicates that, as $\bar{q}$ increases, the dipole–Born theory fails first due to the breakdown of the first-order perturbation theory and then, at larger $\bar{q}$, due to the poor description of the interaction of shorter range than the dipole. For molecules with small D, however, the dipole–Born cross section is good only at $\bar{q}$ much smaller than ~ 1 a.u., which shows the greater importance of the short-range interaction than the second- and higher-order terms in the dipole interaction for such molecules.

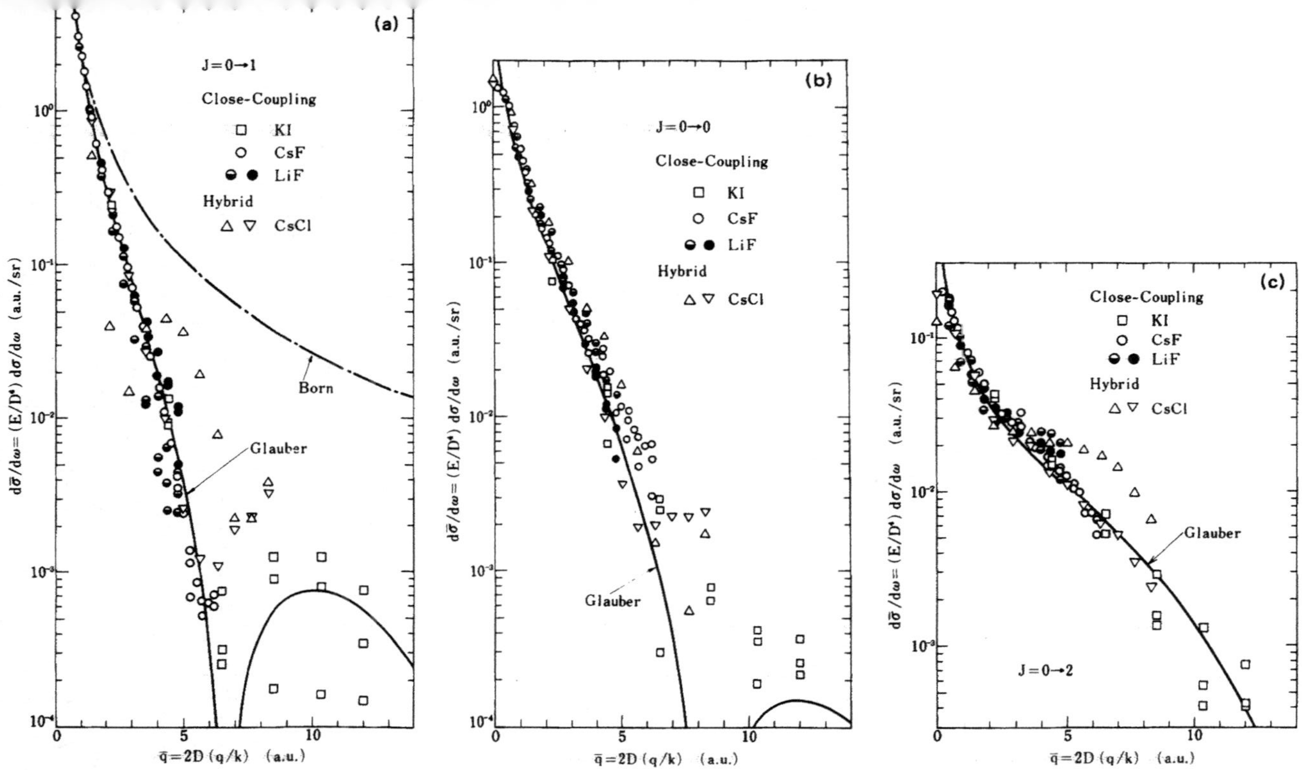

FIGURE 14

Reduced differential cross section $d\bar{\sigma}/d\omega = (E/D^4)d\sigma/d\omega$ plotted versus reduced momentum transfer $\bar{q} = 2D(q/k) \cong 4D \sin(\theta/2)$, D being the dipole moment, q the momentum transfer, E the collision energy, and k the wave number of electron. Close-coupling calculations with the dipole potential and a shorter-range potential of a few different choices: □, KI (6.74 eV, $\theta \leqq 90°$);[7] ○, CsF (1 eV, $\theta \leqq 60°$);[133] ●, LiF (5.44 eV, $\theta \leqq 55°$);[7] ◒, LiF (20 eV, $\theta \leqq 55°$).[7] Hybrid S-matrix calculations: ▽, CsCl (5 eV, $\theta \leqq 60°$);[134] △, CsCl (20 eV, $\theta \leqq 60°$).[134] —·—, dipole–Born approximation. ———, dipole–Glauber approximation. (a) Rotational transition $J = 0 \rightarrow 1$. (b) $J = 0 \rightarrow 0$. (c) $J = 0 \rightarrow 2$. (Taken from Shimamura.[29])

The summed Glauber reduced cross section in Fig. 13 appears to connect smoothly near $\bar{q} \cong 1$ a.u. the Born cross section (149) for smaller $\bar{q}$ and the classical-perturbation cross section (145) for larger $\bar{q}$. This is similar to the case with a central potential c/r^2 discussed in Section 5.3.1. In contrast with the central potential, however, the anisotropic potential (144) has two classical trajectories contributing to any single classical deflection angle. The quantum-mechanical waves associated with these two trajectories interfere with each other, giving rise to oscillations in the Glauber reduced cross section around the classical-perturbation result (145). Although obscured to some extent by the steep curves in Fig. 13, these oscillations stand out clearly in the ratio of the Glauber to classical-perturbation curves; close-coupling results also show these oscillations clearly.[(29)] Dickinson observes this interference effect in a simpler "primitive semiclassical" approach.[(129)] Allan and Dickinson study further details of the interference effect on azimuthal- as well as polar-angle distribution in transitions of linear rotators with both J and M specified.[(135)] They also extend the primitive semiclassical approach and combine it with the sudden approximation (45) to derive an approximation to the Glauber cross section.[(136)]

5.3.3. Division into Distinct Partial-Wave Regions

Section 5.3.2 has discussed the difference between distant collisions and closer collisions by distinguishing between different regions of the scattering angle θ. Almost equivalently, one may distinguish between different regions of the angular momentum l of the incident electron. The latter way of looking at the process affords both further insight into the dynamics and procedures for more accurate treatments of the collision problem than those outlined in Section 5.3.2.

The high centrifugal barriers restrict the high partial waves to the large-r region. Only low-l waves can penetrate to the molecular electron cloud. Therefore, as mentioned in Section 4.2.3, the division of all partial waves into groups is essentially the same as the division of the configuration space discussed in Section 4. Roughly speaking, the compound-system region ($r < a$) in Fig. 8 corresponds to a low-l region, the body-frame potential-scattering region ($a < r < a_F$) to a medium-l region, and the laboratory-frame potential-scattering region ($a_F < r$) to a high-l region. Based on this recognition, Clark and Siegel[(137)] and Siegel *et al.*[(134,138,139)] propose and use an approximate hybrid S-matrix method. This method employs the continuum multiple-scattering method (Chapter 7, Section 3.6), a fixed-nuclei multicenter-potential approach in the body frame, for the calculation of low-l elements of the S matrix. For medium values of l (or l'), the body-frame fixed-nuclei approach is still made, but the interaction between the molecule and the electron is approximated by the point-dipole potential (144). For high partial waves only glancing collisions occur, and the electron and the molecule can exchange only a single quantum of angular momentum through the weak dipole interaction at large r. The laboratory-frame S-matrix elements in the dipole–Born approximation are used for high l (or l').

Collins and Norcross[7] also use a three-part factorization of partial waves in the most elaborate version of their methods. They divide the laboratory-frame region of l into two subregions, and use the dipole–Born approximation in the higher-l subregion and the laboratory-frame close-coupling equations with the multipole expansion (128) of the potential in the lower-l subregion. They regard the body-frame region of l as a single whole, and do not subdivide it. There, they solve the close-coupling equations with the single-center multipole expansion of the potential.

It has long been customary to employ the dipole–Born approximation for treating the many high partial waves that contribute to scattering by dipolar molecules.[140,141] This procedure is efficient, because literature exists on the closed forms of both the S-matrix element for any partial wave and the scattering amplitude, which is the sum of the contributions from all partial waves.[140,141] In other words, one may apply a closure formula in the dipole–Born approximation to the infinite sum over the partial waves; for low and intermediate partial waves, for which the dipole–Born approximation is invalid, corrections are made by calculating the difference between the partial cross sections (or the S-matrix elements) in the dipole–Born approximation and in a more realistic approximation. This calculation for low and intermediate partial waves may be made in the body-frame sudden or adiabatic-nuclei approximation according to the arguments made so far in this subsection. Norcross and Padial discuss this idea in detail and suggest its extension to a general multipole potential by the name "multipole-extracted adiabatic-nuclei (MEAN)" approximation.[31] In passing, the study of electron–atom collisions is also based often on the division into distinct partial-wave regions and on the treatment of each region by a separate method (see, e.g., Refs. 142 and 143); closure formulas in the Born approximation appear often useful (e.g., Ref. 142).

Mittleman and von Holdt[144] and Mittleman *et al.*[145] solve the problem of scattering by the dipole potential (144), which arises in the intermediate-l region in the hybrid S-matrix method. Clark[146] and Clark and Siegel[137] consider how the results of Refs. 144 and 145 may be cast into a form convenient for use in the hybrid-method calculations.

The partial-wave expansion of the wave function

$$\Psi(\mathbf{r}') = \sum_{l\lambda} x^{-1}\psi_{l\lambda}(x)Y_{l\lambda}(\theta',\varphi') \qquad (x = kr) \tag{150}$$

substituted into the Schrödinger equation gives coupled equations

$$\sum_{l}[(d^2/dx^2 + 1)\delta_{l'l} - V_{l'l}^{\lambda}x^2]\,\psi_{l\lambda}(x) = 0 \tag{151a}$$

with

$$V^{\lambda}_{l'l} = V^{\lambda}_{ll'} = l(l+1)\delta_{l'l} - (2\mu eD/\hbar^2)\langle Y_{l'\lambda}|\cos\theta'|Y_{l\lambda}\rangle$$
$$= l(l+1)\delta_{l'l} - (2\mu eD/\hbar^2)[(l_>^2 - \lambda^2)/(2l+1)(2l'+1)]^{1/2}\delta_{l',l\pm 1}, \tag{151b}$$

where $l_>$ is the larger of l and l'. Note that Eq. (151a) is independent of the energy just like Eq. (134). Truncate the infinitely large tridiagonal matrix $\mathbf{V}^{\lambda}$ to a finite size, and diagonalize it for each λ as

$$\mathbf{U}^{\lambda}\mathbf{V}^{\lambda}(\mathbf{U}^{\lambda})^{\dagger} = \mathbf{L}^{\lambda}(\mathbf{L}^{\lambda} + \mathbf{1}). \tag{152}$$

Here, $\mathbf{L}^{\lambda}$ is a diagonal matrix with diagonal elements L^{λ}_N ($N = 0,1,2,\ldots$). Of the two possible choices of L^{λ}_N in general we adopt the value that approaches N, which has the same significance of the angular-momentum quantum number as l, as the dipole moment D tends to zero. The unitary matrix $\mathbf{U}$ transforms the spherical harmonics into the dipole angular functions

$$\Theta_{N\lambda}(\theta',\varphi') = \sum_l U^{\lambda}_{Nl}(\theta',\varphi'). \tag{153}$$

Correspondingly, $\mathbf{U}^{\lambda}$ transforms the channel functions $\psi_{l\lambda}(x)$ into a new set of channel functions $\tilde{\psi}_{N\lambda}(x)$ satisfying the uncoupled equations

$$[d^2/dx^2 + 1 - L^{\lambda}_N(L^{\lambda}_N + 1)/x^2]\,\tilde{\psi}_{N\lambda}(x) = 0. \tag{154}$$

These equations have the same form as Eq. (134). Because L^{λ}_N is always positive for any intermediate or large angular momenta (see below), the solution of Eq. (154) that satisfies the boundary condition at the origin is $x^{1/2}J_{\nu}(x)$ with $\nu = L^{\lambda}_N + \frac{1}{2} > 0$. The phase shift is $\delta^{\lambda}_N = (N - L^{\lambda}_N)\pi/2$ [see Eq. (134a)], and the corresponding element of the diagonal S matrix in the dipole-angular-function representation is $\exp(2i\delta^{\lambda}_N)$. The usual S matrix in the body frame is obtained by the unitary transformation back to the spherical-harmonic representation as

$$\mathbf{S}^{\lambda} = (\mathbf{U}^{\lambda})^{\dagger}\exp(2i\boldsymbol{\delta}^{\lambda})\mathbf{U}^{\lambda}. \tag{155}$$

The laboratory-frame S matrix follows from the frame transformation (94).

In the laboratory-frame region the rotational Hamiltonian H_r is important and the channel energies are nondegenerate. Therefore, the transformation (152) introduces a nondiagonal energy matrix that couples equations similar to Eq. (154). Chandra neglects the off-diagonal elements of the transformed energy matrix and solves the resultant decoupled equations analytically.[78] This procedure, being still simple, can improve over the dipole–Born approximation.

If the dipole moment D in Eq. (151b) is large ($>D_c^\lambda$), one or more lowest eigenvalues of $\mathbf{V}^\lambda$ are too small for L_N^λ to be real. Then, a particle with low N falls to the center, as is the case noted in Section 5.3.1, and $\tilde{\psi}_{N\lambda}$ is ill defined.[144,145] The critical dipole moment D_c^λ is 1.625 D (0.639 a.u.) for $\lambda = 0$, and 9.64 D (3.79 a.u.) for $\lambda = 1$.[144,147] Above the critical dipole moment the wave function $\psi_{l\lambda}(x)$ for any l is undefined, because it is a linear combination of $\tilde{\psi}_{N\lambda}(x)$ for all N including the lowest one ($N = 0$). This phenomenon stems from the nonphysical singularity of the point-dipole potential at the origin. Therefore, we may and should neglect the low-N components, which have small coefficients anyway, when calculating $\psi_{l\lambda}(x)$ with an intermediate or high l for an overcritical dipole moment. For the same reason the S matrix (155) with low-N elements of $\boldsymbol{\delta}^\lambda$ neglected may be used also for an overcritical D.

The phenomenon of falling to the center is closely related to the bound-state energy spectrum for the point-dipole potential. Garrett[2,148] reviews the studies of the scattering and bound-state problems associated with overcritical D for fixed and rotating point and finite dipoles; see Refs. 2, 148, and 149 for further references.

Another closely related subject is the threshold behavior of the cross sections for electron scattering by a dipolar molecule and for photodetachment of an electron attached to a dipolar molecule.[150–153] Mathematically difficult problem as it is, this subject is yet to be solved exactly, but has seen recent progress; see the brief review by Garrett[2] and references therein.

5.4. *Charge–Quadrupole Interaction*

The multipole interaction of the longest range between an electron and a quadrupolar molecule (with vanishing dipole moment) is the quadrupole interaction, which takes a form [see Eqs. (128a) and (129)]

$$(\hbar^2/\mu)V = (\hbar^2/\mu)r^{-3}\,\bar{V}_{20}(R_T)Y_{20}(\hat{\mathbf{r}}') = -eQP_2(\cos\theta')/r^3 \tag{156}$$

for a linear molecule and for a symmetric-top molecule having an n-fold symmetry axis with $n \geqq 3$.

The differential and integral cross sections in the Born approximation for scattering by the quadrupole potential (156) follow from Eqs. (133a,b) as

$$d\sigma^{B1}(vJK\mathbf{k} \to v'J'K'\mathbf{k}')/d\omega = (k'/k)A \tag{157a}$$

and

$$\sigma^{B1}(vJKk \to v'J'K'k') = 4\pi(k'/k)A, \tag{157b}$$

where

$$A = (4/45)\,(\mu e/\hbar^2)^2 Q_{v'v}^2 \{C(J2J';K0K')\}^2 \tag{158a}$$

$$= \begin{cases} \dfrac{4}{45}\left(\dfrac{\mu e}{\hbar^2}\right)^2 Q_{v'v}^2 \delta_{K'K} \dfrac{\{3K^2 - J(J+1)\}^2}{J(J+1)(2J+3)(2J-1)} & (J = J' \neq 0) \\[2ex] \dfrac{4}{15}\left(\dfrac{\mu e}{\hbar^2}\right)^2 Q_{v'v}^2 \delta_{K'K} \dfrac{K^2(J_>^2 - K^2)}{J_>(J_>^2 - 1)(2J+1)} & (|J - J'| = 1, J \neq 0, J' \neq 0) \\[2ex] \dfrac{2}{15}\left(\dfrac{\mu e}{\hbar^2}\right)^2 Q_{v'v}^2 \delta_{K'K} \dfrac{(J_>^2 - K^2)\{(J_> - 1)^2 - K^2\}}{J_>(J_> - 1)(2J_> - 1)(2J+1)} & (|J - J'| = 2) \\[2ex] 0 & (|J - J'| \geq 3; J = J' = 0; |J - J'| = 1, J_> = 1). \end{cases} \tag{158b}$$

Here, $J_> = \max(J,J')$ and $Q_{v'v} = (\chi_{v'}|Q|\chi_v)$. The Clebsch–Gordan coefficient in Eq. (158a) has led to the selection rules that $|\Delta J| \leq 2$, that $K = K'$, that $J \neq 0$ if $J = J'$, and that $K \neq 0$ (and hence, $J \neq 0$ and $J' \neq 0$) if $|\Delta J| = 1$.

The angular distribution (157a) is isotropic. There is no forward peaking of the differential cross sections, as opposed to scattering by the dipole potential. The rotational sudden approximation makes no difference to the cross sections.

The region of the configuration space where the quadrupole potential dominates the collision dynamics is usually small. As r decreases, the strong potentials of shorter range than the quadrupole soon take over the latter. Thus the study of scattering by the quadrupole potential alone is less important than the case of the dipole potential. Close to the threshold of the rotational excitation allowed by the quadrupole–Born formulas (158), however, distant collisions appear important and so does the quadrupole potential (156).

Take, for example, scattering of an electron by a homonuclear diatomic molecule, such as H_2 and N_2, in a $^1\Sigma$ state. The transitions with $|\Delta J| = 2$ are by far the strongest among the inelastic processes at very low energies in the absence of a resonance (see Table 2 in Section 3.3.2 and Fig. 17 in Section 6.3). Close to the threshold of an excitation $J \to J + 2$, the wave number k' in the outgoing channel is small. Only the outgoing s wave is appreciable in the region of the potential, because of the behavior $\psi_{l'} \cong k'^{1/2} r j_{l'}(k'r) \cong k'^{-1/2} \times (k'r)^{l'+1}/(2l' + 1)!!$ of the partial-wave functions at small $k'r$; scattering in the outgoing channel is weak close to the threshold. The interaction potential between an electron and a homonuclear $^1\Sigma$ diatomic molecule couples the outgoing s wave only with the incoming waves with even l (Section 4.2.2),

$s \to s$ coupling between different rotational channels being excluded. Therefore, the $d \to s$ contribution determines the threshold behavior of the cross section. The K matrix element behaves as $\propto k'^{l'+1/2} = k'^{1/2}$, which is the case also in the absence of the quadrupole potential.[154,155] The low-energy incoming d wave is prevented by the centrifugal barrier from penetrating to the small-r strong-interaction region, experiencing only negligible distortion. The outgoing s wave experiences only weak scattering, because k' is very small. The Born approximation suffices for determining the threshold behavior.[154]

By partial-wave analysis of their model-potential close-coupling calculations Lane and Geltman confirm for the transition $J = 0 \to 2$ of H_2 the dominance of the $d \to s$ contribution up to about 0.02 eV above threshold.[48] They also observe good accuracy of the quadrupole–Born approximation in this energy range. At higher energies the outgoing p wave, which couples with the incoming p wave, takes over, and the $p \to p$ contribution remains the principal contributor throughout the entire energy range considered by them, i.e., up to about 20 eV.[48] At small energies ($\lesssim 10^{-1}$ eV) the centrifugal barrier restricts both incoming and outgoing p waves in the large-r quadrupole-potential region where the distortion of the waves is almost negligible. This makes the quadrupole–Born approximation still reliable in this energy range. Chang's partial-wave analysis of the quadrupole–Born cross section also shows the shift of the dominance from the $d \to s$ to the $p \to p$ partial cross section as the energy increases from the threshold.[156]

Figure 6 illustrates the validity of the Born approximation for the integral cross section near the threshold of excitation.[33] The Born differential cross section, however, is less reliable, the close-coupling results by Feldt and Morrison showing anisotropic angular distribution even close to the threshold.[33]

The elastic cross sections $J \to J$ for H_2, on which the effect of the short-range interaction is strong, are much larger than the inelastic cross sections (Section 3.3.1 and Fig. 11a). Lane and Geltman find that elastic scattering is primarily of the $s \to s$ character at low energies, the $p \to p$ component becoming comparable above about 4 eV.[48]

5.5. *Interaction of Shorter Range*

Inclusion of terms higher than the dipole and the quadrupole in the multipole expansion (128a) provides an accurate representation of the electrostatic potential at short distances. Importance of these higher terms in the potential scattering region of Fig. 8 depends on the case. They are unimportant, e.g. for highly dipolar molecules. A reliable but not always necessary method is to solve close-coupling equations including an increasing number of multipole potential terms, and to observe the convergence of the K matrix elements or other physical quantities. This method is useful numerically, when the single-center expansion (128a) is used also in the compound-system region of Fig. 8.[7]

As the incident electron approaches the molecule, the electronic state is (virtually) excited to various levels. In other words the molecular charge distribution polarizes. An effect of this on the channels associated with the ground electronic state is the addition of a polarization potential V_{pol}, which is nonlocal and depends on the collision energy, in general. The asymptotic form of V_{pol} is local and independent of the energy. It decreases as $\sim r^{-4}$ and involves a spherical term and a term proportional to P_2 (cos θ') for a linear molecule [Eq. (24) of Chapter 1 or Eq. (15) of Chapter 7].

The polarization potential at short distances is difficult to determine. Often used is a model polarization potential that is formed by modifying the asymptotic form using the cutoff factor $1 - \exp[-(r/r_c)^6]$. Having an ambiguity in the choice of the empirical cutoff parameter r_c, this phenomenological model potential may hamper the physical interpretation of otherwise physically clear-cut theory. When only the large-r part of the potential is important, however, the cutoff polarization potential is useful, because then the cross section hardly depends on r_c.

Allowance for the polarization potential in the Born approximation [see Eq. (126) of Chapter 1][157] appears to improve the quadrupole–Born cross section for rotational excitation of quadrupolar molecules at low energies. An example on H_2 is seen in Fig. 6. Because $Q > 0$ for H_2, the contribution from the polarization potential monotonically increases the Born cross section as the energy increases. This is in contrast with the $e + N_2$ collisions, in which the negative quadrupole moment of N_2 makes the quadrupole and polarization contributions partly cancel, leading to a decrease in the rotational excitation cross section above $\sim 10^{-2}$ eV.[157]

Further discussion on the polarization potential is found in Chapter 1, Section 3.1 and in Chapter 7, Section 3.5.

6. Physics in the Compound-System Region

6.1. Multicenter Electrostatic Field

The potential field in the region of the molecular electron cloud has singularities at the positions of all nuclei. These singularities are stronger for heavier nuclei with larger nuclear charges. Convergence of the brute-force single-center expansion (128a) of the multicenter potential is slow in the compound-system region for all molecules but the lightest. For example, Morrison and Collins report calculations, on one hand, on H_2 including only three multipole terms $L = 0$, 2, and 4 in Eq. (128a) and, on the other hand, on N_2 including terms L up to $L_{\max} = 28$.[158] Collins and Norcross observe that $L_{\max}$ of 36 to

45 are necessary for the alkali halides LiF, LiCl, NaF, and NaCl.[7] The highest multipole term L_{max} to be retained in Eq. (128a) is determined by the contribution from the nuclei to the electrostatic potential. That from the molecular electrons may be truncated at somewhat lower L.[7,158]

The necessity of so many terms for convergence of the single-center expansion reflects inappropriateness of the concept of multipole potentials in the compound-system region. A similar situation occurs in quantum chemical calculations of molecular structures or bound-state energies, for which multicenter description of the wave functions is much more appropriate and useful than the single-center expansion. Quantum chemical approaches with some modifications may be useful for studying the dynamics of the compound system in electron–molecule collisions.

The R-matrix method is one of the possible approaches; instead of tracing a particle from $r = 0$ to $r = a$ by solving the close-coupling equations we deal with the compound system as a whole by calculating the Hamiltonian matrix in terms of multicenter integrals. These integrals are similar to those occurring in quantum chemistry, but are over the finite compound-system region.

Multicenter integrals over a finite region are difficult to compute in general. This had prompted the development of a class of computational methods, called L_2 (or L^2) methods. In these methods the basis functions are square-integrable (or L_2 functions) just like those of the bound-state basis functions. Quantum chemical program packages may be easily adapted for scattering calculations. The compound-system region is vaguely defined, in effect, as the region spanned by the basis functions.

Another way of avoiding the multicenter integrals is to divide the compound-system region into spherical subregions centered at nuclei and interstitial subregions, and to assume a simple (often spherically symmetric or constant) model potential in each subregion. The continuum multiple-scattering method (CMSM) for electron–molecule collisions is based on this idea of taking approximate account of the multicenter field. The CMSM is an extension of an approximate method of calculating structures of large molecules and solids.

Chapter 7 contains a more detailed account of the R-matrix method as applied to electron–molecule collisions (Section 3.8), of L_2 methods (Sections 3.7 and 3.9), and of the CMSM (Section 3.6).

6.2. *Indistinguishability of All Electrons*

Another important feature of the compound-system region is that all electrons in the system are indistinguishable and strongly correlated with each other in this region. The motion of each electron hinges on the motion of the other electrons. Nevertheless, the truncated close-coupling expansion of the total wave function in terms of given target wave functions presupposes particular states of motion of the "target" electrons, and forces the "incident" electron to move in the given field of the "target," thus distinguishing between the two kinds of electrons. An artificial way to allow for the indistinguishability between the two is to antisymmetrize the total wave function. This procedure introduces into the close-coupling equations a nonlocal potential, called the exchange potential, which represents the effect of the exchange between the "incident" and the "target" electrons, and which is derived in Chapter 1, Section 6.6. The "incident" electron at any position in the compound-system region feels the nonlocal potential that depends on the motion of the electron itself *throughout* the region. This precludes division into subregions in the presence of the nonlocal potential. Chapter 7, Sections 3.1–3.4 discuss in detail how one may treat the exchange potential in actual computations. Often an approximate local form replaces the nonlocal form (Chapter 7, Section 3.4 and Chapter 1, Section 3.2).

The strange nonlocal potential is a result of the antisymmetrization after distinguishing the "incident" electron from the "target" electrons. A more natural way to treat all electrons on a par would be provided by the R-matrix method with electron configurations carefully chosen to represent the indistinguishability and the strong correlation effect. Another possibility is the hyperspherical-coordinate approach.[69,159,160] The radial variable R in this approach is related to the moment of inertia $\Sigma m_i r_i^2$ of the system by $(\Sigma m_i)R^2 = \Sigma m_i r_i^2$. The hyperradius R represents the size of the system, and includes symmetrically all particles having the same mass. It is possible to determine a complete set of functions of other (angle) variables that defines scattering channels and that leads to coupled equations in R without a nonlocal potential. Solving these equations from $R = 0$ outward, we may trace the evolution of the system from the condensation limit to the fragmentation limit. Unfortunately, neither the hyperspherical-coordinate method nor the R-matrix method in the form briefly suggested in this paragraph has yet been applied to electron–molecule collisions. Most *ab initio* calculations are in the static (or electrostatic) approximation or in the static-exchange approximation (which includes the ground-state electrostatic potential and the corresponding exchange potential) with or without polarization potential. In some cases an electronically excited real state (Chapter 7, Section 5) or pseudostate (Chapter 7, Section 3) of the target is closely coupled with the ground state.

These *ab initio* calculations sometimes yield eigenphase sums that rapidly increase with the energy by about π. This is a resonance phenomenon (Chapter 3) caused by capture of the incident electron by the molecular field or, precisely speaking, by that field of the target which the particular approximation made in the calculations introduces. Some of the calculated resonances are real. Others are spurious or mere artifacts of the approximation, and disappear if the approximation is improved on. Physics of real resonances is fully discussed in other chapters of this volume. Here, attention is called to the ubiquitousness and diversity of spurious resonances found in electron–atom collisions, which have been studied more extensively than electron–molecule collisions; one should be careful of spurious resonances also in the latter collisions.

Spurious resonances have long been known to occur frequently in variational calculations, and a variety of methods of removing them have been proposed.[161,162] Spurious resonances of similar nature often creep also into the R-matrix method[163,164] and into a modified close-coupling method involving square-integrable basis functions.[164,165] The close-coupling method including pseudostates (Chapter 1, Section 3.1) and similar R-matrix calculations produce spurious Feshbach resonances.[164,166,167] Adjustable parameters in model polarization potentials can artificially and easily shift the positions of molecular shape resonances.[168] Spurious resonances occur, naturally, also in electron–molecule collisions; examples have been found in calculations in a static-plus-local-exchange-potential approximation.[7,169] These facts suggest that calculated results should be carefully examined and the possibility of a spurious resonance be convincingly excluded, before any resonancelike feature is assumed to be a real resonance.

6.3. Representation in Terms of Adiabatic Angular Basis Set

We have discussed the hybrid S-matrix method in Section 5.3.3. In the intermediate-l region we have studied scattering by the dipole potential by diagonalizing the effective potential that is the sum $(\hbar^2/\mu)\ \{(\hat{\mathbf{l}}^2/2r^2) + V(\mathbf{r}')\}$ of the centrifugal and dipole terms. The same r^{-2} dependence of the dipole potential as the centrifugal potential has made the diagonalization independent of r and necessary only once. This is no longer the case with the low-l, small-r region where the potential is more complex than the dipole. And yet the diagonalization of the effective potential at each r provides us with a deeper physical insight into the collision dynamics.[170]

Let $(\hbar^2/\mu)V(\mathbf{r}')$ be any local potential that describes the interaction between an electron and a linear molecule to a good approximation. We may find such a potential even in the compound-system region by use of an approximate local form for the exchange potential. The partial-wave expansion (150) of the wave function in terms of spherical harmonics casts the Schrödinger equation into the form of coupled equations (151a) with the effective potential $V^{\lambda}_{l'l}/x^2$ replaced by

$$k^{-2}V^{\lambda}_{l'l}(r) = k^{-2}\langle Y_{l'\lambda}(\theta',\varphi')|(\hat{\mathbf{l}}^2/r^2) + 2V(\mathbf{r}')|Y_{l\lambda}(\theta',\varphi')\rangle. \qquad (159)$$

Diagonalization of this matrix

$$\mathbf{U}^{\lambda}(r)\mathbf{V}^{\lambda}(r)(\mathbf{U}^{\lambda}(r))^{\dagger} = \tilde{\mathbf{V}}^{\lambda}(r) \qquad (160)$$

defines a set of adiabatic potential curves $\tilde{V}_N^\lambda(r)$ and a set of adiabatic angular functions

$$\Theta_{N\lambda}(\theta',\varphi';r) = \sum_l U_{Nl}^\lambda(r) Y_{l\lambda}(\theta',\varphi') \tag{161}$$

adapted to the anisotropic molecular potential; cf. Eq. (153).

The functions $\Theta_{N\lambda}$ are nearly equal to the r-independent spherical harmonics $Y_{N\lambda}$ when $\hat{\mathbf{l}}^2/r^2 \gg 2V(\mathbf{r}')$, i.e., at very small r, at large r for nondipolar molecules, and at all r for large l. For dipolar molecules the $\Theta_{N\lambda}$ approach, as r increases, the dipole angular functions of Section 5.3.3, which are also independent of r.

The channel functions $\tilde{\psi}_{N\lambda}(r)$ defined by

$$\Psi(\mathbf{r}') = \sum_{N\lambda} r^{-1} \tilde{\psi}_{N\lambda}(r) \Theta_{N\lambda}(\theta',\varphi';r) \tag{162}$$

satisfy equations

$$\left[\frac{d^2}{dr^2} + k^2 - \tilde{V}_N^\lambda(r)\right] \tilde{\psi}_{N\lambda}(r) + \sum_{N'} \left\{ 2\left[\frac{d\mathbf{U}^\lambda(r)}{dr}(\mathbf{U}^\lambda(r))^\dagger\right]_{N'N} \frac{d\tilde{\psi}_{N'\lambda}(r)}{dr} + \left[\frac{d^2\mathbf{U}^\lambda(r)}{dr^2}(\mathbf{U}^\lambda(r))^\dagger\right]_{N'N} \tilde{\psi}_{N'\lambda}(r) \right\} = 0. \tag{163}$$

Unlike Eqs. (154) these equations are dynamically coupled by the kinetic energy operator where and only where the unitary matrix $\mathbf{U}^\lambda$ or the adiabatic angular functions $\Theta_{N\lambda}$ depend appreciably on r. The use of $\Theta_{N\lambda}$ condenses the whole effect of the molecular anisotropy into the dynamic coupling confined to low partial waves (low N) and to small (but not too small) and medium r. For highly dipolar molecules the coupling is confined almost exclusively in the compound-system region.

Le Dourneuf *et al.* apply this procedure to both dipolar and nondipolar diatomic molecules.[170] They study scattering of electrons by model local potentials of these molecules at equilibrium nuclear separation; approximate exchange and polarization potentials are included. The following first explains their results on scattering by $N_2(^1\Sigma)$.

The adiabatic potentials $\tilde{V}_N^0(r)$, shown in Fig. 15a for the σ_g symmetry ($\lambda = 0, N = 0,2,4,\ldots$) in $e + N_2$, depart appreciably from the centrifugal potential only for the lowest three partial waves $N = 0$, 2, and 4. These are the only partial waves that may be appreciably distorted at low energies ($\lesssim 40$ eV). Figure 15b shows the sharp concentration of the adiabtic angular function $\Theta_{00}(\theta',\varphi';r)$ near the nuclei, i.e., around $r \cong r_N$ and around $\theta' \cong 0$ and $\theta' \cong \pi$. The eigenphase

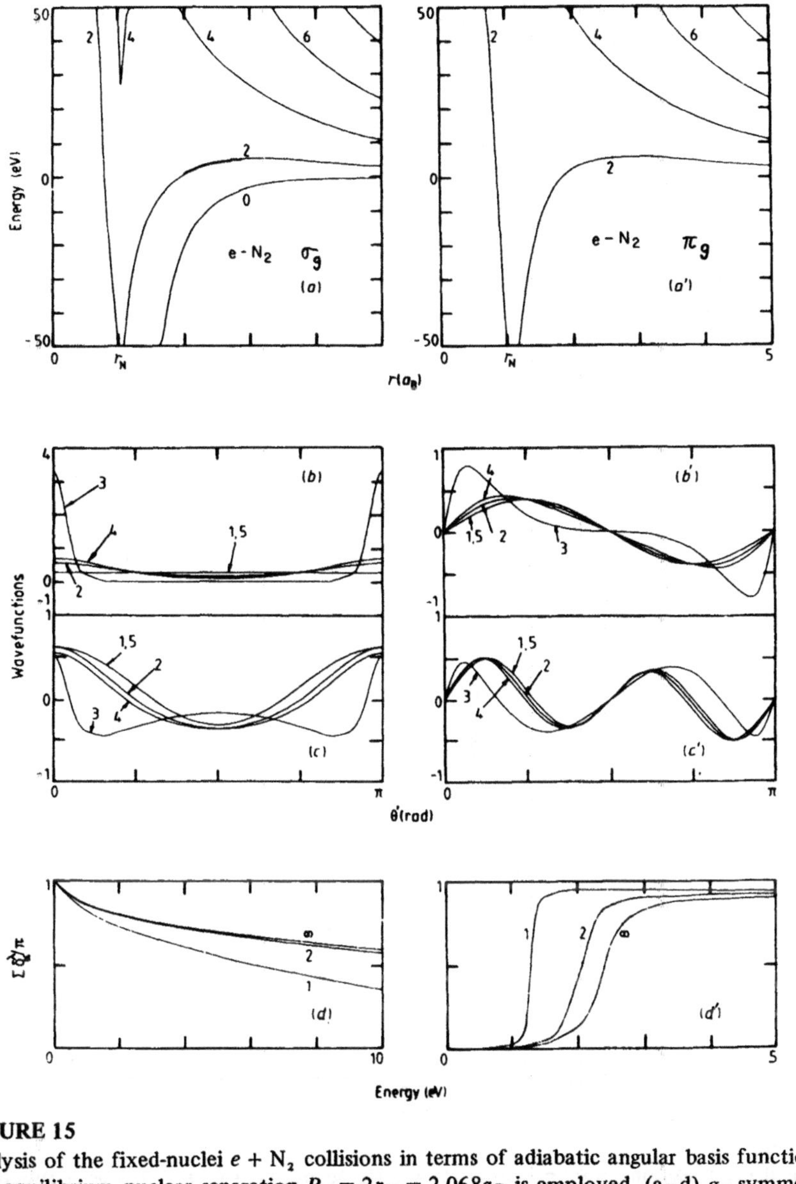

FIGURE 15

Analysis of the fixed-nuclei $e + N_2$ collisions in terms of adiabatic angular basis functions. The equilibrium nuclear separation $R_e = 2r_N = 2.068a_B$ is employed. (a–d) σ_g symmetry ($\lambda = 0, N = 0,2,4, \ldots$). (a'–d') π_g symmetry ($\lambda = 1, N = 2,4,6, \ldots$). (a, a') Adiabatic potentials $\tilde{V}_N^\lambda(r)$ in eV versus r in atomic units (a_B). (b,c,b',c') Adiabatic angular functions $\Theta_{N\lambda}(\theta', \varphi'; r)$ versus θ'. The numbers attached to the curves signify: 1, $r = 0$; 2, $r = 0.7a_B$; 3, $r = r_N = 1.034a_B$; 4, $r = 2a_B$; 5, $r = 5a_B$. (b) $(N,\lambda) = (0,0)$. (c) $(N,\lambda) = (2,0)$. (b') $(N,\lambda) = (2,1)$. (c') $(N,\lambda) = (4,1)$. (d,d') Eigenphase sum $\sum_\alpha \delta_\alpha^\lambda$ in units of π. The number of adiabatic channels $\tilde{\psi}_{N\lambda}$ included in each calculation is indicated for each curve. (Reproduced from Le Dourneuf *et al.*[170])

sum decreases monotonically as the energy increases up to 10 eV; see Fig. 15d. Note that inclusion of only two lowest partial waves $\tilde{\psi}_{00}$ and $\tilde{\psi}_{20}$ accurately reproduces the exact eigenphase sum for the present model.

Figures 15a′–15d′ are for the π_g symmetry ($\lambda = 1, N = 2,4,6, \ldots$) in e + N_2. In this case only the lowest adiabatic potential $\tilde{V}_2^1(r)$ departs appreciably from the centrifugal potential. The potential $\tilde{V}_2^1(r)$ has a deep well and a barrier inside which an electron with the energy of about 1.3 eV is temporarily trapped. This gives rise to a potential or shape resonance that manifests itself as a sudden rise of the eigenphase sum of Fig. 15d′. The dynamic coupling with the upper channels ($N \geqq 4$) shifts the resonance upward only by about 1 eV.

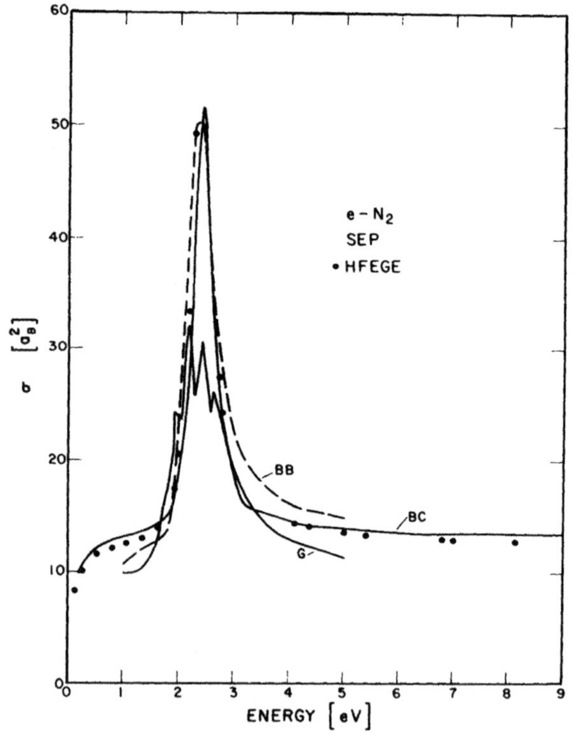

FIGURE 16

Integral cross section in atomic units for $e + N_2$ scattering. Curve G having a structure: Absolute total-cross-section measurements by Golden.[171] Others are computed static-exchange-polarization cross sections for electronically elastic scattering by the fixed-nuclei ground state N_2. BC: Burke and Chandra;[168] they orthogonalize the scattering orbitals to all molecular orbitals to mock the exchange potential (Chapter 7, Section 3.4); $^2\Sigma_{g,u}$, $^2\Pi_{g,u}$. BB: Buckley and Burke,[172] nonlocal exchange, $^2\Sigma_{g,u}$, $^2\Pi_{g,u}$, $^2\Delta_{g,u}$. •: Morrison and Collins,[158] local-exchange approximation, $^2\Sigma_{g,u}$, $^2\Pi_{g,u}$, $^2\Delta_{g,u}$. (Reproduced from Morrison and Collins.[158])

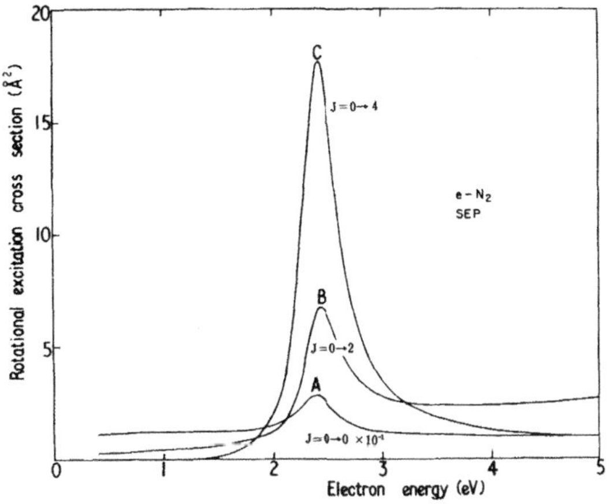

FIGURE 17
Computed integral cross sections for rotational transitions $J \to J'$ of N_2 in collisions with electrons. Static-exchange-polarization calculations in the fixed-nuclei rotational-sudden approximation. A cutoff polarization potential is used, and the exchange potential is mocked by orthogonalizing the scattering orbitals to the molecular orbitals (Chapter 7, Section 3.4). A: $J = 0 \to 0$ reduced by a factor of ten. B: $J = 0 \to 2$. C: $J = 0 \to 4$. (Reproduced from Burke and Chandra.[168])

We find this $^2\Pi_g$ resonance in Fig. 11a as a hump in the cross section for the $e + N_2$ collisions. The experimental results in Fig. 16 reveal a structure in this hump.[171] This structure is introduced by the coupling with the nuclear motion (Chapter 3, Section 5 and Chapter 7, Sections 4.3 and 4.4), and is not reproduced by the static-exchange-polarization calculations, included in Fig. 16, for a fixed nuclear distance.[158, 168, 172] Chandra and Temkin, e.g., allow for the nuclear motion by closely coupling the vibrational channels, and reproduce the experimentally observed structure.[23, 53, 173]

The lowest ($N = 2$) partial wave in the π_g symmetry has mainly the d-wave ($l = 2$) character. A d wave can directly couple the rotational state $J = 0$ with both states $J = 2$ and 4, of which the state $J = 4$ has a larger statistical weight. This explains the greater cross section for the transition $J = 0 \to 4$ than that for $J = 0 \to 2$ in the resonance region (Fig. 17). By contrast, the $e + H_2$ collisions show only a weak p-wave resonance, and the d wave is nearly negligible below 10 eV. A p wave directly couples $J = 0$ with $J = 2$, but not with $J = 4$. The transition $J = 0 \to 4$ requires a second-order coupling, which is weak for the $e + H_2$ collisions at low energies. Therefore, the cross section for $J = 0 \to 4$ is extremely small for $e + H_2$ (Section 3.3.1). The $J = 0 \to 4$ cross section for the $e + N_2$ collisions is also very small off the $^2\Pi_g$ resonance below about 1 eV. There, the relations (62a) and (62b) on p. 122 are applicable.

The differential cross section for the $e + N_2$ collisions in the static-exchange-polarization approximation, shown in Fig. 18, has a shape at the $^2\Pi_g$ resonance different from that off the resonance. The double-minimum shape at 2.3 eV is consistent with the strong d-wave $[P_2(\cos\theta)]$ contribution.

Figure 19 on $e +$ HCl scattering of the σ symmetry ($\lambda = 0, N = 0,1,2,\ldots$) exemplifies the application of the adiabatic angular basis functions to a heteronuclear diatomic molecule.[170] Wells and kinks occur in the adiabatic potentials in Fig. 19a at two different radial distances r_{Cl} and r_H that correspond to the

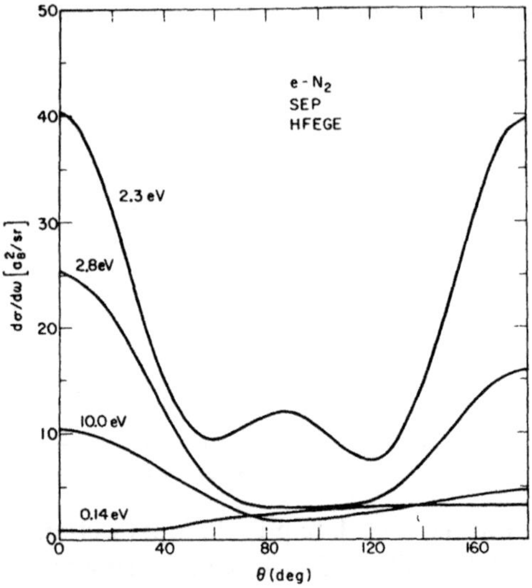

FIGURE 18
Computed differential cross sections for electronically elastic scattering of electrons by the ground-state N_2 molecules with their nuclei held fixed. The computations include the symmetries ${}^2\Sigma_{g,u}$, ${}^2\Pi_{g,u}$, and ${}^2\Delta_{g,u}$, and use the electrostatic, a local-exchange, and a cutoff polarization potentials. (Reproduced from Morrison and Collins.[158])

positions of the two nuclei. The adiabatic angular function $\Theta_{00}(\theta',\varphi';r)$ in Fig. 19b is condensed near the nucleus of Cl ($r \cong r_{\text{Cl}}, \theta' \cong \pi$) and near that of H($r \cong r_{\text{H}}, \theta' \cong 0$). As r becomes large, the $\Theta_{00}(\theta',\varphi';r)$ approaches the dipole angular function $\Theta_{00}(\theta',\varphi')$ of Section 5.3.3, just as any $\Theta_{N\lambda}(\theta',\varphi';r)$ approaches $\Theta_{N\lambda}(\theta',\varphi')$.

Because the coupled equations (163) approach the uncoupled dipole equations (154) asymptotically, the physical solutions of the former equations have the same asymptotic forms as those of the latter except for the phases. The phase difference $\bar{\delta}_N^\lambda$ between the two functions of the same channel (N, λ) represents the net effect of the interaction of shorter range than the dipole, and may be called the dipole-extracted phase shift. We may also define a dipole-extracted S matrix and dipole-extracted eigenphases $\bar{\delta}_\alpha^\lambda$ in the standard manner.

The dipole-extracted eigenphase sum in Fig. 19e exhibits a rapid resonance-like variation with the energy already in the three-channel approximation. This rapid variation is due to the lowest eigenchannel, as seen in Fig. 19f. In the three-channel approximation the lowest eigenchannel ($\alpha = 0$) is mainly of the $N = 0$ character, but the $N = 2$ component is mixed in at the energies of the rapid variation of the eigenphase (Fig. 19g). The second lowest eigenchannel ($\alpha = 1$) is almost completely of the $N = 1$ character below 10 eV (Fig. 19h).

The resonancelike behavior is observed also by Gianturco and Thompson,[174] by Collins *et al.*,[175] and by Padial *et al.*[176] in the static approximation with the effects of exchange and polarization taken into account approximately. Being absent without exchange or without polarization, the resonancelike behavior is a result of a strong cooperative effect involving both exchange and

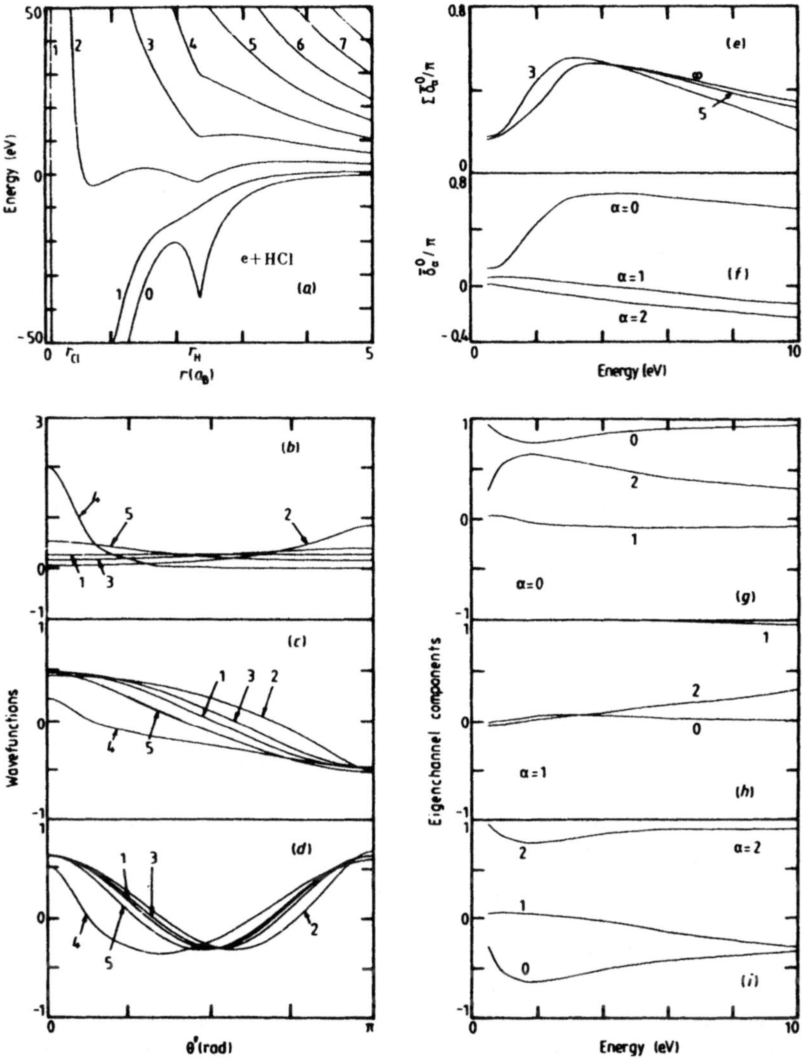

FIGURE 19

Analysis of the fixed-nuclei e + HCl collisions of the σ symmetry in terms of adiabatic angular basis functions ($\lambda = 0, N = 0,1,2, \ldots$). The distance r_{H} of the H nucleus from the center of mass is $2.36a_B$, and the distance r_{Cl} of the Cl nucleus is $0.06a_B$. (a) Adiabatic potentials $\widetilde{V}_N^{\lambda}(r)$ in eV versus r in atomic units (a_B). (b–d) Adiabatic angular functions $\Theta_{N\lambda}(\theta',\varphi';r)$ versus θ'. The numbers attached to the curves signify: 1, $r = 0$; 2, $r = r_{\mathrm{Cl}} = 0.06a_B$; 3, $r = 1a_B$; 4, $r = r_{\mathrm{H}} = 2.36a_B$; 5, $r = 5a_B$. (b) $N = 0$. (c) $N = 1$. (d) $N = 2$. (e) Dipole-extracted eigenphase sum $\Sigma\tilde{\delta}_{\alpha}^{\lambda}$ in units of π. The number of adiabatic channels $\widetilde{\psi}_{N\lambda}$ included in each calculation is indicated for each curve. (f) Dipole-extracted eigenphases $\tilde{\delta}_{\alpha}^{\lambda}$ in the three-channel approximation. (g–i) Adiabatic-channel components $\widetilde{\psi}_{N\lambda}$ ($N = 0$, 1,2) in each eigenchannel α. (g) $\alpha = 0$. (h) $\alpha = 1$. (i) $\alpha = 2$. (Reproduced from Le Dourneuf *et al.*[170])

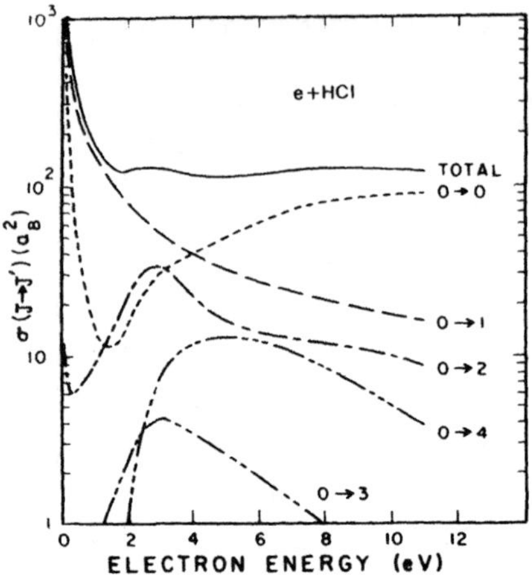

FIGURE 20
Computed integral cross sections in atomic units for rotational transitions $J = 0 \to J'$ of HCl in collisions with electrons. Static, local-exchange, cutoff-polarization approximation. Total: Sum over the final rotational state J'. (Reproduced from Padial *et al.*[176])

polarization, and is sensitive to the cut-off parameter in the model polarization potential.[176]

The rapid phase variation of the lowest eigenchannel with large $N = 0$ and 2 components results in a marked hump in the cross section for the transfer of two units of angular momentum and in a slight hump in the cross section summed over the final rotational state (Fig. 20). Rohr and Linder observe in their crossed-beam experiments a similar broad but clear hump in the cross sections for vibrational excitation around 2–3 eV.[177] Figure 20 shows that the dipole–Born-allowed transition $J = 0 \to 1$ is dominant only below ~2 eV, the $J = 0 \to 0$ cross section surpassing the $J = 0 \to 1$ cross section above 4 eV.[176] Figure 21 compares the rotationally summed differential cross sections computed by Padial *et al.*[176] with the rotationally unresolved, vibrationally elastic differential cross sections measured by Rohr and Linder.[177]

The adiabatic angular basis functions that diagonalize the operator $(\hbar^2/\mu) \times \{(\hat{\mathbf{l}}^2/2r^2) + V(\mathbf{r}')\}$ provide both a visual representation of the dynamics and rapid convergence of the anisotropy-adapted partial-wave expansion, provided that the electron–molecule collision is modeled by elastic scattering by a local potential.[170] When the coupling between the motion of the incident electron and the molecular rotation is important, an appropriate adiabatic basis set diagonalizes an operator $B\hat{\mathbf{J}}^2 + (\hbar^2/\mu)\{(\hat{\mathbf{l}}^2/2r^2) + V\}$ including the rotational energy, as is argued by Mullaney and Truhlar;[178] see also Clark.[146] Le Dourneuf *et al.*'s basis set[170] is a simplification of that by Mullaney and Truhlar.[178] Incidentally, the latter basis set may be more usefully applied to atom–molecule and ion–molecule collisions, in which the relative motion is slower than in electron–molecule collisions at the same energy.[179] The adiabatic basis set that diagonalizes $B\hat{\mathbf{J}}^2 + (\hbar^2/\mu)V$, i.e., another simplification of the Mullaney–Truhlar basis set, is particularly useful for studying low-energy ion–molecule collisions.[180]

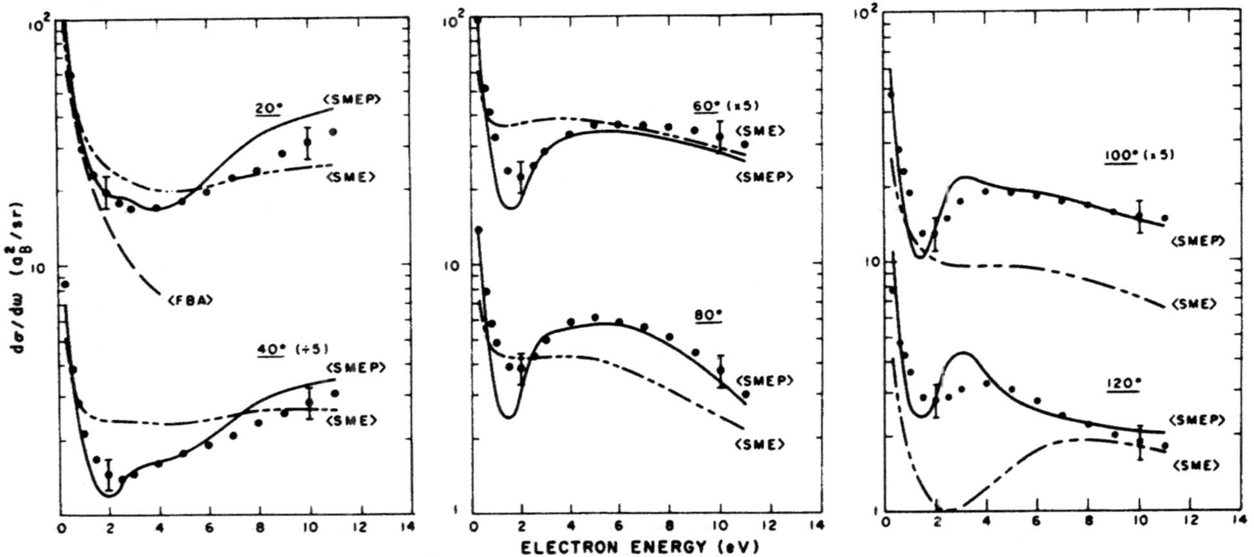

FIGURE 21

Rotationally summed differential cross sections for e + HCl scattering at several scattering angles versus the collision energy. ⟨SMEP⟩, static model-(local-)exchange polarization calculation. ⟨SME⟩, static model-(local-)exchange calculation. ⟨FBA⟩, first-Born-approximation calculation with the dipole and quadrupole interactions. ●, crossed-beam measurements (rotationally unresolved and vibrationally elastic).[177] Renormalized to the SMEP calculations by a single factor that is independent of both energy and angle. (Reproduced from Padial *et al.*[176])

Appendix: Summary of Formulas

1. Rotation matrix elements
   Definition: Eqs. (30), (31) — Coordinate rotation: Eq. (32)
   Integral formula: Eq. (60)

2. Spherical harmonics
   Definition: Eq. (34) — Neumann expansion of r_{12}^{-1}: Eq. (127)
   Integral formula: Eq. (60) — Rayleigh expansion of plane wave: Eq. (132)
   Clebsch–Gordan series: Eq. (84)

3. Clebsch–Gordan coefficients

	Quantum mechanical:	Classical:
Definition and expression:	Eqs. (35), (37), (63)	Eqs. (8), (18)
Special case:	Eq. (39)	Eq. (20b)
Asymptotic formula:	Eq. (64)	Eq. (12)
Orthonormality:	Eq. (38)	Eqs. (4), (20)
Interrelation:	Eqs. (19), (36), (40)	Eq. (19)

4. Rotational energy
   Quantum mechanical: Eqs. (28), (33) — Classical: Eqs. (1), (16), (17)

5. Cross-section relations

	Quantum mechanical:	Classical:
Linear dependence		
Linear rotator:	Eq. (62)	Eq. (5)
Special case:	Eqs. (62a,b) (p. 122)	
Symmetric top:	Eq. (62)	Eq. (22)
Spherical top:	Eq. (67)	Eq. (26)
Dipolar rotator:	Eq. (133b) (Born, Glauber, general; study Section 5.3.2.)	
Asymptotic formula		
Linear rotator:	Eqs. (13)–(15) with Eq. (65) ($c_1 = 1/2$)	Eqs. (13)–(15) ($c_1 = 1/2$)
Spherical top:	Eqs. (13)–(15) with Eq. (65) modified [see below Eq. (67)] ($c_1 = 1$)	Eqs. (13)–(15) ($c_1 = 1$)
Detailed balance		
Linear rotator and Symmetric top:	Eq. (23)	Eq. (23)
Spherical top:	Eq. (27)	Eq. (27)

Sum rules

Linear rotator and Spherical top:	Eqs. (68), (69) [see remark below Eq. (69)]	Eqs. (6), (11)
Symmetric top:	Eqs. (68), (69)	Eqs. (24), (25)

6. Sudden approximation

Wave function: Eqs. (43), (44), (92) Scattering amplitude: Eqs. (45), (51), (95)

Cross section: Eqs. (96), (97) Nonsudden index: Eq. (66)

Analysis of energy-loss spectra: Eqs. (74a,b) with or without Eq. (64)

7. Frame transformation

	Laboratory frame:	Body-fixed frame:
Angular basis function:	Eq. (75)	Eq. (76)
Wave-function expansion:	Eq. (79)	Eq. (88)
Close-coupling equations:	Eq. (80)	Eq. (89)
Asymptotic form:	Eq. (81)	
Scattering amplitude:	Eq. (83)	
Cross sections:	Eqs. (85), (87)	

Frame transformation and wave-function matching: Eqs. (77), (90), (91), (94)

8. R matrix

	Single channel:	Multichannel:
Definition:	Eq. (99)	Eq. (119)
Expansion in terms of R-matrix states:	Eq. (100)	
Basis-set expansion (variationally optimized R matrix):	Eqs. (112), (114)	Eq. (125)
K matrix in terms of R matrix:	Eq. (104)	Eq. (121)
Internal wave function:	Eq. (115)	
Generalized R matrix for more than two regions:	Eqs. (117), (118)	

9. Other formulas

Reduced cross section and reduced momentum transfer

Scattering by cr^{-2}: Eqs. (136), (137), (138)

Highly dipolar molecule: Eqs. (126), (145), (149)

Single-center expansion of electrostatic field: Eqs. (128a), (130), (130a)

Multipole moment: Eqs. (128b), (129)

Born approximation

General multipole: Eq. (133) Dipole: Eqs. (146), (147)

Quadrupole: Eqs. (157), (158)

Eikonal approximation: Eqs. (139), (141), (143)

Dipole angular function: Eq. (153) Channel equations: Eq. (154)

Adiabatic angular function: Eq. (161) Channel equations: Eq. (163)

References and Notes

1. D.W. Norcross and L.A. Collins, Recent developments in the theory of electron scattering by highly polar molecules, *Adv. At. Mol. Phys.* **18**, 341–397 (1982).
2. W.R. Garrett, Electron interactions with polar molecules, in: *Physics of Electronic and Atomic Collisions,* XIIth Intern. Conf. Phys. Electron. Atom. Coll., Gatlinburg, Tennessee, July 1981 (S. Datz, ed.), pp. 65–77, North-Holland, Amsterdam (1982).
3. P.G. Burke, Theory of electron molecule collisions, in: *Quantum Dynamics of Molecules* (R.G. Woolley, ed.), pp. 483–547, Plenum Press, New York (1980).
4. N.F. Lane, The theory of electron–molecule collisions, *Rev. Mod. Phys.* **52**, 29–119 (1980).
5. P.G. Burke, Theory of low energy electron–molecule collisions, *Adv. At. Mol. Phys.* **15**, 471–506 (1979).
6. N. Lane, The state of the theory, in: *Electron–Molecule Scattering* (S.C. Brown, ed.), Chap. 5, pp. 147–183, John Wiley & Sons, New York (1979).
7. L.A. Collins and D.W. Norcross, Electron collisions with highly polar molecules: Comparison of model, static, and static-exchange calculations for alkali-metal halides, *Phys. Rev. A* **18**, 467–498 (1978).
8. Y. Itikawa, Electron scattering by polar molecules, *Phys. Rept.* **46**, 117–164 (1978).
9. S.S. Bhattacharyya and A.S. Dickinson, Inter-relations between classical cross sections for rotational transitions in sudden atom–molecule collisions, *J. Phys. B* **12**, L521–L524 (1979).
10. P.J. Brussaard and H.A. Tolhoek, Classical limits of Clebsch–Gordan coefficients, Racah coefficients and $D^l_{mn}(\phi,\theta,\psi)$ functions, *Physica* **23**, 955–971 (1957).
11. A.R. Edmonds, *Angular Momentum in Quantum Mechanics,* Princeton University Press, Princeton, New Jersey (1957). See the third printing with corrections, second edition (1974); note especially that Section 4.1 discussing the symmetric-top functions has been revised. See also M. Bouten, On the rotation operators in quantum mechanics, *Physica* **42**, 572–580 (1969); and A.A. Wolf, Rotation operators, *Am. J. Phys.* **37**, 531–536 (1969).
12. I. Shimamura, Moments of the spectra for rotational transitions induced by collisions or by external perturbations, *Z. Phys. A* **309**, 107–117 (1982).
13. H. Goldstein, *Classical Mechanics,* Addison-Wesley, Reading, Massachusetts (1950).
14. M.E. Rose, *Elementary Theory of Angular Momentum,* John Wiley & Sons, New York (1957).
15. U. Fano and G. Racah, *Irreducible Tensorial Sets*, Academic Press, New York (1959).
16. I. Shimamura, Sum rules for the rotational structure in the molecular transition spectrum, *J. Phys. B* **15**, 93–100 (1982).
17. S.I. Drozdov, The scattering of fast neutrons by nonspherical nuclei. I, *Sov. Phys. JETP* **1**, 591–592 (1955) [Transl. of: *Zh. Eksp. Teor. Fiz.* **28**, 734–735 (1955)].
18. D.M. Chase, Adiabatic approximation for scattering processes, *Phys. Rev.* **104**, 838–842 (1956).
19. S.I. Drozdov, Scattering of fast neutrons by nonspherical nuclei. III, *Sov. Phys. JETP* **3**, 759–761 (1956) [Transl. of: *Zh. Eksp. Teor. Fiz.* **30**, 786–788 (1956)].
20. G.F. Chew and M.L. Goldberger, The scattering of elementary particles by complex nuclei–A generalization of the impulse approximation, *Phys. Rev.* **87**, 778–782 (1952).
21. S. Altshuler, Theory of low-energy electron scattering by polar molecules, *Phys. Rev.* **107**, 114–117 (1957).
22. Yu.D. Oksyuk, Excitation of the rotational levels of diatomic molecules by electron impact in the adiabatic approximation, *Sov. Phys. JETP* **22**, 873–881 (1966) [Transl. of: *Zh. Eksp. Teor. Fiz.* **49**, 1261–1273 (1965)].

23. N. Chandra and A. Temkin, Hybrid theory and calculation of $e-N_2$ scattering, *Phys. Rev. A* **13**, 188–203 (1976).
24. B.H. Choi and R.T. Poe, Vibrational and rotational transitions in low-energy electron–diatomic-molecule collisions. II. Hybrid theory and close-coupling theory: An $l_{z'}$-conserving close-coupling approximation, *Phys. Rev. A* **16**, 1831–1840 (1977).
25. E. Gerjuoy and B.K. Thomas, Applications of the Glauber approximation to atomic collisions, *Rep. Prog. Phys.* **37**, 1345–1431 (1974).
26. M. Shugard and A.U. Hazi, Theory of electron–molecule scattering: Comments on the adiabatic-nuclei approximation, *Phys. Rev. A* **12**, 1895–1902 (1975).
27. R.J. Glauber, High-energy collision theory, in: *Lectures in Theoretical Physics*, Vol. 1 (W.E. Brittin and L.G. Dunham, eds.), pp. 315–414, Interscience, New York (1959).
28. E.S. Chang and A. Temkin, Rotational excitation of diatomic molecular systems. II. H^+_2, *J. Phys. Soc. Jpn.* **29**, 172–179 (1970).
29. I. Shimamura, Systematics in the cross sections for excitation of molecules by electrons, in *Symposium on Electron-Molecule Collisions* (I. Shimamura and M. Matsuzawa, eds.), pp. 13–30, University of Tokyo, Tokyo (1979).
30. R.K. Nesbet, Energy-modified adiabatic approximation for scattering theory, *Phys. Rev. A* **19**, 551–556 (1979).
31. D.W. Norcross and N.T. Padial, The multipole-extracted adiabatic-nuclei approximation for electron–molecule collisions. *Phys. Rev. A* **25**, 226–238 (1982).
32. T.N. Rescigno, A.E. Orel, A.U. Hazi, and B.V. McKoy, Ab initio study of vibrational excitation of HF by low-energy electrons, *Phys. Rev. A* **26**, 690–693 (1982).
33. A.N. Feldt and M.A. Morrison, Breakdown of the adiabatic-nuclear-rotation approximation for near-threshold $e-H_2$ collisions, *J. Phys. B* **15**, 301–308 (1982).
34. F. Linder and H. Schmidt, Rotational and vibrational excitation of H_2 by slow electron impact, *Z. Naturforsch.* **26a**, 1603–1617 (1971).
35. R.W. Crompton, D.K. Gibson, and A.I. McIntosh, The cross section for the $J = 0 \rightarrow 2$ rotational excitation of hydrogen by slow electrons, *Aust. J. Phys.* **22**, 715–731 (1969).
36. S. Hara, Rotational excitation of H_2 by slow electrons, *J. Phys. Soc. Jpn.* **27**, 1592–1597 (1969).
37. E.S. Chang and A. Temkin, Rotational excitation of diatomic molecules by electron impact, *Phys. Rev. Lett.* **23**, 399–403 (1969).
38. U. Fano and D. Dill, Angular momentum transfer in the theory of angular distributions, *Phys. Rev. A* **6**, 185–192 (1972).
39. R. Goldflam, D.J. Kouri, and S. Green, On the factorization and fitting of molecular scattering information, *J. Chem. Phys.* **67**, 5661–5675 (1977).
40. V. Khare, On the l_z-conserving energy sudden approximation for atom–diatom scattering, *J. Chem. Phys.* **68**, 4631–4640 (1978).
41. D.J. Kouri, Rotational excitation II: Approximation methods, in: *Atom–Molecule Collision Theory* (R.B. Bernstein, ed.), Chap. 9, pp. 301–358, Plenum Press, New York (1979).
42. E.U. Condon and G.H. Shortley, *The Theory of Atomic Spectra*, Cambridge University Press, Cambridge (1935).
43. I. Shimamura, State-to-state rotational transition cross sections from unresolved energy-loss spectra, *Chem. Phys. Lett.* **73**, 328–333 (1980).
44. I. Shimamura, Partial-sum rules for and asymmetry between rotational transitions $J \pm \Delta J \leftarrow J$. *Phys. Rev. A* **28** 1357–1362 (1983).
45. S.F. Wong and L. Dubé, Rotational excitation of N_2 by electron impact: 1–4 eV, *Phys. Rev. A* **17**, 570–576 (1978).
46. S. Green, Rotational excitation of symmetric top molecules by collisions with atoms. II. Infinite order sudden approximation, *J. Chem. Phys.* **70**, 816–829 (1979).

47. D.W. Norcross, Application of the adiabatic-nuclei approximation to energy-loss cross sections for collisions with molecules, *Phys. Rev. A* **25**, 764–772 (1982).
48. N.F. Lane and S. Geltman, Rotational excitation of diatomic molecules by slow electrons: Application to H_2, *Phys. Rev.* **160**, 53–67 (1967).
49. S.K. Srivastava, R.I. Hall, S. Trajmar, and A. Chutjian, Pure rotational excitation of H_2 at electron impact energies of 3 to 100 eV, *Phys. Rev. A* **12**, 1399–1401 (1975).
50. S.F. Wong and G.J. Schulz, Rotational and vibrational excitation of H_2 by electron impact at 4.5 eV: Angular distributions, *Phys. Rev. Lett.* **32**, 1089–1092 (1974).
51. I. Shimamura, Energy loss by slow electrons and by slow atoms in a molecular gas, *Phys. Rev. A* **23**, 3350–3353 (1981).
52. I. Shimamura, Rotational-state dependence of the collisional pumping rate for astrophysical masers, *Progr. Theor. Phys. (Kyoto)* **68**, 178–190 (1982).
53. N. Chandra and A. Temkin, Tabulation of hybrid theory calculated e–N_2 vibrational and rotational cross sections, NASA Technical Note, NASA TN D-8347 (1976).
54. L.H. Beard, D.J. Kouri, and D.K. Hoffman, Exact factorization relations, consistency conditions, and the reduction to the ES approximation for the T-matrix, *J. Chem. Phys.* **76**, 3623–3629 (1982).
55. V.N. Ostrovsky and V.I. Ustimov, The connections between the rotational transition cross sections in diatomic molecules, *J. Phys. B* **14**, 1139–1156 (1981).
56. A.D. Buckingham, B.J. Orr, and J.M. Sichel, Angular distribution and intensity in molecular photoelectron spectroscopy I. General theory for diatomic molecules, *Phil. Trans. R. Soc. Lond. A* **268**, 147–157 (1970).
57. J.M. Sichel, Angular distribution and intensity in molecular photoelectron spectroscopy II. Information contained in rotational fine structure, *Mol. Phys.* **18**, 95–100 (1970).
58. H. Ehrhardt and F. Linder, Rotational excitation of H_2 by slow electrons in a beam experiment, *Phys. Rev. Lett.* **21**, 419–421 (1968).
59. G. Joyez, J. Comer, and F.H. Read, Resonant rotational excitation of H_2 by electron impact, *J. Phys. B* **6**, 2427–2440 (1973).
60. E.S. Chang and S.-F. Wong, Isotope effects in molecular scattering by electrons, *Phys. Rev. Lett.* **38**, 1327–1330 (1977).
61. J. Comer and M. Harrison, Observation of the effects of rotational transitions in the resonant scattering of electrons from N_2, *J. Phys. B* **6**, L70–L72 (1973).
62. H. Tanaka, L. Boesten, and I. Shimamura, Experimental determination of state-to-state rotational transition cross sections: $e+N_2$, VIIth Intern. Conf. Atom. Phys., Cambridge, Mass., Book of Abstracts, pp. 43–44 (1980).
63. K. Jung, Th. Antoni, R. Müller, K.-H. Kochem, and H. Ehrhardt, Rotational excitation of N_2, CO and H_2O by low-energy electron collisions, *J. Phys. B* **15**, 3535–3555 (1982).
64. F.H. Read, Branch structure in rotational excitation by resonant electron molecule collisions, *J. Phys. B* **5**, 255–264 (1972).
65. C. Bottcher, The formalism of electron–molecule scattering, *Chem. Phys. Lett.* **4**, 320–322 (1969).
66. U. Fano, Quantum defect theory of l uncoupling in H_2 as an example of channel-interaction treatment, *Phys. Rev. A* **2**, 353–365 (1970): Erratum, *Phys. Rev. A* **15**, 817 (1977).
67. E.S. Chang and U. Fano, Theory of electron–molecule collisions by frame transformations, *Phys. Rev. A* **6**, 173–185 (1972).
68. P.G. Burke, I. Mackey, and I. Shimamura, R-matrix theory of electron–molecule scattering, *J. Phys. B* **10**, 2497–2512 (1977).

69. U. Fano, Unified treatment of collisions, *Phys. Rev. A* **24**, 2402–2415 (1981).
70. W.G. Harter, C.W. Patterson, and F.J. da Paixao, Frame transformation relations and multipole transitions in symmetric polyatomic molecules, *Rev. Mod. Phys.* 50, 37–83 (1978).
71. A.M. Arthurs and A. Dalgarno, The theory of scattering by a rigid rotator, *Proc. R. Soc. A* **256**, 540–551 (1960).
72. A. Temkin and F.H.M. Faisal, Adiabatic theory of rotational excitation of non-Σ states, *Phys. Rev. A* **3**, 520–521 (1971).
73. N. Chandra, Rotational excitation in e^-–N_2 scattering, *J. Phys. B* **8**, 1338–1348 (1975).
74. P.G. Burke and A.-L. Sinfailam, Electron–molecule interactions II. Scattering by closed-shell diatomic molecules, *J. Phys. B* **3**, 641–659 (1970).
75. N. Chandra, Low-energy electron scattering from CO. II. *Ab initio* study using the frame transformation theory, *Phys. Rev. A* **16**, 80–108 (1977).
76. M. Le Dourneuf, Vo Ky Lan, and B.I. Schneider, Vibrational and rotational excitation of molecules by electrons: New developments of the frame transformation theory using the *R*-matrix and the variable phase methods, in *Symposium on Electron–Molecule Collisions* (I. Shimamura and M. Matsuzawa, eds.), Appendix, University of Tokyo, Tokyo (1979).
77. Vo Ky Lan, M. Le Dourneuf, and B.I. Schneider, Rotational excitation of diatomic polar molecules by slow electron impact. Practical implementation of the frame transformation theory, in: *Electronic and Atomic Collisions,* Abstracts of Contributed Papers, XIth Intern. Conf. Phys. Elec. Atom. Coll., Kyoto, 1979 (K. Takayanagi and N. Oda, eds.), pp. 290–291, North-Holland, Amsterdam (1979).
78. N. Chandra, Low-energy electron scattering from CO: III. Analytic method for outer region in frame-transformation theory, *J. Phys. B* **15**, 4465–4476 (1982).
79. C.A. Weatherford and R.J.W. Henry, Frame-transformation approximation for low-energy e^-–$^1\Sigma^+$ systems using the noniterative integral equation method, *Can. J. Phys.* **55**, 442–451 (1977).
80. E.P. Wigner and L. Eisenbud, Higher angular momenta and long range interaction in resonance reactions, *Phys. Rev.* **72**, 29–41 (1947).
81. A.M. Lane and R.G. Thomas, *R*-matrix theory of nuclear reactions, *Rev. Mod. Phys.* **30**, 257–353 (1958).
82. R.F. Barrett, B.A. Robson, and W. Tobocman, Calculable methods for many-body scattering, *Rev. Mod. Phys.* **55**, 155–243 (1983).
83. P.G. Burke and W.D. Robb, The *R*-matrix theory of atomic processes, *Adv. At. Mol. Phys.* **11**, 143–214 (1975).
84. I. Shimamura, *R*-matrix theories, in *Electronic and Atomic Collisions*, invited Papers and Progress Reports, The Xth Intern. Conf. Phys. Electron. Atom. Coll., Paris, July 1977 (G. Watel, ed.), pp. 213–230, North-Holland, Amsterdam (1978).
85. C. Bloch, Une formulation unifiée de la théorie des réactions nucléaires, *Nucl. Phys.* **4**, 503–528 (1957).
86. A.M. Lane and D. Robson, Optimization of nuclear resonance reaction calculations, *Phys. Rev.* **178**, 1715–1724 (1969).
87. I. Shimamura, *R*-matrix theory of atomic continuum processes, *J. Phys. B* **10**, 2597–2618 (1977).
88. B.I. Schneider, *R*-matrix theory for electron–molecule collisions using analytic basis set expansions. II. Electron–H_2 scattering in the static-exchange model, *Phys. Rev. A* **11**, 1957–1962 (1975).
89. B.I. Schneider, The *R*-matrix method: Applications to electron–molecule collisions,

in *Electronic and Atomic Collisions*, Invited Papers and Progress Reports, The Xth Intern. Conf. Phys. Electron. Atom. Coll., Paris, July 1977 (G. Watel, ed.), pp. 257–269, North-Holland, Amsterdam (1978).

90. I. Shimamura, Variational R-matrix propagation method—A new numerical method for solving close-coupling equations, in *Report of the CECAM Workshop on Electron Molecule Collisions*, pp. 69–80, CECAM, Université Paris-Sud, Orsay (1977).
91. B.I. Schneider and R.B. Walker, The coupled channel R-matrix propagation method, *J. Chem. Phys.* **70**, 2466–2470 (1979).
92. K.L. Baluja, P.G. Burke, and L.A. Morgan, R-matrix propagation program for solving coupled second-order differential equations, *Comput. Phys. Commun.* **27**, 299–307 (1982).
93. M. Danos and W. Greiner, Eigenchannel theory of nuclear reactions, *Phys. Rev.* **146**, 708–712 (1966).
94. R.F. Barrett, L.C. Biedenharn, M. Danos, P.P. Delsanto, W. Greiner, and H.G. Wahsweiler, The eigenchannel method and related theories for nuclear reactions, *Rev. Mod. Phys.* **45**, 44–108 (1973).
95. U. Fano and C.M. Lee, Variational calculation of R matrices. Application to Ar photoabsorption, *Phys. Rev. Lett.* **31**, 1573–1576 (1973).
96. M.J. Seaton, Quantum defect theory I. General formulation, *Proc. Phys. Soc.* **88**, 801–814 (1966).
97. M.J. Seaton, Quantum defect theory II. Illustrative one-channel and two-channel problems, *Proc. Phys. Soc.* **88**, 815–832 (1966).
98. M.J. Seaton, Quantum defect theory, *Rep. Prog. Phys.* **46**, 167–257 (1983).
99. U. Fano, Unified treatment of perturbed series, continuous spectra and collisions, *J. Opt. Soc. Am.* **65**, 979–987 (1975).
100. U. Fano, Evolution of quantum-defect methods, *Comments At. Mol. Phys.* **10**, 223–230 (1981); *ibid.* **13**, 157–169 (1983).
101. R.C. Slater, M.G. Fickes, W.G. Becker, and R.C. Stern, Scattering of alkali halides by electrons. I. CsF, *J. Chem. Phys.* **60**, 4697–4709 (1974).
102. W.G. Becker, M.G. Fickes, R.C. Slater, and R.C. Stern, Scattering of alkali halides by electrons. II. CsCl, *J. Chem. Phys.* **61**, 2283–2289 (1974).
103. R.C. Slater, M.G. Fickes, W.G. Becker, and R.C. Stern, Scattering of alkali halides by electrons. III. KI, *J. Chem. Phys.* **61**, 2290–2293 (1974).
104. M.R.H. Rudge, S. Trajmar, and W. Williams, Electron scattering by highly polar molecules. I. KI, *Phys. Rev. A* **13**, 2074–2086 (1976).
105. L. Vušković, S.K. Srivastava, and S. Trajmar, Electron scattering by highly polar molecules II. LiF, *J. Phys. B* **11**, 1643–1652 (1978).
106. L. Vušković and S.K. Srivastava, Electron scattering by highly polar molecules III. CsCl, *J. Phys. B* **14**, 2677–2685 (1981).
107. S.K. Srivastava, H. Tanaka, and A. Chutjian, Elastic scattering of intermediate energy electrons by HCN, *J. Chem. Phys.* **69**, 1493–1497 (1978).
108. K. Rohr, Interaction mechanisms and cross sections for the scattering of low-energy electrons from HBr, *J. Phys. B* **11**, 1849–1860 (1978).
109. H. Tanaka, S.K. Srivastava, and A. Chutjian, Absolute elastic differential electron scattering cross sections in the intermediate energy region. IV. CO, *J. Chem. Phys.* **69**, 5329–5333 (1978).
110. S.K. Srivastava, A. Chutjian, and S. Trajmar, Absolute elastic differential electron scattering cross sections in the intermediate energy region. I. H_2, *J. Chem. Phys.* **63**, 2659–2665 (1975).
111. T.W. Shyn, W.E. Sharp, and G.R. Carignan, Angular distribution of electrons elastically scattered from CO_2, *Phys. Rev. A* **17**, 1855–1861 (1978).

112. T.W. Shyn and G.R. Carignan, Angular distribution of electrons elastically scattered from gases: 1.5–400 eV on N_2. II, *Phys. Rev. A* **22**, 923–929 (1980).
113. T.W. Shyn and W.E. Sharp, Angular distributions of electrons elastically scattered from H_2, *Phys. Rev. A* **24**, 1734–1740 (1981).
114. D.E. Stogryn and A.P. Stogryn, Molecular multipole moments, *Mol. Phys.* **11**, 371–393 (1966).
115. O.H. Crawford, Scattering of low-energy electrons from polar molecules, *J. Chem. Phys.* **47**, 1100–1104 (1967).
116. Y. Itikawa, Electron-impact rotational excitation of a symmetric-top molecule: Application to NH_3, *J. Phys. Soc. Jpn.* **30**, 835–842 (1971).
117. R.C. Stabler, Rotational excitation of molecular ions by slow electrons, *Phys. Rev.* **131**, 679–683 (1963).
118. R.F. Boĭkova and V.D. Ob"edkov, Rotational and vibrational excitation of molecular ions by electrons, *Sov. Phys. JETP* **27**, 772–774 (1968) [Transl. of: *Zh. Eksp. Teor. Fiz.* **54**, 1439–1444 (1968)].
119. S.-I. Chu and A. Dalgarno, Rotational excitation of CH^+ by electron impact, *Phys. Rev. A* **10**, 788–792 (1974): Erratum, *Phys. Rev. A* **12**, 725 (1975).
120. S.-I. Chu, Rotational excitation of symmetric-top molecular ions by electron impact, *Phys. Rev. A* **12**, 396–405 (1975).
121. A.S. Dickinson and J.M. Muñoz, Rotational excitation of polar molecular ions by slow electrons, *J. Phys. B* **10**, 3151–3163 (1977).
122. L.D. Landau and E.M. Lifshitz, *Quantum Mechanics*, 3rd ed., Pergamon Press, Oxford (1977).
123. R.G. Newton, *Scattering Theory of Waves and Particles*, 2nd ed., Springer-Verlag, New York (1982).
124. N.F. Mott and H.S.W. Massey, *The Theory of Atomic Collisions*, 3rd ed., Oxford University Press, Oxford (1965).
125. W.M. Frank, D.J. Land, and R.M. Spector, Singular potentials, *Rev. Mod. Phys.* **43**, 36–98 (1971).
126. E.A. Mason, J.T. Vanderslice, and C.J.G. Raw, Quantum effects in small-angle molecular-beam scattering, *J. Chem. Phys.* **40**, 2153–2164 (1964).
127. R.J. Cross, Jr., Classical small-angle scattering from anisotropic potentials, *J. Chem. Phys.* **46**, 609–618 (1967).
128. W.R. Gentry, Ion–dipole scattering in classical perturbation theory, *J. Chem. Phys.* **60**, 2547–2553 (1974).
129. A.S. Dickinson, Differential cross sections for electron scattering by strongly polar molecules, *J. Phys. B* **10**, 967–981 (1977).
130. E.A. Gislason and J.G. Sachs, Differential cross sections from classical perturbation scattering theory. I. Ion–dipole scattering, *Chem. Phys.* **25**, 155–163 (1977).
131. K. Takayanagi, Low-energy electron scattering from strongly polar molecules, *Progr. Theor. Phys. (Kyoto)* **52**, 337–338 (1974).
132. O. Ashihara, I. Shimamura, and K. Takayanagi, Low-energy electron scattering from strongly polar molecules. I. Glauber approximation, *J. Phys. Soc. Jpn.* **38**, 1732–1741 (1975).
133. Y. Itikawa, Low-energy electron scattering from strongly polar molecules. III. Close-coupling calculation for CsF, *J. Phys. Soc. Jpn.* **41**, 619–624 (1976).
134. J. Siegel, J.L. Dehmer, and D. Dill, Effects of long- and short-range forces in e–CsCl scattering, *J. Phys. B* **14**, L441–L446 (1981).
135. R.J. Allan and A.S. Dickinson, Electron scattering by state-selected polar molecules, *J. Phys. B* **13**, 3493–3503 (1980).
136. R.J. Allan and A.S. Dickinson, A semiclassical approximation for sudden S-matrix

elements in rotational excitation and simple methods for electron scattering by strongly polar molecules, *J. Phys. B* **14**, 1675–1694 (1981).
137. C.W. Clark and J. Siegel, Electron–polar-molecule scattering at intermediate values of *J*: A closed-form treatment, *J. Phys. B* **13**, L31–L37 (1980).
138. J. Siegel, J.L. Dehmer, and D. Dill, Differential cross sections for *e*–LiF scattering, *J. Phys. B* **13**, L215–L219 (1980).
139. J. Siegel, J.L. Dehmer, and D. Dill, Hybrid calculation of electron–polar-molecule scattering: Integrated and momentum-transfer cross sections for LiF, *Phys. Rev. A* **23**, 632–640 (1981).
140. Y. Itikawa, Rotational transition in polar molecules by electron collisions. II. Differential cross sections, *J. Phys. Soc. Jpn.* **27**, 444–452 (1969).
141. O.H. Crawford and A. Dalgarno, The scattering of thermal electrons by carbon monoxide, *J. Phys. B* **4**, 494–502 (1971).
142. D.G. Thompson, The elastic scattering of slow electrons by neon and argon, *Proc. R. Soc. (London) A* **294**, 160–174 (1966).
143. W.C. Fon, P.G. Burke, and A.E. Kingston, Elastic scattering of electrons by atomic hydrogen, *J. Phys. B* **11**, 521–530 (1978).
144. M.H. Mittleman and R.E. von Holdt, Theory of low-energy-electron scattering by polar molecules, *Phys. Rev.* **140**, A726–A729 (1965).
145. M.H. Mittleman, J.L. Peacher, and B.F. Rozsnyai, Rotational excitation of polar molecules by electrons, *Phys. Rev.* **176**, 180–186 (1968).
146. C.W. Clark, Electron scattering from diatomic polar molecules. II. Treatment by frame transformations, *Phys. Rev. A* **20**, 1875–1889 (1979).
147. O.H. Crawford, Bound states of a charged particle in a dipole field, *Proc. Phys. Soc.* **91**, 279–284 (1967).
148. W.R. Garrett, Scattering by a dipolar system: Divergence of cross sections at the critical moment for a point dipole rotor, *Phys. Rev. A* **23**, 1737–1744 (1981).
149. W.R. Garrett, Critical binding of electron–dipole rotor systems; electronically excited states, *J. Chem. Phys.* **73**, 5721–5725 (1980).
150. T.F. O'Malley, Effect of long-range final-state forces on the negative-ion photodetachment cross section near threshold, *Phys. Rev.* **137**, A1668–A1672 (1965).
151. I.I. Fabrikant, Scattering of slow electrons by polar molecules, *Sov. Phys. JETP* **44**, 77–82 (1976) [Transl. of: *Zh. Eksp. Teor. Fiz.* **71**, 148–158 (1976)].
152. I.I. Fabrikant, Threshold behavior of the cross sections for scattering of electrons by polar molecules, *Sov. Phys. JETP* **46**, 693–697 (1977) [Transl. of: *Zh. Eksp. Teor. Fiz.* **73**, 1317–1324 (1977)].
153. P.C. Engelking, Strong electron–dipole coupling in photodetachment of molecular negative ions: Anomalous rotational thresholds, *Phys. Rev. A* **26**, 740–745 (1982).
154. J.N. Bardsley and R.K. Nesbet, Threshold behavior of inelastic-scattering cross sections, *Phys. Rev. A* **8**, 203–206 (1973).
155. E.P. Wigner, On the behavior of cross sections near thresholds, *Phys. Rev.* **73**, 1002–1009 (1948).
156. E.S. Chang, Study of e–H_2 scattering near the rotational threshold, *Phys. Rev. A* **2**, 1403–1406 (1970).
157. A. Dalgarno and R.J. Moffett, The rotational excitation of molecular nitrogen by slow electrons, *Proc. Nat. Acad. Sci. India* **A33**, 511–521 (1963).
158. M.A. Morrison and L.A. Collins, Exchange in low-energy electron–molecule scattering: Free-electron-gas model exchange potentials and applications to e–H_2 and e–N_2 collisions, *Phys. Rev. A* **17**, 918–938 (1978).
159. Yu. F. Smirnov and K.V. Shitikova, The method of *K* harmonics and the shell model, *Sov. J. Part. Nucl.* **8**, 344–370 (1977).

160. U. Fano, Correlations of two excited electrons, *Rep. Prog. Phys.* **46**, 97–165 (1983).
161. J. Callaway, The variational method in atomic scattering, *Phys. Rep.* **45**, 89–173 (1978).
162. R.K. Nesbet, *Variational Methods in Electron–Atom Scattering Theory*, Plenum Press, New York (1980).
163. P.G. Burke, K.A. Berrington, and C.V. Sukumar, Electron–atom scattering at intermediate energies, *J. Phys. B* **14**, 289–305 (1981).
164. A.E. Kingston and H.R.J. Walters, Electron–atom scattering at intermediate energies, *Comments At. Mol. Phys.* **11**, 177–191 (1982).
165. P.G. Burke and A.J. Taylor, Correlation in the elastic and inelastic S-wave scattering of electrons by H and He^+, *Proc. Phys. Soc.* **88**, 549–562 (1966).
166. P.G. Burke and J.F.B. Mitchell, Electron scattering by atomic hydrogen using a pseudo-state expansion IV. The convergence of the S-state expansion at intermediate energies, *J. Phys. B* **6**, 320–328 (1973).
167. W.C. Fon, K.A. Berrington, P.G. Burke, and A.E. Kingston, Elastic scattering of electrons by atomic hydrogen at intermediate energies, *J. Phys. B* **14**, 1041–1051 (1981).
168. P.G. Burke and N. Chandra, Electron–molecule interactions III. A pseudo-potential method for e^-–N_2 scattering, *J. Phys. B* **5**, 1696–1711 (1972).
169. L.A. Collins, W.D. Robb, and D.W. Norcross, Low-energy collisions of electrons with highly polar molecules: Orthogonalization and model exchange potentials, *Phys. Rev. A* **20**, 1838–1840 (1979).
170. M. Le Dourneuf, Vo Ky Lan, and J.M. Launay, Accurate representation of molecular electrons by a few potential-adapted partial waves, *J. Phys. B* **15**, L685–L690 (1982).
171. D.E. Golden, Low-energy resonances in e^-–N_2 total scattering cross sections: The temporary formation of N_2^-, *Phys. Rev. Lett.* **17**, 847–848 (1966).
172. B.D. Buckley and P.G. Burke, The scattering of low-energy electrons by diatomic molecules, *J. Phys. B* **10**, 725–739 (1977).
173. N. Chandra and A. Temkin, Hybrid theory calculation of simultaneous vibration-rotation excitation in e–N_2 scattering, *Phys. Rev. A* **14**, 507–511 (1976).
174. F.A. Gianturco and D.G. Thompson, Theoretical evidence for short-lived resonances in electron scattering by HCl and HF molecules, *J. Phys. B* **10**, L21–L26 (1977).
175. L.A. Collins, R.J.W. Henry, and D.W. Norcross, Electron collisions with polar molecules: Exchange and polarisation in elastic scattering by HCl, *J. Phys. B* **13**, 2299–2307 (1980).
176. N.T. Padial, D.W. Norcross, and L.A. Collins, Vibrationally elastic scattering of electrons by HCl, *Phys. Rev. A* **27**, 141–148 (1983).
177. K. Rohr and F. Linder, Vibrational excitation of polar molecules by electron impact I. Threshold resonances in HF and HCl, *J. Phys. B* **9**, 2521–2537 (1976), and private communications to the authors of Ref. 177 (see Ref. 177).
178. N.A. Mullaney and D.G. Truhlar, Rotationally and orbitally adiabatic basis sets for electron–molecule scattering, *Chem. Phys. Lett.* **58**, 512–517 (1978).
179. N.A. Mullaney and D.G. Truhlar, The use of rotationally and orbitally adiabatic basis functions to calculate rotational excitation cross sections for atom–molecule collisions, *Chem. Phys.* **39**, 91–104 (1979).
180. K. Takayanagi, Low energy ion molecule collisions, in: *Physics of Electronic and Atomic Collisions*, Invited Papers of the XIIth Intern. Conf. Phys. Electron. Atom. Coll., Gatlinburg, Tennessee, July 1981 (S. Datz, ed.), pp. 343–354, North-Holland, Amsterdam (1982).

3

VIBRATIONAL EXCITATION OF MOLECULES BY SLOW ELECTRONS

ARVID HERZENBERG

1. Introduction

This chapter deals with the excitation of vibration in a molecule by the impact of an electron with an energy of a few electron volts, rarely as high as 30 eV. The field has been reviewed so many times in the last 10 years[1–11] that another paper-by-paper review hardly seems appropriate. What does seem appropriate is an elementary exposition of the physical concepts, particularly resonances and virtual states, which lie behind much of the discussion. These concepts are understood very well by the very few people who actually use them in calculations, but unfortunately different people speak and write about them in ways which at first sight seem to be describing different physical objects. Lengthy discussion usually leads to the conclusion that everyone is talking about the same physics, but it is at present difficult for the newcomer to the field to see this because there is a lack of elementary introductions. This chapter, then, is an attempt at such an elementary introduction. The reader who wants to know more about current theoretical developments will find it helpful to go to the published reviews, starting with the most recent, by Lane[1] and Burke.[2] The reader who wants to know how the experiments are done will find a good starting point in the papers[11] on the state of the experimental art at the first and second Schulz Symposia, by Linder (in 1977) and Read (in 1979).

In this article, the orientation of the molecule will be regarded as fixed in the laboratory during a collision, because the projectile electrons will usually be sufficiently fast to pass the molecule in a time short compared with a rotational period. (See Chapter 2, Section 3.2.) A very slow electron, below about 0.1 eV, offers special difficulties for the theory, because the tail of the potential from the molecule may be rotating at a *linear* velocity sufficiently high to be important, even though the *angular* velocity of the molecule is small; this problem has an extensive literature.[12] See also Chapter 2, Section 5.3.

At first sight, one might expect vibrational excitation to be improbable,

ARVID HERZENBERG • Department of Applied Physics, Yale University, New Haven, Connecticut 06520

because little energy is transferred in the collision of an electron with a free atom; the electron tends to bounce off like a fly off an elephant. For example when an electron of 5 eV collides with an atom of atomic weight 5 and kinetic energy 0.2 eV (about the kinetic energy of an atom in the vibrational ground state of a small molecule), the energy lost by the electron is only of order 1/50 eV. It is therefore surprising to see, in Fig. 1, the large cross sections for the excitation of a single vibrational quantum by an electron in H_2, N_2, N_2O, O_2, and HCl. The differential cross section reaches values as large as a few times a_B^2 in isolated peaks of the order of 1 eV wide. ($a_B = 0.52 \times 10^{-8}$ cm.) Between the peaks, much smaller magnitudes are typical.

It might be thought that the large observed cross sections would emerge as soon as one did a quantum-mechanical calculation. The differential cross sections per unit solid angle for the scattering of an electron of initial momentum $\hbar \mathbf{k}_i$ to a final momentum $\hbar \mathbf{k}_f$ by a molecule whose state changes from ψ_i to ψ_f is (in atomic units, which we shall use throughout this chapter, where $\hbar = 1, m = 1, e = 1$, with m and e the mass and charge of the electron)

$$\frac{d\sigma_{f\leftarrow i}}{d\hat{\mathbf{k}}_f} = \frac{v_f}{v_i} |T_{fi}(\mathbf{k}_f, \mathbf{k}_i)|^2, \tag{1}$$

where T_{fi} is the scattering amplitude, given by

$$T_{fi}(\mathbf{k}_f, \mathbf{k}_i) = \left(-\frac{1}{2\pi}\right)(\Phi_f|V|\Psi_i). \tag{2a}$$

V is the interaction potential. Ψ_i is the exact solution of the Schrödinger equation, having an asymptotic form

$$\Psi_i = e^{i\mathbf{k}_i \cdot \mathbf{r}} \psi_i + \text{outgoing waves}; \tag{2b}$$

Φ_f is

$$\Phi_f = e^{i\mathbf{k}_f \cdot \mathbf{r}} \psi_f; \tag{2c}$$

ψ_i and ψ_f are the initial and final states of the target; they are normalized so that

$$\begin{aligned}(\psi_i|\psi_j) = \delta_{ij} &\equiv 1 \qquad \text{if } i = j,\\ &\equiv 0 \qquad \text{if } i \neq j;\end{aligned} \tag{2d}$$

$\mathbf{r}$ is the position of the projectile electron. The spin state of the projectile is to be

FIGURES 1a–j:
Observed cross sections for vibrational excitation of molecules by electrons.

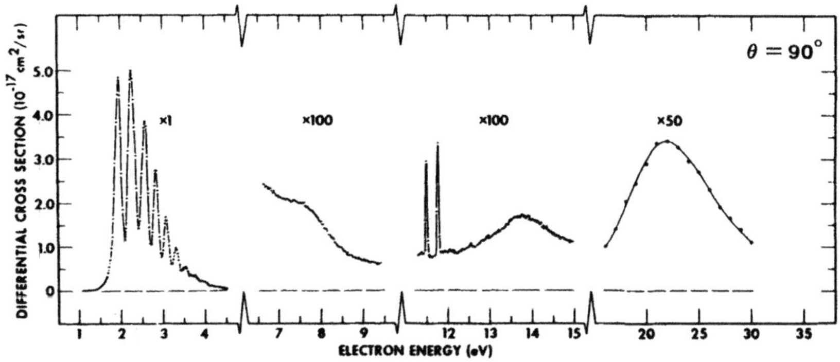

FIGURE 1a
An overview for $e + N_2$ $(v = 0 \rightarrow v = 1)$ at impact energies up to 30 eV. If one integrates over scattering angles, the height of the bump at 2–3 eV becomes $\approx 2 \times 10^{-16}$ cm², and the broad bump at 20–25 eV $\approx 10^{-17}$ cm². If both bumps are integrated over energy, the 20–25 eV bump is about four times smaller than the 2–3 eV bump. (From Ref. 5.)

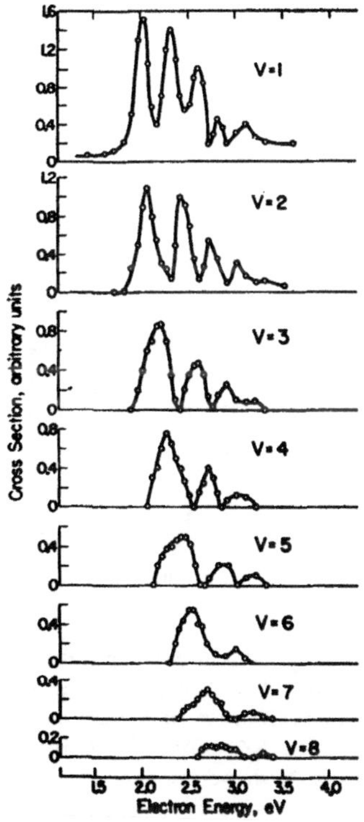

FIGURE 1b
$e + N_2$; excitations $v = 0 \rightarrow v \neq 0$. The widths of the subpeaks are real, and not due to experimental resolution. The positions of the subpeaks are shown in Fig. 14. The measurements were done at a scattering angle of 72°. (From ref. 74.)

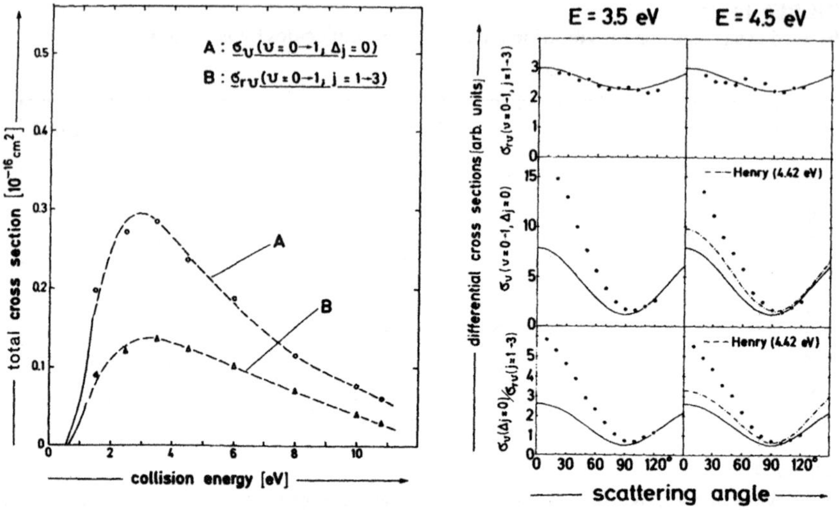

FIGURE 1c

$e + H_2$; excitation $v = 0 \rightarrow v = 1$ with resolution of the rotational states. The diagram on the right shows the angular distributions: experiment, · · · (the variation of the angular distribution with energy is very weak between 2 and 6 eV); calculation by Henry[75] with allowance for exchange and polarization, and inclusion of three vibrational levels –·–·–; pure resonance model[76] ———. (See Section 4; from Ref. 77.)

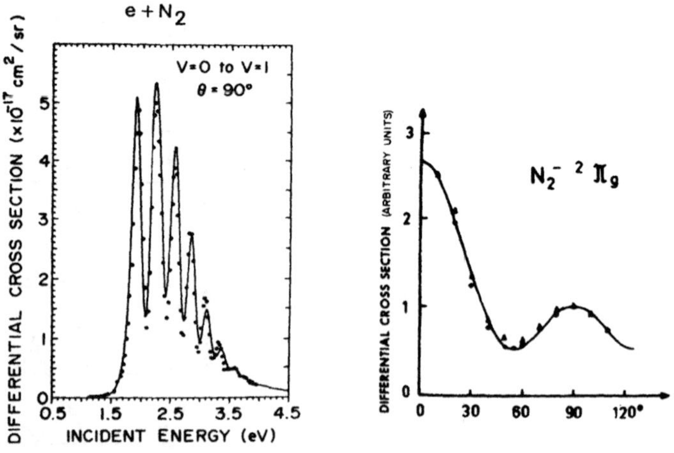

FIGURE 1d

$e + N_2$; absolute magnitudes and angular distributions for $v = 0 \rightarrow v = 1$. Left: Experiment,[78] . . . ; semiempirical resonance theory (see Section 5) ———. (From Ref. 49.) For the significance of the agreement between the observed and calculated absolute values, see Section 5.4, point (4). Right: Experiment: circles and triangles. Continuous curve: theory for a Π_g shape resonance dominated by $l = 2$ outside the target. The variation of the angular dependence over the resonance region is small. (From Ref. 79.)

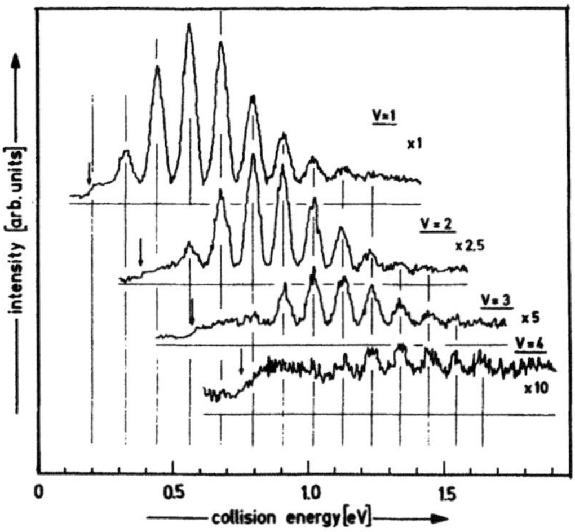

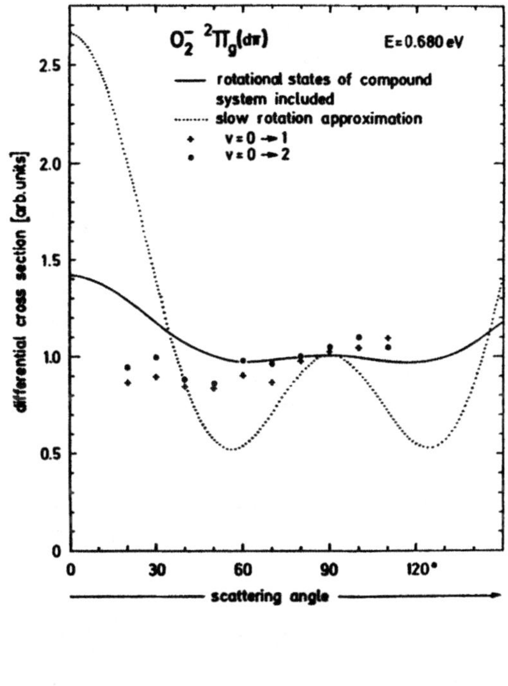

FIGURE 1e
$e + O_2$; relative cross sections for $v = 0 \to 1, 2, 3, 4$. Note the coincidence of the peaks in the final channels, in contrast to Figs. 1b and 14. The diagram corresponds to a scattering angle of 60°. (From Ref. 80.)

included in ψ_i and ψ_f. The initial and final speeds of the electron are v_i and v_f. The functions Ψ_i and Φ_f should be antisymmetrized between the projectile, scattered, and target electrons, but we shall generally omit the antisymmetrization to avoid making the formulas too cumbersome. (The antisymmetrization rarely introduces important qualitative physical features.)

The simplest possible calculation estimates the cross section by ignoring

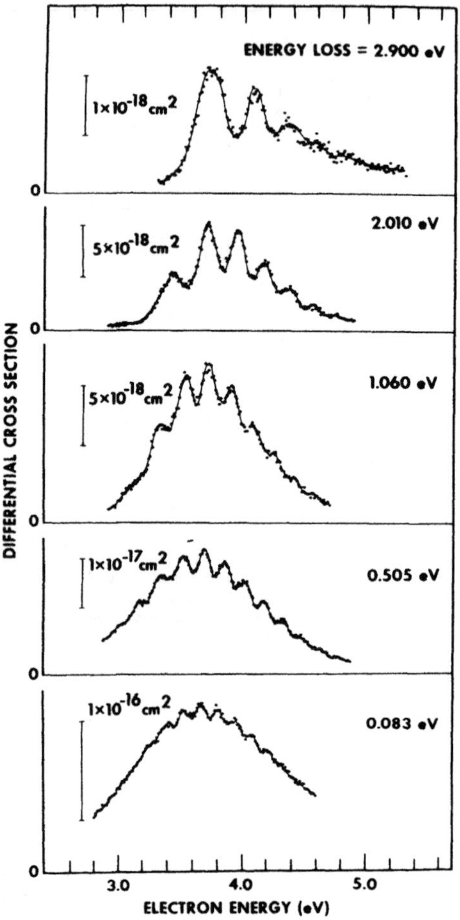

FIGURE 1f
$e + CO_2$; a variety of vibrational excitations in a triatomic molecule with three normal modes of vibration. Compare Fig. 18a for a theoretical fit. (From Ref. 81.) The scattering angle is 90°. The levels excited are, starting from the bottom; (0 1 0), (3 0 0), (6 1 0), (12 1 0), (18 1 0).

the distortion of the wave function, that is in the Born approximation, where one sets $\Psi_i \to \Phi_i$ in Eq. (2a). Then, one obtains[13]

$$T_{fi}(\mathbf{k}_f, \mathbf{k}_i) \to -\frac{2}{q^2} \int d\mathbf{r} e^{-i\mathbf{q}\cdot\mathbf{r}} (\psi_f | \rho(\mathbf{r}) | \psi_i), \tag{3a}$$

where

$$\mathbf{q} \equiv \mathbf{k}_f - \mathbf{k}_i, \tag{3b}$$

and where $\rho(\mathbf{r})$ is the charge density of the molecule:

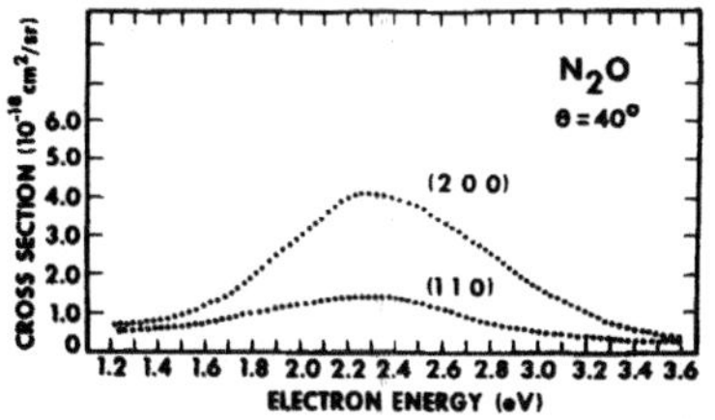

FIGURE 1g
$e + N_2O$; two excitation channels in a triatomic molecule. Note the absence of substructure in the energy dependence, in contrast with Fig. 1f. See Fig. 11 for a theoretical fit. (From Ref. 47.)

$$\rho(\mathbf{r}) = \sum_{\alpha} e_\alpha \delta(\mathbf{r} - \mathbf{r}_\alpha), \tag{4}$$

α running over all the particles in the molecule. $\mathbf{r}_\alpha$ is the position and e_α the charge of the αth particle.

If the molecule has a dipole moment, its potential has a weak tail ($\propto 1/r^2$) which can scatter an electron of a few volts, where $|\mathbf{k}_i| \approx a_B^{-1}$, through small angles. Then $|\mathbf{q}| \ll a_B^{-1}$, and for small molecules whose linear dimensions are a few a_B, one has $|\mathbf{q}\cdot\mathbf{r}| \ll 1$ for the positions $\mathbf{r}$ which contribute to the integral in Eq. (3a); one may replace $e^{-i\mathbf{q}\cdot\mathbf{r}} \to 1 - i\mathbf{q}\cdot\mathbf{r}$. The first term leads to the integral $(\psi_f|\int d\mathbf{r}\,\rho(\mathbf{r})|\psi_i) = (\psi_f|Q|\psi_i)$, where Q is the total electric charge of the molecule. Since Q is a constant of the motion, one has $(\psi_f|Q|\psi_i) = 0$ for inelastic

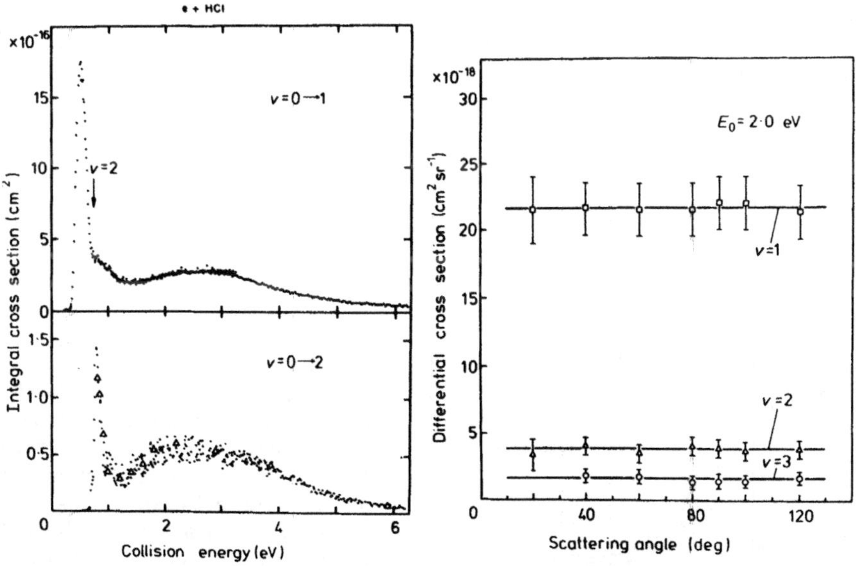

FIGURE 1h
$e + HCl$; vibrational excitation near threshold. Note the threshold peak and the flat angular distribution, which is characteristic of all energies below 4 eV. (From Ref. 82.)

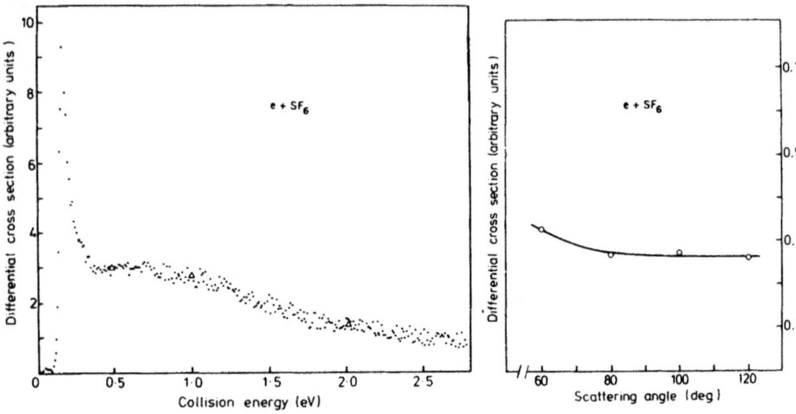

FIGURE 1i
$e + SF_6$; excitation of one quantum of the breathing vibration near threshold. (In the left figure, the scattering angle is 80°. In the right figure, the impact energy is 1 eV.) The symmetry of the molecule precludes a dipole moment, demonstrating that a long-range potential is not essential for such threshold peaks. (From Ref. 83.)

scattering ($\psi_i \neq \psi_f$), which is our concern in this chapter. The term linear in **q** in Eq. (3) leads to

$$T_{fi}(\mathbf{k}_f, \mathbf{k}_i) \to \frac{2i\mathbf{q}\cdot\mathbf{d}_{fi}}{q^2}, \tag{5}$$

where

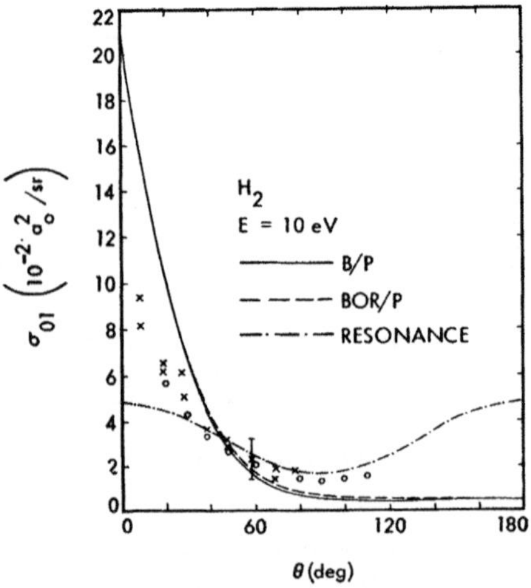

FIGURE 1j
$e + H_2; v = 0 \to v = 1$ at different scattering angles at an impact energy above the broad resonance in Fig. 1c. Experiment: × and ○. Calculation in Born approximation with polarization ——— and – – –. Pure resonance model for the resonance in Fig. 1c: —·—. This diagram hints at the presence of both resonant and nonresonant contributions. (From Ref. 84.)

$$\mathbf{d}_{fi} \equiv (\psi_f | \int d\mathbf{r}\, \mathbf{r}\rho(\mathbf{r}) | \psi_i) \tag{6}$$

is the transition dipole moment. [The origin of $\mathbf{r}$ in the integral in Eq. (6) is arbitrary, because a change of origin leads only to the addition of a term proportional to $(\psi_f | Q | \psi_i)$.]

One may average $|T_{fi}|^2$ over the orientation of the molecule, because the collision is usually so fast that the molecule does not have time to rotate. Let the average be $\overline{|T_{fi}|^2}$. One obtains from Eq. (5) for a given $\mathbf{q}$, $\overline{|T_{fi}|^2} = \frac{4}{3}|\mathbf{d}_{fi}|^2/q^2$. Thus the differential cross section is peaked in the forward direction. The total cross section is obtained by integrating Eq. (1) over all directions of the scattered electron, which yields

$$\sigma_{f\leftarrow i} = \int d\hat{\mathbf{k}}_f \frac{d\sigma_{f\leftarrow i}}{d\hat{\mathbf{k}}_f} = \frac{4\pi}{3} \frac{|\mathbf{d}_{fi}|^2}{k_f k_i} \ln\left(\frac{1+\epsilon}{1-\epsilon}\right)$$

or

$$\sigma_{f\leftarrow i} = \frac{8\pi}{3} \frac{|\mathbf{d}_{fi}|^2}{k_i^2} \ln\left(\frac{2k_i^2}{\omega}\right), \qquad \text{if } \omega \ll \tfrac{1}{2}k_i^2,$$

where ω is the excitation energy transferred to the target, and $\epsilon \equiv 2k_f k_i/(k_f^2 + k_i^2)$.

For vibrational transitions without change of electronic state, the magnitude of $\mathbf{d}_{fi}$ comes from the product of the charge associated with each nucleus and its surrounding cloud of electrons, and the vibrational amplitude of the nuclei. The vibrational amplitudes are[14] of order $(m/M)^{1/4}$, M being a typical nuclear mass. Consider a molecule in which there is some internal charge transfer between atoms, so that there is about a single net electronic charge associated with each vibrating nucleus; the molecule could be either polar, or nonpolar like CO_2 or SF_6. Then $|\mathbf{d}_{fi}|$ is typically of order 10^{-1} for each normal mode, if $M \approx 10^4\, m$.

The total cross section for the excitation of a single vibrational quantum of 0.2 eV becomes $0.3 \times 10^{-16}\ \text{cm}^2$ for an incident energy of 5 eV.

In molecules with little or no internal transfer of charge between atoms, such as H_2 or N_2, each nucleus is surrounded by a cloud of electrons of the same charge. The transition dipole moment is then very small, so that the molecule has no fluctuating electric field extending outside the transition charge density $(\psi_f | \rho(\mathbf{r}) | \psi_i)$; an electron can excite vibration only by interacting at a short range of order a_B. Although it is hardly realistic to ignore the distortion of the wave function in such a case, let us do it nevertheless, just to bring out the need to take distortion into account. We approximate the contribution to $\rho(\mathbf{r})$ from the nucleus at $\mathbf{r}_N$ and its associated electrons by the function $Z\delta(\mathbf{r}-\mathbf{r}_N)-s(|\mathbf{r}-\mathbf{r}_N|)$, the first term representing the nucleus, and the second the cloud of electrons distributed in a spherically symmetric fashion. The equality of the charges requires that $\int d\mathbf{r}\, s(|\mathbf{r}-\mathbf{r}_N|)=Z$. For the purposes of our estimate, we suppose all the atoms to be identical. The expression (3a) becomes

$$T_{fi}(\mathbf{k}_f, \mathbf{k}_i) \longrightarrow -\frac{2F(\mathbf{q})}{q^2}\sum_N (\psi_f|e^{-i\mathbf{q}\cdot\mathbf{r}_N}|\psi_i), \tag{7}$$

where

$$F(\mathbf{q}) = \int d\mathbf{r}\, e^{-i\mathbf{q}\cdot\mathbf{r}}[Z\delta(\mathbf{r})-s(|\mathbf{r}|)]. \tag{8}$$

A convenient approximation for $s(|\mathbf{r}|)$ is

$$s(|\mathbf{r}|) = Z\frac{e^{-r/r_0}}{4\pi r_0^2 r}, \tag{9}$$

where r_0 is a measure of the radius of the electronic cloud. In accordance with the Thomas–Fermi model,[15] take $r_0=0.79Z^{-1/3}$, which corresponds to half the electrons lying within a distance $1.33Z^{-1/3}$ from the nucleus.

Equations (8) and (9) then lead to

$$F(\mathbf{q}) = Z\frac{r_0^2q^2}{1+r_0^2q^2}. \tag{10}$$

Only an integral over the nuclear coordinates remains to be done to evaluate the matrix element $(\psi_f|e^{-i\mathbf{q}\cdot\mathbf{r}_N}|\psi_i)$ in Eq. (7). Write $\mathbf{r}_N=\mathbf{r}_N^{(0)}+\delta\mathbf{r}_N$, where $\mathbf{r}_N^{(0)}$ is the equilibrium position of nucleus N. $\delta\mathbf{r}_N$ is of order $(m/M)^{1/4}\approx 0.1$ if the molecule starts in its vibrational ground state $i=0$. $|\mathbf{q}|$ is 0.86, for an electron of energy 5 eV scattered through 90°. Thus $|\mathbf{q}\cdot\delta\mathbf{r}_N|\ll 1$, so that one can approximate the exponential by $e^{-i\mathbf{q}\cdot\mathbf{r}_N^{(0)}}(1-i\mathbf{q}\cdot\delta\mathbf{r}_N)$. The contribution from the first term vanishes for inelastic scattering, where $\psi_f\neq\psi_0$, so that $(\psi_f|e^{-i\mathbf{q}\cdot\mathbf{r}_N}|\psi_0)\cong -i\mathbf{q}\cdot(\psi_f|\delta\mathbf{r}_N|\psi_0)e^{-i\mathbf{q}\cdot\mathbf{r}_N^{(0)}}$, and, from Eqs. (7) and (10),

$$T_{f0}(\mathbf{k}_f, \mathbf{k}_0) \longrightarrow \left[\frac{2iZr_0^2}{1+r_0^2q^2}\right]\mathbf{q}\cdot\sum_N (\psi_f|\delta\mathbf{r}_N|\psi_0)e^{-i\mathbf{q}\cdot\mathbf{r}_N^{(0)}} \tag{11}$$

To estimate the differential cross section according to Eq. (1), we ignore coherence between different nuclei, so that $|\Sigma_N(\psi_f|\delta\mathbf{r}_N|\psi_0)e^{-i\mathbf{q}\cdot\mathbf{r}_N^{(0)}}|^2$ becomes $\Sigma_N|(\psi_f|\delta\mathbf{r}_N|\psi_0)|^2$. If ψ_f differs from ψ_0 only by a single vibrational quantum in one vibrational mode, the expression $\Sigma_N|\cdots|^2$ becomes $1/(2\omega M)$, where ω is the excitation energy, and M the mass of a nucleus. Since ω is of order $M^{-1/2}$ atomic units,[14] $\Sigma_N|(\psi_f|\delta\mathbf{r}_N|\psi_0)|^2$ is of order $(1/2\sqrt{M})$. If, for example, $Z = 7$, then $r_0 = 0.79/7^{1/3} = 0.41$. If the electron starts with 5 eV and is scattered through 90°, then $|\mathbf{q}| = 0.86$. If one finally averages over the orientation of the molecule relative to $\mathbf{q}$, the differential cross section becomes, from Eqs. (1) and (11),

$$\frac{d\sigma_{f\leftarrow 0}}{d\hat{\mathbf{k}}_f} \approx 0.5 \times 10^{-2} a_B^2.$$

The total cross section, integrated over all final directions, becomes[16] of order 10^{-18} cm^2.

Thus for small molecules with finite transition dipole moments, the no-distortion picture leads to vibrational excitation cross sections of order 10^{-17} cm^2 at impact energies of a few electron volts, with an angular distribution strongly peaked in the forward direction. This is somewhat smaller in magnitude than the maxima in Fig. 1, and completely different from the angular distributions in Fig. 1, which diverge to a much smaller extent from spherical symmetry. For molecules without a transition dipole moment, the vibrational excitation cross sections are even smaller, of order[16] 10^{-18} cm^2, about two orders of magnitude below those in Fig. 1. To understand these failures of the no-distortion picture, we next consider some of the distortions which might occur.

2. Virtual States and Resonances

2.1. Introduction

The simplest examples of distortions of wave functions occur in the scattering of an electron by a spherically symmetric potential well. The failings of this case as a model for a real electron near a real molecule are obvious: molecules are not spherically symmetric, and they possess the dynamical degrees of freedom of the electrons and of the vibrations and rotations of the nuclei. The neglect of the vibrations and rotations is not serious, provisionally, because the electrons inside a molecule are almost always fast compared with the nuclei, so that they see the nuclei as almost static. We shall see in Sections 3 and 4 how the nuclear dynamics may be introduced at a second stage, after the electronic wave function has been determined with a picture of stationary nuclei. To neglect the electronic degrees of freedom, one has to try to represent the effects of electronic relaxation by an optical potential; at low incident energies, this reduces to polarization and exchange potentials. In spite of these gross simplifications, the case of the single electron in a spherical potential well is worth looking at because it is simple and soluble.

The Hamiltonian is, in atomic units (i.e., $\hbar = 1$, electron mass = 1, electron charge = 1)

$$H = -\tfrac{1}{2}\nabla^2 + V, \tag{12}$$

where $V = V(r)$ is the potential well. The solution of the Schrödinger equation at energy E

$$(H - E)\Psi(\mathbf{r}) = 0 \tag{13}$$

behaves as $r \to \infty$ like

$$\Psi(\mathbf{r}) \to e^{i\mathbf{k}\cdot\mathbf{r}} + T(\theta)\frac{e^{ikr}}{r}, \tag{14}$$

where $\mathbf{k}$ is the initial momentum; the direction of the scattered electron may be described by polar angles (θ, ϕ), with $\mathbf{k}$ as the polar axis; an arbitrary plane which contains the polar axis defines the origin of ϕ. The amplitude $T(\theta)$ of the wave scattered in the direction (θ, ϕ) is

$$T(\theta) = -\frac{1}{2\pi}(\Phi_f|V|\Psi), \tag{15a}$$

where $\Phi_f = \exp(i\mathbf{k}_f\cdot\mathbf{r})$, $\mathbf{k}_f$ being the final momentum with polar coordinates (k, θ, ϕ). One may expand

$$T(\theta) = \sum_{l=0}^{\infty} T_l Y_{l0}(\theta). \tag{15b}$$

If one expands

$$\Psi(\mathbf{r}) = \sum_{l=0}^{\infty} \frac{R_l(r)}{r} Y_{l0}(\theta), \tag{16}$$

then the radial wave function $R_l(r)$ satisfies

$$\left[-\frac{1}{2}\frac{d^2}{dr^2} + V_{\text{eff}}(r) - E\right] R_l(r) = 0, \tag{17a}$$

where

$$V_{\text{eff}}(r) \equiv V(r) + \frac{l(l+1)}{2r^2} \tag{17b}$$

may be called the "effective potential." The boundary conditions are

$$R_l(0) = 0, \tag{18}$$

and

$$R_l(r) \xrightarrow[r\to\infty]{} [4\pi(2l+1)]^{1/2} i^l r j_l(kr) + T_l e^{ikr}, \tag{19}$$

where the first term on the right comes from the expansion of $\exp(i\mathbf{k}\cdot\mathbf{r})$ in spherical harmonics.[18]

We shall suppose V to be of short range, so that $V(r) = 0$ for $r > r_0$, r_0 being of the order $r_0 \approx 2a_B$. The "effective potential" V_{eff} in Eq. (17) is plotted in Fig. 2 for $r > r_0$ for $l = 0, 1, 2$, and 3.

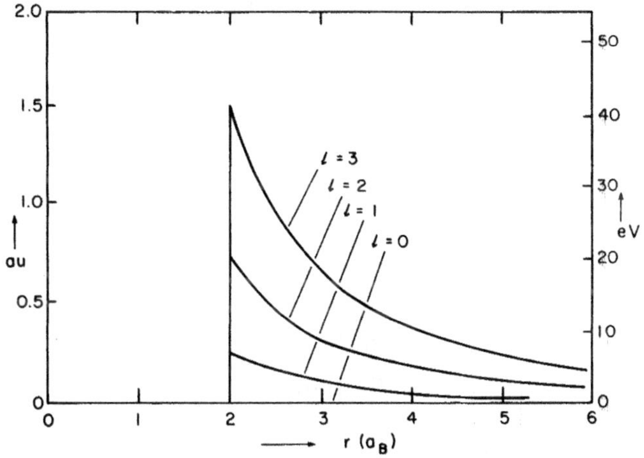

FIGURE 2
The centrifugal barrier for an electron near a small molecule. The interaction has been omitted at distances smaller than $2a_B$, where it is dominated by short-range effects which are generally attractive. The curves shown contain no allowance for polarization; the addition of a static polarization potential corresponding to an electron near an N_2 molecule would reduce the potential at $r = 2a_B$ by 0.39 a.u. = 10.7 eV. The polarization correction falls off with distance much more rapidly than the centrifugal potential shown: r^{-4} as compared with r^{-2}. (1 a.u. = 27.2 eV; $a_B = 0.52 \times 10^{-8}$ cm.)

The solutions of Eq. (17) depend on whether $V_{\text{eff}}(r)$ has a barrier higher than the energy of the projectile when $r > r_0$. Since there is no barrier at all for $l = 0$, we have to consider this case separately from $l > 0$.

2.2. Potential Well Scattering with $l = 0$

Let the effective potential be $V_{\text{eff}} = -U$ for $r < r_0$; for $r > r_0$, we take $V_{\text{eff}} = 0$, corresponding to $l = 0$. For the sake of illustration, we shall take U to be of order 1 a.u. (=27.2 eV) and $r_0 = 2a_B$. The radial factor in the waves function $R_0(r)$ is shown in Fig. 3 for three values of U. The energy E is zero in each case, so that the solution of Eq. (17a) for $r > r_0$ is $R_0(r) = c(r - a)$, where c and a are constants. a is called the "scattering length." There are three possibilities:

(i) When $r > r_0$, $dR_0/dr > 0$, and $a < 0$. There is no bound state for $U \approx 1$ a.u.

(ii) When $r > r_0$, $dR_0/dr < 0$, and $a > 0$. The potential possesses a bound state, because a slight reduction in E allows R_0 to join smoothly to a decreasing exponential in $r > r_0$.

(iii) At $r = r_0$,

$$\left[\frac{r}{R_0}\frac{dR_0}{dr}\right] \ll 1,$$

so that R_0 is almost constant for $r > r_0$ ($|a| \to \infty$). There is then a very weakly bound state when

$$\left[\frac{r}{R_0}\frac{dR_0}{dr}\right]_{r=r_0} < 0;$$

if

$$\left[\frac{r}{R_0}\frac{dR_0}{dr}\right]_{r=r_0} > 0,$$

one calls this case[19] a "virtual state."

To calculate the scattering cross section near $E = 0$, note that, for $r > r_0$, $R_0(r) = (4\pi)^{1/2}(\sin kr)/k + T_0 e^{ikr}$; this follows from Eq. (16) and the formula $Y_{00} = 1/(4\pi)^{1/2}$. Thus in the limit $k \to 0$, $R_0(r) \to (4\pi)^{1/2} r + T_0$, so that $c = (4\pi)^{1/2}$ and $T_0(k = 0) = -(4\pi)^{1/2} a$. The scattering cross section becomes, from Eqs. (14) and (15b), $\sigma = 4\pi a^2$ at $k = 0$. Therefore $\sigma(k = 0)$ can be very large if the potential has either a very weakly bound state or a virtual state.

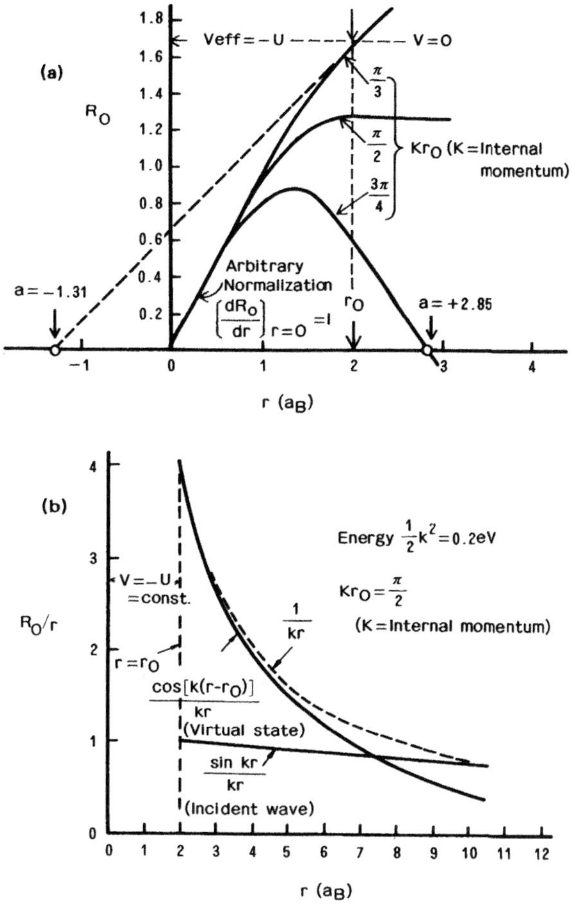

FIGURE 3
The radial factor in an s wave function of an electron near a spherically symmetric potential well V_{eff} of radius $2a_B$ and constant depth. (a) Radial factor $R_0(r)$ in the sense of Eq. (16) for three different well depths. The normalization is arbitrary, and chosen only for convenience in drawing the picture. The intercept of the asymptote of R_0 with the horizontal axis is the "scattering length" a. (b) $R_0(r)/[r(4\pi)^{1/2}]$ at an energy of 0.2 eV outside the well. The curve $\cos[k(r-r_0)]/kr$ corresponds to $dR_0/dr = 0$ at $r = r_0$, the normalization being fixed by the incoming wave part of the incident plane wave. The s-wave part of the incident plane wave is shown as $(\sin kr)/kr$. The label "virtual state" refers to the curve $\cos[k(r-r_0)]/kr$.

The magnitude of R_0 inside the well is important for later applications to vibrational excitation. The normalization is arbitrary in Fig. 3a, chosen simply so that one can get all cases on the same picture. However, in the scattering problem with the asymptotic form specified by Eq. (14), the normalization comes from the $l = 0$ part of the incident wave $e^{i\mathbf{k}\cdot\mathbf{r}}$, i.e., from $(\sin kr)/kr$, whose incoming part is $-e^{-ikr}/2ikr$. One finds, in cases (i) and (ii)

$$\left[\frac{R_0(r)}{r}\right]_{r=r_0} Y_{00} = 1 - \frac{a}{r_0}.$$

In case (iii), if

$$\left[\frac{dR_0(r)}{dr}\right]_{r=r_0} = 0,$$

one has

$$\frac{R_0(r_0)}{r_0} Y_{00} = \frac{ie^{-ikr_0}}{kr_0}$$

Thus if a is of order r_0, the $l = 0$ component of Ψ has a modulus of order unity at $r = r_0$. However, if there is either a very weakly bound state or a virtual state, so that $(dR_0/dr)_{r_0} = 0$, then the $l = 0$ component of Ψ has modulus $(1/kr_0)$. For example, for $E = 0.2$ eV, $r_0 = 2a_B$, the probability densities are, for the incident wave

$$\left(\frac{\sin kr_0}{kr_0}\right)^2 \cong 1,$$

while for $(dR_0/dr)_{r=r_0} = 0$,

$$|\Psi(r = r_0)|^2 \cong \left(\frac{1}{kr_0}\right)^2 = 17.$$

Thus the probability density at the target can be enhanced by an order of magnitude over the incident wave if there is a very weakly bound state or a virtual state, and if the impact energy is below about 0.5 eV. Figure 3b shows the spherically symmetric part of the wave function, i.e., $[R_0(r)/r(4\pi)^{1/2}]$, for $r > r_0$ for the case $[dR_0/dr]_{r=r_0} = 0$; the $l = 0$ part of the incident wave $\exp(i\mathbf{k}\cdot\mathbf{r})$ is shown for comparison. We have also drawn the curve $(1/kr)$; comparison with $[R_0(r)/r(4\pi)^{1/2}]$ shows that the enhancement of the wave function over the incident wave extends out to about $6a_B$.

2.3. Potential-Well Scattering with $l > 0$

We next consider an effective potential in Fig. 2 which possesses a barrier in $r > r_0$ due to the centrifugal potential from an angular momentum $l > 0$. In $r < r_0$, we shall suppose that $V_{\text{eff}}(r) = -U$, where U is constant, positive, and of the order of 1 a.u. According to Eq. (17a), the radial wave function has the asymptotic form

$$R_l(r) \longrightarrow A_l[e^{-ikr} - S_l(k)e^{+ikr}], \tag{20}$$

where one can show from Eq. (19) that

$$A_l = (-1)^{l+1}[4\pi(2l+1)]^{1/2}/2ik,$$

and where

$$\tfrac{1}{2}k^2 = E. \tag{21}$$

One must have

$$|S_l| = 1, \tag{22}$$

because the outgoing flux must equal the incoming flux.

Let us construct R_l by starting from the incoming wave e^{-ikr}. The momentum inside the well, to be denoted by K, is given by $\frac{1}{2}K^2 - U = E$. Let $-R_b$ and T_b denote the amplitude reflection and transmission coefficients of the barrier for waves incident from $r > r_0$, so that a wave e^{-ikr} from $r > r_0$ gives rise to a transmitted wave $e^{-iK(r-r_0)} \cdot T_b e^{-ikr_0}$ in $r < r_0$, and a reflected wave $-R_b e^{+ik(r-r_0)} \cdot e^{-ikr_0}$ in $r > r_0$. Similarly, let a wave e^{+iKr} incident on the barrier from $r < r_0$ give rise to a wave $T_b' e^{+ik(r-r_0)} e^{iKr_0}$ transmitted into $r > r_0$, and a wave $-R_b' e^{-iK(r-r_0)} \cdot e^{iKr_0}$ reflected back into $r < r_0$. The origin acts as a reflector, according to the boundary condition (18), so that a wave e^{-iKr} from $r > 0$ gives rise to a reflected wave $-e^{+iKr}$.

The wave transmitted into $r < r_0$ from the incoming part of Eq. (20) is $A_l e^{-iK(r-r_0)} \cdot T_b e^{-ikr_0}$. The reflection at the origin adds a wave $-A_l e^{+iKr} \times e^{iKr_0} T_b e^{-ikr_0}$, and the reflection at the barrier another wave $-A_l e^{-iK(r-r_0)} \times T_b e^{-ikr_0}(-R_b' e^{2iKr_0})$. Adding successive waves reflected at $r = 0$ and $r = r_0$, one obtains for the wave function in $0 < r < r_0$

$$(e^{-iKr} - e^{+iKr}) A_l T_b e^{ir_0(K-k)}[1 + (+R_b' e^{2iKr_0}) + (+R_b' e^{2iKr_0})^2 + \cdots]$$

$$= A_l \cdot I \sin Kr, \tag{23a}$$

where

$$I = -\left[\frac{2iT_b e^{ir_0(K-k)}}{1 - R_b' e^{2iKr_0}}\right]. \tag{23b}$$

This amplitude can become very large if all the multiply reflected components are in phase, that is if $R_b' e^{2iKr_0}$ is real and positive. This condition is called "resonance." The energy at which a resonance occurs is determined by the optical path $2Kr_0$ for the round trip inside the well. At a maximum of the internal amplitude as a function of energy, one has $(1 - R_b' e^{2iKr_0}) = 1 - |R_b'|$. Because of the conservation of probability flux at a reflection, $K|R_b'|^2 + k|T_b'|^2 = K$, so that $1 - |R_b'|^2 = (k/K)|T_b'|^2$. If the barrier is a good reflector, then $|R_b'| \cong 1$, so that $1 - |R_b'|^2 \cong 2(1 - |R_b'|)$.

At a maximum with respect to energy, the amplitude I becomes

$$|I| = \frac{2|T_b|}{1 - |R_b'|} \cong \frac{2|T_b|}{\frac{1}{2}(1 - |R_b'|^2)} = \frac{4|T_b|}{(k/K)|T_b'|^2} = \frac{4}{|T_b'|}. \tag{24}$$

The last step comes from the relation $T_b' = KT_b/k$. Thus, the stronger the barrier, the larger the enhancement of the amplitude inside the well.

Figure 4 shows R_l near the potential well for a range of energies straddling a resonance.

The amplitude S_l of the outgoing wave in Eq. (20) comes from the wave reflected outwards when the incoming wave hits the barrier, plus the waves transmitted through the barrier from the inside, that is from the part $e^{iKr}/2i$ of $\sin Kr$. With the aid of Eq. (23b), one finds

$$S_l(k) = R_b e^{-2ikr_0} + \frac{T_b T_b' e^{2ir_0(K-k)}}{1 - R_b' e^{2iKr_0}}. \tag{25}$$

2.4. Singularities in Potential-Well Scattering

The enhancement of the wave function shown in Figs. 3 and 4 is connected with singularities of the outgoing wave as a function of energy. We shall need these singularities in the next section.

If $l = 0$, so that there is no centrifugal barrier, then the asymptotic form of the wave function is $R_0(r) = (i/2k)[e^{-ikr} - S_0 e^{+ikr}]$, the coefficient of the term e^{-ikr} being determined by the incident plane wave. In terms of the phase shift,[17] one may write $S_0 = e^{2i\delta} = (\cos\delta + i\sin\delta)/(\cos\delta - i\sin\delta)$. Multiplying numerator and denominator by $k/\sin\delta$, and using the expansion[20] $k\cot\delta = -1/a + O(k^2)$, one finds that

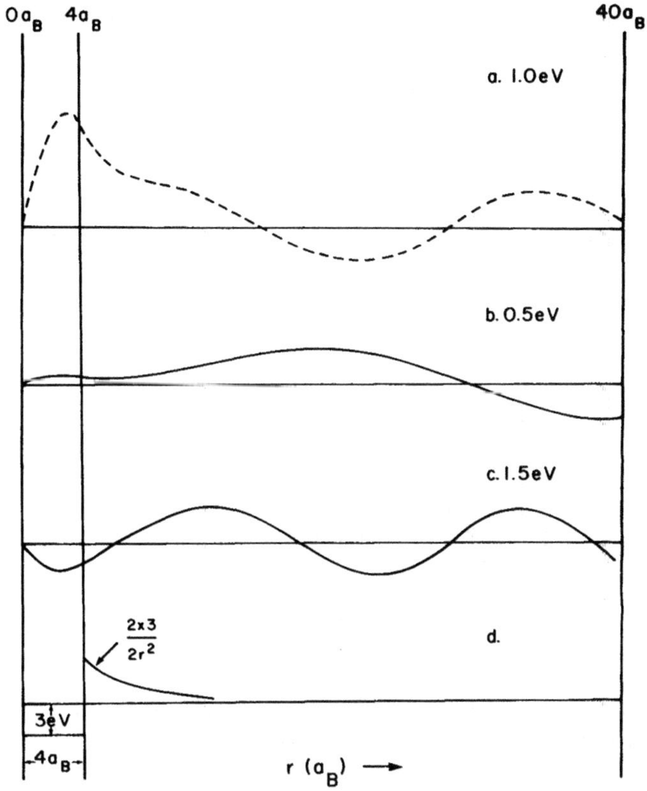

FIGURE 4
Radial factor R [in the sense of Eq. (16)] for an electron near the potential shown in Fig. 4d. (The repulsive barrier corresponds to angular momentum $2\hbar$.) For the well parameters shown, there is a resonance at energy 1.0 eV (= 0.0368 a.u.). The amplitudes outside the barrier are chosen to be the same in Figs. 4a, 4b, 4c, which correspond to energies 1.0, 0.5, 1.5 eV. Note the enhancement of the internal amplitude in $r < 4a_B$ at 1.0 eV, and the diminution at the same distances at 0.5 and 1.5 eV.

$$S_0 = \frac{-1/a + O(k^2) + ik}{-1/a + O(k^2) - ik}.$$

Therefore when a is so large that the $O(k^2)$ term may be neglected near $|k| = a^{-1}$, S_0 has a pole at $k = i/a$, on the imaginary axis close to the origin of the k plane. For the weakly bound state, where $a > 0$, the pole is on the positive imaginary axis. If the potential is just not strong enough to possess a bound state, then $a < 0$ and the pole is on the negative imaginary axis; this pole *is* the "virtual state." As we have already seen in Section 2.2, the pole forces the wave function to have a very small logarithmic derivative just outside the potential well, which combines with the factor $1/r$ in the three-dimensional wave function to produce the enhancement shown in Fig. 3b.

If $l > 0$, so that there is a centrifugal potential barrier, then both I [Eq. (23b)] and S_l [Eq. (25)] will have a pole at a resonance energy where

$$1 - R'_b e^{2iKr_0} = 0. \tag{26}$$

Such a resonance energy will be denoted by W_n, and the associated value of K by K_n, such that

$$W_n = \tfrac{1}{2}K_n^2 - U; \tag{27}$$

U is the depth of the effective potential inside the well introduced in Section 2.3. The subscript n distinguishes the roots from one another. Since $|R'_b| < 1$ for real energies (because the amplitude reflected from a barrier cannot exceed the amplitude of the incident wave), Eq. (26) cannot have roots for K corresponding to real energies. Following convention, we write

$$W_n = E_n - \frac{i}{2}\Gamma_n, \tag{28}$$

where E_n and Γ_n are both real. To find the complex roots of Eq. (26), write $R'_b = |R'_b| \exp(i\alpha)$, so that Eq. (26) becomes $|R'_b| e^{i(2K_n r_0 + \alpha)} = e^{i \cdot 2n\pi}$ with n an integer which denotes the number of wavelengths in the optical path for a double passage across the potential well, taking into account the phase change α at reflection. $|R'_b|$ and α will be regarded as constants over the energy important at a resonance. Thus $i(2K_n r_0 + \alpha - 2n\pi) = -\ln|R'_b| = -\ln[1-(1-|R'_b|)] \cong 1 - |R'_b|$, if the barrier is a good reflector so that $|1-|R'_b|| \ll 1$. Taking real parts, one has $-2\,\mathrm{Im}(K_n r_0) = 1 - |R'_b|$. It follows from Eqs. (27) and (28) that $\Gamma_n = -\mathrm{Im}(K_n^2) = -2\,\mathrm{Re}\,K_n \cdot \mathrm{Im}\,K_n = \mathrm{Re}\,K_n \cdot (1 - |R'_b|)/r_0$. Because $|1 - |R'_b|| \ll 1$, one may replace $(1 - |R'_b|)$ by $(1 - |R'_b|)(1 + |R'_b|)/2 = (1 - |R'_b|^2)/2$, so that finally

$$\Gamma_n = \frac{\mathrm{Re}\,K_n}{2r_0}(1 - |R'_b|^2). \tag{29}$$

Equation (29) has a simple interpretation. Since $\mathrm{Re}\,K_n$ is the velocity of the electron inside the potential well, $\mathrm{Re}\,K_n/(2r_0)$ is the frequency of collisions with the barrier, while the factor $(1 - |R'_b|^2)$ is the probability that the electron should be transmitted through the barrier at each collision. Thus Γ_n is the rate at which the particle leaks from the well through the barrier if it has been trapped inside.

It will be convenient to expand the denominators $(1 - R'_b e^{2iKr_0})$ in Eqs. (23b) and (25) about their zeros at $E = W_n$, so that

$$(1 - R'_b e^{2iKr_0}) \longrightarrow (E - W_n)\left[\frac{d}{dE}(-R'_b e^{2iKr_0})\right]_{E=W_n}.$$

If we continue to regard R'_b as constant in the neighborhood of the resonance, then

$$\frac{d}{dE}(-R'_b e^{2iKr_0}) = -2ir_0 \frac{dK}{dE}(R'_b e^{2iKr_0}).$$

Setting $R'_b e^{2iKr_0} = 1$ at $E = W_n$, and putting $dK/dE = 1/K$, one has $(1 - R'_b e^{2iKr_0}) \rightarrow (E - W_n)[-2ir_0/K]_{E=W_n}$, so that in Eqs. (23)

$$I \longrightarrow \frac{kT'_b e^{ir_0(K-k)}}{r_0(E - W_n)}; \tag{30}$$

we have used the relation $T'_b = KT_b/k$. Similarly, Eq. (25) may be rewritten

$$S_l(k) \cong R_b e^{-2ikr_0} - \frac{kT'^2_b}{2ir_0} \frac{e^{2ir_0(K-k)}}{E - W_n}. \tag{31}$$

2.5. Time Delay

The enhancement of the wave function inside the well enclosed by the barrier for $l > 0$, illustrated in Fig. 4, suggests that a particle can become trapped at a resonance. This phenomenon is important in vibrational excitation of molecules by electrons because it lengthens the time during which a temporarily trapped electron can push on the nuclei with its electric field.

Consider a wave packet,

$$\psi(z, t) = \int_0^\infty dE\, a(E) e^{i(kz - Et)}, \tag{32}$$

where z is the distance of the electron along the direction of propagation. The packet consists of plane waves of momentum k traveling in the positive z direction, superposed in such a way that ψ interacts with a potential well for only a short time. ψ satisfies the time-dependent Schrödinger equation for a free electron provided that k and E are related by

$$k(E) = +(2E)^{1/2}; \tag{33}$$

the positive square root must be taken to ensure that the electrons travel in the $+z$ direction, as they do from a real electron gun.

In a wave packet corresponding to electrons from a real gun, the coefficients $a(E)$ in Eq. (32) superpose waves in a range of energies of width γ, say, near a central energy E_0. A convenient form for analytic discussion is

$$a(E) = \frac{\gamma}{(E-E_0)^2 + \gamma^2} \tag{34}$$

with $\gamma > 0$. Appendix D shows that Eq. (32) becomes approximately

$$\psi(z, t) \cong \pi e^{-\gamma|z/v_g - t|} e^{i[k(E_0)z - E_0 t]}, \tag{35}$$

where

$$v_g = \left(\frac{dE}{dk}\right)_{k=k_0} \tag{36}$$

is the group velocity. According to Eq. (35), the width of the wave packet in the z direction is of order $2v_g/\gamma$, while the wave packet as a whole moves at speed v_g and passes over the origin ($z = 0$) in a time of order $1/\gamma$.

If there is a resonance within the band of energies straddled by the wave packet, there may be an important build-up of the wave function within the well, as is described by Eqs. (23a) and (23b). If the resonance is associated with a particular angular momentum $l(>0)$, the resonant amplitude $A_l I \sin Kr$ in Eqs. (23a) and (23b) arises from a component of e^{ikz}, in Eq. (32), given by the first term on the right in Eq. (19). According to Eq. (20), the corresponding incoming wave is asymptotically $A_l Y_{l0} e^{-ikr}/r$, with A_l as defined after Eq. (20); this corresponds to $1/r$ times the term e^{-ikr} in Eq. (20). If one superposes contributions from different E according to Eq. (32), the resonant amplitude inside the well becomes

$$Y_{l0} \cdot A_l \int_0^\infty dE\, a(E) I(E) \sin Kr\, e^{-iEt}.$$

According to Eq. (30), the integrand may be written

$$\left[\frac{a(E)e^{-iEt}}{E - W_n}\right] \cdot \left[\frac{kT_b' e^{ir_0(K-k)} \sin Kr}{r_0}\right].$$

The second factor $[\cdots]$ varies little over the wave packet, and may therefore be regarded as a constant. This is because $Kr_0 \approx \pi$ at the center of the wave packet if the wave function inside the well contains about half a cycle, while the range ΔK of values satisfies $(\Delta K/K) = (\Delta k/k)(k^2/K^2)$, where Δk and k are, respectively, the range of momenta and the central momentum of the incident wave packet. Therefore if the wave packet is somewhat larger than a wavelength (i.e., $\Delta k/k \lesssim 1$), and has a low energy (so that $k^2/K^2 \ll 1$), then one has $\Delta K/K \ll 1$. Therefore the time dependence of the amplitude inside the well is determined by the integral

$$\int_0^\infty dE\, \frac{a(E)e^{-iEt}}{E - W_n}. \tag{37}$$

The integrand has poles at $E = E_0 \pm i\gamma$ from the incident wave packet, and a pole at $E = W_n$ in the lower half plane from the resonance. The integral may be evaluated by deforming the contour of integration. For $t < 0$ (Fig. 5a), one obtains a contribution from the pole at $E_0 + i\gamma$, proportional to $e^{-\gamma|t|} \times e^{-iE_0 t}$, and a contribution from the positive imaginary axis. The latter comes from the neighborhood of $E = 0$, and is thus determined by the tail of near-zero-energy electrons produced by the idiosyncrasies of the electron gun; this contribution is not characteristic of resonance scattering and should be ignored. For $t > 0$, one deforms the contour as in Fig. 5b, and obtains contributions from the two poles at $E_0 - i\gamma$ and W_n, proportional to $e^{-\gamma|t|}e^{-iE_0 t}$ and $e^{-(\Gamma_n/2)t}e^{-i\mathrm{Re}W_n t}$, respectively. (There is also a contribution from the negative imaginary axis which should again be ignored because it is determined by the near-zero-energy electrons from the gun.)

The two contributions proportional to $e^{-\gamma|t|}$ last as long as the contact with the incident wave packet. If one chooses $\gamma \gg \Gamma_n/2$, then the time of order $1/\Gamma_n$, for which the contribution $e^{-\Gamma_n t/2}$ from the resonance lasts, is much longer than the duration $1/\gamma$ of the contact with the wave packet. There is a significant time delay at the resonance.

So far we have considered a shape resonance with $l > 0$. If $l = 0$, there is the possibility of scattering without a barrier in the presence of a weakly bound state or of a virtual state, where the scattered wave outside the potential has the form $(-i/2k)(S_0 - 1)e^{+ikr} = -a/(1 + ika)e^{+ikr}$, with the expression $S_0 = (1 - ika)/(1 + ika)$ discussed in Section 2.4. If the energy is so low that $kr \ll 1$ just outside the potential well, the time dependence of the scattered wave just outside the well is given by the integral

$$\int_0^\infty dE\, a(E)e^{-iEt}\left(\frac{-a}{1 + ika}\right). \tag{38}$$

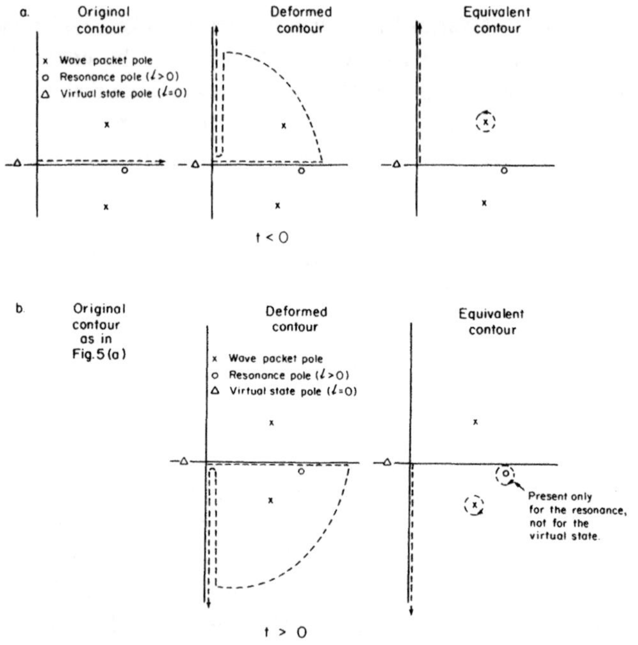

FIGURE 5
The integrations in Eqs. (37) and (38) in the complex E plane. (a) $t < 0$; (b) $t > 0$.

The pole at $(ika + 1) = 0$ occurs at $E = \frac{1}{2}k^2 = -1/2a^2$, on the negative real E axis associated with either the positive or negative imaginary k axis, depending on whether $a > 0$ or $a < 0$. Thus if one evaluates the integral for $t < 0$ by deforming the contour as in Fig. 5a, or for $t > 0$ as in Fig. 5b, one obtains contributions only from the poles at $E_0 \pm i\gamma$ from $a(E)$ in the incident wave packet. Unlike the case of $l > 0$, there is no contribution from the pole in the scattering amplitude. Thus there is no time delay associated with either the weakly bound S state or the virtual state.

The only property of a resonance on which the time delay depends is the pole at $E = W_n$. Thus the time delay is not restricted to resonances in a potential well enclosed by a barrier, as in the examples of Sections 2.3 and 2.4, but would occur for any pole in the scattering amplitude just below the real E axis.

2.6. *General Resonances*

The trapping of an electron at a resonance is so important for vibrational excitation that one has to extend the theory beyond the simple potential-well model. In this section we regard the nuclei in a molecule as fixed, and set up a formalism sufficiently general to treat the dynamics of many interacting electrons. The extension to moving nuclei is given in Sections 3, 4, and 5.

Figure 4a shows a resonance, that is a state which is "almost bound" within a barrier, in the sense that the amplitude outside the barrier is very small compared with the amplitude inside. It represents a wave function which leaks out slowly through the barrier, while an incident wave compensates for the outgoing probability flux. If one enclosed the potential well in a large spherical reflector, the wave leaking out would be reflected back, and there would be a stationary state with a distinctive wave function: a strong central core with weak tails extending to the mirror. We shall call such a state a "boxed-in resonance." It will be denoted by b and normalized so that $\langle b|b\rangle = 1$.

Although we introduced this picture of a resonance with a potential barrier, the picture would work with any other trapping mechanism which gives rise to an almost-bound state. The mechanism could just as well be binding of a particle to an excited state of the target, in such a way that there is only a weak coupling to lead to autoionization ("Feshbach mechanism,"[21] of which many examples are discussed in Chapter 5, Section 2). Or it could be any other trapping mechanism still to be discovered. Whatever the mechanism, a large spherical reflector would yield a stationary state consisting of a strong central core with weak tails extending out to the reflector.

The radius of the stabilizing mirror has to be large enough not to interfere with the geometric structure of the inner region where the amplitude is large. However, the mirror should be no larger than is necessary to satisfy this criterion, because when the Hamiltonian has added to it a potential representing the mirror, there will be many other eigenstates in addition to b, and it is convenient to keep the spacing of their energy levels as large as possible by keeping the enclosing box small.

For the purpose of illustration of the general formalism, we go back to the s wave of a single particle at $\mathbf{r}$ in a spherically symmetric potential $V(r)$. The potential is shown in Fig. 6a, where $V(r) = -U(0 < r < \rho_1)$, $V(r) = +B(\rho_1 < r < \rho_2)$, and $V(r) = 0(r > \rho_2)$; U is the well depth, and B the barrier height, both positive. The reflecting mirror is introduced by adding to the potential a term

$$v = V_0\theta(r - \rho), \tag{39a}$$

where ρ is the radius of the mirror, and V_0 a large positive constant with the dimension of energy. θ is defined as $\theta(x) = 1$ for $x > 0$, $\theta(x) = 0$ for $x < 0$. The complete Hamiltonian is now $H = -\frac{1}{2}\partial^2/\partial r^2 + V + v$. If $V_0 \to \infty$, the effect of v is to make the wave function vanish at $r = \rho$.

The eigenvalues of H are plotted in Fig. 6d as ρ varies from 7 to 35 atomic units. It is seen that, for the parameter values given in the caption, there is an

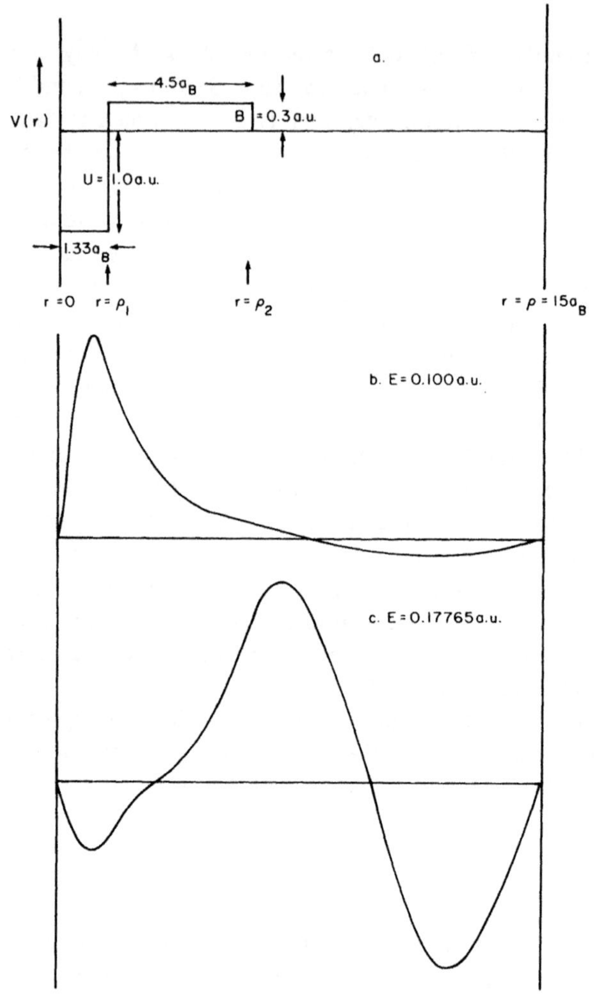
a.
V(r)
$4.5a_B$
B = 0.3 a.u.
U = 1.0 a.u.
$1.33a_B$
r = 0
$r = \rho_1$
$r = \rho_2$
$r = \rho = 15a_B$
b. E = 0.100 a.u.
c. E = 0.17765 a.u.

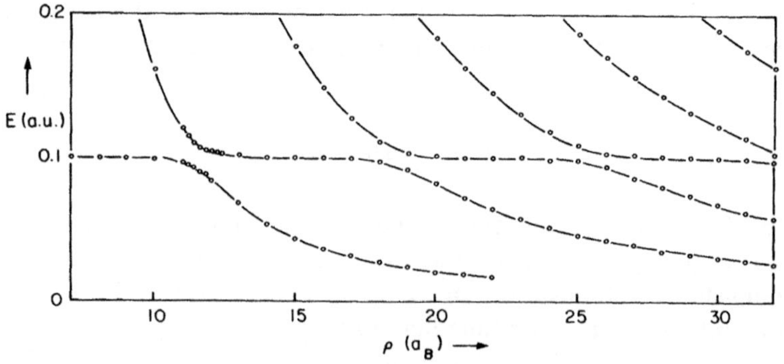
0.2
E (a.u.)
0.1
0
10
15
20
25
30
ρ (a_B)

eigenvalue close to 0.1 a.u. at all values of ρ except near points where the horizontal $E = 0.1$ comes close to intersecting an eigenvalue curve which varies with ρ. The wave functions belonging to the stable and to one of the variable eigenvalues are shown in Figs. 6b, 6c. Evidently the stable eigenvalue belongs to a boxed-in resonance, while the variable eigenvalues represent states in which the particle is mainly outside the potential well and barrier. The stability of the stable root against changes of ρ comes from the small magnitude of the tail of the wave function which reaches the mirror, in comparison with the large amplitude inside the well.

The notion of the boxed-in resonance can be generalized for a many-electron system without difficulty. The mirror potential then has to act on all the electrons, so that in place of Eq. (39a), one has

$$v = V_0 \sum_{i=1}^{N} \theta(r_i - \rho), \tag{39b}$$

where r_i is the distance of the ith electron from the mass-center of the molecule, and V_0 has the same meaning as before. The energy of a resonance can be calculated by searching for eigenvalues of the Hamiltonian with the added term v which are stable against changes of the mirror radius. (The idea of using stability of eigenvalues to distinguish resonant and nonresonant states was introduced by Taylor.[22] The method has usually been used not with an enclosing mirror, but with a basis set of functions which extend to progressively larger distances.) A boxed-in-resonance of a many-electron system satisfies

$$(H_e + v - E_0)b = 0. \tag{40}$$

$H_e = H_e(\mathbf{r}; \mathbf{R})$ is the Hamiltonian of all the electrons; their coordinates are denoted by $\mathbf{r}$; $\mathbf{R}$ stands for the coordinates of the nuclei, which for the moment are being regarded as fixed. The eigenvalue E_0 varies with $\mathbf{R}$.

FIGURE 6

The calculation of a "boxed-in resonance" for a one-electron model. (a) The potential. In this diagram, the enclosing box has a radius $\rho = 15a_B$. (b) A stationary eigenstate of the potential in (a) at energy 0.100 a.u. This state is concentrated inside the well in $r < 1.33a_B$, and corresponds to a quasistationary state in the absence of the large box. (c) A stationary state in the potential in (a), at energy 0.17765 a.u. This state is mainly outside the well in $r < 1.33a_B$, and corresponds to scattering without trapping in the absence of the large box. (d) The variation of the lowest few eigenvalues of a one-electron Hamiltonian with the potential in $0 < r < (1.33 + 4.5)a_B$ when the radius ρ of the large box is varied. The states shown in Figs. 6b and 6c may be found at $\rho = 15a_B$. Note that there is an eigenvalue of $0.100a_B$ at almost all box radii; it corresponds to a state of the type shown in Fig. 6b, with a peak inside the well, and a weak tail extending to the enclosing box.

If a target, and an electron shot at it, have a total energy E close to E_0, it is convenient to write the wave function Ψ_i

$$\Psi_i = \xi_i b + \Psi_i^P, \tag{41}$$

where $\xi_i = \xi_i(\mathbf{R}; E)$, and where Ψ_i^P is orthogonal to b:

$$\langle b | \Psi_i^P \rangle = 0. \tag{42}$$

The suffix i stands for all the quantum numbers of the initial state. The details of the construction (41) are described in Appendix A. In the example in Section 2.3, the term $\xi_i b$ would represent the large amplitude inside the well described by Eq. (23) close to a resonance.

Ψ_i^P may be separated into two parts

$$\Psi_i^P = \xi_i b^{(+)} + \Phi_i^P, \tag{43}$$

where the term $\xi_i b^{(+)}$ describes the resonant part of the outgoing wave [like the second term on the right in Eq. (25)], and Φ_i^P the nonresonant part [like the first term on the right in Eq. (25)]. $b^{(+)}$ [see Eq. (A.7b)] is a tail of outgoing waves which arise from b when the confining mirror v is removed. Φ_i^P [see Eq. (A.7a)] is sometimes called the "potential-scattering" part of the wave function; in the example of Section 2.3, it contains the incoming wave and the part of the outgoing wave which comes from reflection of the incoming wave at the outside of the barrier. Both $b^{(+)}$ and Φ_i^P are orthogonal to b:

$$\langle b | b^{(+)} \rangle = \langle b | \Phi_i^P \rangle = 0 \tag{44}$$

[see Eq. (A.9)]. If one denotes the sum of b and $b^{(+)}$ by

$$S = b + b^{(+)}, \tag{45}$$

then Eq. (41) becomes

$$\Psi_i = \Phi_i^P + \xi_i S. \tag{46}$$

$S[\equiv S(\mathbf{r}; \mathbf{R}; E)]$ contains the nuclear configuration $\mathbf{R}$ as parameter. Appendix A shows that [Eq. (A. 12)]

$$\xi_i = \frac{\langle b | H_{el} | \Phi_i^P \rangle}{E - \hat{W}}, \tag{47}$$

where $\hat{W}[\equiv \hat{W}(\mathbf{R};E)]$ is defined as [Eq. (A. 13)]

$$\hat{W} \equiv \langle b|H_e|S\rangle. \tag{48}$$

With Eqs. (45) and (40), one finds from Eq. (48)

$$\hat{W} = E_0 + \Delta - \frac{i}{2}\Gamma, \tag{49a}$$

where Δ and Γ are real and defined by

$$\Delta - \frac{i}{2}\Gamma \equiv -\langle b|v|b\rangle + \langle b|H_e|b^{(+)}\rangle. \tag{49b}$$

The term $-(i/2)\Gamma$ in Eq. (49a) is a consequence of the fact that $b^{(+)}$ has to be complex to be an outgoing wave.

The terms (49b) are small corrections to $\hat{W}$. The term $\langle b|v|b\rangle$ vanishes in the limit $V_0 \to \infty$, because both the magnitude of b at the mirror surface and the distance b extends into the mirror turn out to vary as $V_0^{-1/2}$; therefore with another factor V_0 from the integrand, one finds $\langle b|v|b\rangle \propto V_0^{-1/2} \to 0$ as $V_0 \to \infty$. The term $\langle b|H_e|b^{(+)}\rangle$ will be much smaller than $\langle b|H_e|b\rangle$ if $|b^{(+)}| \ll |b|$ in the volume occupied by b; this condition has to be satisfied if the leak through the trapping mechanism which gives rise to $b^{(+)}$ is sufficiently small for the physical picture of a resonance to make sense. Since the energy E appears in $\hat{W}$ only in the small terms $\Delta - (i/2)\Gamma$, it follows that $\hat{W}$ varies little with E.

If $\hat{W}$ varies only little with E, the expression (47) for ξ_i must have a pole at $E = W$, where W is the root of the equation

$$W = \hat{W}(\mathbf{R};\, W). \tag{50}$$

A physical interpretation of W is derived in Eq. (A.18); W turns out to be the eigenvalue in the equation

$$(H_e - W)\tilde{S}(\mathbf{r};\mathbf{R}) = 0 \tag{51a}$$

where

$$\tilde{S}(\mathbf{r};\mathbf{R}) \equiv S(\mathbf{r};\,\mathbf{R};\,W). \tag{51b}$$

From the definition of S in Eq. (45) and Eqs. (51), it follows that W is the complex energy of an eigenfunction of H_e which behaves as an outgoing wave when any particle moves off to infinity. Such energies were introduced by

Siegert.[23] This interpretation of W shows that the pole in the wave function is independent of the reflecting mirror one introduced to define b.

The imaginary part $-\Gamma/2$ of W is shown in Appendix A to be negative, so that the positive sign of Γ_n for the potential-well resonance in Eq. (29) was a particular case of a more general result true for any kind of trapping mechanism. It follows that the time delay at a resonance derived in Section 2.5, which depended only on the existence of a pole in the scattering amplitude below the real E axis, exists independently of the kind of trapping mechanism.

3. Vibrational Excitation. I: Wave-Function Enhancement without Trapping

In this section, we discuss the process of Section 2.2, which does not depend on trapping. Here there is an enhancement of the wave function at low energy, but no time delay (according to Section 2.5). Scattering with a time delay and with trapping is discussed in Sections 4 and 5.

The mechanism of the virtual state or very weakly bound state of Section 2.2 may be illustrated with the following schematic model. A spherical breathing oscillator has a radius R which oscillates about an equilibrium value r_0. The Hamiltonian of the oscillator is

$$H_{\text{osc}} = \frac{P^2}{2M} + \frac{M\omega^2}{2}(R - r_0)^2, \tag{52}$$

where P is the momentum conjugate to $(R - r_0)$, M is the effective mass, and ω the natural angular frequency. For the sake of illustration, we use values corresponding to the HCl molecule: $M = 0.97 M_p$, where M_p is the mass of the proton, and $\omega = 0.37$ eV. For r_0, take $2a_B$. The vibrational amplitude in the ground state becomes $\Delta' \equiv |\overline{(R - r_0)^2}|^{1/2} = 0.14 a_B$.

An extra electron at $\mathbf{r}$ sees a potential well

$$V(r; R) = -U\theta(R - r), \tag{53}$$

where $\theta(x) = 1$ if $x > 0$ and $\theta(x) = 0$ if $x < 0$; U is an energy close to $(\pi^2/8r_0^2)$, which corresponds to the appearance of a bound s state. Because $\Delta' \ll r_0$, we use the approximation linear in $(R - r_0)$,

$$V(r; R) = -U\theta(r_0 - r) - (R - r_0)U\delta(r_0 - r), \tag{54}$$

remembering that $[d\theta(R - r)/dR]_{R=r_0} = \delta(r_0 - r)$.

The complete Hamiltonian is

$$H = H_{osc} + V(r;R) + K, \tag{55}$$

where K is the kinetic energy of the electron. The Schrödinger equation $(H-E)\Psi = 0$ may be solved with the expansion

$$\Psi = \sum_v \frac{\psi_v(r)}{r} X_v(R), \tag{56}$$

where $[H_{osc} - (v+\frac{1}{2})\omega]\, X_v(R) = 0$, and where $\psi_v(r) = \delta_{v0} \sin(k_0 r)/k_0 + T_v e^{ik_v r}$ when $r \to \infty$, with $\frac{1}{2}k_v^2 = E - (v+\frac{1}{2})\omega$. If $E < (v+\frac{1}{2})\omega$, that root for k_v has to be chosen which yields a ψ_v decreasing as $r \to \infty$. Only the component of Ψ with angular momentum zero is to be taken into account.

If one retains only the terms $v = 0$ and $v = 1$, one finds, if $k_0 r_0 \ll 1$,

$$\sigma(v = 0 \to v = 1) = 4\pi \frac{k_1}{k_0} \left| \frac{2U\Delta'(1 + 2K_0 r_0 \cot K_0 r_0)}{W_0 W_1 - (2U\Delta')^2} \right|^2, \tag{57}$$

where $\frac{1}{2}K_v^2 = \frac{1}{2}k_v^2 + U$, and $W_v = ik_v - K_v \cot(K_v r_0)$. The quantities $K_0 \cot(K_0 r_0)$ and $K_1 \cot(K_1 r_0)$ are the logarithmic derivatives of ψ_0 and ψ_1 just inside r_0. They are generally of order a_B^{-1}, which leads to a very small value of $\sigma(v=0 \to v=1)$, of order 10^{-17}–10^{-18} cm^2. Values of the order 10^{-15} cm^2 occur near threshold if the logarithmic derivatives are $\ll a_B^{-1}$, i.e., if there is either a virtual state or a very weakly bound state in the potential well.

The energy dependence calculated from Eq. (57) is shown in Fig. 7 for $K_0 r_0 = \pi/2$, when a new bound state is about to be born. Both in magnitude and shape, the calculated threshold peak looks very much like those observed in HCl and SF_6(see Figs. 1h and 1i).

Calculations more realistic than the schematic model have been done on vibrational excitation near threshold in CO_2[(24)] and HCl.[(25)] Both calculations use models which are to some extent fitted to experiment. The calculation[(24)] on $e + CO_2$ was done by Morrison and Lane, who used a fixed axis model in the impulse approximation for the symmetric-stretch vibration (see Appendix C). The electron-scattering amplitude for fixed nuclei, i.e., T^F in Eq. (C.9a), was calculated with a potential obtained from *ab initio* Hartree–Fock wave functions, with the addition of an exchange potential from a free-electron-gas approximation (Chapter 1, Section 3.2 and Chapter 7, Section 3.4), and a polarization potential cutoff at a short distance from the molecule (Chapter 1, Section 3.1 and Chapter 7, Section 3.5), selected so as to reproduce a strong observed shape resonance[(26)] at 3.8 eV. The amplitude for scattering with the excitation of one quantum of vibration in the symmetric stretch mode was obtained from Eq. (C.9a) with harmonic oscillator wave functions for the initial

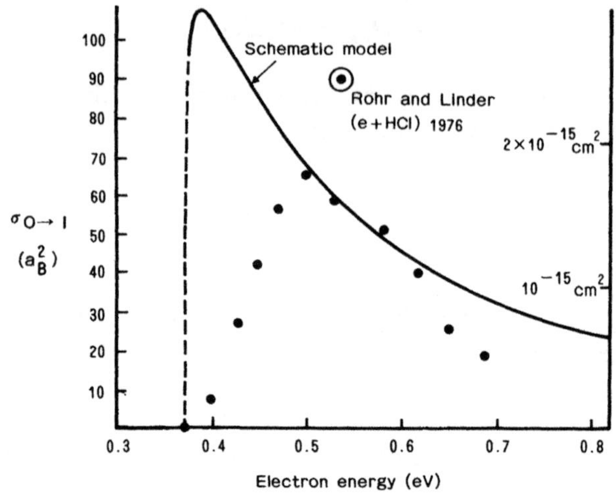

FIGURE 7
Vibrational excitation near threshold of the schematic model discussed in Section 3; on the assumption that the well depth and radius are such that a new bound *s* state is on the verge of being born. The experimental points obtained by Rohr and Linder on HCl (Ref. 82) are shown for comparison.

and final nuclear state. If one denotes this amplitude by $T^{\mathrm{imp}}_{v'=1\leftarrow v=0}$, one obtains the differential cross section from Eq. (1), by multiplying $|T_{v'\leftarrow v}|^2$ by the ratio $(v_{v'}/v_v)$ of the final to the initial velocity of the electron. The result is shown in Fig. 8, as the dotted curve; the drop to zero at threshold is due to the factor $v_{v'=1}/v_{v=0}$. The full curve shows the result if one sets $(v_1/v_0)=1$. In these calculations, the cutoff in the polarization potential was tuned separately at each internuclear separation to place the lowest shape resonance at the energy obtained in a semiempirical "boomerang" calculation. (This term is explained in Section 5.) If one fixes the polarization cutoff at the value required by the equilibrium separation in the CO_2 ground state, one obtains the dashed curve. Although 12 (or more) partial waves belonging to different angular momenta were used in the calculation of the fixed-nuclei amplitude T^F in Eq. (C.9a), only the components of symmetry σ_g were found to contribute significantly to vibrational excitation near threshold.

Vibrational excitation in HCl within 0.5 eV of threshold was treated[25] by Dubé and me, also with a fixed-axis model. Outside the core of the molecule, more than $2a_B$ from the mass center, we expanded the wave function in solutions of the one-electron Schrödinger equation in the dipole potential from the molecule. One finds that at each of the low energies in question for the incoming and scattered electrons, a single such solution permits the electrons to reach the molecule: a solution of symmetry σ without angular nodes. The dipole moment of HCl is so small that this solution turns out to be an almost perfect *s* wave.

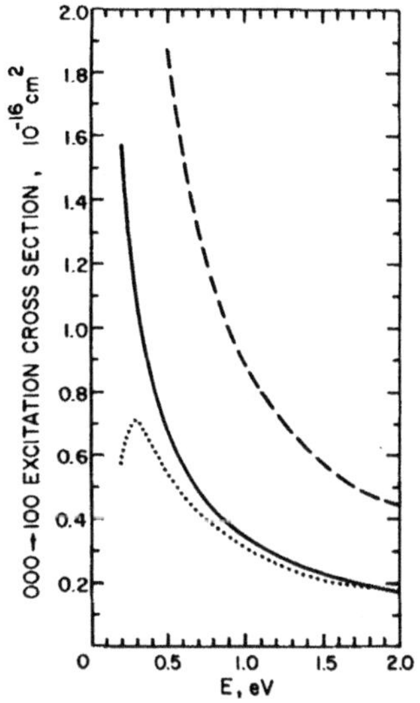

FIGURE 8
$e + CO_2$; calculated cross sections for excitation of the vibrational symmetric stretch mode. Curve · · ·, polarization cut off adjusted separately for each internuclear distance; curve ———, with the factor (final velocity/initial velocity) replaced by unity: curve – – –, polarization cutoff fixed at the radius corresponding to the equilibrium-separation of the nuclei. (From Ref. 24.)

The boundary condition at the molecular surface, at a distance of $2a_B$ from the mass-center, is fixed by the remark that inside the molecule, the potential is so strong that the electron is fast compared with the nuclei, so that the electronic wave function follows the nuclei adiabatically. Therefore the logarithmic derivative at the molecular surface is determined by the nuclear separation; moreover, since the nuclear vibrational amplitudes are small compared with a_B, one may expand the logarithmic derivative in the form $f(R) = f_0 + f_1 Q$, where f_0 and f_1 are constants, and Q is the displacement of the separation from equilibrium. One expands the wave function outside the molecule in the form $\sum_v X_v(Q)\,(\psi_{0v}(r)/r)$, where X_v is the vth vibrational wave function and ψ_{0v} the electronic wave function corresponding to the dipole potential at the energy corresponding to the vth vibrational channel. ψ_{0v} can be expressed as a superposition of ingoing and outgoing waves, whose amplitudes can be determined in terms of f_0 and f_1 by fitting the logarithmic derivative at the molecular surface. The value turns out to be between values one would expect from a structure ClH and a structure Cl^-H^+. The calculated and experimental cross sections are shown in Fig. 9.

The calculations on threshold excitation of CO_2 and HCl evidently have much in common with one another and with the schematic spherical breathing model. All three calculations produce threshold spikes in the cross section for vibrational excitation (Figs. 7, 8, 9), of similar shape and order of magnitude; the result obtained by Morrison and Lane on CO_2 (dotted curve in Fig. 8) has a peak about an order of magnitude below the other two for $v = 0 \rightarrow v = 1$ but it appears from the dashed curve in Fig. 8 that a minor change in the polarization potential would raise the $e + CO_2$ cross section appreciably. In each of the three

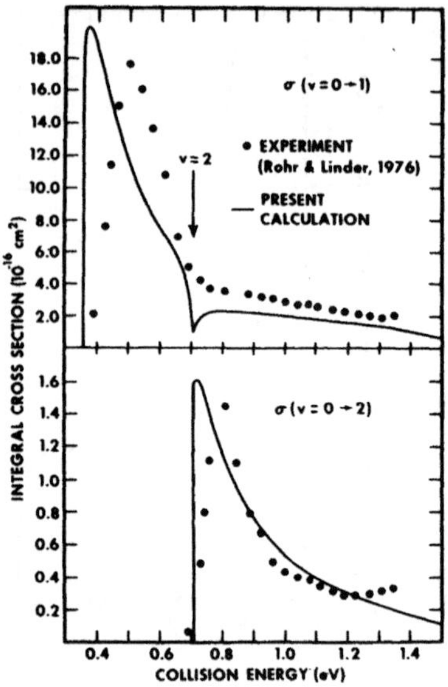

FIGURE 9
e + HCl; calculated (——) and experimental (· · ·) vibrational excitation near threshold. The model used for the calculation had two adjustable parameters. (From Ref. 25.)

calculations, the wave function of the extra electron has a strong s-wave component outside the molecule. In the case of e + HCl, we verified that the electronic wave function outside the molecule falls off as $1/r$ near threshold, in the way characteristic of either a barely bound state or a virtual state. Thus all three calculations are consistent with a model in which there is either a barely bound state or a virtual state, with a strong s-wave component.

An attempt at an *ab initio* calculation of the electronic states of HCl^- has been made by Taylor, Goldstein, and Segal.[27] They use the "stabilization method," where one calculates the eigenvalues of the Hamiltonian matrix with a basis set of wave functions of finite spatial extent, and picks out those eigenvalues which remain stable where the basis set is enlarged. Among other stable roots, they find two of Σ^+ symmetry which appear to follow the ground electronic state of HCl, at binding energy zero for the extra electron at nuclear separations between 0.7 and 1.5 Å. The calculation does not tell one whether these states are to be interpreted as trapped compound states, or as virtual states. Nesbet[28] has suggested that the calculation produces *two* states of zero energy for the extra electron because that is the closest the approximation can get to reality, where the single pole in the scattering amplitude discussed in Section 2.4 for the case $l = 0$ passes from the upper half of the imaginary k axis when $a > 0$ down to the lower half of the binding is weakened so that $a < 0$; numerical experiments[29] show that an $l = 0$ pole moving down the imaginary axis eventually meets another moving up. Nesbet's suggestion is that the two zero-energy states for the extra electron found by Taylor *et al.* are to be associated with two such poles. Nesbet points out that similar threshold peaks have been obtained in electronic excitation in e + He scattering.[30]

4. Vibrational Excitation. II: Electron Scattering at a Shape Resonance; Impulse Limit

If there is a potential barrier around a molecular target, an electron may be trapped temporarily in a resonance, and the associated time delay and amplitude enhancement may lead to a large cross section for vibrational excitation. In Sections 4 and 5, we consider shape resonances, where the trapping occurs in the potential well presented by the lowest electronic state, and the target electrons are not affected dynamically, except through a minor polarization. We suppose the energy to be too low for the excitation of higher electronic levels. One has to distinguish between short-lived resonances, where the time delay is too small for the nuclei to relax, and longer-lived resonances, where the nuclei relax before the extra electron is reemitted. In the first case (discussed in Section 4), the "impulse approximation" or "adiabatic approximation" may be used. (See Appendix C.) In the second (discussed in Section 5), one has to treat the relaxation of the nuclei in the presence of electronic autoionization. (See Appendix B.)

In this section (Section 4), we consider trapping by a mechanism so weak that the time delay $\hbar/\Gamma$ is short compared with the time for a significant displacement of the nuclei. If the temporary negative ion has an electronic energy surface with a minimum, then a "significant displacement" can be interpreted as the slide of the nuclei to the bottom of the well, which takes a quarter of a vibrational cycle; if ω^- is the angular frequency of the nuclear vibration in the negative ion, this time is $\pi/(2\omega^-)$. If the negative ion does not have a minimum in its electronic energy surface, then a "significant displacement" becomes a dissociation in which a bond is broken. However, the time required by the nuclei to slide down the potential surface to dissociation is of the same order as the time they would take to slide to the bottom of the well if there were a minimum; thus one can still take the time $(\pi/2\omega^-)$, with some typical value of ω^-, as the time for a significant displacement.

If we take ω^- to be of order 0.3 eV, we can call a resonance short-lived if $\Gamma \gg 0.2\,\text{eV}$ [in order that $(\hbar/\Gamma) \ll (\pi/2\omega^-)$]. In applying this criterion, two points have to be watched. First, the appropriate value ω^- depends somewhat on the nuclear masses, and the slope and the nature of the potential surface. Second, Γ generally varies with the nuclear configuration, so that one has to be sure that Γ is large enough at all the configurations likely to occur before one labels a resonance as short-lived.

The integral cross section for vibrational excitation is, according to Eq. (1),

$$\sigma_{f\leftarrow i} = \frac{v_f}{v_i} \int d\hat{\mathbf{k}}_f |T_{fi}(\mathbf{k}_f, \mathbf{k}_i)|^2, \qquad (58)$$

where v_i and v_f are the initial and final velocities of the bombarding electron, and T_{fi} is the scattering amplitude for an electron with initial momentum $\mathbf{k}_i$ and final momentum $\mathbf{k}_f$. Since the collision time is short compared with the natural periods of the target, Eq. (58) may be calculated in the impulse approximation described in Appendix C, so that [see Eq. (C.9a)]

$$T_{fi} \cong \int d\mathbf{R}\, \mathscr{R}_f^*(\mathbf{R}) T_{fi}^F(\mathbf{R}) \mathscr{R}_i(\mathbf{R}), \tag{59}$$

where T_{fi}^F is the scattering amplitude of an electron scattered elastically by a molecule with fixed nuclei from an initial momentum $\mathbf{k}_i$ to a final momentum $\mathbf{k}_f$. The functions $\mathscr{R}_i$ and $\mathscr{R}_f$ are the initial and final vibrational wave functions of the molecule, and $\mathbf{R}$ stands for the vibrational coordinates. The orientation of the molecule is being regarded as fixed. (Since $|\mathbf{k}_f| \neq |\mathbf{k}_i|$ for an inelastic scattering, T^F has to be calculated off the energy shell.) The name "adiabatic approximation" for Eq. (59) is also in common use.

For T^F, we adopt the amplitude for scattering through a resonance when the nuclei are fixed, given in Eq. (A.21a) in Appendix A; in place of Eq. (58), we obtain

$$\sigma_{f\leftarrow i} = \frac{v_f}{v_i} \cdot \frac{1}{(2\pi)^2} \int d\hat{\mathbf{k}}_f \left| \int d\mathbf{R}\, \mathscr{R}_f^*(\mathbf{R}) \frac{\langle \Phi_f | V_f | b \rangle^F \langle b | V_i | \Phi_i \rangle^F}{E - \hat{W}(\mathbf{R})} \mathscr{R}_i(\mathbf{R}) \right|^2 \tag{60}$$

The superscript F on the matrix elements signifies that the nuclei are to be kept fixed. The energy E in the denominator is the sum of the energy of the incident electron and the electronic energy of the molecule at the equilibrium configuration in the initial state; since this electronic energy is stationary at equilibrium, its dependence on configuration $\mathbf{R}$ may be neglected in comparison with the dependence of $\hat{W}(\mathbf{R})$. The resonance energy $\hat{W}$ is as defined in Eqs. (48) and (49) for fixed nuclei.

One may simplify the expression (60) by treating the matrix elements $\langle \cdots \rangle^F$ as constants corresponding to the equilibrium configuration $\mathbf{R}_0$, and taking them outside the integral, because they vary with a scale of a_B which is large compared with the extent of the vibrational wave functions $\mathscr{R}_i$ and $\mathscr{R}_f$. The variation of the matrix elements with $\mathbf{R}$ would only add a correction to their values at equilibrium, which have to be nonzero because they are related to the width by the relation (A.24a) in Appendix A

$$\frac{1}{4\pi} \int d\hat{\mathbf{k}}_f |\langle \Phi_f | V_f | b \rangle^F|^2 \cong \frac{\pi \Gamma_0}{k_0}, \tag{61a}$$

$$\frac{1}{4\pi} \int d\hat{\mathbf{k}}_0 |\langle b | V_i | \Phi_i \rangle^F|^2 \cong \frac{\pi \Gamma_0}{k_0}, \tag{61b}$$

where Γ_0 is the width at the equilibrium of the target, and k_0 the momentum of the electron emitted there. In using Eqs. (61), we are ignoring the dependence of the matrix elements on the initial and final energies of the electron, which appear in Φ_f and Φ_i, because we are interested in energy differences of the order of vibrational spacings, which are usually small compared with the electronic energies over which the integrals vary appreciably. To use Eq. (61b), one has to average $\sigma_{f\leftarrow i}$ over the incident direction $\hat{\mathbf{k}}_i$; this is permissible because the integration over $\hat{\mathbf{k}}_f$ makes $\sigma_{f\leftarrow i}$ independent of $\hat{\mathbf{k}}_i$. One finally obtains, in place of Eq. (60), for the short-lived resonances,

$$\sigma_{f\leftarrow i} \cong \frac{v_f}{v_i} \cdot \frac{\pi\Gamma_0^2}{k_0^2} \left| \int d\mathbf{R} \frac{\mathscr{R}_f^*(\mathbf{R})\mathscr{R}_i(\mathbf{R})}{E - \hat{W}(\mathbf{R})} \right|^2. \tag{62}$$

One has to retain the dependence of the denominator on **R** [unlike the matrix elements in the numerator in Eq. (60)] since E cancels Re $\hat{W}$ at the resonance, so that the real part of the denominator is not just a correction to the value at equilibrium.

To evaluate the remaining integral in Eq. (62), we note the fact that $\hat{W}(\mathbf{R})$ varies with a scale of order a_B, which is much larger than the width of the vibrational wave functions. If, for example, there is only a single vibrational coordinate R, one may approximate $\hat{W}(R) \rightarrow \hat{W}(R_0) + S\cdot(R-R_0)$, where R_0 is the equilibrium, and $S = [\partial\hat{W}/\partial R]_{R=R_0}$, or the force felt by the nuclei because of the extra electron. If, moreover, one wants only the one-quantum transitions, one can use the linear approximation $[E-\hat{W}(R)]^{-1} = [E-\hat{W}(R_0)]^{-1} + [E-\hat{W}(R_0)]^{-2}\cdot S\cdot(R-R_0)$ to turn Eq. (62) into

$$\sigma_{f\leftarrow i} \cong \frac{v_f}{v_i} \cdot \frac{\pi\Gamma_0^2}{k_0^2} \cdot \frac{|S|^2}{|E-\hat{W}(R_0)|^4} \left| \int dR\, \mathscr{R}_f^*(R)(R-R_0)\mathscr{R}_i(R) \right|^2 \tag{63}$$

for the one-quantum transitions. For the transition $v=0\rightarrow v=1$ and for harmonic oscillator wave functions, one obtains $|\int dR\,\mathscr{R}_f^*(R)(R-R_0)\mathscr{R}_i(R)|^2 = (2M\omega)^{-1}$, where M is the reduced nuclear mass in units of the electron mass, and ω the vibrational quantum in atomic units. For the maximum cross section at the energy at the center of the resonance, one finds

$$\text{for } v = 0 \rightarrow v = 1, \qquad \sigma_{f\leftarrow i}(\max) = \frac{v_f}{v_i} \cdot \frac{4\pi}{k_0^2} \left[\frac{|S|^2}{\Gamma_0^2 \cdot \frac{1}{2}M\omega} \right]. \tag{64}$$

The parameter $|S|^2/(\Gamma_0^2 \cdot \frac{1}{2}M\omega)$ has a simple physical interpretation. Since $(\hbar/\Gamma)$ is the time delay, and S the force exerted by the extra electron, $(\hbar S/\Gamma)$ is the vibrational impulse the nuclei receive in the collision. The mean vibrational momentum in the ground state is $(\frac{1}{2}M\omega)^{1/2}$. Since $\hbar = 1$ in atomic units, $[|S|^2/(\Gamma_0^2 \cdot \frac{1}{2}M\omega)]^{1/2}$ is the ratio of the impulse to the mean momentum.

As an example of orders-of-magnitude, a very broad shape resonance in $e + H_2$ scattering peaks at about 3 eV of incident energy when the H_2 target is in its ground state. (See Figs. 1c and 10.) The cross-section maximum for the excitation $v = 0 \rightarrow 1$ is observed to be 0.4×10^{-16} cm^2. According to Eq. (64), this is consistent with a value $(|S|/\Gamma_0) \cong 0.5 a_B^{-1}$, much smaller than the mean momentum of $3a_B^{-1}$ in the vibrational ground state. With such a small delivery of momentum to vibrations, the cross sections for the excitation of higher vibrational states fall off rapidly.

Sometimes a molecule possesses enough symmetry to make possible a decomposition of T^F in Eq. (59) into independent components belonging to different symmetries which make additive contributions to T^F. For example, in e–H_2 collisions, there is a rotational axis and a center of symmetry, so that the wave function in a fixed-nuclei model can be split into components of definite angular momentum about the axis, and definite parity. Each component is characterized by a phase shift $\delta(R, E)$, depending on the nuclear configuration and the energy, and appearing in T^F in a factor $(e^{2i\delta} - 1)$. The one-quantum transitions may be calculated by replacing

$$(e^{2i\delta} - 1) \rightarrow 2i(R - R_0)\left(\frac{d\delta}{dR}\right)_{R_0} e^{2i\delta(R_0)} + (e^{2i\delta(R_0)} - 1)$$

so that the measure of the relative importance of the different components is the derivative $(d\delta/dR)_{R_0}$.

The connection between the phase shift and Eq. (62) may be made by taking the factor Γ_0^2 inside the integral, so that one gets a factor $\Gamma_0/(E - \hat{W})$ in the integrand. If one treats Γ as a constant in $\hat{W}$, then $\hat{W} = \text{Re}\,\hat{W} - (i/2)\Gamma_0$, so that

$$\frac{\Gamma_0}{E - \hat{W}} = \left(\frac{\Gamma_0}{E - \hat{W}} + i\right) - i = i\left(\frac{1 + i\tan\delta}{1 - i\tan\delta} - 1\right) = i(e^{2i\delta} - 1),$$

where $\tan\delta = \Gamma_0/2/(\text{Re}\,\hat{W} - E)$. Thus if one retains only the symmetry component of T^F associated with the resonance, one obtains in place of Eq. (62)

$$\sigma_{f\leftarrow i} \cong \frac{v_f}{v_i} \cdot \frac{\pi}{k_0^2} \left| \int d\mathbf{R}\, \mathscr{R}_f^*(\mathbf{R})(e^{2i\delta(\mathbf{R})} - 1)\mathscr{R}_i(\mathbf{R}) \right|^2. \qquad (62a)$$

The integral now has the form used by Chang in Ref. 32, by Chang and Fano,[33] and by Temkin and his collaborators.[34] If one retains more than a single symmetry component, as these authors often do, then one has to add several terms inside the modulus sign in Eq. (62a). In the limit of a short collision time, the one-quantum transitions predominate, and Eq. (62a) gets replaced by (for a single vibrational degree of freedom)

$$\sigma_{f\leftarrow i} \cong \frac{v_f}{v_i}\cdot\frac{4\pi}{k_0^2}\left(\frac{d\delta}{dR}\right)^2_{R_0}\left|\int dR\mathscr{R}_f^*(R)(R-R_0)\mathscr{R}_i(R)\right|^2. \quad (62b)$$

Good examples of resonant vibrational excitation in the impulse limit are hard to find, because it is not easy to satisfy the condition that the lifetime be much shorter than a vibrational period and yet longer than a nonresonant collision time. For example, for a molecule with a vibrational quantum of 0.2 eV, one has $1/\omega \approx 3\times 10^{-14}$ sec. On the other hand, the nonresonant collision time is somewhat longer than $\lambdabar/v$, where v is the velocity of the incident electron and $\lambdabar$ its reduced wavelength; for an impact energy of 2 eV, one has $\lambdabar/v \approx 1.5\times 10^{-14}$ sec, little different from $1/\omega$. Nevertheless, there are examples of resonance peaks in electron–molecule scattering which are most simply interpreted as belonging to the impulse limit because they possess no vibrational structure; the simplest interpretation of the absence of structure is that the lifetime is too short for the nuclei to vibrate. (See also Section 5.2.) One such example is a resonance observed in $e + N_2O$ at an impact energy of ~ 2.3 eV (See Fig. 1g). As a word of caution, one should add that in polyatomic molecules, an electron-resonance profile without a vibrational structure could come also from several vibrational modes with many overlapping vibrational levels; no examples of this kind have been analyzed.

For the simplest examples of electron–molecule resonances, one naturally looks to e–H_2 scattering. (The e–H_2 problem has a very large literature; for a complete scholarly review, the reader is recommended to consult Lane's article.[1]) The simplest resonance should come from the transient occupation of the lowest vacant antibonding orbital by the scattered electron. This orbital is, roughly, a superposition of two atomic 1s orbitals centered on the two nuclei, with symmetry σ_u and a nodal plane which bisects the nuclear axis. An electron can enter or leave this orbital through a p wave with zero component of angular momentum about the nuclear axis.

Observations of vibrational excitation $v = 0 \rightarrow 1$ in $e + H_2$ scattering (see Fig. 10) show a structureless peak between 3 and 6 eV. Early analyses of this region, which were reviewed by Massey,[35] showed that the Born approximation would lead to a cross section much smaller than that observed, so that a compound state becomes an interesting hypothesis. Schulz and Asundi[36] therefore started to study dissociative attachment, which would indicate whether a compound state was present, but found only a very weak peak, with a cross section of order 10^{-21} cm^2. Presumably because of this small magnitude, the

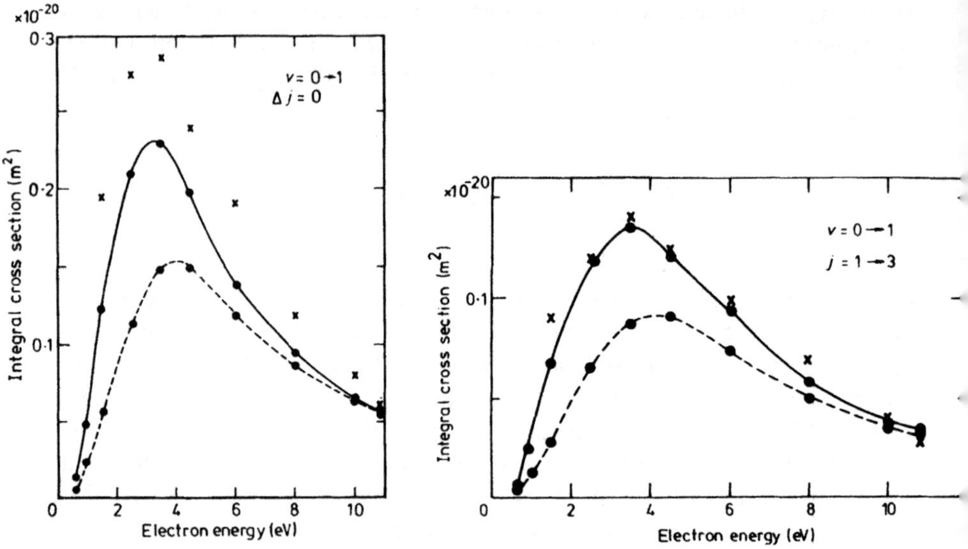

FIGURE 10
$e + H_2$; observed (X X X) and *ab initio* calculated cross sections for excitation $v = 0 \rightarrow v = 1$, for different rotational transitions, Static-exchange (– –·– –). Including polarization (—●—●—●—) (From Ref. 42).

idea of a compound state gained acceptance only slowly; however, the existence of a compound state was put beyond doubt when Allan and Wong[37] showed experimentally that the dissociative attachment near ~3 eV has a cross section as large as 10^{-16} cm^2 if the target molecule is vibrationally excited, and Wadehra and Bardsley[38] showed theoretically that the observations were consistent with a σ_u resonance. (See Chapter 4, Section 5.1.)

An attempt[39] to interpret the 3–6-eV peak in vibrational excitation in terms of a σ_u resonance met with qualitative success already in 1966. A compound state had been suggested earlier by Chen and Magee,[40] who proposed that the coupling between the extra electron and the nuclear velocity should capture or eject an electron, but the magnitude of the cross section again came out much too small. The later[39] compound state models neglected this mechanism, and relied on adiabatic capture and decay by leakage through a centrifugal barrier. However, the shape resonance due to the lowest σ_u orbital is not sufficiently dominant near the equilibrium separation of the nuclei to account with precision for vibrational excitation in e–H_2 scattering near 3 eV. The difficulty of providing an unambiguous physical interpretation was emphasized by Takayanagi,[41] who showed that a distorted wave calculation could get close to the observed magnitude of the cross section. It seems that although the resonance picture gives one a useful qualitative picture of e–H_2 scattering near 3 eV, a precise account of the observations can be obtained only by mixing so much direct scattering with the resonance that much of the physical simplicity is lost. Very recently, vibrational excitation in e–H_2 scattering below 10 eV has been accounted for very well by an *ab initio* calculation by Klonover and Kaldor,[42] using a *T*-matrix expansion due to Rescigno, McCurdy, and McKoy[43] (see Chapter 7, Section 3.9); the theoretical curve in Fig. 10 comes from this calculation. See also Fig. 8 of Chapter 7, Section 3.9.

The reader is referred to an interesting series of papers[44] by Lane, Henry, and Chang for a description of the structure of the wave function in e–H_2 scattering below 10 eV. A paper by Chang[32] published in 1974 is particularly useful as an exposition of the strengths and weaknesses of the resonance model. A paper by Temkin and Sullivan[45] should also be consulted, to show that the e–H_2 problem is still not finished.

Because e–H_2 scattering does not provide a clear-cut example of an electron–molecule resonance in the impulse limit, Dubé and I analyzed[46] a structureless resonance observed[47] in $e + N_2O$ at an impact energy of 2.3 eV. This work is an application of Eq. (62) to a polyatomic molecule with three different vibrational modes. The fit of observations shown in Fig. 11 was obtained with four adjustable parameters, representing the width and the slopes of the potential surface at equilibrium. The absolute magnitude comes out of the calculation because one can get the width from the energy dependence in Eq. (62), and that determines the absolute magnitude through the factor Γ_0^2.

5. Vibrational Excitation. III: Long-Lived Resonances and Nuclear Relaxation

5.1. The Nuclear Wave Equation

If a resonance has a lifetime long enough for the nuclei to slide down the electronic energy surface, one has to take into account the nuclear relaxation which we ignored in Section 4. In this section, we consider resonances sufficiently long-lived to vibrate a little, but we ignore dissociation; that is the subject of Chapter 4. For the sake of simplicity, we restrict ourselves to diatomic molecules, although little modification is needed to extend the theory to polyatomic molecules. The decay rate has to satisfy $\Gamma \lesssim \omega^-$, where ω^- is the angular frequency of the nuclear vibration in the negative ion. (A typical value for ω^- in a small molecule might be $\omega^- \approx 0.3$ eV.)

Because a resonance with $\Gamma \lesssim \omega^-$ can vibrate, even if only through one or two cycles, the observed cross sections show vibrational structure, unlike the impulse resonances in Section 4. Examples among those in Fig. 1 are $e + N_2 \rightarrow N_2^-(^2\Pi_g)$ at an impact energy of 2.3 eV, and $e + CO_2 \rightarrow CO_2^-(^2\Pi_u)$ at 3.8 eV.

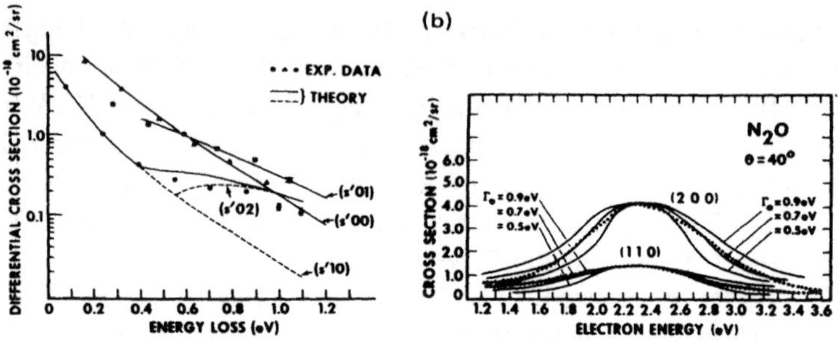

FIGURE 11
A fit of $e + N_2O$ in the impulse approximation. (a) Maxima of the energy loss peaks. The small circles, triangles, and squares are experimental, from Ref. 47. The corresponding calculated points have been joined by the continuous curves. The dashed curves correspond to the calculation of two different modes of excitation which are not resolved in the experiments; they have been summed and joined by a single continuous curve. (From Ref. 46.) (Scattering angle 40°.) (b) Energy dependence. · · ·, experimental results. The continuous curves correspond to calculations with three different values of the decay width at the center of the resonance. (From Ref. 46.)

Moreover, the nuclear wave function becomes so severely distorted during the lifetime of the resonance that many vibrational levels are excited if the energy is high enough.

The electronic resonant wave function may be approximated by a boxed-in compound state $b(\mathbf{r}; \mathbf{R})$, as defined in Section 2.6. The electronic coordinates are represented by $\mathbf{r}$, and the nuclei by $\mathbf{R}$. The complete scattering wave function Ψ_i at an energy close to the resonance has the form

$$\Psi_i(\mathbf{r}, \mathbf{R}) = \xi_i(\mathbf{R}, E) b(\mathbf{r}, \mathbf{R}) + \Psi_i^p(\mathbf{r}, \mathbf{R}), \tag{65}$$

where the suffix i stands for all the quantum numbers defining the initial state. Ψ_i^p contains everything orthogonal to b with respect to the electrons:

$$\langle b | \Psi_i^p \rangle_e = 0, \tag{66}$$

where the suffix e signifies an integration over the coordinates of the electrons while the nuclei are held fixed. The function ξ_i is the wave function of the nuclei in the resonance. In the short-lived resonances discussed in Section 4, ξ_i was a slightly modulated version of the initial vibrational state $\mathscr{R}_i(\mathbf{R})$; it appears as the function $\langle b|V_i|\Phi_i\rangle^F / [E - \hat{W}(\mathbf{R})] \mathscr{R}_i(\mathbf{R})$ in the integral in Eq. (60). However, in the long-lived resonances, the relaxation makes ξ_i look very different from the initial vibrational state.

The key to the calculation of ξ_i is the Born–Oppenheimer approximation, which rests on the assumption that ξ_i carries most of the nuclear momentum in the transient negative ion, because ξ_i varies with **R** much more rapidly than b does. For example, for a nuclear kinetic energy of 1 eV, which is typical for a resonance at an energy of a few electron volts in electron–molecule scattering, the reduced wavelength of ξ_i is of order $0.1\,a_B$, whereas b varies on the scale of $1\,a_B$ characteristic of valence orbitals. In the Born–Oppenheimer approximation, whenever the nuclear kinetic energy acts on the wave function, a derivative of b with respect to **R** is neglected in comparison with a derivative of ξ_i.

Considerable confusion has been caused in the literature by suggestions that the coefficient b multiplying the nuclear function in Eq. (65) should be the complete wave function (41) for an electron scattered by a molecule with fixed nuclei, instead of just the normalized wave function b of the resonant core we defined in Section 2.6. Because of the form of the coefficient (47) in the fixed-nuclei model, the complete wave function (41) has a sharp kink when **R** varies so as to sweep Re $\hat{W}$ through E, and therefore does not satisfy the condition that the coefficient of ξ_i in Eq. (65) should vary slowly with **R**. The difficulty is removed by identifying b in Eq. (65) with the boxed-in compound state, corresponding to the physical picture where the electronic wave function follows the nuclei adiabatically within the trapping region. The abruptly varying coefficient (47) of the fixed-nuclei model then gets replaced by the rapidly oscillating function ξ_i of the nuclei in Eq. (65). As the Born–Oppenheimer approximation requires, the rapid variation with respect to R appears in the nuclear wave function, and not in the electronic factor.

To calculate the wave function, substitute Eq. (65) into the Schrödinger equation, and project parallel and orthogonally to b with respect to the electrons. After some calculations, which are explained in detail in Appendix B, one obtains, with the use of the Born–Oppenheimer approximation, the equation[49]

$$[K + W(\mathbf{R}) - E]\,\xi_i(\mathbf{R}) = -\langle b|H_e|\Phi_i^P\rangle_e. \tag{67}$$

Here K is the kinetic energy of the nuclei; $W(\mathbf{R})$ is the Siegert energy corresponding to a resonance for nuclei fixed at **R**, as defined in Eq. (50); E is the total energy; H_e is the electronic Hamiltonian defined for fixed **R**; Φ_i^P is a potential scattering wave function which asymptotically contains the incident wave, and which is constructed so that $\langle b|\Phi_i^P\rangle_e = 0$. The subscript e signifies that the integration is over the coordinates of the electrons, while the nuclei are held fixed.

The approximations behind Eq. (67) are explained in Appendix B. One approximation is to drop terms of the order $\lambda\!\!\!^{-}_N/a_B$, where $\lambda\!\!\!^{-}_N$ is a nuclear reduced wavelength; ($\lambda\!\!\!^{-}_N/a_B$ is 0.025 in the example of a nuclear kinetic energy of 1 eV, and a nuclear mass of 10 proton masses). A second approximation is to drop the energy of coupling between the nuclear velocities v_N and the electrons in the transient negative ion, which is of order v_N/a_B, in comparison with the spacing ($\equiv \Delta E_e$) of electronic energy levels in a box just big enough to enclose the quasistationary state. In the example of the nuclear kinetic energy of 1 eV, and the nuclear mass of 10 proton masses, one has $v_N/a_B \approx 0.05$ eV; this is much smaller than the expected value of the spacing, $\Delta E_e \gtrsim 1$ eV.

The justification (in Appendix B) of the approximations behind Eq. (67) is more delicate than the justification of the Born–Oppenheimer approximation in the stationary states of molecules which do not emit electrons. In the case of Eq. (67), one has to consider the tail of outgoing waves from the transient negative ion, in addition to the structure of the wave function inside the quasistationary state. Moreover, for the stationary states of molecules, it is reasonable to suppose that the electronic wave function follows the nuclei adiabatically wherever it is not negligibly small. By contrast, in scattering through a quasistationary state, there is no adiabatic following when an electron is outside the transient negative ion, even though the electronic wave function is not vanishingly small there; the electrons follow the nuclei adiabatically only when they are within the transient negative ion, or within the residual molecule after one electron has been emitted.

Equation (67) is a wave equation for ξ_i in which $W(\mathbf{R})$ is a local complex potential, and the matrix element on the right is a source.

The literature contains much discussion whether the complex potential is local. Our statement is that the potential is local because $\lambda\!\!\!^{-}_N/a_B \ll 1$, and $\Delta E_e \gg v_N/a_B$, as we explained above [see also after Eq. (B.30a)]. These conditions imply that the electronic wave function follows the nuclei adiabatically when all the electrons are close to the molecule. A formulation of the theory by Bardsley,[50] which does not emphasize the adiabatic following when all the electrons are within the molecule, leads to a complex potential which is generally nonlocal. Bardsley shows that his potential becomes local if the energy of the incident electron outside the molecule is so high that several vibrational states may be excited. The question whether all one needs for a potential to be local is electronic velocities which are large inside the molecule, or whether the energies outside have also to be large ($\gtrsim 1$ eV), has still not been finally settled. Analyses of experiments with, in part, the purpose of determining whether a nonlocal potential is necessary have been made by Parlant and Fayard,[51] and by Hazi, Rescigno, and Kurila.[52]

Once ξ_i has been obtained by solving Eq. (67), the scattering cross sections may be calculated from [see Appendix B, Eq. (B.41a)]

$$\sigma_{f\leftarrow i} = \frac{v_f}{v_i}\frac{1}{16\pi^3}\left|\int dR\,\mathscr{R}_f^*(R)\tilde{\zeta}(R, k_f)\tilde{\xi}_i(R)\right|^2, \tag{68}$$

where v_i and v_f are the initial and final velocities of the electron.

Here $\tilde{\xi}_i(R)$ is defined in Eq. (B.36b) and $\mathscr{R}_f$ is the final vibrational state;

$$\tilde{\xi}(R, k_f) \equiv \left| \int d\hat{\mathbf{k}}_f |\langle \Phi_f^F | \mathrm{V} | b \rangle_e|^2 \right|^{1/2},$$

where Φ_f^F is the electronic factor in the final state, with fixed nuclei. Expression (68) corresponds to a case with just a single electronic resonance, with angular momentum zero about the molecular axis; if there are two degenerate resonances with equal and opposite angular momenta, the expression (68) has to be multiplied by two.

An approximation made during the derivation is that only a single orbital angular momentum contributes to the outgoing wave, higher angular momenta being inhibited by a centrifugal potential barrier.

5.2. Solutions of the Nuclear Wave Equation—Qualitative Features

A qualitative discussion of Eqs. (67) and (68) is enough for an understanding of several important features of resonant electron–molecule scattering. The source term on the right of Eq. (67) is nonzero only within the Franck–Condon width of the initial vibrational state, of the order of $2(M\omega)^{-1/2}a_B$ in magnitude if the target is initially in its lowest vibrational state. (M is the nuclear reduced mass in electron masses, and ω the vibrational quantum of the neutral target in atomic units.) For example, for $M \approx 10^4$ and $\omega \approx 10^{-2}$ a.u. ≈ 0.3 eV, the Franck–Condon width is $0.2a_B$. This is typically only a tenth of the nuclear separations, so that the term on the right is often zero over most of the range of separations allowed at a given E.

When the right-hand side of Eq. (67) is zero, the WKB approximation may be used because the wavelength is usually small; at a kinetic energy of 1 eV, and $M \approx 10^4$, the reduced wavelength is only $0.04a_B$.

The wave number is complex: $[2M(E - W)]^{1/2} \cong k_0 + i\Gamma/2v$, where $k_0 \equiv [2M(E - \mathrm{Re}\, W)]^{1/2}$, and v is the velocity of the nuclei. (The expression is valid only if $\Gamma \ll Mv^2$.) Therefore the solutions of the homogeneous equation are damped waves, falling in amplitude by a factor e in a distance $(\Gamma/2v)^{-1}$. For example, if $\Gamma = 0.1$ eV and $M = 10^4$, then at a kinetic energy of 1 eV one has $(\Gamma/2v)^{-1} = 1.5a_B$, just of the order of a typical nuclear separation. Physically, the decay represents the reduction of the probability $|\xi|^2$ of the compound state b as the extra electron is emitted. If the nuclear separation increases by ΔR in a time $\Delta\tau = \Delta R/v$, then $|\xi|^2$ decreases by a factor $\exp[-\Gamma\Delta\tau] = |\exp[-(\Gamma/2v)\Delta R]|^2$.

If the potential $W(R)$ contains a well (i.e., is not purely repulsive), then the solution of Eq. (67) is a wave emitted by the source on the right of Eq. (67), plus reflected waves generated at the turning points. This process is illustrated in Fig. 12b, which shows just one wave packet in each direction. [The

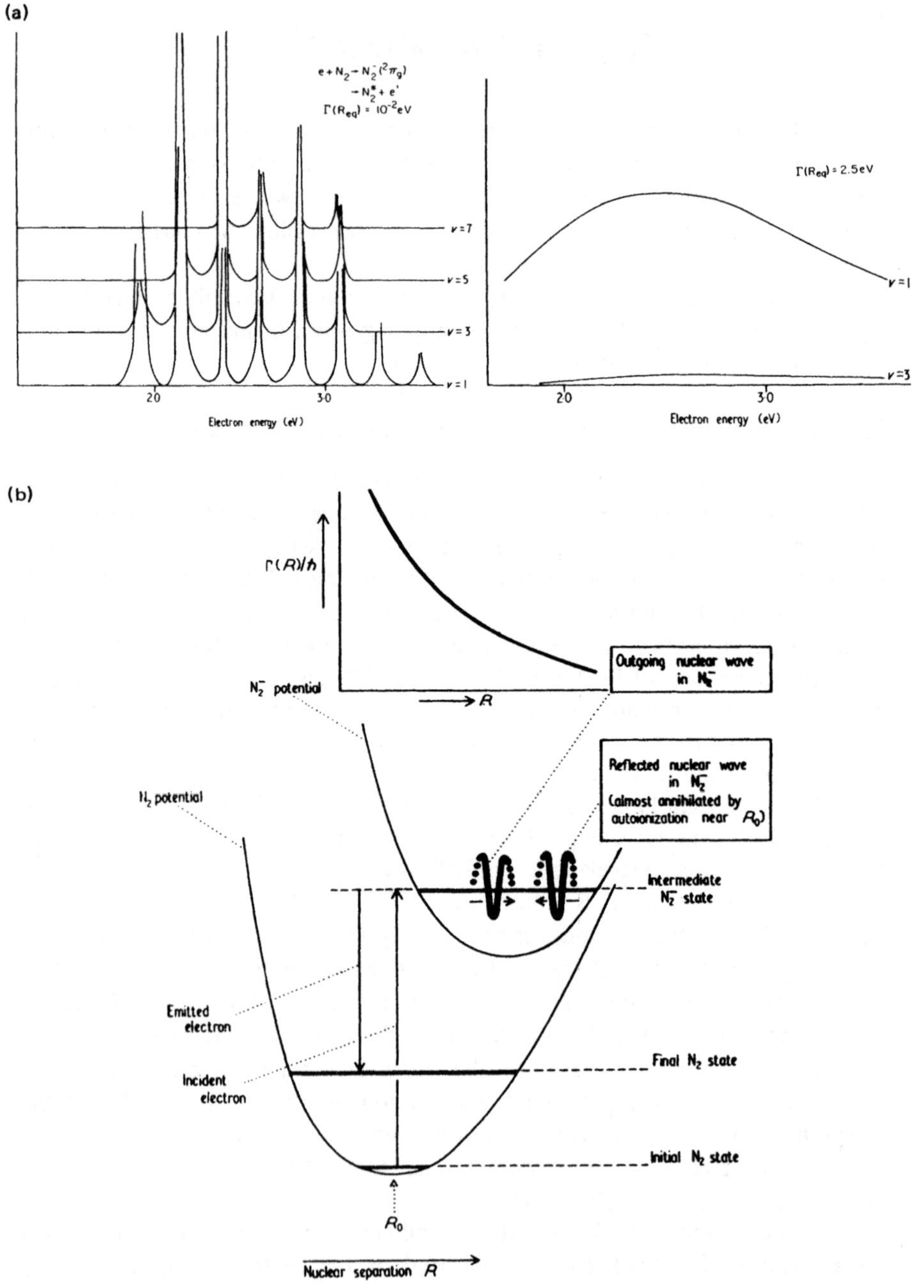
(a)
e + N2 → N2-(2πg)
→ N2* + e'
Γ(Req) = 10-2eV
ν = 7
ν = 5
ν = 3
ν = 1
2·0
3·0
Electron energy (eV)
Γ(Req) = 2.5eV
ν = 1
ν = 3
2·0
3·0
Electron energy (eV)
(b)
Γ(R)/ħ
R
N2- potential
N2 potential
Outgoing nuclear wave in N2-
Reflected nuclear wave in N2-
(almost annihilated by autoionization near R0)
Intermediate N2- state
Emitted electron
Incident electron
Final N2 state
Initial N2 state
R0
Nuclear separation R

wave in Fig. 12b actually illustrates the time-dependent picture corresponding to Eq. (67), which is obtained by taking the time Fourier transform.]

The character of the superposition of waves in Fig. 12b depends on how many reflected waves are important; this number is of order ω^-/Γ. If $\Gamma/\omega^- \ll 1$, then there are many reflected waves, and the amplitude becomes large only if they are all in phase; the negative ion then has vibrational levels, with each level slightly broadened by autoionization; this case may be called a "compound molecule." If $\Gamma/\omega^- \gg 1$, then the wave emanating from the source on the right of Eq. (67) is damped out by autoionization before any reflections arise; one then has one of the short-lived resonances discussed in Section 4, and one may speak of the "impulse limit." Figure 12b illustrates an intermediate case, in which a single reflected wave is important; the obvious name for this case is the "boomerang model." The transition between the two extremes is illustrated in Fig. 12a.

If vibrational structure can be seen in the cross sections, as in several examples in Fig. 1, one can often tell by inspection what the order of magnitude of Γ/ω^- is. If the peaks in different vibrational channels appear at the same incident energy, as in the example of O_2 at impact energies below 1 eV, then there are true vibrational levels; therefore there must be many reflected waves, and $\Gamma/\omega^- \ll 1$. Since ω^- in O_2 is of order 0.1 eV,[53] one must have $\Gamma \ll 0.1$ eV. In the case of N_2 at impact energies around 2 eV, there are peaks which appear in the different channels at different energies, so that at almost all energies there are some channels with a large cross section (Fig. 14). Therefore the nuclear wave function ξ_i within the resonance band cannot fluctuate violently in amplitude, and there cannot be many reflections, because that would lead to constructive interference only at discrete vibrational energies, with little amplitude between. On the other hand, there have to be some reflected waves, for otherwise there would be interference in $\xi_i(\mathbf{R})$ to produce the oscillations in the cross sections which one observes as the energy varies. If there has to be more than one wave, but not many more than one, then there must be about one reflected wave in ξ_i. This single reflected wave interferes with the wave

FIGURE 12
Qualitative results from a complex-potential model for a resonance in electron–molecule scattering. (a) If the electron-decay width in the model which yields Figs. 16 and 17 is *reduced* by a factor 60, one obtains the *left* curve in Fig. 12a. It shows a vibrational spectrum for the negative ion, with narrow peaks at the same incident energies for all the final channels. Compare Fig. 1e for $e + O_2$. The *right* curve in Fig. 12a is obtained if one *increases* the electron-decay width in the model used for Figs. 16 and 17 by a factor 4. Compare Fig. 1c for $e + H_2$. (From Ref. 66.) (b) A schematic picture of the wave function of the nuclei in the $e + N_2$ resonance at 2–3 eV of electron impact energy. The magnitude of the electron-decay width here corresponds to the "boomerang model," where only a single outgoing and returning nuclear wave survive before the negative ion dies from autoionization. (From Ref. 66.)

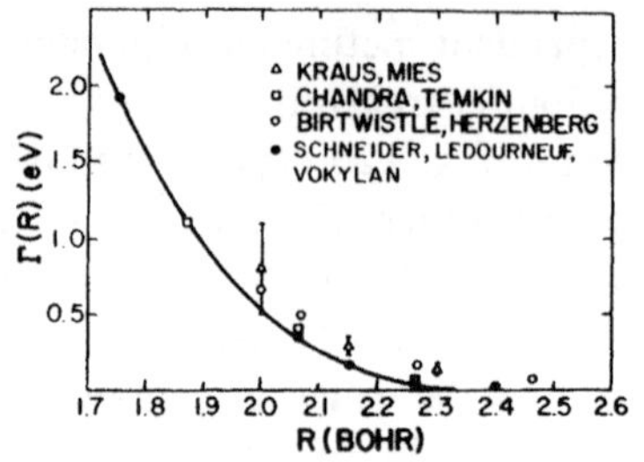

FIGURE 13
Electron decay width of the e–N_2 resonance in Figs. 1b and 1d as a function of nuclear separation. The width decreases with increasing separation because the energy of the emitted electron decreases (see Fig. 15), and tunnelling through the centrifugal potential barrier becomes more difficult (see Fig. 2 for $l = 2$). (From Ref. 64.)

emitted by the souce on the right in Eq. (67) to form a standing wave whose nodes drift slowly as the energy changes.[54] In Eq. (68), the overlap of this standing wave with the final vibrational state oscillates as the energy changes. (One can actually explain the shift of the observed peaks with vibrational channel in terms of this picture.[54]) Calculations[54,55,63–66] show that Γ is not constant in the N_2^- resonance, because the penetrability of the centrifugal barrier increases with the energy of the emerging electron; thus Γ rises towards low separations (Fig. 13). Therefore the probability of autoionization becomes very large at the end of the first vibrational cycle, when the nuclear wave packet is momentarily stationary just where Γ is largest.

5.3. The Complex Potential Surface

The complex potential $W(\mathbf{R})$ has to be known before the wave equation (67) can be solved for the nuclear wave function ξ_i in the transient negative ion. The Born–Oppenheimer picture for the electronic wave function close to the molecule implies that calculations with fixed nuclei are sufficient to determine $W(\mathbf{R})$. If W is known, the "feeding term" $\langle b|H_e|\Phi_i^P\rangle_e$ on the right of Eq. (67) is also determined by Eq. (A.24a) and the approximation just before Eq. (A.21a).

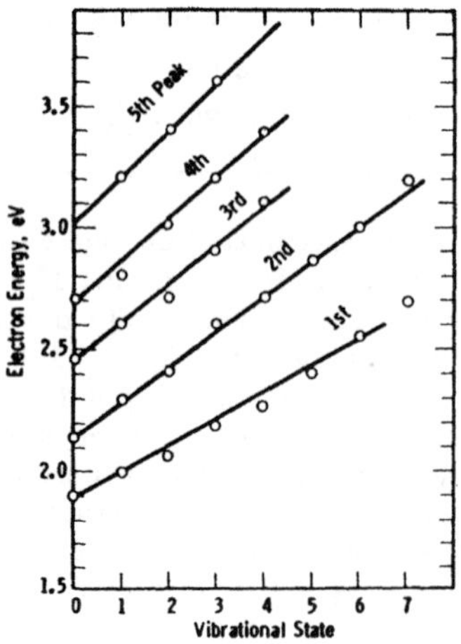

FIGURE 14
Observed variation of the energies of the cross-section peaks in Fig. 1b with the final vibrational channel. These shifts are characteristic[54] of the boomerang wave function in Fig. 12b. (From Ref. 74.)

The potential surface with the most extensive literature is the shape resonance observed in $e + N_2$ scattering at an impact energy of about 2.3 eV when the target molecule is initially in its lowest vibrational state. The electronic configuration of this resonance was suggested to be $(N_2\text{core})\cdot(\pi_g 2p)$ by Gilmore[56] in 1965; this suggestion was discussed further by Bardsley, Mandl, and Wood.[57] As a test of this assignment, the angular distribution expected for the scattering of a π_g electron was given[58] by Bardsley and Read, and by O'Malley and Taylor; the predicted expression agrees well with the observed angular distribution of electrons which have excited one or more quanta of vibrations.[59] The example of this resonance will suffice as an illustration. (Potential surfaces have been calculated also for other cases; examples are quoted in Refs. 60 and 61.)

A direct way of obtaining a complex resonance energy $W(\mathbf{R})$ is to calculate the scattering of an electron by a molecule with fixed nuclei near the resonance and to fit the resonant phase shift δ with the expression $\tan \delta = (\Gamma/2)/(\text{Re } W - E)$; see Section 4 near Eq. (62a). Many calculations for $e + N_2$ scattering with fixed nuclei have been done over the last ten years, particularly by Burke,[62] in collaboration with Sinfailam, Chandra, and Buckley, by Chandra and Temkin,[63] and by several other groups. A list and description of these calculations is given in Section C.1 of Lane's recent review.[1] The results shown in Fig. 15 due to Chandra and Temkin,[63] and to Schneider, Le Dourneuf, and Vo Ky Lan[64] were obtained in this way (as preliminaries, one should add, to calculations with relaxing nuclei).

In Fig. 15, the curve due to Chandra and Temkin contains one adjustable parameter (in a polarization potential), which is used to tune the center of the resonance to the observed energy. The curve due to Schneider *et al.* is completely *ab initio*, without any adjustable parameters at all. (The method used to obtain that curve–from the R matrix–is described in Chapter 7, Section 4.4.) An important feature of the calculation in Ref. 64 is the use of a wave function for the resonance which is optimized for all $N + 1$ electrons, where N is the number of electrons in the neutral target. The idea of allowing all $N + 1$ electrons to relax simultaneously in a resonance calculation was introduced by Schneider and Hay;[61] the physical reason is the time delay associated with resonance scattering. This one simple device has removed a large part of the need to add a parametric "correlation energy," which had previously been customary in otherwise *ab initio* electron–molecule calculations at a resonance.

The first near-*ab initio* calculation of the electronic energy of the $^2\Pi_g$ resonance in N_2 was done by Krauss and Mies;[65] at that time a negative "correlation energy" of 0.7 eV had still to be added to get the resonance energy to agree with experiment.

The imaginary parts of the complex energies from these calculations are shown in Fig. 13. Since the $^2\Pi_g$ resonance decays by emitting a π_g electron with

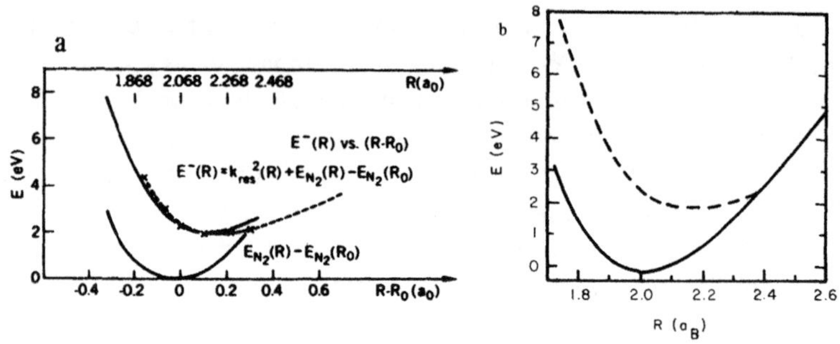

FIGURE 15
The electronic energy of N_2, and the real part of the electronic energy of the N_2^- resonance in Figs. 1b and 1d. (a) Upper full curve (for N_2^-) from a near-*ab initio* calculation.[63] Dashed curve (for N_2^-) from the fit[66] of the cross sections in Figs. 1d and 17. Lower full curve for neutral N_2. (From the first item in Ref. 63.) (b) Dashed curve (for N_2^-) from a fully *ab initio* calculation; full curve for neutral N_2. (Extracted from Ref. 64.)

a minimum angular momentum $l = 2$, Γ becomes small at small emission energies where the penetration of the centrifugal potential barrier becomes difficult. An approximate expression for Γ is[55] $\Gamma(R) \propto k(R)^{2l+1}$, where $k(R)$ is the wave number of the electron emitted at separation R, and l is the smallest angular momentum contributing to the outgoing wave. This expression explains the rise of Γ toward low R where the potential curves of the neutral molecule and negative ion move apart.

A very high precision in the complex potential is needed for the vibrational excitation cross sections to agree with experiment. According to Eq. (67), the scattering amplitude is proportional to the overlap integral of two rapidly oscillating wave functions, the final vibrational state $\mathscr{R}_f(R)$ and the nuclear wave function $\xi_i(R)$ of the negative ion. The nodes of these two functions must be located to an accuracy much better than a reduced wavelength, which is $0.04a_B$ for a nuclear kinetic energy of 1 eV with a reduced mass of 7 proton masses. This precision is at least an order of magnitude more severe than that needed in calculations of electronic excitation at energies of about 10 eV.

5.4. Semiempirical Calculations

Much progress may be made even without an *ab initio* complex potential, if one guesses a reasonable shape for the complex function $W(R)$ and adjusts parameters in it until the calculated cross sections agree with experiment. This procedure is less arbitrary than appears at first sight, because of the great richness of structure in the cross sections at any one electron–molecule resonance when the nuclei relax. [This structure has its origin in the effect of the rapid oscillation of ξ_i and the vibrational wave functions $\mathscr{R}_f$ on Eq. (68).] For example, at the e–N_2 resonance at $\sim$2.3 eV with a target initially in its lowest vibrational state (see Fig.1b), there are over 40 different peaks in the energy dependence of the vibrational excitation cross sections in all the different channels. Each peak has its own energy, width, and intensity. If one succeeds in accounting for all these peaks with even a half-dozen adjustable parameters, the success is hardly likely to be accidental.

A severe test for such a calculation is to compare the predictions of a parametrized model with measurements which are not known at the time the parameters are chosen (and which are not trivially related to some known result, as for example by time reversal). This has become possible recently with the $N_2^-(^2\Pi_g)$ shape resonance, where advances in experimental technique have led to measurements[48] of vibrational excitation of an already vibrationally excited molecule; the parametrized theory passed this test[67] with flying colors (Fig. 16).

Although one would always prefer to have a completely *ab initio* calculation, there is a number of reasons why several groups[68, 69] have published calculations with parametrized models:

(1) The calculations yield a large return in physical understanding for a small effort. [To use Eqs. (67) and (68), one need usually only solve a single second-order ordinary differential equation for each energy and incident channel, and do one quadrature for each cross section (Fig. 17).]

(2) Cross sections showing vibrational structure need the very high precision of calculation to which we alluded in Section 5.3. This precision is much easier to achieve by carefully tuning an adjustable parameter than by getting an *ab initio* calculation right. It may in the future become customary to start from

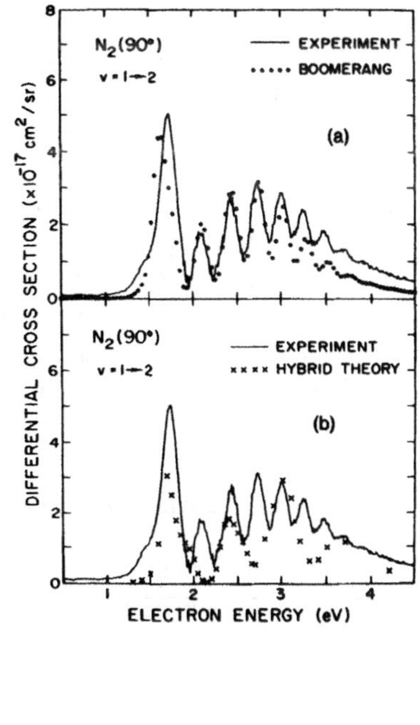

FIGURE 16
A test of a semiempirical model. The points (· · ·) in Fig. 16a were calculated for the transition $v = 1 \rightarrow 2$ with a model whose parameters had previously been fixed by fitting to transition $v = 0 \rightarrow v' \neq 0$ (see Fig. 17). The experimental curve in Figs. 16a and 16b is due to Wong, Michejda, and Stamatovic (to be published).[48] The diagram in Fig. 16b illustrates the difficulty of achieving a similar precision in some *ab initio* calculations. The points (X X X) calculated from the "hybrid theory" are obviously qualitatively correct, but fail to get the number and positions of the peak right. (From Ref. 49.)

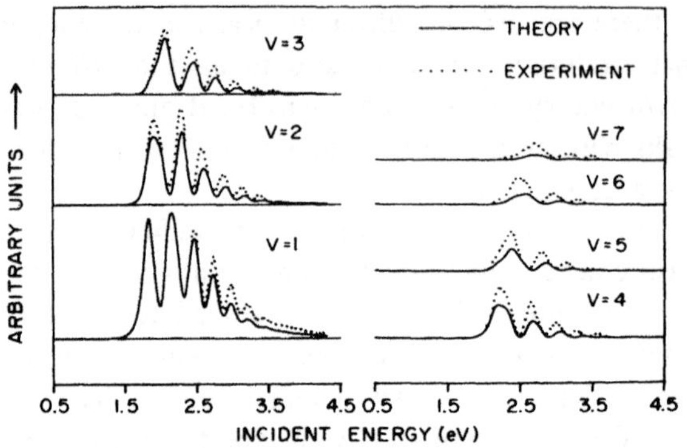

FIGURE 17
A fit of a model with six adjustable parameters to observations on $e + N_2$ for $v = 0 \rightarrow v \neq 0$. Compare Fig. 16a. (From Ref. 49.)

a fairly rough *ab initio* calculation of complex energies and electronic matrix elements, and to correct the *ab initio* results by fine tuning of adjustable parameters to get the vibrational structure right. This procedure would put the correct identification of an electronic resonance beyond doubt, without asking the electronic *ab initio* calculation to have a precision of $10^{-2} a_B$ in the location of vibrational nodes. The procedure is illustrated by Fig. 18.

(3) Resonance cross sections for changes of vibrational state, starting from vibrationally excited targets, are often needed for calculations to optimize gas lasers. The easiest way to get these cross sections is to use a model with parameters calibrated on laboratory observations of vibrational excitation starting from the ground state. Some fits and predictions in this spirit at an $e + CO$ resonance are shown in Fig. 19.

(4) Calculations with adjustable parameters are particularly suited to getting the absolute magnitude of vibrational excitation cross sections right; these absolute magnitudes are often the most important aspect in applications. The absolute magnitude is determined largely by $\tilde{\zeta}$ in Eq. (68), and the feeding term in Eq. (67); both are related to Γ by Eq. (A.24a) and the approximations just before Eq. (A.21a). Γ is largely determined by the energy dependence of the cross sections, without regard to absolute magnitude, because Γ determines the relative magnitudes of successive nuclear waves in the negative ion. The procedure is described in detail in Ref. 49. The calculation of an absolute cross section in this way is illustrated in Fig. 1d.

These semiempirical calculations all rely on numerical solution of Eq. (67). Yelets and Kazansky[(100)] have shown that results which are quite as satisfactory can be obtained very simply by solving Eq. (67) in the WKB approximation.

5.5. *Ab Initio Calculations in the Born–Oppenheimer Approximation*

The first solutions of the nuclear wave equation (67) for resonances sufficiently long-lived to show vibrational structure were obtained by adjusting parameters in $W(\mathbf{R})$ until the calculated vibrational excitation cross sections

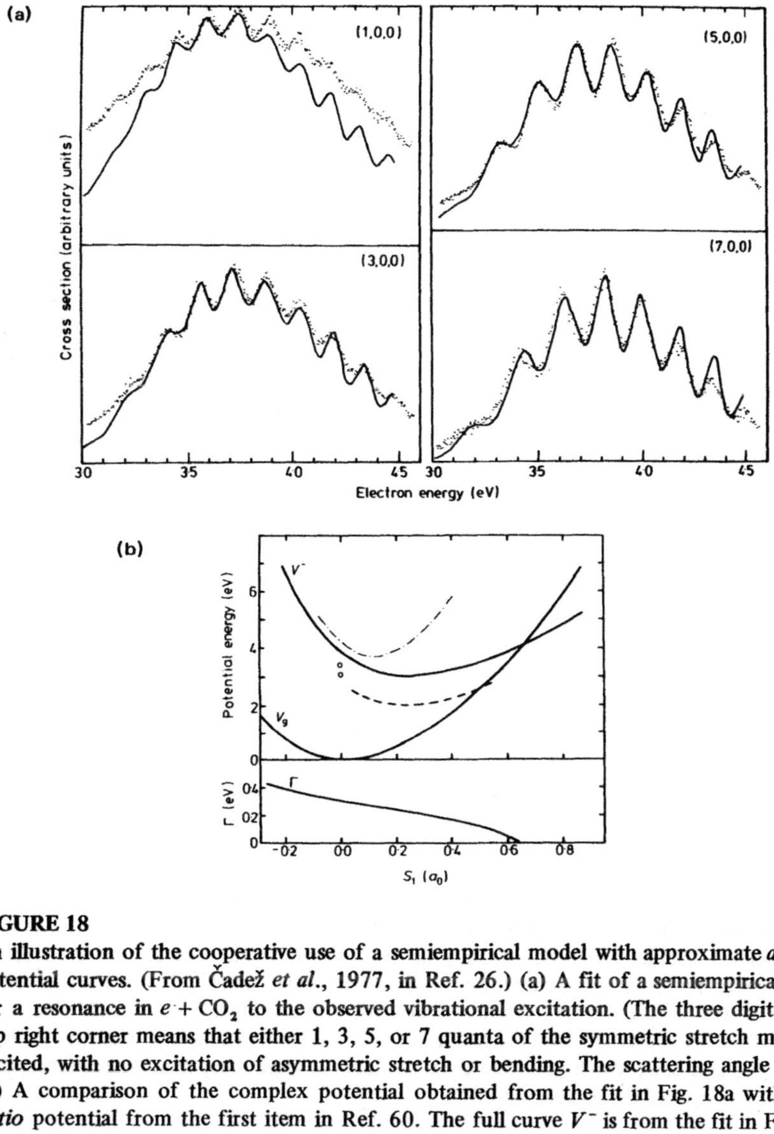

FIGURE 18

An illustration of the cooperative use of a semiempirical model with approximate *ab initio* potential curves. (From Čadež *et al.*, 1977, in Ref. 26.) (a) A fit of a semiempirical model for a resonance in $e + CO_2$ to the observed vibrational excitation. (The three digits in the top right corner means that either 1, 3, 5, or 7 quanta of the symmetric stretch mode are excited, with no excitation of asymmetric stretch or bending. The scattering angle is 20°.) (b) A comparison of the complex potential obtained from the fit in Fig. 18a with an *ab initio* potential from the first item in Ref. 60. The full curve V^- is from the fit in Fig. 18a; the curve —·—·— is an *ab initio* potential from Claydon *et al.*, and the curve — — — an *ab initio* potential from Pacansky *et al.*, Ref. 60. The curve for Γ is from the semiempirical fit.

agreed with experiment.[49,66,68–71] In the case of the shape resonances $N_2^-(^2\Pi_g)$ and $CO_2^-(^2\Pi_u)$, the approximate *ab initio* calculations of W which were available[60,65] turned out to be in fair agreement with the potentials W obtained by parameter adjustment. (The *ab initio* and fitted potential curves are compared in Figs. 13, 15, and 18.)

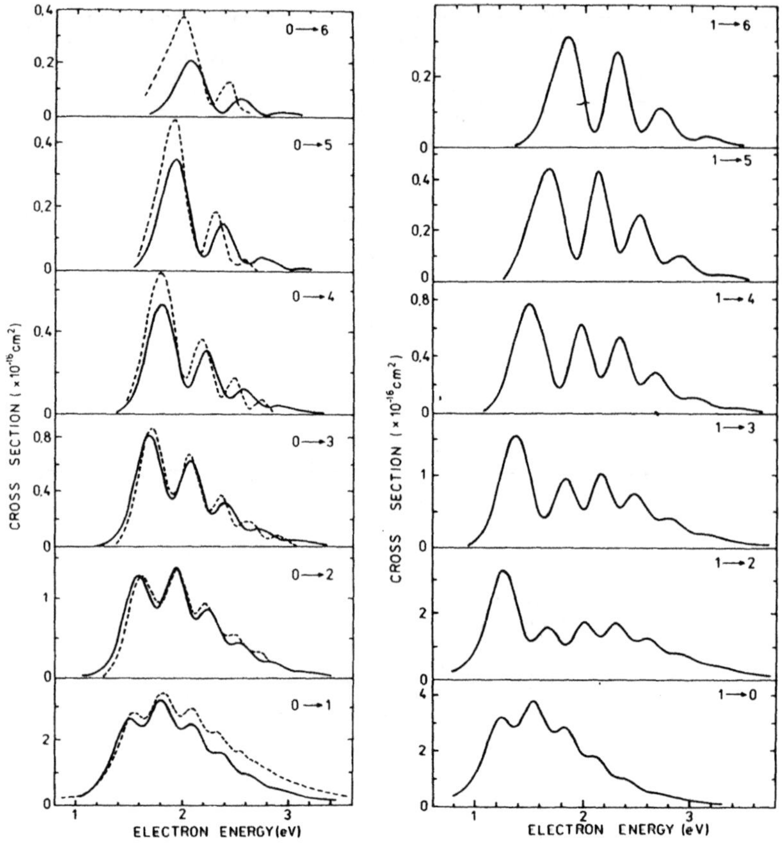

FIGURE 19

An example of the use of a semiempirical resonance model calibrated on measurements of collisions starting from the target ground state; the model is used to predict cross sections for collisions starting from vibrationally excited states. In the diagram on the left, the dashed curves represent experiments on the vibrational excitations of CO near the resonance at about 2 eV; the full curves are from calculations with a semiempirical model fitted to the experiments. The curves in the diagram on the right are from calculations with the calibrated semiempirical model, which yield the cross sections for vibrational transitions starting from vibrationally excited CO. (From Ref. 85.)

The discrepancies between the *ab initio* and fitted curves in Figs. 13, 15, and 18 suggest that the *ab initio* calculations lack important correlation effects. A large part of these correlations should be included by Schneider *et al.*'s technique of optimizing the transient negative ion wave function with fixed nuclei with respect to all $N + 1$ electrons.

A test of Eq. (68) with the potential for $N_2^-(^2\Pi_g)$ calculated by Schneider *et al.*'s technique (the details of that calculation are described in Chapter 7,

Section 4.4) has been made by Hazi *et al.*,[52] using the complex potential calculated by Schneider, Le Dourneuf, and Vo Ky Lan.[64] The theory is now completely *ab initio*. Hazi *et al.* report that their results agree with those described in the next paragraph, so that the methods are consistent.

Schneider, Le Dourneuf, and Vo Ky Lan[64] calculated the scattering at the $N_2^-(^2\Pi_g)$ resonance in a way different from that of Hazi *et al.* They did *ab initio* calculations by the R-matrix technique (see Chapter 7, Section 4.4), which uses real rather than complex potentials. The coupling of the trapped state to the continuum, which the imaginary part of a complex W would represent, is introduced by the use of more than a single Born–Oppenheimer wave function in a volume of molecular dimensions. Between them, these wave functions represent both the trapped state and the continuum in the neighborhood of the molecule. In the calculations published so far,[64] each of these electronic wave functions is a product of the same N_2-core function and a function $F_k(r)$ for the extra electron. The orbitals of the N_2-core are taken from a self-consistent field calculation for the $N_2^-(^2\Pi_g)$ state optimized with respect to all $N+1$ electrons. The functions F_k are eigenfunctions of a one-electron Hamiltonian, and have vanishing radial derivatives at a distance of $8a_B$ from the mass center of the molecule. The associated eigenvalues $E_k(R)$ are found to have stable minima; Schneider, Le Dourneuf, and Vo Ky Lan use $E_k(R)$ as the potential in a nuclear wave equation to yield a set of stationary states, $\theta_{kq}(R)$, which are numbered by q. The complete wave function may be written $\Sigma_{kq}\theta_{kq}(R)\Psi_k(\mathbf{R})\alpha_{kq}$, where $\Psi_k(\mathbf{R})$ is the product of F_k and the N_2 core function for nuclear coordinates $\mathbf{R}$. The α_{kq} are numerical coefficients fixed by the R-matrix technique so as to match the wave function to spherical Bessel and Neumann functions outside the sphere of radius $8a_B$; long-range potentials outside the sphere are neglected. Schneider, Le Dourneuf, and Vo Ky Lan find that only the $l=2$ partial wave is important asymptotically. The resulting vibrational cross sections are shown in Fig. 11 of Chapter 7, Section 4.4 and are discussed there.

5.6. *Other Calculations*

One way of taking the relaxation of the nuclei at an electron resonance into account is to expand the wave function in a series $\Sigma_v X_v(\mathbf{R})F_v(\mathbf{r})$, where X_v is a vibrational state of the neutral target and F_v a function of the electronic coordinates. If the resonance indicates by its vibrational structure that its lifetime is comparable to a vibrational period, then it is a good approximation to ignore rotation of the molecule, which occurs with a period about 100 times as long as the vibrational period of a small molecule at room temperature. Rotational excitation can then be treated by the impulse approximation (Chapter 2, Section 3.2). Calculations like this have been named "hybrid" by Temkin,[63] who introduced them. An important aspect of the hybrid method is that it does not use the Born–Oppenheimer approximation, so that the functions F_v are treated as if they were all different. Within the Born–Oppenheimer approximation, one would replace F_v by $F(\mathbf{r}; \mathbf{R})\alpha_v$, where the function F is the same for all v, and α_v is a numerical coefficient which depends on energy; there would be only a single electronic function to calculate. Therefore calculations within the Born–Oppenheimer approximation use much less computer time than the hybrid method. As a consequence, the hybrid method does not presently yield such good agreement with experiment as the *ab initio* Born–Oppenheimer calculations of Schneider, Le Dourneuf, and Vo Ky Lan,[(64)] or of Hazi *et al.*[(52)] See also Chapter 7, Section 4.2 for the description of the hybrid theory.

Some authors have explored new ways of calculating vibrational excitation at a resonance in the Born–Oppenheimer approximation. Nesbet[(72)] shows how one can take nuclear relaxation into account by modifying the element $\langle v'|S(\epsilon, \mathbf{R})|v\rangle$ of the S matrix in the impulse approximation (Chapter 7, Section 4.1); ϵ is the energy of the incident electron, and $\mathbf{R}$ stands for the nuclear coordinates. v and v' are the initial and final nuclear vibrational states. Nesbet replaces ϵ by $\epsilon - H_n$ where H_n is the operator for the nuclear energy, including the kinetic energy. He shows how the matrix elements of the modified S can be calculated from the complex potential $W(\mathbf{R})$ for the negative ion, and demonstrates good agreement with experiment when using the parametrized potential from Ref. 54. He calls this method the energy-modified adiabatic approximation.

Also using the Born–Oppenheimer approximation, Domcke and Cederbaum[(73)] obtain elegant analytic expressions for the cross sections from Eq. (68) by replacing $W(\mathbf{R})$ by an expression containing terms of order R^0, R^1, and R^2; they replace Eq. (67) by an equation with a non-Hermitian Hamiltonian to describe the transient negative ion.

6. Current Developments

So far in this chapter, we have discussed the collisions of electrons with molecules under several restrictions:

1. The impact energies were low, generally below 5 eV, so that electronic excitation was impossible; therefore the electronic structure of the target was polarized, but not disrupted by excitation.
2. The targets were simple, usually diatomic and occasionally triatomic.
3. The fixed-axis approximation used neglected rotational effects.

We shall close with a few brief comments on recent advances to lift these restrictions.

6.1. Electronic Excitation

Huetz *et al.*[86] have studied a resonant state of N_2 in which the electronic structure of the N_2 target is disrupted, having a single π_u hole in the $(\cdots \pi_u^4 \sigma_g^2)$ electronic core of N_2, and two additional π_g electrons above the core, with overall symmetry $^2\Pi_u$. The resonance appears in the impact of 8.5–9.5 eV electrons on N_2 molecules in their $^1\Sigma_g^+$ ground state $(\cdots \pi_u^4 \sigma_g^2)$, and decays predominantly to the $A^3\Sigma_u^+$ electronically excited state $(\cdots \pi_u^3 \sigma_g^2 \pi_g)$. However, there is also some decay to the electronic ground state, which is observed to lead to vibrational excitation with a final v up to $v = 25$! (See Chapter 5, Section 2.7, Fig. 18). The excitation of the $A^3\Sigma_u^+$ state through the $^2\Pi_u$ resonance was first discussed by Čadež and Fayard,[89] with a modification of the boomerang model; Huetz *et al.*[86] have further modified the model to describe the vibrational excitation of the electronic ground state with satisfactory results. For further details see Chapter V, Section 2.7.

6.2. High Excitation Energies

Figure 1a shows observations on vibrational excitation in $e + N_2$ at impact energies up to 30 eV. The broad bump between 20 and 25 eV was first reported by Pavlovic *et al.*[96] A more recent example of a similar high-energy bump in $e + CO_2$ is shown in Fig. 20a.[87] The case of $e + CO_2$ (and others!) have been studied by[88] Dehmer, Dill, and their collaborators in the vibrational impulse approximation with a model where the incident electron sees a potential due to the electrons of the target frozen in their ground state (for further details of the model see Chapter 7, Section 3.6); a very satisfactory example of such a calculation on $e + CO_2$ is shown in Fig. 20b. Results of similar calculations on $e + N_2$ are shown in Chapter 7, Section 3.6, Fig. 4.

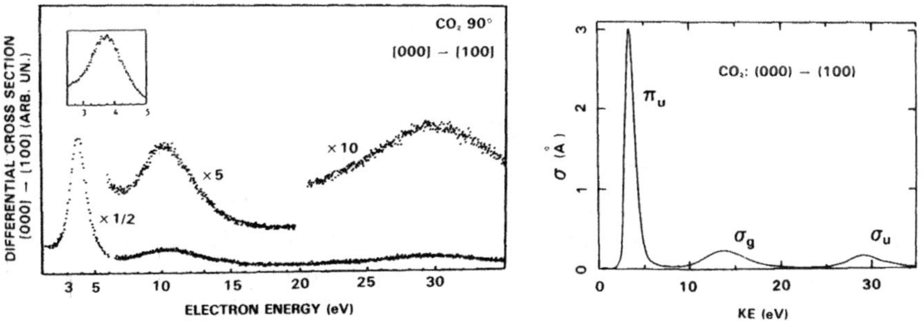

FIGURE 20
$e + CO_2$; excitation of one symmetric stretch vibrational quantum at impact energies up to 35 eV. (a) Observations (from Ref. 87). There are three resonances at ~4, 10, and 30 eV. (b) Calculated cross sections, with a one-electron model in the impulse approximation (from Ref. 88).

At impact energies above 5–10 eV, one expects electronic excitation to occur (the case described in Section 6.1 is an example). What effect this has on the high-energy bumps in vibrational excitation is still an open question.

6.3. More Complicated Targets

Recent work on electron impact on molecules more complicated than diatomics and triatomics has been reviewed by Jordan, Michejda, and Burrow.(90) An example is(91) $e + C_6H_6$ (benzene) where a shape resonance is observed due to the temporary occupation of the lowest vacant π-electron orbital. The fact that the charge density of the extra electron has the sixfold symmetry of the molecule leads to interesting selection rules for vibrational excitation.(91)

6.4. Rotational Effects

Changes of rotational state in electron–molecule scattering can usually be analyzed in the rotational impulse approximation, with wave functions calculated with a fixed-axis model. This is described in detail in Chapter 2, Section 3.2. Individual rotational states in an electron–molecule collision were resolved first for the $e + H_2$ collisions by Ehrhardt and Linder in 1968,(92) in the region of the low $^2\Sigma_u^+$ shape-resonance peak at $\sim$3 eV; the rotational impulse approximation works quite well, even if one ignores the nonresonant part of the scattering amplitude.(98, 76) The extension to a comparison of $e + H_2$ and $e + D_2$ was observed and discussed by Chang and Wong.(93)

In molecules with nuclei more massive than H, for example N_2, the larger moment of inertia pushes the rotational levels so close together that individual rotational levels can usually not be observed in electron scattering; however, it may still be possible to detect a broadening of electron lines by rotational excitation.(94) One recent application of this broadening has been(95) to separate the vibrationally elastic scattering in $e + N_2$ into resonant and nonresonant parts. The extreme tails of the rotationally broadened electron lines are dominated by resonant scattering, because near the $^2\Pi_g$ shape resonance in $e + N_2$, the incoming and outgoing electrons both supply two units of angular momentum, so that the largest rotational changes in resonant scattering are $\Delta J = \pm 4$; this is much larger than the changes of angular momentum which nonresonant scattering can produce in $e + N_2$ at an energy of a few volts. See also Chapter 2, Section 3.3, which includes description of the analysis of the broadening due to rotational excitation.

Chapter 2 discusses many other aspects of rotational transitions that occur simultaneously with vibrational excitation.

Acknowledgments

Such contributions to this material as I have been able to make would not have been possible without stimulating interaction with so many colleagues that one could not possibly mention them all. First, there were the experimenters, George Schulz and many others, without whose maddening insistence that mechanisms should be explicable in terms of simple pictures, much that now seems clear would not have become so. My gratitude goes also to the theorists whose cooperative efforts finally persuaded a not always enthusiastic community that it really was not all that complicated. My collaborators Dick Abram, Norman Bardsley, Dominique Billy, David Birtwistle, John Briggs, Louis Dubé, Jean-Pierre Gauyacq, and Franz Mandl know best how much I owe them for the fact that things were actually finished. This work was partly supported by grant N00014-75-C-0559 from ARPA, monitored by the Office of Naval Research.

Appendix A: Resonance Scattering of Electrons by a Target with Fixed Nuclei

Resonance scattering of electrons appeared in atomic physics in the early 1960s. The theoretical approaches which were then developed for the fixed-nuclei problem were reviewed by Smith[21] in 1966. In electron–molecule scattering, one has to handle nuclei which may move, but very slowly compared with the electrons. Appendices A and B of this chapter aim at that problem.

Let an isolated resonance[21] turn into the boxed-in compound state $b(\mathbf{r}, \mathbf{R})$, in the sense of Section 2.6, when the system of particles is enclosed in a potential box. The box potential will be denoted by $v(\mathbf{r}; \mathbf{R})$, where $\mathbf{r}$ stands for the coordinates of the electrons, and $\mathbf{R}$ for the coordinates of the nuclei. In this appendix, the $\mathbf{R}$ are regarded as fixed. If $H_e(\mathbf{r}; \mathbf{R})$ is the Hamiltonian for the electrons, then b satisfies

$$(H_e + v - E_0)b = 0, \tag{A.1}$$

where $E_0 = E_0(\mathbf{R})$ is the eigenvalue which corresponds to the boundary condition $b \to 0$ if one of the electrons moves off to infinity.

In Fig. 6b, b is illustrated by the curve with the strong core and the weak tail extending to the box. The separation of the wave function into the strong core and the weak tail is the characteristic which makes it a resonance, that is, a state which is almost bound. Because of the relatively small amplitude in the tail, the exact position of the box represented by v is not very important, provided only that the box is large enough not to interfere with the structure of the central core in b.

Let $\Psi_i(\mathbf{r}, \mathbf{R})$ be the wave function in a scattering problem at an energy E

near E_0, defined as in Eq. (2b), with the suffix i standing for all the quantum numbers in the initial state. Ψ_i satisfies

$$(H_e - E)\Psi_i = 0. \tag{A.2}$$

Without loss of generality, one may write

$$\Psi_i = \xi_i b + \Psi_i^P, \tag{A.3}$$

where $\xi_i = \xi_i(\mathbf{R}; E)$, and where $\Psi_i^P\,[= \Psi_i^P(\mathbf{r}, \mathbf{R}; E)]$ satisfies

$$\langle b|\Psi_i^P\rangle = 0, \tag{A.4}$$

where the integration runs over the electronic coordinates while the nuclei are held fixed. The function Ψ_i^P contains the incoming and the scattered waves.

Projecting Eq. (A.2) onto b, one has

$$\langle b|(H_e - E)|\Psi_i\rangle = 0,$$

or, with Eq. (A.1a),

$$(E_0 - E)\xi_i - \langle b|v|b\rangle\xi_i = -\langle b|H_e|\Psi_i^P\rangle. \tag{A.5}$$

To project Eq. (A.2) orthogonally to b, introduce the operator

$$P = |b\rangle\langle b|,$$

so that

$$(1 - P)|b\rangle = 0.$$

P depends on $\mathbf{R}$ because the ket $|b\rangle$ and the bra $\langle b|$ do. The projection of Eq. (A.2) orthogonal to b is

$$(1 - P)(H_e - E)\Psi_i = 0,$$

or, with Eq. (A.3),

$$(1 - P)(H_e - E)\Psi_i^P = -\xi_i(1 - P)H_e b. \tag{A.6}$$

The solution of Eq. (A.6) is made complicated by the condition (A.4a). The simplest way to satisfy Eq. (A.4) is to omit the term $PE\Psi_i^P$ in Eq. (A.6), replacing Eq. (A.6) by

$$[(1-P)H_e - E]\,\Psi_i^p = -\xi_i(1-P)H_e b. \qquad \text{(A.6')}$$

By projecting Eq. (A.6') onto b, and noting that $\langle b|(1-P) = 0$, one finds that Eq. (A.4a) is satisfied unless $E = 0$. Since there is no reason why this isolated case should be particularly important, we shall ignore it.

To solve Eq. (A.6'), introduce a solution Φ_i^p of the homogeneous equation

$$[(1-P)H_e - E]\,\Phi_i^p = 0, \qquad \text{(A.7a)}$$

and a solution $b^{(+)}$ of the inhomogeneous equation

$$[(1-P)H_e - E]\,b^{(+)} = -(1-P)H_e b. \qquad \text{(A.7b)}$$

Φ_i^p is defined by the demand that if one electron moves off to infinity, then Φ_i^p is to behave like a product of the initial target state and the incident plane wave, plus an outgoing scattered wave; we shall refer to Φ_i^p as the "potential scattering function." The function $b^{(+)}$ is defined by the demand that if one electron moves off to infinity, $b^{(+)}$ is to behave like a sum of products of an outgoing wave and a state of the residual molecule. We shall refer to $b^{(+)}$ as the "outgoing part." One may now rewrite the solution Ψ_i^p of Eq. (A.6') in the form

$$\Psi_i^p = \Phi_i^p + \xi_i b^{(+)}. \qquad \text{(A.8)}$$

By projecting Eqs. (A.7a) and (A.7b) onto b, it follows that

$$\langle b|\Phi_i^p\rangle = 0 \quad \text{and} \quad \langle b|b^{(+)}\rangle = 0, \qquad \text{(A.9)}$$

unless $E = 0$; as with Ψ_i^p, we shall ignore this possibility. Condition (A.9a) enables us to modify Eqs. (A.7a) and (A.7b) by replacing the operator $(1-P)H_e$ on the left by the Hermitian operator $(1-P)H_e(1-P)$, because the added terms $(1-P)H_e P\Phi_i^p$ and $(1-P)H_e P b^{(+)}$ vanish.

It is convenient to define a "decaying state" by

$$S \equiv b + b^{(+)}. \qquad \text{(A.10)}$$

S depends on E, because $b^{(+)}$ does. Then with Eqs. (A.3a), (A.8a), and (A.10a), one may write

$$\Psi_i = \Phi_i^p + \xi_i S; \qquad \text{(A.10')}$$

i.e., Ψ_i is the sum of a potential scattering function and a decaying state.

To determine ξ_i, rewrite Eq. (A.5) in the form

$$[\langle b|H_e|b\rangle - E]\,\xi_i = -\langle b|H_e|(\Phi_i^P + \xi_i b^{(+)})\rangle \tag{A.11}$$

by using Eq. (A.8) and the relation $\langle b|H_e|b\rangle = E_0 - \langle b|v|b\rangle$, which follows by projecting Eq. (A.1) onto b. With Eqs. (A.10) and (A.11), one now has

$$\xi_i = \frac{\langle b|H_e|\Phi_i^P\rangle}{E - \hat{W}}, \tag{A.12}$$

where

$$\hat{W} = \hat{W}(\mathbf{R}; E) \equiv \langle b|H_e|S\rangle. \tag{A.13}$$

$\hat{W}$ will generally be complex because of the complex outgoing wave $b^{(+)}$ in S.

If the binding mechanism which holds the core together in the resonance in Fig. 6b is strong, the term on the right of Eq. (A.7b), which acts as the source of $b^{(+)}$, will be small. In other words, the coupling term $(1-P)H_e b$ between b and the part of function space orthogonal to b will be small. The outgoing part $b^{(+)}$ will then be very small compared with b in the central core, so that $\hat{W}$ may be written as the sum of E_0 and a small correction:

$$\hat{W} = E_0 + \left(\Delta - \frac{i}{2}\Gamma\right), \tag{A.14}$$

where, according to Eq. (A.1),

$$\Delta - \frac{i}{2}\Gamma \equiv -\langle b|v|b\rangle + \langle b|H_e|b^{(+)}\rangle. \tag{A.15}$$

By definition, Δ is the real part of the right-hand side, and $-\Gamma/2$ the imaginary part. Therefore, ξ_i has a pole close to E_0 at $E = W$, where W is a root of the equation

$$W - \hat{W}(\mathbf{R}; W) = 0. \tag{A.16}$$

Clearly, W depends only on $\mathbf{R}$ and not on E. Since $\hat{W}$ is generally complex, so is W.

To see the physical significance of W, consider the equation satisfied by the decaying state S, which can be constructed by rearranging Eq. (A.7b). Taking the term on the right over to the left, and using Eq. (A.10), one has

$$(1-P)H_e S - E b^{(+)} = 0.$$

Write $PH_eS = b\hat{W}$, with the aid of Eq. (A.13), and add $(E - \hat{W})b^{(+)}$ to both sides to get

$$[H_e - \hat{W}(\mathbf{R}; E)]\, S(\mathbf{r}, \mathbf{R}; E) = [E - \hat{W}(\mathbf{R}; E)]\, b^{(+)}(\mathbf{r}, \mathbf{R}; E). \qquad \text{(A.17)}$$

If E is set equal to the root W of Eq. (A.16), the term on the right vanishes, and one obtains

$$[H_e - W(\mathbf{R})]\, \tilde{S}(\mathbf{r}; \mathbf{R}) = 0, \qquad \text{(A.18)}$$

where

$$\tilde{S}(\mathbf{r}; \mathbf{R}) \equiv S(\mathbf{r}; \mathbf{R}; W). \qquad \text{(A.19)}$$

According to Eq. (A.18a), $\tilde{S}$ is a "Siegert state,"[23] having the eigenvalue W. Equation (A.7b) tells us that if one of the electrons moves off to infinity, $\tilde{S}$ behaves like a sum of products of a residual molecular state and an outgoing wave, with total energy W.

Worth noting is that Eq. (A.18) does not contain the box potential v which was necessary to define the boxed-in compound state b. Therefore W is a property of H_e, and independent of any devices one might introduce to define the first approximation S to $\tilde{S}$.

It follows from Eq. (A.17) that $\hat{W}$ is a good approximation to W if $E \cong \text{Re}\, \hat{W}$. According to Eqs. (A.10), (A.13), and (A.14), the factor $(E - \hat{W})$ in Eq. (A.17) becomes $-i\, \text{Im}\, \langle b|H_e|b^{(+)}\rangle = i\Gamma/2$, which will be very small compared with $\hat{W}$ if the coupling between b and the continuum is weak. The term $(E - \hat{W})b^{(+)}$ in Eq. (A.17) is therefore very small compared with $\hat{W}S\, [= \hat{W}(b + b^{(+)})]$ inside the compound state because

$$|(E - \hat{W})| \ll |\hat{W}| \quad \text{and} \quad |b^{(+)}| \ll |b|.$$

Therefore, if $E \cong \text{Re}\, \hat{W}$, the term on the right of Eq. (A.17) may be neglected inside the molecule. Since S joins onto the outgoing part $b^{(+)}$ belonging to energy E at the molecular surface, S satisfies the definition[31] of a "Kapur–Peierls state," and $\hat{W}$ becomes the corresponding Kapur–Peierls energy.

The scattering amplitude through a resonance is obtained by substituting Eqs. (A.10′) and (A.12) into the general expression (2a), so that

$$T_{f\leftarrow i} = \left(-\frac{1}{2\pi}\right) \frac{\langle \Phi_f | V_f | S \rangle \langle b | H_e | \Phi_i^P \rangle}{E - \hat{W}}, \qquad \text{(A.20)}$$

where V_f is the interaction between the departing electron and the rest.

The numerator in Eq. (A.20) lacks symmetry between the initial and final states. A more symmetric form may be obtained by noting that the resonant part of the scattered wave is contained in $b^{(+)}$, and expanding $b^{(+)}$ [in Eq. (A.7b)] in terms of the potential scattering functions from Eq. (A.7a) which asymptotically is a product of the final target state and an incoming scattered wave; such functions will be denoted by $\Phi_f^{p(-)}$. These functions are useful because asymptotically, $b^{(+)}$ must come entirely from the plane wave components. After some algebra, one obtains an expression for $T_{f\leftarrow i}$ in which the factor $\langle\Phi_f|V|S\rangle$ in Eq. (A.20) is replaced by $\langle\Phi_f^{p(-)}|H_e|b\rangle$, so that symmetry between the initial and final states is regained.

For practical applications, it is convenient to approximate the exact result (A.20a) by

(1) replacing $\langle\Phi_f|V_f|S\rangle \to \langle\Phi_f|V|b\rangle$,

(2) replacing $\langle b|H_e|\Phi_i^p\rangle \to \langle b|V_i|\Phi_i\rangle$,

where V_i is the interaction between the incident electron and the rest; then

$$T_{f\leftarrow i} \cong \left(-\frac{1}{2\pi}\right)\frac{\langle\Phi_f|V_f|b\rangle\langle b|V_i|\Phi_i\rangle}{E-\hat{W}}. \tag{A.21}$$

The justification for the first replacement is that most of the integral $\langle\Phi_f|V_f|S\rangle$ comes from the region where all particles are together, because V_f decreases as the departing electron moves out, and in the inner region $|b^{(+)}| \ll |b|$. The justification for the second replacement comes from approximating $\Phi_i^p \to (1-P)\Phi_i$, i.e., taking into account Eq. (A.9), but ignoring polarization effects in Φ_i^p. One then has

$$\langle b|H_e|\Phi_i^p\rangle \cong \langle b|(H_e - \langle b|H_e|b\rangle)|\Phi_i\rangle;$$

if the initial energy is close to the resonance (i.e., $E \cong \langle b|H_e|b\rangle$), then $(H_e - \langle b|H_e|b\rangle)|\Phi_i\rangle \cong V_i|\Phi_i\rangle$, which leads to the second replacement.

The sign of Γ in Eq. (A.14) is important in Section 2.5. Multiply Eq. (A.18) by $\tilde{S}^*$, and subtract the complex conjugate equation. Integrate over a volume a little larger than the core of the compound state, to obtain

$$\frac{\left(-\frac{1}{2}\right)\sum_n \int_{\mathrm{IR}} [\tilde{S}^*\nabla_n^2\tilde{S} - \tilde{S}\nabla_n^2\tilde{S}^*]\,d\mathbf{r}}{\int_{\mathrm{IR}} d\mathbf{r}\,|\tilde{S}|^2} = W - W^* = -i\Gamma,$$

where the suffix n runs over all the $(Z+1)$ electrons, and $d\mathbf{r}$ stands for the $3(Z+1)$-dimensional element of volume of all their coordinates. The suffix IR on the integrals means that the integration volume should be cutoff outside the core of the compound state. A partial integration may be done in each term in the numerator, to yield

$$\Gamma = \frac{\left(-\frac{i}{2}\right) \sum_n \int_{S_n} dS_n \left[\tilde{S}^* \frac{\partial \tilde{S}}{\partial r_n} - \frac{\partial \tilde{S}^*}{\partial r_n} \tilde{S}\right]}{\int_{\mathrm{IR}} d\mathbf{r} |\tilde{S}|^2},$$

where dS_n stands for a product of volume elements for all electrons except the nth, and a surface element for electron n. On the surface for the nth electron, write

$$\tilde{S} = \sum_\alpha \frac{e^{ik_\alpha r_n}}{r_n} S_{\alpha n} \psi_\alpha(\not{n}) Y_{J(\alpha)}(\hat{\mathbf{r}}_n), \tag{A.22}$$

where $\psi_\alpha(\not{n})$ stands for the αth eigenstate of the Hamiltonian of all particles except the nth, and $Y_{J(\alpha)}(\hat{\mathbf{r}}_n)$ for a normalized angular momentum eigenfunction for particle n. $\frac{1}{2}k_\alpha^2$ is the energy of the nth electron when the rest are in state $\psi_\alpha(\not{n})$. $S_{\alpha n}$ is a numerical coefficient. The states $\psi_\alpha(\not{n})$ are orthonormal, i.e., $(\psi_\alpha(\not{n}) Y_{J(\alpha)} | \psi_\beta(n) Y_{J(\beta)}) = \delta_{\alpha\beta}$, where the integration runs over all electronic coordinates except the radial coordinate r_n. Remembering that $dS_n = d\mathbf{r}(\not{n}) d\hat{\mathbf{r}}_n r_n^2$, where $d\mathbf{r}(\not{n})$ is a volume element for all electrons except the nth, and $d\hat{\mathbf{r}}_n$ an element of solid angle for electron n, one finds

$$\Gamma = \frac{1}{2} \sum_{\alpha n} (k_\alpha + k_\alpha^*) |\tilde{S}_{\alpha n}|^2 \Big/ \int_{\mathrm{IR}} d\mathbf{r} |\tilde{S}|^2 > 0. \tag{A.23}$$

The sign of the inequality comes from the fact that $\tilde{S}$ has to be a decaying state, so that the real part of k_α has to be positive.

A useful relation between the matrix element $\langle \Phi_f | V_f | S \rangle$ in Eq. (A.20) and the width Γ follows because the factor $(-1/2\pi)\langle \Phi_f | V_f | S \rangle$ in Eq. (A.20) is the amplitude emerging from the resonant state. It is therefore identical with $Y_{J(\alpha)}(\hat{\mathbf{k}}_f) S_{\alpha n}$ for the same electron leaving in direction $\hat{\mathbf{k}}_f$ from a residual target in state α. If only the electronic ground state is accessible energetically, and we accordingly replace the suffix α by 0, we get, from Eq. (A.23),

$$\Gamma \cong (\mathrm{Re}\, k_0) \cdot |S_{0n}|^2 \cdot \int d\hat{\mathbf{k}}_f |Y_{J(0)}(\hat{\mathbf{k}}_f)|^2 = (\mathrm{Re}\, k_0) \cdot \int d\hat{\mathbf{k}}_f \left| -\frac{1}{2\pi} \langle \Phi_f | V_f | S \rangle \right|^2,$$

or

$$\overline{|\langle\Phi_f|V_f|S\rangle|^2} = \frac{\pi\Gamma}{\mathrm{Re}\,k_0}, \tag{A.24}$$

where the horizontal bar in $\overline{|\langle\cdots\rangle|^2}$ denotes an average over the direction of the outgoing electron. The denominator in Eq. (A.23) has been dropped, being near equal to $\langle b|b\rangle = 1$. For a resonance of physical importance, one can replace Re k_0 by the momentum of the entering electron at the center of the resonance. It has been emphasized by Hazi *et al.*[52] that this replacement is not permissible at very low energies, where the dependence of Φ_f on energy may become very strong.

Appendix B: Resonance Scattering of an Electron by a Molecule with Relaxing Nuclei

In this appendix, we consider the scattering of an electron by a molecule whose nuclei relax during the interaction. We suppose that there is an almost-bound state, in the sense of Appendix A, in which all the electrons (including the projectile) are confined to a volume of molecular dimensions from which one of them may escape by leaking past some imperfect trapping mechanism. For the sake of simplicity, we assume that the interaction of the projectile with the molecule is negligible outside the trapping volume, and that the orientation of the molecule is fixed. Also for the sake of simplicity, we restrict ourselves to diatomic molecules.

Corresponding to each configuration $\mathbf{R}$ of the nuclei, one may define a boxed-in resonance $b(\mathbf{r}, \mathbf{R})$ for the electrons (see Fig. 6b), to approximate the almost trapped state, the nuclei being regarded as fixed for the purpose of this definition. Exactly as in Appendix A [Eq. (A.1a)], the state b satisfies

$$[H_e + v - E_0]\,b = 0, \tag{B.1}$$

where H_e, v, and b are all defined for the same fixed $\mathbf{R}$. b will be normalized so that

$$\langle b|b\rangle_e = 1, \tag{B.2}$$

the suffix e signifying an integration over the electrons while the nuclei are held fixed. Without any approximation, the complete wave function may now be written, analogously to Eq. (A.3),

$$\Psi_i(\mathbf{r}, \mathbf{R}) = \xi_i(\mathbf{R})b(\mathbf{r}, \mathbf{R}) + \Psi_i^p(\mathbf{r}, \mathbf{R}), \tag{B.3}$$

where ξ_i and Ψ_i^p have still to be determined; they now contain nuclear wave functions varying with $\mathbf{R}$, instead of belonging to a fixed $\mathbf{R}$, as in Eq. (A.3); for the moment, we have only the restriction

$$\langle b|\Psi_i^p\rangle_e = 0. \tag{B.4}$$

The suffix i stands for all the quantum numbers in the initial state.

The function ξ_i is the wave function of the nuclei in the transient negative ion whose electronic state is approximated by b. The function Ψ_i^p must contain two parts:

$$\Psi_i^p = \Phi_i^p + B_i; \tag{B.5}$$

B_i is the outgoing wave due to the decay of the negative ion. Φ_i^p is the remainder, including the initial plane wave of the projectile, multiplied by the initial state of the target, as well as outgoing scattered waves. To satisfy Eq. (B.4), we shall force Φ_i^p and B_i to satisfy

$$\langle b|\Phi_i^p\rangle_e = \langle b|B_i\rangle_e = 0. \tag{B.6}$$

The functions ξ_i, Φ_i^p, and B_i have to be constructed so as to satisfy the Schrödinger equation

$$(H-E)\Psi_i = 0, \tag{B.7a}$$

where

$$H = H_e + K, \tag{B.7b}$$

K being the nuclear kinetic energy. It is convenient to split Eq. (B.7a) into two separate equations by projecting for the electrons parallel to b and orthogonally to b:

$$\langle b|(H-E)|\Psi_i\rangle_e = 0 \tag{B.8}$$

and

$$(1-P)(H-E)\Psi_i = 0. \tag{B.9}$$

The operator P is defined by

$$(Pf)(\mathbf{r}, \mathbf{R}) \equiv b(\mathbf{r}, \mathbf{R}) \int d\mathbf{r}' b^*(\mathbf{r}', \mathbf{R}) f(\mathbf{r}', \mathbf{R}), \tag{B.10}$$

where f is an arbitrary function of **r** and **R**.

Equations (B.8) and (B.9) will be solved in the adiabatic approximation, which rests on the assumption that nuclear wave functions vary with **R** much more rapidly than electronic wave functions. Therefore whenever a product of a nuclear and an electronic wave function is differentiated with respect to **R**, the derivative of the electronic factor will be dropped. Thus one may approximate, in Eq. (B.8)

$$\langle b|K|\Psi_i\rangle_e \to K\langle b|\Psi_i\rangle_e \tag{B.11}$$

and in Eq. (B.9)

$$(1-P)K\Psi_i \to K(1-P)\Psi_i = K\Psi_i^p, \tag{B.12}$$

because the derivatives with respect to **R** of the function b are small compared with the derivatives of the nuclear wave functions in Ψ_i. Equations (B.8) and (B.9) may now be replaced by

$$(K-E)\langle b|\Psi_i\rangle_e + \langle b|H_e|\Psi_i\rangle_e = 0 \tag{B.13}$$

and

$$(K-E)(1-P)\Psi_i + (1-P)H_e\Psi_i = 0. \tag{B.14}$$

Equation (B.13) may be rewritten with the aid of Eqs. (B.2), (B.3), (B.4), (B.5), and (B.6),

$$(K-E)\xi_i + \langle b|H_e|(\xi_i b + B_i)\rangle_e = -\langle b|H_e|\Phi_i^p\rangle_e. \tag{B.15a}$$

In Eq. (B.14), one has $(1-P)\Psi_i = \Psi_i^p = B_i + \Phi_i^p$, so that

$$(K-E)(B_i + \Phi_i^p) + (1-P)H_e(B_i + \Phi_i^p) = -(1-P)H_e\xi_i b. \tag{B.16}$$

To solve this, we demand that Φ_i^p satisfy the homogeneous equation

$$(K-E)\Phi_i^p + (1-P)H_e\Phi_i^p = 0, \tag{B.17}$$

and B_i the inhomogeneous equation

$$[K-E+(1-P)H_e]B_i = -(1-P)H_e\xi_i b. \tag{B.18a}$$

The solutions of Eqs. (B.17) and (B.18) satisfy Eq. (B.6). To prove this, project both equations on b, and use the Born–Oppenheimer approximation to write

$$\langle b|K|\Phi_i^p\rangle_e \to K\langle b|\Phi_i^p\rangle_e,$$

$$\langle b|K|B_i\rangle_e \to K\langle b|B_i\rangle_e.$$

One obtains, since $\langle b|(1-P)=0$,

$$(K-E)\langle b|\Phi_i^p\rangle_e = 0, \qquad \text{(B.19)}$$

$$(K-E)\langle b|B_i\rangle_e = 0. \qquad \text{(B.20)}$$

Equations (B.19) and (B.20) are wave equations without sources, for the functions $\langle b|\Phi_i^p\rangle_e$ and $\langle b|B_i\rangle_e$, respectively. Therefore if these functions vanish as $R\to\infty$, they vanish everywhere. (Recall that we are restricting ourselves to diatomic molecules.)

As any one electron moves off to infinity, Φ_i^p is to behave like a product of the initial plane wave and the initial state of the target, plus outgoing waves. B_i is to behave like an outgoing wave.

While the function Φ_i^p may be determined from the single equation (B.17), the functions ξ_i and B_i have to be found by a simultaneous solution of Eqs. (B.15) and (B.18). Recall that the direction $\hat{\mathbf{R}}$ of the nuclear axis is kept fixed, so that only the magnitude R is a variable in $\mathbf{R}=R\hat{\mathbf{R}}$. The nuclear kinetic energy is $K=-(1/2M)\partial^2/\partial R^2$.

It is convenient to construct ξ_i and B_i from equations in which the source term on the right of Eq. (B.18) is replaced by a delta function $\delta(R-R')$; R is the nuclear variable in the equation, and R' the separation of the nuclei in the source. Let the solutions analogous to ξ_i and B_i be $g\equiv g(R, R', \hat{\mathbf{R}})$ and $G\equiv G(\mathbf{r}, R, R', \hat{\mathbf{R}})$; the modified equations become

$$(K-E)g+\langle b|H_e|b\rangle_e g+\langle b|H_e|G\rangle_e = -\delta(R-R'), \qquad \text{(B.15b)}$$

$$[(K-E)+(1-P)H_e]\,G = -(1-P)H_e gb, \qquad \text{(B.18b)}$$

where $Kg\equiv-(1/2M)\partial^2 g/\partial R^2$, $KG\equiv-(1/2M)\partial^2 G/\partial R^2$. The solutions of (B.15a) and (B.18a) are (with $\mathbf{R}=R\hat{\mathbf{R}}$)

$$\xi_i(R\hat{\mathbf{R}}) = \int g(R,R',\hat{\mathbf{R}})dR'\rho(R',\hat{\mathbf{R}}) \qquad \text{(B.21)}$$

$$B_i(\mathbf{r},R\hat{\mathbf{R}}) = \int G(\mathbf{r},R,R',\hat{\mathbf{R}})dR'\rho(R',\hat{\mathbf{R}}), \qquad \text{(B.22)}$$

where $\rho(R,\hat{\mathbf{R}})$ is the value of $\langle b|H_e|\Phi_i^p\rangle_e$ at $R=R\hat{\mathbf{R}}$.

Except at the point $R = R'$, the modified equations (B.15b) and (B.18b) are homogeneous in g and G. By integrating Eq. (B.15b) through the interval $(R' - \epsilon) < R < (R' + \epsilon)$, where ϵ is positive and infinitesimal, one finds that

$$\lim_{\epsilon \to 0} \left[\left(\frac{\partial g}{\partial R} \right)_{R'+\epsilon} - \left(\frac{\partial g}{\partial R} \right)_{R'-\epsilon} \right] = 2M. \tag{B.23}$$

The function G appears in Eq. (B.15b) in the integral $\langle b | H_e | G \rangle_e$, so that one needs to know G only in the volume occupied by the boxed-in resonance. Within this volume, it is convenient to define a new function $\hat{G} \equiv \hat{G}(\mathbf{r}, R, R', \hat{\mathbf{R}})$ so that

$$G(\mathbf{r}, R, R', \hat{\mathbf{R}}) = g(R, R', \hat{\mathbf{R}}) \hat{G}(\mathbf{r}, R, R', \hat{\mathbf{R}}). \tag{B.24}$$

The kinetic energy term in Eq. (B.18b) becomes

$$KG = -\frac{1}{2M} \frac{\partial^2 g}{\partial R^2} \hat{G} - \frac{1}{M} \frac{\partial g}{\partial R} \frac{\partial \hat{G}}{\partial R} - \frac{1}{2M} g \frac{\partial^2 \hat{G}}{\partial R^2}. \tag{B.25}$$

Let the nuclear momentum be k_N, so that g varies with respect to R with a characteristic distance k_N^{-1}, whose order of magnitude is such that $k_N a_B \gg 1$. (A value $k_N a_B \approx 40$ would correspond to a nuclear kinetic energy in the transient negative ion of the order of 1 eV, for a nuclear mass equal to 10 protons.) If the Born–Oppenheimer approximation were valid in the transient negative ion, then $\hat{G}$ would be an electronic wave function which varies with respect to R with a characteristic distance a_B, so that the three terms on the right of Eq. (B.25a) would be of relative order $(k_N a_B)^2 : k_N a_B : 1$, the first being by far the largest. Therefore, for the moment, we drop the second and third terms on the right of Eq. (B.25a), so that in place of Eq. (B.18b), we have

$$\left[-\left(E - \frac{Kg}{g} \right) + (1 - P) H_e \right] \hat{G} = -(1 - P) H_e b. \tag{B.26a}$$

A comparison of Eq. (B.26a) with Eq. (A.7b) in Appendix A shows that

$$\hat{G} = b^{(+)} \left(\mathbf{r}, \mathbf{R}, E - \frac{Kg}{g} \right), \tag{B.26b}$$

where $Kg/g \equiv Kg(R, R', \hat{\mathbf{R}})/g(R, R', \hat{\mathbf{R}})$. In other words, $\hat{G}$ is the tail of outgoing waves of an electronic quasistationary state, the energy available to the electrons being $E-(Kg/g)$, that is the energy left over when the local nuclear kinetic energy Kg/g is subtracted from the total energy E. (It is worth noting that the division by g in the ratio Kg/g usually offers no difficulty, because the damping of g by autoionization makes g a complex wave, whose real and imaginary parts do not vanish at the same points.) Equation (B.15b) becomes, with Eqs. (B.26b), (A.10), and (A.13)

$$(K-E)g+\hat{W}\left(\mathbf{R};E-\frac{Kg}{g}\right)g = -\delta(R-R'). \qquad \text{(B.27a)}$$

To simplify Eq. (B.27), write

$$E-\frac{Kg}{g} = W', \qquad \text{(B.28a)}$$

so that Eq. (B.27) becomes

$$-W'+\hat{W}(\mathbf{R},W') = -\frac{\delta(R-R')}{g}. \qquad \text{(B. 27b)}$$

Comparison with Eq. (A.16) shows that, except at $R=R'$, one has

$$W' = W(\mathbf{R}), \qquad \text{(B.29a)}$$

where $W(\mathbf{R})$ is the electronic Siegert energy from Eq. (A.16). Note that the dependence of W' on R', being still present in Eq. (B.27b), has now disappeared.

Using Eq. (B.29) in Eq. (B.28), and multiplying by g, we get $[K+W(\mathbf{R})-E]\,g(R,R',\hat{\mathbf{R}})=0$ everywhere except at $R=R'$. We denote by g_- the solution of this equation in the region $R<R'$ which satisfies $g_-(0,R',\hat{\mathbf{R}})=0$, and by g_+ the solution in $R>R'$ which behaves as $R\to\infty$ either like a bound state or like an outward traveling wave, depending on the energy available. The complete solution of Eq. (B.27a) is obtained by setting $g=g_\pm$ in $R\gtrless R'$, assigning multiplicative constants to g_+ and g_- such that $g_+(R',R',\hat{\mathbf{R}})=g_-(R',R',\hat{\mathbf{R}})$ and such that [to satisfy Eq. (B.23)] $(\partial g_+/\partial R)_{R=R'}-(\partial g_-/\partial R)_{R=R'}=2M$. The function g now satisfies

$$[K+W(\mathbf{R})-E]\,g(R,R',\hat{\mathbf{R}}) = -\delta(R-R'). \qquad \text{(B.28b)}$$

The function $\hat{G}$ can now be written, except at $R=R'$,

$$\hat{G} = b^{(+)}(\mathbf{r},\mathbf{R};W(\mathbf{R})), \qquad \text{(B.28c)}$$

if one substitutes $Kg/g = E - W(\mathbf{R})$ from Eq. (B.28b) into Eq. (B.26b). Because this expression for $\hat{G}$ is independent of R', one has

$$B_i(\mathbf{r}, R\hat{\mathbf{R}}) = \xi_i(R\hat{\mathbf{R}})b^{(+)}(\mathbf{r}, \mathbf{R}; W(\mathbf{R})), \tag{B.29}$$

as one can see by substituting Eqs. (B.24) and (B.28c) into Eq. (B.22), and using Eq. (B.21).

An equation for ξ_i is obtained if one multiplies Eq. (B.28b) by $\rho(R', \hat{\mathbf{R}})$, and integrates over R'. Then the function g on the left of Eq. (B.28b) is replaced by ξ_i [according to Eq. (B.21)], while the term on the right becomes

$$-\int \delta(R - R')dR'\rho(R', \hat{\mathbf{R}}) = \rho(R, \hat{\mathbf{R}}) \equiv \langle b|H_e|\Phi_i^P\rangle_e.$$

Therefore ξ_i satisfies

$$[K + W(\mathbf{R}) - E]\,\xi_i(\mathbf{R}) = -\langle b|H_e|\Phi_i^P\rangle_e. \tag{B.30}$$

Equation (B.30) is discussed in Section 5, where it is quoted as Eq. (67). It shows that ξ_i is the solution of a wave equation with a source $-\langle b|H_e|\Phi_i^P\rangle$, and with a local complex potential given by the Siegert energy $W(\mathbf{R})$.

We have still to justify the crucial approximation in the derivation of Eq. (B.30), namely, the dropping of the last two terms in Eq. (B.25). It is only necessary to estimate the dropped terms with the first approximation (B.26b) for $\hat{G}$. This first approximation says that $\hat{G}$ is indeed an electronic wave function, so that our previous estimate of the relative magnitudes of all terms in Eq. (B.25) is correct. The larger of the two neglected terms, $-(1/M)(\partial g/\partial R)(\partial G/\partial R)$, is of the order of magnitude $v_N a_B^{-1}\, g\hat{G}$, where v_N is the nuclear velocity (since $-i\partial/\partial R$ is the nuclear momentum operator). For the values $k_N a_B \approx 40$ and $M = 10$ proton masses, which we quoted above, one has $v_N a_B^{-1} \approx 0.05$ eV; this has to be small compared with the terms retained, which are given by the square brackets in Eq. (B.26a). The first term in the square brackets, $(E - Kg/g)$, being the difference of the total energy E and the local kinetic energy Kg/g of the nuclei, is the energy of the electrons, and therefore approximately equal to the energy E_0 of the electronic resonance. The second term in the square brackets, namely, $(1-P)H_e\hat{G}$, may be decomposed into contributions from eigenstates of $(1-P)H_e(1-P)$, because $\hat{G}$ is defined to be orthogonal to b. All eigenvalues may contribute, except E_0, which is excluded by the projection operators $(1-P)$. Therefore the square brackets in Eq. (B.26a) are of the order of ΔE_e, the spacing of the electronic energy levels in a box that is just big enough to contain the resonant state. Since the box will have to be of atomic dimensions, $\Delta E_e \gtrsim 1$ eV; therefore $\Delta E_e \gg v_N a_B^{-1}$, so that the neglected terms in Eq. (B.25a) are smaller than the terms retained by at least one order of magnitude.

It is important that this justification of the Born–Oppenheimer approximation comes from a consideration of what happens when all the electrons are together in the box used to define b. If any electron passes out of the box, the Born–Oppenheimer approximation may break down. In other words, we claim that the electrons follow the nuclei adiabatically in the quasistationary state, but any electron which leaves the resonance need no longer follow adiabatically.

To calculate cross sections for vibrational excitation, use Eqs. (B.3), (B.5), and (B.29) in Eqs. (2a) and (1), and retain only the resonant term $\xi_i(b + b^{(+)})$ in the wave function Ψ_i. The scattering amplitude becomes

$$T_{fi}(\mathbf{k}_f, \mathbf{k}_i) \cong \left(-\frac{1}{2\pi}\right)(\Phi_f|V|\xi_i(b + b^{(+)}))$$

$$= \left(-\frac{1}{2\pi}\right)\int d\mathbf{R}\,\mathscr{R}_f^*(\mathbf{R})\langle\Phi_f^F(\mathbf{r}, \mathbf{R})|V|b\rangle_e \xi_i(\mathbf{R}), \qquad \text{(B.31)}$$

where we have written

$$\Phi_f(\mathbf{r}, \mathbf{R}) = \mathscr{R}_f(\mathbf{R})\Phi_f^F(\mathbf{r}, \mathbf{R}), \qquad \text{(B.32)}$$

as in Appendix C, and dropped the small term $b^{(+)}$. $\mathscr{R}_f(\mathbf{R})$ is the vibrational wave function in the final molecular state, and $\Phi_f^F(\mathbf{r}, \mathbf{R})$ the electronic wave function, with the nuclei held fixed. The reason for splitting up the integrand in Eq. (B.31a) in this way is that $\mathscr{R}_f^*(\mathbf{R})$ and $\xi_i(\mathbf{R})$ are nuclear wave functions, and therefore vary with $\mathbf{R}$ with a wavelength of order $0.1\,a_B$, much more rapidly than the factor $\langle\cdots\rangle_e$, which contains only electronic wave functions and therefore varies with the much larger characteristic length a_B.

In calculating ξ_i from Eq. (B.30), one may replace the term on the right by $-\langle b|V|\Phi_i\rangle_e$, by approximating $\Phi_i^P \cong (1 - P)\Phi_i$ [as in the transition from Eq. (A.20) to Eq. (A.21) in Appendix A]. A factor $\mathscr{R}_i(\mathbf{R})$ for the initial vibrational wave function may be extracted by writing

$$-\langle b|V|\Phi_i\rangle_e = -\langle b|V|\Phi_i^F\rangle_e \mathscr{R}_i(\mathbf{R}),$$

where

$$\Phi_i(\mathbf{r}, \mathbf{R}) = \mathscr{R}_i(\mathbf{R})\Phi_i^F(\mathbf{r}, \mathbf{R}).$$

The function $\Phi_i^F(\mathbf{r}, \mathbf{R})$ is the electronic wave function in the initial state, calculated for fixed nuclei.

If one writes

$$\langle b|V|\Phi_i^F\rangle_e \equiv \zeta'(\mathbf{R}, \mathbf{k}_i), \tag{B.33}$$

one may split off a factor for the dependence on the direction $\hat{\mathbf{k}}_i$ by defining

$$\zeta'(\mathbf{R}, \mathbf{k}_i) \equiv \tilde{\zeta}(R, k_i)A(\mathbf{R}, \mathbf{k}_i), \tag{B.34a}$$

where

$$\begin{aligned}\tilde{\zeta}(R, k_i) &= [\textstyle\int d\hat{\mathbf{k}}_i|\zeta'(\mathbf{R}, \mathbf{k}_i)|^2]^{1/2} \\ &\cong |4\pi^2\Gamma(\mathbf{R})/k_i|^{1/2}\end{aligned} \tag{B.34b}$$

according to Eq. (61b). This expression may be further simplified if (B.33) is dominated by the component of $e^{i\mathbf{k}_i\cdot\mathbf{r}}$ with the lowest angular momentum with which electrons can enter or leave the compound state, because components of higher angular momentum have too large a centrifugal barrier to overcome. Then one may replace

$$e^{i\mathbf{k}_i\cdot\mathbf{r}} \to 4\pi i^l j_l(k_i r) Y_{lm}^*(\hat{\mathbf{k}}_i) Y_{lm}(\hat{\mathbf{r}}), \tag{B.35a}$$

where l is the smallest angular momentum in question. m is the projection on the molecular axis, but because we have consistently considered only a single resonance, we have to suppose that $m = 0$; only minor changes in the formalism would be necessary to treat resonances which appear as a degenerate pair with $m = \pm|m|$. (The necessary details are given in Ref. 49.) $Y_{lm}(\hat{\mathbf{r}})$ and $Y_{lm}(\hat{\mathbf{k}}_i)$ are spherical harmonics referred to the molecular frame. With (B.35a), one gets

$$A(\mathbf{R}, \mathbf{k}_i) \to Y_{l0}(\hat{\mathbf{R}}, \hat{\mathbf{k}}_i), \tag{B.35b}$$

where we have inserted $\hat{\mathbf{R}}$ to remind us to refer to the molecular axis. With Eqs. (B.33), (B.34a), and (B.35b), Eq. (B.30) may be replaced by

$$[K + W(\mathbf{R}) - E]\,\tilde{\xi} = -\tilde{\zeta}(R, k_i)\mathscr{R}_i(\mathbf{R}), \tag{B.36a}$$

where

$$\tilde{\xi}(R) Y_{l0}(\hat{\mathbf{R}}, \hat{\mathbf{k}}_i) \equiv \xi_i(R, \hat{\mathbf{R}}), \tag{B.36b}$$

R being the separation of the two nuclei in our diatomic molecule. Remember that we are treating the direction of the nuclear axis as fixed, so that the rotational part of the nuclear kinetic energy is being dropped. The factor Y_{l0} in Eq. (B.36b) describes the electron, not the nuclei.

In the expression (B.31) for the scattering amplitude, the matrix element $\langle\cdots\rangle_e$ may be rewritten in a way analogous to Eqs. (B.33), (B.34b), and (B.35b). Write

$$\langle \Phi_f^F(\mathbf{R})|V|b\rangle_e \equiv \zeta(\mathbf{R}, \mathbf{k}_f), \tag{B.37}$$

so that by comparison with Eq. (B.33),

$$\zeta'(\mathbf{R}, \mathbf{k}_f) = \zeta^*(\mathbf{R}, \mathbf{k}_f). \tag{B.38}$$

By an argument similar to that which led to (B.35b), one finds

$$\zeta(\mathbf{R}, \mathbf{k}_f) \cong \tilde{\zeta}(R, k_f) Y_{l0}^*(\hat{\mathbf{R}}, \hat{\mathbf{k}}_f). \tag{B.39}$$

The asterisk may be dropped because Y_{l0} is real.

To evaluate the amplitude (B.31a) in the fixed-axis model, note that the vibrational wave functions $\mathscr{R}_f(\mathbf{R})$ and $\mathscr{R}_i(\mathbf{R})$ depend only on the magnitude R of $\mathbf{R}$, and not on the direction $\hat{\mathbf{R}}$, so that the integral $\int d\mathbf{R}\ldots$ becomes an integral $\int dR \ldots$ over the interval $0<R<\infty$. Substituting for $\langle\cdots\rangle_e$ in Eq. (B.31) from Eqs. (B.37) and (B.39), and using Eq. (B.36b), one finds

$$T_{fi}(\mathbf{k}_f, \mathbf{k}_i) = \left(-\frac{1}{2\pi}\right) \int dR\, Y_{l0}^*(\hat{\mathbf{R}}, \hat{\mathbf{k}}_f) \mathscr{R}_f^*(R) \tilde{\zeta}(R, k_f) \tilde{\xi}(R) Y_{l0}(\hat{\mathbf{R}}, \hat{\mathbf{k}}_i).$$

We shall give the expression for $\sigma_{f\leftarrow i}$, the integral of the differential scattering cross section over all final directions $\hat{\mathbf{k}}_f$, averaged over the direction of the molecular axis $\hat{\mathbf{R}}$. This last average may be replaced by an average over the initial direction $\hat{\mathbf{k}}_i$, so that

$$\sigma_{f\leftarrow i} = \frac{v_f}{v_i} \cdot \frac{1}{4\pi} \int d\hat{\mathbf{k}}_i\, d\hat{\mathbf{k}}_f |T_{fi}(\mathbf{k}_f, \mathbf{k}_i)|^2 \tag{B.40}$$

$$= \frac{v_f}{v_i} \cdot \frac{1}{16\pi^3} \left| \int dR\, \mathscr{R}_f^*(R) \tilde{\zeta}(R, k_f) \tilde{\xi}(R) \right|^2 \tag{B.41}$$

with the aid of the normalization integrals

$$\int d\hat{\mathbf{k}}_i |Y_{l0}(\hat{\mathbf{k}}_i)|^2 = \int d\hat{\mathbf{k}}_f |Y_{l0}^*(\hat{\mathbf{k}}_f)|^2 = 1.$$

Equation (B.40a) applies if there is only a single resonant electronic state with $m = 0$. If there were two degenerate resonances with $m = \pm|m|$, one would have to multiply Eq. (B.41) by two.

Appendix C: The Impulse Approximation

If the collision time of an electron with a molecule is short compared with the periods of the motions of the nuclei, one can express the scattering amplitudes to a good approximation in terms of scattering amplitudes calculated for fixed nuclei. This is called the "impulse approximation," or the "adiabatic approximation." A derivation of the approximation which leads to a precise interpretation of the term "short collision time" was given in 1952 by Chew and Goldberger.[97] Applications to electron–molecule scattering[98] are usually traced to an independent later derivation by Chase.[99] This appendix gives a brief sketch of the Chew–Goldberger derivation, mainly because of its useful expression for the error in the approximation. Other aspects of the impulse approximation are discussed in Chapter 2, Section 3.2.

The Chew–Goldberger derivation[97] was originally intended to be used in the scattering of elementary particles by complex nuclei, a problem which needs a notation rather different from electron–molecule scattering. Here we shall establish the relation between the calculations in the two problems, and then leave the reader to get the details from Chew and Goldberger's article.

The Hamiltonian may be written

$$H = H_0 + V, \tag{C.1}$$

where H_0 is the kinetic energy of the incident electron plus the Hamiltonian of the target, and V is the interaction between the target and projectile.

H_0 can be split up again:

$$H_0 = H_{e0} + K, \tag{C.2}$$

where H_{e0} is the electronic Hamiltonian of the target and the kinetic energy of the projectile; K is the kinetic energy of the nuclei. (Chew and Goldberger use the symbol K for our H_{e0}, and U for our K.) The problem is to calculate scattering as if the nuclei were stationary during the collision, that is by ignoring K during the collision, just retaining K to define the initial and final rotational and vibrational states.

The initial and final states Φ_i and Φ_f of the target and projectile are defined in Eq. (2c); they satisfy

$$(H_0 - E_i)\Phi_i = 0, \qquad (H_0 - E_f)\Phi_f = 0. \tag{C.3}$$

The exact solution of the Schrödinger equation, Ψ_i, satisfies

$$(H-E_i)\Psi_i = 0, \tag{C.4}$$

and the boundary condition (2b) when any electron is at a large distance from the target. The differential cross section per unit solid angle is given by Eqs. (2a) and (1).

To bring in the fixed-nuclei approximation, introduce a complete set of basis vectors $|\mathbf{R}\rangle$ in which the positions of the nuclei are sharp; the basis vectors satisfy $\int|\mathbf{R}>d\mathbf{R}<\mathbf{R}| = I$, where I is the unit operator, and $\langle\mathbf{R}|\mathbf{R}'\rangle=\delta(\mathbf{R}-\mathbf{R}')$; the values $\mathbf{R}$ range over all the nuclear coordinates. One can now write $|\Phi_{i\text{ or }f}\rangle = \int|\mathbf{R}>d\mathbf{R}<\mathbf{R}|\Phi_{i\text{ or }f}\rangle$, where $\langle\mathbf{R}|\Phi_{i\text{ or }f}\rangle=\langle\mathscr{R}_{i\text{ or }f}(\mathbf{R})|\Phi^F_{i\text{ or }f}(\mathbf{R})\rangle$; the factor $\mathscr{R}_{i\text{ or }f}(\mathbf{R})$ is the initial or final nuclear wave function, and $\Phi^F_{i\text{ or }f}(\mathbf{R})$ is defined by

$$[H_{e0}(\mathbf{R})-E_{i\text{ or }f}(\mathbf{R})]\,|\Phi^F_{i\text{ or }f}(\mathbf{R})\rangle = 0. \tag{C.5}$$

The energy $E_{i\text{ or }f}(\mathbf{R})$ is the electronic energy of the target plus the kinetic energy of the extra electron, defined for nuclei at $\mathbf{R}$. A complete scattering state for the fixed-nuclei picture is $\Psi^F_i(\mathbf{R})$; it behaves asymptotically like

$$|\Psi^F_i(\mathbf{R})\rangle = |\Phi^F_i(\mathbf{R})\rangle + \text{outgoing electronic waves},$$

and satisfies

$$[H_{e0}(\mathbf{R})+V(\mathbf{R})-E_i(\mathbf{R})]\,|\Psi^F_i\rangle = 0. \tag{C.6}$$

(Remember that $H_{e0}+V=H_e$, the complete electronic Hamiltonian.) The scattering amplitude for fixed nuclei $[\equiv T^F_{fi}(\mathbf{R})]$ is

$$T^F_{fi}(\mathbf{R}) = -\frac{1}{2\pi}\langle\Phi^F_f(\mathbf{R})|V(\mathbf{R})|\Psi^F_i(\mathbf{R})\rangle. \tag{C.7}$$

(We have written $\mathbf{R}$ in $|\Phi^F_i(\mathbf{R})\rangle$, $|\Phi^F_f(\mathbf{R})\rangle$, and $|\Psi^F_i(\mathbf{R})\rangle$ to remind us that the states are defined for fixed $\mathbf{R}$; they are, of course, functions of the electronic coordinates.)

The complete scattering amplitude may now be written,

$$T_{fi} = T^{\text{imp}}_{fi} + (T_{fi}-T^{\text{imp}}_{fi}), \tag{C.8}$$

where T^{imp}_{fi} is the impulse approximation, which is defined as[97]

$$T_{fi}^{\text{imp}} = \int d\mathbf{R}\, \mathscr{R}_f^*(\mathbf{R})\, T_{fi}^F(\mathbf{R})\, \mathscr{R}_i(\mathbf{R}), \tag{C.9}$$

whereas in Eq. (C.7), T_{fi}^F is the scattering amplitude for fixed nuclei. The correction in Eq. (C.8) turns out to be[97]

$$T_{fi} - T_{fi}^{\text{imp}} = \left(-\frac{1}{2\pi}\right)\langle\Phi_f|V\left\{\frac{1}{E_i + iy - H_0 - V}\,[K, \omega^{(+)}]\,|\Phi_i\rangle\right\}, \tag{C.10}$$

where

$$\omega^{(+)} \equiv 1 - i\int_{-\infty}^{0} dx\, e^{yx}\, e^{iH_e x}\, V e^{-iH_{e0}x}$$

is the operator which turns Φ_i^F into Ψ_i^F:

$$\omega^{(+)}|\Phi_i^F(\mathbf{R})\rangle = |\Psi_i^F(\mathbf{R})\rangle;$$

$\omega^{(+)}$ is defined for fixed $\mathbf{R}$. y is an infinitesimal real positive quantity.

The correction (C.10) has a simple physical interpretation. The factor $1/(E_i + iy - H_0 - V)$ is the collision time, so that the factor $\{\cdots\}$ is the change in the scattering state for the electrons which comes from the displacements of the nuclei during the collision. An estimate of Eq. (C.10) may be made by going to lowest order in K, i.e., replacing H_0 in the denominator by H_{e0}. This approximation is particularly useful at a resonance, because $H_{e0} + V = H_e$, the complete electronic Hamiltonian, so that the factor $1/(E_i + iy - H_0 - V)$ becomes $1/(i\Gamma/2)$, where $1/\Gamma$ is the resonant time delay.

Appendix D: Wave Packets

The integral (32) with the expression (34) for $a(E)$ gets most of its magnitude from a region of width 2γ centered on E_0. Because of the convergence factor $a(E)$, one may extend the lower limit to $-\infty$ if $\gamma \ll E_0$, that is, if the frequency spread is small compared with the central frequency. In the factor $\exp[i(kz - Et)]$, write $k_0 = (2E_0)^{1/2}$, expand $kz = k_0 z + (E - E_0)(dk/dE)_{k_0} z + \cdots$, and drop the higher terms; we shall return to them in a moment. The integral (32) may now be written

$$\frac{1}{2i} \exp\,[i(k_0 z - E_0 t)] \int_{-\infty}^{\infty} dE \left[\frac{1}{E-(E_0+i\gamma)} - \frac{1}{E-(E_0-i\gamma)}\right] e^{i(E-E_0)x},$$

where $x \equiv z/v_g - t$, with $v_g \equiv (dE/dk)_{k=k_0}$. The integral can be evaluated by closing the contour above the real E axis if $x > 0$, below if $x < 0$. The result (35) is obtained if one uses the residue theorem on the poles at $E_0 \pm i\gamma$.

The next term in the expansion of kz turns out to be $\frac{1}{8}(\gamma/E_0)^2 k_0 z$. If the wave packet passes the target at $t \cong 0$, then Eq. (35) says that the approximation should be correct out to $|z| = v_g/\gamma$. Since $v_g = (dE/dk)_{k_0} = k_0$, one must have

$$\left|\frac{1}{8}\left(\frac{\gamma}{E_0}\right)^2 k_0 \frac{v_g}{\gamma}\right| = \frac{1}{4}\frac{\gamma}{E_0} \ll 1.$$

This will be true if the width in frequency is small compared with the central frequency.

References and Notes

1. N.F. Lane, The theory of electron–molecule collisions, *Rev. Mod. Phys.* **52**, 29 (1980).
2. P.G. Burke, Theory of low energy electron–molecule collisions, *Adv. At. Mol. Phys.* **15**, 471 (1979).
3. D.E. Golden, Resonances in electron atom and molecule scattering, *Adv. At. Mol. Phys.* **14**, 1 (1978).
4. B.L. Moiseiwitsch, Recent progress in atomic collision theory, *Rep. Prog. Phys.* **40**, 843 (1977).
5. G.J. Schulz, in *Principles of Laser Plasmas* (G. Bekefi, ed.), John Wiley and Sons, New York (1976).
6. Sir Harrie Massey, *Negative Ions*, 3rd ed., Cambridge University Press, Cambridge (1976).
7. K. Takayanagi, The theory of low energy electron–molecule scattering, in *Invited Lectures, Review Papers, and Progress Reports*, IX International Conf. on the Physics of Electronic and Atomic Collisions (ICPEAC), Seattle, 1975 (J.S. Risley and R. Geballe, eds.), p. 219.
8. F.H. Read, in *Atoms, Molecules, and Lasers*, Trieste, Italy, 17 January–10 April, 1973; IAEA, Vienna (1974), p. 567.
9. G.J. Schulz, Resonances in electron impact on diatomic molecules, *Rev. Mod. Phys.* **45**, 423 (1973).
10. Reviews before 1973 (a more extensive list of references is given in Ref. 1.): D.E. Golden, N.F. Lane, A. Temkin, and E. Gerjuoy, Low energy electron–molecule scattering experiments and the theory of rotational excitation, *Rev. Mod. Phys.* **43**, 642 (1971); H.S. Taylor, Models, interpretations, and calculations concerning resonant electron scattering processes in atoms and molecules, *Adv. Chem. Phys.* **18**, 91 (1970); H.S.W. Massey, E.H.S. Burhop, and H.B. Gilbody, in *Electronic and Ionic Impact Phenomena*,

Vol. II, Oxford University Press, London, 1969 (the material in this volume is brought up to date in Vol. V); J.C.Y. Chen, Theory of transient negative ions of simple molecules, *Adv. Radiat. Chem.* **1,** 245 (1969); J.N. Bardsley and F. Mandl, Resonant scattering of electrons by molecules, *Rep. Prog. Phys.* **31,** 471 (1968).

11. First Schulz Symposium (Yale, 1977), *see*: *Electron–Molecule Scattering* (S. Brown, ed.), John Wiley and Sons, New York (1979). Second Schulz Symposium (Yale, 1979), *see*: F.H. Read, Electron–atom scattering – experiments, *Phys. Scr.* **27,** 103 (1983). The other three papers from the same symposium on electron–atom scattering are in *Comments. At. Mol. Phys.* **14,** 45–101 (1983); (W.R. Bennett, "Gas Lasers"; L. Spruch, "Theory"; and W.D. Watson, "Astrophysics").
12. Y. Itikawa, Electron scattering by polar molecules, *Phys. Rep.* **46,** 117 (1978); W.R. Garrett, Low-energy electron scattering by polar molecules, *Mol. Phys.* **24,** 465 (1972); See also Ref. 35.
13. H.A. Bethe, On the theory of the passage of fast corpuscular rays through matter, *Ann. Phys.* (*Leipzig*) **5,** 325 (1930); quoted by N.F. Mott and H.S.W. Massey, *The Theory of Atomic Collisions,* 3rd ed., Oxford, University Press, London (1965), p. 477.
14. The vibrational amplitude and energy in a molecule is estimated by, for example, L.I. Schiff, *Quantum Mechanics,* 3rd ed., McGraw-Hill, New York (1968), Sec. 49, p. 477.
15. For the Thomas–Fermi model, see for example L.D. Landau and E.M. Lifshitz, *Quantum Mechanics,* 3rd ed., Pergamon Press, Oxford (1977), Sec. 70, p. 259 (translated from the Russian by J.B. Sykes and J.S. Bell).
16. H.S.W. Massey, Excitation of molecular vibration by impact of slow electrons, *Trans. Far. Soc.* **31,** 556 (1935).
17. N.F. Mott and H.S.W. Massey, *The Theory of Atomic Collisions,* 3rd ed., Oxford, University Press, London (1965), § II.1.
18. Reference 17, p. 22.
19. J.R. Taylor, *Scattering Theory,* John Wiley & Sons, New York (1972); J.M. Blatt and V.F. Weisskopf, *Theoretical Nuclear Physics,* John Wiley & Sons, New York (1952), p. 68.
20. Reference 17, p. 46.
21. Some of the key papers in resonance-scattering theory which have influenced Appendix A are: H. Feshbach, Unified theory of nuclear reactions, Ann. Phys. (N. Y.) **5,** 357 (1958); *ibid.* **19,** 287 (1962); T.F. O'Malley and S. Geltman, Compound-atom states for two-electron systems, *Phys. Rev.* **137,** A1344 (1965); W.H. Miller, Resonances in the scattering of electrons from atoms, *Phys. Rev.* **152,** 70 (1966); K. Smith, Resonant scattering of electrons by atomic systems, *Rep. Prog. Phys.* **29,** 373 (1966); "Boxed-in-resonances" have recently been discussed by C.H. Maier, L.S. Cederbaum, and W. Domcke, A spherical-box approach to resonances, *J. Phys. B* **13,** L119 (1980).
22. H.S. Taylor, in Ref. 10; H.S. Taylor and R. Yaris, The rotated-coordinate method and its relation to the stabilization method, *Comments At. Mol. Phys.* **9,** 73 (1980).
23. A.J.F. Siegert, On the derivation of the dispersion formula for nuclear reactions, *Phys. Rev.* **56,** 750 (1939).
24. M.A. Morrison and N.F. Lane, Threshold structure in the vibrational excitation of CO_2 by low-energy electrons, *Chem. Phys. Lett.* **66,** 527 (1979).
25. L. Dubé and A. Herzenberg, Vibrational excitation of polar molecules by slow electrons: HCl, *Phys. Rev. Lett.* **38,** 820 (1977).
26. M.J.W. Boness and G.J. Schulz, Vibrational excitation of CO_2 by electron impact, *Phys. Rev. Lett.* **21,** 1031 (1968); Vibrational excitation in CO_2 via the 3.8-eV resonance, *Phys. Rev. A* **9,** 1969 (1974); I. Cadez, M. Tronc, and R.I. Hall, Oscillations in electron impact cross sections of CO_2 between 3 and 5 eV, *J. Phys. B* **7,** L132 (1974); I. Cadez, F. Gresteau, M. Tronc, and R.I. Hall, Resonant electron impact excitation of

CO_2 in the 4 eV region, *J. Phys. B* **10**, 3821 (1977); C. Szmytkowski and M. Zubek, Calculation of the cross sections for resonant vibrational excitation of CO_2 by electron collision, *J. Phys. B* **10**, L31 (1977); C. Szmytkowski, M. Zubek, and J. Drewko, Calculations of cross sections for vibrational excitation and de-excitation of CO_2 by electronic collisions, *J. Phys. B* **11**, L371 (1978); W. Domcke and L.S. Cederbaum, Simple formula for the vibrational structure of resonances in electron–molecule scattering, *J. Phys. B* **10**, L47 (1977); Theory of the vibrational structure of resonances in electron–molecule scattering, *Phys. Rev. A* **16**, 1465 (1977).

27. H.S. Taylor, E. Goldstein, and G.A. Segal, Resonance states of HCl^- and electron–HCl scattering processes, *J. Phys. B* **10**, 2253 (1977).
28. R.K. Nesbet, Interpretation of low-energy electron–HCl scattering phenomena, *J. Phys. B* **10**, L739 (1977).
29. H.M. Nussenzveig, The poles of the *S*-matrix of a rectangular potential well or barrier, *Nucl. Phys.* **11**, 499 (1959).
30. P.G. Burke, A.J. Taylor, J.W. Cooper, and S. Ormonde, The scattering of electrons by helium atoms, Proc. Vth ICPEAC, p. 376, Leningrad, 1967; P.G. Burke, J.W. Cooper, and S. Ormonde, Low-energy scattering of electrons by helium, *Phys. Rev.* **183**, 245 (1969); H. Ehrhardt, L. Langhans, and F. Linder, Die differentiellen Anregungsfunktionen der $n = 2$-Zustände von Helium, *Z. Phys.* **214**, 179 (1968).
31. P.L. Kapur and R. Peierls, The dispersion formula for nuclear reactions, *Proc. Roy. Soc. A* **166**, 277 (1938).
32. E.S. Chang, Comment on rotational and vibrational excitation of H_2 by electron impact, *Phys. Rev. Lett.* **33**, 1644 (1974).
33. E.S. Chang and U. Fano, Theory of electron–molecule collisions by frame transformations, *Phys. Rev. A* **6**, 173 (1972).
34. E.S. Chang and A. Temkin, Rotational excitation of diatomic molecules by electron impact, *Phys. Rev. Lett.* **23**, 399 (1969); A. Temkin and E.C. Sullivan, Rotational-vibrational coupling in the theory of electron–molecule scattering, *Phys. Rev. Lett.* **33**, 1057 (1974).
35. H.S.W. Massey *et al.*, in Ref. 10.
36. G.J. Schulz and R.K. Asundi, Isotope effect in the dissociative attachment in H_2 at low energy, *Phys. Rev.* **158**, 25 (1967).
37. M. Allan and S.F. Wong, Effect of vibrational and rotational excitation on dissociative attachment in hydrogen, *Phys. Rev. Lett.* **41**, 1791 (1978).
38. J.M. Wadehra and J.N. Bardsley, Vibrational- and rotational-state dependence of dissociative attachment in e–H_2 collisions, *Phys. Rev. Lett.* **41**, 1795 (1978).
39. J.N. Bardsley, A. Herzenberg, and F. Mandl, Electron resonances of the H_2^- ion, *Proc. Phys. Soc., London* **89**, 305 (1966); Vibrational excitation and dissociative attachment in the scattering of electrons by hydrogen molecules, *Proc. Phys. Soc., London* **89**, 321 (1966).
40. J.C.Y. Chen and J.L. Magee, Excitation of molecular vibration by slow electron impact, *J. Chem. Phys.* **36**, 1407 (1962).
41. K. Takayanagi, Vibrational excitation of hydrogen molecules by slow electrons, *J. Phys. Soc. Jpn.* **20**, 562 (1965); Excitation of molecular vibration by slow electrons, *ibid.* **20**, 2297 (1965).
42. A. Klonover and U. Kaldor, Zero-point vibration in low-energy electron–molecule scattering theory, *J. Phys. B* **12**, L61, 323 (1979).
43. T.N. Rescigno, C.W. McCurdy, and V. McKoy, Discrete basis set approach to nonspherical scattering, *Chem. Phys. Lett.* **27**, 401 (1974); Discrete-basis-set approach to nonspherical scattering. II, *Phys. Rev. A* **10**, 2240 (1974); Low-energy e^-–H_2 elastic cross sections using discrete basis functions, *Phys. Rev. A* **11**, 825 (1975).

44. N.F. Lane and R.J.W. Henry, Polarization potential in low-energy electron–H_2 scattering, *Phys. Rev.* **173**, 183 (1968); R.J.W. Henry and N.F. Lane, Polarization and exchange effects in low-energy electron–H_2 scattering, *Phys. Rev.* **183**, 221 (1969); *Phys. Rev. A* **4**, 410 (1971); R.J.W. Henry, Vibrational and rotational excitation of molecular hydrogen by electron impact, *Phys. Rev. A* **2**, 1349 (1970); R.J.W. Henry and E.S. Chang, Rotational-vibrational excitation of H_2 by slow electrons, *Phys. Rev. A* **5**, 276 (1972).
45. A. Temkin and E.C. Sullivan, in Ref. 34.
46. L. Dubé and A. Herzenberg, Resonant electron–molecule scattering: The impulse approximation in N_2O, *Phys. Rev. A* **11**, 1314 (1975).
47. R. Azria, S.F. Wong, and G.J. Schulz, Vibrational excitation in N_2O via the 2.3-eV shape resonance, *Phys. Rev. A* **11**, 1309 (1975).
48. S.F. Wong, J. Michejda, and A. Stamatovic, unpublished. Their results for $e + N_2$, $v = 1 \rightarrow 2$, were published by permission in Ref. 49, Fig. 4.
49. L. Dubé and A. Herzenberg, Absolute cross sections from the "boomerang model" for resonant electron–molecule scattering, *Phys. Rev. A* **20**, 194 (1979).
50. J.N. Bardsley, Configuration interaction in the continuum states of molecules, *J. Phys. B* **1**, 349, 365 (1968).
51. G. Parlant and F. Fiquet-Fayard, Comparison of $\Gamma(R)$ theory (local width) and $\Gamma(E)$ theory (energy dependent width) for resonant excitation. Application to $O_2^-\ {}^2\Pi_g$, in IX ICPEAC, Seattle, 1975, Book of Abstracts, p. 279.
52. A.U. Hazi, T.N. Rescigno, and M. Kurila, Cross-sections for resonant vibrational excitation of N_2 by electron impact, *Phys. Rev. A* **23**, 1089 (1981).
53. The data on the potential surface and vibrational spectrum of the electronic ground state of O_2^- are assembled by Schulz (see Ref. 9, Fig. 69, p. 471).
54. A. Herzenberg, Oscillatory energy dependence of resonant electron–molecule scattering, *J. Phys. B* **1**, 548 (1968).
55. J.M. Blatt and V.F. Weisskopf, in Ref. 19, Chapter 8, Section 5.
56. F.R. Gilmore, Potential energy curves for N_2, NO, O_2 and corresponding ions, *J. Quant. Spectrosc. Radiat. Transfer* **5**, 369 (1965).
57. J.N. Bardsley, F. Mandl, and A.R. Wood, The structure of the 2 eV N_2^- resonance, *Chem. Phys. Lett.* **1**, 359 (1967).
58. J.N. Bardsley and F.H. Read, Predicted angular distribution for resonant scattering of electrons by molecules, *Chem. Phys. Lett.* **2**, 333 (1968); T.F. O'Malley and H.S. Taylor, Angular dependence of scattering products in electron–molecule resonant excitation and in dissociative attachment, *Phys. Rev.* **176**, 207 (1968).
59. H. Ehrhardt and K. Willmann, Die Winkelabhängigkeit der Resonanzstreuung niederenergetischer Elektronen an N_2, *Z. Phys.* **204**, 462 (1967).
60. C.R. Claydon, G.A. Segal, and H.S. Taylor, Theoretical interpretation of the electron scattering spectrum of CO_2, *J. Chem. Phys.* **52**, 3387 (1970); J. Pacansky, U. Wahlgren, and P.S. Bagus, SCF *ab initio* ground state energy surfaces for CO_2 and CO_2^-, *J. Chem. Phys.* **62**, 2740 (1975) (see also Ref. 27 [HCl^-]); M. Krauss and W.J. Stevens, Electronic structure of the negative ion of HCl, *J. Chem. Phys.* **74**, 570 (1981).
61. B.I. Schneider and P.J. Hay, Elastic scattering of electrons from F_2; An *R*-matrix calculation, *Phys. Rev. A* **13**, 2049 (1976); T.N. Rescigno and C.F. Bender, The stability of the F_2^- ion: A model for dissociative attachment, *J. Phys. B* **9**, L329 (1976).
62. P.G. Burke and A.L. Sinfailam, Electron–molecule interactions, II: Scattering by closed-shell diatomic molecules, *J. Phys. B* **3**, 641 (1970); P.G. Burke and N. Chandra, Electron–molecule interactions, III: A pseudo-potential method for e^-–N_2 scattering, *J. Phys. B* **5**, 1696 (1972); B.D. Buckley and P.G. Burke, The scattering of low-energy electrons by diatomic molecules, *J. Phys. B* **10**, 725 (1977).

63. N. Chandra and A. Temkin, Hybrid theory and calculation of $e-N_2$ scattering, *Phys. Rev. A* **13,** 188 (1976); Hybrid theory calculation of simultaneous vibration-rotation excitation in $e-N_2$ scattering, *Phys. Rev. A* **14,** 507 (1976); Hybrid theory calculation of electron–N_2 scattering at 5 and 10 eV, *J. Chem. Phys.* **65,** 4537 (1976); see also B.H. Choi and R.T. Poe, Vibrational and rotational transitions in low-energy electron–diatomic-molecule collisions, *Phys. Rev. A* **16,** 1831 (1977).
64. B.I. Schneider, M. Le Dourneuf, and Vo Ky Lan, Resonant vibrational excitation of N_2 by low-energy electrons: An *ab initio* *R*-matrix calculation, *Phys. Rev. Lett.* **43,** 1926 (1979); see also B.I. Schneider, M. LeDourneuf, and P.G. Burke, Theory of vibrational excitation and dissociative attachment: An *R*-matrix approach, *J. Phys. B* **12,** L365 (1979).
65. M. Krauss and F.H. Mies, Molecular-orbital calculation of the shape resonance in N_2^-, *Phys. Rev. A* **1,** 1592 (1970).
66. D.T. Birtwistle and A. Herzenberg, Vibrational excitation of N_2 by resonance scattering of electrons, *J. Phys. B* **4,** 53 (1971).
67. See Ref. 49, and A. Herzenberg, Strong excitation of nuclear motions in electron–molecule collisions, in *Invited Papers and Progress Reports,* X ICPEAC, Paris, 1977 (G. Watel, ed.), North-Holland, Amsterdam (1978), p. 1.
68. I.M. Cadez *et al.* (1977), in Ref. 26. A. Huetz, I. Cadez, F. Gresteau, R.I. Hall, D. Vichon, and J. Mazeau, Formation and decay of the first $^2\Pi_u$ state of N_2^-, *Phys. Rev. A* **21,** 622 (1980).
69. C. Szmytkowski and M. Zubek, Calculation of the cross sections for resonant vibrational excitation of CO_2 by electron collision, *J. Phys. B* **10,** L31 (1977); Calculation of resonant vibrational excitation of CO by scattering of electrons, *J. Phys. B* **10,** L27 (1977); C. Szmytkowski *et al.* (1978), in Ref. 26; M. Zubek and C. Szmytkowski, Electron impact vibrational excitation of CO in the range 1–4 eV, *Phys. Lett. A* **74,** 60 (1979).
70. A. Herzenberg and F. Mandl, Vibrational excitation of molecules by resonance scattering of electrons, *Proc. R. Soc., London, A* **270,** 48 (1962).
71. J.C.Y. Chen, Theory of subexcitation electron scattering by molecules. II. Excitation and de-excitation of molecular vibration, *J. Chem. Phys.* **40,** 3513 (1964).
72. R.K. Nesbet, Energy-modified adiabatic approximation for scattering theory, *Phys. Rev. A* **19,** 551 (1979).
73. W. Domcke and L.S. Cederbaum, in Ref. 26.
74. G.J. Schulz, Vibrational excitation of N_2, CO, and H_2 by electron impact, *Phys. Rev.* **135,** A988 (1964).
75. R.J.W. Henry, in Ref. 44.
76. R.A. Abram and A. Herzenberg, Rotational excitation of H_2 by slow electrons, *Chem. Phys. Lett.* **3,** 187 (1969) (see also Ref. 98).
77. F. Linder and H. Schmidt, Rotational and vibrational excitation of H_2 by slow electron impact, *Z. Naturforsch. A* **26a,** 1603 (1971).
78. The dots in Fig. 1d are absolute experimental measurements by Wong, Michejda, and Stamatovic which were previously published in the same diagram by permission in Ref. 49.
79. H. Ehrhardt, L. Langhans, F. Linder, and H.S. Taylor, Resonance scattering of slow electrons from H_2 and CO angular distributions, *Phys. Rev.* **173,** 222 (1968).
80. F. Linder and H. Schmidt, Experimental study of low energy $e-O_2$ collision processes, *Z. Naturforsch. A* **26a,** 1617 (1971).
81. M.J.W. Boness and G.J. Schulz (1974), in Ref. 26.
82. K. Rohr and F. Linder, Vibrational excitation of polar molecules by electron impact I. Threshold resonances in HF and HCl, *J. Phys. B* **9,** 2521 (1976); see also J.P. Ziesel,

I. Nenner, and G.J. Schulz, Negative ion formation, vibrational excitation, and transmission spectroscopy in hydrogen halides, *J. Chem. Phys.* **63,** 1943 (1975).

83. K. Rohr, Vibrational excitation of SF_6 by low-energy electron impact, *J. Phys. B* **10,** 1175 (1977).

84 S. Trajmar, D.G. Truhlar, J.K. Rice, and A. Kupperman, Electron scattering by H_2 with and without vibrational excitation. III. Experimental and theoretical study of inelastic scattering, *J. Chem. Phys.* **52,** 4516 (1970).

85. M. Zubek and C. Szmytkowski, in Ref. 69.

86. A. Huetz *et al.,* in Ref. 68.

87. M. Tronc, R. Azria, and R. Paineau, Shape resonances in vibrational excitation of molecules: [100] symmetric stretch mode excitation in CO_2, *J. Phys. (Paris) Lett.* **40,** L323 (1979).

88. J.L. Dehmer and Dan Dill, Shape-resonance enhanced nuclear motion effects in electron–molecule scattering and molecular photoionization, in *Invited Papers and Progress Reports,* XI ICPEAC (N. Oda and K. Takayanagi, eds.), North-Holland, Amsterdam (1980), p. 195.

89. I. Cadez and F. Fayard, The calculation of cross sections for some resonant excitations of N_2, in VIII ICPEAC, Belgrade, 1973 (B.C. Cobic and M.V. Kurepa, eds.), p. 454.

90. K.D. Jordan, J.A. Michejda, and P.D. Burrow, Electron transmission studies of the negative ion states of substituted benzenes in the gas phase, *J. Am. Chem. Soc.* **98,** 7189 (1976).

91. S.F. Wong and G.J. Schulz, Vibrational excitation in benzene by electron impact via resonances: Selection rules, *Phys. Rev. Lett.* **35,** 1429 (1975).

92. H. Ehrhardt and F. Linder, Rotational excitation of H_2 by slow electrons in a beam experiment, *Phys. Rev. Lett.* **21,** 419 (1968).

93. E.S. Chang and S.F. Wong, Isotope effects in molecular scattering by electrons, *Phys. Rev. Lett.* **38,** 1327 (1977).

94. F.H. Read, Branch structure in rotational excitation by resonant electron molecule collisions, *J. Phys. B* **5,** 255 (1972); J. Comer and M. Harrison, Observation of the effects of rotational transitions in the resonant scattering of electrons from N_2, *J. Phys. B* **6,** L70 (1973).

95. S.F. Wong and L. Dubé, Rotational excitation of N_2 by electron impact: 1–4 eV, *Phys Rev. A* **17,** 570 (1978).

96. Z. Pavlovic, M.J.W. Boness, A. Herzenberg, and G.J. Schulz, Vibrational excitation in N_2 by electron impact in the 15–35 eV region, *Phys. Rev. A* **6,** 676 (1972).

97. G.F. Chew and M.L. Goldberger, The scattering of elementary particles by complex nuclei – A generalization of the impulse approximation, *Phys. Rev.* **87,** 778 (1952).

98. Y.D. Oksyuk, Excitation of the rotational levels of diatomic molecules by electron impact in the adiabatic approximation, *Sov. Phys. JETP* **22,** 873 (1966); R.A. Abram and A. Herzenberg, Ref. 76; E.S. Chang and A. Temkin, Rotational excitation of diatomic molecules by electron impact, *J. Phys. Soc. Jpn.* **29,** 172 (1970).

99. D.M. Chase, Adiabatic approximation for scattering processes, *Phys. Rev.* **104,** 838 (1956).

100. I.S. Yelets and A.K. Kazansky, *JETP* **82,** 450 (1982).

4

DISSOCIATION OF MOLECULES BY SLOW ELECTRONS

R.N. COMPTON and J.N. BARDSLEY

1. Introduction

At a first glance it might appear unlikely that the nuclei within a molecule can be separated in a collision with such a light particle as an electron. However, according to the Born–Oppenheimer principle, the effective potential controlling the motion of the nuclei is given by the eigenvalues of the electronic Hamiltonian. Thus, a change in the electronic structure of the molecule can have a profound effect on the nuclear motion, and an electronic transition will often be accompanied by vibrational excitation or dissociation of the molecule.

It is useful to divide the various dissociative processes into three groups that we will call dissociative excitation, dissociative ionization, and dissociative capture. Each of these three processes can be viewed, following the Born–Oppenheimer ideas, in terms of an electronic transition in which the nuclei do not move, followed by dissociation of the nuclei along a modified potential.

In dissociative excitation the initial step involves the electronic excitation of the target molecule, either through direct or exchange scattering. Dissociation will be a likely outcome if the excited state has a potential curve with a minimum at a relatively large value of R.

With respect to electron-impact ionization, the removal of one of the target electrons from a stable molecule almost always leads to a more weakly bound electronic state. This is especially true if the ejection of the electron leaves the molecular ion in an excited state. For example, the excited electronic states of H_2^+ have potential curves that are either completely repulsive or have very weak attractive regions at extremely large R. Hence ionization of a molecule often leads to dissociation.

R.N. COMPTON • Oak Ridge National Laboratory, Oak Ridge, Tennessee 37831. J.N. BARDSLEY • Department of Physics, University of Pittsburgh, Pittsburgh, Pennsylvania 15260. Research sponsored in part by the Office of Health and Environmental Research, U.S. Department of Energy under Contract No. W-7405-eng-26 with the Union Carbide Corporation and in part by the Division of Theoretical Physics, National Science Foundation Grant No. PHY8105074.

In dissociative capture, which includes dissociative attachment of electrons to neutral molecules and dissociative recombination in electron collisions with positive ions, the incident electron becomes temporarily attached to the target molecule. This leads to an unstable intermediate state which can decay by electronic emission or dissociation. The probability of dissociation will be large if the lifetime of this state against electron emission is of the same order as, or longer than, the vibrational period ($\sim 10^{-14}$ s), so that the nuclear motion can be disturbed significantly by the presence of the extra electron.

Dissociative processes can have important effects in ionized gases containing molecules. For example, in the ionosphere and in many laboratory afterglows, dissociative recombination is the dominant reaction leading to charge neutralization. Although dissociative attachment does not lead to the removal of charge, the increase in the mass of the particle carrying the negative charge results in a decrease in the mobility. Thus even small amounts of attaching molecules such as F_2 and NF_3 can have a significant effect on the conduction of electrons through a gas.

Dissociation into neutral fragments is important, not only because of the energy loss associated with the process, but because, since the fragments are usually more reactive than the parent molecule, interesting secondary processes may ensue. These processes are often aided by the internal energy or superthermal kinetic energy of the dissociating fragments.

In some applications molecular dissociation results from indirect processes or multiple collisions. For example, in a hydrogen plasma, following electronic excitation of the $B\,^1\Sigma_u^+$ or $C\,^1\Pi_u$ states of H_2, a radiative transition back to the ground electronic state may leave the nuclei in a continuum state. This process is particularly efficient in H_2 because of the significant difference between the equilibrium nuclear separation in the ground and excited electronic states. Alternatively, dissociation may result from collisions of a molecule with two or more electrons. Vibrational excitation followed by dissociative attachment is important in plasmas containing molecules such as H_2, HCl, and HF, for which the attachment rate is small if the molecules are unexcited.

The experimental study of dissociative processes is simplest for those processes in which charged particles disappear, such as dissociative recombination, or in which one of the fragments is charged. The detection of neutral products is more difficult, especially if the fragments are in the ground electronic state. Fortunately many processes lead to electronically excited products.

Excited states with low principal quantum number which are allowed to emit to a lower state can be detected optically. For molecular hydrogen, Balmer emission has been investigated for $n = 3$–11 to $n = 2$. Fragments which have been excited to states with large principal quantum number ($n \gtrsim 15$) are arbitrarily designated as "high-Rydberg atoms" and have long radiative lifetimes and must be detected in another manner. The high-Rydberg atoms have been detected by one of four basic methods:

1. Collisional ionization:

$$A^{**} + M \rightarrow A^{+} + M + e,$$

$$\rightarrow A^{+} + M^{-}.$$

2. Penning ionization:

$$A^{**} + M \rightarrow A + M^{+} + e.$$

3. Resonance ionization near a metal surface.
4. Ionization in a strong electric field.

The last method is particularly useful since absolute cross sections can be obtained along with an approximate measurement of the distribution of states with principal quantum number [See Schiavone *et al.*[1]]. Also processes 1, 2, and 3 can occur for certain metastable states, whereas field ionization is only appropriate to detecting high-Rydberg fragments. Metastable states (i.e., state with lifetimes $\gtrsim 10^{-4}$ s) can be detected by Auger emission of an electron from a surface provided that the excitation energy is above the work function of that surface. In some cases, the most notable of which is $H(2\,^2S)$, the fragment can be induced to emit by applying an external electric field to quench the metastable state, thereby inducing photon emission which can be detected with a photomultiplier.

The kinetic energy distribution of excited fragments has been studied in a number of instances. Since the high-Rydberg and metastable fragments are long-lived, the resulting velocity distribution of the excited fragment can be easily measured by standard time-of-flight techniques. For H_2, observations of anticrossing in Zeeman sublevels of excited ($n = 3$ or 4) H atoms induced by the motional electric fields provide information on the mean velocity of the excited fragments.[2,3] It is also found that the Balmer linewidths produced in electron impact excitation of H_2 far exceed the thermal Doppler widths.[4,5] The "wings" of the Balmer profile correspond to the emission by "fast" excited H atoms and provide information on the velocity distribution of these fast atoms. Study of the Doppler profiles of atomic radiation was also crucial in the demonstration that dissociative recombination is the dominant mechanism for recombination of electrons with molecular ions.[6]

Most of the theoretical techniques that have been applied to dissociative processes are adapted from the basic techniques discussed by other authors in this volume. At high energies, dissociative excitation and dissociative ionization can be treated by the various forms of the Born approximation (Chapter 1, Section 6.5). The nuclear motion is described within the sudden, or adiabatic-nuclei approximation, in which the transition probability for the excitation or ejection of the electron is computed as a function of the internuclear separation, and is folded between the initial and final nuclear wave function. At lower incident energies, the close coupling, *R*-matrix, or algebraic-variational methods can be applied for electrons scattered by the molecule with fixed-nuclei, still working within the adiabatic-nuclei approximation. This approach is described more fully in Chapter 1, Section 6, in Chapter 3, Section 5.5, and in Chapter 7, Section 4.

Dissociative capture processes have traditionally been described by resonant scattering theory, which is discussed in Chapter 3 and summarized in Chapter 7, Section 4.3. The incident electron is captured to form a resonant state which is stabilized by dissociation. These resonant states can also play a role in dissociative excitation or dissociative ionization, e.g.,

$$
\begin{aligned}
e + H_2 &\to H_2^{-**} \to H_2(^3\Sigma_u) + e \\
&\to H + H + e, \\
e + H_2 &\to H_2^{**} + e \\
&\to (H + H^+ + e) + e, \\
e + H_2 &\to H_2^{**} + e \\
&\to H + H + e.
\end{aligned}
$$

The development of resonant scattering theories has led to semiempirical models which have greatly facilitated the interaction between theorists and experimenters in the analysis of dissociative processes. Due to the significant advances that have been made in quantum chemistry during the past few years, one can now calculate the properties of the resonant states from first principles and compare with the parameters assumed in the empirical models.

It will be apparent from the examples of dissociative attachment discussed in this review that the lifetime of the resonant states against autodetachment can be very short, so that the postulation of a discrete negative ion state should be questioned. The development of alternative theoretical techniques for the treatment of dissociative capture would be very worthwhile, and the recent extensions of *R*-matrix theory are a first step in this direction. (See Chapter 7, Section 4.4.)

2. Dissociation of Molecules into Ground Electronic State Fragments

We consider first the particularly difficult problem of detecting the neutral ground state fragments resulting from electron impact dissociation of neutral molecules. The detection problem arises from the fact that the fragments are uncharged and possess no internal electronic energy which can readily be used to detect the fragments. Another difficulty arises from the fact that two or more neutral fragments are also produced from dissociative excitation and ionization reactions. Hence, one must somehow subtract these reaction channels from the total dissociative reaction channels in order to deduce the cross section for producing electronic ground state fragments. The neutral fragments can also be left vibrationally or rotationally excited; however, there are no experimental studies reported to date on this particularly difficult problem. The few experiments performed thus far detect only ground electronic state products without regard to ro-vibronic excitation. The future uses of laser-induced fluorescence techniques could be very fruitfully applied to probing the final vibrational and rotational state distributions. Laser-induced fluorescence (LIF) has been used[7] to detect the nascent rotational distribution of $N_2^+(X\,^2\Sigma_g^+)$ formed through electron impact.

A number of techniques have been used to detect the ground-state neutrals. The most direct method is to somehow remove the fragments and observe the pressure change in a closed gas system. As early as 1927, Glocker, Baxter, and Dalton,[8] and Hughes and Skellet[9] detected the dissociation of H_2 by noting the decrease in gas pressure as the resulting H atoms were trapped on the surrounding walls of their apparatus. Corrigan,[10] Winters,[11–13] and Teloy[14] have independently used active "getters" to more effectively trap the active atomic fragments. The experiment itself is basically simple. A known volume of gas is enclosed in a tube with an electron source, a pressure gauge, and a freshly evaporated getter. In some experiments the ionization gauge itself acts as the electron source, the getter absorbs gas emitted from the walls as well as decomposition fragments but does not react with the gas itself. The rate at which gas molecules leave the gas phase is designated dN/dt. All fragments (including ion fragments) are assumed to be absorbed on the walls or at the getter. The dissociation cross section is then defined through the relationship

$$\sigma = \frac{1}{n_e l}\frac{V}{P}\frac{dP}{dt},$$

where P and V are the pressure and volume of the system and n_e is the number of electrons per second traveling a length l through the tube. Note that one does not need to determine the absolute pressure, only the ratio of the change in pressure to total pressure. Thus, an ion gauge is quite adequate for determining the relative pressure.

Since molecular dissociation can occur at the hot filament it is necessary to correct for this effect. The rate of change of pressure with the electron beam off (or at an energy below which dissociation occurs) establishes the contribution due to dissociation at the filament. It turns out that the contribution due to dissociation at the filament and electron impact dissociation are comparable.

Low work function filaments (e.g., LaB_6 covered Rh) are used in order to keep the filament temperature low for a given electron beam current. Winters[11] estimates that the cross sections derived in this method should be accurate to $\sim\pm 20\%$. Contributions to the error are $V(\pm 2\%)$, $I(\pm 9\%)$, $n_e(\pm 3\%)$, and $(1/P) \times (dP/dt)$ $(\pm 6\%)$.

Winters[11] has discussed some of the precautions necessary in applying this technique. It is clear that many of these difficulties limit this technique to only certain gases. The primary assumption in these experiments is that the pressure drop in a closed system is directly proportional to the number of fragmentation events. The following effects would invalidate this assumption:

1. Some gases themselves adsorb to the walls of the apparatus. Notable among these are H_2O, HCl, and CO. In fact, almost any polar molecule would adsorb to some extent. The amount of gas adsorbed is a function of the pressure. Thus, when the pressure decreases, previously adsorbed molecules enter the gas phase making the measured cross section smaller than the correct value.
2. Electron impact on an absorbed surface can enhance adsorption or desorption depending upon the type of adsorbed molecules.
3. In certain systems, fragments in the gas phase can recombine with adsorbed fragments to create a gas phase molecule thereby reducing the measured cross section. This turned out to be important in the case of N_2.[11]
4. Undissociated ions, or undissociated, long-lived, excited neutrals can react with the surface and leave the gas phase causing the measured dissociation cross section to be too large.

All of the above criteria have been met or corrected for the cases of H_2,[10] N_2,[11,14] CH_4,[12] and C_2H_6 (ethane),[13] but will severely limit the use of this technique to many other gases.

Another method which has been employed to measure dissociative excitation cross sections is to use a second electron beam and a mass spectrometer to detect the fragments. This method was introduced by Beck and Osberghaus[15] and successfully used by others.[16–28] These so-called dual-chamber experiments have provided much new information on the type of fragments produced, but because of the many uncertainties due to collection efficiencies, detection sensitivities, expected anisotropies in angular distribution, and the like, these experiments have not provided satisfactory cross-section measurements. These experiments have provided a wealth of information about the types of neutrals produced but add little new information about the details of the collision phenomenon itself. Such information is of practical use in fields such as radiation chemistry but is not of direct interest to our discussion of the physics of electron impact dissociation.

Marx *et al.*[29] and Polak and Slovetsky[30] have used electron spin resonance (ESR) techniques to detect the free radicals produced by low-energy electron impact dissociation of some hydrocarbons. The free radicals are trapped in a solid matrix and the characteristic ESR spectrum is used to detect and identify the radicals. The strength of the signal is proportional to the number of radicals produced. Marx *et al.*[29] have used this method to study the production of *t*-butyl radicals and neopentyl radicals from electron impact excitation of neopentane. The neopentane molecule has the same structure as methane with the hydrogens replaced by methyl groups (CH_3). The *t*-butyl radical is missing one CH_3 and the neobutyl radical is missing one H atom from a methyl group. The production of both radicals is first order in pressure up to 10^{-3} torr and nearly second order from 10^{-3} to 10^{-1} torr. For electron energies below the ionization potential the radicals can only be produced from excited states (or temporary negative ion states). If the electron energy is above the ionization potential, the radicals can originate from dissociative ionization processes or from superexcited states. As in the case of the dual-chamber experiments, very little quantitative information has been obtained from these experiments.

Butkovskaya *et al.*[31] have employed the dual-chamber experiment to study the production of H_2 and CH_4 in the collisions of electrons having energies below the ionization potential of propane. The sensitivity to H_2 production was particularly good since H_2^+ production from propane itself by electron impact is particularly weak. This permits improvement of the signal-to-noise ratio in determining the H_2 fragment. They failed to detect methane as a dissociation product and placed an upper limit of 10^{-17} cm^2 on its production cross section. The cross section for H_2 production peaked at 6.5 eV with a cross section of $3 \pm 2 \times 10^{-17}$ cm^2. This maximum is close to the cross section for negative ion production. Owing to the close proximity in peak energy they attributed the H_2 production to the overall mechanism

$$e + C_3H_8 \rightarrow (C_3H_8)^- \rightarrow (C_3H_6 + e) + H_2.$$

The electron may or may not be bound to the C_3H_6. Much more work is needed to examine the effects of compound negative ion states in molecular dissociation leaving neutral products and a free electron. This could be a very important mechanism at energies below the ionization threshold. We will discuss this aspect further in Section 5 on dissociative electron attachment.

Corrigan[10] has published total cross sections for dissociation of molecular hydrogen using the method of atomic hydrogen absorption and pressure measurement. Atomic hydrogen, produced by electron impact excitation of H_2, was completely absorbed by a coating of molybdenum trioxide on the walls of a glass tube. The cross section was obtained from the known electron current, and the change in pressure for a given time of irradiation. For energies above 20 eV the cross sections are inaccurate due to the formation of H_2^+ which can be trapped by either the MoO_3 or by the ion–molecule reaction with H_2 producing H_3^+ and H. Subtraction of the known ionization cross section from the total cross section gives an estimate of the total cross section for dissociation of H_2.

Cross sections for excitation of H_2 to the antibonding $b\,^3\Sigma_u^+$ state have been calculated by many different techniques. A comparison of these most recent calculations[32] with measured cross sections of Corrigan is shown in Fig. 1. Agreement between experiment and theory appears good although the measured cross section contains contributions from other dissociation mechanisms which we explore below.

The total cross section for dissociation of H_2 contains many contributions other than direct excitation to the $b\,^3\Sigma_u^+$ state. One important contribution is the decay of the $a\,^3\Sigma_g^+$ state to the $b\,^3\Sigma_u^+$ antibonding state. The optical emission from the $a\,^3\Sigma_g^+$ state to the $b\,^3\Sigma_u^+$ state is often used as a continuum light source between 1600 and 5000 Å.

Rescigno *et al.*[33] have calculated differential and total cross sections for the excitation of $H_2(X\,^1\Sigma_g^+)$ to $H_2(a\,^3\Sigma_g^+)$ for incident electrons between 13 and 20 eV. More recently, Böse[34] has measured relative optical excitation functions for $a\,^3\Sigma_g^+ \rightarrow b\,^3\Sigma_u^+$ light emission in H_2. Energy resolution of the electron gun was 20 to 100 meV and was sufficient to observe small structures in the excitation function due to negative ion resonances. The light was detected at right angles to the electron beam and the photomultiplier was sensitive in the wavelength region from 1900 to 6000 Å with a detection efficiency of ~13%. The radiation detected from threshold to ~13 eV is due entirely to the $a \rightarrow b$ transition, however, at higher energies, radiation from cascade transitions and $e\,^3\Sigma_u^+ \rightarrow a\,^3\Sigma_g^+$ are present but estimated to contribute less than 20% of the cross section. The structures in the region of 13.5 to 14.5 eV are due to decay of resonant H_2^- states into the $a\,^3\Sigma_g^+$ state. In general, the contribution of negative ion states to the dissociation cross section of H_2 is expected to be minor. Thus, we see that the cross section for dissociation of molecular hydrogen is dominated by direct excitation to the $b\,^3\Sigma_u^+$ state and theoretical calculations are accurate to within a factor of 2 in predicting the magnitude and shape of the cross section.

Winters[12] has determined the total cross section for dissociation of methane (CH_4). The onset occurs at the energy of the first excited singlet state of CH_4 (~80 000 cm^{-1} or 10 eV). The data shows that at ~80 eV, dissociation into neutral fragments is the same as dissociation into ionic fragments. Aarts *et al.*[35] and Donohue *et al.*[36] have investigated the optical emission spectra for CH_4 and only radiation from fragments could be identified. The cross sections for dissociation into excited fragments which emit are small ($< 10^{-17}\,cm^2$)[35] and do not make a significant contribution at any energy. This prompted Winters[12] to conclude that the neutral fragments are primarily

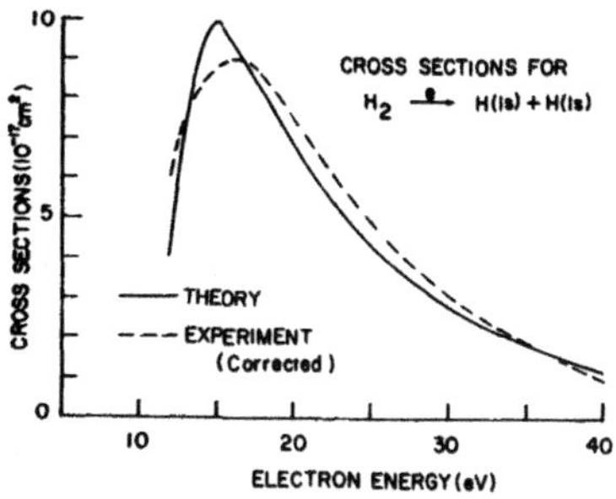

FIGURE 1
Theoretical cross section for the dissociative process $H_2 \xrightarrow{e} H(1s) + H(1s)$ of Chung and Lin[32] along with the experimental values of Corrigan[10] corrected as described in Ref. 32.

ground state or long-lived excited species. Schiavone *et al.*[37] have determined cross sections for the production of long-lived Rydberg states in CH_4 which are also $< 10^{-17}\,cm^2$ (see Section 2.3). Dissociative attachment is also a very minor process.

Winters[13] has measured the total cross section for dissociation of ethane (C_2H_6) and the data are shown in Fig. 2 compared with data of methane and hydrogen. A maximum of $7.65 \times 10^{-16}\,cm^2$ is reached at ~ 80 eV and is slightly higher than the total ionization cross section of Rapp and Golden[38] at that energy. Winters[13] also measured the dissociation cross section for deuterated ethane and consistently obtained a small isotope effect ($\sigma C_2H_6 = 1.09\sigma C_2D_6$) but was uncertain as to whether the difference was real or an experimental artifact. The dissociation cross section was compared 19 times, always with the result $\sigma C_2H_6 > \sigma C_2D_6$. Platzman[39] has offered a general explanation for such

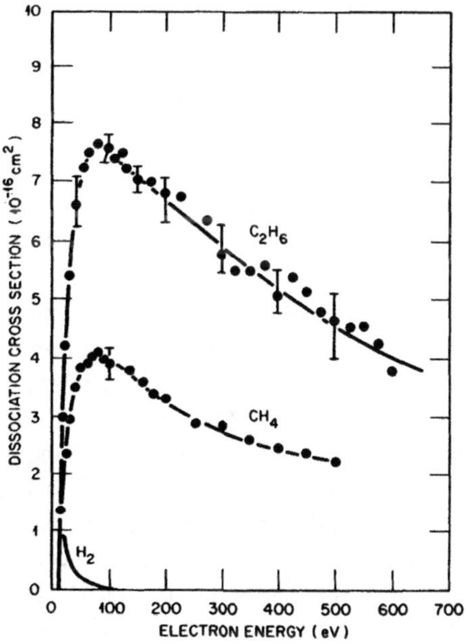

FIGURE 2
Comparison of the experimental cross sections for dissociation of H_2,[10] CH_4,[10] and C_2H_6[13] (ethane) as a function of electron energy.

isotope effects which has been used by Vroom and de Heer[40] to explain observed isotope effects for dissociative excitation of hydrogen (H_2, D_2) and methane (CH_4, CD_4), by assuming that preionization and dissociation are competing paths for highly excited states (sometimes referred to as "superexcited states"). Since dissociation depends upon the velocity and hence reduced mass of the dissociating fragments, a greater dissociation cross section would be observed for C_2H_6, as is observed. Alternately, the ionization cross section for C_2D_6 should be greater than that for C_2H_6. Such an effect would be more accurately determined than the corresponding dissociation isotope effect.

At this point it may be useful to make brief mention of the concept of a superexcited state. Platzman[41] pointed out that a large part of the oscillator strength of most molecules lies above the lowest ionization potential and is associated with electronically excited states that decay by autoionization and dissociation. The role of such states in electron collisions with simple hydrocarbon molecules and H_2O was reviewed in 1975 by de Heer,[42] but it is only recently that theoretical techniques[43,44] have become available to probe the spectra of these states and to study the competition between autoionization and predissociation in their decay.

Because of the lack of spherical symmetry in molecular wave functions, the calculation of inelastic scattering cross sections is much more difficult than for atoms. Most early calculations used forms of the Born approximation. A good example of this type of calculation is the study of the excitation of the $b\,^3\Sigma_u^+$ and $B\,^1\Sigma_u^+$ states of H_2 by Fliflet and McKoy.[45] They use a distorted-wave approximation in which both the incoming and outgoing continuum functions are calculated in the field of the target ground state. (See Chapter 1, Section 6.5.3, Fig. 10, and Chapter 7, Section 5.2.)

For low-energy electrons the R-matrix, variational, or close-coupling methods are potentially more accurate. Chung and Lin[32] have performed a series of two-state calculations on excitation to five states, including the repulsive $b\,^3\Sigma_u^+$ state (see Fig. 1). Weatherford[45a] has examined the treatment of exchange in such calculations. He shows that the total cross section can be computed with reasonable accuracy using a semiclassical exchange approximation, provided that the continuum orbital is constrained to be orthogonal to target orbitals. However, this approximation leads to a significant decrease in the differential cross section at angles above 120°.

In each of these calculations, the nuclei are assumed to be fixed at the equilibrium separation, and the probability of electronic excitation is evaluated only at this single value of R. This is almost certainly adequate for providing a reliable estimate of the cross section for dissociative excitation from the ground vibrational state of the molecule. However, in order to test the sensitivity of the cross section to the initial vibrational state, the calculations should be repeated at several values of R, preferably larger than the equilibrium value.

Further work is also needed to resolve the discrepancy, of a factor of 3, between the cross sections obtained by the distorted-wave and close-coupling methods around 15 eV, and to study the effect of resonant states of H_2^- around 10 eV.

The distorted-wave method has also been applied[46] to dissociation of F_2 following excitation of the lowest $^3\Pi_u$ state. A low-energy resonance is observed, which, however, disappears when allowance is made for target polarization effects. This indicates that neither the standard distorted-wave nor two-state close-coupling methods may be adequate for low-energy collisions by polarizable molecules.

3. Dissociation into Excited Fragments

3.1. Dissociation of Molecules into Fragments Which Emit Radiation

As early as 1925, Blackett and Franck[47] observed Balmer radiation from dissociative electron impact excitation of molecular hydrogen. Since that time there have been numerous studies of dissociative excitation of $H_2(D_2)$ leading to the emission of Lyman ($n \geq 2 \rightarrow n = 1$) and Balmer ($n \geq 3 \rightarrow n = 2$) radiation. Table 1 lists these studies since 1950. Many measurements of atomic cross sections have been reported and a critical review of the experiments prior to 1975 has appeared.[70]

In discussing the experimental results, it is important to emphasize the problems associated with measuring the Lyman and Balmer emission cross sections. Hydrogen atoms in a given state of n can exist in l sublevels ($l \leq n-1$) which are slightly different in energies due to relativistic effects.[71] The energy separation of the l sublevels becomes smaller as n increases. If these levels are subjected to an external electric field, the Stark effect can mix the various l states and lead to a new set of Stark levels which are slightly shifted relative to the unperturbed levels. The strength of the electric field necessary to completely mix the l sublevels can be very small for large n and l. For example, if a hydrogen atom with $n = 3$, $l = 1$ or 2, and j (total angular momentum including spin) $= 3/2$, an electric field of 2 V/cm will completely mix the l sublevels. In many cases the excited atoms will possess kinetic energy from the dissociation process. If a magnetic field is present, the atom experiences a "motional" electric field which can further mix the l sublevels and alter the observed emission. Wilcox and Lamb,[51] for example, have found that magnetic fields of less than 50 G increase the H_α intensity due to mixing of the fine-structure (l) states of the excited atom.

Weaver and Hughes[57] and Khayrallah and Smith[61,69] clearly demonstrated that l-mixing can seriously affect the decay of excited hydrogen atoms. They determined the lifetime of the excited H atoms by studying the time

TABLE 1
Catalog of Studies of Dissociative Electron Impact Excitation of H_2, D_2, and HD to Produce Lyman (*L*) and Balmer (*H*) Radiation

Molecule	Energy range (eV)	Type of radiation	Reference	Year
H_2		H_α	Lamb and Retherford[48]	1950
H_2			Fite and Brackmann[49]	1958
H_2		H_α	Lamb and Sanders[50]	1960
H_2		H_α	Wilcox and Lamb[51]	1960
H_2	0–400		Williams, Martinez, and Dunn[52]	1967
H_2			de Heer *et al.*[53]	1967
H_2, D_2	0–1000		Burrows and Dunn[54]	1968
H_2, D_2	50–6000		Vroom and de Heer[55]	1969
			Ott, Kauppila, and Fite[56]	1970
H_2	20–400		Weaver and Hughes[57]	1970
H_2	0–400	L_α	Mumma and Zipf[58]	1971
			McConkey and Donaldson[59]	1972
H_2			Walker and St. John[60]	1974
H_2			Khayrallah and Smith[61,69]	1975, 1977
H_2, D_2	10–540		Khayrallah[62]	1976
H_2	0–400	L_α, H_α, β, γ, δ	Freund, Schiavone, and Bader[63]	1976
H_2, D_2	0–2000		Möhlmann, de Heer, and Los[64]	1977
H_2, HD, D_2	0–100	H_α, H_β	Karolis and Harting[65]	1978
H_2	0–2000	H_α	Möhlmann and de Heer[66]	1979
	100	$H_\alpha - n, \theta, L$		
H_2	11–16	L_α	Böse and Linder[67]	1979
D_2	17–100	H_β	Higo and Ogawa[68]	1981

dependence of the Balmer-α emission in a pulsed electron impact experiment. The observed lifetime of the excited state was 22 ns instead of the known 5.4-ns lifetime. The longer lifetime is attributed to Stark mixing of other l states, most likely $3d$.

A further complication arises when one attempts to relate the measured emission cross section to the actual cross section for formation of excited hydrogen atoms. Balmer radiation consists of transition from upper levels with $n \geqslant 3$ and $0 \leqslant l \leqslant n-1$ to the lower levels $n = 2$ and $l = 0$ or 1. The excited (n, l) state can radiate to all lower n states with the selection rule $\Delta l = \pm 1$ and thus the Balmer emission cross section is always less than the total cross section for forming the excited atom. This also means that for increasing n, the number of decay channels of an (n, l) state also increases and an even smaller fraction of excited states is detected via Balmer radiation. For example, only 12% of the $4\,^2P$, 58% of the $4\,^2S$, 75% of the $4\,^2D$, and more of the $4\,^2F$ excited atoms emit Balmer radiation.

The measurement of photoemission cross sections for excited H atoms is further complicated by the fact that vibrational bands of molecular emission [Werner

$(C^1\Pi_u - X^1\Sigma_g^+)$ and Lyman $(B^1\Sigma_u^+ - X^1\Sigma_g^+)$] fall in the wavelength region of the Lyman-α line. Most early experiments could not remove these artifacts, and Fite and Brackmann,[49] for example, called their excitation functions "the cross section for countable ultraviolet photons (CUV)." Their CUV cross section was measured relative to the emission of Lyman-α from direct electron excitation of hydrogen atoms, which in turn was normalized to the Born approximation above 300 eV. Carriére and de Heer[72] later determined that ~25% of the radiation detected by Fite and Brackmann[49] was molecular radiation. Finally, collisional quenching of the H(2*s*) can occur which would contribute to the Lyman-α radiation. Pressures lower than 10^{-3} torr are required to exclude appreciable contributions of H(2*s*) to this collision induced emission.

A critical review of the experimental measurements of the cross section for dissociative excitation of Lyman-α radiation by electron impact on H_2 has been presented in the paper by Mumma and Zipf.[58] The threshold for production of $H(^2P)$ atoms is 14.674 eV. Mumma and Zipf[58] also detected a small structure in the Lyman-α cross section at ~27 eV. The feature appears as a distinct break in a "Bethe" plot and the energy happens to be close to the threshold for simultaneous ionization and excitation of H_2 fragments, i.e., $H^+ + H(^2P)$, at 28.27 eV. If this is true, it may provide evidence that doubly excited states may be contributing to the production of $H(^2P)$ atoms. We will see below that structure is also seen in the Balmer-α and Balmer-β cross section which can be attributed to excitation of repulsive molecular ionic states which dissociate to $H^+ + H(3l)$ or $H(4l)$. The contribution of doubly excited states in the production of photons from H_2 will be discussed in Section 4.

Böse and Linder[67] have employed an electron–photon coincidence technique to study electron impact excitation of H_2 and D_2 in the threshold energy region (11–16 eV). Measurements of coincidence between the threshold electrons and VUV photons (1120–1300 Å) provided evidence that triplet states are formed above 14.7 eV which predissociate into H(1*s*) + H(2*p*) atoms. The detected radiation is most certainly Lyman-α. Böse and Linder[67] used the Kronig selection rules for predissociation, especially that $g \nleftrightarrow u$ and $+ \nleftrightarrow -$ to identify the triplet states which are predissociating. The cross section for excitation of these predissociating triplet states is estimated from data of Weingartshofer *et al.*[73] to be in the order of 10^{-17} cm^2. Provided that the branching ratio for predissociation is approximately unity, these processes are expected to contribute to the cross section for dissociation of H_2 into two neutral atoms (see Section 2).

The dissociation of H_2 into excited atoms leading to Balmer $\alpha(H_\alpha)$ $(n = 3 \rightarrow n = 2)$ and Balmer $\beta(H_\beta)$ $(n = 4 \rightarrow n = 2)$ has been thoroughly studied (see Table 1). Much of the effort has been devoted to struggling with the difficult and sometimes subtle experimental artifacts which affect the measurements. The most recent studies by Karolis and Harting[65] discuss the earlier work in detail and present cross sections for H_α and H_β emission for H_2 and D_2. They were careful to eliminate magnetic and electric fields and to correct for molecular radiation near the Balmer lines. However, their measurements are relative cross sections with particular emphasis upon the determination of isotope effects. Figure 3 shows the data points of Karolis and Harting[65] normalized to the absolute values of Vroom and de Heer[55] which are included as solid lines for comparison. A significant isotope effect is observed. Similar isotope studies have been published earlier and the results are summarized in Table 2. At high energies the cross section ratios H_α/D_α and H_β/D_β are in reasonable agreement (± 10%). However, at low energy significant differences are seen between the data of Karolis and Harting[65] and Möhlmann *et al.*[64] versus that of Burrows and Dunn (see Ref. 65) and Khayrallah.[62] The former observe a dramatic decrease in the hydrogen $H_{\alpha,\beta}$ emission cross section as the energy increases, whereas the latter authors find essentially no change in the isotope effect. Since all the measurements agree at high energies, the difference cannot be due to errors in pressure measurements or other types of discrimination. The main differences occur near threshold where Stark mixing of l substates is expected to be most

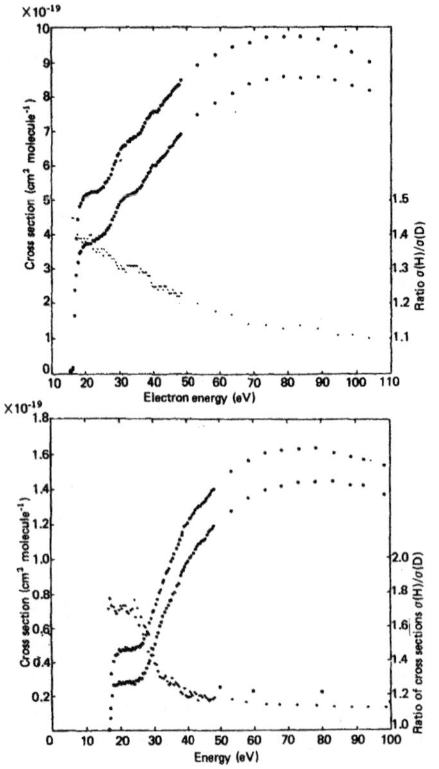

FIGURE 3
Cross section for Balmer α (top) and Balmer β (bottom) production, normalized to the Vroom and de Heer[55] cross section for hydrogen at 500 eV. Values for the ratio between cross sections are also plotted [o, hydrogen; •, deuterium; +, ·, ratio]. Taken from Karolis and Harting.[65]

TABLE 2
(a) H_α/D_α Cross-Section Ratios. (b) H_β/D_β Cross-Section Ratios

Electron energy (eV)	Karolis and Harting[65]	Burrows and Dunn[54]	Khayrallah[62]	Möhlmann *et al.*[64]
		(a)		
20	1.41	1.12	1.08	1.50
25	1.36	1.14	1.08	1.50
35	1.31	1.12	1.07	1.32
45	1.25	1.12	1.06	1.24
60	1.18	1.11	1.06	1.21
100	1.12		1.08	1.19
		(b)		
20	1.70	1.22	1.18	(2.2)
25	1.65	1.20	1.15	2.11
35	1.26	1.12	1.11	1.79
45	1.20	1.11	1.11	1.46
60	1.14		1.11	1.30
100	1.12			1.23

important. Karolis and Harting[65] suggest that the magnetic collimation in the case of Burrows and Dunn[54] may have similarly affected their results; however, the experiments of Khayrallah were not performed in the presence of a magnetic field. One fact seems well established: All of the studies agree that the cross section for producing $H_{\alpha,\beta}$ is greater than that for producing $D_{\alpha,\beta}$.

The observed isotope effect can be understood in a number of quite different ways and the exact explanation is not yet clarified. Vroom and de Heer used the observation of an isotope effect to support the existence of so-called "superexcited" states (discussed in Section 2). "Superexcited" states are states which lie above the first ionization potential and can decay by autoionization, radiation, or dissociation. All of these decay modes may be in competition. The time required for a molecule to dissociate is isotope dependent whereas the autoionization lifetime is relatively isotope independent. Thus, the lighter isotope is expected to give a larger cross section. Accordingly, Vroom and de Heer[55] attributed the larger cross section for $H_{\alpha,\beta}$ emission compared with $D_{\alpha,\beta}$ to competition between the various decay modes of the "superexcited" states.

Karolis and Harting[65] have analyzed the cross sections and isotope effects shown in Fig. 3 in some detail. Owing to the pronounced "break" in the cross section and ratio of $H_{\alpha,\beta}/D_{\alpha,\beta}$ at ~25 eV they divide their discussion to the region from 16–23 eV and above 23 eV. We briefly summarize their analysis. In the energy range between 16 and 23 eV H(3l) or H(4l) atoms can be produced by (1) excitation of the bound $(1s\sigma_g)(n''l)$ states which correlate to H(s) + H($n'l$) to energies above the dissociation energy followed by either direct dissociation or predissociation via curve crossing with adiabatic $(2p\sigma_u)(nl)$ states or with the diabatic repulsion states; (2) direct excitation into vibrational levels of $(1s\sigma_g)(n''l)$ states followed by predissociation. Autoionizing transitions to the $1s\sigma_g$ state of H_2^+ or radiative transition downward and predissociation into the diabatic states may yield ion pairs rather than excited atoms (see Hazi[74]). The observed isotope dependence of the dissociation cross sections may result from unequal excitation cross sections for the intermediate states $(1s\sigma_g)(n''l)$ or by unequal likelihood of competing processes taking place, or both. Further detailed speculation can be found in the paper by Karolis and Harting.[65]

The thresholds observed at 26 and 35 eV are explained as direct excitation of doubly excited adiabatic or diabatic molecular states. The adiabatic states $^3\Pi_g$, $^3\Sigma_u^+$, and $^1\Sigma_g^+$ lie below the $2p\sigma_u$ repulsive states of H_2^+ (see Bottcher and Docken[75] and Bottcher[76]) and dissociate almost entirely to $H + H^*$, $H^+ + H$, and $H^+ + H^-$ (see Hazi[74]). There is a considerable body of evidence that excitation of molecular hydrogen for energies above ~26 eV results in excited $H(2s)$ atoms (see Section 3.2), Rydberg atoms (see Section 3.3), and protons (see Section 4). We shall also see in the next section that Freund *et al.*[77] have measured the kinetic energies of fast excited H atoms which emit Balmer radiation for excitation energies above 26 eV and conclude that such atoms come from doubly excited states leading to $H(1s) + H(nl)$.

The fact that the cross sections exhibit such a small isotope effect provides further strong evidence that doubly excited states are responsible for the majority of the L_α and H_β emission. The near equality of the cross sections further implies that if doubly excited states are involved, dissociation must dominate over autoionization as a much larger isotope effect would be observed. As we shall see, in the case of dissociative attachment (i.e., $e + H_2 \rightarrow H_2^- \rightarrow H^- + H$), autodetachment dominates over dissociation and a large isotope effect is observed (σH_2 can be as large as $1000\sigma D_2$ depending on the energy).

Information on the kinetic energy distribution of excited hydrogen atoms has come from measurements of the Doppler profile of the Balmer lines and from the shapes of anticrossing signals induced by a "motional" electric field.

Emission from a single Balmer "line" consists of a superposition of seven components. Due to Doppler motion of the hydrogen atom, these components are smeared out. Glass-Maujean *et al.*,[78–81] Freund *et al.*,[77] and Ito *et al.*[83,84] have found that the Balmer linewidths far exceed the thermal Doppler widths and actually exhibit structure. At low energies a single Doppler broadened peak is observed corresponding to the production of "slow" excited atoms. As the electron energy is increased wings appear corresponding to "fast" excited atoms. Freund *et al.*[77] measured the mean velocity of the "fast" atom to be ~6.1 eV (velocity = 3.5×10^6 cm/s) and was the same for all levels for $n = 3$ to 6. The identity of the states is deduced from a study of the energy dependence of the excitation cross section of the central peak ("slow atoms") and the wings ("fast atoms"). The "slow atoms" result from dissociated Rydberg states and the fast atoms result from excitation of doubly excited states.

Glass-Maujean *et al.* have described a powerful new technique for determining the velocity distribution of excited H atoms based on the anticrossing effect (see Wieder and Eck[82] and Julien, Glass-Maujean, and Descoubes[81]). The nearly degenerate l sublevels of an excited H atom can be shifted into coincidence with an external magnetic field. The sublevels are said to "cross". The two sublevels can be made to repel or "anticross" with the further application of a small electrostatic field. Transitions occur between these perturbed levels, which tends to equalize the populations of the sublevels. Therefore, the radiation emitted from these sublevels is changed and the change can be detected as an anticrossing signal. If the excited atom is born with kinetic energy in a magnetic field, the atom immediately experiences a motional electric field of magnitude $\mathbf{v} \times \mathbf{B}$. The anticrossing signal strength versus velocity can be calculated to fit with the experimental measurements in order to deduce the mean excited atom velocity. This method also provides information on the relative production rates of the various levels excited. For example, Glass-Maujean[80] has determined that the relative population rates of the $3S$, $3P$, and $3D$ levels have been measured for the "slow" H atoms. There is good agreement between the conclusions of both the anticrossing and Doppler profile measurements (see Glass-Maujean[80]).

Many groups have investigated the collision of electrons with hydrogen-containing molecules resulting in excited hydrogen atoms with $n \geqslant 3$. Möhlmann and de Heer[66] have recently studied Balmer radiation for n as high as 11, i.e., $(n = 11, l \pm 1) \rightarrow (n = 2, l)$ for HCl, H_2O, H_2S, and CH_4 and compared these results with all others before them. Table 3 gives the molecule, electron energy range, and n level observed.

The important information to result from these studies is that (1) optically forbidden (i.e., dipole forbidden) excitations of the parent molecule dominate in the formation of $H(n = 3)$ and (2) the cross section for production of Balmer radiation can be approximated by the empirical relationship $\sigma(n \rightarrow 2) = an^{-b}$, where a and b are parameters characteristic of the type of parent molecule. Values of a and b are presented in Table 4. It is known that the intensities of spectral lines of a given series are proportional to $(n^*)^{-3}$ where $n^* = n - \delta$ (δ = quantum defect) and, for large n, $(n^*)^{-3}$ approaches $(n)^{-3}$. The fact that the variation of σ is closer to n^{-5} than to n^{-3} can be attributed to one or more of a number of factors. First, we must remember that the emission cross section is ***not*** the cross section for formation of $H(n)$. Second, the role of autoionization is no doubt important in explaining the difference between the expected n^{-3} and observed n^{-5} dependence. The high $H(n)$ states most probably result from excitation of high lying molecular Rydberg states which lie in the continuum

TABLE 3
Previous Studies of Balmer Radiation Produced by Electrons Incident on Various Molecules

Molecule	Electron energy	Balmer level $n' \to n = 2$	Ref.
CH_4, CD_4, C_2H_4, C_2H_6, C_6H_6	50–6000	3–6	84
H_2O	50–6000	3–6	85
CH_4, C_2H_4, C_2H_2	0–1000	3–6	86
C_2H_6	100	3, 4, 5	
C_6H_6	0–1000	3–6	87
H_2O	0–1000	3–6	88
14 hydrocarbons up to C_8H_{18}	0–2000	3	89
Chlorinated methanes	0–2000	3–8	90
CH_4, C_2H_4, CH_3OH	0–2000	4–10	91
HCl, HBr, H_2O, H_2S, NH_3, CH_4	0–2000	3	
HCl, HBr, H_2O, H_2S, NH_3, CH_4	100	3–11	66

(i.e., superexcited states). Since the probability of autoionization increases with increasing energy, it is clear that a growing competition between autoionization and dissociation into neutral fragments would favor autoionization. Finally, as we shall see below, internal conversion of electronic to vibrational energy increases with internal excitation energy and may contribute to the deviation from the n^{-3} dependence.

The prefactor, a, is primarily a measure of the size of the cross section, which depends upon the molecular structure of the ground and excited states of the molecule. In this connection, it is worth noting that Möhlmann and de Heer[(66)] have shown that the emission cross sections for H_α as well as CH($A^2\Delta \to X^2\Pi$) radiations *decrease* remarkably with an increasing number of atoms in the molecule for the series CH_4, C_2H_2, C_2H_4, C_2H_6, cyclo-C_3H_6, C_3H_8, C_4H_{10}, cyclo-C_6H_{12}, C_5H_{12}, cyclo-C_5H_{10}, neo-C_5H_{12}, C_6H_{14}, and C_8H_{18}. In general, the cross section for energy transfer from an incident projectile (electron) to a molecule will increase with an increasing number of valence electrons (i.e., atoms) in the molecule. In fact, the total cross sections for inelastic electron scattering is known to increase strongly in going from CH_4 to C_8H_{18}.

Since the initial excitation probability actually increases with a number of

TABLE 4
Empirically Determined a and b Which Fit the Relationship $\sigma(n \to 2) = an^{-b}$ at 100 eV

	H_2	HCl	HBr	H_2O	H_2S	NH_3	CH_4
a	5.46	18.05	17.71	11.48	16.23	6.21	5.83
b	5.74	5.38	5.02	5.26	5.07	4.57	4.76

atoms, another channel or channels must compete more favorably with dissociation into excited fragments as the size of the molecule increases. The internal conversion of electronic to vibrational energy is believed reasonable. Internal conversion processes are directly related to the density of vibrational levels at the energy of the excited electronic state. The density of vibrational states is approximately proportional to a factor $E^N/\Pi h\nu_i$, where E is the excess energy above the ground state, and N is the number of degrees of freedom with frequencies ν_i ($i = 1$ to N). Thus, this vibrational density of states increases rapidly with the number of atoms [$N = 3 \times$ (number of atoms) $- 6$] and the probability of internal conversion is expected to increase greatly in going from CH_4 to C_8H_{18}. The vibrationally excited molecule will also dissociate but can also be left in a lower electronic state. Therefore, the initially excited states will yield increasing amounts of H and CH fragments in the ground state and fewer in the excited state as the number of atoms in the molecule increases.

Donohue, Schiavone, and Freund[(92)] have made a detailed study of the optical emission from excited fragments produced by electron impact on methane, ethylene, and methanol. They find that superexcited states dominate the production of excited fragments below ~25 eV, and for incident energies above 30 eV doubly excited states play an important role. The onsets for emission are 1 to 8 eV above the thermodynamic limits which implies that there is considerable excess energy available. At high energies (> 100 eV) optically forbidden transitions (perhaps two-electron excitations) tend to dominate. Donohue *et al.*[(92)] studied emission from excited C, CH, OH, and CH^+ fragments as well as H atoms.

Isotope effects (i.e., H–D substitution) in the emission cross section of the Balmer lines for methane,[(85)] methanol,[(93)] acetonitrile,[(94)] and HCl[(95)] have been studied. In all cases the emission from the deuterated species is less than that of the hydrogenated species. The isotope effect for HCl and DCl is, for example, $\sigma_H/\sigma_D = 1.20 \pm 0.05$, 1.15 ± 0.04, and 1.07 ± 0.05 for the H_β, H_γ, and H_δ lines, respectively. Hydrogen chloride seems to be a simple case in which dissociation from a single repulsive state gives rise to excited hydrogen atoms. The isotope effect was attributed to competition between autoionization and dissociation of the "superexcited" molecule.

Orient and Srivastava[(96)] have given a recent review along with new data on the cross sections for L_α emission by electron impact on methane. Appreciable differences among the various experimental data are found.

3.2. Metastable Fragments

Electron impact dissociation of molecular hydrogen resulting in metastable H(2*s*) atoms has been well studied experimentally.[(55, 97–106)] The metastable H(2*s*) [or D(2*s*)] atoms were detected either by direct electron ejection from the entrance of an electron multiplier[98–101, 103] or by using a strong, well-localized electric field to mix the 2*s* and 2*p* states thereby causing emission of Lyman-α photons which can be detected with a photomultiplier.[97, 102] In all experiments, the velocity distribution of the H(2*s*) [or D(2*s*)] atoms was determined by the time-of-flight method. Subsequent to some of the experimental studies, Hazi and Wiemers[107] have pointed out that it is important to account for effects such as thermal motion of the target gas and electron–molecule momentum transfer. Hazi and Wiemers[107] find, for example, that the calculated velocity distribution is sensitive to the angle of detection. Although these effects are important, they appear to be less than the differences between the experimental data, as we will see.

Leventhal, Robiscoe, and Lea[97] first showed that dissociation of H_2 resulted in two distinct groups of H(2*s*) atoms – one "slow" group (~0.3 eV) and one "fast" group (~4.1 eV). Subsequent studies[98–102, 106] confirm this observation. All of the data show an unexpected cutoff on the low-energy side of the slow group of 2*s* atoms. It was first[97] suggested that the deficiency in the low-energy tail of the "slow" metastables could be accounted for if those states which mainly contribute to the production of slow metastables possessed a potential maxima at large internuclear separation. The fact that all of the data show a low-energy cutoff in the "slow" H(2*s*) distribution appears somewhat controversial [see the comments by Glass-Maujean[80]]. This effect was not observed in the cross-section measurements of Vroom and de Heer and may be due to a lack of sensitivity in the time-of-flight methods. The fact that photodissociation of H_2 results in an H(2*s*) distribution which is peaked at zero energy prompted Glass-Maujean to offer an explanation. Atoms with zero kinetic energy can be produced through optically allowed transitions which have large cross sections at threshold; if optically forbidden transitions are much more important, they would mask the optically allowed ones in most of the experiments except at threshold where their cross sections are small.

Later studies by Czarnik and Fairchild[102] and Ryan *et al.*[105] showed pronounced structure in the energy distribution of the "slow" group of H(2*s*) atoms. The random thermal motion of the H_2 target gas in previous studies had smeared out the fine structure in the H(2*s*) energy distribution. Czarnik and Fairchild[102] cooled their hydrogen to 50 K and Ryan *et al.*[105] used a capillary hole array to collimate the target gas in order to reduce the thermal velocity component of the H_2 gas. The spectra of Czarnik and Fairchild[102] could be decomposed into six or seven independent peaks each having a shape attributable to the thermal motion of the target molecules and momentum transfer to them by the scattered electrons. The structures arise from predissociation of bound vibrational levels as the energies of the peaks correlated well with peaks in the optical absorption spectra, photodissociation spectra, and photoionization spectra. For H_2 and D_2, the primary predissociation processes appear to be predissociation of the D state via the B' state. The predissociation structures observed for HD were more prominent and were not as easily interpreted. Ryan *et al.*[105] estimate for H_2 that at 100-eV electron energy ~58% of the "slow" H(2s) fragments results from direct dissociation of the B' state and ~42% result from predissociation of the D state via the B' state.

The excitation function for the "slow" H(2s) atoms indicates that singlet–triplet transitions prevail at threshold while for energies greater than 25 eV singlet–singlet transitions dominate.[100,101] The threshold is measured to be 14.6 ± 0.3 eV as compared with the minimum energy requirement of 14.6 eV. We will see later that the states like $(1\sigma_u)(2\sigma_g)\,^3\Sigma_u$ which could dissociate into H(2s) + H(1s) are significant in other collision processes. However, they probably autoionize too rapidly to be important in the production of metastable atoms. This has been confirmed for the $(1\sigma_u)(2\sigma_g)\,^1\Sigma_u$ state in calculations by Kirby, Guberman, and Dalgarno[108] which suggest that only 3% of the molecules formed do not decay by autoionization. The low-lying $^1\Sigma_g$ states are also known to autoionize rapidly, but further work on the triplet states and Π states is needed.

It was initially assumed[97] that the fast metastable atoms are produced through the dissociation of the doubly excited states, such as $(1\sigma_u)(2\sigma_g)\,^3\Sigma_u$, that lie below the first excited state of H_2^+. These states could be formed in collisions of electrons with energy above ~26 eV and dissociate into H(1s) + H(2s). The kinetic energy distribution of the fragments would be expected to peak around 8 eV. The observed threshold depends somewhat on the kinetic energy of the dissociating atoms. Metastable atoms with kinetic energy of about 2 eV are first seen near 29 eV,[101,102] whereas the threshold for the production of 3.8 eV atoms is ~32 eV. These figures suggest that the dissociation limit of the molecular states excited during the collision is ~25 eV, corresponding to two H atoms in $n = 2$ levels. From the angular distribution of the dissociating metastable H atoms and the shape of the excitation function, Misakian and Zorn[100,101] suggest that the dominant state is $(1\pi_u)(2\sigma_g)\,^1\Pi_u$ dissociating to H(2s) + H(2p). Hazi and Wiemers[107] have calculated the velocity distribution of the "fast" H(2s) atoms which result from dissociation of the $^1\Pi_u$ autoionizing state in both H_2 and D_2 and compared the results with the experimental data available at that time. They used the potential energy curves and autoionization widths for the doubly excited $^1\Pi_u$ state calculated by Bottcher and Docken.[75] They conclude that for H_2, 10% of the molecules excited to the $^1\Pi_u$ state autoionize and 90% dissociate to H(2s). For D_2, the branching is 14% and 86%, respectively. Autoionization increases for the heavier D_2 since the nuclei travel slower.

Carnahan and Zipf[104] have determined the relative differential cross sections for production of 2s metastables at 90° to the electron beam direction. For H_2, HD, and D_2 the fraction of fast 2s signals to total 2s intensity at 100 eV was 0.12 ± 0.02. The total absolute differential cross section for the production of H(2s) due to electron impact on H_2 is $d\sigma(100\,\text{eV})/d\Omega = 4.5 \times 10^{-19}\,\text{cm}^2/\text{sr}$. In the dissociation of the heteronuclear HD molecule, it was found that the ratio of fast H(2s) to D(2s) was about 1 to 1, while the same ratio comparing Rydberg atoms was nearly 2 to 1. We will comment upon this interesting observation in the next subsection.

Clampitt[109] has studied the production of metastable $H(2^2S)$ atoms upon collisions of a number of hydrogen containing molecules. The data are shown in Table 5. It is suggested that the excited atoms result from dissociation of "superexcited" states as discussed in Section 2 (see also Ref. 41).

Two groups[110–112] have studied the production of metastable CO and N_2 fragments from electron impact dissociation of CO_2 and N_2O. Freund and Klemperer[110,111] used a detector for the metastables based upon the Auger process by which the electronic excitation energy of the metastable molecule is transferred to an electron in the conduction band of a metal detector which was cesium in their case. This surface was also sensitive to stray light and thermionic emission at room temperature. In order to detect higher-lying metastable states, a less sensitive heated tantalum wire was used which was warmed by direct resistance heating to a temperature just below that for thermionic emission. Clampitt and Newton[112] detected the metastable species by electron ejection from a cesium-antimonide photocathode of a 931 Å photomultiplier tube. The time-of-flight method was used in both experiments to determine the velocity of the metastable fragments. The original data for CO_2 of both groups were originally contradictory as to the observed thresholds and conclusions; however, the later study of Freund[111] resolved this discrepancy and we will discuss only these results. Freund's[111] data show that CO_2 is dissociated by electron impact to give two metastable fragments, $CO(a\,^3\Pi)$ and $O(^5S)$. The experimental dissociation limit for production of $CO(a^3\,\Pi)$ lies at 11.65 ± 0.3 eV, which agrees with the 11.45 eV energy of

$$e + CO_2 \rightarrow CO(a\,^3\Pi) + O(^3P) + e.$$

TABLE 5
Metastable H(2s) Atoms Produced by Electron Impact as Reported by Clampitt[98]

Process	Threshold (eV)	Kinetic energy (eV)
$H_2O \rightarrow H(2\,^2S) + OH$	~19	3.7
$NH_3 \rightarrow H(2\,^2S) + NH_2$	~15	0.2
$NH_3 \rightarrow H(2\,^2S) + NH_2$	~22	2.8
$C_2H_2 \rightarrow H(2\,^2S) + C_2H$	~19	1.6
$C_3H_8 \rightarrow H(2\,^2S) + C_3H_7$	~22	2.7
$C_2H_4 \rightarrow H(2\,^2S) + C_2H_3$	~19	1.7

The excitation function for production of $CO(a\,^3\Pi)$ exhibits weak breaks and structures can be attributed to dissociation processes involving higher triplet states of CO and excited states of O. The cross section for formation of $CO(a\,^3\Pi)$ peaks at ~27 eV and the magnitude is estimated to be of the order of $\sim 4 \times 10^{-17}\,\mathrm{cm}^2$. The cross section was accurately measured relative to the cross section for direct excitation of CO to the $a\,^3\Pi$ state $[\sigma(CO_2) = 0.4 \pm 0.2\sigma(CO)]$. A more accurate measurement of either cross section would be of interest to interpreting the production of $CO(a\,^3\Pi)$ in the atmosphere of Mars. The so-called Cameron bands of $CO(a\,^3\Pi \rightarrow X\,^1\Sigma)$ are prominent in the ultraviolet emission spectrum from the atmosphere of Mars.[113]

Freund and Klemperer[110] were able to make absolute identification of the metastable species by (1) time-of-flight measurements, (2) energetic thresholds, (3) minimum energy requirements of their detector, and (4) electric deflection techniques. The last piece of evidence is rather clever and deserves further comment. A strong electric quadrupole field will deflect a polar molecule leaving nonpolar molecules unaffected. In the case of CO_2 the metastables were deflected leaving only CO^* as a possible candidate. The fragment from N_2O showed no deflection, eliminating all possibilities except N_2^*.

The threshold for production of N_2^* is 8.7 eV, which is close to the threshold of 8.84 eV corresponding to the production of $N_2(B\,^3\Pi) + O(^3P)$.

3.3. High Rydberg Fragments

Atoms with one electron excited to a state of large principal quantum number can be considered to a good approximation as hydrogenic and we can use the theory of Bethe and Salpeter[71] to discuss some of the pertinent properties of these high-Rydberg states. High-Rydberg states are very fragile; the total energy required to remove an electron from the hydrogen atom in the ground state ($n = 1$) is 13.6 eV and for $n = 20$ the binding energy is only 0.034 eV. The Rydberg electron occupies an orbital with a mean radius of

$$\langle r \rangle = 0.26\,[3n^2 - l(l+1)]\ \text{Å},$$

which for $l = 0$ and $n = 20$ gives $\langle r \rangle = 156$ Å. Thus, the geometric cross section for such an atom is large ($7.6 \times 10^{-12}\,\mathrm{cm}^2$) and collisional ionization cross section of this order of magnitude has been found. Also, the radiative lifetime for a state with large n and fixed l is proportional to n^3 and lifetimes are found to be correspondingly large ($\sim 10^{-5}$ to 10^{-4} s for $n = 20$). The fact that the atoms are long-lived and fragile makes it possible to produce them in one region of space, let them drift under their own motion to another region of space, and efficiently detect them by one of the four methods outlined in Section 1. Since fragment ions of the same mass as the Rydberg atoms are produced, it is important to be able to physically or temporally separate the two.

Kupriyanov[114,115] was the first to show that long-lived highly excited atoms result from electron impact excitation of high-Rydberg states of H_2, D_2, CO, O_2, N_2, H_2O, and C_2H_4 molecules. These excited atoms are thought to be primarily high-Rydberg atoms. The mass spectra of the ions formed in the second chamber were investigated as functions of the gas pressure (3×10^{-7} to 3×10^{-4} torr) and electron current in the first chamber. In order to suppress ions from the first chamber, the second chamber was biased 22 V positive. A second plate between the two chambers was biased to 38 V in order to deflect electrons and negative ions. The electron energy extended over the range from 5 to 100 eV with a current of $\sim 200\,\mu A$. The Rydberg atoms were detected by all of the methods discussed in Section 1. Three important results emerged from this pioneering work: (1) the threshold for production of high-Rydberg fragments (O^{**}, N^{**}, C^{**}) corresponds to the threshold for the formation of singly charged molecular ions; (2) a second threshold occurs at the energy corresponding to the doubly charged molecular ions; (3) it was found that the cross sections were about the same for O_2, N_2, and CO, i.e.,

$$\sigma(N^{**}) \approx \sigma(O^{**}) \approx \sigma(C^{**} + O^{**}),$$

and Kupriyanov[115] further noted that the dissociative ionization cross sections are also about the same, i.e.,

$$\sigma(N^{+}) \approx \sigma(O^{+}) \approx \sigma(C^{+} + O^{+})$$

as are the total ionization cross sections $\sigma(N_2^+) \approx \sigma(O_2^+) \approx \sigma(CO^+)$.

These results provided the basis for what Freund *et al.*[116–119] have called the "core-ion model of high-Rydberg dissociation." The model is centered around the assumption that the excited electron in a high-Rydberg molecule is almost completely nonbonding since it will have a small probability density at the core. The potential energy curves of the high-Rydberg molecule will be parallel with those of the ion core and will lie less than ~ 0.1 eV lower in energy for both bound and repulsive states. Thus, dissociation is governed entirely by the potential energy surfaces of the ion core. The Rydberg electron and the ion core can be viewed as uncoupled until the distance between the separating nuclei is comparable to the mean radius of the Rydberg orbital, $\langle r \rangle$. Until the nuclei reach the vicinity of $\langle r \rangle$, the electron is merely a spectator to the dissociation. As the nuclei continue to separate, the molecular Rydberg electron sometimes remains bound to one of the ionic dissociation fragments into a high-Rydberg orbital. Thus, the core-ion model qualitatively accounts for the observations of Kupriyanov.[114,115] A further important prediction is that the translational energy of that high-Rydberg atom is expected to be the same as that which the atomic ion acquires from dissociation of the molecular core ion. Freund[116] was the first to confirm this prediction for the case of O^{**} from O_2. Freund[116] used a time-of-flight technique to measure the kinetic energy distribution of high-Rydberg oxygen atoms produced by electron impact on O_2. The efficient collisional ionization process

$$O^{**} + SF_6 \rightarrow O^+ + SF_6^-$$

was used to detect the high-Rydberg O^{**} fragments. It was found that the translational energy distribution of O^{**} agreed exactly with the known kinetic energy distribution of O^+ ions from dissociative ionization of O_2. Freund and co-workers have subsequently studied dissociative excitation leading to high-Rydberg fragments for N_2,[117] CO,[118] methane,[119] ethylene,[119] and ethane,[119] and generally confirm the predictions of the core-ion model that the kinetic energy distributions of the high-Rydberg fragment could be very similar to that of the corresponding ion fragment. In all these later experiments, field ionization was used to detect the high-Rydberg atoms.

Zipf and co-workers[104,120–122] have also studied the translational spectroscopy of "metastable" fragments using a copper-beryllium electron multiplier as a detector of the "metastable" atoms. The signal contains contributions from true metastables and high-Rydberg fragments. They also successfully interpreted their results in terms of the core-ion model for high-Rydberg dissociation.

Carnahan and Zipf[104] have studied the production of Rydberg states from HD and find twice as many Rydberg atoms of H as of D. This indicates that one should not view the process in terms of the incoherent excitation of molecular states of HD since the individual electronic states have definite parity and thus would dissociate into H^* or D^* with equal probability. They also measured the rate of production of metastable states, and found equal production of $H(2s)$ and $D(2s)$.

In a recent study Schiavone *et al.*[123] have presented measurements of the principal quantum number distributions, effective radiative lifetimes, and excitation cross sections including both the excitation functions and absolute cross sections for 13 molecules. They studied the dependence of the high-Rydberg signal on the experimental variables such as beam current, background gas pressure, and time of flight of the high-Rydberg fragments to the field ionization detector. The distribution with n was determined from the known electric field in the analyzer. Specifically, the electric field was set to ionize and remove high-Rydberg fragments having principal quantum numbers greater than 15, 20, or 80. The time-of-flight distributions were then subtracted to obtain energy distributions for $n \approx 21–80$, $n \approx 16–20$, or $n \approx 15$. Schiavone *et al.*[124] have earlier studied the production of high-Rydberg states when rare gas atoms are excited by electron impact and made some interesting comparisons. First, the n distributions peak at lower n values than those for the rare gases. Second, the fragments appear to form in high-angular-momentum (*l*) states as a result of dissociation as opposed to rare gas atoms which form initially in low-*l* high-Rydberg states. Also, the energy dependence of the excitation function exhibits a slow rise with a peak near 100 eV (resembling a dissociative process) and does not display the step function threshold characteristic of the rare gases.

Table 6 presents cross sections for production of high-Rydberg fragments for electrons of 100 eV incident on 13 molecules. The data of Schiavone *et al.*(123) and Zipf *et al.*(120, 121) are included where the estimated error in cross section is ± 50% for Schiavone *et al.*(123) and a factor of 2 for Zipf *et al.*(120, 121) The cross sections for excitation of the rare gases(125) are also included for comparison. The experiments were sensitive to high-Rydberg atoms with $15 \leqslant n \leqslant 80$ and the detection was made at right angles to the electron beam. In calculating the cross sections, the angular distribution was assumed to be isotropic.

All excitation functions for the data of Schiavone *et al.*(123) show a maximum somewhere between 85 and 150 eV. The implication of this is that the primary dissociation process involves initial excitation by electric dipole-allowed transitions to high-Rydberg molecular states.

Finally, we comment upon the production of high-Rydberg N^{**} fragments from nitrogen. Both Smyth *et al.*(117) and Wells *et al.*(121) find that the N^{**} cross

TABLE 6

Cross Sections for Formation of High-Rydberg Atoms ($15 \lesssim n \lesssim 80$) at 100 eV

Molecule → fragment	Cross section ($\times 10^{-4}$ Å^2)	
	Schiavone *et al.*(119,123)	Finn *et al.*(120)
$H_2 \rightarrow H^{**}$	2.2	11
$D_2 \rightarrow D^{**}$	1.8	8.4
$CH_4 \rightarrow C^{**}$	2.5	
$N_2 \rightarrow N^{**}$	19	134
$CO \rightarrow C^{**}; O^{**}$	18	
$CO_2 \rightarrow O^{**}$	35	
$CH_4 \rightarrow H^{**}$	36	100
$C_2H_6 \rightarrow H^{**}$	66	
$C_3H_8 \rightarrow H^{**}$	56	
$C_6H_{14} \rightarrow H^{**}$	37	
$C_2H_4 \rightarrow H^{**}$	57	
$C_3H_4 \rightarrow H^{**}$	70	
$C_3H_6 \rightarrow H^{**}$	29	
$C_4H_6 \rightarrow H^{**}$	53	
$He \rightarrow He^{**}$	2.3	
$Ne \rightarrow Ne^{**}$	1.7	
$Ar \rightarrow Ar^{**}$	4.5	
$Kr \rightarrow Kr^{**}$	4.7	
$Xe \rightarrow Xe^{**}$	8.5	

section exhibits a sharp resonancelike behavior with a threshold of ~24.1 eV corresponding to fragment production of $N(^4S) + N^{**}$. At higher energies some structures occur in the kinetic energy distributions and high kinetic energies (up to 30 eV) were observed. Wells *et al.*[121] pointed out that the high kinetic energies of the fragment atoms (up to 30 eV) and the large internal excitation energy (14.5 eV) of the high-Rydberg atoms may lead to chemical reactions with atmospheric gases not considered previously. Also the high kinetic energies could contribute to the heat input in the upper atmosphere, especially under disturbed conditions such as auroras.

4. Dissociative Ionization

The dissociation of molecules into fragment positive ions is a much studied subject. Literally thousands of papers have been written on fragment ion production, energy thresholds, cross sections, etc. We will briefly discuss here the recent data on ion kinetic energy distributions and angular distributions where such studies reveal something about the collision dynamics. Unfortunately, the experimental data on this subject are oftentimes in wide disaccord. So much so, that a conclusive review at this time is not warranted. We will only cite some of the recent studies in H_2, O_2, and N_2.

Electron impact dissociation of H_2 into a proton and hydrogen atoms was first observed by Bleakney[126] and Lozier[127] over 50 years ago. Since that time, many studies have been reported on measurements of thresholds, cross sections, angular distributions, and kinetic energy distributions of the resulting protons. In some cases, deuterium was studied because of experimental efficacy or to purposefully examine isotope effects. Table 7 summarizes the studies since 1960. The large number of studies is due to the fundamental nature of the problem and to the considerable disagreement among the experimental observations. There are many experimental obstacles associated with such studies. One major problem is that the cross section for H^+ production is small ($\sim 10^{-18}$ cm^2 at 40 eV). The ratio of H^+ to H_2^+ increases approximately linearly from zero at ~18 eV (electron energy) to only 0.015 at 25 eV.[132] This together with the fact that the kinetic energy distribution extends from ~0 to ~10 eV, results in a small number of ions detected at a given energy. Also, since one is dealing with light ions at low energy, the transmission efficiency of a given energy analyzer is very important.

The first measurements of proton kinetic energy distribution were performed by Dunn and Kieffer[128] in 1963. They concluded that the proton kinetic energy distributions above ~2 eV could be described by Franck–Condon transitions from the $H_2\ {}^1\Sigma_g^+$ to the $(2p\sigma_u)\ {}^2\Sigma_u^+$ repulsive state of H_2^+. Like measurements by Kieffer and Dunn[129] and van Brunt and Kieffer[130] presented

TABLE 7
Summary of Experimental Studies of Proton (Deuteron) Production from Hydrogen (Deuterium)

Ref.	Authors (year)	Species	Electron energy range (eV)	Energy analysis	Peaks observed (eV, Lab)	Angular range	Mass analysis
128	Dunn and Kieffer (1963)	H_2					
129	Kieffer and Dunn (1967)	H_2 D_2	150 only	Magnetic	~8, ~6 (shoulder)	55° only	60° magnetic sector
130	van Brunt and Kieffer (1970)	H_2 D_2	50–300	Magnetic	~6, ~8	20°–160° 200°–340°	60° magnetic sector
131 132	Crowe and McConkey (1973)	H_2	26–300	Time-of-flight (TOF)	~2, 4, 7.5 shoulder at 6	40°–140°	Quadrupole
133	Stockdale *et al.* (1975)	D_2	26–300	TOF	1, 2, 5, 8		Quadrupole
134	Köllmann (1975)	H_2		90° cylindrical analyzer	~8	35°–145°	60° magnetic sector
135	Köllmann (1978)	H_2	23–100	90° cylindrical analyzer	See text, broad peak	35°–145°	60° magnetic sector
136	van Wingerden *et al.* (1979)	D_2	8000	TOF	See text	4π (total)	None
137	Johnson and Franklin (1980)	H_2	30–75	TOF	4, 6 shoulder ~2, ~7.5	30°–150°	Quadrupole
138	Burrows *et al.* (1980)	H_2 D_2 HD	30–100	TOF	~4, ~7 shoulder at ~2	90° only	Retardation (see Ref. 138)
139	Landau, Hall, and Pichou (1981)	H_2	20–100	127° cylindrical analyzer	~1.5, 4, 7	30°–120°	None

evidence that other excitation processes were contributing to the H^+ production in addition to the $^2\Sigma_u^+$ state. The angular distributions measured by these authors contain an isotropic and an anisotropic component. Selection rules first derived by Dunn[140] predict that for a transition between Σ_g^+ and Σ_u^+ states, the protons should exhibit a $\cos^2\theta$ distribution going to zero at 90°. van Brunt and Kieffer[130] also demonstrated the importance of including molecular recoil and the ejected electron momentum in describing the dissociation. However, such effects are insufficient to explain the observed isotropic component. In order to rationalize these results, the authors[129,130] proposed that autoionization from high-lying Rydberg states of H_2 may contribute significantly to the production of H^+ (D^+) by dissociative ionization.

In 1973, Crowe and McConkey[131,132] reported experimental evidence for very well resolved peaks at 2, 4, and ~7.5 eV with shoulders at ~6 and ~8 eV on the 7.5-eV peak. Their data revealed the existence of many new states lying between the $^2\Sigma_g^+$ and $^2\Sigma_u^+$ states of H_2^+ in the Franck–Condon region. The remarkable difference between the data of Crowe and McConkey[131,132] and the earlier data[128–130] is the appearance of many sharp peaks where only one had been previously observed (see Fig. 4). Since the publication by Crowe and McConkey, others have examined the energy distributions for H^+ (D^+) from H_2 (D_2). The experiments of Stockdale *et al.*,[133] Johnson and Franklin,[137] and Landau *et al.*[139] obtain energy spectra in reasonable agreement with that of Crowe and McConkey,[131,132] although the peaks are not as well resolved. In contrast, the proton kinetic energy spectra of Köllmann[134,135] and Burrows *et al.*[138] show only one major peak (aside from transitions to the $^2\Sigma_g^+$ state) with a possible shoulder on the high-energy side of the major peak. These data agreed reasonably well with the earlier experiments.[128–130]

In reviewing the experiments described above, the present authors became aware of a potential artifact in the experiments of Crowe and McConkey,[131,132] Stockdale *et al.*,[133] and Johnson and Franklin.[137] We note from Table 8 that these groups employed an *in line* quadrupole mass filter to detect the protons or deuterons. Since an rf field can effectively ionize high-Rydberg states, it is conceivable that some of the "proton" signal is due to rf field ionization of

FIGURE 4
Proton energy distributions obtained for 50-eV incident electrons on H_2. Dots are Crowe and McConkey[131,132] data taken at 27° to the incident electron beam; dashed line is data of van Brunt and Kieffer[130] taken at 23°. Arrows indicate structure observed by Crowe and McConkey in the wings of the 8 eV peak.

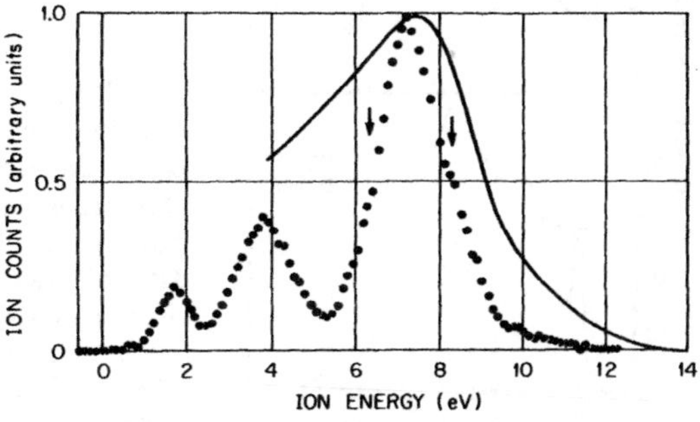

TABLE 8
Autoionization Widths (in eV) for the $^1\Sigma_g\,(1\sigma_u)^2$ State of H_2

Authors	$R = 1.4a_B$	$R = 2.0a_B$	$R = 2.5a_B$
Bottcher[75]	0.97	0.98	0.44
Robb[147]	0.65	1.33	1.8
Hazi[149]	0.65	1.25	1.68
Takagi[148]	0.75	1.62	1.98

high-Rydberg states.[141] Figure 5 shows the D^+ kinetic energy distributions of Burrows *et al.*[138] and Stockdale *et al.*[133] along with the Rydberg atom data of Carnahan and Zipf.[104] One readily sees that the disagreement in the D^+ data at high energy could be reconciled by assuming that the data of Stockdale *et al.*[133] are partly contaminated by rf field ionization of high-Rydberg atoms. Further experiments are needed to determine the exact effects of high-Rydbergs on the data of Crowe and McConkey,[131,132] Stockdale *et al.*,[133] and Johnson and Franklin;[137] however, the data must be affected to some degree.

RF-field ionization of high-Rydberg H atoms cannot account for all the low-energy structures since the data of Landau *et al.*[139] also exhibit the low-energy structures in the H^+ distributions and are not subject to rf field ionization.

Although the experiments are not in satisfactory agreement, they all require autoionization processes in addition to transitions to the $^2\Sigma_g^+$ and $^2\Sigma_u^+$ states to explain the data. In addition, dissociative photoionization studies[142] and experiments on dissociative ionization of H_2 by He^+ ions at a few MeV[143] require the existence of such H_2 autoionizing states. Further evidence for such states is found from experiments in which high-Rydberg H atoms are produced (see Section 3.3).

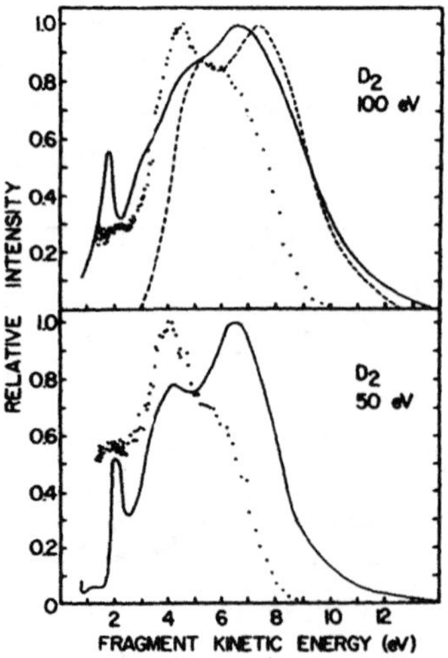

FIGURE 5
Kinetic-energy distributions of D^+ fragments from D_2 at the indicated electron-impact energies. The observation angle was 90°. The dotted curves are the data of Ref. 138. The solid line curves are the 90° data of Stockdale *et al.*[133] taken at the same electron energies. The dashed-line curve is the Rydberg-atom data of Carnahan and Zipf[104] taken with 100-eV electrons and 90° observation angle.

There have been several calculations[74–76,144–146] of the potential curves and lifetimes of the doubly excited states of $H_2(1\sigma_u)(nl\lambda)$ that form a Rydberg series leading to the first excited state of H_2^+. The lowest state $(1\sigma_u)^2$ $^1\Sigma_g^+$ is of special interest with respect to dissociative recombination of H_2^+ and electron-impact dissociation of H_2, in which it appears at scattering energies in the range 24–32 eV. There is general agreement concerning the position of this state, but the autoionization width has been more controversial. Nevertheless, as can be seen from Table 8, there is a growing consensus on both the magnitude and R dependence of the width. Note that as R approaches the stabilization point, close to 2.7 a.u., at which the potential curve crosses that of the ground state of H_2, the width should not go continuously to zero, as suggested in some of the earlier papers.

The results of Robb[147] and by Takagi and Nakamura[148] are derived from calculations of the phase shifts for electrons scattered by H_2^+ molecules with fixed nuclei, using the close-coupling and algebraic-variational methods, respectively. The doubly excited states appear as resonances and the position and width can be obtained by fitting to a Fano profile. Hazi[149] uses the Stieltjes-imaging technique[150] to compute the strength of the configuration interaction between the doubly excited state and the electron-scattering continuum. The matrix element can be obtained as a function of electron energy and nuclear separation. The values given in Table 8 correspond to the energy at which the resonance would occur if the nuclei were fixed. The earlier calculation by Bottcher and Docken[75] using perturbation theory appears to give the correct order of magnitude, but a poor R dependence for the width.

The lowest doubly excited state that can be formed by photon absorption by the ground state of H_2 is the $(1\sigma_u)(2\sigma_g)$ $^1\Sigma_u$ state. This state leads to a peak in photon-impact and electron-impact excitation of H_2 around 27 eV. A comparison between the various calculations with the experimental data on proton production has recently been made by Kirby *et al.*[151] The comparison suggests either that the most recent calculations of the width are too large, or that the competition between autoionization and dissociation cannot be adequately described using a complex potential with a local width. Further study of the distribution of vibrational states of H_2^+ formed after the autoionization of this state would be useful.

These doubly excited states have repulsive potential curves which can lead directly to dissociation, either with or without the emission of an electron. However, there are many electronically excited states with attractive potential curves that undergo predissociation through interactions with other states. This predissociation leads to narrow structures in the kinetic energy distribution of the dissociation fragments.

Study of predissociating states in H_2 has led to interesting comparisons between theory and photo absorption experiments.[152–155] The development of multichannel quantum defect theory[156–158] has led to comprehensive theoretical treatments of both autoionization and predissociation. The presence of the dissociation channel leads to significant deviations from the propensity rules that have been so successful in describing autoionization alone.

The energy and angular distribution of O^+ ions from dissociative ionization of O_2 has been studied by Kieffer *et al.*,[159] Stockdale and Deleanu,[160,161] and Schopman and Locht.[162,163] Figure 6 compares the data for ion production[160] with that for high-Rydberg atoms[164] and dissociative photoionization.[165] All measurements are consistent with the suggestion that the large peak at ~0.8 eV is due to predissociation of the $B\,^2\Sigma_g^-$ state of O_2^+. Vibrational structure found in this peak[161] is also consistent with this picture. The fact that the high-Rydberg oxygen atoms have the same energy dependence as the O^+ is also consistent with the core-ion model of dissociation discussed in Section 3.3.

Studies of the production of N^+ and N^{++} from dissociative ionization of N_2 do not show the same degree of agreement as for the case of O_2. Köllmann[166] has presented the most recent studies and has reviewed the earlier work of Kieffer and van Brunt,[167] Crowe and McConkey,[168] and Deleanu and Stockdale.[169] The latter two groups also studied N^{++} from N_2.

5. Dissociative Attachment

The formation of negative ions through dissociative attachment of electrons to molecules has been studied extensively over the last 20 years, using both electron beam and swarm techniques. Some of the experimental data on simple molecules have been explained qualitatively in terms of resonant scattering theory, as can be seen, for example, in the earlier reviews by Bardsley and Mandl,[170] Schulz,[171] and Compton and Huebner.[172] In this section we will discuss several examples to show how some recent data have supported the interpretation in terms of resonant scattering, whereas other data are as yet unexplained. The section will be subdivided by molecular species, beginning at the simplest diatomic molecules and working towards large molecular clusters.

5.1. H_2 and D_2

It is ironic that these molecules, for which *ab initio* calculations are simplest and for which there is a great deal of very interesting experimental data, are the ones for which the application of the traditional theory of dissociative attachment is most in doubt because of the very short lifetime of the intermediate molecular negative ions.

There are two states of H_2^- that dissociate adiabatically to $H + H^-$ in the ground state. The lowest $X\,^2\Sigma_u^+$ state is stable against electron emission only for $R \gtrsim 3a_B$, and the $B\,^2\Sigma_g^+$ state becomes unstable around $5.3a_B$.[173] The molecular

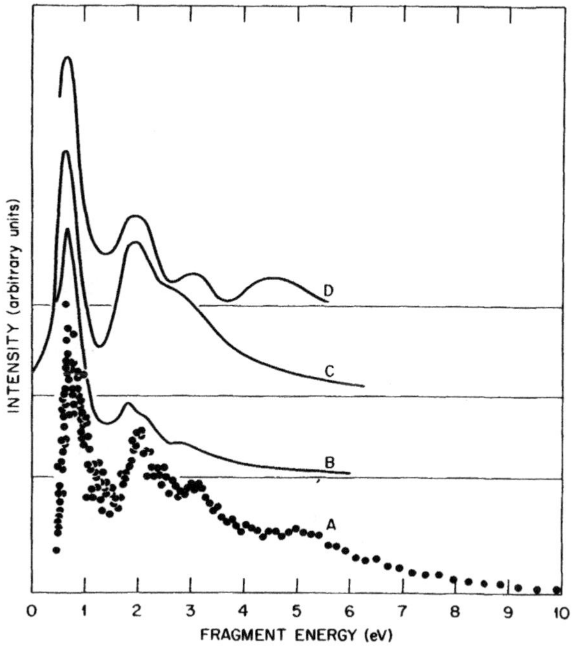

FIGURE 6
Energy spectra of oxygen fragments from dissociation of O_2. (A) O^+ detected at 50 eV electron energy. (B) O^+ detected at 35 eV electron energy. (C) High-Rydberg O atoms at 39 eV electron energy. (D) O^+ detected at 34.6 eV photon energy. Taken from Ref. 160.

structure of the ground $X\,^2\Sigma_u^+$ state is $(1\sigma_g)^2(1\sigma_u)$, and for $R < 3a_B$ only the weak p-wave centrifugal barrier holds the third electron in the vicinity of the nuclei. If a resonant state exists, its lifetime must be extremely short. Most authors have assumed that the potential curve for H_2^- can be extended to smaller R and that for $R < 3a_B$ it lies above that of the ground state of H_2. However, Bottcher (private communication) has suggested that the state may become a virtual state, as discussed in Chapter 3 of this volume, with an energy whose real part is below that of the H_2 ground state.

The interpretation of the sharp threshold peak in attachment to H_2 (see Chapter 1, Section 2.4, Fig. 2) in terms of resonant scattering theory has been strengthened through the analysis,[174,175] of the recent experiment[176] on the temperature dependence of the attachment cross sections. Allan and Wong[176] passed an electron beam through a trochoidal electron monochromator and studied the collisions with the hydrogen molecules in an irridium oven at temperatures between 300 and 1600 K. The negative ions were collected through a quadrupole mass filter and channeltron. From the lowering of the attachment threshold as the oven temperature is increased, Allan and Wong were able to deduce the threshold values for the cross section for dissociative attachment to

H_2 and D_2 in excited vibrational and rotational levels. They showed that the threshold cross section for attachment to D_2 molecules in the $v = 5$ state is almost six orders of magnitude larger than the corresponding value for the ground state. Theoretical estimates of these cross sections for various vibrational and rotational states were obtained independently by Wadehra and Bardsley,[174,175] using a semiempirical resonance model with parameters chosen to fit previous experimental data involving ground-state molecules only. The enhancement in the cross section for excited states arises because the intermediate H_2^- can be formed at larger values of R. There is then a greater probability that the H_2^- ion can dissociate without emission of the third electron.

Although the dependence on the initial vibrational state was expected, at least qualitatively, there had been some debate about the effect of rotational excitation following the prediction by Chen and Peacher[177] of a very large enhancement. The theoretical cross sections for H^- production have been used by Bacal and Hamilton[178] in their explanation of the anomalously high negative ion production in the H^- sources that are being developed for injection into Tokamak devices.

In spite of this recent success of resonant scattering theory there remains some uncertainty concerning the size of the isotope effect in the attachment cross section, both near threshold and around 11 eV. A new measurement of the ratio of H^- and D^- production cross sections would be worthwhile. Also, further theoretical study of the narrow peaks that are seen[179] near 12 and 15 eV would be valuable, since it would help us to understand curve-crossing phenomena involving autoionizing states. Studies of the angular distributions of the dissociation fragments[180] have already proved valuable in the interpretation of the lower structures.

5.2. N_2 and NO

Hiraoka *et al.*[181] have reported the detection of N^- ions with lifetimes greater than 10^{-6} s from electron collisions in nitric oxide, and in mixtures of N_2 and O_2. The appearance potential of the N^- ion indicated that the ion was a ^1D state, although the ^1S symmetry could not be excluded. Autodetachment from either ^1D or ^1S into $N(^4S^0) + e$ would be very slow because stable if it were below the ^2D state of N. However, Klots and Compton (unpublished) have repeated these experiments using a time-of-flight mass spectrometer and were unable to confirm the observation.

A short-lived autodetaching state of N^- has been produced in e–N_2 and e–NO collisions by Spence and Burrow[182] and by Mazeau *et al.*[183] Studies of the energy distribution of the inelastically scattered electrons reveal the existence of an autoionizing state of N^- at an energy of 0.07 ± 0.02 eV above that of the ground state of N, with a width of 16 ± 5 meV (Chapter 5, Section

2.7, Fig. 19). The short lifetime of this state and its low energy strongly suggest that it is the ground ^3P state of N^-.

Huetz *et al.*[184] have performed an extensive study of the production of $N^-(^3P)$ in electron collisions with N_2. The cross section arises from an apparent threshold at 9.83 eV, and peaks near 10.1 eV. At a scattering angle of 90°, the maximum differential cross section is $\sim 2 \times 10^{-20}$ cm^2/sr. The maximum integrated cross section measured by Spence and Burrow[182] is $(2.4 \pm 1.2) \times 10^{-18}$ cm^2, which seems to be somewhat higher than one would expect from these differential measurements. Huetz *et al.*[185] have also studied the enhancement in the dissociation cross section due to vibrational excitation in the initial N_2 molecules, using a microwave discharge. Their results suggest that in going from $v = 0$ to 4 the cross section increases only by a factor of 4, which is much less than is found in H_2 and D_2. See also Chapter 5, Section 2.7.

Huetz *et al.*[184–186] have analyzed these results, and related measurements of vibrational excitation and electronic excitation in e–N_2 collisions, using a complex potential model for the $A\,^2\Pi_u$ state of N_2^-. They assume that the potential curve lies below its parent $A\,^3\Sigma_u^+$ state of N_2 between 1.7 and 3.3 Å and at smaller separations is a shape resonance in the field of the parent state. The choice of the parameters describing the resonance near its equilibrium separation was based on an earlier empirical analysis by Cadez and Fayard.[187] Although this model gives a good fit to the cross sections for vibrational excitation and electronic excitation of the $A\,^3\Sigma_u$ state and of the energy dependence of the "attachment" cross section, the magnitude of the latter cross section is badly underestimated. This suggests that the rate of decay of the resonance into its parent is overestimated whereas the coupling to the ground state is too weak.

Although calculations of the energies of the N^- states are not yet sufficiently accurate to establish the exact value of the affinity of the ground state, it does seem likely that the lowest $^1D^e$ and $^1S^e$ states lie well below their parents ($^2D^0$ and $^2S^0$) and that they are metastable against autodetachment.[188] In view of the failure of attempts[189] to detect N^- ions produced by double electron capture by fast N^+ ions, further studies of the production of metastable N^- ions by electron impact on diatomic molecules would seem worthwhile.

5.3. O_2

The major feature observed in dissociative attachment to the ground $X\,^3\Sigma_g^-$ state of O_2 is a bell-shaped curve with a width of 1.7 eV and a maximum value of about 1.3×10^{-18} cm^2 at an energy of 6.5 eV. The angular distribution of the O^- ions, observed by van Brunt and Kieffer,[190] suggests that this peak is due to a $^2\Pi_u$ state of O_2^- that leads to ground-state fragments $O^-(^2P) + O(^3P)$. This assignment is supported by the calculation of Krauss *et al.*[191] and Das *et al.*[192] In experiments with oxygen gas heated to around 2000 K, Henderson *et al.*[193] and Spence and Schulz[194] found the peak to occur at lower energy and noticed a significant lowering of the apparent threshold for the reaction from the expected 3.64 eV. O'Malley[195] argued convincingly that this indicates that the cross section for attachment is enhanced significantly by vibrational excitation in the target molecule.

Studies of attachment to the metastable $^1\Delta_g$ state of O_2 have been reported by Burrow[196] and by Belic and Hall.[197] The $^2\Pi_u$ resonance is again seen, with a maximum attachment cross section of $\sim 4 \times 10^{-18}$ cm^2. Two higher resonances are also seen which lead to excited $O(^1D)$ and $O(^1S)$ atoms.

5.4. F_2, Cl_2, Br_2, and I_2

Interest in attachment to these molecules, and to the hydrogen-halides to be discussed in Section 5.5, has been stimulated by the realization that the formation of the upper laser state in rare-gas halide lasers proceeds predominantly through the recombination of positive rare-gas ions and negative halogen ions that are produced by dissociative attachment.

Although most of the experiments on attachment to F_2 have involved swarm techniques, a beam measurement has been made by Chantry[198] (Chapter 7, Section 4.3, Fig. 9). The results by the two methods have been compared by Schneider and Brau.[199] The results are consistent for mean energies above 0.5 eV. The beam measurements for energies above 0.1 eV can be fitted very well[200] by assuming that the ground $^2\Sigma_u$ state of F_2^- is stable only for $R > 2.6a_B$ and leads to a *p*-wave shape resonance at smaller R. The calculation employed a nonlocal width and gave a cross section peaking at 6×10^{-16} cm^2 at 0.07 eV and decreasing to $\sim 3.2 \times 10^{-17}$ cm^2 at 1 eV. However, the cross sections measured in the beam experiment continue to rise as the electron energy is decreased below 0.1 eV, reaching a value well above 10^{-15} cm^2.

Hall[201] had previously analyzed the swarm data in terms of a $^2\Sigma_u$ resonance, but with a local width. His results are consistent with Chantry's cross sections at energies between 0.2 and 1 eV, but again are lower at energies below 0.2 eV. Although Hall's results contain a peak at zero energy, his theory is inappropriate at very low energies since it violates the Wigner threshold laws.

The existence of a zero-energy peak would imply a significant contribution to the cross section from the capture of *s*-wave electrons. This would be of great theoretical interest since the lowest resonant state of F_2^- that contains an *s*-wave component is the $^2\Sigma_g$ state, as discussed below. Further experimental study of this very difficult regime would be worthwhile.

The theoretical models predict larger attachment cross section for F_2 molecules in the $v = 1$ and $v = 2$ states, but the enhancement is not so great as found in H_2 and HCl. For example, in the nonlocal-width theory at 0.5 eV the ratio $\sigma_{DA}(v=1)/\sigma_{DA}(v=0)$ is close to 2. These ratios are qualitatively compatible with the small increase in the attachment rates observed by Trainor and Jacob[202] when the gas temperature is raised from 300 to 500 K. Since both theoretical models are chosen to fit the attachment data for ground-state molecules and experimental studies of the attachment rates to excited states are difficult, the models should be tested by looking at another collision process, such as vibrational excitation. The nonlocal width model leads to smaller cross sections than those obtained by Hall.[201] Both theories predict the attachment cross section to be greater than the vibrational excitation cross sections.

The great progress that has been achieved in the last few years in *ab initio* calculations can be judged from the recent studies by Hazi *et al.*[203] of dissociative attachment and vibrational excitation for F_2. Both the potential curves for F_2 and F_2^- and the width of the F_2^- state have been computed and the resultant cross sections are in reasonable agreement with the semiempirical fits of Bardsley *et al.*[200] The only significant difference from experiment is in the attachment cross section near zero energy (see Chapter 7, Section 4.3, Fig. 9).

The beam measurements of Chantry[198] on F_2 show further structure at higher energies, with a shoulder at 3 eV and a broad peak between 5 and 8 eV. In Cl_2 three similar peaks, near 0, 2.5, and 5.5 eV have been observed by Tam and Wong[204] and by Kurepa and Belic,[205] who also reported a very small shoulder around 10 eV.

There are six states of the diatomic halogen negative ions that dissociate into ground-state fragments. These have symmetries ${}^2\Sigma_u^+$, ${}^2\Pi_g$, ${}^2\Pi_u$, and ${}^2\Sigma_g^+$, with the ${}^2\Pi$ states being split into $J = 1/2, 3/2$ levels by spin–orbit forces. In F_2 and Cl_2 it is tempting to ascribe the first three observed structures to the ${}^2\Sigma_u$, ${}^2\Pi_g$, and ${}^2\Pi_u$ states since all those could lead to shape resonances. The ${}^2\Sigma_g^+$ state could perhaps lead to a broad Feshbach resonance similar to the $B\,{}^2\Sigma_g^+$ state of H_2^-. However, as pointed out by Tam and Wong,[204] the ${}^2\Sigma_u$ state may be stable against electron emission in Cl_2 and the heavier molecules Br_2 and I_2.

In the experiment by Tam and Wong the ions were detected in a direction at right angles to the incident beam. If the finite acceptance angle of the detector, the rotation of the molecule during dissociation, and the effect of spin–orbit forces are all ignored, then one would expect that the formation of the ${}^2\Sigma_u^+$ and ${}^2\Pi_g$ resonance would result in no dissociating ion in such a direction. However, in view of these qualifications, further study of the angular distributions of the ions and further theoretical analysis is needed before any firm conclusions can be made concerning the identification of the peaks. Calculations of the negative ion potential curves are difficult. Rescigno and Bender[206] have carried out large configuration interaction calculations for F_2 and F_2^-, and find that the ground state of F_2^- becomes unstable near the equilibrium separation in F_2. Further work on excited states and on the heavier halogen molecules would be worthwhile.

Kurepa *et al.*[207] have recently reported the observation of six peaks between 0 and 8.5 eV in dissociative attachment to Br_2. They tentatively assign these peaks to the six lowest states of Br_2^- with the symmetries given above. However, once again the lowest structure seems to peak at zero energy. As was pointed out above, a nonzero cross section at zero energy implies the capture of an *s*-wave electron, and only the highest ($^2\Sigma_g^+$) state can have an *s*-wave component. Thus, the zero-energy peaks in each of the halogens must be regarded as special threshold effects, perhaps enhanced by the $^2\Sigma_g^+$ resonance or perhaps unconnected with the resonances. Ion pair production is seen in *e*—Br_2 collisions between 10 and 40 eV with a maximum cross section of 5.3×10^{-18} cm^2 around 18 eV.

In a pioneering swarm experiment in I_2, Truby[(208)] observed the thermal attachment rate to increase from 0.9×10^{-10} to 4.2×10^{-10} cm^3 s^{-1} when the whole gas was heated from 253 K to 467 K. With the gas temperature fixed at 298 K selective heating of the electrons from an average energy of 0.04 eV up to 0.27 led to an increase in the attachment rate from 1.8×10^{-10} cm^3 s^{-1}. Shipsey[(209)] was able to explain the effect of gas heating by assuming a *p*-wave resonance ($^2\Sigma_u$ or $^2\Pi_u$) with a potential curve crossing that of the I_2 ground state just outside the Franck—Condon region. An alternative model, with a constant $\Gamma(R)$, was suggested by Birtwistle and Modinos,[(210)] but they claim that neither model is able to explain the effect of selective heating of the electrons. The observation by Tam and Wong[(204)] of two separate peaks in the attachment cross section, near 0 and 0.9 eV, perhaps explains the failure of these models, which both assume a single resonance, to explain all of Truby's data.

Brooks *et al.*[(211)] have recently studied the temperature dependence of electron attachment in an electron swarm experiment with a mixture of I_2(1%) and N_2(99%) between 35 and 110°C. By changing the strength of the electric field, they were able to vary the average energy of the electron between 0.5 and 2.5 eV. They found a definite minimum in the cross section near 1 eV, confirming that more than one resonant state is important. However, the minimum in their cross section is very close to the energy at which Tam and Wong find a peak.

5.5. Hydrogen Halides

The cross sections for electron attachment to HCl and HBr[(212, 213)] show sharp peaks near the threshold energies of 0.82 and 0.40 eV, respectively. The rate of decrease of the cross section at energies just above these attachment thresholds appears to change abruptly at the thresholds for vibrational excitation. This has been interpreted by Fiquet-Fayard[(214, 215)] as being due to the additional competition between attachment and vibrational excitation. If the temporary negative ion can decay through the emission of *s*-wave electrons, the slope of the attachment cross section should be discontinuous at the vibrational excitation thresholds.

The appearance of this structure suggests that the HCl^- and HBr^- states decay mostly through the electron emission and that the survival factor is small. Thus one must expect a significant enhancement in the attachment cross section for vibrationally excited molecules. Experiments by Allan and Wong[(216)] on attachment to hot HCl molecules appear to confirm this. An analysis by Bardsley *et al.*,[(200)] based on an empirical resonance model, suggests that the presence of rotationally excited molecules is also important. They find that for each vibrational state the threshold cross section increases as J increases, with particularly large changes between $J = 9$ and 10, and between $J = 18$ and 19. These abrupt changes arise because the number of competing vibrational excitation channels is reduced at these values. However, this resonance model does not predict the threshold peaks in vibrational excitation very accurately. The exact nature of the intermediate state of HCl^- that leads to dissociative attachment is not clear, and further exploration of alternative models would be worthwhile.

Two peaks are observed in the production of H^- from HCl. The first has a steep onset near 6.9 eV, whereas the second rises less rapidly and has a maximum at 9.2 eV. It is interesting that the steep rise at 6.9 eV is well above the threshold for the formation of H^-, so that it is not the usual vertical-onset phenomenon. Taylor *et al.*[(217)] have suggested that it arises because of a maximum in the potential curve for the dissociating HCl^- state. Azria *et al.*[(218)] have observed fine structure in the second peak which they attribute to the predissociation of Feshbach resonances in HCl^-.

At higher energies one can produce excited H atoms along with a negative halogen ion. For example, in e–HF scattering, Abouaf and Teillet-Billy[(219)] have seen several peaks between 12.5 and 15.5 eV, which appear to be correlated with structures seen in transmission experiments by Spence and Noguchi.[(220)] Thus predissociation of Feshbach resonances appears in at least two of the hydrogen halides.

Azria *et al.*[(221)] have recently studied the branching ratio for the processes

$$e + HX \rightarrow HX^- \begin{cases} \nearrow H^- + X(^2P_{1/2}) \\ \searrow H^- + X(^2P_{3/2}), \end{cases}$$

where X is Cl, Br, or I. They see contributions arising from both $^2\Sigma$ and $^2\Pi$ resonant states of the negative ions, and cannot explain their results in terms of completely adiabatic motion. There appears to be significant diabatic coupling between the resonant states which varies significantly with the velocity of the dissociation fragments. Theoretical analysis of these experiments would be very interesting, since we do not yet know exactly how nonadiabatic couplings depend on the widths of the interacting molecular states.

5.6. *Triatomic Molecules*

Dissociative attachment in triatomic molecules provides further examples of several of the features that we have discussed above. For example, the first direct observation of an enhancement, by several orders of magnitude, in an attachment cross section due to vibrational excitation in the molecules was made on N_2O by Chantry.[222] It was suggested[223,224] that excitation of the bending mode leads to the capture of electrons with very low energy, and that in the bent configuration the ground state of N_2O^- is stable against electron emission. This suggestion was confirmed by the quantal calculations by Hopper *et al.*[225] Studies of the angular distributions of the O^- produced from N_2O have also been made,[224] giving information on the partial waves that are involved in the electron capture step.

The enhancement of the cross section for attachment to excited states and the accompanying lowering of the threshold energy lead to problems in using dissociative attachment in the determination of electron affinities. In cases where the cross section has a steep onset, the threshold energy can often be deduced more accurately from the position of maximum slope than from the lowest energy at which attachment is observed.

The cross section for O^- production from CO_2 has a prominent peak between 4 and 5 eV. Abouaf *et al.*[226] have observed structure in this peak, with a spacing of 0.12 eV, similar to that observed in backscattering experiments by Burrow and Sanche,[227] in transmission measurements by Sanche and Schulz,[228] and in vibrational excitation by Boness and Schulz.[229] In this latter experiment the exact positions of the cross section peaks depend on the final vibrational state, showing that the vibrational spectrum of the CO_2^- ion is not fully developed.

This structure can be understood in terms of Herzenberg's boomerang model[230,231] (Chapter 3, Section 5.2) and the CO_2^- potential surfaces described by Claydon *et al.*[232] In the linear symmetry configuration the ground state of CO_2^- has symmetry $^2\Pi_u$. On bending this state is split, giving a lower $^2A_1$ state and a higher $^2B_1$ state. If CO_2^- is formed by the attachment of an electron to a CO_2 molecule in its equilibrium configuration, there is an initial force leading to symmetric stretching of the C–O bands. However, the symmetric linear configuration represents a saddle point for CO_2^- and the ground state is unstable with respect to an asymmetrical stretch and, in the case of the $^2A_1$ state, to bending. At energies above 3.99 eV, the CO_2^- state is unstable both through electron emission and through dissociation via the asymmetric stretch mode, so that the vibrations in the symmetric stretch mode do not lead to a well-developed level spectrum.

In e–CO_2 collisions at higher energies, O^- production is seen again around 8, 12, and 17 eV, whereas O_2^- appears, with a smaller cross section, between 11 and 14 eV, and C^- is produced between 16 and 20 eV.[232–234] The production of negative ions from attachment to the similar molecules OCS and CS_2 has also been studied experimentally.[235,236] As in the case of CO_2, evidence was found for vibrational excitation in the neutral fragments.

Attachment to H_2O is of special interest because there are three dissociation channels with thresholds near 4 eV. OH^- ions can be produced at an energy of 3.3 eV or more, O^- ions require at least 3.6 eV, and 4.4 eV is needed to give H^- ions. A valuable experimental and theoretical analysis has been recently published by Jungen, Vogt, and Staemmler,[237] with a tabulation of earlier experimental work. The attachment proceeds mainly through three resonances of the H_2O^- ions, which appear at first sight to be fine examples of the classic Feshbach resonances, with two electrons in the lowest unoccupied orbital outside the positive ion H_2O^+ in the $^2B_1$, $^2A_1$, and $^2B_2$ states. The binding energy of the two outer electrons is around 6 eV in each of the resonant states. Jungen *et al.*[237] have carried out a CI calculation which shows that there is a strong correlation between the two outer electrons and that the character of their orbitals is intermediate between the Rydberg and valence type. The symmetry assignment for the lowest resonance is supported by the angular distribution of the dissociation fragments as measured by Trajmar and Hall.[238]

All three dissociation channels have been observed near the lowest two resonant states, but the peaks in the cross section for H^- production are lower by 0.5 eV than the corresponding peaks in O^- and OH^- production (6.5 vs. 7.0 eV and 8.6 vs. 9.1 eV). Only O^- and OH^- have been observed from the third resonance at 11.8 eV. The CI calculations suggest that there may be a barrier in the potential surface that hinders the formation of O^- and OH^- fragments, which might explain the different peak positions.

Tronc *et al.*[239] have observed some very interesting isotope effects in the formation of H^- or D^- from H_2O, HDO, and D_2O through the lowest resonance. The ratio $\sigma(H^-/H_2O)/\sigma(D^-/D_2O)$ is 1.25. The relatively small value of this ratio suggests the dissociation proceeds with only moderate competition from autodetachment. However, when comparing the production of H^- and D^- from the same isotope HDO, they find $\sigma(H^-/HDO)/\sigma(D^-/HDO) = 3.55$. Although the D^- ions will separate more slowly than the H^- ions it seems unlikely that one can attribute this larger ratio to the increased competition from autodetachment. However, it was demonstrated by Fiquet-Fayard and her colleagues,[240,241] using both a quantum theory analysis in terms of wave packets and a classical study with Monte-Carlo selection of trajectories, that dissociation of a triatomic molecule from the neighborhood of a saddle point in a repulsive potential surface leads predominantly to the separation of the lighter atom, and that isotope ratios up to 5 can be obtained without any competition from detachment. Very similar isotope effects have been observed in H_2S, HDS, and D_2S,[239] for which the lower two resonances appear at 5.35 and 7.65 eV.

For simple molecules one can study details of the dissociation cross sections by measuring the energy loss suffered by the scattered electron. For example, in H_2 there is a gap in the spectrum for discrete excitations between 4.5 and 11.3 eV, and the observation of an energy loss between these limits is evidence for a dissociative event. This technique also gives information about the kinetic energy of the dissociation fragments without detection of the neutral atoms.

Let E_i and E_r be the energy of the incident and residual (or scattered) electron, and ΔE be the energy loss ($\Delta E = E_i - E_r$). The kinetic energy distribution can be obtained by studying the scattering cross section as a function of E_r, for fixed E_i. Alternatively, one can fix ΔE and vary E_i. Huetz and Mazeau[242] have performed such an experiment for H_2, with ΔE between 4.6 and 5.7 eV. They observed several narrow peaks with incident energy between 11 and 16 eV. By comparing the results obtained at different values of ΔE, it becomes clear that these peaks can arise in two ways. Some peaks, occurring at fixed values of E_i, are due to the formation of resonant states of H_2^- and can be correlated with similar structures observed in vibrational excitation. Other peaks appear at fixed values of E_r. These are attributed to the formation of doubly excited states of H^- which decay by emitting electrons with a specific energy.

$$e + H_2 \rightarrow H + H^{-*} \rightarrow H + (H + e).$$

The energy of the emitted electrons agrees well with the positions of resonances observed in e–H scattering.

This process can be regarded as dissociative attachment leading to quasi-stable states of the negative ion, and has been thoroughly analyzed in the thesis of Huetz.[243] Similar processes in N_2 and NO are discussed in the section on dissociative attachment.

5.7. *Halogenated Methanes*

Chlorofluoromethane (CCl_xF_y, where $x + y = 4$) molecules have been of considerable recent interest because the decomposition products of some members of this class are suspected of altering the equilibrium of the stratospheric ozone layer. Although photodecomposition is generally believed to be the main decomposition pathway, it has been shown that slow electrons are $\sim 10^4$ times more effective at decomposition than photons of 7.5 eV.[244] Photodecomposition is possible at lower photon energies when the chlorofluoromethanes are adsorbed on surfaces such as grains of sand.[245, 246] It has been suggested that weakly bound electrons on these surfaces are responsible for this effect.[244] Electron attachment studies of these compounds are also important because they are most easily detected in the environment using the electron capture detector.

Verhaart, van Sprang, and Brongersma[246] have studied the neutral products produced by low-energy (< 0.5 eV) electron impact on CCl_4, $CFCl_3$, and CF_2Cl_2. The neutral chemistry induced by the free radicals produced was found to be much more specific as compared to reactions induced by photons or high-energy electrons (discharges). Wall effects, unfortunately, complicate the interpretations.

Pejčev, Kurepa, and Čadež[247] have recently determined absolute cross sections for dissociative electron attachment to dichloro-difluoromethane, CF_2Cl_2. The cross section has a maximum value of $1.8 \times 10^{-16}\,cm^2$ at ~zero energy with subsidiary peaks at 0.6 and 3.5 eV. CF_2Cl_2 represents a potential source of chlorine atoms in the stratosphere.

Dissociative attachment to the halomethanes is generally discussed in terms of a vertical transition from the ground surface (or in some cases from vibrationally excited surfaces due to "hot bands") to a repulsive surface leading to product ions and neutral radicals.[248] The shape and magnitude of the cross section are then mainly determined by the relative spacing (and shape) of the two (or more) respective potential energy surfaces. In this connection, Peyerimhoff and Buenker[249] have constructed *ab initio* potential energy curves for $CFCl_3$ and CF_2Cl_2. The results of these calculations give quantitative support to this interpretation of the dissociative attachment phenomena in halocarbons. The calculated transition energies agreed well with experiment.[245]

Verhaart, van Sprang, and Brongersma[246] have studied the electron transmission spectra along with dissociative attachment for $CFCl_3$, CF_2Cl_2, CF_3Cl, and CF_4. They observed a sharp vibrational progression for CF_3Cl and CF_4 at energies below 2 eV. Both spectra were very similar. Cl^- is also observed from CF_3Cl in this energy range. The authors argue that "dissociative negative ion states do not live long enough to exhibit well-defined vibrational structure, or even to be observed at all in transmission." Based on these observations they suggest that there are two different resonances in the 0–2-eV region, one Rydberg and one valence type, the Rydberg probably being at the lower energy.

6. Dissociative Attachment to Ultracold Molecules and Molecular Clusters

The cross section for electron attachment to molecules can be especially sensitive to thermal effects. Henderson, Fite, and Brackmann[193] found, for example, that the maximum in the cross section for the process $e + O_2 \rightarrow O^- + O$ shifts to lower energy and increases in magnitude upon heating the O_2. Also, the energy onset for O^- from CO_2 is temperature dependent; at room temperature the threshold is 0.25 eV below the expected thermodynamic threshold.[250] The most dramatic effect of internal energy upon electron attachment cross sections is found in the case of N_2O. For N_2O, at room temperature, a separate O^- peak appears at ~0.5 eV.[234] Chantry[234] attributes this dramatic temperature effect to thermal excitations of the bending modes of N_2O. Such excitations would yield good Franck–Condon overlap with the bent configuration of the N_2O^- ground electronic state.

Klots and Compton[251,252] have largely eliminated these artificial thresholds by employing sonic nozzle expansions to adiabatically cool the gas. For example, Fig. 7 shows the O^- signal from the reaction $e + N_2O \rightarrow O^- + N_2$ as a function of electron energy. The prominent peak at ~0.5 eV in the room-temperature data is absent in the nozzle experiment. Similarly, the onset for O^- from CO_2 shown in Fig. 8 occurs at the proper thermodynamic threshold. Electron beam experiments employing sonic nozzle expansions offer a number of advantages: (1) rotational (~1 K), vibrational (~20–50 K) and translational (~1 K) cooling; (2) translational cooling greatly reduces the Doppler profile from the energy dependence of a resonance or threshold; (3) increased molecular density for a given collision chamber pressure.

Van der Waals polymers can also be produced in sonic expansions. Klots and Compton have studied electron attachment to polymers of CO_2,[251,252] N_2O,[252] and H_2O.[253] In the case of CO_2, the most intense cluster ions occur with the general formula $O^-(CO_2)_n$ with $0 \leqslant n \leqslant 6$ (sec Fig. 8). Long-lived ions with charge-to-mass ratios corresponding to $(CO_2)_n^-$ with $2 \leqslant n \leqslant 6$ were detected. Carbon dioxide is linear whereas CO_2^- is bent at about 135°. For bond angles less than ~140°, the CO_2^- curve lies below that for CO_2. The experimental vertical[254] and adiabatic[255] electron affinity of CO_2 are −3.8

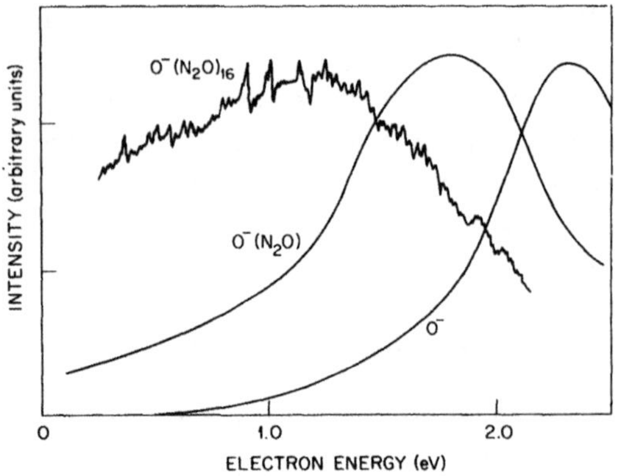

FIGURE 7
Negative ions resulting from electron collisions with a supersonic beam of nitrous oxide as a function of electron energy. Taken from Ref. 252.

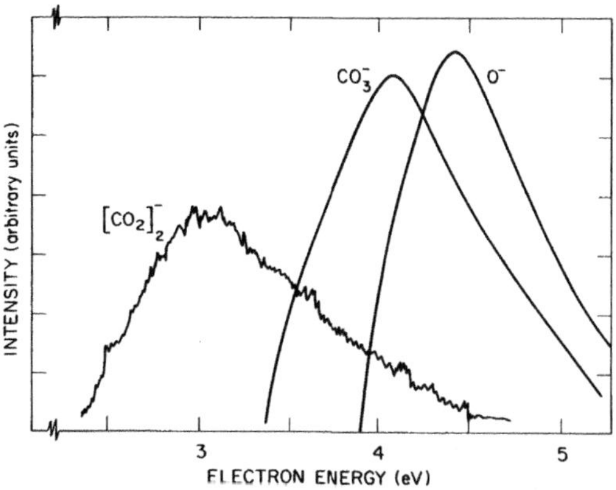

FIGURE 8
Negative ions resulting from electron collision with a supersonic beam of carbon dioxide as a function of electron energy. Taken from Ref. 252.

and -0.6 ± 0.2 eV, respectively. Linear CO_2^- autodetaches in $\sim 10^{-15}$ s. However, due to poor Franck–Condon factors between the bent CO_2^- and linear CO_2, the lifetime of ground state CO_2^- is $\sim 90\,\mu s$. The $(CO_2)_n^-$ ions appear to be stable ($\tau > 10^{-3}$ s). Rossi and Jordan[256] have employed the Hartree–Fock method to predict that the adiabatic electron affinity of $(CO_2)_2$ is small but positive. Their most stable configuration consisted of a bent CO_2 "solvated" by a nearly linear CO_2 (C_s symmetry). If this turns out to be true, these experimental results will provide the first observation of a solvated electron in the gas phase. Solvation energies can be large [e.g., > 0.6 eV in the case of $(CO_2)_2^-$] and could provide stability for a large number of anions. Klots and Compton originally began these cluster experiments in order to search for the "hydrated" electron in the gas phase. Thus far, they have been unsuccessful.[253] The large dipole moment of $(H_2O)_2$ (> 2.4 D) assures us that $(H_2O)_2^-$ is bound. However, the dipole moment of $(H_2O)_3$ is probably very small. Thus, unless $(H_2O)_2^-$ can be further solvated by water molecules any connection with the hydrated electron is tenuous. In this connection, Armbruster *et al.*[257] have recently reported the first observation of free hydrated electrons, i.e., the hydrated electron in the gas phase. By expanding both electrons and steam from a nozzle into a vacuum, they observe $(H_2O)_n^-$ with $n \geqslant 8$ in a quadrupole mass spectrometer.

7. Dissociative Recombination

In gases at low densities electrons can combine with positive ions in a number of different ways, e.g.,

$$e + AB^+ \rightarrow AB + h\nu \qquad \text{Radiative,}$$

$$e + AB^+ \rightarrow AB^{**} \rightarrow AB^* + h\nu \qquad \text{Dielectronic,}$$

$$\left.\begin{array}{l} e + AB^+ \rightarrow A + B \\ e + AB^+ \rightarrow A + B^* \\ e + AB^+ \rightarrow A^+ + B^- \end{array}\right\} \quad \text{Dissociative.}$$

Radiative recombination cross sections are small since the time an electron spends in the local vicinity of the ion is small (10^{-15} s) compared to the radiative lifetime (10^{-8} s). Cross sections for such processes have not been directly measured; however, applying detailed balancing to photoionization data yields cross sections of $\sim 10^{-20}$ cm^2. The cross sections are expected to exhibit roughly a $1/E$ energy dependence.

Dielectronic recombination is a resonant-type process in which the incident electron simultaneously excites an electron of the ion and is captured into a high Rydberg state of the resultant neutral atom. Such multiply excited states lie in the continuum and autoionization is possible. In the event that the excited ion core has a small radiative lifetime, and the Rydberg state of the captured electron is sufficiently high so that it does not interact significantly with the core electron, radiation will occur before autoionization results in an excited neutral and a free electron. Cross sections for these processes can be large (10–100 Å^2); however, the width of the resonances are so narrow (millielectron volts) that they are difficult to detect. No experiments have yet been performed on this important process.

Radiative and dielectronic recombination can occur in the case of atomic or molecular ions. For most molecular ions the primary process is dissociative recombination. Here the electron is initially captured into neutral states which lie in the continuum. Assuming that the electron is captured into an antibonding state of the molecule, the nuclei begin to separate while autoionization is occurring. Since the potential energy varies with the separation of the nuclei, the potential energy of the dissociating molecule may cross below the ionic surface. At this point, autoionization becomes likely and dissociation will be essentially complete. Thus, the cross section depends upon two factors: (1) the initial capture probability, and (2) the time required to get to the crossing point compared to the autoionization lifetime. Autoionization times are $\sim 10^{-14}$ s compared to dissociation times of 10^{-15} to 10^{-18} s and therefore dissociative recombination cross sections can be large and are known to approach 10^{-13} cm^2 at low energy. Because the cross sections can be large over a broad energy range, dissociative recombination is easily observable and has been widely studied. Bates and Massey[(258)] were the first to show that dissociative recombination rates could be very large. Three basic experimental methods have been employed to measure these rates: (1) electron–ion recombination in plasmas, (2) electron-beam–ion collisions, and (3) electron-beam–ion trap experiments. Each of the methods have their advantages and disadvantages which we will discuss below.

7.1. Plasmas

The first measurements on dissociative recombination processes came from rate coefficients in laboratory plasmas. The first laboratory measurements on dissociative recombination were made in 1949 by Biondi and Brown.[259] Biondi and his associates have provided the most extensive and productive results of these studies.[260–264] The experimental method involves creating a plasma in a microwave cavity. The decrease of the number density of the electrons is measured as a function of time after the microwave discharge is turned off. The decrease in positive ion intensity is measured simultaneously. From the decay curves of electron density or ion intensity they are able to determine the rates of electron–ion recombination at various temperatures under conditions such that the electron and ion temperatures are equal to the gas temperature: $T_e = T_i = T_{gas}$. They are also able to heat the microwave cavity, and thus obtain recombination rates in which the gas temperature (and the electron and ion temperature) are varied. Ambient temperatures up to 450 K have been reached. Alternatively the electrons can be heated selectively through absorption of microwave radiation. The recombination rate is a product of the cross section and the relative collision velocity (approximate electron velocity) averaged over a Maxwellian velocity distribution at $T = T_{gas}$. Cunningham and Hobson[265] used a gas-driven shock to control the internal vibrational energy as well and established the importance of internal energy in the recombination process. With the exception of He_2^+, recombination rates between 10^{-7} and 10^{-5} $cm^3\,s^{-1}$ (300 K) have been reported. The primary advantage of the plasma technique is that absolute rates can be obtained over a rather wide temperature range. The disadvantages are (1) the energy range is small and the energy resolution is poor, (2) only those species which can be produced in a plasma are accessible, and (3) the internal energy states cannot be controlled. As we shall see, all methods suffer from the latter defficiency.

7.2. Colliding Beams

The groups of Peart and Dolder,[266–269] McGowan *et al.*,[270,271] and Dunn *et al.*[272–274] have developed the colliding beams to a rather high degree of sophistication. In these experiments, an electron beam collides with an ion beam and one observes the appearance of neutrals,[266–271] the disappearance of the ions, or the production of photons from excited dissociation fragments.[272–274]

The basic differences between the methods used by the three groups are the angles at which the electron and ion beams collide. If an electron of mass m and laboratory energy E_e collides with an ion with mass M and laboratory energy E_i at an angle θ, then the relative energy of the interaction is

$$E = \frac{M}{M+m}\left[E_e + m\left(\frac{E_i}{M}\right) - 2\left(\frac{m}{M}E_eE_i\right)^{1/2} \cos\theta\right]. \tag{1}$$

For the case of electron collisions with H_2^+, if the electron and ion energies are 20 and 10 keV, respectively, the interaction energy varies from 18.06 eV for a merged-beam configuration ($\theta = 0$) to 19.42 eV for an inclined-beam geometry ($\theta = 45°$), to 22.72 eV for the case in which the two beams are orthogonal. Thus using the merged- or inclined-beam method, quite low collision energies can be attained. Note, however, that the relative energy is very dependent upon the angle of intersection, i.e., $dE/d\theta \propto [(m/M)E_eE_i]^{1/2} \sin\theta$. Thus, good energy resolution may be lost unless the angle between the two beams is well defined.

The method of Peart and Dolder[(266–269)] employs beams inclined at 10°. The background pressure was 2×10^{-10} torr, even with both beams on. The H_2^+ or D_2^+ ion beam was produced by electron impact ionization and was assumed to have a distribution of vibrational states given by von Busch and Dunn.[(274)] The ion energy was usually 60 keV, although 40- and 50-keV beams were used to attain the lower interaction energies. If we let the capital letters I and V represent the current and velocity of the electron beam and if R represents the rate of detection of atom pairs, then the recombination cross section can be written

$$\sigma_r = \frac{R}{I}\frac{e^2vVF}{(v^2 + V^2 - 2vV\cos\theta)^{1/2}} \sin\theta, \tag{2}$$

where F is a beam overlap factor and is calculated from measured current-density distribution and v is the ion velocity. The relative energy was varied between 0.3 and 0.4 eV by changing the electron energy. The resolution depends primarily upon the angular spread of the two beams (≈ 1°). The estimated resolution ranged from 0.15 to 0.3 eV and the energy scale is accurate to 0.2 eV. The cross sections are believed to be accurate to ± 10%.

McGowan *et al.*[(270,271)] employ a merged-beam method where essentially zero relative velocity between the two beams can be obtained, i.e., the two beams are traveling collinear. As in the case of Dolder *et al.*,[(266–269)] the neutrals are detected with a surface barrier detector. The detector surface area was large (300 mm^2) in order to accommodate the spatial spread of the dissociative products. The cross section in this case is given by

$$\sigma = \frac{Re^2F^1}{IL}\left|\frac{vV}{v - V}\right|, \tag{3}$$

where L is the path length over which the two beams interact and F^1 is a beam overlap factor or the so-called "form factor." The measured cross section is symmetric about $v = V$. This also allows one to determine the absolute zero of energy. The electron beam is brought into coincidence with the ion beam by the use of a trochoidal electron spectrometer. The trochoidal spectrometer, although capable of high-energy resolution (20 meV), was operated at ~0.1 eV and 15 to 30 μA intensity. Equation (1) shows that high-energy resolution is obtained at low center-of-mass energies. The laboratory electron energies were above 40 eV. The ion beam was produced in a 450-keV van de Graaff accelerator. At this ion energy the equivalent electron energy required to produce zero center-of-mass energy is 120 eV. The energy resolution depends (primarily) upon the angular spread of the two beams from 180° and to a lesser extent upon the energy resolution of each of the two beams. The angular spread $\Delta\theta$ about $\theta = 0$ is difficult to assess. They make the valid assumption that in the center-of-mass energy below 0.01 eV the center-of-mass energy, E_{cm}, is a function of θ and $\Delta\theta$ only. Making the further assumption that $\theta \propto 1/E_{cm}$, they fit the shape of the measured cross section at the extreme low energy (on both sides of zero). This led to a best fit of $\Delta\theta = 1.5°$. For the case that $\theta = 0$ the energy spread is

$$\Delta E \sim E_{cm}(\Delta\theta) - E_{cm}(0) \sim \left(\frac{m}{M} E_i E_e\right)^{1/2} (\Delta\theta)^2 . \tag{4}$$

For $\Delta\theta = 1.5°$, the energy spread is 0.08 eV at $E_{cm} = 0$. The total energy spreads including uncertainties in ΔE_e and ΔE_i are 0.02 eV for energies near zero and $\leqslant 0.04$ eV for energies of ~1 eV. In support of this analysis the structures observed in the experimental data (see Fig. 9) are less than 0.05 eV.

The authors[(270, 271)] considered the effect of electron spiraling on the magnitude of the measured cross sections and found this to be a negligible effect. They used an experimentally derived angle of merging to estimate the radius pitch of the spiral. For 100-eV electrons moving in a 25-G magnetic field, the spiraling diameter is 0.3 mm and the pitch is 8.5 cm.

Peart and Dolder[(268)] have pointed out that if electrons follow a spiral path, θ is no longer zero and good resolution, which is a prime advantage of merged beams, would be lost. Now since the electrons are emitted from a hot filament ($T \sim 1200°$C) the electrons must have at least a small transverse spread in velocity corresponding to the thermal contribution. Thus if the electron beam is accelerated to 120 eV, the angular spread of these transverse components of the beam is ~0.02 radians. This would correspond to a maximum center-of-mass energy spread of 0.1 eV. We conclude that in the experiments of McGowan *et al.*,[(270, 271)] the transverse components must be "skimmed" off by the entrance apertures due to the large radius of curvature of the spiraling electrons.

Dunn and colleagues[(272, 273)] have crossed orthogonal beams of electrons and D_2^+ ions to determine the absolute cross sections for forming excited D

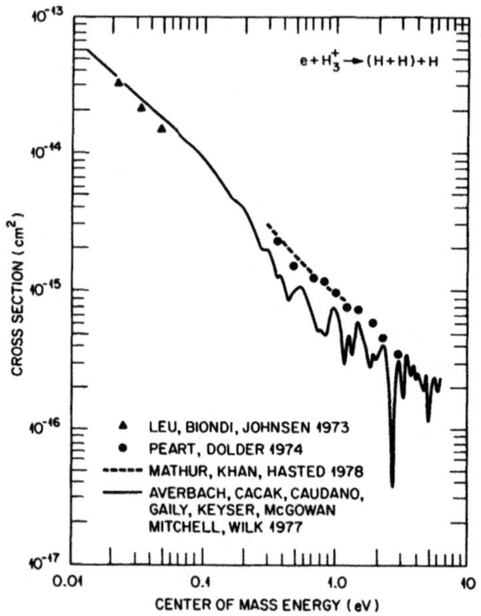

FIGURE 9
Dissociative recombination cross sections for $e + H_3^+ \rightarrow (H + H) + H$ as a function of electron energy.[264, 270, 269, 278]

atoms from dissociative recombination. The experimental procedure is to collide a mass-analyzed beam of D_2^+ ions of known vibrational population distribution[273] at right angles with a magnetically confined beam of electrons of variable energy. The photon flux is measured into a well-defined cone along the third orthogonal direction. Background pressures are typically 10^{-9} torr. The fact that the motion of the excited D atoms extends along the ion beam direction beyond the field of view of the detector posed some substantial correction of the observed count rate to obtain cross sections. The measurements were further complicated by the presence of a fixed magnetic field of 2×10^{-2} T which was used to confine the electron beam. The motion of the fast excited D atom experienced a motional electric field of sufficient magnitude (~3000 V/m for a 500-eV D_2^+) to cause Stark mixing of the Zeeman-shifted $D(n = 4)$ states. The estimated errors due to photon detection and electron spiraling lead to an uncertainty in the measured cross sections of 14% and 8% at electron energies of 0.6 and 7 eV, respectively.

7.3. *Ion Trap*

Trapped-ion techniques have been developed by Dunn[274,277] and colleagues and more recently by Mathur, Khan, and Hasted.[278] The method basically consists of directing an electron beam of variable energy into a cloud of trapped ions and recording the decrease in the number of trapped ions. The rate at which ions disappear is related to the recombination rate.

Dunn *et al.*[274–276] use a static system consisting of a quadrupole ion trap of the Penning type. The ions are contained in the r–θ plane by use of a large homogeneous magnetic field (11.7 kG). Containment along the Z axis is achieved by applying an electrostatic potential of the form $V = K(r^2 - 2Z^2)$, which creates a harmonic potential well along the Z axis. The residence time of the ions ranges from minutes to hours depending upon the ionic species. The upper limit is determined by reactions with background gas. These trapping times are sufficiently long that most vibrationally and rotationally excited ions and all electronically excited ions will have relaxed to the ground state. Some notable exceptions are homonuclear diatomic molecules which have infrared radiative lifetimes of four months. Coulomb collisions between trapped ions establish a Maxwellian distribution in ~0.02 s. Since the storage times are orders of magnitude larger than this, the ions reach a definite temperature. The oscillatory axial motion of the ions causes an alternating image current to flow from one end cap of the quadrupole to the other. If a resistor is inserted in this current path, the Joule heating losses will cause the ions to approach the noise temperature of the resistor. The collective motion of the ions is then damped and the ions are eventually cooled. Fast ions escaping the trap will also act to cool the ions. One of the major advantages of the ion trap is that ground-state ions can be studied. This is to be contrasted with the colliding-beam experiments where the ion ro-vibrational temperature is determined by the Franck–Condon factors of the ionization event. The useful ion population is limited by the space charge to about 10^4 ions or an ion density of about $10^6/\text{cm}^3$. The absolute cross sections are determined in a straightforward manner. One first measures the natural decay period, τ_n, for the ions without the electron beam. Electrons of a well-defined energy are then introduced into the trap until a measurable number of ions is lost by recombination. The cross section is calculated from

$$\sigma = (e/I)A\,[(I/t_e)\ln(N_1/N_2) - 1/\tau_n]\,, \tag{5}$$

where I is the electron current and N_1 and N_2 are the number of ions present before and after introducing the electrons for a time t_e. The factor A is the electron-beam overlap factor with the ion cloud. Simple geometric arguments give A to within a factor of 2 or 3, but a better method is to use the more accurate recombination rates for plasma data to determine A. The most accurate recombination rates are for O_2^+ (Ref. 265), which give a value of A which is about 1.9 times the value estimated from simple geometric calculations.

Mathur, Khan, and Hasted[278] have introduced an ion-trap experiment in which an intense electron beam (15 μA V$^{-3/2}$) provides a space-charge trap for the ions of the order of 1 eV. A molecular beam is injected into the hollow electron beam and the positive ions produced by electron-impact ionization oscillate in the space-charge potential for periods of up to 1 s. A second beam of electrons whose energy can be varied is introduced axially into the confined ion trap and the destruction of the trapped ions is monitored in a mass spectrometer as a function of the axial-electron-beam energy. The cross sections derived from this method are not absolute and are normalized to the inclined-beam measurements of Peart and Dolder(268) at 1 eV. The measurements were performed at sufficiently low pressures ($\sim 10^{-7}$ torr) so that ion–molecule reactions such as $H_2^+ + H_2 \rightarrow H_3^+ + H$ are a negligible contribution to the destruction rate compared with recombination. The H_2^+ and H_3^+ data were independent of hollow-beam electron density or of gas pressure, so that collisional relaxation of vibrational states is small.

7.4. *Theoretical Calculations and Experimental Results*

The theoretical basis for the study of dissociative recombination was laid in 1950 by Bates,(279) who realized that during the collision of an electron with an ion in its ground electronic state the interelectronic Coulomb force can lead to the formation of a doubly excited state of the neutral molecule with a potential curve sufficiently repulsive that dissociation into neutral fragments is very likely. The later development of resonant scattering theory(280, 281) showed that the Bates calculation of the recombination rate was correct except for the neglect of the competition between autoionization and dissociation in the decay of the doubly excited state.

During the last 15 years it has become evident that the excited vibrational and rotational levels of molecular Rydberg states of high n are subject to both autoionization and predissociation. Thus it is possible that these states can also act as intermediates in dissociative recombination through the two-step process

$$\begin{aligned} e + AB^*(v_i) &\rightarrow AB^*(n \gtrsim 7, v = v_i + 1) \\ &\rightarrow AB^{**} \rightarrow A + B^*. \end{aligned} \qquad (6)$$

The first stage involves the formation of the Rydberg level through the inverse process to autoionization and the second involves the predissociation of the Rydberg state through the doubly excited repulsive state AB^{**}. It is clear that if experiments could be performed with low-energy ($\sim$100 meV) electron beams with good energy resolution, this indirect recombination mechanism would lead to structure in the cross section, since it is important near the discrete levels of the Rydberg state.

The development of multichannel quantum defect theory has led to an elegant description of dissociative recombination, including both the direct and indirect mechanisms. Giusti(282) has shown that Rydberg states with $n < 7$ may also influence the recombination cross sections and that many of the structures will appear as dips in the cross section. This point has also been made by O'Malley,[283] who considers the Rydberg states as resonances in the atom–atom continuum that forms the final state. The interference between the direct and indirect processes can then be treated by a modification of Fano's theory of configuration interaction.[284]

In spite of the 30 years that have elapsed since the publication of the paper by Bates, theoretical progress on dissociative recombination has been very slow and it is debatable whether any completely reliable *ab initio* calculations have yet been published. In this review we will survey the calculations that have been performed on three ions, H_2^+, CH^+, and NO^+, and then briefly discuss recombination in polyatomic ions.

7.4.1. e–H_2^+

The calculation of doubly excited states of the H_2 molecule was discussed in Section 4. Although there are an infinite number of resonances that can be formed in e–H_2^+ collisions with energies between 5 and 15 eV, the $^1\Sigma_g^+(1\sigma_u)^2$ state lies well below the others and is dominant for collisions of very slow electrons with ions in low vibrational states. Giusti *et al.*[285] have calculated the recombination cross sections for such collisions. The width of the $^1\Sigma_g^+$ resonance was obtained from analysis of the two-state close-coupling calculations of e–H_2^+ scattering at fixed R by Robb[286] and the Stieltjes-imaging calculations by Hazi. The width is so large (~ 1–2 eV) that there is severe competition between autoionization and dissociation in the decay of the doubly excited state. The amount of autoionization depends strongly on the number of vibrational excitation channels open and the recombination cross section is discontinuous, falling significantly at each vibrational excitation threshold. Thus at low energies in the resonant theory the autoionization must be described through a nonlocal decay width. Above about 1.5 eV there are enough open vibrational channels that the local-width approximation becomes valid. The second major feature of the calculation is that for incident energies below 1 eV the cross section for recombination of ions in the $v = 0$ state is very small. The cross section for $v = 1$ is greater by more than one order of magnitude, and a further increase by a factor of ~ 2 is seen in the cross section for $v = 2$. For the higher vibrational states the calculated cross section is of a similar order of magnitude, being $\sim 2 \times 10^{-15}$ cm^2 at 0.1 eV. However, for these states the higher resonances may also be important.

The calculations have been extended to higher energies and to collisions with higher vibrational states through the inclusion of estimates of the potential curves and widths of the excited state. The cross sections for recombination with specific vibrational levels show oscillations as a function of energy, because of the variations in the Franck–Condon factors associated with the dominant resonances.

Comparison of these theoretical results with the measured values is difficult because of the uncertainty regarding the vibrational states present in the experimental studies. At energies below the first vibrational excitation threshold at 0.3 eV, the experimental results are close to those predicted by theory for a mixture of many vibrational states, but at higher energies the experimental values are larger by a factor of 2–3. This suggests that the theoretical model may overestimate the competition from autoionization, or that some significant resonant states have been excluded. The sensitivity of the calculated values to the assumed resonance width has been studied. However, a good comparison between theory and experiment will only be possible when a clean experiment is performed with low-energy electrons and ions in a specific vibrational state.

These calculations also suggest that indirect recombination may lead to an increase in the recombination cross section for ground-state ions at a few specific energies. Significant enhancement is possible only because the cross section for direct recombination is very small for these ions. For vibrationally excited ions the indirect process leads predominantly to dips in the cross section which may have been observed by Auerbach *et al.*[(270)] (see Fig. 9).

The earlier calculations of recombination cross sections[(145,287)] appear to have been based on inadequate representations of the resonance width and contain some surprising resonancelike structures in the cross section for direct dissociative recombination,[(145)] for which no explanation is offered. Furthermore, the large difference in the predicted recombination rates for the rotational levels $J = 0$ and $J \geqslant 1$ (within the ground vibrational state) is difficult to understand.

Zhdanov and Chibisov[(288)] have made an interesting comment regarding the relative importance of doubly excited states of the second Rydberg manifold, $(1\sigma_u)(nl\lambda)$, with large n. Although the probability of forming such states is small ($\propto n^{-3}$), the competition from autoionization is less. Although the complete neglect of autoionization by these authors has been criticized,[(289)] their main argument is valid. For recombination at moderate electron energies (5–10 eV) and with high vibrational levels of the ion the resonant states with large n will contribute significantly to the cross section.

The determination of the atomic states produced in dissociative recombination is not yet complete. Through studies of the atomic radiation produced by the recombination of $e + D_2^+$, Dunn and his colleagues[(272,273)] have measured the cross sections for the formation of the $2p$ state of D and of the states with $n = 4$. In each case the contribution is only a small fraction ($\lesssim$ 10%) of the total recombination cross section. Peart and Dolder[(290)] have shown that for H_2^+, ion-pair formation

$$e + H_2^+ \rightarrow H^+ + H^- \quad (7)$$

accounts for only about 1% of the total cross section for electron capture. A full theoretical analysis is under way, but simple arguments suggest that while formation of the $n = 3$ levels might be important when they are energetically accessible, there will be a broad distribution among many final states, extending to the high Rydberg levels that have been observed in other dissociative processes.

7.4.2. e–CH^+

The order of magnitude of the cross section for recombination of thermal electrons with ground state CH^+ ions has been debated since 1972, but is still in doubt. In their analyses of the abundance of CH^+ and CH in interstellar clouds, Solomon and Klemperer(291) and Frisch(292) were forced to the conclusion that the recombination rate cannot greatly exceed $10^{-9}\,cm^3\,s^{-1}$, which is over two orders of magnitude smaller than is found for most diatomic ions at low temperature. There is theoretical evidence for an anomalously low rate for H_2^+ under these conditions and experimental evidence for a low rate for He_2^+. The astrophysical analyses suggest that CH^+ is a third ion with a very small recombination coefficient.

Early quantal calculations, by Bardsley and Junker(293) and by Krauss and Julienne,(294) supported a large rate, but later, calculations by Giusti-Suzor and Lefebvre-Brion(295) were compatible with a small rate. In all of the states of interest, in both CH^+ and CH, the inner four electrons fill the orbitals 1σ and 2σ. In the dominant configuration of the ground $X^1\Sigma$ state of CH^+, the outer two electrons occupy the 3σ orbital. The valence configuration $(3\sigma)(1\pi)$ gives rise to a $^3\Pi$ state, with an excitation energy of about 1.1 eV and a $^3\Pi$ state at 3.0 eV.(296) The early calculations suggested that one of the $^2\Pi$ states of CH with valence configuration $(3\sigma)\,(1\pi)\,(4\sigma)$ has a repulsive potential curve which crosses that of the ground state of CH^+ within the Franck–Condon region. However, Giusti-Suzor and Lefebvre-Brion(295) suggest that the potential curve for the lowest state of this kind lies well below the CH^+ curve and is only weakly coupled to the e–CH^+ continuum. Their conclusion seems reasonable, in view of the low excitation energies for the Π states of CH^+ as calculated by Green *et al.*(296) The significance of the second state with this orbital character is still unclear.

Mitchell and McGowan(297) have measured the recombination cross section for CH^+ ions by the merged-beam technique. The ions were produced by the dissociation of methane gas and were identified by mass analysis. Various buffer gases were introduced into the ion source in an attempt to quench any electronic or vibrational excitation in the CH^+ ions that are produced in the dissociation process. It was found that the addition of at least 5% of O_2 to the methane leads to a reduction of 30% in the observed recombination cross section and that no further reduction was seen when inert gases were added to this mixture. The cross section was determined to be $(5.1 \pm 0.5) \times 10^{-14}\,cm^2$ at 0.01 eV, with an energy variation between 0.0044 and 1 eV that is consistent with a power law $E^{-(0.89 \pm 0.11)}$. This leads to a recombination rate of $\sim 3 \times 10^{-7}$ $cm^3 s^{-1}$, which is considerably larger than the upper limit set by the astrophysical models.

Some attempts have been made to solve this problem by discovering new production mechanisms for CH^+. Stecher and Williams(298) suggested the ion–molecule reaction

$$C^+ + H_2 \rightarrow CH^+ + H. \quad (8)$$

However, this reaction is endoergic by 0.4 eV for ground-state H_2 and one must postulate the existence of significant amounts of vibrationally excited H_2. Elitzur and Watson(299) have shown that this process could be important in interstellar clouds that are subjected to shock waves that may be associated, for example, with the expansion of HII regions or stellar winds. The radiative association reaction

$$C^+ + H \rightarrow CH^+ + h\nu \quad (9)$$

is also being reexamined(300) since there exist many low-energy resonances that have been seen in the inverse photoabsorption process.(301) However, it seems unlikely that the resonant enhancement in the rate of this process can be enough to provide a complete solution to this problem.

7.4.3. e–N_2^+, NO^+, and O_2^+

For an understanding of the earth's ionosphere, one needs to know the recombination rates for the ions N_2^+, NO^+, and O_2^+ for temperatures between 150 and 2000 K, and to know the states of the atomic fragments. Indeed the recombination process is so important in the upper atmosphere that attempts have been made to deduce the recombination rates from measurements of ion concentrations as measured from satellites.(302–304) However, our ability to infer reaction rates from such measurements is not universally accepted, and there has been spirited debate concerning some of the conclusions.(305,306) Dalgarno(307) has recently reviewed the atmospheric analyses and we concentrate here on the information obtained from laboratory measurements.

The early afterglow measurements by Kasner(308) and by Mehr and Biondi(261) for N_2^+ ions gave results that differ both in magnitude and in temperature dependence. The cross sections measured by Mul and McGowan,(271) using the merged-beam technique, support the temperature dependence ($\sim T_e^{-0.4}$) observed by Mehr and Biondi, but are higher (by ~30%) in magnitude. When compared to Kasner's results, the merged-beam data agree well in magnitude but not in temperature dependence. In view of the uncertainty in regard to the number of excited ions in each experiment, one should perhaps not expect exact agreement. However, for the atmospheric ions one does not expect the very large differences in the cross sections for different vibrational levels that might exist for H_2^+ and He_2^+ and have been seen in dissociative attachment to some neutral molecules.

The recombination rate at translational temperatures between 700 and 2700 K was measured by Cunningham and Hobson[(309)] by the shock-tube technique. Their results agree very well with the afterglow measurements by Mehr and Biondi.[(261)] The ions in the plasma were not identified, but it is likely that N_2^+ was the dominant ion.

Kasner and Biondi[(310)] first demonstrated that the recombination rate for N_4^+ ions is very high. More recently Whitaker *et al.*[(311)] performed a more accurate experiment over a range of electron temperature between 300 and 5600 K. The measured rate was 1.4×10^{-6} $(T_e/300)^{-0.41}$. Unfortunately, the rate for N_3^+, which may be important in ionized air at atmospheric pressure, has not yet been measured.

For NO^+ the merged-beam cross sections in Ref. 271 are in reasonable agreement with those measured by Walls and Dunn.[(275)] At energies between 0.04 and 0.2 eV the trapped-ion data fall more rapidly, varying as $E^{-1.5}$ as compared to $E^{-1.0}$ in the merged-beam data. The slower energy dependence is expected theoretically if there is little competition between autoionization and dissociation after the capture of the incident electron, but the faster dependence cannot be ruled out when autoionization is significant. The recombination rate obtained from the cross sections of Mul and McGowan[(271)] is approximately 60% of the rate obtained in an afterglow by Huang *et al.*[(312)] The recombination rate obtained from the trapped-ion data seem consistent with the afterglow data below 600 K but then falls off more rapidly.

Although the disparity in these results could be attributed to the presence of excited vibrational levels in the afterglow experiments, and perhaps also in the merged-beam experiment, it is difficult to understand the particular nature of the difference in the temperature dependence. Since the ion trap should contain only ground-state ions, it appears that the temperature dependence of recombination to the ground state is steeper than that of the excited states. No convincing theoretical interpretation has yet been given for such behavior.

By analysis of the predissociation of the Rydberg states of NO, Bardsley[(280)] showed that half of the observed recombination can be attributed to the formation of the $B\,^2\Pi$ and $B'\,^2\Delta$ states of NO, both of which dissociate to $N(^2D) + O(^3P)$. A later calculation by Lee[(313)] predicts significant contribution from the $A'\,^2\Sigma^+$ and $I\,^2\Sigma^+$ states. Because of an avoided crossing between potential curves for these states, the exact branching ratio for production of $N(^2D)$ or $N(^4S)$ has not been calculated, but it seems likely that the excited state $N(^2D)$ will be produced in at least 60% of the collisions.

For oxygen the dimer-ion data of Walls and Dunn[275] and the afterglow measurements of Mehr and Biondi[261] are in good agreement. The cross section falls steadily from $2 \times 10^{-15}\,\text{cm}^2$ at 0.1 eV to $\sim 2 \times 10^{-16}\,\text{cm}^2$ at 1 eV, with a recombination rate of $2 \times 10^{-7}\,\text{cm}^3\,\text{s}^{-1}$ at 300 K. At around 1.5 eV Walls and Dunn see a very pronounced structure in the cross section, with a peak value of $7 \times 10^{-16}\,\text{cm}^2$. This peak could be due to the formation and predissociation of one of the lowest members of the Rydberg series converging to the $a\,^4\Pi_u$ state of O_2^+. One possible intermediate state would be

$$^3\Pi_g(1\sigma_g)^2(1\sigma_u)^2(2\sigma_g)^2(2\sigma_u)^2(3\sigma_g)^2(1\pi_u)^3(1\pi_g)^2(3\sigma_u). \qquad (10)$$

Unfortunately, there is no spectroscopic information on this state, since it cannot be formed in an allowed transition from either the ground state of O_2 or the two low-lying metastable states.

The merged-beam data of Mul and McGowan[271] show the same energy dependence ($\sim E^{-1}$) from 0.01 to 1 eV, with a slightly higher magnitude. The shock-tube data of Cunningham and Hobson[314] seem also to be consistent with a temperature dependence of $T^{-0.63}$ between 600 and 2500 K. For energies below 1 eV, the current situation is very satisfactory, but further examination of the 1.5-eV peak would be worthwhile.

The recombination rate for the O_4^+ ion has been measured by Kasner and Biondi[315] to be $\sim 2.3 \times 10^{-6}\,\text{cm}^3\,\text{s}^{-1}$ at 205 K.

One important feature of dissociative recombination in O_2^+ is the yield of metastable state $O(^1S)$. Zipf[316,317] has shown that the rate of formation is sensitive to the vibrational state of the recombining ion. This suggests that more than one resonant state of O_2 is active in the recombination. An impressive theoretical study of the relevant states of O_2 has been carried out by Guberman,[318] confirming that both metastable states ^1S and ^1D can be formed in the recombination. A thorough discussion of this question, and of its atmospheric significance, has been presented by Bates and Zipf.[319]

7.4.4. Inert-Gas Ions

Recombination of the inert gas ions has been studied in both in afterglows and in shock tubes. A difference in the apparent temperature dependence of the rates obtained by the two methods for Ne_2^+, Ar_2^+, and Kr_2^+ has been interpreted by O'Malley, Cunningham, and Hobson.[320] They conclude that the rate of recombination of vibrationally excited ions is considerably lower than that of ground-state ions. This does not appear to be the case for other ions, and is the opposite of the situation usually found for dissociative attachment. A possible explanation of the behavior has been put forward,[321] and is worthy of further examination.

Studies by Shiu *et al.*[322,323] of the radiation emitted following recombination of Ar_2^+, Kr_2^+, and Xe_2^+ ions suggest that almost all energetically allowed final states are populated. However, some states do not have emission lines within the range of their detectors and so further work would be useful.

There is still considerable uncertainty regarding the rate of dissociative recombination of He_2^+ ions. Theory suggests that the rate is small for ions in the ground vibrational state but that it may be large for vibrationally excited ones. The experimental situation is unclear, since most measurements involve a mixture of two-body and three-body processes. A critical review of the experimental situation has recently been given by Bates.[324]

7.4.5. Complex Ions

We have already seen that the recombination coefficients for dimer ions, such as N_4^+, O_4^+, and $NO \cdot NO^+$, appear to be greater than those for the corresponding diatomic ions. Using the afterglow technique with a mass spectrometer to identify the ions and with microwave heating to raise the electron temperature, Biondi and his colleagues[325,262] have studied the recombination of electrons with the clustered ions $NH_4^+ \cdot (NH_3)_n$ and $H_3O^+ \cdot (H_2O)_n$. The recombination rates are large and they appear to vary slowly with electron temperature.

The magnitude of the recombination rate can be understood in terms of indirect dissociative recombination, since there are many vibrational degrees of freedom which can absorb energy from the incident electron, leading to temporary electron capture. The subsequent behavior of the complex is unclear. Perhaps the extra electron cascades down through the Rydberg manifold until it reaches a valence orbit around the central ion. The polarization forces holding the external molecules will then be weakened and the potential surface within the central H_3O or NH_4 molecule will probably be repulsive. Alternatively the core ion may undergo normal dissociative recombination, through the formation of a doubly excited state, and be split.

The most puzzling feature of the experimental results is the weak dependence of the recombination rate upon electron temperature. This implies that the recombination cross sections vary inversely with velocity and are of the order of $10^{-13}\,cm^2$ at energies around 0.5 eV. In spite of studies of the indirect recombination process by Smirnov[326] and by Bottcher,[327] and the extended discussion of the direct process by Bates,[324] no quantitative explanation of this phenomenon has yet emerged.

With regard to simpler polyatomic ions, there is considerable interest in determining the fragmentation patterns for astrophysically significant ions such as H_3O^+, CH_3^+, NH_4^+, and $HCNH^+$. Herbst[328] has applied statistical arguments to this problem. He finds that several break-up channels are usually important and that the attachment of a single hydrogen atom is prominent for each of these ions.

The large recombination coefficients and rapid formation rates associated with most polyatomic ions have led to serious difficulties in the design of experiments to measure specific recombination rates. For example, in a molecular hydrogen plasma the dominant ion is often H_5^+ (see, e.g., Trainor[329]). However, careful studies of the smaller ions H_2^+ and H_3^+ have been performed.[268–270,278] Theoretical studies on H_3^+ recombination are under way and a thorough comparison of theory and experimental results for this ion should be particularly beneficial.

In Fig. 9, we compare the experimental data on electron–H_3^+ recombination for the merged ion beams, plasma, and trapped-ion experiments. The experiments are in agreement as to the magnitude and general energy dependence of the cross section; however, the data of Auerbach *et al.* show considerable structure between 0.5 and 10 eV which is not evident in the other data. Clearly more experimental and theoretical work is desirable.

References and Notes

1. J.A. Schiavone, D.E. Donohue, D.R. Herrick, and R.S. Freund, Electron-impact excitation of helium; Cross sections, *n* and *l* distributions of high Rydberg states, *Phys. Rev. A* **16**, 48 (1977).
2. L. Julien, M. Glass-Maujean, and J.P. Descoubes, On the dissociation of the H_2 molecule following electron impact and leading to the obtention of atomic hydrogen in the $n = 3$ levels, *J. Phys. B* **6**, L196 (1973).
3. M. Glass-Maujean, Dissociative excitation of H_2 by electron impact, *Comments At. Mol. Phys.* **7**, 83 (1977).
4. M. Glass-Maujean, Abstracts, *Fourth International Conference on Atomic Physics* (Heidelberg, 1974), p. 260.
5. R.S. Freund, J.A. Schiavone, and D.F. Brader, Dissociative excitation of H_2: Spectral line shapes and electron impact cross sections of the Balmer lines, *J. Chem. Phys.* **64**, 1122 (1976).
6. T.R. Connor and M.A. Biondi, Dissociative recombination in neon: Spectral line-shape studies, *Phys. Rev. A* **140**, 778 (1965).
7. J. Allison, T. Kondow, and R.N. Zare, Laser-induced fluorescence measurement of the nascent rotational distribution of $N_2^+(X\ ^2\Sigma_g^+)$ formed by electron impact on N_2, *Chem. Phys. Lett.* **64**, 202 (1979).
8. G. Glocker, W.P. Baxter, and R.H. Dalton, The activation of hydrogen by electron impact, *J. Am. Chem. Soc.* **49**, 58 (1927).
9. A. Hughes and A.M. Skellet, Dissociation of hydrogen by electrons, *Phys. Rev.* **30**, 11 (1927).
10. S.J.B. Corrigan, Dissociation of molecular hydrogen by electron impact, *J. Chem. Phys.* **43**, 4381 (1965).
11. H.F. Winters, Ionic adsorption and dissociation cross sections for nitrogen, *J. Chem. Phys.* **44**, 1472 (1966).

12. H.F. Winters, Dissociation of methane by electron impact, *J. Chem. Phys.* **63,** 3462 (1975).
13. H.F. Winters, Dissociation of ethane by electron impact, *Chem. Phys.* **36,** 353 (1979).
14. E. Teloy, *Transactions of the Third International Vacuum Congress* (H. Adam, ed.), p. 613, Pergamon Press, New York (1967).
15. D. Beck and O. Osberghaus, The mass spectrum of the fragments from section collisions with molecules, *Z. Phys.* **160,** 406 (1960).
16. D. Beck, Neutral fragments from hydrocarbons under electron impact, *Discuss. Faraday Soc.* **36,** 56 (1964).
17. D. Beck and A. Niehaus, Reactions of excited butane molecules, *J. Chem. Phys.* **37,** 2705 (1962).
18. H. Genzel and O. Osberghaus, Dissociation of some chromatic molecules by electron impact, *Z. Naturforsch. A* **22,** 331 (1967).
19. C.E. Melton, Higher transmission and dual electron beam ion sources for mass spectrometry, *J. Sci. Instrum.* **43,** 927 (1966).
20. C.E. Melton, The mass spectrometer as a radiolytic and catalytic laboratory, *Int. J. Mass Spectrom. Ion Phys.* **1,** 353 (1968).
21. C.E. Melton, Study by mass spectrometry of the decomposition of ammonia by ionizing radiation in a wide-range radiolysis source, *J. Chem. Phys.* **45,** 4414 (1966).
22. C.E. Melton and P.S. Rudolph, Radiolysis of methane in a wide-range radiolysis source of a mass spectrometer: I. Individual and total cross sections for the production of positive ions, negative ions, and free radicals by electrons, *J. Chem. Phys.* **47,** 1771 (1967).
23. C.E. Melton, Radiolysis of water vapor in a wide range radiolysis source of a mass spectrometer, *J. Phys. Chem.* **74,** 582 (1970).
24. A. Niehaus, Excitation and dissociation of molecules by electron impact, *Z. Naturforsch. A* **22,** 690 (1967).
25. F.W. Lampe and A. Niehaus, Ionization potentials of free radicals formed by electron impact, *J. Chem. Phys.* **49,** 2949 (1968).
26. R.A. Saunders, J.T. Larkins, and F.E. Saalfield, Detection of the neutral fragments produced in an electron impact, *Int. J. Mass Spectrom. Ion Phys.* **3,** 203 (1969).
27. F.J. Preston, M. Tsuchiya, and H.J. Svec, A mass spectrometer for the detection of neutral species formed by electron impact, *Int. J. Mass Spectrom. Ion Phys.* **3,** 323 (1969).
28. J.R. Reecher, G.D. Flesch, and J.J. Svec, A mass spectrometer for the characterization of neutral species produced in the electron bombardment of gases, *Int. J. Mass Spectrom. Ion Phys.* **19,** 351 (1976).
29. R. Marx, G. Mauclaire, and M. Wallart, Formation of radicals in the gas phase by slow electron impact, *J. Chem. Phys.* 1680 (1971).
30. L.S. Polak and D.I. Slovetsky, Production of free radicals from electron impact collisions with hydrocarbons, *Int. J. Radiat. Phys. Chem.* **8,** 257 (1976).
31. N.I. Butkovskaya, M.N. Larichev, I.O. Leipunskii, I.I. Morozov, I. Yu. Razuvaen, and V.L. Tal'rose, *Dokl. Akad. Nauk SSSR* No. **233,** 398 (1977).
32. S. Chung and C.C. Lin, Application of the close-coupling method to excitation of electronic states and dissociation of H_2 by electron impact, *Phys. Rev. A* **17,** 1874 (1978).
33. T.N. Rescigno, C.W. McCurdy, Jr., V. McKoy, and C.F. Bender, Low energy electron-impact excitation of the hydrogen molecule. *Phys. Rev. A* **13,** 216 (1976).
34. N. Böse, Excitation function of $a\,^3\Sigma_g^+ \rightarrow b\,^3\Sigma_u^+$ light emission in H_2 measured by monoenergetic electrons, *J. Phys. B* **11,** L83 (1978).

35. J.F.M. Aarts, C.I.M. Beenakker, and F.J. de Heer, Radiation from CH_4 and C_2H_4 produced by electron impact, ***Physica (Utrecht)*** **53,** 32 (1971).
36. D.E. Donohue, J.A. Schiavone, and R.S. Freund, Molecular dissociation by electron impact: Optical emission from fragments of methane, ethylene, and methanol, *J. Chem. Phys.* **67,** 769 (1977).
37. J.A. Schiavone, S.M. Tarr, and R.S. Freund, High-Rydberg atomic fragments from electron-impact dissociation of molecules, *J. Chem. Phys.* **70,** 4468 (1979).
38. D. Rapp and P. Golden, Total cross section for ionization and attachment in gases by electron impact. I. Positive ionization, *J. Chem. Phys.* **43,** 1464 (1965).
39. R.L. Platzman, Superexcited states of molecules, *Radiat. Res.* **17,** 419 (1962).
40. D.A. Vroom and F.J. de Heer, Production of excited hydrogen atoms by impact of fast electrons on some simple hydrocarbons, *J. Chem. Phys.* **50,** 580 (1969).
41. R.L. Platzman, Superexcited states of molecules, and the primary action of ionizing radiation, *Vortex* **23,** 372 (1962).
42. F.J. de Heer, Superexcited states of molecules produced by electron impact, *Int. J. Radiat. Phys. Chem.* **7,** 137 (1975).
43. H. Nakamura, Branching ratio for preionization and predissociation of a superexcited state of a diatomic molecule, *Chem. Phys. Lett.* **33,** 151 (1975).
44. A. Giusti-Suzor, Application of quantum defect theory to preionization, predissociation, and dissociative recombination, in *Physics of Electronic and Atomic Collisions, Invited Papers of Twelfth International Conference on the Physics of Electronic and Atomic Collisions.* Gatlinburg, Tennessee, July 1981 (S. Datz, ed.), p. 381, North-Holland, Amsterdam (1982).
45. A.W. Fliflet and V.B. McKoy, Distorted-wave-approximation cross sections for excitation of the $b\ ^3\Sigma_u^+$ states of H_2 by low-energy electron impact, *Phys. Rev. A* **21,** 1863 (1980).
45a. C.A. Weatherford, Excitation of the $b\ ^3\Sigma_u^+$ state of H_2 by electron impact: Semiclassical exchange potentials at intermediate energies, *Phys. Rev. A* **22,** 2519 (1980).
46. A.W. Fliflet, V. McKoy, and T.N. Rescigno, Dissociation of F_2 by electron impact excitation of the lowest $^3\Pi_u$ electronic state, *Phys. Rev. A* **21,** 788 (1980).
47. P.M.S. Blackett and J. Franck, Excitation spectrum of hydrogen by electron collisions, *Z. Phys.* **34,** 389 (1925).
48. W.E. Lamb, Jr., and R.C. Retherford, Fine structure of the hydrogen atom, *Phys. Rev.* **79,** 549 (1950).
49. W.L. Fite and R.T. Brackmann, Collisions of electrons in the hydrogen atoms. II. Excitation of L_α radiation, *Phys. Rev.* **112,** 1151 (1958).
50. W.E. Lamb, Jr., and T.M. Sanders, Fine structure of short-lived state of hydrogen by a microwave-optical method. I., *Phys. Rev.* **119,** 1901 (1960).
51. L.R. Wilcox and W.E. Lamb, Jr., Fine structure of short-lived states of hydrogen by a microwave-optical method. II., *Phys. Rev.* **119,** 1915 (1960).
52. E.R. Williams, J.V. Martinez, and G.H. Dunn, Dissociative excitation of hydrogen by electron impact, *Bull. Am. Phys. Soc.* **12,** 233 (1967).
53. F.J. de Heer, H.R. Moustafa Moussa, and M. Inokuti, Production of Lyman-α radiation in collisions of electrons on molecular hydrogen (0.05–3 keV), *Chem. Phys. Lett.* **1,** 484 (1967).
54. K. Burrows and G.H. Dunn, Isotope effects in the dissociative excitation of H_2 and D_2, *Bull. Am. Phys. Soc.* **13,** 215 (1968).
55. D.A. Vroom and F.J. de Heer, Production of excited atoms by impact of fast electrons on molecular hydrogen and deuterium, *J. Chem. Phys.* **50,** 580 (1969).
56. W.R. Ott, C.E. Kauppila, and W.L. Fite, Polarization of Lyman-α radiation emitted in electron collisions with hydroxyl atoms and molecules, *Phys. Rev. A* **1,** 1089 (1970).

57. L.D. Weaver and R.H. Hughes, Production of $n = 3$ and $n = 4$ states of atomic hydrogen by electron impact, *J. Chem. Phys.* **52,** 2299 (1970).
58. M.J. Mumma and E.C. Zipf, Dissociative excitation of vacuum ultraviolet emission features by electron impact on molecular gases. I. H_2 and O_2, *J. Chem. Phys.* **55,** 1661 (1971).
59. J.W. McConkey and F.G. Donaldson, Vacuum ultraviolet excitation of H(2*p*) by electron impact on H_2, *Can. J. Phys.* **50,** 2211 (1972).
60. J.D. Walker, Jr., and R.M. St. John, Design of high intensity atomic hydrogen sources and determination of Balmer cross sections, *J. Chem. Phys.* **61,** 2394 (1974).
61. G. Khayrallah and S.J. Smith, Production of excited hydrogen atoms ($n = 3, 4, 5$) in dissociative excitation of molecular hydrogen by electron impact, in *Proceedings of Ninth International Conference on Physics of Electronic and Atomic Collisions,* p. 808 (1975).
62. G.A. Khayrallah, Electron impact dissociation of molecular hydrogen and deuterium: Production of atomic hydrogen and deuterium α, β, and γ Balmer lines, *Phys. Rev. A* **13,** 1989 (1976).
63. R.S. Freund, J.A. Schiavone, and D.F. Bader, Dissociative excitation of H_2: Spectral line shapes and electron impact cross sections of the Balmer lines, *J. Chem. Phys.* **64,** 1122 (1976).
64. G.R. Möhlmann, F.J. de Heer, and J. Los, Emission cross sections of Balmer α, β, γ radiation for electrons (0–2000 eV) on H_2 and D_2, *Chem. Phys.* **25,** 103 (1977).
65. C. Karolis and E. Harting, Electron impact dissociation cross sections in hydrogen and deuterium, leading to Balmer alpha and beta emission, *J. Phys. B* **11,** 357 (1978).
66. G.R. Möhlmann and F.J. de Heer, Production of Balmer radiation by electron impact (0–2000 eV) on small hydrogen-containing molecules, *Chem. Phys.* **40,** 157 (1979).
67. N. Böse and F. Linder, Threshold excitation in H_2 and D_2 by electron impact and predissociation of triplet states measured by electron–photon coincidence, *J. Phys. B* **12,** 3805 (1979).
68. M. Higo and T. Ogawa, Translational energy distribution of excited deuterium atoms produced by controlled electron impact on D_2, *Chem. Phys.* **56,** 15 (1981).
69. G.A. Khayrallah and S.J. Smith, The cross section for production of the 3*s* state of atomic hydrogen in dissociative excitation of H_2 by electrons, *Chem. Phys. Lett.* **48,** 289 (1977).
70. D.E. Gerhart, Comprehensive optical and collisional data for radiation action. I. H_2, *J. Chem. Phys.* **62,** 821 (1975).
71. H.A. Bethe and E.E. Salpeter, *Quantum Mechanics of One and Two Electron Atoms,* Springer-Verlag, Berlin (1975).
72. J.D. Carriére and F.J. de Heer, Cross sections for Lyman-α radiation, *J. Chem. Phys.* **56,** 2993 (1972).
73. A. Weingartshofer, E.M. Clark, J.K. Holmes, and J.W. McGowan, Resonances and their effects above and below the electron impact ionization threshold of H_2, *J. Phys. B* **8,** 1552 (1975).
74. A.U. Hazi, Distribution of final state resulting from autoionization of the $^1\Sigma_g$ state of H_2 and D_2, *J. Chem. Phys.* **60,** 4358 (1974).
75. C. Bottcher and K. Docken, Autoionizing states of the hydrogen molecules, *J. Phys. B* **7,** L5 (1974).
76. C. Bottcher, Dissociative ionization of the hydrogen molecule, *J. Phys. B* **7,** L352 (1974); *Proc. R. Soc. London, Ser. A* **340,** 301 (1974).
77. R.S. Freund, J.A. Schiavone, and D.F. Baden, Dissociative excitation of H_2: Spectral line shapes and electron impact cross sections of the Balmer lines, *J. Chem. Phys.* **64,** 1122 (1976).

78. M. Glass-Maujean, L. Julien, and Ti Dohnalik, Stark-induced anticrossings in hydrogen: Experimental study, measurement of the $^2P_{3/2}-^2D_{3/2}$ Lamb shifts in H($n = 3$ and 4), *J. Phys. B* **11,** 421 (1978).
79. M. Glass-Maujean and J.P. Descoubes, Calculations of Stark-mixing signals resulting from second-order coupling in hydrogenic atoms, *Opt. Commun.* **4,** 365 (1972).
80. M. Glass-Maujean, Electron impact dissociation of H_2 and D_2 studied from the anticrossing signals and by the Doppler-profile technique, *J. Phys. B* **11,** 431 (1978).
81. L. Julien, M. Glass-Maujean, and J.P. Descoubes, On the dissociation of the H_2 molecule following electron impact and leading to the obtention of atomic hydrogen in the $n = 2$ levels, *J. Phys. B* **6,** L196 (1973).
82. H. Wieder and T.G. Eck, Anticrossing signals in resonance fluorescence, *Phys. Rev.* **153,** 103 (1967).
83. K. Ito, N. Oda, Y. Hatano, and T. Tsuboi, Doppler profile measurements of Balmer-α radiation by electron impact on H_2, *Chem. Phys.* **17,** 35 (1976).
84. K. Ito, N. Oda, Y. Hatano, and T. Tsuboi, The electron energy dependence of the Doppler profiles of the Balmer-α emission from H_2, D_2, CH_4, and other simple hydrocarbons by electron impact, *Chem. Phys.* **21,** 203 (1977).
85. D.A. Vroom and F.J. de Heer, Production of excited hydrogen atoms by impact of fast electrons on some simple hydrocarbons, *J. Chem. Phys.* **50,** 573 (1969).
86. D.A. Vroom and F.J. de Heer, Production of excited hydrogen atoms by impact of fast electrons on water vapor, *J. Chem. Phys.* **50,** 1883 (1969).
87. C.I.M. Beenakker and F.J. de Heer, Dissociative excitation of some aliphatic hydrocarbons by electron impact, *Chem. Phys.* **7,** 130 (1975).
88. C.I.M. Beenakker and F.J. de Heer, Dissociative excitation of benzene by electron impact, *Chem. Phys. Lett.* **29,** 89 (1974).
89. C.I.M. Beenakker, F.J. de Heer, H.B. Krop, and G.R. Möhlmann, Dissociative excitation of water by electron impact, *Chem. Phys.* **6,** 445 (1975).
90. G.R. Möhlmann and F.J. de Heer, Emission cross section of H_α and CH($A^2\Delta \rightarrow X^2\Pi$) radiation for electron impact on hydrocarbons, *Chem. Phys.* **19,** 233 (1977).
91. G.R. Möhlmann, K.K. Bhutani, K.H. Shima, and F.J. de Heer, to be published.
92. D.E. Donohue, J.A. Schiavone, and R.S. Freund, Molecular dissociation by electron impact: Optical emission from fragments of methane, ethylene, and methanol, *J. Chem. Phys.* **67,** 769 (1977).
93. I. Fujita, M. Hatada, T. Ogawa, and K. Hirota, The emission spectra of hydrogen atoms produced from methanol by electron impact, *Bull. Chem. Soc. Jpn.* **44,** 1751 (1971).
94. I. Tokue, I. Nishiyama, and K. Kuchitsu, Isotope effects for the emission impurities of the hydrogen Balmer lines produced in the electron impact dissociation of CH_3CN and CD_3CN, *Chem. Phys. Lett.* **35,** 69 (1975).
95. T. Ogawa, M. Higo, M. Toyoda, and N. Ishibashi, Isotope effects in the emission cross section of the Balmer lines (β, γ, δ) produced by controlled electron impact on HCl and DCl, *Chem. Lett. Chem. Soc. Jpn.* p. 493 (1978).
96. O.J. Orient and S.K. Srivastava, Cross section for *L*-α emission by electron impact on methane, *Chem. Phys.* **54,** 183 (1981).
97. M. Leventhal, R.T. Robiscoe, and K.R. Lea, Velocity distributions of metastable H atoms produced by dissociative excitation of H_2, *Phys. Rev.* **158,** 49 (1967).
98. R. Clampitt, H(2*s*) atoms produced by dissociative excitation of molecules, *Phys. Lett. A* **28,** 581 (1969).
99. R. Clampitt and A.S. Newton, Metastable species produced by electron impact excitation of N_2, H_2, N_2O, and CO_2, *J. Chem. Phys.* **50,** 1997 (1969).
100. M. Misakian and J.C. Zorn, Final states in the dissociative excitation of molecular hydrogen, *Phys. Rev. Lett.* **27,** 174 (1971).

101. M. Misakian and J.C. Zorn, Dissociative excitation of molecular hydrogen by electron impact, *Phys. Rev. A* **6,** 2180 (1972).
102. J.W. Czarnik and C.E. Fairchild, New data on the velocity distribution of H(2*s*) atoms produced by dissociative excitation of H_2, *Phys. Rev. Lett.* **26,** 807 (1971).
103. M.J. Mumma and E.C. Zipf, Dissociative excitation of vacuum ultraviolet emission features by electron impact on molecular gases. I. H_2 and O_2, *J. Chem. Phys.* **55,** 1661 (1971).
104. B.L. Carnahan and E.C. Zipf, Dissociative excitation of H_2, HD, and D_2 by electron impact, *Phys. Rev. A* **16,** 991 (1977).
105. S.R. Ryan, J.J. Spezeski, O.F. Kalman, W.E. Lamb, Jr., L.C. McIntyre, Jr., and W.H. Wing, Time-of-flight study of H(2*S*) and D(2*S*) produced by electron impact on H_2, D_2, and HD: Evidence for predissociation, *Phys. Rev. A* **19,** 2192 (1979).
106. J.J. Spezeski, O.F. Kalman, and L.C. McIntyre, Jr., Time-of-flight study of H(2*s*) and D(2*s*) produced by electron impact on H_2 and D_2: Fast peaks, *Phys. Rev. A* **22,** 1906 (1980).
107. A.U. Hazi and K. Wiemers, Velocity distribution of H(2*s*) resulting from the electron impact dissociation of H_2, *J. Chem. Phys.* **66,** 5296 (1977).
108. K. Kirby, S. Guberman, and A. Dalgarno, Resonant dissociative photoionization of H_2, *J. Chem. Phys.* **70,** 4635 (1979).
109. R. Clampitt, H(2*s*) atoms produced by dissociative excitation of molecules, *Phys. Lett.* **28A,** 581 (1969).
110. R.S. Freund and W. Klemperer, Molecular beam time-of-flight measurements for the study of metastable and repulsive electronic states, *J. Chem. Phys.* **47,** 2897 (1967).
111. R.S. Freund, Dissociation of CO_2 by electron impact with the formation of metastable CO($a\ ^3\Pi$) and O(^5S), *J. Chem. Phys.* **55,** 3569 (1971).
112. R. Clampitt and A.S. Newton, Metastable species produced by electron excitation of N_2, H_2, N_2O, and CO_2, *J. Chem. Phys.* **50,** 1997 (1969).
113. C.A. Barth, W.G. Fastie, C.W. Hord, J.B. Pearce, K.K. Kelly, A.I. Stewart, G.E. Thomas, P.G. Anderson, and O.F. Raper, Mariner 6: Ultraviolet spectrum of Mars's upper atmosphere, *Science* **165,** 1004 (1969).
114. S.E. Kupriyanov, Detection of HeH^+ ions and long-lived highly excited molecular and atomic states, *Zh. Eksp. Teor. Fiz.* **49,** 467 (1965) [*Sov. Phys. JETP* **21,** 311 (1965)].
115. S.E. Kupriyanov, Formation of highly excited atoms in electron collisions with CO, O_2, and N_2, *Zh. Eksp. Teor. Fiz.* **55,** 460 (1968) [*Sov. Phys. JETP* **28,** 240 (1969)].
116. R.S. Freund, Dissociation by electron impact of oxygen into metastable quintet and long-lived high-Rydberg atoms, *J. Chem. Phys.* **54,** 3125 (1971).
117. K.C. Smyth, J.A. Schiavone, and R.S. Freund, Dissociative excitation of N_2 by electron impact: Translational spectroscopy of long-lived high-Rydberg fragment atoms, *J. Chem. Phys.* **59,** 5225 (1973).
118. K.C. Smyth, J.A. Schiavone, and R.S. Freund, Dissociative excitation of CO by electron impact: Translational spectroscopy of long-lived high-Rydberg fragment atoms, *J. Chem. Phys.* **60,** 1358 (1974).
119. J.A. Schiavone, D.E. Donohue, and R.S. Freund, Molecular dissociation by electron impact: High-Rydberg fragments from methane, ethylene, and ethane, *J. Chem. Phys.* **67,** 759 (1977).
120. T.G. Finn, B.L. Carnahan, W.C. Wells, and E.C. Zipf, Dissociation of CH_4 and CD_4 by electron impact: Production of metastable and high-Rydberg hydrogen and carbon fragments, *J. Chem. Phys.* **63,** 1596 (1975).
121. W.C. Wells, W.L. Borst, and E.C. Zipf, Translational spectroscopy of metastable

fragments produced by dissociative excitation of atmospheric gases by electron impact. I. Nitrogen, *Phys. Rev. A* **14,** 695 (1976).
122. W.C. Wells and E.C. Zipf, Translational spectroscopy of metastable fragments produced by dissociative excitation of chlorine, *J. Chem. Phys.* **66,** 5828 (1977).
123. J.A. Schiavone, S.M. Tarr, and R.S. Freund, High-Rydberg atomic fragments from electron-impact dissociation of molecules, *J. Chem. Phys.* **70,** 4468 (1979).
124. J.A. Schiavone, D.E. Donohue, D.R. Herrick, and R.S. Freund, Electron-impact excitation of helium: Cross sections, *n* and *l* distributions of high-Rydberg states, *Phys. Rev. A* **16,** 48 (1977).
125. J.A. Schiavone, S.M. Tarr, and R.S. Freund, Electron-impact excitation of the rare-gas atoms to high-Rydberg states, *Phys. Rev. A* **20,** 71 (1979).
126. W. Bleakney, The ionization of hydrogen by single electron impact, *Phys. Rev.* **35,** 1180 (1930).
127. W.W. Lozier, A study of the velocities of H^+ ions formed in hydrogen by dissociation following electron impact, *Phys. Rev.* **36,** 1285 (1930).
128. G.H. Dunn and L.J. Kieffer, Dissociative ionization of H_2: A study of angular distributions and energy distribution of resultant fast protons, *Phys. Rev.* **132,** 2109 (1963).
129. L.J. Kieffer and G.H. Dunn, Dissociative ionization of H_2, *Phys. Rev.* **158,** 61 (1967).
130. R.J. van Brunt and L.J. Kieffer, Angular distribution of protons and deuterons produced by dissociative ionization of H_2 and D_2 near threshold, *Phys. Rev. A* **2,** 1293 (1970).
131. A. Crowe and J.W. McConkey, Experimental evidence for new dissociation channels in electron-impact ionization of H_2, *Phys. Rev. Lett.* **31,** 192 (1973).
132. A. Crowe and J.W. McConkey, Dissociative ionization by electron impact. I. Protons from H_2, *J. Phys. B* **6,** 2088 (1973).
133. J.A.D. Stockdale, V.E. Anderson, A.E. Carter, and L. Deleanu, Dissociative ionization of molecules by electron impact. I. Apparatus and kinetic energy distributions of D^+ ions from D_2, *J. Chem. Phys.* **63,** 3886 (1975).
134. K. Köllmann, Dissociative ionization of H_2, N_2, and CO by electron impact – measurement of kinetic energy, angular distributions, and appearance potentials, *Int. J. Mass Spectrom. Ion Phys.* **17,** 261 (1975).
135. K. Köllmann, Energetic protons from autoionizing states of H_2, *J. Phys. B* **11,** 339 (1978).
136. B. van Wingerden, Ph. E. van der Leeuw, F.J. de Heer, and M.J. van der Wiel, Partial oscillator strengths (25–40 eV) for repulsive states of D_2 (H_2), *J. Phys. B* **12,** 1559 (1979).
137. J.P. Johnson and J.L. Franklin, Kinetic energies and angular distributions of H^+ ions produced by electron impact on H_2, *Int. J. Mass Spectrom. Ion Phys.* **33,** 393 (1980).
138. M.D. Burrows, L.C. McIntyre, Jr., S.R. Ryan, and W.E. Lamb, Jr., Dissociative ionization of H_2, D_2, and HD using electron-impact excitation, *Phys. Rev. A* **21,** 1841 (1980).
139. M. Landau, R.I. Hall, and F. Pichou, Proton production from H_2 by electron impact excitation of autoionizing states near threshold, *J. Phys. B* **14,** 1509 (1981).
140. G.H. Dunn, Anisotropies in angular distributions of molecular dissociation products, *Phys. Rev. Lett.* **8,** 62 (1962).
141. The four poles of a quadrupole mass spectrometer are connected to both AC and DC voltages. The rf voltage, *V*, and dc voltage, *U*, are given by

$$V = 7.22mf^2r_0^2$$

$$U = 1.2mf^2r_0^2,$$

where f is the rf frequency (megacycles/s), m is the mass (amu), and r_0 is the radius tangent to the four poles (cm). For $m = 23$, $f = 1.6$ and $r_0 = 3$, $V = 2826$ V, and $U = 635$ V. These are rather large voltages; however, a given Rydberg atom will not see the full voltage and resulting field since the pole voltages alternate plus, minus, plus, minus.
142. S. Strathdee and R. Browning, Dissociative photoionization of H_2 at 26.9 eV, *J. Phys. B* **9**, L505 (1976); Dissociative photoionization of H_2: Proton kinetic energy spectra, *J. Phys. B* **12**, 1789 (1979).
143. R.M. Wood, A.K. Edwards, and M.F. Steven, Dissociative ionization of H_2 and D_2 produced by bombardment with fast He^+ ions, *Phys. Rev. A* **15**, 1433 (1977).
144. T.F. O'Malley, Some diabatic (quasistationary) states of small diatomic systems – projected atomic orbitals, *J. Chem. Phys.* **51**, 322 (1969).
145. C. Bottcher, Dissociative recombination of the hydrogen molecular ion, *J. Phys. B* **9**, 2899 (1976).
146. A.U. Hazi, Comment on the dissociative ionization of H_2, *J. Phys. B* **8**, L262 (1975).
147. W.D. Robb, private communication (1980).
148. H. Takagi and H. Nakamura, Elastic scattering of electrons from H_2^+ and studies of related dynamic processes, in *Abstracts of Contributed Papers, Twelfth International Conference on the Physics of Electronic and Atomic Collisions*, Gatlinburg, Tennessee, June 1981, p. 457.
149. A.U. Hazi, private communication (1981).
150. A.U. Hazi, Stieltjes moment method for calculating resonance widths, in *Electron–Molecule and Photon–Molecule Collisions* (T. Rescigno, V. McKoy, and B. Schneider, eds.), p. 281, Plenum Press, New York (1979).
151. K. Kirby, T. Uzer, A.C. Allison, and A. Dalgarno, Dissociative photoionization of H_2 at 26.9 eV and 30.5 eV, *J. Chem. Phys.* **75**, 2820 (1981).
152. F. Fiquet-Fayard and O. Gallais, Predissociation of H_2 and D_2 ($D\ ^1\Pi_u$): Comparison of calculated and experimental line widths, *Chem. Phys. Lett.* **16**, 18 (1972).
153. P. Borrell, P.M. Guyon, and M. Glass-Maujean, H_2 and D_2 photon impact predissociation, *J. Chem. Phys.* **66**, 818 (1977).
154. M. Glass-Maujean, J. Breton, and P.M. Guyon, Accidental predissociation of the $4p\pi\ ^1\Pi_u^+$ state of H_2, *Phys. Rev. Lett.* **40**, 181 (1978).
155. J. Durup, On isotope effects in the predissociation of HD, *J. Phys. (Paris)* **39**, 941 (1978).
156. R. Colle, MQDT analysis of predissociation. I. A two-channel problem with semi-classical scattering parameters, *J. Chem. Phys.* **74**, 2910 (1981).
157. M. Raoult and Ch. Jungen, Calculation of vibrational preionization in H_2 by multi-channel quantum defect theory: Total and partial cross sections and photoelectron angular distributions, *J. Chem. Phys.* **74**, 3388 (1981).
158. H. Takagi and H. Nakamura, Autoionization of highly excited Rydberg states of diatomic molecules, *J. Chem. Phys.* **74**, 5808 (1981).
159. L.J. Kieffer, G.M. Lawrence, and J.M. Slater, in *Proceedings of Seventh International Conference on the Physics of Electronic and Atomic Collisions*, Amsterdam (1971), p. 574.
160. J.A.D. Stockdale and L. Deleanu, Dissociative ionization of O_2 and N_2 by electron impact; N^+ and O^+ kinetic energies and angular distributions, *Chem. Phys. Lett.* **22**, 204 (1973).
161. J.A.D. Stockdale and L. Deleanu, Vibrational structure in kinetic energy spectra of O^+ ions from electron impact dissociative ionization of O_2: Predissociation of the $B\ ^2\Sigma_g^-$ state of O_2^+, *Chem. Phys. Lett.* **28**, 588 (1974).
162. J. Schopman and R. Locht, The observation of predissociation in the oxygen molecular ion by low-energy electron impact, *Chem. Phys. Lett.* **26**, 596 (1974).

163. R. Locht and J. Schopman, The dissociative ionization in oxygen, *Int. J. Mass Spectrom. Ion Phys.* **15**, 361 (1964).
164. R.S. Freund, Dissociation by electron impact of oxygen into metastable quintet and long-lived high-Rydberg atoms, *J. Chem. Phys.* **54**, 3125 (1971).
165. P.H. Doolittle, R.I. Schoen, and K.E. Schubert, Dissociative photoionization of O_2, *J. Chem. Phys.* **49**, 5108 (1968).
166. K. Köllman, Dissociative ionization of H_2, N_2, and CO by electron impact: Measurement of kinetic energy, angular distributions, and appearance potentials, *Int. J. Mass Spectrom. Ion Phys.* **17**, 261 (1975).
167. L.J. Kieffer and R.J. van Brunt, Energetic ions from N_2 produced by electron impact, *J. Chem. Phys.* **46**, 2728 (1967).
168. A. Crowe and J.W. McConkey, Dissociative ionization by electron impact. II. N^+ and N^{++} from N_2, *J. Phys. B* **6**, 2108 (1973).
169. L. Deleanu and J.A.D. Stockdale, Dissociative ionization of molecules by electron impact. II. Kinetic energy and angular distributions of N^+ and N^{++} ions from N_2, *J. Chem. Phys.* **63**, 3898 (1975).
170. J.N. Bardsley and F. Mandl, Resonant scattering of electrons by molecules, *Rep. Prog. Phys.* **31**, 471 (1968).
171. G.J. Schulz, Resonances in electron impact on diatomic molecules, *Rev. Mod. Phys.* **45**, 423 (1973).
172. R.N. Compton and R.H. Huebner, Collisions of low-energy electrons with molecules: Threshold excitation and negative-ion formation, in *Radiation Chemistry* (M. Burton and J.L. Magee, eds.), Vol. 2, p. 281, Wiley-Interscience, New York (1970).
173. J.N. Bardsley and J.S. Cohen, Variational calculations of resonant states of H_2^-, *J. Phys. B* **11**, 3645 (1978).
174. J.M. Wadehra and J.N. Bardsley, Vibrational- and rotational-state dependence of dissociative attachment in e–H_2 collisions, *Phys. Rev. Lett.* **41**, 1795 (1978).
175. J.N. Bardsley and J.M. Wadehra, Dissociative attachment and vibrational excitation in low-energy collisions of electrons with H_2 and D_2, *Phys. Rev. A* **20**, 1398 (1979).
176. M. Allan and S.F. Wong, Effect of vibrational and rotational excitation on dissociative attachment in hydrogen, *Phys. Rev. Lett.* **41**, 1791 (1978).
177. J.C.Y. Chen and J.L. Peacher, Survival probability in dissociative attachment, *Phys. Rev.* **163**, 103 (1967).
178. M. Bacal and G.W. Hamilton, H^- and D^- production in plasmas, *Phys. Rev. Lett.* **42**, 1538 (1979).
179. M. Tronc, R.I. Hall, C. Schermann, and H.S. Taylor, Interference in dissociative attachment cross sections of H_2 around 14 eV, *J. Phys. B* **12**, L279 (1979).
180. M. Tronc, F. Fiquet-Fayard, C. Schermann, and R.I. Hall, Differential cross sections and angular distributions of H^- from dissociative electron attachment to H_2 between 3.75 eV and 13 eV, *J. Phys. B* **10**, 305 (1977).
181. H. Hiraoka, R.K. Nesbet, and L.W. Welsh, Experimental observations of the negative atomic nitrogen ion, *Phys. Rev. Lett.* **39**, 130 (1977).
182. D. Spence and P.D. Burrow, Resonant dissociation of N_2 by electron impact, *J. Phys. B* **12**, L179 (1979).
183. J. Mazeau, F. Gresteau, R.I. Hall, and A. Huetz, Energy and width of $N^-(^3P)$ from observation of its formation by dissociative attachment to N_2 and NO, *J. Phys. B* **11**, L557 (1978).
184. A. Huetz, F. Gresteau, and J. Mazeau, Dissociative attachment in N_2, *J. Phys. B* **13**, 3275 (1980).
185. A. Huetz, F. Gresteau, R.I. Hall, and J. Mazeau, Initial vibrational state dependence

of resonant excitation and dissociative attachment in electron–N_2 scattering, *J. Chem. Phys.* **72**, 5297 (1980).

186. A. Huetz, I. Cadez, F. Gresteau, R.I. Hall, D. Vichon, and J. Mazeau, Formation and decay of the first $^2\Pi_u$ state of N_2^-, *Phys. Rev. A* **21**, 622 (1980).
187. I. Cadez and F. Fayard, The calculation of cross sections for some resonant excitation, *VIII ICPEAC Abstracts* (B.C. Cobic and M.V. Kurepa, eds.), p. 454, Institute of Physics, Belgrade (1974).
188. L.D. Thomas and R.K. Nesbet, Low-energy electron scattering by atomic nitrogen, *Phys. Rev. A* **12**, 2369 (1975).
189. B. Hird and S.P. Ali, Existence of the negative ion of atomic nitrogen, *Phys. Rev. Lett.* **41**, 540 (1978).
190. R.J. van Brunt and L.J. Kieffer, Angular distribution of O^- from dissociative electron attachment to O_2, *Phys. Rev. A* **2**, 1899 (1970).
191. M. Krauss, D. Neumann, A.C. Wahl, G. Das, and W. Zemke, Excited electronic states of O_2^-, *Phys. Rev. A* **7**, 69 (1973).
192. G. Das, A.C. Wahl, W.T. Zemke, and W.C. Stwalley, Accurate *ab initio* potential curves for the $X\,^2\Pi_g, A\,^2\Pi_u, a\,^4\Sigma_u^-$ and $^2\Sigma_u^-$ states of the O_2^- ion, *J. Chem. Phys.* **68**, 4252 (1978).
193. W.R. Henderson, W.L. Fite, and R.T. Brackmann, Dissociative attachment of electrons to hot oxygen, *Phys. Rev.* **183**, 157 (1969).
194. D. Spence and G.J. Schulz, Temperature dependence of dissociative attachment in O_2 and CO_2, *Phys. Rev.* **188**, 280 (1969).
195. T.F. O'Malley, Calculation of dissociative attachment in hot O_2, *Phys. Rev.* **155**, 59 (1967).
196. P.D. Burrow, Dissociative attachment from the $O_2(a\,^1\Delta_g)$ state, *J. Chem. Phys.* **59**, 4922 (1973).
197. D.S. Belic and R.I. Hall, Dissociative electron attachment to metastable oxygen $(a^1\Delta_g)$, *J. Phys. B* **14**, 365 (1981).
198. P.J. Chantry, private communication (1978).
199. B.I. Schneider and C.A. Brau, Dissociative attachment of electrons by F_2, *Appl. Phys. Lett.* **33**, 569 (1978).
200. J.N. Bardsley, C. Derkits, and J.M. Wadehra, Dissociative attachment in HCl, DCl, and F_2, to be published.
201. R.J. Hall, Dissociative attachment and vibrational excitation of F_2 by slow electrons, *J. Chem. Phys.* **68**, 1803 (1978).
202. D.W. Trainor and J.H. Jacob, Electron dissociative attachment rate constant for F_2 and NF_3 at 300 and 500°K, *Appl. Phys. Lett.* **35**, 920 (1979).
203. A.U. Hazi, A.E. Orel, and T.N. Rescigno, *Ab initio* study of dissociative attachment of low-energy electrons to F_2, *Phys. Rev. Lett.* **46**, 918 (1981).
204. W.C. Tam and S.F. Wong, Dissociative attachment of halogen molecules by 0–8 eV electrons, *J. Chem. Phys.* **68**, 5626 (1978).
205. M.V. Kurepa and D.S. Belic, Dissociative attachment of electrons to chlorine molecules, *Chem. Phys. Lett.* **49**, 608 (1977).
206. T.N. Rescigno and C. Bender, The stability of the F_2^- ion: A model for dissociative attachment, *J. Phys. B* **9**, L329 (1976).
207. M.V. Kurepa, D.S. Babic, and D.S. Belic, Electron–bromine-molecule total ionisation and electron attachment cross sections, *J. Phys. B* **14**, 375 (1981).
208. F.K. Truby, Temperature dependence of attachment in I_2 vapor, *Phys. Rev.* **188**, 508 (1969).
209. E.J. Shipsey, Thermal-energy dissociative attachment of I_2; detection of the curve crossing point for the experimental measurements, *J. Chem. Phys.* **52**, 2274 (1970).

210. D.T. Birtwistle and A. Modinos, Dissociative attachment in iodine, *J. Phys. B* **11,** 2949 (1978).
211. H.L. Brooks, S.R. Hunter, and K.J. Nygaard, Temperature dependence of the electron attachment coefficient in iodine, *J. Chem. Phys.* **71,** 1870 (1979).
212. J.P. Ziesel, I. Nenner, and G.J. Schulz, Negative ion formation, vibrational excitation, and transmission spectroscopy in hydrogen halides, *J. Chem. Phys.* **63,** 1943 (1975).
213. R. Azria, L. Roussier, R. Paineau, and M. Tronc, Attachment electronique dissociatif sur HCl et DCl, *Rev. Phys. Appl.* **9,** 469 (1974).
214. F. Fiquet-Fayard, Theoretical investigation of dissociative attachment in HCl and DCl, *J. Phys. B* **7,** 810 (1974).
215. F. Fiquet-Fayard, Theoretical problems in the interpretation of dissociative attachment experiments, *Vacuum* **24,** 533 (1974).
216. M. Allan and S.F. Wong, Dissociative attachment from vibrationally and rotationally excited HCl and HF, *J. Chem. Phys.* **74,** 1687 (1981).
217. H.S. Taylor, E. Goldstein, and G.A. Segal, Resonance states of HCl^- and electron–HCl scattering processes, *J. Phys. B* **10,** 2253 (1977).
218. R. Azria, M. Tronc, Y. LeCoat, and D. Simon, Differential cross section for negative ion formation in HCl, *XI ICPEAC Abstracts,* p. 360, Kyoto (1979).
219. R. Abouaf and D. Teillet-Billy, Predissociations in the dissociative attachment processes leading to C^-/CO and F^-/FH, *XI ICPEAC Abstracts,* p. 356, Kyoto (1979).
220. D. Spence and T. Noguchi, Feshbach resonances associated with Rydberg states of the hydrogen halides, *J. Chem. Phys.* **63,** 505 (1975).
221. R. Azria, Y. LeCoat, and M. Tronc, Spin–orbit splitting and dynamical coupling for differential H^- cross section in hydrogen halides, *XII ICPEAC Abstracts,* p. 411, Gatlinburg, Tennessee (1981).
222. P.J. Chantry, Formation of N_2O^- via ion–molecule reactions in N_2O, *J. Chem. Phys.* **51,** 3380 (1969).
223. J.N. Bardsley, Negative ions of N_2O and CO_2, *J. Chem. Phys.* **51,** 3384 (1969).
224. M. Tronc, F. Fiquet-Fayard, C. Schermann, and R.I. Hall, Angular distributions of O^- from dissociative electron attachment to N_2O between 1.9 and 2.9 eV, *J. Phys. B* **10,** L459 (1977).
225. D.G. Hopper, A.C. Wahl, R.L.C. Wu, and T.O. Tiernan, Theoretical and experimental studies of the N_2O^- and N_2O ground state potential energy surfaces. Implications for the $O^- + N_2 \rightarrow N_2O + e$ and other processes, *J. Chem. Phys.* **65,** 5474 (1976).
226. R. Abouaf, R. Paineau, and F. Fiquet-Fayard, Dissociative attachment in NO_2 and CO_2, *J. Phys. B* **9,** 303 (1976).
227. P.D. Burrow and L. Sanche, Elastic scattering of low-energy electrons at 180° in CO_2, *Phys. Rev. Lett.* **28,** 333 (1972).
228. L. Sanche and G.J. Schulz, Electron transmission spectroscopy: Resonances in triatomic molecules and hydrocarbons, *J. Chem. Phys.* **58,** 479 (1973).
229. M.J.W. Boness and G.J. Schulz, Vibrational excitation in CO_2 via the 3.8 eV resonance, *Phys. Rev. A* **9,** 1969 (1974).
230. A. Herzenberg, Oscillatory energy dependence of resonance electron–molecule scattering, *J. Phys. B* **1,** 548 (1968).
231. W. Domcke and L.S. Cederbaum, A simple formula for the vibrational structure of resonances in electron–molecule scattering, *J. Phys. B* **10,** L47 (1977).
232. C.R. Claydon, G.A. Segal, and H.S. Taylor, Theoretical interpretation of the electron scattering spectrum of CO_2, *J. Chem. Phys.* **52,** 3387 (1970).
233. D. Spence and G.J. Schulz, Dissociative electron attachment in CO_2, *J. Chem. Phys.* **60,** 216 (1974).

234. P.J. Chantry, Temperature dependence of dissociative attachment in N_2O, *J. Chem. Phys.* **51**, 3369 (1969).
235. J.P. Ziesel, G.J. Schulz, and M. Milhand, S^- formation by dissociative attachment in OCS and CS_2, *J. Chem. Phys.* **62**, 1936 (1975).
236. R. Abouaf and F. Fiquet-Fayard, Dissociative attachment in SO_2 and OCS: Evidence for vibrational excitation of the SO, O_2, and CO fragments, *J. Phys. B* **9**, L323 (1976).
237. M. Jungen, J. Vogt, and V. Staemmler, Feshbach resonances and dissociative electron attachment of H_2O, *Chem. Phys.* **37**, 49 (1979).
238. S. Trajmar and R.I. Hall, Dissociative attachment in H_2O and D_2O: Energy and angular distribution of H^- and D^- fragments, *J. Phys. B* **7**, L458 (1974).
239. M. Tronc, S. Gowsand, R. Azria, and F. Fiquet-Fayard, Effects isotopiques dans la formation de H^- et D^- par attachement dissociatif sur H_2S, HDS, D_2S et H_2O, HDO, D_2O, *J. Phys. (Paris)* **34**, 381 (1973).
240. F. Fiquet-Fayard, M. Sizun, and H. Abgrall, Intramolecular isotope effect in dissociations, *Chem. Phys. Lett.* **37**, 72 (1976).
241. S. Goursand, M. Sizon, and F. Fiquet-Fayard, Energy partitioning and isotope effects in the fragmentation of triatomic negative ions: Monte Carlo scheme for a classical trajectory study, *J. Chem. Phys.* **65**, 5453 (1976).
242. A. Huetz and J. Mazeau, Observation of unstable H^- from the dissociation of H_2^-, *J. Phys. B* **14**, L591 (1981).
243. A. Huetz, Excitation and resonant dissociation of diatomic molecules by electron impact, Thesis, Université-Paris VI (1981).
244. E. Illenberger, H.V. Scheunemann, and H. Baumgärtel, Negative ion formation in CF_2Cl_2, CF_3Cl, and $CFCl_3$ following low-energy (0–10 eV) impact with near monoenergetic electrons, *Chem. Phys.* **37**, 21 (1979).
245. P. Ausloos, R.E. Rebbert, and L. Glasgow, Photodecomposition of chloromethanes adsorbed on silica surfaces, *J. Res. Natl. Bur. Stand. A* **82**, 1 (1977).
246. G.J. Verhaart, H.A. Van Der Hart, and H.H. Brongersma, Electron chemistry. Chemical reactions in CCl_4, $CFCl_3$, and CF_2Cl_2 induced by low-energy electrons, *Chem. Phys.* **1**, 389 (1980).
247. V.M. Pejčev, M.V. Kurepa, and I.M. Čadež, Total ionization and electron attachment cross sections of CCl_2F_2 by electron impact, *Chem. Phys. Lett.* **63**, 301 (1979).
248. W.E. Wentworth, R. George, and H. Keith, Dissociative thermal electron attachment to some aliphatic chloro, bromo, iodo, compounds, *J. Chem. Phys.* **51**, 1791 (1969).
249. S.D. Peyerimhoff and R.J. Buenker, Potential curves for dissociative electron attachment of $CFCl_3$, *Chem. Phys. Lett.* **65**, 434 (1979).
250. G.J. Schulz and D. Spence, Temperature dependence of the onset for negative-ion formation in CO_2, *Phys. Rev. Lett.* **22**, 47 (1969).
251. C.E. Klots and R.N. Compton, Electron attachment to carbon dioxide clusters in a supersonic beam, *J. Chem. Phys.* **67**, 1779 (1977).
252. C.E. Klots and R.N. Compton, Electron attachment to van der Waals polymers of carbon dioxide and nitrous oxide, *J. Chem. Phys.* **69**, 1636 (1978).
253. C.E. Klots and R.N. Compton, Electron attachment to van der Waals polymers of water, *J. Chem. Phys.* **69**, 1644 (1978).
254. M.J. Boness and G.J. Schulz, Vibrational excitation in CO_2 via the 3.8-eV resonance, *Phys. Rev. A* **9**, 1969 (1974).
255. R.N. Compton, P.W. Reinhardt, and C.D. Cooper, Collisional ionization of Na, K, Cs by CO_2 and CS_2: Molecular electron affinities, *J. Chem. Phys.* **63**, 3821 (1975).
256. A.R. Rossi and K.D. Jordan, Comment on the structure and stability of $(CO_2)_2^-$, *J. Chem. Phys.* **70**, 4422 (1979).

257. M. Armbruster, H. Haberland, and H.-G. Schindler, Negatively charged water clusters, or the first observation of free hydrated electrons, *Phys. Rev. Lett.* **47,** 323 (1981).
258. D.R. Bates and H.S.W. Massey, The basic reactions in the upper atmosphere II. The theory of recombination in the ionized layers, *Proc. R. Soc. London Ser. A* **192,** 1 (1947).
259. M.A. Biondi and S.C. Brown, Measurement of electron–ion recombination, *Phys. Rev.* **76,** 1697 (1949).
260. C.S. Weller and M.A. Biondi, Recombination attachment and ambipolar diffusion of electrons in photo-ionized NO afterglows, *Phys. Rev.* **172,** 198 (1968).
261. F.J. Mehr and M.A. Biondi, Electron temperature dependence of recombination of O_2^+ and N_2^+ ions with electrons, *Phys. Rev.* **181,** 264 (1969).
262. M.T. Leu, M.A. Biondi, and R. Johnsen, Measurements of the recombination of electrons with $H_3O^+ \cdot (H_2O)_n$-series ions. *Phys. Rev. A* **7,** 292 (1973).
263. M.T. Leu, M.A. Biondi, and R. Johnsen, Measurements of recombination of electrons with HCO^+ ions, *Phys. Rev. A* **8,** 420 (1973).
264. M.T. Leu, M.A. Biondi, and R. Johnsen, Measurements of recombination of electrons with H_3^+ and H_5^+ ions, *Phys. Rev. A* **8,** 413 (1973).
265. A.J. Cunningham and R.M. Hobson, Experimental measurements of dissociative recombination in vibrationally excited gases, *Phys. Rev.* **185,** 98 (1969).
266. B. Peart and K.T. Dolder, Collisions between electrons and H_2^+ ions. IV. Measurement of cross sections for dissociative ionization, *J. Phys. B* **6,** 2409 (1973).
267. B. Peart and K.T. Dolder, Measurements of cross sections for the dissociative recombination of D_2^+ ions, *J. Phys. B* **6,** L359 (1973).
268. B. Peart and K.T. Dolder, Collisions between electrons and H_2^+ ions: V. Measurements of cross sections for dissociative recombination, *J. Phys. B* **7,** 236 (1974).
269. B. Peart and K.T. Dolder, Measurements of dissociative recombination of H_3^+ ions, *J. Phys. B* **7,** 1948 (1974).
270. D. Auerbach, R. Cacak, R. Caudano, T.D. Gaily, C.J. Keyser, J. Wm. McGowan, J.B.S. Mitchell, and S.F.J. Wilk, Merged electron–ion beam experiments. I. Method and measurements of (e–H_2^+) and (e–H_3^+) dissociative-recombination cross sections, *J. Phys. B* **10,** 3797 (1977).
271. P.M. Mul and J. Wm. McGowan, Merged electron–ion beam experiments. III. Temperature dependence of dissociative recombination for atmospheric ions NO^+, O_2^+, and N_2^+, *J. Phys. B* **12,** 1591 (1979).
272. R.A. Phaneuf, D.H. Crandall, and G.H. Dunn, Production of D* ($n = 4$) from electron–D_2^+ dissociative recombination, *Phys. Rev. A* **11,** 528 (1975).
273. M. Vogler and G.H. Dunn, Dissociative recombination of electrons and D_2^+ to yield D(2p), *Phys. Rev. A* **11,** 1983 (1975).
274. F. von Busch and G.H. Dunn, Photodissociation of H_2^+ and D_2^+: Experiment, *Phys. Rev. A* **5,** 1726 (1972).
275. F.L. Walls and G.H. Dunn, Measurement of total cross sections for electron recombination with NO^+ and O_2^+ using ion storage techniques, *J. Geophys. Res.* **79,** 1911 (1974).
276. R.A. Heppner, F.L. Walls, W.T. Armstrong, and G.H. Dunn, Cross-section measurements for electron–H_3O^+ recombination, *Phys. Rev. A* **13,** 1000 (1976).
277. R.D. DuBois, J.B. Jeffries, and G.H. Dunn, Dissociative recombination cross sections for NH_4^+ ions and electrons, *Phys. Rev. A* **17,** 1314 (1978).
278. D. Mathur, S.V. Khan, and J.B. Hasted, Dissociative recombination in low-energy e–H_2^+ and e–H_3^+ collisions, *J. Phys. B* **11,** 3615 (1978).

279. D.R. Bates, Dissociative recombination, *Phys. Rev.* **78**, 492 (1950).
280. J.N. Bardsley, The theory of dissociative recombination, *J. Phys. B* **1**, 365 (1968).
281. C. Bottcher, Theory of dissociative recombination and related processes, *Proc. R. Soc. London Ser. A* **340**, 301 (1974).
282. A. Giusti, A multichannel quantum defect approach to dissociative recombination, *J. Phys. B* **13**, 3867 (1980).
283. T.F. O'Malley, Rydberg levels and structure in dissociative recombination cross sections, *J. Phys. B* **14**, 1229 (1951).
284. U. Fano, Effects of configuration interaction on intensities and phase shifts, *Phys. Rev.* **124**, 1866 (1961).
285. A. Giusti-Suzor, J.N. Bardsley, and C. Derkits, Dissociative recombination in low-energy $e{-}H_2^+$ collisions, *Phys. Rev. A* **28**, 682 (1983).
286. D. Robb, private communication.
287. K. Rai Dastidar and T.K. Rai Dastidar, Dissociative recombination of H_2^+, HD^+, and D_2^+ molecular ions, *J. Phys. Soc. Jpn.* **46**, 1288 (1979).
288. V.P. Zhdanov and M.I. Chibisov, Dissociative recombination of electrons with molecular ions involving the formation of highly excited atoms, *Zh. Eksp. Teor. Fiz.* **74**, 75 (1978) [*Sov. Phys. JETP* **47**, 38 (1978)].
289. C. Derkits, J.N. Bardsley, and J.M. Wadehra, Dissociative recombination in $e + H_2^+$ collisions, *J. Phys. B* **12**, L529 (1979).
290. B. Peart and K.T. Dolder, Collisions between electrons and H_2^+ ions. VI. Measurements of cross sections for the simultaneous production of H^+ and H^-, *J. Phys. B* **8**, 1570 (1975).
291. P.M. Solomon and W. Klemperer, The formation of diatomic molecules in interstellar clouds, *Astrophys. J.* **178**, 389 (1972).
292. P. Frisch, Abundances of interstellar CH and CH^+ radicals, *Astrophys. J.* **173**, 301 (1972).
293. J.N. Bardsley and B.R. Junker, Dissociative recombination of CH^+ ions, *Astrophys. J.* **183**, L135 (1973).
294. M. Krauss and P.S. Julienne, Dissociative recombination of $e + CH^+$ $(X\,^1\Sigma^+)$, *Astrophys. J.* **183**, L139 (1973).
295. A. Giusti-Suzor and H. Lefebvre-Brion, The dissociative recombination of CH^+ ions, *Astrophys. J.* **214**, L101 (1977).
296. A.E.S. Green, P.S. Bagus, A.D. McLean, and Y. Yoshimine, Calculated potential energy curves for CH^+, *Phys. Rev. A* **5**, 1614 (1972).
297. J.B.A. Mitchell and J.W. McGowan, The dissociative recombination of CH^+ $X\,^1\Sigma^+$ $(v = 0)$, *Astrophys. J.* **222**, L77 (1978).
298. T.P. Stecher and D.A. Williams, CH and CH^+ formation in ion–molecule reactions, *Astrophys. J.* **177**, L141 (1972).
299. M. Elitzur and W.D. Watson, Formation of CH^+ in interstellar shocks, *Astrophys. J.* **222**, L141 (1978).
300. T. Uzer and A. Dalgarno, private communication.
301. P.C. Cosby, H. Helm, and J.T. Moseley, Observation of predissociated levels of CH^+ $(A\,^1\Pi_1)$, *Astrophys. J.* **235**, 52 (1980).
302. D.G. Torr, M.R. Torr, J.C.G. Walker, A.O. Nier, L.H. Brace, and H.C. Brinton, Recombination of O_2^+ in the ionosphere, *J. Geophys. Res.* **81**, 5578 (1976).
303. D.G. Torr, M.R. Torr, J.C.G. Walker, L.H. Brace, H.C. Brinton, W.B. Hanson, J.H. Hoffman, and M. Oppenheimer, Recombination of NO^+ in the atmosphere, *Geophys. Res. Lett.* **3**, 209 (1976).

304. N. Orsini, D.G. Torr, H.C. Brinton, L.H. Brace, W.B. Hanson, J.H. Hoffman, and A.O. Nier, Determination of the N_2^+ rate coefficient in the ionosphere, *Geophys. Res. Lett.* **4,** 431 (1977).
305. M.A. Biondi, Objections to the $N_2^+ + e^-$ dissociative recombination coefficients inferred from analysis of atmosphere explorer measurements, *Geophys. Res. Lett.* **5,** 661 (1978).
306. D.G. Torr and N. Orsini, The effect of N_2^+ recombination on the aeronomic determination of the charge exchange rate coefficient of $O^+(^2D)$ with N_2, *Geophys. Res. Lett.* **5,** 657 (1978).
307. A. Dalgarno, Atomic physics from atmospheric and astrophysical studies, *Adv. At. Mol. Phys.* **15,** 37 (1979).
308. W.H. Kasner, Study of the temperature dependence of electron–ion recombination in nitrogen, *Phys. Rev.* **164,** 194 (1967).
309. A.J. Cunningham and R.M. Hobson, Dissociative recombination at elevated temperatures. IV. N_2^+ dominated afterglows, *J. Phys. B* **5,** 2328 (1972).
310. W.H. Kasner and M.A. Biondi, Electron–ion recombination in nitrogen, *Phys. Rev.* **137,** A317 (1965).
311. M. Whitaker, M.A. Biondi, and R. Johnsen. Electron temperature dependence of dissociative recombination of electrons with $N_2^+ \cdot N_2$ dimer ions, *Phys. Rev. A* **24,** 743 (1981).
312. C.-M. Huang, M.A. Biondi, and R. Johnsen, Variation of electron–NO^+ ion recombination coefficient with electron temperature, *Phys. Rev. A* **11,** 901 (1975).
313. C.M. Lee, Multichannel dissociative recombination theory, *Phys. Rev. A* **16,** 109 (1977).
314. A.J. Cunningham and R.M. Hobson, Dissociative recombination at elevated temperatures. III. O_2^+ dominated afterglows, *J. Phys. B* **5,** 2320 (1972).
315. W.H. Kasner and M.A. Biondi, Temperature dependence of the electron–O_2^+ ion recombination coefficient, *Phys. Rev.* **174,** 139 (1968).
316. E.C. Zipf, The $O(^1S)$ state: Its quenching and formation by the dissociative recombination of vibrationally excited O_2^+ ions, *Geophys. Res. Lett.* **6,** 881 (1979).
317. E.C. Zipf, A laboratory study of the dissociative recombination of vibrationally excited O_2^+ ions, *J. Geophys. Res.* **85,** 4232 (1980).
318. S.L. Guberman, Potential curves for dissociative recombination of O_2^+, *Int. J. Quant. Chem. Symp.* **13,** 531 (1979).
319. D.R. Bates and E.C. Zipf, The $O(^1S)$ quantum yield from O_2^+ dissociative recombination, *Planet. Space Sci.* **28,** 1081 (1980).
320. T.F. O'Malley, A.J. Cunningham, and R.M. Hobson, Dissociative recombination at elevated temperatures. II. Comparison between theory and experiment in neon and argon afterglows, *J. Phys. B* **5,** 2126 (1972).
321. A.J. Cunningham, T.F. O'Malley, and R.M. Hobson, On the role of vibrational excitation in dissociative recombination, *J. Phys. B* **14,** 773 (1981).
322. Y.J. Shiu, M.A. Biondi, and D.P. Sipler, Dissociative recombination in xenon: Variation of the total rate coefficient and excited state production with electron temperature, *Phys. Rev. A* **15,** 494 (1977).
323. Y.J. Shiu and M.A. Biondi, Dissociative recombination in krypton (and argon): Dependence of the total rate coefficient and excited-state production with electron temperature, *Phys. Rev A* **16,** 1817 (1977); *ibid.* **17,** 868 (1978).
324. D.R. Bates, Aspects of recombination, *Adv. At. Mol. Phys.* **15,** 235 (1979).
325. C.M. Huang, M.A. Biondi, and R. Johnsen, Recombination of electrons with $NH_4^+ \cdot (NH_3)_n$ series ions, *Phys. Rev. A* **14,** 984 (1976).

326. B.M. Smirnov, Recombination of an electron and a complex ion, *Sov. Phys. JETP* **45**, 731 (1977).
327. C. Bottcher, Dissociative recombination of large molecular ions, *J. Phys. B* **11**, 3887 (1978).
328. I. Herbst, What are the products of polyatomic ion–electron dissociative recombination reactions?, *Astrophys. J.* **222**, 508 (1978).
329. D.W. Trainor, Dissociative recombination of electrons with hydrogen ions, *Chem. Phys. Lett.* **55**, 361 (1978).

5

MOLECULAR SPECTROSCOPY BY ELECTRON SCATTERING

R.I. HALL and F.H. READ

1. Introduction

The previous chapters have been primarily concerned with the theoretical description of electron–molecule scattering processes and with the appropriate comparisons between theory and experiment. The emphasis has tended to be on *reaction mechanisms*, rather than on the *spectroscopy* of the molecular states involved in the reactions. In the present chapter we take a different and complementary approach, and deal mainly with the use of electron–molecule collisions as a spectroscopic tool for investigating the electronic structure of simple molecules. We shall see that there is considerable information to be gained from electron scattering spectroscopy which is not available from optical experiments.

In Section 2 we consider the spectroscopy of negative-ion resonances. These resonances appear as sharp changes in electron–molecule scattering cross sections at low incident electron energies (typically 1–10 eV). An example is shown in Fig. 1, taken from the work of Sanche and Schulz.[1] This spectrum shows the yield of electrons that pass through a sample of NO gas without being elastically or inelastically scattered. As the incident electron energy is varied the yield of transmitted electrons fluctuates rapidly. This yield is differentiated before being displayed in the spectrum, to make the fluctuations more visible. The analysis of this spectrum will be discussed in Section 2.1.

The structures which appear in electron scattering spectra such as that shown in the figure are caused by the formation of molecular negative ions M^- which quickly decay (with a lifetime typically in the range 10^{-15} to 10^{-12} s) by any of the routes that are energetically accessible. This can be illustrated by writing

---

R.I. HALL • Centre National de la Recherche Scientifique, and Laboratoire de Physique et Optique Corpusculaires, Université Pierre et Marie Curie, Paris, France. **F.H. READ** • Department of Physics, University of Manchester, Manchester, England.

$$e + M_{n_i v_i J_i} \rightarrow M^{-*}_{n_r v_r J_r} \rightarrow \begin{cases} e + M_{n_i v_i J_i}, & \text{elastic scattering,} \\ \text{or } e + M_{n_f v_f J_f}, & \text{inelastic scattering,} \\ \text{or } e + e + M^+, & \text{ionization,} \\ \text{or } e + A + B, & \text{molecular dissociation,} \\ \text{or } A^- + B, & \text{dissociative attachment,} \\ \text{etc.,} & \end{cases} \quad (1)$$

where M^{-*} is the temporary negative ion and n, v, and J represent the set of electronic, vibrational, and rotational quantum numbers of the molecule. When inelastic scattering occurs at least one of the quantum numbers n_f, v_f, and J_f is different from the initial set n_i, v_i, and J_i.

Resonances may appear in any of the accessible decay routes. The incident electron energy at which a resonance occurs is given by the difference between the energy of the initial state of the target molecule and that of the negative ion. The observed energy width of a resonance depends on the natural decay width of the negative ion and the width introduced by the finite resolution of the electron energy analyzers. Therefore the energies, and sometimes also the widths, of the negative-ion states come directly from the experimental measurements. Other spectroscopic properties of interest are the symmetries and potential energy functions of these short-lived molecules. The experimental energy resolution and theoretical interpretation have now reached the stage at which most or all of these properties can be deduced for resonant states of several simple molecules.

In Section 3 we consider the use of the electron energy-loss technique to

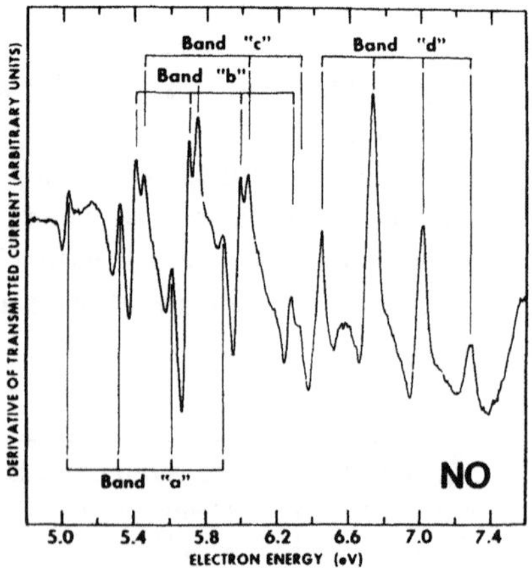

FIGURE 1
Derivative of the electron current transmitted through a gas cell containing NO, as a function of the incident electron energy. Four series of resonances can be seen. (Sanche and Schulz.[1]).

obtain spectroscopic information about neutral molecules. The relevant scattering process is either resonance-aided inelastic scattering [see Eq. (1)] or direct (i.e., nonresonant) inelastic scattering,

$$e + M_{n_i v_i J_i} \rightarrow e + M_{n_f v_f J_f}. \tag{2}$$

The usual experimental practice is to keep the incident electron energy fixed, while the energy loss of the scattered electrons is scanned. Figure 2 shows a spectrum obtained in this way by Lassettre *et al.*(2) The incident electron energy is 70 eV and the scattering angle is zero. The dominant transition is the Schumann–Runge continuum of O_2, extending from about 7 to 9 eV, and the peaks at higher energies represent transitions to discrete excited states.

It is convenient to divide the discussion of Section 3 into four subsections, according to the range of incident electron energy and of electron energy loss. In the first subsection we shall deal with high incident energies ($\gtrsim$ a few hundred electron volts) and low energy losses ($\lesssim 50$ eV), and shall see that the spectra essentially reproduce photoabsorption spectra. At higher values of the energy loss, dealt with in the following subsection, inner-shell excited states of molecules can be studied, often at a higher energy resolution than is available in the analogous photoabsorption experiments, thus allowing the states and their natural decay widths to be studied in greater detail. Furthermore, optically forbidden transitions show up much more strongly than in the analogous photoabsorption experiments. After that we shall consider intermediate incident energies (~20 to 100 eV), for which the electron energy-loss technique allows the study of optically forbidden valence and Rydberg transitions. This results in valuable information on the spin and spatial symmetries of molecular excited states, and complements the limited range of transitions that can be studied by

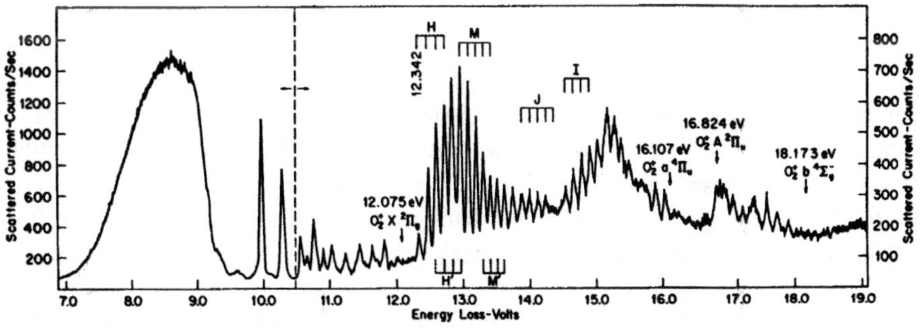

FIGURE 2
Electron energy-loss spectrum of oxygen, taken at an incident electron energy of 70 eV and zero scattering angle (Lassettre *et al.*(2)).

optical methods. Finally, at near-threshold incident energies (i.e., within about 0.1 eV of the threshold energy), optically forbidden transitions often appear even more strongly, showing the existence of states that are not found in optical absorption spectra. Resonance effects are often of importance in the near-threshold range, but not in general at the high-energy and intermediate-energy ranges.

There have been several previous books and reviews which have a partial overlap with the coverage of the present chapter. Amongst these are the books of Hasted,(3) McDaniel,(4) Massey and Burhop,(5) Christophorou,(6) Sevier,(7) and Massey(8) and the reviews of Lassettre,(9) Trajmar *et al.*,(10) Inokuti,(11) Dillon,(12) Hasted,(13) Schulz,(14) Lassettre and Skerbele,(15) Read,(16) Brion,(17) Williams,(18) Golden,(19) and Celotta and Huebner.(20)

The present chapter is intended to be an introduction to the subject, and no attempt has been made to give a comprehensive account of all the work that has been carried out in this field. Such detailed accounts can be found in some of the books and reviews listed above. Our aim is rather to describe the more important aspects of electron-impact spectroscopy, illustrating our account with a few specific examples of experimental results. To keep the length of the chapter within reasonable bounds we have also omitted discussion of the topic of binary (e, $2e$) experiments, in which coincidences between scattered and ejected electrons are used to deduce information about the scattering mechanism and also about the electronic structure of molecules. Recent reviews of this topic have been given by McCarthy(21) and Giardini-Guidoni *et al.*(22)

2. Resonance Spectroscopy of Molecules

Resonances occur in electron–molecule scattering and perturb the various cross sections at specific impact energies. The resonances can be considered as quasistationary states of the negative ion M^- formed by the electron and the target molecule M.

A schematic illustration of the energies and configurations of states of the molecular positive ion M^+, the neutral molecule M, and the negative ion M^- is shown in Fig. 3. Two types of resonance state of M^- are illustrated, namely, Feshbach and shape resonances. In the present section we shall largely be concerned with the spectroscopy of Feshbach resonances: shape resonances (at least those associated with the ground state) have already been treated in detail in Chapter 3.

We see from Fig. 3 that the configuration of a Feshbach resonance is that of an excited state of M, usually referred to as the *parent state*, with an additional electron in an excited orbital. Alternatively the resonance configuration may be viewed as that of a state (often the ground state) of M^+, referred to

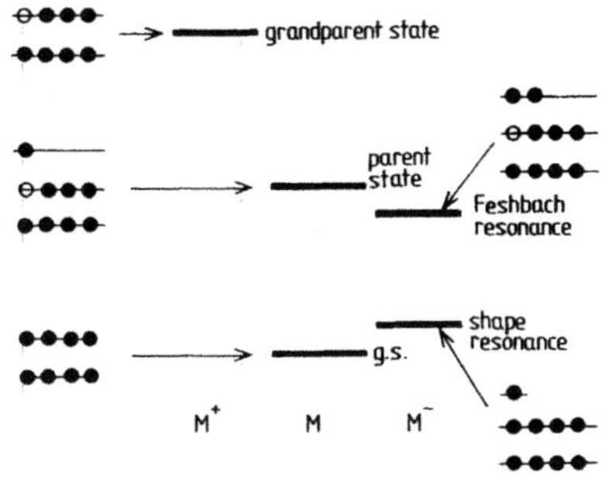

FIGURE 3
Schematic illustration of the energies and configurations of related states of the molecules M^+, M, and M^-. Solid and open circles represent occupied and unoccupied orbitals, respectively.

as the *grandparent state*, with the addition of two electrons in excited orbitals. We shall see later that the "grandparent model" is usually the more useful, and that the two excited electrons normally have the same principal quantum number n and possess a high degree of mutual correlation.

Feshbach resonances are generally long-lived compared to a typical vibrational period (10^{-14} s), as their decay involves the rearrangement of the orbitals of both the excited electrons. Thus the temporary negative ion can make many vibrations before decaying. The nuclear and electronic parts of the wave function of the resonance therefore have forms similar to those of bound neutral molecular states, but the time-dependent part has the form

$$\psi(t) \propto \exp(-iWt/\hbar), \tag{3}$$

where W is the complex energy

$$W = E_r - \tfrac{1}{2} i\Gamma. \tag{4}$$

The probability density of the resonant state therefore has the time dependence

$$|\psi(t)|^2 \propto \exp(-\Gamma t/\hbar) = \exp(-t/\tau), \tag{5}$$

where Γ is the resonance width and τ the resonance lifetime.

Because the lifetimes of Feshbach resonances tend to be long, their resonance widths Γ are correspondingly small (typically less than 20 meV), and so the resonances appear as narrow structures in the cross sections of the elastic and inelastic processes represented in Eq. (1). As we shall see in more detail later, the energy dependence of a scattering cross section in the neighborhood of an isolated sharp resonance can be represented by the Breit–Wigner formula

$$\sigma(E) = \left| A + \frac{B}{E - E_r + \frac{1}{2} i\Gamma} \right|^2. \tag{6}$$

Here A represents direct (i.e., nonresonant) scattering, and E_r is the resonance energy that appears in Eq. (4). Interference between the direct and resonant processes gives rise to the well-known Beutler–Fano forms. In electron–molecule scattering a special situation exists which is not found in electron scattering from atoms, nor in nuclear reactions. It is nearly always possible to find an inelastic channel to which the resonance decays where the direct process is weak and can be neglected. In this case we have

$$\sigma(E) \propto \frac{\Gamma_{\text{in}}\Gamma_{\text{out}}}{(E-E_r)^2 + (\frac{1}{2}\Gamma)^2} \qquad (7)$$

where Γ is the total width and Γ_{in} and Γ_{out} are partial widths. The resonance then appears as a Lorentzian peak centered at E_r with a width of Γ at half height.

Shape resonances differ from Feshbach resonances in one important respect: They lie above their parent state in energy (see Fig. 3). The extra electron is temporarily retained by the effective potential between it and the parent state M, provided that this potential has a local maximum (forming an effective barrier) at a separation of the order of a few angstroms. Such a barrier can be created by the combination of an attractive potential and a centrifugal repulsive potential that is proportional to $l(l+1)/r^2$, where $l\hbar$ is the angular momentum of the electron. The resonance can then be pictured as decaying by the electron tunneling out through the potential barrier. These resonances usually have short lifetimes compared to a vibrational period because they decay preferentially and easily to their parent state. Shape resonances therefore appear in excitation cross sections near their threshold as broad peaks with little or no vibrational structure. Those associated with the ground state (also known as single-particle resonances) are particularly prominent in electron scattering by molecules as they lead to strong vibrational and rotational excitation (see Chapter 3). Shape resonances which have excited states as parents are also known and are called core-excited shape resonances. They have been observed for both valence and Rydberg states. The spectroscopy of this type of resonance will also be discussed here. A third category of sharp structures in the cross sections can also be defined whose energy is the same as that of the parent state. These are called virtual states rather than resonances and are rarer and less well known than the other two types (see, however, Chapter 3, Section 3).

The spectroscopy of resonances as it will be described in this chapter consists of, first, considering resonances of quasibound states of negative ions as mentioned above and, second, characterizing these states by their different parameters [i.e., potential energy curve (real part), width, configuration, symmetry, branching ratios, coupling mechanisms, etc]. This type of spectroscopy is now an important part of electron-impact spectroscopy and has developed because of the improvements in electron scattering techniques that have come about in recent years. Many instruments now exist which can produce quasi-monochromatic electron beams and certain types are also able to analyze the scattered electrons with sufficient resolution and sensitivity to be able to identify numerous electronic, vibrational, and, sometimes, rotational excitation processes.

The following sections are not intended as a comprehensive review or catalogue of all resonant processes occurring in molecules. All this is contained in the books and reviews mentioned in the introduction. Instead resonant spectroscopy is presented from an experimental point of view. Experimental results are used to demonstrate typical resonance phenomena and we indicate how observations in specific cases can be interpreted with a view to obtaining knowledge about the resonant states and their associated processes. This account will be restricted to resonances associated with excited electronic states of molecules, i.e., core-excited resonances, as those associated with ground states are discussed in Chapter 3. This choice then further restricts us to diatomic molecules, since at the present time very little is known about core-excited resonances in polyatomic molecules.

2.1. *The Grandparent Model and Feshbach Resonances*

As stated above, Feshbach resonances are narrow, typically less than 20 meV wide, and show up as sharp structure in many cross sections lying energetically below excited states. Their narrowness implies that the resonances represent only a small localized perturbation to the total cross section and for this reason are not easily observable. These resonances are most easily detected by observing some bulk cross section such as the total, metastable production, or photon production cross section, under conditions of high sensitivity and resolution. A particularly fruitful technique for detecting small sharp features, which has revealed a multitude of resonances in both atoms and molecules, is that used by Sanche and Schulz.[1] It is basically a transmission technique in which the derivative of the electron current that has passed through a gas cell is measured. A typical spectrum obtained by these authors for NO is that already shown in Fig. 1. This spectrum displays a complex series of peaks and dips which can be divided into four vibrational progressions. The problem that we now want to discuss is how these progressions can be identified in terms of temporary negative-ion states.

The grandparent model, mentioned above, has revealed itself to be invaluable for obtaining information on the configurations of resonant states. This model was first introduced by Simpson and Fano[23] to help in the classification of the resonances observed in elastic scattering of electrons by helium and neon atoms. Sanche and Schulz[1] used the model to interpret molecular resonances, following the work of Weiss and Krauss,[23a] who noted from their calculations on a limited number of molecules that it is the Rydberg states and not the valence states that have positive electron affinities and can bind an extra electron. Thus Feshbach resonances can be considered as two Rydberg electrons bound in the field of a positive-ion core. This core can be in its ground state or in an excited state and is the grandparent state of the resonance. Adding one Rydberg electron to the core forms the parent state mentioned above and adding a second one forms the negative-ion state.

Rydberg electrons play virtually no role in the binding of the nuclei and therefore the nuclear motion of the positive-ion core in the grandparent state is almost unpertubed by the addition of the two extra electrons. Thus it was postulated that the vibrational spacings should be the same for the positive and negative ions. Indeed Sanche and Schulz(1) verified that the four series in NO shown in Fig. 1 have the same vibrational spacings as the $NO^+X^1\Sigma^+$ ground state. On the other hand when the Franck–Condon overlap factors between the vibrational wave functions of the target molecule and positive ion are used to predict the relative cross sections of the resonance vibrational levels, the agreement with the measured values is only qualitative. The reason for this will be discussed in Section 2.2, where we shall see that with a suitable modification the relative cross sections are still consistent with the grandparent model. It was noticed also that the energy with which the two electrons are bound to the core (about 4.1 eV) is essentially independent of the core state. This is illustrated clearly when the energies of the resonant structures of NO are compared to the state energies of NO^+ after shifting the resonant scale by 4.1 eV (Fig. 4). Each group of resonances corresponds to an NO^+ grandparent state. This convincingly demonstrates the validity of the grandparent model in NO.

For Feshbach resonances to exist the binding energy of the two Rydberg electrons to the core must be large enough to cause the resonance energy to lie below that of the parent state. Otherwise if the resonance is unbound with respect to the parent state it falls into the category of core-excited shape reson-

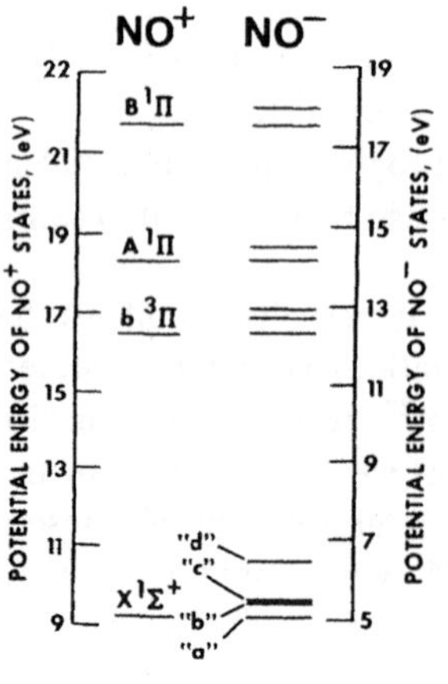

FIGURE 4
Energy-level diagram of NO^+ grandparent states compared with the energy-level diagram of NO^- resonances observed in a transmission experiment. The two energy scales have been displaced by 4.23 eV, the binding energy of the lowest member of series *a* with respect to its grandparent $X^1\Sigma^+$ state of NO^+. Each state of NO^+ gives rise to a Rydberg series of NO^- states. (Sanche and Schulz(1)).

ances: it then couples strongly to its parent and is probably too broad to be detected in transmission-type experiments. It turns out that only certain combinations of Rydberg orbitals are sufficiently bound and guidelines have been established from a survey of many atoms and molecules which indicate what the likely characteristics of the orbitals are for them to produce Feshbach resonances (see, for example, Spence[(24)]). The first of these guidelines is that the principal quantum number n must be the same for both electrons. Second, the angular momenta of the two electrons must be the same or little different. In other words, Feshbach resonances will exist for configurations such as ns^2, np^2, nd^2, and also, but with a smaller affinity with respect to the parent state, for $nsnp$ configurations. The underlying reason for this is that in order for the two outer electrons to be sufficiently bound they must screen each other from the charge of the ion core as little as possible, and this can happen only when the orbitals are strongly correlated. For instance two s electrons with the same n tend to lie on opposite sides of the molecule at approximately the same distance from the core, resulting in a low mutual screening (see, for example, Read[(25)]). However, for the electrons with different values of n the inner electron screens the core much more effectively and the system will not have a sufficiently large binding energy.

The grandparent model implies that the two Rydberg electrons $nl_1\lambda_1$ and $nl_2\lambda_2$ which are far from the core couple together to form a state with well-defined total spin S and orbital angular momentum component Λ. This state then couples to the s and λ of the core state forming the negative ion state with total spin S_t and total angular momentum component Λ_t. This is then a $\lambda\Lambda S$ coupling scheme and is analogous to the jLS scheme used for atomic resonances.[(26)]

This model can now be taken further and made more quantitative. The mean effective binding energy ϵ of a single Rydberg electron $nl\lambda$ can be expressed as

$$\epsilon(nl\lambda) = \frac{R}{(n - \delta_{nl\lambda})^2}, \tag{8}$$

where R is the Rydberg energy (13.605 eV) and $\delta_{nl\lambda}$ is the quantum defect. The quantum defect is required as a correction to the principal quantum number because a Rydberg electron does not move in a pure Coulomb field produced by a point charge, but penetrates into the core of the molecule where it experiences a different field. Thus $\delta_{nl\lambda}$ is a strong function of l because the core penetration depends strongly on the angular momentum. It depends also on λ, but more weakly, and is nearly independent of n.

When two electrons are bound in the field of a positive-ion core, as in a Feshbach resonance, we can make the assumptions (i) that each electron has the same quantum defect as if it were alone, and (ii) that the two electrons provide a

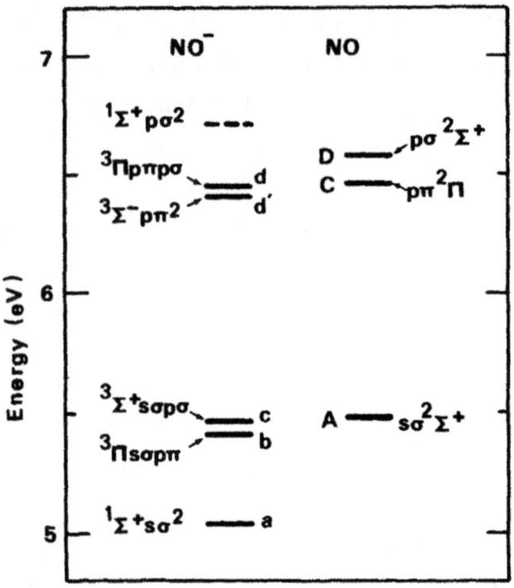

FIGURE 5
Energy-level diagram of the lowest members of the Feshbach resonance series of NO^-, indicating their configuration and symmetry, compared to the lowest member of their parent Rydberg states of NO (Gresteau *et al.*[28]).

mutual and partial screening of the ionic core. The core charge of Ze is then reduced to an effective value $(Z-S)e$, where S is a screening parameter. The total binding energy of a Feshbach resonance can then be written[27]

$$\begin{aligned}\epsilon(nl_1\lambda_1 nl_2\lambda_2) &= \frac{R(1-S)^2}{(n-\delta_{nl_1\lambda_1})^2} + \frac{R(1-S)^2}{(n-\delta_{nl_2\lambda_2})^2} \\ &= (1-S)^2\,[\epsilon(nl_1\lambda_1)+\epsilon(nl_2\lambda_2)]\,. \end{aligned} \tag{9}$$

If we again use NO as an example we can obtain the values of the screening parameter corresponding to the various combinations of molecular orbitals with $n=3$. Figure 5 shows an energy-level diagram of the $v=0$ levels of the NO^- and NO states associated with the $(1\pi)^4\ {}^1\Sigma^+$ ground state of NO^+. The resonance energies are those measured by Gresteau *et al.*[28] The negative-ion configurations are the outcome of theoretical calculations of Lefebvre-Brion[29] confirmed and amended by the further experiments of Gresteau *et al.*[30] The information shown in Fig. 5 is listed in Table 1 for convenience along with the values of S calculated using Eq. (9). These values of S are typical for such configurations, as shown by the results of Brunt *et al.*[27] who obtained the values of S for many configurations and core states in four diatomic molecules made up of second-row atoms (CO, N_2, NO, and O_2). Furthermore these values are also very similar to those obtained for the equivalent configurations in a large number of atoms (Read[25]). Thus it appears that the screening parameters depend only on l and λ of the Rydberg electrons, and are independent of the excitation energy of the core or of whether the core is an atomic or molecular ion. This behavior of S appears to indicate the general validity of the grandparent model.

TABLE 1

For the Five Feshbach Resonance Series of NO^-: the Configuration ($nl_1\lambda_1\, nl_2\lambda_2$); Energy of the Lowest Member $E(v=0)$; Symmetry of the Resonance State; Binding Energy $\epsilon(nl_1\lambda_1 nl_2\lambda_2)$ of the Feshbach Resonance to the NO^+ Ground State; Binding Energy $\epsilon(nl_1\lambda_1)$, $\epsilon(nl_2\lambda_2)$ of the Parent Rydberg State(s) to the NO^+ Ground State; the Screening Parameter S[a]

Series	Configuration $nl_1\lambda_1 nl_2\lambda_2$	$E(v=0)$ (eV)	Resonance symmetry	$\epsilon(nl_1\lambda_1 nl_2\lambda_2)$ (eV)	$\epsilon(nl_1\lambda_1)$, $\epsilon(nl_2\lambda_2)$ (eV)	Screening constant S
a	$(3s\sigma)^2$	5.04	$^1\Sigma^+$	4.22	3.81	0.26
b	$3s\sigma 3p\pi$	5.41	$^3\Pi$	3.85	3.81, 2.79	0.24
c	$3s\sigma 3p\sigma$	5.46	$^3\Sigma^+$	3.81	3.81, 2.68	0.23
d'	$(3p\pi)^2$	6.41	$^3\Sigma^-$	2.85	2.79	0.29
d	$3p\pi 3p\sigma$	6.45	$^3\Pi$	2.81	2.79, 2.68	0.28

[a]From Gresteau *et al.*[28, 30]

A Feshbach resonance which might be expected to exist in NO^- but which has not been clearly observed experimentally, is that with a $p\sigma^2\ ^1\Sigma^+$ configuration based on the NO^+ ground state. Analogous configurations are known to exist for other diatomic molecules, and Brunt *et al.*[27] find that the screening parameter is 0.31. The energy of this configuration in NO^- can be predicted using Eq. (9) and is found to occur at 6.71 eV (dashed line in Fig. 5). This resonance is therefore *above* its $p\sigma\ ^2\Sigma^+$ parent state and so belongs to the shape-resonance category. It should thus be broader than the other series, which explains why it is less visible, and it should decay preferentially to the parent state.

It is interesting to note that the binding energy for the pair of Rydberg electrons in the above-mentioned molecules is always of the order of 4 eV. This constancy was at one time almost considered to be a rule. However, when one looks at the trend of $\epsilon((ns\sigma)^2)$ values for a column of the periodic table a strong variation appears. This is illustrated in Fig. 6, which indicates the binding energies of the lowest Feshbach resonance of the molecular halogens, measured by Spence.[31] These binding energies could also be calculated using Eq. (9), but neither the Rydberg states nor their quantum defects are known. However, as suggested by Spence,[31] molecular quantum defects are very similar to those of the corresponding atoms, differing usually by less than 10%. The binding energies using the atomic quantum defects are added to the figure (crosses) and we see that they are in reasonable agreement apart from an overestimation of the value F_2. The general trend of the calculated values is correct, however, and simply demonstrates the variation of the quantum defect with n. As n increases the number of closed shells inside the Rydberg electrons increases and the penetration of the electron into the core diminishes due to orthogonality with inner-shell electrons having the same value of l. As a result δ, instead of increasing by one

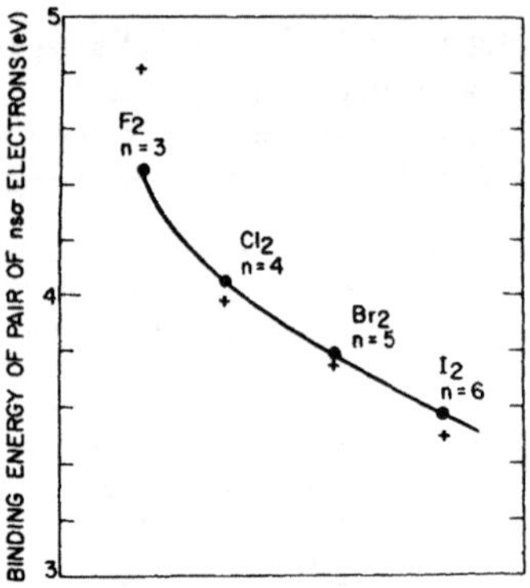

FIGURE 6

Binding energies of the lowest pair of $ns\sigma$ electrons to ground-state positive-ion cores for the molecular halogens (solid circles). The binding energy decreases monotonically by almost 1 eV as the molecular size increases from fluorine to iodine (Spence[31]). The crosses are the values of the binding energy obtained using Eq. (9).

with each value of n, increases by slightly less than this. In other words, as the molecule gets bigger more and more shells are filled, and the Rydberg electrons are forced into more distant orbitals and consequently are less bound.

Another useful method of analyzing complex Feshbach resonance structure is by isoelectronic comparison. This can be applied when the energy levels are known for the neutral or ionic species which have the same number of electrons as the resonance states. A good example of this is given by Spence and Noguchi,[32] who used the known energies and configurations of Feshbach resonances in the noble gases to help unravel the features observed in the cross sections of the isoelectronic hydrogen halides. These authors noticed that the structure observed in the transmission spectra of the hydrogen halides and the rare gases is very similar and can be divided into the same number of groups. The rare gas structure can be interpreted with the aid of the grandparent model and a configuration assigned to each group. These assignments are then carried over to the hydrogen halides thus allowing a relatively easy identification of the parent and grandparent states of all the observed structures.

An interesting result of these observations in the hydrogen halides which is in sharp contrast to the case of the molecular halogens is that the binding energy of the lowest Feshbach resonance $(ns\sigma)^2$ is constant when going from HCl to HI. This would indicate that the Rydberg $ns\sigma$ orbitals are characteristic of the hydrogen atom and are not perturbed by substitution of the halogen atom.

2.2. Coupling Mechanisms of Feshbach Resonances. I: Electronic Coupling

In this section we consider how the processes of creation and decay of a Feshbach resonance are affected by the initial and final energy of the electron and by the vibrational motion of the resonance.

Because the lifetime of a Feshbach resonance is much longer than a vibrational period, the nuclear wave function of the resonance is well represented by that of a stationary state. When the vibrational excitation energy is high, the nuclei can make excursions that are large compared to those in the initial state of the target molecule. As a result the resonance can decay outside the Franck–Condon region of the initial state to vibrational channels that are not accessible by direct excitation. The resonance process then constitutes the only contribution to the corresponding vibrational excitation cross sections. This purely resonant situation is found very frequently in molecules, as discussed in Chapter 3. Resonances can lead to excitation of many vibrational quanta whereas direct processes are limited to at most one or two. It is nearly always possible when studying a Feshbach resonance to find a decay channel where the direct contribution is small or negligible. This is illustrated by Fig. 7, obtained by Gresteau *et al.*,[28] which shows the differential cross section for excitation of the $v = 3$ level of NO. The vibrational progressions produced by the various NO^- Feshbach series discussed in the previous section clearly dominate this cross section, while the contribution from the direct excitation process is comparatively very weak, even although the vibrational level is one of low excitation energy.

The cross section for resonant excitation of a molecule in the absence of direct scattering can be represented by a Breit–Wigner formula which has the following form (Fiquet-Fayard[33]):

$$\sigma \propto \frac{1}{E_i} \frac{\gamma(E_i)|\langle\chi_0|\chi_r\rangle|^2 \gamma(E_f)|\langle\chi_r|\chi_f\rangle|^2}{(E_i - E_r)^2 + (\frac{1}{2}\Gamma)^2}, \tag{10}$$

where χ_0, χ_r, and χ_f are the initial-, resonance-, and final-state vibrational wave functions, respectively. The partial widths Γ_{in} and Γ_{out} that appear in Eq. (2) have been expressed here as a product of a Franck–Condon factor and a width $\gamma(E)$ that represents the coupling between the states. This coupling is purely electronic: it depends on the energy of the electron (initial E_i or final E_f) and is

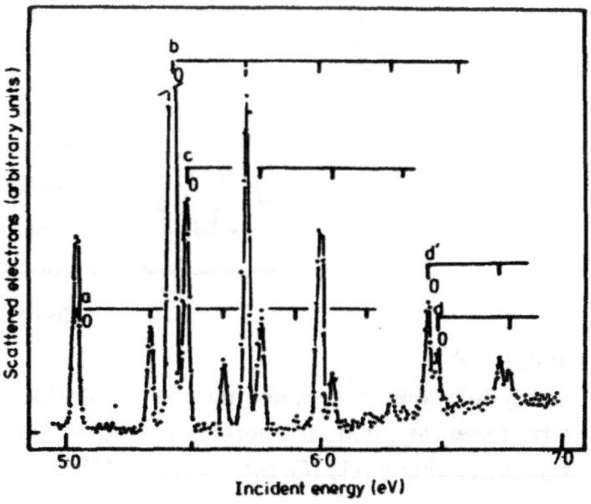

FIGURE 7
Differential excitation cross section of the $v = 3$ vibrational level of the NO ground state observed at a 140° scattering angle. The Feshbach resonance series of NO^- are indicated. (Gresteau *et al.*[28])

assumed to be independent of the internuclear distance (R). This is known as the $\Gamma(E)$ approximation, as opposed to the $\Gamma(R)$ approximation, which is more appropriate for short lifetime resonances and will be discussed in Section 2.7. The Born–Oppenheimer approximation is of course implicit in Eq. (10).

The variation of the electronic width Γ with electron energy is a characteristic of resonant scattering and depends specifically on the resonant state and the states with which it is reacting. Thus the resonant intensity observed for a particular final level depends not only on the Franck–Condon factors but also on the energy of the vibrational level, as this determines the final electron energy.

This effect is nicely illustrated by the NO^- Feshbach resonances and their decay to vibrational levels of the ground state. An example of an NO resonance spectrum is shown in Fig. 7. Up to four vibrational levels of the Feshbach series in Fig. 7 have substantial intensity. This is because the resonances are built on

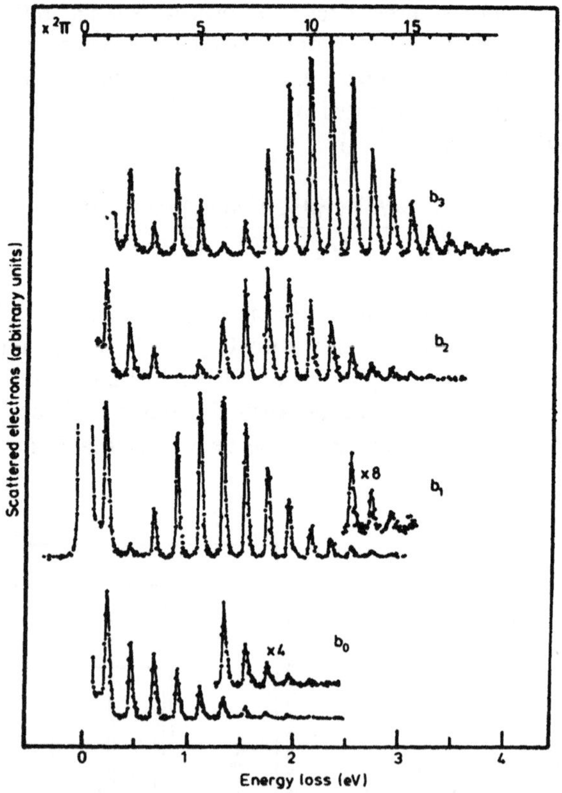

FIGURE 8
Energy-loss spectrum showing the excitation of high vibrational levels of the NO ground state, taken at incident energies corresponding to the vibrational levels of series b and at a 140° scattering angle (Gresteau *et al.*[28]).

the NO^+ ground state, which is formed from NO by the removal of an antibonding valence electron. The equilibrium internuclear distance of NO^+ is thus smaller (1.06 Å) than that of NO (1.15 Å). Hence the vibrational wave functions of the resonance levels have large overlaps with those of high vibrational levels of the ground state. In fact noticeable resonance intensities are observed for excitation of NO levels up to $v = 18$. This can be seen from Fig. 8, which presents energy-loss spectra taken at incident energies corresponding to the vibrational levels (with quantum numbers r) of series b, which is the most intense of the five series in NO. The general shape of the vibrational intensity envelopes roughly represents the Franck–Condon factors between the resonance and the ground state.

For the experimental results illustrated in Figs. 7 and 8 the instrumental resolution is approximately 20 meV, and is therefore broad compared to the resonance widths, which are estimated to be less than 10 meV. Thus when $E_i = E_r$ the resonance is formed over its entire profile. In this case the expression for the observed vibrational intensities in the spectra is obtained by integrating σ [Eq. (10)] over the energy E_i, which gives an expression of the following form (Parlant and Fiquet-Fayard[34]):

$$I \propto \frac{\gamma(E_r)|\langle\chi_0|\chi_r\rangle|^2 \gamma(E_f)|\langle\chi_r|\chi_f\rangle|^2}{E_r\Gamma_r} . \tag{11}$$

Now if we again assume that the grandparent model is valid then the potential curve of the NO^- resonance is the same as that of NO^+ (except for a vertical shift) and all of the Franck–Condon factors in Eq. (11) can be determined. Gresteau *et al.*[28] show that the variation of the electronic coupling γ with energy can then be deduced from the above equation by using the observed relative vibrational intensities for the different resonance levels. The result is shown in Fig. 9. This $\gamma(E)$ function is shown to be the same for all the resonance levels and depends only on the electron energy (initial or final). This observation thus validates the assumption implicit in Eq. (10) that γ is purely

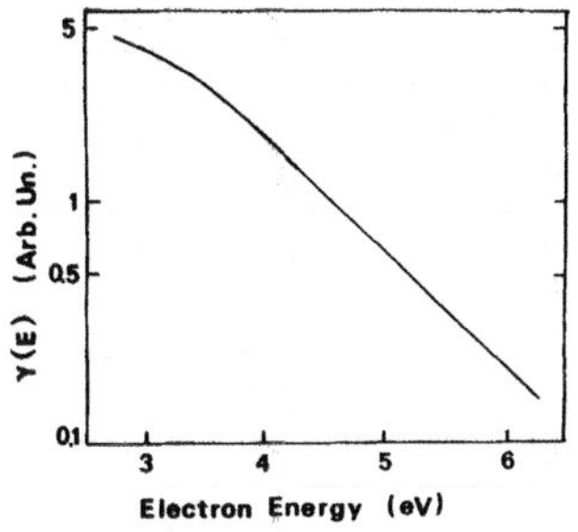

FIGURE 9
Relative variation of the electronic coupling with electron energy for the Feshbach resonance series b of NO^- on a semilog scale (Gresteau *et al.*[28]).

electronic and depends only on the electron energy. The electronic coupling is in fact a strong function of electron energy in the case in point, as Fig. 9 indicates. When the electron energy doubles (3 to 6 eV) the electronic coupling changes by a factor of 20.

In the above analysis one can question the validity of the grandparent model and the use of the NO^+ potential curve for the NO^- resonance state. As it turns out, if we make the reasonable assumption that $\gamma(E)$ is smoothly varying, then strong constraints are placed on the Franck–Condon factors, particularly near the minima in the intensity envelopes of Fig. 8. The results of the analysis indicate that the internuclear equilibrium separation of NO^- is the same as that of NO^+ to within 1%, which is a very strong argument in favor of the grandparent model for this molecule and resonant state.

When using this model to compare resonance level intensities, as seen, for example, in a transmission experiment, with Franck–Condon factors to the negative ion state, the variation of $\gamma(E)$ must be taken into account. The formation intensity I_r is obtained[28] from Eq. (11) by summing over all the final states, giving

$$I_r \propto \frac{\gamma(E_r)|\langle\chi_0|\chi_r\rangle|^2}{E_r} \,. \tag{12}$$

Table 2 shows the large variation of $\gamma(E)$ over the energy range of the series b levels taken from the curve of Fig. 9, together with the calculated values of I_r. The latter values agree well with the observed intensities of Sanche and Schulz,[1] whereas the Franck–Condon factors are very different.

Also listed in Table 2 are the relative values of the total widths Γ_r obtained by Gresteau *et al.*[28] The variation of Γ_r with vibrational energy is much smaller than for $\gamma(E)$ but is dependent on it. If $\Gamma(E)$ were to be energy independent then Γ_r would also be invariant. In fact Γ_r increases with resonant level r, which

TABLE 2

For Each of the Four Vibrational Levels (r) of the Feshbach Resonance Series b of NO^-, (i) Energy (E_r), (ii) the Electronic Coupling $\gamma(E)$ Taken from Fig. 9, (iii) the Franck–Condon Factors $|\langle X_0|X_r\rangle|^2$, (iv) the Resonance Intensities, Calculated and Measured (Sanche and Schulz[1]), and (v) the Total Width Γ_r[a]

Resonance level (r)	E_r (eV)	$\gamma(E)$ (eV)	$\lvert\langle\chi_0\vert\chi_r\rangle\rvert^2$	I_r calculated	I_r Sanche and Schulz	Γ_r (eV)
0	5.41	1.38	0.48	0.70	0.57	0.92
1	5.70	1	1	1	1	1
2	5.99	0.74	0.92	0.65	0.64	1.08
3	6.29	0.57	0.48	0.25	0.24	1.16

[a]Gresteau *et al.*[28]

can be understood in the following way. Γ_r is the sum of the partial widths to the various vibrational levels of the ground state, and each of these partial widths has the form $\gamma(E_f)|\langle\chi_r|\chi_f\rangle|^2$. As r increases, the partial widths to higher levels increase not only because the Franck–Condon factors are more favorable but also because $\gamma(E_f)$ increases as the final electron energy (E_f) diminishes. Thus the summed partial widths for higher values of r represent a stronger coupling to the ground electronic state and so Γ_r is larger.

No other Feshbach series in NO, or in any other molecule, has been studied in this detailed way, but an appraisal of the relative intensities of series a and d of NO^- measured by Sanche and Schulz[(1)] reveals strong deviations from the Franck–Condon factors to the NO^+ ground state, with the differences having the same trend as for series b. This would then imply, if the grandparent model holds for these states, that $\gamma(E)$ also decreases as the resonance level energy increases, although the exact behavior is not necessarily the same as that of series b.

2.3. Coupling Mechanisms of Feshbach Resonances. II: Dynamic Coupling

The above information on the electronic coupling was obtained by observing the decay of an NO^- Feshbach resonance series to the ground electronic state of NO. Continuing to use NO as an example, we can now look at this decay to the Rydberg states of NO. This reveals another type of process, having its origins in a breakdown of the Born–Oppenheimer approximation due to dynamic coupling between nuclear and electronic motion.

As demonstrated above, the grandparent model is valid for NO^-, which implies that the five Feshbach series (see Fig. 5) have identical potential curves except for vertical shifts in energy, as do the associated three Rydberg parent states. If we assume that the Born–Oppenheimer approximation holds and that the coupling is purely electronic then the only allowed transitions between the resonance and Rydberg states are those in which the vibrational quantum number (v) does not change ($\Delta v = 0$). All other transitions, with $\Delta v \neq 0$, would have a zero intensity because the nuclear wave functions are orthogonal.

In fact transitions with $\Delta v = 1$ are also observed, as can be seen in the cross sections for vibrational excitation of the $A\,^2\Sigma^+$ Rydberg state (Fig. 10), obtained by Gresteau *et al.*[(30)] The resonant structure corresponds to $\Delta v = 0$ transitions from series d together with $\Delta v = 1$ transitions from series b and c. $\Delta v = 0$ transitions are not possible for the latter two series as they lie below their analogous Rydberg level. All the transitions that have been unambiguously observed are displayed in Fig. 11. Series a is not shown as both the $v = 0$ and $v = 1$ levels are below the lowest vibrational level of the $A\,^2\Sigma^+$ Rydberg state (Fig. 5).

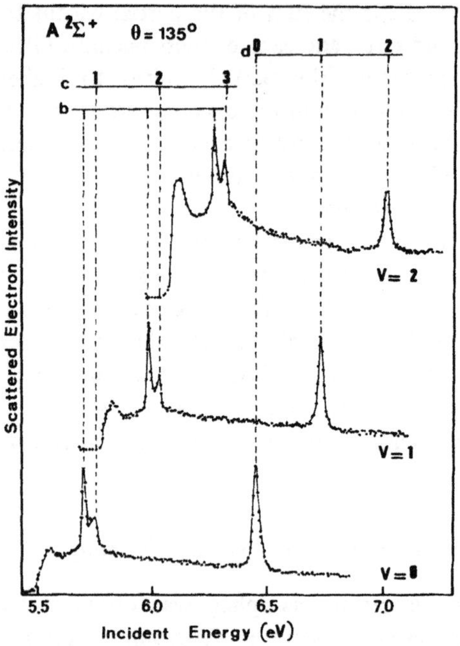

FIGURE 10
Differential excitation cross sections of the $v = 0$, 1, and 2 vibrational levels of the $A\ ^2\Sigma^+$ state of NO, observed at 135°. The peaks correspond to transitions from the vibrational levels of the indicated Feshbach resonance series of NO^-. (Gresteau *et al.*[30])

This situation is analogous to that occurring in autoionization transitions, which also proceed through an interaction between a quasibound state and a continuum. There $\Delta v = 1$ transitions are observed under similar circumstances. This phenomenon can be explained in terms of the nuclear kinetic energy operator T_n which is neglected in the Born–Oppenheimer approximation. The term which leads to the observed effects can be written as

$$M = \left\langle \chi_r \left| \frac{d}{dR} \right| \chi_f \right\rangle \left\langle \phi_r \left| \frac{d}{dR} \right| \phi(\text{Ryd} + e) \right\rangle , \tag{13}$$

where ϕ_r is the electronic wave function of the resonant state and $\phi(\text{Ryd} + e)$ that of the Rydberg state and an outgoing electron. R is the internuclear separation. The first (nuclear) part of M is nonzero for $v = \pm 1$, which thus accounts for the existence of these transitions. However, the second part of M must also be nonzero for the transition to be observed. Now the effect of the operator d/dR is such[28] that this part is nonzero only for transitions where a change of one orbital occurs. Thus the resonance can couple only to a state with which it has all its orbitals in common except one, i.e., to a parent state.

We are now in a position to discuss the reason why the transitions shown in Fig. 11 are observed. Both series b and c couple to their A-state parent through the radial coupling mechanism by ejecting a $p\sigma$ and $p\pi$ electron, respectively. In

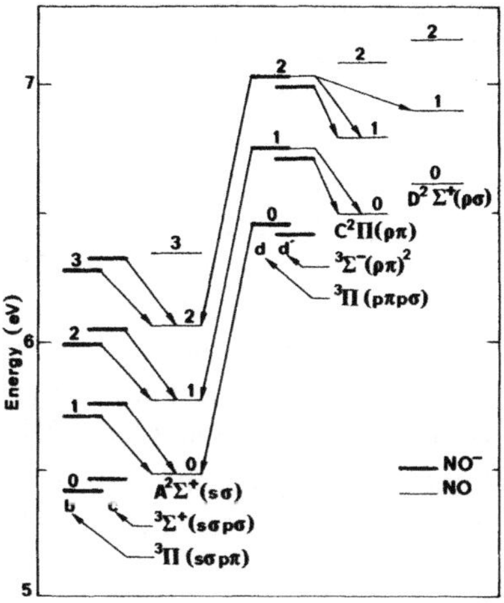

FIGURE 11
Energy-level diagram of the vibrational levels of NO and NO^- states, showing the experimentally observed transitions (Gresteau *et al.*[30]).

the same way, series d' decays to its C-state parent, while series d is seen in the cross sections of both the C and D states as it has a $p\sigma$ or $p\pi$ orbital in common with one or the other of these parent states. Series d decays into the A state also, but with $\Delta v = 0$ only; the absence of $\Delta v = 1$ transitions indicates that the resonant state has no Rydberg orbital in common with the A state.

It is important to notice also that series d' is not observed in the A-state cross sections, although $\Delta v = 0$ and 1 transitions are both energetically allowed. The reason for this is the existence of a selection rule arising from the conservation, during the collision, of the reflection symmetry of the electronic wave function in a plane containing the internuclear axis. The rule is strict[35] and applies when the resonant and final molecular states both have Σ symmetry, and is expressed as

$$\Sigma^+ \rightarrow \Sigma^+, \qquad \Sigma^- \rightarrow \Sigma^-, \qquad \Sigma^+ \not\leftrightarrow \Sigma^-. \tag{14}$$

The complete absence of series d in the $A\ ^2\Sigma^+$ channel indicates therefore that this resonance series has Σ^- symmetry.

We see that all the above-mentioned rules have been helpful in determining or confirming the shown symmetries and configurations. Thus we have demonstrated another technique of resonance spectroscopy which can be employed when the circumstances allow, although for the time being this is the only example.

2.4. *Potential Curves of Feshbach Resonances in* H_2

In the previous sections we have assumed that the potential curves of the resonances can be established by using the grandparent model. We have seen that this is a good procedure for NO where the Rydberg orbitals have 14 valence electrons binding the nuclei in the NO^+ core. However, this is not always the case. In particular the model is inapplicable for H_2. The reason for this is that the H_2^+ binding is due to only one electron and so the addition of one or two further electrons to form Rydberg states and Feshbach resonances is sufficient to change the internuclear separation. Much complex structure resulting from Feshbach resonances has been observed in many channels of this molecule, and information is available on the potential curves of the lowest states.

In the same way as for NO^-, the H_2^- Feshbach resonances have been observed in the vibrational channels of the ground state: a sample of the spectra recorded by Comer and Read[36] is shown in Fig. 12. The vibrational wave functions of the H_2 ground state are known and they can be used together with the observed intensities and Eq. (11) to determine the potential curves of the

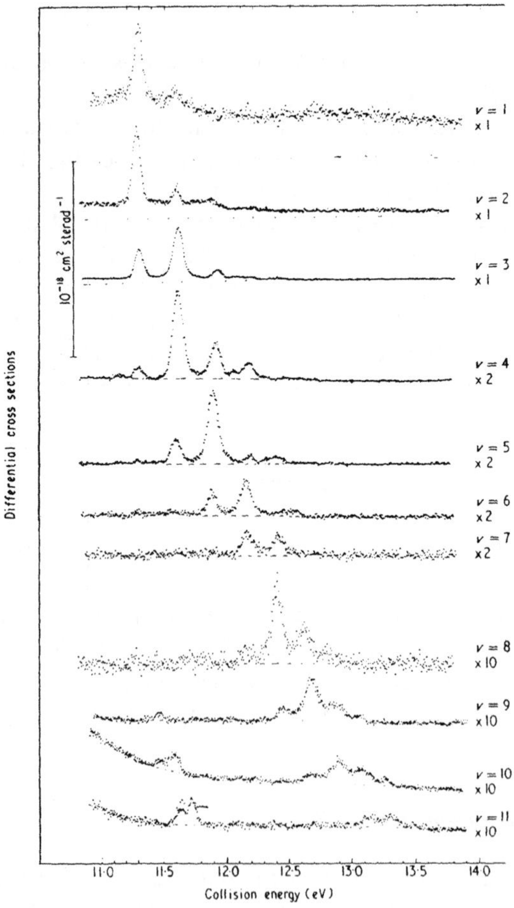

FIGURE 12
Energy dependence of the absolute differential cross sections for vibrational excitation of the H_2 ground electronic state observed at a scattering angle of 85°. For each curve the cross-section scale should be multiplied by the factor indicated. The zero lines are shown broken (Comer and Read[36]).

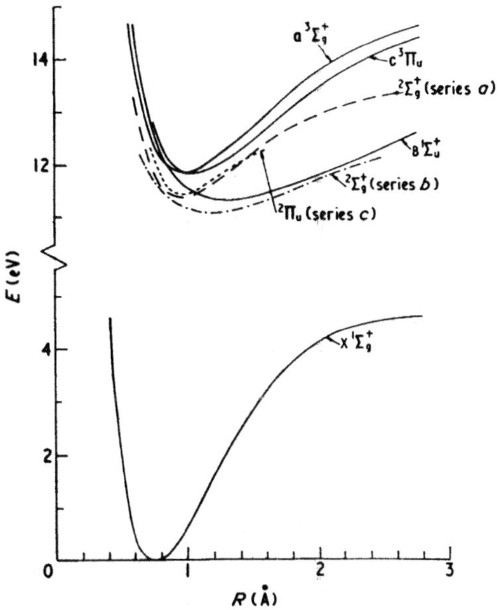

FIGURE 13

Potential energy curves for the Feshbach resonances of H_2^- (broken curves) and some corresponding states of H_2 (full curves). (Comer and Read[36]).

resonances. But before this can be done we have to face the problem that the function $\gamma(E)$ is not known. An obvious choice is to assume that $\gamma(E)$ is constant. This is probably a reasonable assumption when one considers that the Feshbach series lie over 11 eV above the ground state. Potential curves obtained making this assumption for three resonance series are shown in Fig. 13 (Comer and Read[36]). In a further study, Joyez *et al.*[35] found the lowest member of series *c* to be at 11.19 eV, which implies that the curve of Fig. 13 should be shifted down slightly. These three series are built on the H_2^+ core and, as can be seen, have potential curves which are quite different from each other (and different also from that of H_2^+). This illustrates the point, made above, that in the particular case of H_2 the Rydberg orbitals, at least the lowest ones, influence the binding of the molecule.

The symmetries of the three series are also indicated in Fig. 13. These symmetries were obtained and confirmed by two other techniques of resonance spectroscopy. The first consists of the observation of the angular behavior of resonantly scattered electrons, and the second the observation of rotational cross sections. These methods are described in the following two sections.

2.5. Resonance Symmetry and Differential Cross Sections

A resonance can be represented as a temporary negative ion with well-defined configuration and symmetry. The resonance symmetry acts as a type of angular momentum filter determining which partial waves contained in the incident electron beam can combine with the target molecule to form the resonant state. Let us consider the case of diatomic molecules and take the internuclear axis as the direction of quantization for the incident partial waves. The amplitude of a particular wave in the molecular frame then depends on the orientation of the molecular axis with respect to the incident electron beam. Thus the probability of forming the resonance for a partial wave compatible with the resonance symmetry is determined by the molecular orientation. Similarly the decay probability for ejection of the electron into a particular scattering angle is also dependent on the resonance symmetry and the orientation of the molecule with respect to this scattering angle. Thus the angular behavior of resonantly scattered electrons is characteristic of the resonant-state symmetry, and its observation should help to indicate this symmetry. In fact the resonant state is well characterized by the angular distribution only when one partial wave contributes to the process. Otherwise, if several partial waves are present, different resonant symmetries can often yield similar differential cross sections which experimental measurements and their inherent uncertainties are unable to separate. However, in practice one partial wave often makes a dominant contribution to the process and it is possible to interpret the observations in terms of resonant symmetries, assuming of course that the resonance is observed in a channel free from direct processes.

Consider a typical experiment on a diatomic molecule (M) where the rotational structure is not resolved and the resonant scattering is assumed to be independent of spin, i.e., the reaction

$$\underset{(l_1 m_1)}{e} + \underset{(l_i m_i)}{\mathrm{M}} \rightarrow \underset{(l_r m_r)}{\mathrm{M}^{-*}} \rightarrow \underset{(l_f m_f)}{\mathrm{M}^*} + \underset{(l_2 m_2)}{e}, \tag{15}$$

where the l's are the angular momentum quantum numbers and the m's their projection onto the internuclear axis. Assuming that the ion M^{-*} does not rotate during the molecular lifetime (this is referred to as the impulse approximation) then the differential cross section can be expressed as[37,38]

$$d\sigma(\theta)/d\Omega \sim \sum_k A_k P_k(\cos\theta), \tag{16}$$

where P_k is the Legendre polynomial of order k, and A_k is a product of Wigner $3j$ coefficients. Of interest are the selection rules which govern the resonant reaction. First, there is the obvious rule that the projected angular momentum must be preserved throughout the collision:

$$m_1 + m_i = m_r = m_f + m_2. \tag{17}$$

There is also a selection rule

$$\mathbf{l}_1 + \mathbf{l}_i = \mathbf{l}_r = \mathbf{l}_f + \mathbf{l}_2. \tag{18}$$

Conservation of parity imposes further relationships in the case of homonuclear diatomics. The values of l_1 must be even for g(initial)→g(resonance) and u→u transitions, and odd for g→u and u→g transitions. In both cases the angular behavior is symmetric with respect to the 90° scattering angle.

If we stipulate that there is only one partial wave of appreciable amplitude in the initial and final channels then further two selection rules are given by the above theory.[37, 38] First, k is an even number only, which implies that the angular behavior is symmetric with respect to 90°, even for heteronuclear diatomics. Second, the maximum value of k is given by

$$k_{\max} = \min(2l_1, 2l_2). \tag{19}$$

There are two reasons why only one partial wave is expected to participate in resonant scattering. First, at low energies where most resonant phenomena occur, the amplitude of the various spherical harmonics in the partial-wave expansion of the incident electron wave decreases rapidly with l. Second, in the case of shape resonances where the binding mechanism of the resonance electron depends on the value of l and its associated centrifugal barrier, higher values of l lead to wider barriers thus making penetration less probable. In the case of Feshbach resonances, where Rydberg orbitals characterize the negative-ion state, it can be hoped that these orbitals are similar to atomic orbitals and that they can be adequately represented by a single spherical harmonic, which automatically determines a single spherical harmonic or partial wave in the entrance channel.

It often occurs, as can be seen above, that the resonances are easily observed in the vibrational-level cross sections of the ground state. In such a situation the initial and final states are the same ($m_i = m_f$) and the expression for A_k in Eq. (16) takes the simple form

$$A_k(l, m) = (2k+1)\begin{pmatrix} l & l & k \\ 0 & 0 & 0 \end{pmatrix}^2 \left[\begin{pmatrix} l & l & k \\ m & -m & 0 \end{pmatrix}^2 + p\begin{pmatrix} l & l & k \\ m & m & -2m \end{pmatrix}^2\right], \tag{20}$$

where $l_1 = l_2 = l$ and $m_r = m$, and p takes the values 0.5 for $m = 0$ and 1.0 for $m \neq 0$. The theory of Chang[39] leads to expressions for the differential cross sections similar to those obtained with Eqs. (16) and (20) for homonuclear diatomic molecules, but in the case of heteronuclear molecules the expressions of Chang are somewhat simpler.

Let us see how these equations can be applied to the three H_2^- Feshbach resonances discussed in the previous section. For example, a resonance series of symmetry Π_u would be formed from the ground state of ${}^1\Sigma_g^+$ symmetry by the addition of partial waves having odd l and $m = 1$, i.e., partial waves such as (1, 1), (3, 1), (5, 1), etc. If we take the lowest one then Eqs. (16) and (20) give a differential cross section of the form

$$d\sigma(\theta)/d\Omega \propto 1 + 7\cos^2\theta. \tag{21}$$

This angular behavior is shown in Fig. 14, where it is compared with the observed angular distribution of series c and with the theoretical curves for some other resonant symmetries formed by their lowest allowed partial wave. This comparison led Comer and Read(36) to assign Π_u symmetry to series c, on the assumption that the finite lifetime of the resonances does not alter the angular behavior (21) appreciably. Chang(40) has argued that series c has the symmetry ${}^2\Delta_g$, but recent calculations of the potential energy curves of H_2^- (Buckley and Bottcher(41) and Bardsley and Cohen(42)) do not reveal the presence of such a state in the appropriate energy range. Studies of the rotational structure (see the next section) also support Π_u symmetry for series c.

Observations of angular dependences are particularly useful for the study of shape resonances with ground-state parents, where the lifetimes are generally shorter than the rotational period and where the above-mentioned theory is suitable. However, long-lived resonances, particularly of the Feshbach type, can feasibly live long enough for the molecule to rotate, in which case the theories of Bottcher and Read(43) or Chang(39) are appropriate for interpreting the observations.

At the present time the energy resolution of electron scattering experiments is generally too low to clearly separate rotational structure either in the energy-loss process or in the resonant state. But techniques are steadily improving and when rotational processes are resolved new information will be available on the resonant states. This has already been demonstrated for H_2, which is the

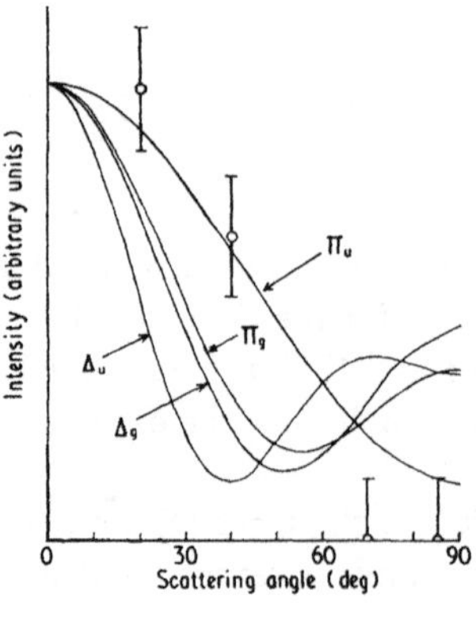

FIGURE 14
Angular distribution for H_2^- series c resonant peaks observed in the ground electronic state $v = 7$ exit channel. The curves are the calculated angular distribution for various symmetries of the resonant state, normalized at 0°. (Comer and Read.(36))

one, and for the time being only, exception to the above statement as far as Feshbach resonances are concerned. Here rotational levels of the ground state are sufficiently well separated for individual rotational channels to be studied. Furthermore, rotational levels of some Feshbach resonances have been detected, which has given further insight into the resonant scattering process, as will be discussed in the next section.

2.6. Resonance Symmetry and Rotational Structure

In the particular case of H_2, cross sections for excitation to various rotational levels of the final molecule can be observed in electron-impact experiments. Also, the lifetimes of the Feshbach series *a* and *c* are sufficiently long for them to have discrete rotational structure. In this case there exist selection rules governing the appearance of resonance rotational levels in the rotational cross sections, thus giving assistance in the identification of the resonant state symmetry.

Let the initial and final rotational quantum numbers of H_2 be $\mathbf{J}_i$ and $\mathbf{J}_f$ and the angular momenta of the initial and final partial waves be $\mathbf{l}_1$ and $\mathbf{l}_2$. If we assume that the electronic symmetry of the molecule does not change during the collision then we must have

$$\mathbf{J}_i + \mathbf{l}_i \to \mathbf{N}_r \to \mathbf{J}_f + \mathbf{l}_f, \tag{22}$$

where $\mathbf{N}_r$ is the total angular momentum of the resonant state, excluding the electron spin.

Nuclear spin statistics and their effect on the scattering also have to be considered.[16] Hydrogen molecules in the ground state ${}^1\Sigma_g^+$ with parallel nuclear spins (orthohydrogen) have only odd values of J_i and those with antiparallel spins (parahydrogen) have even values of J_i. As the nuclear spins are unaffected by the electron collision, then $\Delta J = J_f - J_i$ can only be even for the two molecular species.

We must also consider the selection rule (14) mentioned in Section 2.3 which results from the conservation of reflection symmetry of the electronic wave functions. In the case of H_2, where the initial state has Σ^+ symmetry, this implies that Σ^- resonance states cannot be formed (i.e., the Σ^- rotational sequences associated with Λ doubling in the states with $m \neq 0$).

From the above considerations it can be shown[35] that

$$\begin{aligned} &N_r - J_i \text{ is even for } \Sigma_g \to \Sigma_g, \Sigma_u \to \Sigma_u, \\ &N_r - J_i \text{ is odd for } \Sigma_g \leftrightarrow \Sigma_u \end{aligned} \tag{23}$$

TABLE 3
Values of N_r Which Could Be Observed in the Ground-State Vibrational Cross Sections of H_2 for Different Symmetries of the Resonant State[a]

Transition $J_i \rightarrow J_f$	$^2\Sigma_g^+$ s waves only	$^2\Sigma_g^+$ s and d waves	$^2\Pi_g$ d waves	$^2\Delta_g$ d waves	$^2\Sigma_u^+$ p waves	$^2\Pi_u$ p waves
	Symmetry of resonant state					
$1 \rightarrow 3$	–	1,3	1,2,3	2,3	2	2
$3 \rightarrow 5$	–	3,5	3,4,5	3,4,5	4	4
$1 \rightarrow 5$	–	3	3	3	–	–
$2 \rightarrow 0$	–	0,2	2	2	1	1
$0 \rightarrow 0$	0	0,2	2	2	1	1
$1 \rightarrow 1$	1	1,3	1,2,3	2,3	0,2	1,2
$2 \rightarrow 2$	2	0,2,4	1,2,3,4	2,3,4	1,3	1,2,3
$3 \rightarrow 3$	3	1,3,5	1,2,3,4,5	2,3,4,5	2,4	2,3,4

[a]Joyez *et al.*[35]

transitions between the initial and resonant states. These relations apply to all homonuclear diatomic molecules. The values which N_r can take for various resonant symmetries and for different rotational transitions $J_i \rightarrow J_f$ have been summarized by Joyez *et al.*[35] and are shown in Table 3. This table can now be used to interpret the experimental observations. For example, Fig. 15 shows excitation functions obtained by Joyez *et al.*[35] of various rotational transitions, all having $\Delta v = 1$, in the neighborhood of the $v = 0$ level of resonance series *a*. A distinct double structure can be seen in the excitation functions of the $1 \rightarrow 3$ and $3 \rightarrow 5$ rotational transitions which, when compared to Table 3, indicates that the resonant state can only be Σ_g^+ and that the $l = 2$ partial waves also contribute to the cross section. These authors were able to estimate that the $l = 2$ amplitude is some 4 or 5 times bigger than that of $l = 0$. Thus we are in an unusual situation in which the first allowed partial wave is weaker than the second one. Nevertheless a single partial wave dominates the scattering for reasons related to the form of the resonance orbitals as discussed briefly in the previous subsection.

Joyez *et al.*[35] further extended the analysis of their data in Fig. 15. They noted that the double dip in Fig. 15c is well defined and the minima are separated by 77 meV. Assuming that the dips correspond to the $N_r = 3$ and 5 levels given by Table 3, they deduced the pure vibronic energy of the level E_r and the rotational constant B_r of the resonance, using the relationship

$$E_i = E_r + B_r N_r(N_r + 1) - B_i J_i(J_i + 1), \tag{24}$$

where the ground-state constant B_i is known. The energy location of the rotational structures in the other channels can now be calculated and the resulting

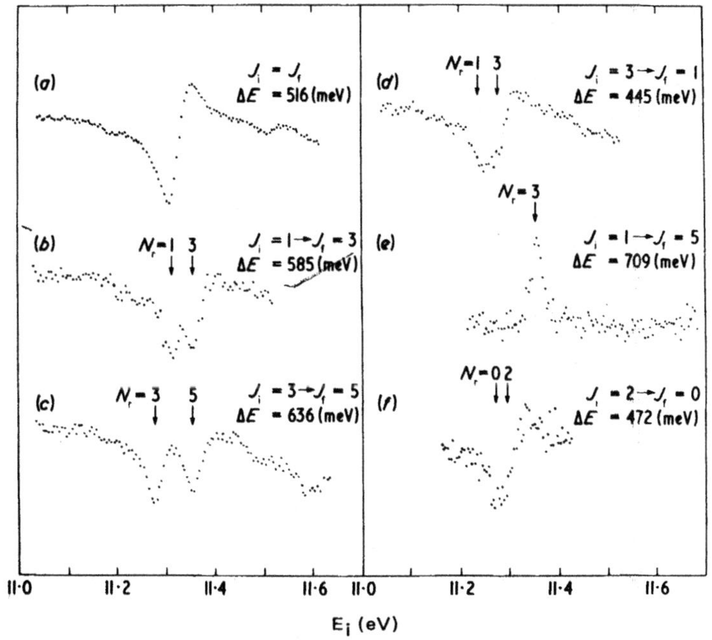

FIGURE 15
Excitation functions for vibrational excitation $\Delta v = 1$ of the ground electronic state of H_2, and for the rotational transitions shown. The scattering angle is 25°. The arrows show the calculated values of N_r (see text). (Joyez *et al.*[35])

energies are indicated by the arrows in the figure. As is clearly demonstrated, the agreement with the observed energies is excellent. The value of B_r thus calculated can be used to deduce the equilibrium internuclear separation, and a value of 0.98 ± 0.02 Å is obtained, which compares favorably with the value of 0.97 ± 0.01 Å obtained by Comer and Read[36] using the technique described in Section 2.4. Other measurements made by Joyez *et al.*[35] indicate that series *c* has only one rotational peak in the channel $J_i = 1 \rightarrow J_f = 3$, which implies (see Table 3) that the symmetry of this series can be $^2\Pi_u$ but cannot be $^2\Delta_g$.

It should be pointed out here that the rotational structure is observed in the presence of direct scattering, which accounts for the variety of resonance profiles shown in Fig. 15. This then illustrates the importance of this type of analysis. The information about the potential curve and symmetry of the resonance can be obtained even in the presence of a large direct signal. Consequently, as soon as the observation of rotational cross sections becomes technically possible in heavier molecules, a wealth of knowledge will immediately become available about the spectroscopic characteristics of long-lived resonances.

2.7. Core-Excited Shape Resonances. I: Valence States

Much less is known about this kind of resonance than about either Feshbach resonances or shape resonances with a ground-state parent. However, there is one case in N_2 where a core-excited shape resonance with $(\pi_u^3\sigma_g^2\pi_g^2)\,^2\Pi_u$ symmetry, whose parent is the lowest valence state $(\pi_u^3\sigma_g^2\pi_g)A\,^3\Sigma_u^+$, is particularly well known and its associated resonant mechanism well understood. This is because the resonance belongs to the special class of resonances which have intermediate lifetimes (of the same order as the vibrational period), and so the experimental observations are amenable to analysis by means of the boomerang model of Birtwistle and Herzenberg(44) (see Chapter 3, Sections 5.2 and 5.4, and Chapter 7, Section 4.3). In this model the resonance is represented by a complex potential whose width depends only on the internuclear distance R. This is referred to as the $\Gamma(R)$ approximation. Both the potential curve and the width of the core-excited shape resonance of N_2 can be extracted from the boomerang model if a reasonable assumption is made about the R dependence of the width.

Intermediate-lifetime resonances are characterized by the production of oscillatory structures in the cross sections for vibrational excitation. The energies of the structures depend on the particular vibrational level being excited. This is illustrated in Fig. 16 (Mazeau *et al.*(45)), which shows the weak oscillations present in the cross sections when the $A\,^2\Pi_u$ state of N_2^-, formed from the $v=0$ level of the N_2 ground state, decays to vibrational levels of its parent state $A\,^3\Sigma_u^+$. The resonant contribution is mainly a broad hump and the secondary oscillations represent only a small part of the cross section, yet the position and amplitude of these oscillations are essential in determining the molecular constants of the resonance.

The vibrational excitation cross section is given in the boomerang model(44) by

$$\sigma_v = \frac{64\alpha\pi^5 m^2}{\hbar^4}\cdot\frac{k_f}{k_i}\left|\int \chi_f^*(R)\,\zeta(R)\,\xi(R)\,dR\right|^2, \tag{25}$$

where $\hbar k_f$ and $\hbar k_i$ are the final and initial electron momenta, respectively, m is the electron mass, $\chi_f(R)$ is the nuclear wave function of the final state, $\xi(R)$ is the nuclear wave function of the resonant state, and $\zeta(R)$ is the electronic matrix element between the resonant state and the final continuum state (consisting of the final molecular state and a free electron). When the initial state is the $v=0$ level of the ground state with nuclear wave function $\chi_0(R)$, and electronic coupling to the resonant state is $\phi(R)$, then $\xi(R)$ satisfies the equation

$$\left[-\frac{\hbar^2}{2m}\frac{d^2}{dR^2} + V^-(R) - \tfrac{1}{2}\,i\Gamma(R) - E\right]\xi(R) = \phi(R)\chi_0(R), \tag{26}$$

where E is the total energy of the system. The electronic energy of the resonance is represented here by a complex energy with a real part $V^-(R)$ and a width $\Gamma(R)$, both of which depend on the internuclear distance. The left-hand side represents the motion of the nuclei with a reduced mass M in the potential $V^-(R)$, damped by the width $\Gamma(R)$. The right-hand side is the feeding term from the initial state.

This shape resonance in N_2 has two decay paths open to it, one back to the ground state $X\,{}^1\Sigma_g^+$ and the other to the $A\,{}^3\Sigma_u^+$ state. $\Gamma(R)$ is then the sum of the two partial widths

$$\Gamma(R) = \Gamma_A(R) + \Gamma_X(R). \tag{27}$$

Since this type of resonance is strongly coupled to its parent state, $\Gamma_X(R)$ can be neglected and $\Gamma(R)$ set equal to $\Gamma_A(R)$ in Eq. (27). We also have the relation[44]

$$[\zeta(R)]^2 = \frac{\hbar^2}{8\pi^2 m}\frac{\Gamma_A(R)}{k(R)} \tag{28}$$

with

$$\frac{\hbar^2}{2m}[k(R)]^2 = V^-(R) - V_A(R), \tag{29}$$

where $V_A(R)$ is the known potential energy curve of the $A\,{}^3\Sigma_u^+$ state.

There is an important unknown quantity in Eq. (26) which must be determined before we can interpret the experimental data: that is the form of $\Gamma(R)$. This can be expressed as an analytic function representing the probability for the electron to penetrate the potential barrier. The height, and hence the penetrability, of the barrier depend on the l value of the scattered electron. This penetration probability depends also on k, thus relating the variation of Γ with R to the variation of Γ with final electron energy. As k tends to zero the effective barrier width gets larger, and so then Γ tends to zero. The form of this function is given by Huetz *et al.*[46]

Equations (25) and (26) can now be used to calculate σ, and by comparing the result with the experimental curves of Fig. 16, $V^-(R)$ and $\Gamma(R)$ can be obtained. The resulting forms[46] of $V^-(R)$ and $\Gamma(R)$ are shown in Fig. 17b: they result in an excellent fit to the experimental data. $V^-(R)$ has its minimum at a larger value of R than its parent A state because of the antibonding nature of the π_g orbital in which the resonant electron is trapped. The $V^-(R)$ curve crosses its parent state to go below it at about 3.1 a.u.: thus the resonance is stable against autodetachment to its parent during part of the nuclear trajectory. In this region the resonance then becomes of the Feshbach type. The width

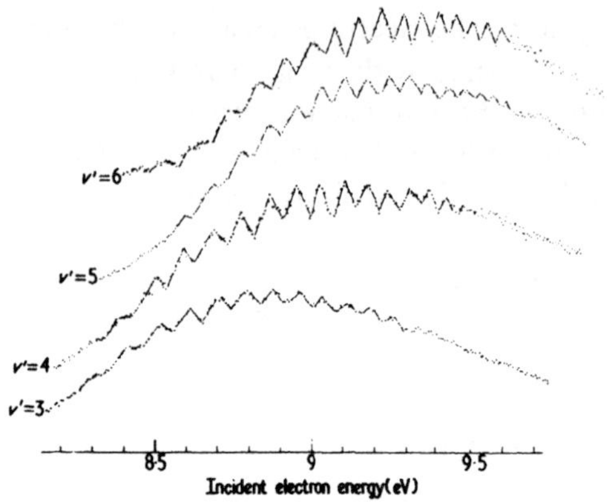

FIGURE 16
Differential cross sections for excitation of the $v = 3$, 4, 5, and 6 vibrational levels of the $A\ ^3\Sigma_u^+$ state of N_2, for a 90° scattering angle. Note the shift in peak positions with vibrational exit channel. (Mazeau *et al.*[45]).

$\Gamma_A(R)$ has a maximum value of 1.1 eV in the Franck–Condon region of the $v = 0$ ground-state level where the $V^-(R)$ and $V_A(R)$ curve separation is at its greatest.

The maximum value of Γ_A corresponds to a lifetime of 5×10^{-15} s, which is some 60 times shorter than the vibrational period of the N_2^- state. Thus the nuclear wave function in the resonant state is very strongly damped. Calculated moduli of the nuclear wave function are shown in Fig. 17a. We see that these wave functions are strongly concentrated where they are formed from the ground state and the main peak has the same form as the ground-state $v = 0$ wave function. Very little of the N_2^- intensity survives to be reflected back from the opposite wall of the potential well (see the lower two functions of Fig. 17a). The main peak in the wave function varies in magnitude only slowly with energy, and so produces the broad peak in the A-state cross sections, while the oscillatory part is strongly energy dependent and so leads to the secondary structure of Fig. 16.

This state of N_2^- also decays to vibrational levels of the ground state up to at least $v = 25$, but with an intensity down by two orders of magnitude compared to the A state. This fact justifies replacing Γ by Γ_A in Eq. (26). Figure 18 shows excitation functions of three high vibrational levels of the ground state. The structure resulting from the $^2\Pi_u$ resonance is the high-frequency part converging to the N_2^- dissociation limit at 9.83 eV (the lower-frequency oscillations above about 9.5 eV are caused by another state of N_2^- dissociating to a higher limit). Here the $A\ ^2\Pi_u$ oscillations are much more spectacular and account for all the observed cross section (although stray electrons provide a large but spurious contribution to the weak scattered signal). This is so because the transition to the high levels takes place at large R and only the oscillating part of the V^- wave functions contributes to the cross section. Huetz *et al.*[46] have also obtained an excellent fit to these ground-state cross sections using the boomerang model and have been able to deduce the form of $\Gamma_X(R)$.

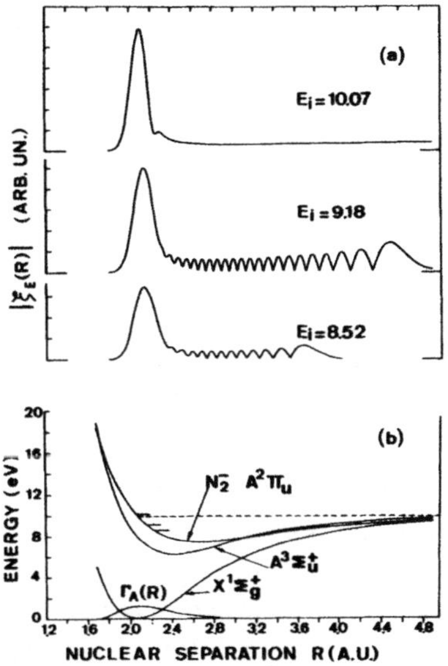

FIGURE 17
(a) Modulus of the nuclear wave function of the $A\ ^2\Pi_u$ state of N_2^- formed from the $v = 0$ level of the N_2 ground state by electron impact at the energies indicated (in eV). (b) Potential curves of the $X\ ^1\Sigma_g^+$ and $A\ ^3\Sigma_u^+$ states of N_2 and the potential curve $V^-(R)$ and partial width $\Gamma_A(R)$ of the $A\ ^2\Pi_u$ state of N_2^-. The dashed line indicates the N_2^- dissociation energy and the three marks show the energies corresponding to the wave functions of (a). The nuclear separation scale is common to both (a) and (b). (Huetz *et al.*[46])

The upper nuclear wave function of Fig. 17a corresponds to an incident energy above the N_2^- dissociation limit, and there is no reflected wave to produce any oscillations. The nuclear wave function is a pure outgoing wave and corresponds to the process of dissociative attachment. Although stable N^- ions do not exist, N_2^- is able to dissociate to unstable $N^-(^3P)$ which subsequently decays by ejecting an electron. By measuring the energy and width of the ejected-electron distribution Mazeau *et al.*[47] were able to determine the energy (70 meV) and width (16 meV) of the ground state of this atomic negative ion. This demon-

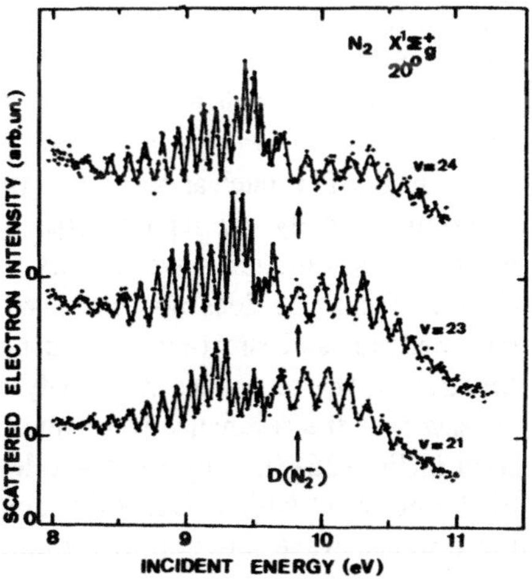

FIGURE 18
Excitation functions of the $v = 21$, 23, and 24 vibrational levels of the N_2 ground state at a 20° scattering angle. The sharp structure converging to the N_2^- dissociation limit is produced by the $^2\Pi_u$ state of the N_2^-. The lower-frequency structure starting at about 9.5 eV results from another state of N_2^-. (Heutz *et al.*[46])

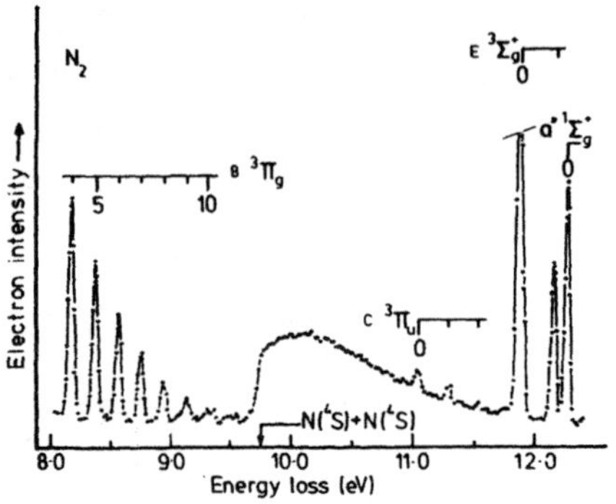

FIGURE 19
Electron yield from electron impact on N_2 at a scattering angle of 135° when the analyzer and detector are set to receive electrons with an energy of 70 meV. Energy losses corresponding to the excitation of various bands of N_2 are indicated. The incident electron energy is obtained by adding 70 meV to the energy loss. (Mazeau *et al.*[47])

strates a new technique of resonance spectroscopy, for obtaining information about resonances in atoms, where instead of forming the atomic resonance states by direct impact they are prepared by fragmentation of molecular resonances and their nature is ascertained by observing the autodetaching electrons.

The probability of forming $N^-(^3P)$, i.e., the dissociative attachment cross section, can be observed by sweeping the incident electron energy and setting an electron detector to collect only electrons with 70 meV energy. Such a recording obtained by Mazeau *et al.*[47] is displayed in Fig. 19. The structureless peak with a rapid onset near the N_2 dissociation limit on the energy-loss scale corresponds to the electron signal from $N^-(^3P)$. Huetz *et al.*[48] showed that N^- was formed by dissociation of the $^2\Pi_u$ state of N_2^- discussed above. Furthermore the form of the peak is observed to change when the electron detector energy is varied by a few tens of meV around the 70 meV value. This effect can be explained as the consequence of the $N^-(^3P)$ ion decaying at large internuclear distances, before complete separation of the nuclei occurs and when it is still perturbed by its atomic partner. This interpretation has given valuable insight into the form of the N_2^- curve at large internuclear separation (see also Chapter 4, Section 5.2).

One further topic in relation to shape resonances which should be discussed here is the effect of the nuclear kinetic energy of the initial state on the resonance phenomena. The nuclear wave function of a long-lived Feshbach resonance is determined solely by its potential curve, and so although the initial target nuclear wave function serves to vary the formation intensity by way of Franck–Condon factors it does not influence the decay processes of the resonance. In intermediate- and short-lived resonances this is not so, since the nuclear wave function of the resonance is strongly dependent on the initial state. In Fig. 17a we saw that the resonance wave function was largely determined by the feeding term in Eq. (26) (i.e., by the $v = 0$ level of the ground state), because evolution of the nuclei at large R is rapidly damped by Γ. On the other hand the effect on the resonant wave function of forming it from an excited state can be observed

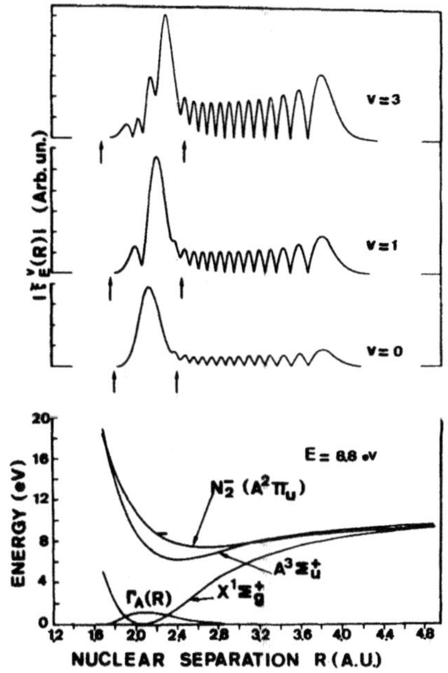

FIGURE 20

The upper part shows the modulus of the nuclear wave function of the $A\ ^2\Pi_u$ state of N_2^- with a total energy (E) of 8.8 eV when it is formed from the indicated vibrational levels (v) of the ground state by electron impact. The vertical arrows indicate the Franck–Condon region of the corresponding ground-state level. The lower part is that already shown in Fig. 17b, reproduced here for convenience of comparison. The small mark on the $V^-(R)$ curve indicates the energy of the system (8.8 eV) corresponding to the wave functions of the upper part of the figure. (Huetz *et al.*[49])

in Fig. 20. These wave functions are those of N_2^- calculated by Huetz *et al.*,[48] all with the same total energy but formed from $v = 1$ and 3 levels of the target molecule as well as that from the $v = 0$ level for comparison. First of all one can notice that the form of the initial target vibrational wave function becomes deformed in the resonant state. For instance, the modulus of the $v = 3$ wave function of the target molecule has four maxima of similar height, but in the resonance, while the four maxima still remain, their intensities are strongly asymmetric. The reason for this is that the target vibrational wave function spans a sufficiently large part of the N_2^- curve for its associated Γ to change considerably over this region. Second, we see from the figure that as the initial v increases the amplitude of the resonance wave function becomes more intense outside the Franck–Condon region. This can be partially attributed to the fact that the increased momentum of the nuclei increases the probability of escaping from the region where Γ is high to the region of larger R where decay is less probable.

Because the negative-ion wave function is strongly modified when the initial vibrational state is changed, the initial state also influences the decay processes of the resonance. The main result of increasing the initial vibrational energy is the enhancement of processes that take place at large R. The oscillations in the A-state cross sections become deeper, as has been verified by Huetz *et al.*[49] Also excitation intensities of high ground-state levels increase and dissociative attachment cross sections become larger (see Chapter 4, Section 5.2).

2.8. Core-Excited Shape Resonances. II: Rydberg States

Core-excited shape resonances probably make a strong contribution to the excitation cross sections of Rydberg states in the threshold region, although this is not well documented. In the few cases that have been studied it appears that a Rydberg state which is the parent of a Feshbach resonance also exhibits resonant structure in its excitation cross section near to threshold. A case in point which demonstrates this is the $b\ ^3\Sigma^+$ state of CO at 10.39 eV which is the lowest Rydberg state in this molecule and is the parent of a well-known Feshbach resonance with its first level at 10.04 eV. The differential cross sections obtained by Mazeau *et al.*[50] (see Fig. 21) reveal complex resonant structures within 2 eV of threshold. The fine structures indicated by the lower arrows result from Feshbach resonances and are not of concern here. What is important are the broad peaks indicated by the upper arrows along with the sharp peak at threshold. This threshold peak has an isotropic angular distribution that indi-

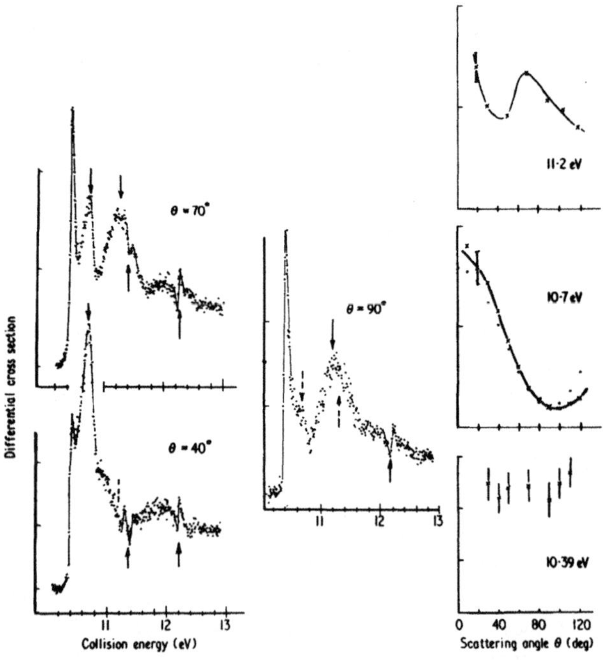

FIGURE 21
Differential excitation functions of the $b\ ^3\Sigma^+$, $v = 0$ level of CO at 40°, 70°, and 90°. The upper arrows indicate shape resonances at 10.7 and 11.2 eV. There is probably a third shape resonance at approximately 12.0 eV. The lower arrows indicate Feshbach resonances. On the right-hand side are the angular behaviors of the scattered electrons at 10.39, 10.7, and 11.2 eV. The points represent a $[1 + 7\cos^2\theta]$ behavior (Π resonance symmetry) normalized to the 10.7 eV curve at 90°. (Mazeau *et al.*[50])

cates a $^2\Sigma^+$ resonant state. This is not of the shape-resonance type as there is no centrifugal barrier when $l = 0$. It is interpreted either as a virtual state, i.e., a state which occurs just at threshold, or as resulting from the overlap of the tail of the lower-lying Feshbach resonance with its parent. The true shape resonances have a maximum occurring at approximately 10.7, 11.2, and perhaps 12.0 eV. They appear to form a sequence which gets broader as they get further from threshold. Also the angular momentum of the scattered electron appears to increase at the same time, as the observed angular distributions indicate. At 10.7 eV the angular distribution corresponds to $l = 1$ and at 11.2 eV the angular behavior indicates that $l = 2$, although higher angular momenta may also be present. This series of shape resonances in CO may well be typical of Rydberg parent states, although this remains to be investigated.

3. Electron Energy-Loss Spectroscopy

In this section of the chapter we deal with the way in which inelastic electron scattering experiments

$$e + \mathrm{M}_{n_i v_i J_i} \rightarrow \mathrm{M}^*_{n_j v_j J_j} + e \tag{30}$$

can be used to determine various *spectroscopic properties* of the molecular excited states M^*. Here, n, v, and J represent the set of electronic, vibrational, and rotational quantum numbers, respectively, of the molecular states. We shall confine our attention to scattering processes in which the electronic state changes ($n_i \neq n_j$, and this change can also be accompanied by vibrational and rotational changes) and in which the incident and scattered electrons have a nonrelativistic velocity. In general the incident electrons are energy selected (see the following chapter for a description of the experimental techniques) so that they have a known mean kinetic energy T and an energy spread ΔT which is much smaller than the excitation energies of the molecular states being excited. The incident electron beam passes through a gas cell, or crosses a molecular beam, and the yield of scattered electrons is measured as a function of the energy loss E_n and the angle of scattering θ.

As mentioned in the introduction to the chapter, this section is subdivided according to the range of incident electron energy and electron energy loss, namely, (i) high incident energy ($\gtrsim$ a few hundred electron volts) and moderate energy loss ($\lesssim 50$ eV), (ii) high incident energy and high energy loss (corresponding to inner-shell excitation), (iii) intermediate incident energy (~ 20 eV to a few hundred eV), and (iv) near-threshold incident energy (within ~ 0.1 eV of the threshold energy). Although this division is a little arbitrary, we shall see that each of these ranges is useful for studying a different subset of the molecular

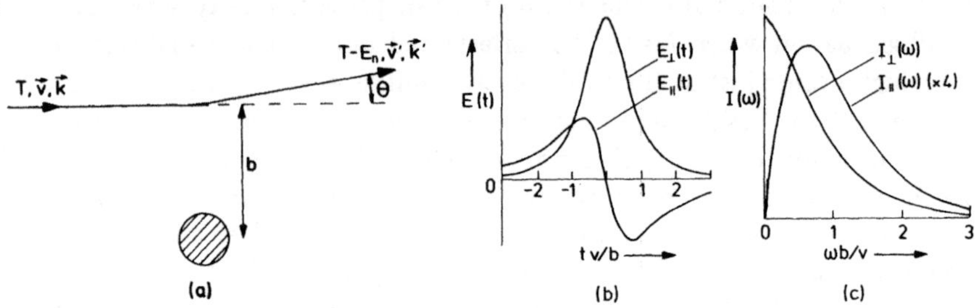

FIGURE 22
(a) Schematic illustration of near-forward inelastic scattering of an electron that has an initial impact parameter b. The electron energy changes from T to $T-E_n$, its velocity from **v** to **v**′, and its momentum from $\hbar\mathbf{k}$ to $\hbar\mathbf{k}'$. (b) Time dependence of the perpendicular and parallel components of the electric field experienced by the molecule if the electron goes past with unchanged velocity or direction. (c) Dependence on the angular frequency ω of the intensity of the Fourier components of the perpendicular and parallel fields.

excited states M*. Before looking at the four ranges in detail, let us start with a simple but useful model of the inelastic scattering process at high incident energy.

In Fig. 22a the motion of an electron as it passes near to a target molecule is represented by a classical trajectory that has an initial impact parameter b. The interaction between the electron and the molecule causes the molecule to be excited to a state of energy E_n, while at the same time the electron energy changes from T to $T-E_n$, its velocity changes from **v** to **v**′, its momentum changes from $\hbar\mathbf{k}$ to $\hbar\mathbf{k}'$, and it is deflected through the angle θ. In this first part of the discussion we shall assume that $E_n \ll T$, and that θ is very small or zero.

The electron creates a varying electric field at the site of the molecule. As a first approximation we can neglect the change in energy and direction of the electron, in which case the components of the electric field perpendicular and parallel to the incident direction are

$$E_\perp(t) = \frac{eb}{(v^2t^2+b^2)^{3/2}}, \tag{31}$$

$$E_\parallel(t) = \frac{evt}{(v^2t^2+b^2)^{3/2}}, \tag{32}$$

where the time t is measured from the time of closest approach. These field components are shown in Fig. 22b. In the perpendicular direction the molecule experiences a symmetric sharp pulse that has the duration $\Delta t \sim \text{few}\,(b/v)$, while in the parallel direction the pulse has a similar duration but is antisymmetric.

The time-dependent fields can be considered as being composed of harmonic components having a range of angular frequencies ω. The amplitude $a(\omega)$ at the frequency ω is given by the Fourier transform of $E(t)$,

$$a(\omega) = \frac{1}{2\pi} \int_{-\infty}^{+\infty} E(t) e^{i\omega t} \, dt. \tag{33}$$

For the fields given by Eq. (31) and (32) the Fourier components are[51]

$$a_{\perp}(\omega) = \frac{e}{\pi b v} x K_1(x), \tag{34}$$

$$a_{\parallel}(\omega) = \frac{e}{\pi b v} x K_0(x), \tag{35}$$

where

$$x = \omega b / v, \tag{36}$$

and where K_0 and K_1 are modified Bessel functions. The intensities at the frequency ω are proportional to $|a(\omega)|^2$ and are given by

$$I_{\perp}(\omega) \propto \frac{1}{b^2 v^2} [x K_1(x)]^2, \tag{37}$$

$$I_{\parallel}(\omega) \propto \frac{1}{b^2 v^2} [x K_0(x)]^2. \tag{38}$$

These intensity components are shown in Fig. 22c. We see that the effect of the electron on the molecule is to create harmonic components having all angular frequencies, but with an effective cutoff in intensity at the frequency

$$\omega_c \sim v/b. \tag{39}$$

The electric field experienced by the molecule is therefore similar to what it would experience if it were immersed for a short time in a field of photons having energies from zero to $\hbar\nu_{\max} \sim \hbar v/b$. We see therefore that the passing electron effectively creates pseudophotons or *virtual photons*. One of these photons can be absorbed by the molecule to give an excited state of energy E_n, provided that

$$E_n \lesssim \frac{\hbar v}{b}. \tag{40}$$

Since energy must be conserved, the kinetic energy of the electron decreases at the same time. We can therefore regard the system of incident plus scattered electrons as a *pseudophoton source.*

If the impact parameter is much larger than the mean radius r of the target molecule the electric field experienced by the molecule is approximately uniform over its spatial extent, as is the effective electric field of a photon beam when the light wavelength λ is much greater than r. The electric field then interacts most strongly with the transition electric dipole moment of the molecule, and the transitions which are excited are predominantly the electric dipole (E1) or *optically allowed* transitions that are excited by photoabsorption. Under these conditions the inelastic electron scattering process simulates the familiar photoabsorption process. Taking r to be of the order of the Bohr radius

$$a_0 = \frac{\hbar^2}{me^2} = 5.292 \times 10^{-11} \text{ m}, \tag{41}$$

where m is the electron mass, and combining the condition $b \gg r$ with the condition (40), we find that the simulation occurs when

$$E_n \ll \frac{\hbar v}{a_0} \,. \tag{42}$$

Expressing v in terms of the incident energy T, and using the Rydberg energy

$$R = \frac{me^4}{2\hbar^2} = 13.602 \text{ eV}, \tag{43}$$

we find that this condition can be expressed in the form

$$T \gg \frac{E_n^2}{4R} \,. \tag{44}$$

A quantity that is useful here and later is the *momentum-transfer vector*

$$\mathbf{K} = \mathbf{k} - \mathbf{k}'. \tag{45}$$

$\hbar\mathbf{K}$ is the linear momentum lost by the electron and gained by the molecule. Expressing the magnitude of $\mathbf{K}$ in terms of k and θ we find that

$$\begin{aligned}(Ka_0)^2 &= (ka_0)^2 + (k'a_0)^2 - 2kk'a_0^2 \cos\theta \\ &= \frac{1}{R} \{(2T - E_n) - 2[T(T - E_n)]^{1/2} \cos\theta\}. \end{aligned} \tag{46}$$

FIGURE 23
Examples of the momentum-transfer vector **K**. E_n has the same magnitude for the three parts of the figure. K becomes larger as θ is increased or T is decreased.

We now want to establish that the condition (42) or (44) is equivalent to the condition

$$ka_0 \ll 1. \tag{47}$$

The way in which K depends on θ and T/E_n is illustrated in Fig. 23. To achieve a small value of Ka_0 it is clearly necessary to have $\theta \ll 1$ and $T \gg E_n$. Under these conditions

$$(Ka_0)^2 \approx \frac{4\bar{T}}{R}\left[\sin^2\frac{\theta}{2} + \left(\frac{E_n}{4\bar{T}}\right)^2\right], \tag{48}$$

where

$$\bar{T} = T - \tfrac{1}{2}E_n. \tag{49}$$

For given values of T and E_n the minimum value of K occurs at $\theta = 0$, and when $T \gg E_n$ it is given by

$$(Ka_0)_{\min} \approx \frac{E_n}{2(R\bar{T})^{1/2}}\,. \tag{50}$$

We see now that conditions (44) and (47) are indeed equivalent for forward scattering. When $\theta \neq 0$ the classical impulsive electric field differs from that shown in Fig. 22b, and so the classically derived condition (44) becomes unreliable, but the condition (47), which as we shall see below has a rigorous quantal foundation, remains valid. This requires that condition (44) be satisfied together with the further condition [obtained from Eq. (48)]

$$\theta^2 \ll R/T. \tag{51}$$

The transitions excited by inelastic electron scattering are then the optically allowed transitions that occur in photoabsorption events.

The classical model is helpful also in obtaining a qualitative understanding of the connections between the polarization and spectral intensity distributions of the electron scattering and photoabsorption processes. Considering forward scattering ($\theta = 0$) events, we see that by detecting only those electrons that are scattered in this direction one is selecting events in which the molecule has interacted with the electron to change its momentum by $\hbar\mathbf{K}$ (in the forward direction) and hence the electron has interacted with the molecule to change its momentum by $-\hbar\mathbf{K}$. The relevant field component for this interaction is therefore $E_{\parallel}(t)$, and the pseudophoton source is effectively polarized in the forward (parallel) direction. For nonforward scattering of the electrons the effective polarization direction changes, as we shall see in more detail in the next section.

To see what information the classical model gives about the spectral intensity distribution we note from Eq. (38) and Fig. 22c that $I_{\parallel}(\omega)$ has a maximum value

$$I_{\parallel,\max}(\omega) \propto \frac{1}{b_m^2 v^2} \tag{52}$$

at the impact parameter

$$b_m \sim v/\omega, \tag{53}$$

and that the intensity drops off sharply for $b \lesssim \frac{1}{2}b_m$ or $b \gtrsim \frac{3}{2}b_m$. The effective impact area is therefore of the order of $2\pi b_m^2$, and the total intensity $I_t(\omega)$ available at ω is approximately proportional to

$$I_{\parallel,\max}(\omega) \times 2\pi b_m^2 \propto 1/v^2 \tag{54}$$

[and we can reach the same conclusion by integrating $I_{\parallel}(\omega)$ over the impact area $2\pi b db$]. Therefore within the limitations of this simple model, and neglecting the high-frequency cutoff in ω, we see that $I_t(\omega)$ does not depend on ω (still assuming that $E_n \ll T$ and $\theta \ll 1$), and that the pseudophoton source is therefore approximately "white", so that the relative intensities seen on an inelastic electron scattering spectrum are approximately the same as those seen in the analogous photoabsorption spectrum. In fact this is only qualitatively correct, as we shall now see.

A more exact treatment is given in Chapter 1, Section 6.5 (see also Inokuti[11]), where it is shown that when $E_n \ll T$ the first Born approximation can be used to derive the differential cross section for inelastic electron scattering. When

$$\theta^2 \ll (E_n/2T)^2 \tag{55}$$

so that Ka_0 has its minimum value [see Eq. (48)], the differential cross section for excitation to the state n is given by

$$\frac{d\sigma}{d\Omega} = \frac{16a_0^2 R^2 T}{E_n^3} f_n^{\text{opt}}. \tag{56}$$

Here f_n^{opt} is the *optical oscillator strength* for the transition from the target state to the excited state n. In optical absorption experiments the absorption strength for photons of energy E_n is directly proportional to f_n^{opt}, and so we see that an electron energy-loss spectrum for forward scattering and for $E_n \ll T$ essentially reproduces the analogous optical absorption spectrum, apart from the kinematic factor E_n^{-3}. Specific examples will be discussed in Section 3.1.

Equation (56) applies not only to excitation of bound states but also to excitation of unbound or continuum states (for which $E_n > I$, I being the ionization energy). The differential cross section and oscillator strength then apply to an energy interval dE of the continuum, rather than to a discrete energy E_n. The electron energy-loss technique is especially suitable for studies of the continuum ($h\nu \gtrsim 20\,\text{eV}$, $\lambda \lesssim 60\,\text{nm}$), since the equivalent photon sources (e.g., synchroton radiation sources) can be difficult to use and expensive to construct. Van der Wiel has coined the description "poor man's synchrotron" for apparatus in which inelastic electron scattering is used to simulate photoabsorption studies. This will be discussed further in Section 3.1.

Turning now to nonforward-scattering processes [in which the condition $\theta^2 \ll (E_n/2T)^2$ is not satisfied], let us start by extending the model depicted in Fig. 22. Large scattering angles occur only when there is a strong interaction between the electron and molecule, which in turn requires a small impact parameter b. When b is comparable with, but still greater than, the molecular size the electric field experienced by the molecule is nonuniform (because it is higher on the side nearer to the electron). The nonuniform components of the field can interact with higher electric moments of the molecule, such as the transition electric quadrupole (E2) moment. In this way *optically forbidden* transitions of the molecule can be excited. By experimentally selecting larger values of θ we tend to select this type of forbidden transitions.

We can also see the origin of the higher-order interactions by expanding the Coulomb potential between the incident electron at position $\mathbf{r}$ and a molecular electron at position $\mathbf{r}_i$ or nucleus of charge Z_α at position $\mathbf{R}_\alpha$. Using a coordinate system in which the center of mass of the molecule is at the origin and the incident electron is on the z axis,

$$\frac{e^2}{|\mathbf{r}-\mathbf{r}_i|} = \frac{e^2}{r} \sum_{l=0}^{\infty} \left(\frac{r_i}{r}\right)^l P_l(\cos\theta_i). \tag{57}$$

where the P_l are Legendre polynomials and it is assumed that $r_i < r$. Summing over the Z electrons and N nuclei in the molecule, we find that the interaction energy is

$$U = \frac{e^2}{r}\left(Z - \sum_{\alpha=1}^{N} Z_\alpha\right) + \frac{e^2}{r^2}\left(\sum_{j=1}^{Z} z_j - \sum_{\alpha=1}^{N} Z_\alpha z_\alpha\right)$$

$$+\frac{e^2}{r^3}\left[\sum_{j=1}^{Z}(3z_j^2 - r_j^2) - \sum_{\alpha=1}^{N} Z_\alpha(3z_\alpha^2 - R_\alpha^2)\right] + \cdots \tag{58}$$

where z_j and r_j are coordinates of the molecular electrons and Z_α and R_α are coordinates of the nuclei. The first term on the right-hand side is identically zero for a neutral molecule, and the second term is the interaction already considered above between the incident electron and the electric dipole moment of the molecule. The third term gives the interaction with a component of the electric quadrupole moment of the molecule. It is much smaller than the dipole term when $r \gg r_i$, but becomes significant when r becomes comparable with r_i.

More exactly, it is shown in Chapter 1, Section 6.5.2 that when T is still high enough for the first Born approximation to be valid the differential cross section can be expressed as

$$\frac{d\sigma}{d\Omega} = \frac{k'}{k} \cdot \frac{4Rf_n(K)}{E_n K^2} \, . \tag{59}$$

The optical oscillator strength which appears in Eq. (56) is replaced here by the *generalized oscillator strength* $f_n(K)$ defined by

$$f_n(K) = \frac{E_n}{R(Ka_0)^2}\left|\int \psi_n^* \sum_{j=1}^{Z} e^{i\mathbf{K}\cdot\mathbf{r}_j}\,\psi_0 d\tau\right|^2 . \tag{60}$$

where ψ_0 and ψ_n are the wave functions of the target and excited states, respectively, and the summation is over the Z electrons in the molecule.

If $(Ka_0) \ll 1$ then $\exp(i\mathbf{K}\cdot\mathbf{r}_j)$ can be expanded in terms of K, leading to an expansion of $f_n(K)$ in even powers of K,

$$f_n(K) = \sum_{\lambda=0}^{\infty} \frac{(Ka_0)^2}{\lambda!} f_n^{(\lambda)}. \tag{61}$$

(The inclusion of other terms, linear and logarithmic in K, due to deviations from the first Born approximation, will be discussed in Section 3.1.)

The quantities $f_n^{(\lambda)}$ are given by

$$f_n^{(\lambda)} = \lambda!\frac{E_n}{R}\sum_{\kappa=1}^{2\lambda+1}\frac{(-)^{\lambda-\kappa+1}}{\kappa!(2\lambda+2-\kappa)!}\,X_n^{(\kappa)}X_n^{(2\lambda+2-\kappa)*}, \tag{62}$$

where

$$X_n^{(\kappa)} = a_0^{-\kappa} \int \psi_n^* \sum_{j=1}^{Z} x_j^{\kappa} \psi_0 d\tau, \tag{63}$$

and x_j is the component of $\mathbf{r}_j$ in the direction $\mathbf{K}$. The first expansion coefficient $f_n^{(0)}$ is the optical oscillator strength mentioned above,

$$f_n^{(0)} = f_n^{\text{opt}} = \frac{E_n}{Ra_0^2} \left| \int \psi_n^* \sum_{j=1}^{Z} x_j \psi_0 d\tau \right|^2 . \tag{64}$$

It has a nonzero value only when the optical selection rules for the transition are obeyed, giving an optically allowed electric dipole transition. When $Ka_0 \ll 1$ only the first term of Eq. (61) is significant. Substituting Eqs. (48) and (61) into Eq. (59), and taking $\theta^2 \ll (E_n/2T)$, we regain Eq. (56).

We see also from Eq. (59) that the differential cross sections for optically allowed transitions are proportional to $1/K^2$ for small K, and so decrease quickly as θ (and hence K) increases. These cross sections therefore tend to be strongly peaked in the forward direction.

The other expansion coefficients $f_n^{(\lambda)}$ in Eq. (61), with $\lambda \geqslant 1$, correspond to mixtures of electric multipole (i.e., electric quadrupole, octupole, etc.) transitions, all of which are optically forbidden. It is also possible to expand $f_n(K)$ in terms corresponding to pure multipole transitions (Read[16]),

$$f_n(K) = \frac{4\pi E_n}{R(Ka_0)^2} \sum_{l=0}^{\infty} M_l, \tag{65}$$

where

$$M_l = \sum_{m=-l}^{+l} |N(l,m)|^2$$

and

$$N(l,m) = \int \psi_n^* \sum_{i=1}^{Z} j_l(Kr_i)\, Y_{lm}(\theta_i \phi_i) \psi_0 d\tau.$$

Here j_l and Y_{lm} are spherical Bessel and spherical harmonic functions, respectively. Each of the $N(l,m)$ has an integrand of definite symmetry with respect to rotations of the electronic coordinates, and so group-theoretical methods can be used to establish which of them vanish (Read and Whiterod[52,53]).

The magnitude of the optically forbidden contributions to the cross section, relative to the optically allowed contributions, can be found from

Eqs. (61) and (62) [and also from Eq. (65), by expanding the j_l]. In the forward direction it is of the order of

$$\frac{(Ka_0)^{2\lambda}}{\lambda!} \approx \frac{1}{\lambda!}\left(\frac{E_n^2}{4RT}\right)^{2\lambda}, \tag{66}$$

which is usually a very small quantity in practice (except for inner-shell excitation, see Section 3.2). The forbidden terms become comparable to the allowed term only when

$$Ka_0 \gtrsim 1. \tag{67}$$

Equation (48) shows that the corresponding scattering angles are given by

$$\theta \gtrsim \left(\frac{R}{T}\right)^{1/2}. \tag{68}$$

Another way of increasing the value of Ka_0 and thus enhancing the ratio of the forbidden to optically allowed components is by reducing the incident energy and so relaxing the condition $E_n \ll T$. For example at $T = 2E_n$, we see from Eq. (46) that Ka_0 changes from $0.4(E_n/R)^{1/2}$ in the forward direction to $2.4(E_n/R)^{1/2}$ in the backward direction. Ka_0 is no longer small in the forward direction, and Eq. (67) is approximately satisfied at all scattering angles. The Born approximation ceases to be valid at incident energies which are this low, but these arguments nevertheless give a correct indication of the observed (and calculated) behavior: the forbidden transitions are greatly enhanced at all scattering angles when T is low.

Finally, let us consider qualitatively those collisions in which the incident electron essentially makes a direct impact with the target molecule. The new process of electron exchange then becomes possible. This process can be pictured classically as that in which the incident electron becomes attached to the target molecule while a target electron is detached to become the outgoing (scattered) electron. The probability of this exchange tends to be highest when the velocity of the incident electron is comparable with the velocity of the target valence electron, which requires that the incident energy T should be of the order of R. This implies in turn that T is near the threshold energy E_n of the reaction. The probability of electron exchange decreases with increasing T. The electrons involved in the reaction are of course indistinguishable, which is not taken into account in this classical model, but the indistinguishability is in one sense removed in those reactions in which the exchange electrons are in different spin states ($m_s = +\frac{1}{2}$ and $-\frac{1}{2}$, respectively). The spin state S of the molecule may then be changed by the scattering event. For example if the target molecule has two electrons in the valence shell, coupled with opposite spin states ($m_1 m_2 = \frac{1}{2} - \frac{1}{2}$) to a singlet state ($S = 0$) of the molecule, and if one of these target electrons (e.g., $m_2 = -\frac{1}{2}$) exchanges with an incident electron of opposite spin state ($m_3 = \frac{1}{2}$), the excited molecule will have both valence electrons in the same spin state ($m_1 m_3 = \frac{1}{2}\frac{1}{2}$) and so will be in triplet state ($S = 1$). This is described as a *spin-exchange process*. It can be represented pictorially as

$$\overset{\uparrow}{1}\,\overset{\downarrow}{2} + \overset{\uparrow}{3} \rightarrow \overset{\uparrow}{1}\,\overset{\uparrow}{3} + \overset{\downarrow}{2}. \tag{69}$$

This type of scattering provides essentially the only way of exciting a singlet–triplet transition, since other mechanisms (such as direct spin-flipping of an electron by magnetic interactions) are comparatively much weaker. The differential cross sections for spin-exchange processes do not show forward peaking, and so the detection of molecular excitation involving a change of spin is enhanced by using low incident energies and observing at large scattering angles. Specific examples will be discussed in Section 3.3.

3.1. Molecular Outer-Shell Spectroscopy at High Incident Electron Energies

In this section we consider the use of the electron energy-loss technique at high incident energies (generally greater than a few hundred eV) and at low energy losses (generally less than about 50 eV). Such studies give information about (i) rotationally and vibrationally excited states, (ii) valence and Rydberg electronic states, and (iii) ionization and dissociation processes within a few eV of their respective thresholds.

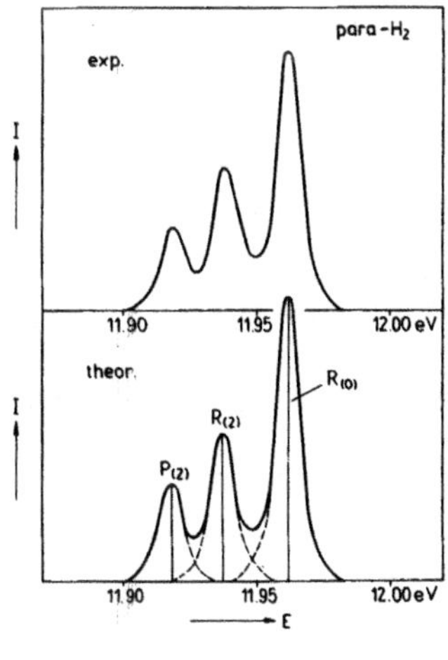

FIGURE 24

Electron energy-loss spectrum of *para*-hydrogen (Boersch *et al.*[54]), showing rotational structure of the Lyman band $v' = 5$ transition. Incident energy = 30 keV, resolution = 7 meV, scattering angle $\leqslant 10^{-4}$ rad. The theoretical curves are obtained by folding the calculated photoabsorption cross section with the apparatus resolution function.

We start with an example of an electron energy-loss spectrum taken by Boersch *et al.*[54] and shown in Fig. 24. The incident energy is 30 keV, the energy resolution is 7 meV, and the angle of scattering is $\leqslant 10^{-4}$ rad. The electron energy loss is scanned from 11.9 to 12.0 eV, and the peaks in the spectrum correspond to the excitation of different rotational (J') levels of a specific electronic (n') and vibrational (v') level of *para*-hydrogen (which has antiparallel proton spins). This energy resolution has seldom been surpassed in the field of electron impact studies on gaseous targets. The resolution was achieved by slowing the electrons down to an energy of a few eV, and then energy selecting with a Wien filter.

To see how closely this and other electron energy-loss spectra correspond to the photoabsorption limit [apart from the factor E_n^{-3} – see Eq. (56)], the quantity Ka_0 is shown in Fig. 25 as a function of E_n/T for various values of θ. To make the figure applicable for all nonrelativistic energies T the left-hand ordinate gives $(Ka_0)(T/R)^{-1/2}$, while the right-hand ordinate gives Ka_0 for the specific incident energy $T = 1$ keV. For the conditions appropriate to Fig. 24, $E_n/T = 4 \times 10^{-4}$, $\theta \leqslant 10^{-4}$, and so we see from Fig. 25 that $(Ka_0)(T/R)^{-1/2} \approx 2.2 \times 10^{-4}$, giving $Ka_0 \approx 10^{-2}$. This implies [Eq. (61)] that the optically forbidden component is negligibly small ($\leqslant 10^{-4}$ of the optically allowed component, assuming that the ratio $f_n^{(1)}/f_n^{(0)}$ is not abnormally large), and so the spectrum corresponds very closely to the photoabsorption spectrum. In practice the generalized oscillator strength $f_n(K)$ differs from the optical oscillator strength $f_n^{\rm opt}$ by about 5% (which is also of the order of the inaccuracies in the experimental measurements) when Ka_0 is in the range from about 0.04 (for example for the $3\,^1P \leftarrow 2\,^1S$ transition in He – see Inokuti[11]) to about 0.03 (for example for the $5\,^1P \leftarrow 1\,^1S$ transition in He). Taking the typical value as

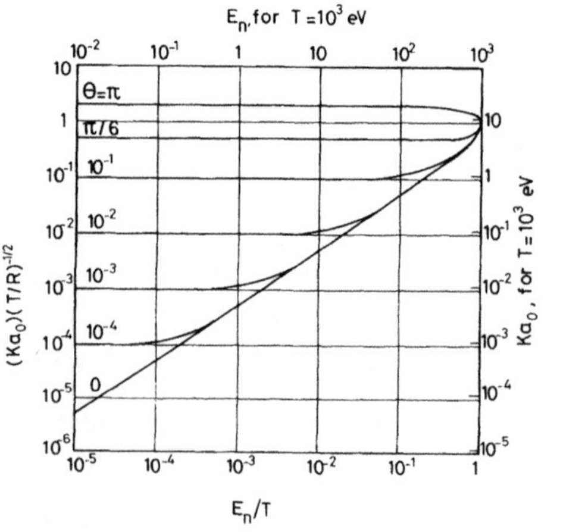

FIGURE 25

Dependence of K on θ, E_n, and T. By using the dimensionless quantities given on the left-hand and bottom axes K can be found for any nonrelativistic conditions. The right-hand and top axes apply to $T = 1$ keV only.

$Ka_0 = 0.1$ [which is also consistent with Eq. (61)] we see from Fig. 25 that for a 5% inaccuracy at $T = 1$ keV we must have $\theta \lesssim 10^{-2}$ and $E_n \lesssim 20$ eV (again assuming that the ratio $f_n^{(1)}/f_n^{(0)}$ is not abnormally large). Figure 25 allows similar estimates to be made for other incident energies.

Although the optical limit is approached most closely at $\theta = 0$, in practice it is not possible to observe at this angle only, since the detector must accept a finite range of angles and there is also necessarily a finite range of angles in the incident beam. If we assume that all scattering angles between 0 and θ_m have the same intrinsic probability per unit solid angle, the resulting experimentally measured cross section is obtained by integrating Eq. (59) over the solid-angle increments $2\pi \sin\theta \, d\theta$. The total observed solid angle is

$$\Delta\Omega = \int_0^{\theta_m} 2\pi \sin\theta \, d\theta = 4\pi \sin^2(\theta_m/2), \tag{70}$$

and the partial cross section for scattering between θ and $\theta + d\theta$ is

$$d\sigma(\theta) = \frac{d\sigma}{d\Omega} \cdot 2\pi \sin\theta \, d\theta = \frac{k'}{k} \frac{4Rf_n^{\text{opt}}}{E_n K^2} 2\pi \sin\theta \, d\theta, \tag{71}$$

where the optically forbidden terms in $f_n(K)$ have been neglected. The partial cross section for the range of θ from 0 to θ_m is then $\Delta\sigma$, where

$$\begin{aligned} \frac{\Delta\sigma}{\Delta\Omega} &= \frac{1}{\Delta\Omega} \int_0^{\theta_m} d\sigma(\theta) \\ &= \frac{k'}{k} \frac{R^2 a_0^2 f_n^{\text{opt}}}{E_n \bar{T} \sin^2(\theta_m/2)} \ln\left\{1 + \left[\frac{4\bar{T}\sin(\theta_m/2)}{E_n}\right]^2\right\}. \end{aligned} \tag{72}$$

For $\theta_m \ll 1$ this becomes

$$\frac{\Delta\sigma}{\Delta\Omega} = \frac{k'}{k} \frac{4R^2 a_0^2 f_n^{\text{opt}}}{E_n \bar{T} \theta_m^2} \ln\left[1 + \left(\frac{2\bar{T}\theta_m}{E_n}\right)^2\right]. \tag{73}$$

This is equivalent to the expressions obtained by Hamnett *et al.*[55] and Huebner *et al.*[56] Putting $\theta_m = 0$ (or less restrictively, $\theta_m \ll E_n/2T$), we regain Eq. (56).

A finite range of scattering angles also affects the average direction of polarization of the excited molecules, as pointed out by Hamnett *et al.*[55] As we saw in the introduction to Section 3, the pseudophotons are effectively polarized in the forward direction when $\theta = 0$. More generally, the direction of polarization is that of the momentum-transfer vector $\hbar\mathbf{K}$ (assuming that the conditions are such that the Born approximation is valid). We see from Fig. 23 that the inclination of $\mathbf{K}$ to the forward direction is ϕ, where

$$\sin\phi = (k/K)\sin\theta. \tag{74}$$

For example, at the value of θ for which $\sigma(\theta)$ has its maximum value, namely,

$$\theta' = \frac{E_n}{2\bar{T}} \tag{75}$$

[which we obtain by differentiating Eq. (71) and assuming that $\theta' \ll 1$], we find that

$$\sin\phi = \frac{(2mT)^{1/2}}{\hbar}\,\frac{\theta'}{\left(\dfrac{2\bar{T}}{R}\right)^{1/2}\dfrac{\theta'}{a_0}} = \frac{1}{\sqrt{2}}, \tag{76}$$

so that the polarization direction is then at 45° to the forward direction. We see therefore that a wide spread of polarization directions exists unless $\theta_m \ll E_n/2T$.

The electron forward-scattering energy-loss technique has been used to simulate photoabsorption spectra by many groups of workers. The aim of the work is usually to measure the optical oscillator strengths f_n^{opt}. A second example, from the work of Geiger and Schroeder,[57] is shown in Fig. 26. Here

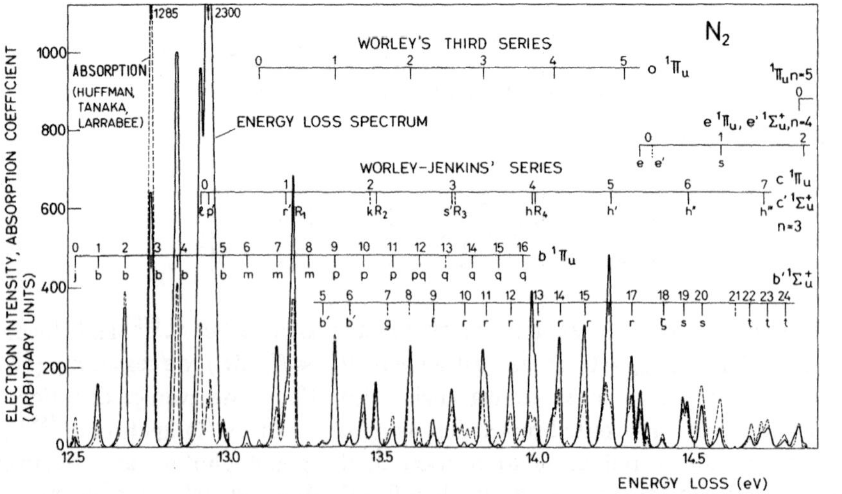

FIGURE 26
Electron energy-loss spectrum of N_2 (Geiger and Schroeder[57]). Incident electron energy 25 keV, resolution 10 meV, scattering angle $\theta < 1 \times 10^{-4}$ rad. The dashed line is a photoabsorption spectrum (Huffman *et al.*[58]).

an electron energy-loss spectrum of N_2 is compared with the photoabsorption spectrum of Huffman *et al.*[58] The comparison shows an advantage possessed by the electron energy-loss technique, manifested by the existence of a discrepancy in the relative intensities of the two spectra in the region between 12.50 and 12.93 eV (this was noticed also in measurements by Lassettre *et al.*[59] at $T =$ 200–500 eV). The discrepancy was first attributed to a failure of the first Born approximation, but it was realized later (Meyer *et al.*,[60] Lawrence *et al.*[61]) that there is a line saturation problem in the original photoabsorption measurements. The relevant lines have a high enough oscillator strength for the photon absorption to saturate at the centers of the lines at the pressures and path lengths used by Huffman *et al.*, but the lines are narrower than the bandwidth of the spectrometer and this saturation was not noticed. This problem does not arise in the electron-impact spectra. Here each absorption line shape is a convolution of the natural line shape with the apparatus energy profile (which may be taken to include the Doppler broadening, see Read[62]), and its area is proportional to the oscillator strength but is independent of the natural width. The question of saturation could arise only if the probability for an incident electron to suffer a particular energy loss in traversing the gas is comparable with unity, which does not happen because the inelastic scattering cross sections are too small at the incident energies used in these experiments. The electron energy-loss technique has been useful in uncovering similar instrumental errors in the photoabsorption spectra of other gases (Inokuti[11]).

Another example of the measurement of optical oscillator strength is shown in Fig. 27, taken from Tan *et al.*[63] They used an incident electron energy of 2.5 keV. This example illustrates the usefulness of the electron energy-loss technique for measurements over a wide range of energy loss, including the continuum. Before the widespread availability of synchrotron radiation sources there was no other convenient way of observing the low-resolution features of an absorption spectrum over such a wide energy range. The electron energy-loss

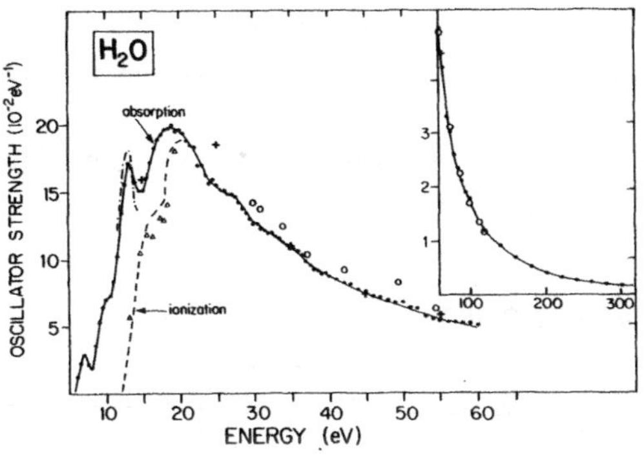

FIGURE 27
Oscillator strength measurements in H_2O, using the inelastic-electron coincidence technique to simulate the photoabsorption spectrum (Tan *et al.*[63]). The full line and points (•) give the total oscillator strength for photoabsorption, and the broken line the part of the oscillator strength corresponding to the production of one or more ions. The other data points correspond to optical measurements (see Tan *et al.*[63] for details).

technique still has the advantage of providing an easier and usually more accurate absolute calibration of an absorption spectrum over very wide energy ranges. The use of oscillator strength sum rules (Kay *et al.*[64]) facilitates this calibration.

Figure 27 illustrates also how the optical oscillator strength can be partitioned according to the products formed by the absorption. The part of the spectrum corresponding to the total absorption oscillator strength is obtained as described above. The result of absorption (of either real or pseudophotons) can be the production of ions and electrons,

$$
\begin{aligned}
h\nu + H_2O \rightarrow\ & H_2O^+ + e, \\
& \text{or } OH + H^+ + e, \\
& \text{or } OH^+ + H^+ + 2e, \\
& \text{etc.,}
\end{aligned}
\tag{77}
$$

or the production of neutral states only,

$$
\begin{aligned}
h\nu + H_2O \rightarrow\ & H_2O^*, \\
& \text{or } H + OH, \\
& \text{etc.}
\end{aligned}
\tag{78}
$$

(often accompanied by photons emitted in the decay of excited neutral fragments). To separate out the part of the yield which corresponds to the production of ions and electrons, Tan *et al.* used a coincidence technique. In this they detected inelastically scattered electrons of known energy (thus simulating photon absorption) in coincidence with either the product electrons or ions (summed over the full range of mass and energy). In a shorthand form these coincidence experiments can be described as $(e, 2e)$ and $(e, e + \text{ion})$ experiments. The result of the measurement of the partial yield due to the production of ions and electrons is shown as the broken line in the figure. The difference between this and the total oscillator strength is the partial oscillator strength for the production of neutrals. Only neutrals can be produced below the first ionization limit, but we see that they are also produced at energy losses that are slightly higher.

This type of coincidence method has been extensively used by van der Wiel and co-workers. It can be represented schematically as

$$e(\text{energy } T) + \text{molecule} \rightarrow \text{collision products} + e(\text{energy } T - E)$$

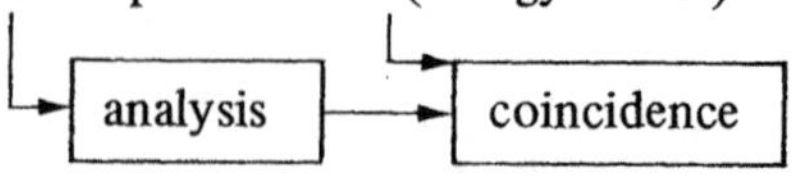

. (79)

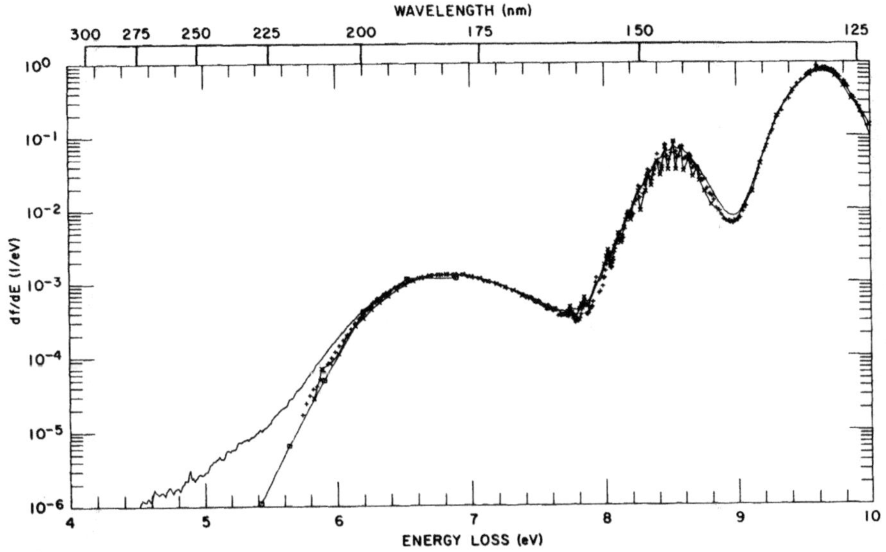

FIGURE 28
Comparison of optical and electron-impact oscillator strengths of N_2O in the 4.10 eV regions (Huebner *et al.*[67]). The solid line shows the electron impact determination, while the specific points represent various optical determinations (see Huebner *et al.*[67] for details).

The technique has proved very useful for studying a wide range of molecular dissociation and ionization processes (see the reviews of Backx and van der Wiel,[65] Brion,[17] and van der Wiel[66]).

There have been many other measurements of optical oscillator strengths by the forward-scattering electron energy-loss technique. Amongst these are the measurements on acetone by Huebner *et al.*,[56] using $T = 100$ eV and $E \leqslant 15$ eV, the measurements on N_2O by Huebner *et al.*,[67] using $T = 100$ eV again and $E \leqslant 14$ eV (Fig. 28 gives one of their results, showing the agreement between the electron impact and photoabsorption results, where the latter exist), the measurements on infrared transitions in N_2O by Kuyatt *et al.*,[68] using $T = 100$ eV and $E \leqslant 0.6$ eV, and the measurement on the chlorofluoromethanes by King and McConkey,[69] using $T = 500$ eV and $E \leqslant 36$ eV. A comprehensive review is given by Celotta and Huebner.[20]

When T, E, and θ are such that Ka_0 is not small, and when perhaps E_n/T is also not small, it is possible to use an extrapolation technique to deduce f_n^{opt} from electron energy-loss spectra. This depends on a knowledge of the way in which the generalized oscillator strength $f_n(K)$ approaches f_n^{opt} at small K. We have seen [Eq. (61)] that when the first Born approximation is applicable ($T \gg E_n$), $f_n(K)$ can be expanded as a series in even powers of K, with f_n^{opt}

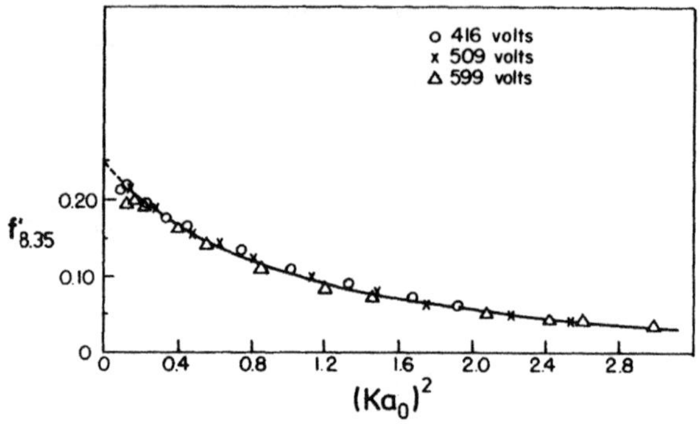

FIGURE 29
Generalized oscillator strengths for the 8.35-eV transition in CO, obtained by the electron energy-loss technique at three different incident electron energies. (Lassettre and Silverman.[71])

as the first term, and so f_n^{opt} is the limiting value of $f_n(K)$ as K goes to zero. Lassettre *et al.*[70] have proven a more general theorem of the limiting oscillator strength, which applies at all incident energies and is independent of the validity of the first Born approximation, namely,

$$\lim_{K \to 0} f_n(K) = f_n^{\text{opt}} \qquad \text{(at all } T\text{)}. \tag{80}$$

Here, $f_n(K)$ is the apparent generalized oscillator strength defined by Eq. (59), rather than by Eq. (60), using the correct differential cross section even when the first Born approximation is invalid. An example of an extrapolation to $K = 0$, taken from the early work of Lassettre and his group (Lassettre and Silverman[71]), is shown in Fig. 29. Here the inelastic scattering cross section is measured at a number of incident energies and scattering angles, and Eq. (59) is used to deduce $f_n(K)$. The fact that the deduced values of $f_n(K)$ depend on $(Ka_0)^2$ but not on T appears to indicate that the first Born approximation is valid for those results, although we see from Eq. (80) that this way of deducing f^{opt} does not depend on whether the Born approximation is valid or not.

Lassettre and his co-workers have made many such determinations of generalized oscillator strengths of valence, Rydberg and continuum levels of many simple molecules, and have used this type of extrapolation to deduce f^{opt}. The impact energy is usually in the range from 300 to 700 eV. Experimental inaccuracies and extrapolation uncertainties lead to errors in f^{opt} that are typically less than 10%, which is often better than that obtainable by optical techniques. The dependence of $f_n(K)$ on K is information which is also useful, for example in comparing with the results of theoretical calculations. Further important information is the dependence of $f(K)$ on K for optically forbidden transitions, for which $f^{\text{opt}} \approx 0$. A comprehensive account of this work, and of the experimental methods used to normalize the cross sections, is given in the review of Lassettre and Skerbele.[15]

Huo[72] has shown that although only even powers of K are present in the expansion of $f(K)$ when the first Born approximation is valid, in practice odd powers of K are always present due to deviations from the Born approximation. In particular, she shows that there is always a linear term,

$$f(K) = f^{\text{opt}} + Kf^{(1)} + K^2 f^{(2)} + \cdots, \tag{81}$$

where

$$f^{(1)} = c_1 T^{-1/2} + c_2 T^{-1} + \cdots. \tag{82}$$

Terms logarithmic in K are also present, the one of lowest order being proportional to $T^{-1/2} K^2 \ln K$. The curve of $f(K)$ versus K^2 therefore has an infinite slope at $K = 0$, since

$$\frac{df}{dK^2} = \frac{c_1}{2KT^{1/2}} + \frac{d_1 \ln K}{T^{1/2}} + \cdots. \tag{83}$$

This infinite slope is not experimentally observable because the smallest physically accessible value of K occurs at $\theta = 0$, and has the nonzero value given by Eq. (50). Nevertheless some care must be taken in extrapolating to find f^{opt}, and Huo recommends that this should be done, where feasible, by plotting $f(K)$ versus K rather than versus K^2. Alternatively the experimental data may be fitted by a semiempirical formula (see Lassettre[73]) which includes a term linear in K at small K. For example Vriens *et al.*[73a] have used the formula

$$f(K) = f^{\text{opt}} (1+x)^{-6} \left(1 + \frac{gx}{1+x}\right), \tag{84}$$

where

$$x = hK^2,$$

and where g and h are empirical constants.

3.2. *Inner-Shell Excitation*

In this section we consider studies in which the excitation energy E_n is high enough ($\geqslant 100$ eV) to cause excitation of an inner-shell electron to a valence or Rydberg orbital of the molecule.

An interesting example is the excitation of a nitrogen K-shell ($1s$) electron in N_2 to the lowest unfilled $2p\pi$ molecular orbital,

$$e + N_2(1s\sigma_g)^2\,(1s\sigma_u)^2\,(2s\sigma_g)^2\,(2s\sigma_u)^2\,(2p\pi_u)^4\,(2p\sigma_g)^2\;{}^1\Sigma_g \rightarrow N_2(1s\sigma_g)^2\,(1s\sigma_u)^1\cdots(2p\sigma_g)^2\,(2p\pi_g)\;{}^1\Pi_u + e. \quad (85)$$

The electron energy-loss spectrum obtained by Wight *et al.*,[74] using $T = 2.5$ keV and forward scattering, is shown in Fig. 30. The energy resolution is about 0.5 eV, but this is sufficient to show the main features of the spectrum. The prominent peak at approximately 401 eV corresponds to the transition given by Eq. (85). It is noticeable that the oscillator strength for this transition is much larger (by a factor of approximately 10) than the sum of the oscillator strengths for the discrete transitions at higher energy. It is noticeable also that the oscillator strength above the ionization limit at 410 eV does not have the normal shape seen in many photoionization continua, but has an additional structure component of large magnitude.

These various features of the oscillator strength distribution of N_2 have been explained by Dehmer and Dill[75] in terms of the shape and strength of the effective potential experienced by the promoted electron. For example when the electron is in the $2p\pi_g$ orbital it has a d-wave ($l = 2$) component of angular momentum, the effective potential of which includes the centrifugal term

$$V^{\text{cent}}(r) = \frac{\hbar^2 l(l+1)}{2mr^2} = \frac{3\hbar^2}{mr^2}. \quad (86)$$

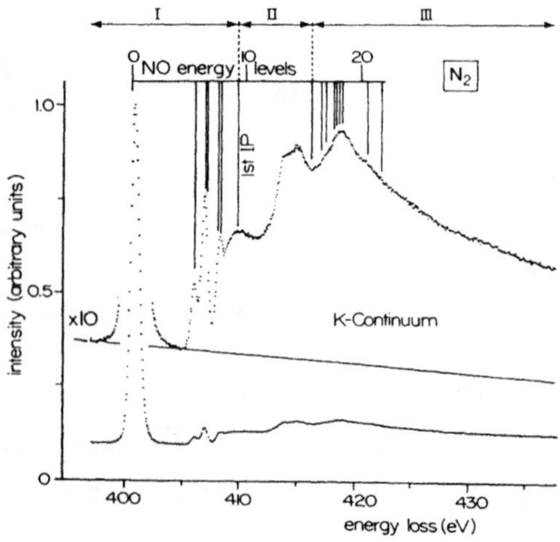

FIGURE 30
Forward-scattering electron energy-loss spectrum of N_2, taken at an incident energy of 2.5 keV and with an energy resolution of 0.5 eV (Wight *et al.*[74]). Atomic $1s$ electrons are promoted to outer molecular orbitals or are ejected into the ionization continuum. Energy levels of the equivalent-core molecule NO are indicated.

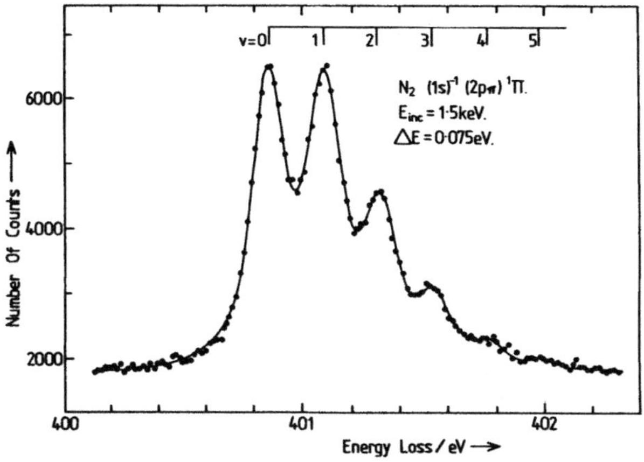

FIGURE 31
Forward-scattering electron energy-loss spectrum of N_2, taken at an incident energy of 1.5 keV and with an energy resolution of 75 meV (King *et al.*[78]). The peaks show the vibrational levels of the inner-shell excited state formed by the promotion of an atomic 1*s* electron to the 2*pπ* molecular orbital. The widths of the peaks correspond to the natural-decay width of the excited state.

This term gives rise to a potential barrier that confines the $2p\pi_g$ orbital and causes it to have a large overlap with the K-shell $1s\sigma_u$ orbital, thus giving the exceptionally high oscillator strength. Brion and his co-workers (see Hitchcock and Brion[76] for references) have used the electron energy-loss technique to study inner-shell excitation spectra of many molecules, and have found many different and curious distributions of oscillator strengths, which lead to a greater understanding of the effective potentials experienced by the outer electrons of these molecules. This and related aspects of inner-shell spectroscopy have been reviewed by Dehmer and Dill.[77]

It is often possible to obtain electron energy-loss spectra which have a higher energy resolution than the corresponding photoabsorption spectra when the states being excited are inner-shell excited states of atoms or molecules. This point is illustrated in Fig. 31, which shows the main $N_2^{K^*}$ peak that appears in Fig. 30, but at an energy resolution of 0.075 eV (King *et al.*[78]). The vibrational structure of the transition is now clearly seen. To obtain this spectrum by the photoabsorption technique it would be necessary to work at the wavelength

$$\lambda = \frac{hc}{E} = 3.1\,\text{nm} \tag{87}$$

(where E is the excitation energy), and with the wavelength spread

$$\Delta\lambda = \lambda\frac{\Delta E}{E} = 0.0006\,\text{nm} \tag{88}$$

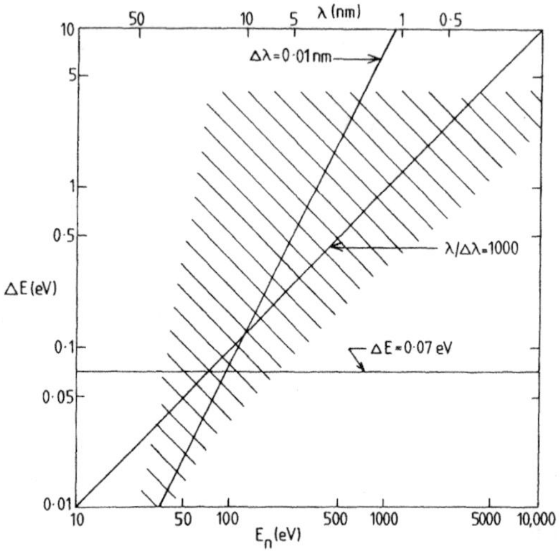

FIGURE 32
The lines show the experimental energy uncertainty ΔE as a function of the excitation energy E_n for two typical operating conditions of an optical spectrometer and a typical resolution of an electron-impact spectrometer. The shaded area indicates the ranges of ΔE and E_n required for measurements of the natural widths of inner-shell excited levels of atoms and molecules.

(where ΔE is the corresponding electron-energy spread). This condition is at present unobtainable.

A comparison of the resolutions obtainable in the electron-impact and photoabsorption techniques is shown in Fig. 32. ΔE is shown as a function of E for a typical operating condition of an electron-impact spectrometer and for two typical operating conditions of an optical spectrometer, namely, one in which $\Delta\lambda$ is constant and another in which $\lambda/\Delta\lambda$ is constant. Because $\Delta\lambda$ is proportional to $\Delta E/E^2$, the electron impact technique has the superior energy resolution at high excitation energies. At present the excitation energy at which the best available resolutions of the two techniques are equal is approximately 150 eV.

Inner-shell excited states usually decay quickly by the process of Auger decay, in which one electron is ejected as another fills the inner-shell vacancy. The lifetimes are typically $\lesssim 10^{-14}$ s, giving natural widths $\Gamma \gtrsim 0.05$ eV. An outcome of the high-energy resolution obtainable in electron energy-loss spectra is the ability to measure these decay widths accurately. For example the separate peaks of Fig. 31 have a width greater than the energy resolution ΔE, and so the value of $\Gamma(0.134 \pm 0.006\,\text{eV})$ can be easily deduced. Figure 32 shows (hatched region) the approximate range of values of Γ and E for inner-shell excited states. The boundaries of this region are not well defined, but the right-hand boundary corresponds to the measured or calculated widths of K-shell excited states (Keski-Rahkonen and Krause[79]), the left-hand boundary represents the minimum E for inner-shell excited states, and the upper boundary represents the position at which the inner-shell states cease to be discrete since Γ becomes greater than their typical spacing. It can be seen from the figure that electron-impact experiments are particularly advantageous for studying the states which have the lowest Γ for a given E (with the exception of the few states in the narrow region below $\Gamma = 0.05$ eV).

TABLE 4
Properties of the Inner-Shell Excited Molecules N_2^{K*} and $C^{K*}O$, and of Their Equivalent-Core Molecule NO[a]

Valence orbital	Molecule	$hc\omega$ (meV)	R_e(nm)	Term value (eV)	Quantum defect	Γ(meV)
	N_2^{K*}	235(2)	0.1162(3)	9.07(3)	0.78(1)	128(6)
$2p\pi$	$C^{K*}O$	273(3)	0.1150(3)	8.65(2)	0.75(1)	100(2)
	NO	236	0.1151	9.26	0.79	
	N_2^{K*}	–	0.1083(4)	3.64(3)	1.07(1)	200(20)
$3s\sigma$	$C^{K*}O$	279(30)	0.1076(10)	3.68(2)	1.08(1)	96(8)
	NO	294	0.1064	3.79	1.11	
	N_2^{K*}	249(3)	0.1068(4)	2.75(3)	0.78(1)	200(20)
$3p\pi$	$C^{K*}O$	301(5)	0.1072(5)	2.72(2)	0.76(1)	96(8)
	NO	295	0.1062	2.78	0.79	

[a]The numbers in brackets give the error in the last significant figure.

The separate peaks of Fig. 31 correspond to excitation to different vibrational levels v of the inner-shell excited molecule. Analysis of this type of spectrum therefore gives information about the geometry and force constants of the excited state. For example the mean energies of the peaks are given by

$$E(v) = E_0 + hc\omega(v+\tfrac{1}{2}) - hc\omega x(v+\tfrac{1}{2})^2,$$

and so $hc\omega$ and $hc\omega x$ can be easily deduced. The mean internuclear separation R_0 of the inner-shell excited molecule can be obtained from the relative areas of the vibrational peaks. To do this King *et al.*(78) assumed that the relative areas are proportional to the Franck–Condon overlap integrals $q_{v'v''}$ between the vibrational wave function of the ground vibrational state ($v'' = 0$) of the target molecule and the vibrational wave functions of the inner-shell excited states (v'). They took the potential functions of both states to be of the Morse form, with that for the target molecule having the values ω, ωx, and R_0 given by Herzberg.(80). The relative peak areas deduced from a least-squares analysis of the spectrum are compared with Franck–Condon factors calculated for different values of R_0 of the excited state, to yield the value giving the best agreement.

We see from this example therefore, how electron-impact spectroscopy provides the values of Γ, ω, ωx, and R_0 of the inner-shell excited N_2^{K*} state. This type of information has been obtained for several other molecules by this technique (see Read and King(81) for a brief review). Table 4 shows a selection of the information obtained on N_2^{K*} and $C^{K*}O$ (a molecule in which a carbon K-shell electron has been excited to an outer valence or Rydberg orbital).

Included in Table 4 is a comparison with the corresponding properties of the ground-state or valence-excited NO molecule. The reason for the similarity in properties of the NO and $C^{K*}O$ molecules is that when a K-shell electron is removed from a C atom the remaining K-shell electron and nucleus have the same total charge as the two K-shell electrons and nucleus of the N atom. The outer electrons of the $C^{K*}O$ and NO molecules therefore move in similar mean fields. NO can therefore be described as the "equivalent-core" molecule.

NO is also the equivalent-core molecule for K-shell excited N_2, provided that we can consider the K-shell vacancy as being localized at one of the nitrogen nuclei. The application of the equivalent-core model to molecules with atoms at equivalent sites (which includes of course homonuclear diatomic molecules) was originally suggested and developed by Prins,[(82)] Nakamura *et al.*,[(83)] and Schwartz.[(84)] The localization of the inner-shell hole in N_2^{K*} has been discussed by King *et al.*[(78)] A localized atomic vacancy [whose wave function is a mixture of those of the $1s\sigma_g$ and $1s\sigma_u$ molecular orbitals – see Eq. (85)] becomes delocalized in a mean time τ_h, where τ_h is related to the energy separation ΔE of the $1s\sigma_g$ and $1s\sigma_u$ orbitals by $\tau_h \Delta E = \hbar$. The value of ΔE has not been measured experimentally, but Cade *et al.*[(85)] have calculated it to be 0.1 eV, which gives $\tau_h \approx 1.3\tau_K$, where τ_K is mean lifetime for Auger decay of the vacancy state. If the inner-shell vacancy is localized at the time of its creation, then the localization is approximately maintained during the existence of the state, even although the two N atoms are indistinguishable. In these circumstances NO is the appropriate equivalent-core molecule, as appears to be borne out by the comparisons shown in the table. It then follows that the inner-shell vacancy should not be described as in Eq. (85), and that the inner-shell excited molecule does not have definite g or u symmetry.

Figure 33 shows another inner-shell energy-loss spectrum for a homonuclear diatomic molecule, Cl_2 (Shaw *et al.*[(86)]). The localization of the inner-shell vacancy is apparent from the fact that the observed spectrum consists of two identical series that are separated in energy and have different relative heights. The energy separation is consistent with the known $(2p_{1/2})-(2p_{3/2})$ energy difference in atomic chlorine, and the relative height is consistent with the statistical weights of the $2p_{1/2}$ and $2p_{3/2}$ levels. More than two series could be expected if the $2p$ vacancy were to have a molecular, rather than atomic, character. The localization of the vacancy implies that the equivalent-core molecule is ArCl, a molecule which is used in ultraviolet lasers, but about which little is known. Shaw *et al.*[(86)] have been able to use the spectrum of Fig. 33 to deduce information about this molecule.

Before leaving the subject of inner-shell excitation by electron impact, let us consider the excitation of dipole-forbidden transitions. We see from Fig. 25 that at the values of T/R and E_n/T typically used in these experiments ($T/R \sim 10^2$ to 10^3, $E_n/T \sim 10^{-2}$ to 10^{-1}), $(Ka_0)_{min}$ is often of the order of unity. For example in high-resolution studies of L-shell excitation in argon (King *et al.*[(87)]) the values of T, E_n, and Ka_0 are 1.5 keV, 250 eV, and 0.90, respectively, and the dipole-forbidden transition $(2p)^{-1} \rightarrow (4p)$ is clearly seen.

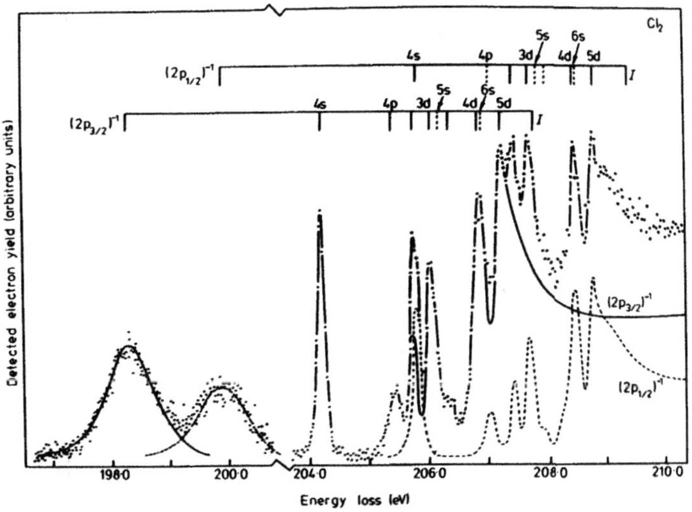

FIGURE 33
Electron energy-loss spectrum of Cl_2 in the region of the L_2 and L_3 ionization edges (Shaw *et al.*[86]). Incident energy 1.5 keV, energy resolution 65 meV. The solid and broken lines show the decomposition into two identical series which are separated in energy and have different relative heights.

The analogous transition $(2p_{3/2})^{-1} \rightarrow (4p)$ in Cl_2 can be seen in the spectrum shown in Fig. 33.

3.3. Electron Scattering Spectroscopy at Intermediate Energies

In the range of incident electron energy from a few eV above threshold to about 100 eV the Born approximation provides in general a rather poor description of the excitation of valence or Rydberg states. This range is therefore not suitable for the simulation of photoabsorption spectra. It is extremely useful, however, for the study of optically forbidden (i.e., electric-dipole forbidden) transitions. These transitions are by definition extremely weak in conventional photoabsorption spectra, and are therefore difficult to observe. Techniques exist (see below) for enhancing the effective absorption cross section, but this usually results in a considerable broadening of the line. Other disadvantages of the optical technique for studying very weak transitions are that small concentrations of impurities can lead to false results and that the presence of "hot" bands (i.e., transitions from thermally populated excited initial states) can mask the transitions being studied. To overcome these difficulties information about the upper states can sometimes be obtained by studying emission spectra.

The electron-impact technique offers a far less restrictive means of studying optically forbidden transitions, since the relevant selection rules are less restrictive than in the photon absorption or emission processes. Of great importance is the fact that it is possible to enhance the differential cross section for forbidden transitions with respect to those for allowed transitions by appropriately choosing the incident energy T and angle of scattering θ. In many cases the type of forbiddenness has been established by studying the dependence of the cross section on T and θ, and in other cases forbidden transitions have been found that are unknown in optical work.

In the past 30 years a considerable amount of work has been carried out in this energy range, much of it stemming from Kupperman, Lassettre, Trajmar, Doering, Kuyatt, Hasted, and their many co-workers. A recent publication by Kupperman *et al.*[88] illustrates the wide-ranging usefulness of the technique for the study of polyatomic molecules. A new power and impetus to the technique has been given recently by Wilden and co-workers, who have used significantly better energy resolutions. In this section we shall illustrate the technique by taking a few specific examples, mostly from the more recent studies. The earlier work is comprehensively described in the reviews of Trajmar *et al.*,[10] Dillon,[12] Hasted,[13] Lassettre and Skerbele,[15] and Celotta and Huebner.[20]

Transitions between molecular states can be optically forbidden (i.e., electric-dipole forbidden) for a variety of reasons, depending on the symmetry of the molecule and the coupling scheme for spin, orbital, and rotational angular momenta (Herzberg[80]). In general the differential cross section for electron-impact excitation via an optically forbidden transition has a dependence on incident energy T and scattering angle θ which is markedly different from that of an allowed transition in the same molecule. For example, as we have seen qualitatively in the introduction to Section 3, allowed transitions have cross sections that are strongly peaked in the forward direction and much smaller in backward scattering, whereas forbidden transitions tend to be more isotropic. More quantitatively, Klump and Lassettre[88a] have deduced the formula

$$\frac{d\sigma}{d\Omega} \propto \frac{(Ka_0)^{2\lambda_m - 2}}{(K^2 a_0^2 + \alpha^2)^{l + l' + \lambda_m + 3}} \tag{89}$$

to represent the approximate dependence on T and θ of the differential cross sections for atomic excitation at high incident energies (when the Born approximation is valid). Here Ka_0 is given by Eq. (46), α is given by

$$\alpha = \left(\frac{I}{R}\right)^{1/2} + \left(\frac{I - E_n}{R}\right)^{1/2}, \tag{90}$$

l and l' are the angular momentum quantum numbers of the initial and final atomic orbitals involved in the transition, and λ_m is the minimum value of λ appearing in the expansion (61) of $f_n(K)$ in powers of $(Ka_0)^2$ (i.e., $\lambda_m = 0$ for an electric-dipole, or optically allowed, transition, $\lambda_m = 1$ for an electric-quadrupole transition, etc). For example the transition $1\,^1S \rightarrow 2\,^1P$ in helium has $l = 0$, $l' = 1$, $\lambda_m = 0$, giving a differential cross section which is approximately proportional to K^{-2} and so is strongly peaked in the forward direction. On the other hand the transition $1\,^1S \rightarrow 2\,^1S$ has $l = l' = 0$, $\lambda_m = 1$, giving the weaker K dependence $(K^2 a_0^2 + \alpha^2)^{-4}$: since $Ka_0 \ll \alpha$ in the forward direction and $Ka_0 \sim \alpha$ in the backward direction the differential cross section still decreases as θ increases, but much less strongly than for the $1\,^1S \rightarrow 2\,^1P$ transition. We see therefore that it is possible to suppress the allowed contributions by observing at large angles of scattering, and in this way many forbidden transitions can be studied that are very difficult to detect in optical work. The dependences on T and θ can also be used in helping to establish the type of forbiddenness. The differences in behavior between different types of forbiddenness are not always marked enough to provide an unambiguous characterization, and so the evidence from electron scattering measurements often has to be combined with other available evidence (e.g., from theoretical calculations of excitation energies).

These points are well illustrated by measurements in N_2. Figure 34 shows an early result of Kupperman *et al.*[89] At a fixed T (35 eV) the ratios of the peak heights in energy-loss spectra are measured at various values of θ. The excitation $X\,^1\Sigma_g^+ \rightarrow b\,^1\Pi_u$ (12.93 eV) is electric-dipole allowed, and is used as a reference. The curve labeled (a) shows the intensity ratio to this reference excitation of the excitation to the $p'\,^1\Sigma_u^+$ (13.21 eV) state. This is also electric-dipole allowed, and the ratio changes by less than a factor of 3 from $\theta = 0$ to 80°. Curve (b) shows the behavior of the excitation to the $^1\Sigma_g^+$ (12.28 eV) state: the gerade symmetry does not change in this process, causing it to be symmetry-forbidden. If the Born approximation were valid one could say that this excitation proceeds predominantly through the monopole term $N(0, 0)$ in Eq. (65), because the initial and final electronic wave functions have an identical symmetry. Expanding the spherical Bessel function for small arguments,

$$j_0(Kr_i) = 1 - \frac{(Kr_i)^2}{6} + \cdots, \tag{91}$$

we see that the first term does not contribute (because of the orthogonality of ψ_0 and ψ_n) and that the second term gives a generalized oscillator strength that is proportional to K^2. This term alone would therefore give a differential cross section [see Eq. (59)] that is independent of K (as long as Ka_0 is small) and is therefore isotropic in this approximation. We see also from this reasoning that $\lambda_m = 1$, and so using Eq. (89) to obtain a better approximation we see that the

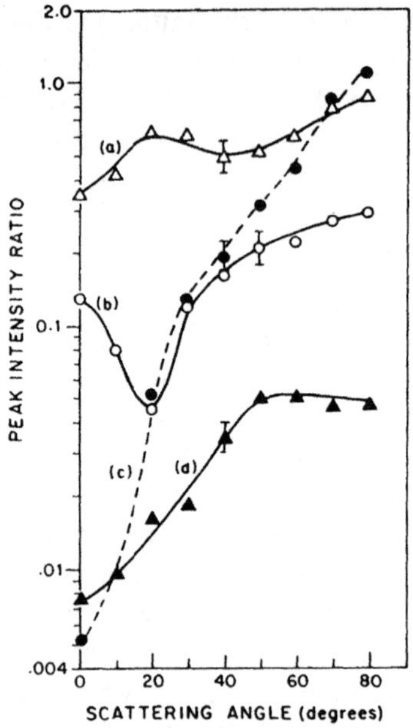

FIGURE 34
Intensity ratios for several transitions in N_2 as a function of scattering angle at an impact energy of 35 eV (Kupperman *et al.*[89]). In all cases the reference transition is the electric-dipole allowed transition to the $b\,^1\Pi_u$ state. The respective upper states are (a) $p'\,^1\Sigma_u^+$ (electric-dipole allowed), (b) $^1\Sigma_g^+$ (g–g symmetry forbidden), (c) $C\,^3\Pi_u$ (spin forbidden), and (d) $E\,^3\Sigma_g^+$ (g–g symmetry forbidden and spin forbidden).

differential cross section is not strongly peaked in the forward direction. On the other hand the allowed excitation is approximately proportional to K^{-2}, and hence it decreases rapidly in going from 0 to 80°. This reasoning thus helps to explain why the cross section for this forbidden excitation is an order of magnitude weaker at 0°, but increases relatively at larger angles. It does not give the cause of the strong dip in curve (b) at 20°.

Curve (c) corresponds to the spin-forbidden excitation of the $C\,^3\Pi_u$ (11.03 eV) state. Here the angular distribution is clearly very different from that of the reference excitation, being much weaker at 0° but of comparable magnitude at 80°. As explained earlier, the roughly isotropic behavior of spin-forbidden excitations arises from the spin-exchange process. Finally, curve (d) corresponds to the spin- and symmetry-forbidden excitation of the $E\,^3\Sigma_g^+$ (11.87 eV) state. It also tends to be more isotropic, but is generally rather small.

In a later study, using an improved energy resolution of 40 meV, Cartwright *et al.*[90] have been able to resolve many more states of N_2 and have measured the angular dependence of differential cross sections. Some of their results are shown in Fig. 35. Four different types of forbidden transitions are illustrated here, together with elastic scattering (which has a θ dependence similar to that of optically allowed transitions). The excitation $X\,^1\Sigma_g^+ \rightarrow A\,^3\Sigma_u^+$ is spin forbidden, and we see that at low energies the angular distribution is approximately isotropic as expected, but that at higher energies a backward scattering peak develops. The excitation $X\,^1\Sigma_g^+ \rightarrow a\,^1\Pi_g$ is symmetry forbidden and in the Born

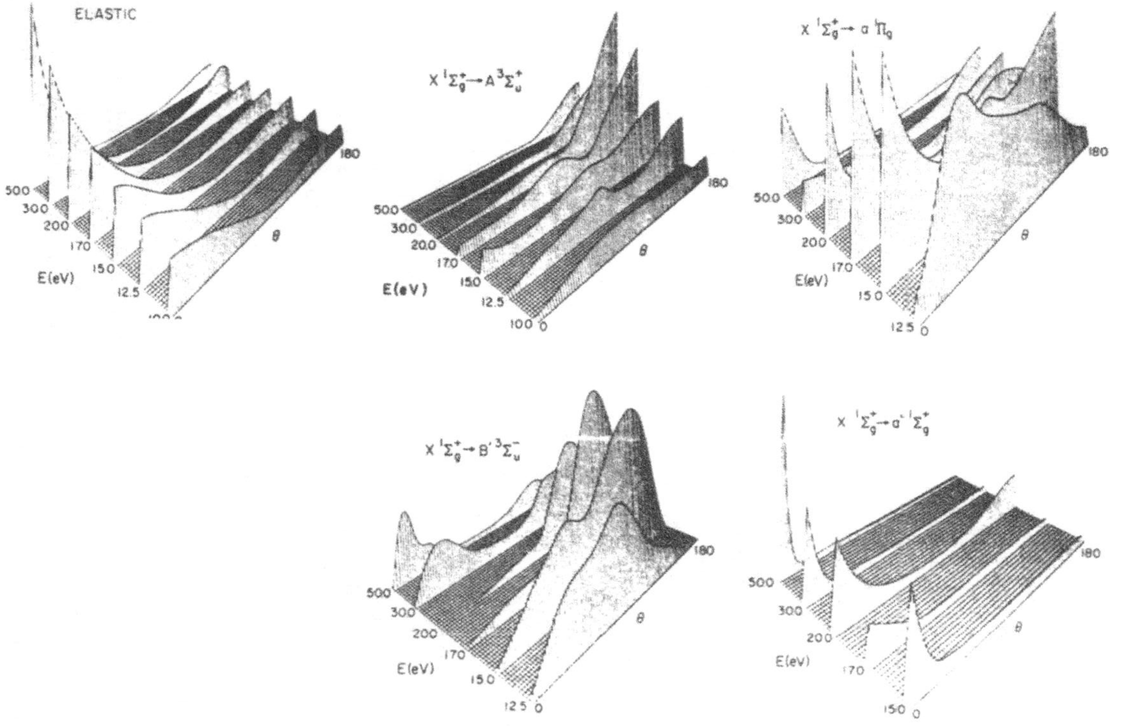

FIGURE 35
Three-dimensional perspective plots of the differential cross sections for electron scattering by N_2. (Cartwright *et al.*[90])

approximation it proceeds through the electric-quadrupole term $N(2, 0)$ in Eq. (65). The spherical Bessel function $j_2(Kr_i)$ is proportional to K^2 at small K, and so $f_n(K) \propto K^2$ and the angular distribution is isotropic (as long as Ka_0 is small) in this approximation. In fact we see that the angular distribution depends strongly on the incident energy, but the forward and backward components are always comparable. The next transition shown, $X\,^1\Sigma_g^+ \rightarrow B'\,^3\Sigma_u^-$, is spin forbidden, but is also symmetry forbidden by the change in reflection symmetry from + to −. Cartwright *et al.*[91] and Goddard *et al.*[92] have shown theoretically that the differential cross section must vanish at 0° and 180° for this type of symmetry change, which is indeed the observed behavior. Finally, we see that the symmetry-forbidden excitation of the Rydberg $a''\,^1\Sigma_g^+$ state has an angular distribution similar to that of elastic or optically allowed transitions. Cartwright *et al.*[90] suggest that this similarity exists because the diffuse nature of the Rydberg orbital causes the excitation to be easily induced by long-range interactions, as is the case in optically allowed excitations (see the introduction to Section 3).

The four different types of forbidden transitions illustrated in Fig. 35 have characteristically different angular distributions, and Cartwright *et al.*[90] find that other examples of forbidden transitions in N_2 fall into this same pattern, each angular distribution being approximately characterized by the type of forbiddenness. But it must be noted that the distributions are in general not sufficiently different from each other to allow a unique identification of an otherwise unknown symmetry.

Further evidence on the type of forbiddenness of a transition can sometimes be obtained by studying the way in which the total or differential cross section depends on the incident energy T. This applies particularly to the identification of spin-forbidden transitions. The spin-exchange process on which such transitions depend occurs most readily when the exchanged (incident and target) electrons have similar velocities, and so the cross sections tend to peak at a few eV above their thresholds. Allowed and symmetry-forbidden transitions on the other hand have cross sections that tend to peak at from about 50 to a few 100 eV above threshold (see, for example, Massey and Burhop[5]). The dependence on energy therefore often gives a reliable identification of spin-forbidden transitions.

The application of these principles to the unraveling of the excited-state structure of a larger molecule is illustrated in Fig. 36, which shows electron energy-loss spectra of benzene (C_6H_6), taken at an energy resolution of 30 meV by Wilden and Comer.[93] When working at near-threshold incident energies it is usual to change the mode of operation of the apparatus so that the analyzer pass-band energy E_s for the scattered electrons is kept constant while the incident electron energy T is varied. The effect of this is that when measuring the yield of the reaction

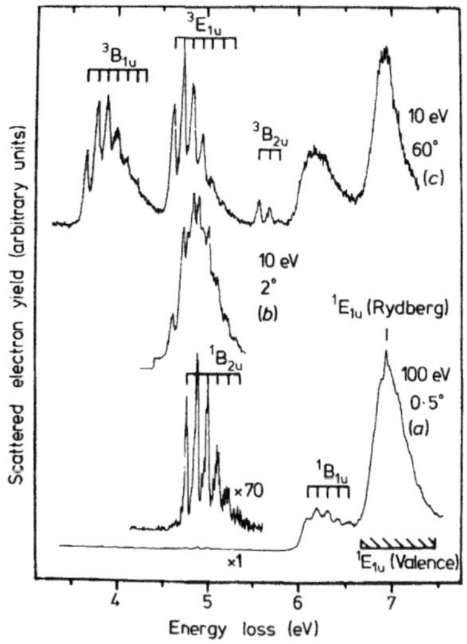

FIGURE 36
Electron energy-loss spectra of benzene (C_6H_6) taken at various incident energies and scattering angles (Wilden and Comer[93]). The energy resolution is 30 meV.

$$e(T) + M(\text{ground state}) \rightarrow M^*(E_n) + e(E_s), \tag{92}$$

a peak is seen when the incident energy is equal to $E_n + E_s$, where E_n is the energy of a molecular excited state. This is known as the *constant residual energy* mode of operation. Each molecular state is then excited at the same energy E_s above its threshold excitation energy. Figure 36 shows spectra taken at residual energies of 10 and 100 eV.

Partly because of its high degree of symmetry (planar-hexagonal) the benzene molecule has long been of intense theoretical and experimental interest (see, for example, Robin[94]), and yet only six valence states of the molecule have been convincingly identified through optical studies. These states are the singlet and triplet states of symmetries B_{1u}, B_{2u}, and E_{1u}, formed by promoting an electron from the highest filled orbital ($1e_{1g}$) to the lowest unfilled orbital ($1e_{2u}$).

In the spectrum labeled (a) in Fig. 36 we see an intense and smooth peak corresponding to the optically allowed excitation of the $^1E_{1u}$ valence state. Superimposed on this is a narrow peak corresponding to the excitation of a Rydberg state that is also of $^1E_{1u}$ symmetry. At lower energies weaker transitions to the $^1B_{1u}$ and $^1B_{2u}$ states can be seen. These would be symmetry forbidden but for the fact that the nuclear vibrational motion introduces a small allowed component into the total electronic-plus-vibrational symmetry of the molecule. This spectrum is taken under conditions (high T, small θ) which simulate the photoabsorption spectrum, and so the transitions appearing here are well known from higher resolution optical studies. In spectrum (b) the values of T and θ have been chosen so that the spin-forbidden transition to the $^3E_{1u}$ state is similar in intensity to the transition to the $^1B_{2u}$ state. In spectrum (c) θ is large enough that the singlet to triplet excitations are dominant below 6 eV.

An optical technique that has been used to study the transition to the lowest triplet state ($^3B_{1u}$) consists of adding a high pressure (typically 100 atm) of O_2 to the absorption cell, so that the spin symmetry of the benzene molecules is perturbed by the magnetic moments of neighboring oxygen molecules (which have $^3\Sigma_g^-$ ground states). The perturbation also broadens the levels, so spoiling the energy resolution, and may also change the spatial symmetry of the molecule (see Wilden and Comer[93] for references and further details). The electron-impact technique does not have these disadvantages.

The transition to the second triplet state ($^3E_{1u}$) is also difficult to study by optical techniques, because of the overlap with the relatively intense transition to the $^1B_{2u}$ state. In the electron energy-loss spectrum it appears clearly, and seven vibrational levels can be identified. The transition to the third triplet state ($^3B_{2u}$) has been observed only in electron-impact spectroscopy (Compton *et al.*,[95] Doering[96,97]). The observed intensity distribution of the vibrational levels is markedly different from that of the other valence transitions, which indicates that the equilibrium geometry of the molecule may be different (i.e., not a plane hexagon) in this state.

Other features in the spectra shown in Fig. 36 and in spectra taken at higher energy losses give a wealth of further information about the excited states of benzene. Other important and simple molecules such as acetylene (Wilden *et al.*[98]) and ethylene (Wilden and Comer[99]) have also been studied in this way. High-resolution electron-impact spectroscopy has clearly become an indispensable aid to molecular structure studies.

3.4. Electron Scattering Spectroscopy at Near-Threshold Energies

The subject of this section is the use of near-threshold electron-impact excitation to obtain spectroscopic information. The definition of the term "near-threshold" is somewhat arbitrary, and there is therefore some overlap with the subject matter of the previous subsection: in general we shall take near-threshold to imply residual (scattered electron) energies of less than a few eV. There is also a partial overlap with the subject matter of Section 2, since resonance mechanisms often play a dominant role in near-threshold excitation processes. We shall therefore make the present subsection rather brief and shall deal only with nonresonance aspects. We shall confine our attention also to high-resolution studies, partly in the interests of brevity and partly because near-threshold electron-impact techniques are usually able to make a significant contribution in the field of molecular spectroscopy only when the electron energy resolution is high. We shall therefore not discuss the earlier low-resolution results, such as those obtained by the trapped-electron or electron scavenger techniques. Compton, Schulz, Brongersma, Hall, Christophorou, Spence, and their various co-workers have been prominent in the field. This work is adequately dealt with in the books by Hasted,[3] McDaniel,[4] and Christophorou,[6] and the reviews by Trajmar *et al.*,[10] Dillon,[12] Hasted,[13] Lassettre and Skerbele,[15] Golden,[19] and Celotta and Huebner.[20] Instead we shall consider only two recent high-resolution results as representative examples of what can be achieved with near-threshold excitation studies.

Figure 37 shows a spectrum of NO taken by Vichon *et al.*[100] at a residual energy of 2 eV, a scattering angle of 140°, and an energy resolution of 40 meV. The experimental conditions have been chosen to enhance the intensities of the quartet transitions relative to the doublet transitions. Nine vibrational levels of the $b\,^4\Sigma^-$ state can be clearly seen. Vibrational levels of the $a\,^4\Pi$ state and of three doublet states are also present. Previous optical studies (see Vichon *et al.*[100] for references and further details) have provided little information on the quartet a and b states, mainly due to the difficulty in observing the spin-forbidden $X\,^2\Pi \rightarrow a\,^4\Pi$ transition in absorption. The vibrational numbering of the a and b states was uncertain, and the vibrational level at 5.72 eV (marked $v = 0$ in Fig. 37) had been thought to be the $v = 1$ level of the $b\,^4\Sigma^-$ state. The error in the vibrational assignment is shown clearly by this spectrum. Vichon *et al.*[100] have used the measured energies of the vibrational levels to obtain the spectroscopic constants T_0, ω_e, and $\omega_e x_e$ and have also used the measured peak

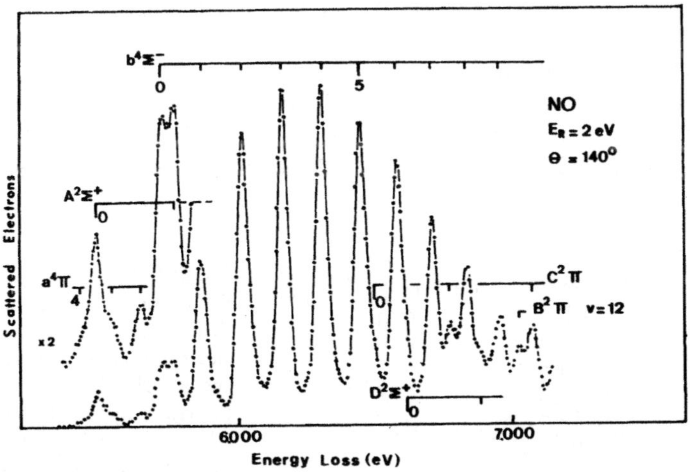

FIGURE 37
Electron energy-loss spectrum of NO, taken with a constant residual energy of 2 eV and at a scattering angle of 140°. (Vichon *et al.*[100])

heights to deduce the Franck–Condon overlap factors (see Chapter 1, Section 6.5.4), and hence to deduce the equilibrium internuclear separation R_e in the $b\,^4\Sigma^-$ state. They have been able to obtain the analogous information for the $a\,^4\Pi$ state from a spectrum taken at a constant residual energy of 6 eV. Spin-forbidden transitions in many other molecules can be studied in this way, the residual energy and scattering angle being chosen in each case to enhance the desired transition relative to others in the same energy region.

Techniques that can truly be described as "threshold" or "near-threshold" are those in which the scattered electrons are detected only when they have a very small energy (in some cases less than about 10 meV). The early ways of achieving this involved the use of electrostatic and magnetic fields to trap low-energy electrons (the trapped-electron technique) or the use of molecules such as SF_6 that have a high attachment cross section for slow electrons (the electron-scavenger technique). These techniques are described in the reviews listed above. The range of residual (scattered electron) energies detected in these techniques was usually from 0 to about 0.1 or 0.2 eV. An improvement in resolution (to 30 meV) was achieved by the Paris group (see, for example, Joyez *et al.*[101] and Mazeau *et al.*[102]) by using a conventional electron energy analyzer tuned to a very low pass-band energy. This has enabled them to obtain high-resolution threshold-excitation spectra of the molecules N_2 and CO.

A technique specifically developed for high-resolution threshold-excitation and threshold-ionization studies is the penetrating-field technique (Cvejanović and Read[103]). The principle of the method is illustrated in Fig. 38. A weak electrostatic field is allowed to penetrate through two windows into an otherwise electrostatically screened target region. Some computed equipotentials are shown in the figure. The electric potential has a saddle point at the center of the region, where the electron and molecular beams intersect, and the resulting quadrupole-type electric field is able to focus low-energy electrons over a wide range of initial angles, causing them to pass out of the windows where they are then reflected at an electrostatic mirror and detected by a channel electron multiplier. Higher-energy electrons are collected over a smaller solid angle, and are not reflected by

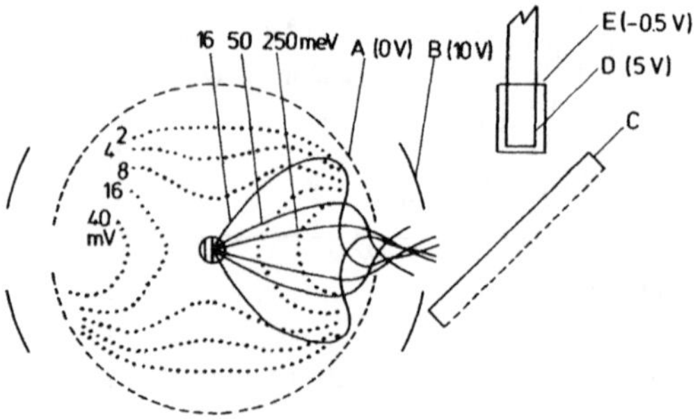

FIGURE 38
Schematic diagram of penetrating-field technique for threshold electron spectroscopy (Cvejanović and Read[103]). A section through the target region, the electrostatic mirror (C), and channel electron detector (D) is shown. The broken lines show computed equipotentials and the full lines computed electron trajectories.

the mirror. In a later version of the technique (Jureta *et al.*[104]) the mirror is replaced by an energy analyzer tuned to a low pass-band energy.

We see that the effect of the penetrating field and external analyzer is to give a very efficient detection efficiency for low-energy electrons, while rejecting higher-energy electrons. The device acts also as a type of electron trap, since if an electron starts with a sufficiently low kinetic energy it is unable to reach the walls of the target region and so can only go out through the windows. For example in Fig. 38 the penetrating field raises the potential of the intersection volume to 12 mV above the potential of the walls, so that an electron having an initial energy less than 12 meV is forced to go through one of the windows, whatever its initial direction of motion. The efficiency with which electrons are extracted through one of the windows and directed to the channel multiplier is therefore 50% from 0 to 12 meV (since half of these electrons will go through the window in question), and falls steeply at higher energies. The range of energies can be controlled by varying the potential on the external electrodes (marked *B* in the figure).

A threshold spectrum obtained using the penetrating-field technique (with the improved analyzer of Jureta *et al.*[104]) is shown in Fig. 39. Several Rydberg series in CO can be clearly seen. The combined effect of the penetrating field and external analyzer results in a cutoff energy of approximately 10 meV for the detected threshold electrons. This very low cutoff value results in a greatly

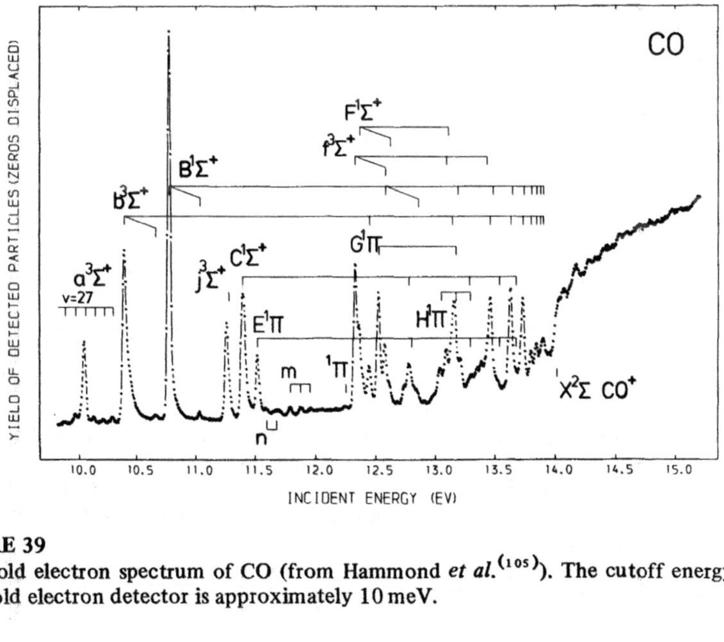

FIGURE 39
Threshold electron spectrum of CO (from Hammond *et al.*[105]). The cutoff energy of the threshold electron detector is approximately 10 meV.

enhanced relative peak height for those states having an excitation cross section that is peaked at or very near to threshold, thus allowing such states to be seen for the first time. The high-energy resolution also allows the excitation energies of many states to be accurately measured for the first time, particularly states of high vibrational quantum number v or high Rydberg principal quantum number n. A detailed analysis of this and other threshold spectra in CO and N_2 has been given by Hammond *et al.*[105] Such studies are valuable in giving additional information about the spectroscopy of small molecules, and will undoubtedly become even more valuable as the attainable energy resolution is improved.

Acknowledgments

We are grateful to Professor M.J. van der Wiel, Dr. J. Comer, and Dr. D.G. Wilden for their helpful comments on the manuscript, and to the authors whose figures we have reproduced, for their permission to do so.

References

1. L. Sanche and G.J. Schulz, Electron transmission spectroscopy: Core excited resonances in diatomic molecules, *Phys. Rev. A* **6**, 69–86 (1972).
2. E.N. Lassettre, A. Skerbele, M.A. Dillon, and K.J. Ross, High-resolution study of electron-impact spectra at kinetic energies between 33 and 100 eV and scattering angles to 16°, *J. Chem. Phys.* **48**, 5066–5069 (1968).
3. J.B. Hasted, *Physics of Atomic Collisions,* Butterworths, London (1964).
4. E.W. McDaniel, *Collision Phenomena in Ionized Gases,* John Wiley and Sons, New York (1964).
5. H.S.W. Massey and E.H.S. Burhop, *Electronic and Ionic Impact Phenomena II,* Oxford University Press, London (1969).
6. L.G. Christophorou, *Atomic and Molecular Radiation Physics,* Wiley-Interscience, New York (1971).
7. K.D. Sevier, *Low-Energy Electron Spectrometry,* John Wiley and Sons, New York (1972).
8. H.S.W. Massey, *Negative Ions,* Cambridge University Press, London (1976).
9. E.N. Lassettre, Inelastic scattering of electrons by atmospheric gases, *Can. J. Chem.* **47**, 1733–1774 (1969).
10. S. Trajmar, J.K. Rice, and A. Kupperman, Electron-impact spectroscopy, *Adv. Chem. Phys.* **18**, 15–90 (1970).
11. M. Inokuti, Inelastic collisions of fast charged particles with atoms and molecules–The Bethe theory revisited, *Rev. Mod. Phys.* **43**, 297–347 (1971).
12. M.A. Dillon, Electron impact, in *Creation and Detection of the Excited State* (A.A. Lamola, ed.), pp. 375–428, Marcel Dekker, New York (1971).
13. J.B. Hasted, Electron scattering spectroscopy, *Contemp. Phys.* **14**, 357–387 (1973).
14. G.J. Schulz, Resonances in electron impact on diatomic molecules, *Rev. Mod. Phys.* **45**, 423–486 (1973).
15. E.N. Lassettre and A. Skerbele. Inelastic electron scattering, in *Methods of Experimental Physics,* Vol. 3 (D. Williams, ed.) pp. 868–951, Academic Press, New York (1974).

16. F.H. Read, Collisions between electrons and molecules, in *Atoms, Molecules and Lasers*, pp. 567–618, International Atomic Energy Agency, Vienna (1974).
17. C.E. Brion, Photoelectron spectroscopy by electron impact, *Rad. Res.* 64, 37–52 (1975).
18. J.F. Williams, The spectroscopy of atomic compound states, in *Progress in Atomic Spectroscopy*, Part B (W. Hanle and H. Kleinpoppen, eds.), pp. 1031–1074, Plenum Press, New York (1978).
19. D.E. Golden, Resonances in electron and molecule scattering, *Adv. At. Mol. Phys.* **14**, 1–87 (1978).
20. R.J. Celotta and R.H. Huebner, Electron impact spectroscopy: An overview of the low-energy aspects, in *Electron Spectroscopy: Theory, Techniques and Applications.* Vol. 3 (C.R. Brundle and A.D. Baker, eds.), pp. 41–125, Academic Press, New York (1979).
21. I.E. McCarthy, Theory of (*e*, 2*e* experiments, in *Coherence and Correlation in Atomic Collisions* (H. Kleinpoppen and J.F. Williams, eds.), pp. 1–12, Plenum Press, New York (1980).
22. A. Giardini-Guidoni, R. Cannilloni, and G. Stefani, Impulsive (*e*, 2*e*) experiments; a tool to test different ionization theories and electronic structure of atoms and molecules, in *Coherence and Correlation in Atomic Collisions* (H. Kleinpoppen and J.F. Williams, eds.), pp. 13–40, Plenum Press, New York (1980).
23. J.A. Simpson and U. Fano, Classification of resonances in the electron scattering cross-section of Ne and He, *Phys. Rev. Lett.* **11**, 158–159 (1963).
23a. A.W. Weiss and M. Krauss, Bound-state calculation of scattering resonance energies, *J. Chem. Phys.* 52, 4363–4368 (1970).
24. D. Spence, Classification of Feshbach resonances in electron–molecule scattering, in *The Physics of Electronic and Atomic Collisions* (J.S. Risley and R. Geballe, eds.), pp. 241–255, University of Washington Press, Seattle (1976).
25. F.H. Read, A modified Rydberg formula, *J. Phys. B* **10**, 449–458 (1977).
26. F.H. Read, J.N.H. Brunt, and G.C. King, The classification of resonances in electron impact on neon, argon, and krypton, *J. Phys. B* **9**, 2209–2219 (1976).
27. J.N.H. Brunt, G.C. King, and F.H. Read, Excitation of carbon monoxide and nitrogen molecules by electron impact below 16 eV: Studies of resonances in the excitation functions of metastable and ultra-violet emitting levels, *J. Phys. B* **11**, 173–192 (1978).
28. F. Gresteau, R.I. Hall, A. Huetz, D. Vichon, and J. Mazeau, Decay of Feshbach resonances in NO: I. Energy dependence of the electronic coupling to the ground state, *J. Phys. B* 12, 2925–2935 (1979).
29. H. Lefebvre-Brion, Nature of the resonant states of NO^-, *Chem. Phys. Lett.* **19**, 456–458 (1973).
30. F. Gresteau, R.I. Hall, A. Huetz, D. Vichon, and J. Mazeau, Decay of Feshbach resonances in NO: II. Dynamic coupling to Rydberg states, *J. Phys. B* **12**, 2937–2945 (1979).
31. D. Spence, Systematics of Feshbach resonances in the molecular halogens, *Phys. Rev. A* **10**, 1045–1052 (1974).
32. D. Spence and T. Noguchi, Feshbach resonances associated with Rydberg states of the hydrogen halides, *J. Chem. Phys.* **63**, 505–513 (1975).
33. F. Fiquet-Fayard, Theoretical problems in the interpretation of dissociative attachment experiments, *Vacuum* **24**, 533–547 (1974).
34. G. Parlant and F. Fiquet-Fayard, The $O_2^{-}\,{}^2\Pi_g$ resonance: Theoretical analysis of electron scattering data, *J. Phys. B* **9**, 1617–1628 (1976).
35. G. Joyez, J. Comer, and F.H. Read, Resonance rotational excitation of H_2 by electron impact, *J. Phys. B* **6**, 2427–2440 (1973).

36. J. Comer and F.H. Read, Potential curves and symmetries of some resonant states of H_2^-, *J. Phys. B* 4, 368–388 (1971).
37. F.H. Read, Angular distribution for resonant scattering of electrons by molecules, *J. Phys. B* 1, 893–908 (1968).
38. D. Andrick and F.H. Read, Angular distributions for the excitation of vibronic states by resonant electron molecule reactions, *J. Phys. B* 4, 389–396 (1971).
39. E.S. Chang, Theory of angular distributions of electrons resonantly scattered by molecules. I. Vibrational and rotational excitation of diatomic molecules, *Phys. Rev. A* 16, 1841–1849 (1977).
40. E.S. Chang, New interpretation of resonances in H_2^-, *Phys. Rev A* 12, 2399–2401 (1975).
41. B.D. Buckley and C. Bottcher, Feshbach projection operator calculations of the resonant states of H_2^-, *J. Phys. B* 10, L635–640 (1977).
42. J.N. Bardsley and J.S. Cohen, Variational calculations of resonant states of H_2^-, *J. Phys. B* 11, 3645–3654 (1978).
43. C. Bottcher and F.H. Read, Angular distributions in resonance scattering of electrons by molecules, *Proc. VIIIth ICPEAC* (Belgrade), pp. 447–488 (1973).
44. D.T. Birtwistle and A. Herzenberg, Vibrational excitation of N_2 by resonance scattering of electrons, *J. Phys. B* 4, 53–70 (1971).
45. J. Mazeau, F. Gresteau, R.I. Hall, G. Joyez, and J. Reinhardt, Electron impact excitation of N_2. I. Resonance phenomena associated with the $A\ {}^3\Sigma_u^+$ and $B\ {}^3\Pi_g$ valence states, *J. Phys. B* 6, 862–872 (1973).
46. A. Huetz, I. Čadež, F. Gresteau, R.I. Hall, D. Vichon, and J. Mazeau, Formation and decay of the first ${}^2\Pi_u$ state of N_2^-, *Phys. Rev. A* 21, 622–628 (1980).
47. J. Mazeau, F. Gresteau, R.I. Hall, and A. Huetz, Energy and width of $N^-({}^3P)$ from observation of its formation by dissociative attachment, *J. Phys. B* 11, L557–560 (1978).
48. A. Huetz, F. Gresteau, and J. Mazeau, Dissociative attachment in N_2, *J. Phys. B* 13, 3275–3284 (1980).
49. A. Huetz, F. Gresteau, R.I. Hall, and J. Mazeau, Initial vibrational state dependence of resonant excitation and "dissociative attachment" in electron–N_2 scattering, *J. Chem. Phys.* 72, 5297–5304 (1980).
50. J. Mazeau, F. Gresteau, G. Joyez, J. Reinhardt, and R.I. Hall, Resonances in electron scattering from CO, *J. Phys. B* 5, 1890–1897 (1972).
51. J.D. Jackson, *Classical Electrodynamics,* John Wiley and Sons, New York (1967).
52. F.H. Read and G.L. Whiterod, Electron impact spectroscopy. I. The determination of the symmetry species of molecular excited states, *Proc. Phys. Soc.* 82, 434–442 (1963).
53. F.H. Read and G.L. Whiterod, Electron impact spectroscopy. III. Calculated cross-sections for inelastic scattering from benzene, *Proc. Phys. Soc.* 85, 71–77 (1965).
54. H. Boersch, J. Geiger, and M. Topschowsky, Rotational structure in the energy loss spectrum of H_2, *Phys. Lett.* 17, 266–267 (1965).
55. A. Hamnett, W. Stoll, G. Branton, C.E. Brion, and M.J. van der Wiel, A kinematic and experimental investigation of the $(e, 2e)$ simulation of photoelectron spectroscopy, *J. Phys. B* 9, 945–957 (1976).
56. R. Huebner, R.J. Celotta, S.R. Mielczarek, and C.E. Kuyatt, Electron energy-loss spectroscopy of acetone vapor, *J. Chem. Phys.* 59, 5434–5443 (1973).
57. J. Geiger and B. Schroeder, Intensity perturbations due to configuration interaction observed in the electron energy-loss spectrum of N_2, *J. Chem. Phys.* 50, 7–11 (1969).
58. R.E. Huffman, Y. Tanaka, and J.C. Larrabee, Absorption coefficients of nitrogen in the 1000–580 Å wavelength range, *J. Chem. Phys.* 39, 910–925 (1963).

59. E.N. Lassettre, F.M. Glaser, V.D. Meyer, and A. Skerbele, Determination of molecular excitation potentials by electron impact. An anomaly in the N_2 spectrum, *J. Chem. Phys.* **42**, 3429–3435 (1965).
60. V.D. Meyer, A. Skerbele, and E.N. Lassettre, Intensity distribution in the N_2 $(b\,^1\Pi_u) \leftarrow X\,^1\Sigma_g^+$ transition, *J. Chem. Phys.* **43**, 3769–3771 (1965).
61. G.M. Lawrence, D.L. Mickey, and K. Dressler, Absolute oscillator strengths of the strongest bands within the dipole-allowed absorption spectrum of nitrogen, *J. Chem. Phys.* **48**, 1989–1994 (1968).
62. F.H. Read, Doppler and other broadening effects in electron scattering experiments, *J. Phys. B* **8**, 1034–1040 (1975).
63. K.H. Tan, C.E. Brion, P.E. van der Leeuw, and M.J. van der Wiel, Absolute oscillator strengths (10–60 eV) for the photoabsorption, photoionization and fragmentation of H_2O, *Chem. Phys.* **29**, 299–309 (1978).
64. R.B. Kay, Ph.E. van der Leeuw, and M.J. van der Wiel, Absolute oscillator strengths for the shape resonance near the *K* edges of N_2 and CO, *J. Phys. B* **10**, 2513–2519 (1977).
65. C. Backx and M.J. van der Wiel, Coincidence measurements with electron impact excitation, in *Vacuum Ultraviolet Radiation Physics* (E. Koch, R. Haensel, and C. Kunz, eds.), pp. 137–153, Pergamon Press, Oxford (1974).
66. M.J. van der Wiel, Inner-shell excitation of molecules by electron impact, in *Electronic and Atomic Collisions* (N. Oda and K. Takayanagi, eds.), pp. 209–218, North-Holland, Amsterdam (1980).
67. R. Huebner, R.J. Celotta, S.R. Mielczarek, and C.E. Kuyatt, Energy absorption by N_2O in the 4 to 14 eV region, *J. Chem. Phys.* **63**, 4490–4494 (1975).
68. C.E. Kuyatt, S.R. Mielczarek, and M.J. Weiss, Intensities of infra-red transitions in N_2O and H_2CO by electron impact spectroscopy, *J. Chem. Phys.* **65**, 3481–3485 (1976).
69. G.C. King and J.W. McConkey, Electron energy-loss spectroscopy in the chlorofluoro-methanes, *J. Phys. B* **11**, 1861–1877 (1978).
70. E.N. Lassettre, A. Skerbele, and M.A. Dillon, Generalized oscillator strength for $1^1S \rightarrow 2^1P$ transition in helium: Theory of limiting oscillator strengths, *J. Chem. Phys.* **50**, 1829–1839 (1969).
71. E.N. Lassettre and S.M. Silverman, Inelastic collision cross-sections of carbon monoxide, *J. Chem. Phys.* **40**, 1256–1261 (1964).
72. W.M. Huo, Limiting slope of the generalized oscillator strengths vs. momentum transfer curve, *J. Chem. Phys.* **71**, 1593–1600 (1979).
73. E.N. Lassettre, Power series representation of generalized oscillator strengths, *J. Chem. Phys.* **43**, 4479–4486 (1965).
73a. L. Vriens, J.A. Simpson, and S.R. Mielczarek, Tests of Born approximations: Differential and total $2^3S, 2^1P$ and 2^1S cross-sections for excitation of He by 100 to 400 eV electrons, *Phys. Rev.* **165**, 7–15 (1968).
74. G.R. Wight, C.E. Brion, and M.J. van der Wiel, *K*-shell energy loss spectra of 2.5 keV electrons in N_2 and CO, *J. Electron Spectrosc.* **1**, 457–469 (1972/3).
75. J.L. Dehmer and D. Dill, Molecular effects on inner-shell photoabsorption *K*-shell spectrum of N_2, *J. Chem. Phys.* **65**, 5327–5334 (1976).
76. A.P. Hitchcock and C.E. Brion, Inner-shell excitation and EXAFS-type phenomena in the chloromethanes, *J. Electron Spectrosc.* **14**, 417–441 (1978).
77. J.L. Dehmer and D. Dill, Connections between molecular photoionization and electron–molecule scattering with emphasis on shape resonances, in *Symposium on Electron–Molecule Collisions* (I. Shimamura and M. Matsuzawa, eds.), pp. 95–104, Univ. of Tokyo, Tokyo (1979).
78. G.C. King, F.H. Read, and M. Tronc, Investigation of the energy and vibrational struc-

ture of the inner-shell $(1s)^{-1}$ $(2p\pi)$ $^1\Pi$ state of N_2 by electron impact with high resolution, *Chem. Phys. Lett.* **52,** 50–54 (1977).

79. O. Keski-Rahkonen and M.O. Krause, Total and partial atomic-level widths, *At. Data Nucl. Data Tables* **14,** 139–146 (1974).
80. G. Herzberg, *Molecular Spectra and Molecular Structure,* Van Nostrand, Princeton (1950).
81. F.H. Read and G.C. King, Inner-shell excited states of molecules, in *Symposium on Electron–Molecule Collisions* (I. Shimamura and M. Matsuzawa, eds.), pp. 155–162, Univ. of Tokyo, Tokyo (1979).
82 M.J.-A. Prins, La structure des discontinuités d'absorption de quelques gaz dans la région intermédiaire (10–100 Å), *Physica* **1,** 1174–1180 (1934).
83. M. Nakamura, M. Sasanuma, S. Sato, M. Watanabe, H. Yamashita, Y. Iguchi, A. Ejiri, S. Nakai, S. Yamaguchi, T. Sagawa, Y. Nakai, and T. Oshio, Absorption structure near the $L_{2,3}$ edge of argon gas, *Phys. Rev. Lett.* **21,** 1303–1305 (1968).
84. W.H.E. Schwartz, X-ray absorption spectroscopy on free molecules, *Angew Chem. Int. Ed.* **13,** 454–465 (1974).
85. D.E. Cade, K.D. Sales, and A.C. Wahl, Electronic structure of diatomic molecules III A. Hartree–Fock wavefunctions and energy quantities for N_2 and N_2^+, *J. Chem. Phys.* **44,** 1973–2003 (1966).
86. D. Shaw, G.C. King, and F.H. Read, An investigation of inner-shell excited states near the $L_{2,3}$ edges of Cl_2 using electron impact with high resolution, *J. Phys. B* **13,** L723–728 (1980).
87. G.C. King, M. Tronc, F.H. Read, and R.C. Bradford, An investigation of the structure near the $L_{2,3}$ edges of argon, the $M_{4,5}$ edges of krypton, and the $N_{4,5}$ edges of xenon, using electron impact with high resolution, *J. Phys. B* **10,** 2479–2495 (1977).
88. A. Kupperman, W.M. Flicker, and O.A. Mosher, Electronic spectroscopy of polyatomics by low-energy variable-angle electron impact, *Chem. Rev.* 79, 77–89 (1979).
88a. K.N. Klump and E.N. Lassettre, Energy transfer cross-sections in relation to generalized oscillator strengths for the donor–acceptor pair benzene–acetone, *J. Chem. Phys.* **68,** 886–895 (1978).
89. A. Kupperman, J.K. Rice, and S. Trajmar, Low-energy, high-angle electron-impact spectroscopy, *J. Phys. Chem.* **72,** 3894–3903 (1968).
90. D.C. Cartwright, A. Chutjian, S. Trajmar, and W. Williams, Electron impact excitation of the electronic states of N_2, *Phys. Rev. A* **16,** 1013–1040 (1977).
91. D.C. Cartwright, S. Trajmar, W. Williams, and D.L. Huestis, Selection rule for Σ^+–Σ^- transitions in electron–molecule collisions, *Phys. Rev. Lett.* **27,** 704–707 (1971).
92. W.A. Goddard, D.L. Huestis, D.C. Cartwright, and S. Trajmar, Group-theoretical selection rules for electron-impact spectroscopy, *Chem. Phys. Lett.* **11,** 329–333 (1971).
93. D.G. Wilden and J. Comer, High resolution electron impact studies of electric dipole forbidden transitions of benzene, *J. Phys. B* **13,** 627–640 (1980).
94. M.B. Robin, *Higher Excited States of Polyatomic Molecules,* Academic Press, London (1975).
95. R.N. Compton, R.H. Heubner, D.W. Reinhardt, and L.G. Christophorou, Threshold electron impact excitation of atoms and molecules: Detection of triplet and temporary negative ions, *J. Chem. Phys.* **48,** 901–909 (1968).
96. J.P. Doering, Electronic energy levels of benzene below 7 eV, *J. Chem. Phys.* **67,** 4065–4070 (1977).
97. J.P. Doering, Low-energy electron-impact study of the first, second, and third triplet states of benzene, *J. Chem. Phys.* **51,** 2866–2870 (1979).

98. D.G. Wilden, J. Comer, and S. Taylor, Energy-loss spectroscopy of the higher excited states of acetylene, *J. Phys. B* **13**, 2849–2857 (1980).
99. D.G. Wilden and J. Comer, Rydberg states of C_2H_4 and C_2D_4: Assignments using the technique of low-energy electron energy-loss spectroscopy, *J. Phys. B* **13**, 1009–1021 (1980).
100. D. Vichon, R.I. Hall, F. Gresteau, and J. Mazeau, Observation of the $a\ ^4\Pi$ and $b^4\Sigma^-$ states of NO by electron-impact spectroscopy, *J. Mol. Spectrosc.* **69**, 341–350 (1978).
101. G. Joyez, R.I. Hall, J. Reinhardt, and J. Mazeau, Low-energy electron spectroscopy of N_2 in the 11.8–13.8 eV energy range, *J. Electron Spectrosc.* **2**, 183–190 (1973).
102. J. Mazeau, C. Schermann, and G. Joyez, Low-energy electron spectroscopy of CO in the 10.600–14.400 eV energy-loss range, *J. Electron Spectrosc.* **7**, 269–279 (1975).
103. S. Cvejanović and F.H. Read, A new technique for threshold excitation spectroscopy, *J. Phys. B* **7**, 1180–1193 (1974).
104. J. Jureta, S. Cvejanović, J.N.H. Brunt, and F.H. Read, Threshold electron spectroscopy of krypton atoms, *J. Phys. B* **11**, L347–351 (1978).
105. P. Hammond, J. Jureta, S.J. Buckman, G.C. King, and F.H. Read, Threshold electron spectroscopy of CO and N_2 using the penetrating field technique, private communication.

6

EXPERIMENTAL TECHNIQUES FOR CROSS-SECTION MEASUREMENTS

SANDOR TRAJMAR and DAVID F. REGISTER

1. Introduction

Instrumentation and experimental techniques for electron-impact cross-section measurements have improved significantly in recent years. This progress has been motivated by the need for electron collision data in connection with the modeling of many practical systems (lasers, plasma devices, planetary ionospheres, chemical systems under the influence of particle bombardment, and so on) and has been aided by advances in vacuum techniques, charged and neutral particle beam technology, progress in electron energy analysis and detection, and computer developments. On one hand the reliability and accuracy of cross-section measurement techniques have reached a stage where a large body of highly accurate (few %) cross sections can now be determined with moderate efforts. On the other hand refinements and sophistication in the experimental techniques make it possible to study and distinguish individual processes which have been averaged over in previous measurements (e.g., individual vibrational–rotational transitions, competing and overlapping dissociation and ionization processes, magnetic sublevel excitation amplitudes, etc.).

The subject of this chapter is electron collision phenomena which may be studied under single-collision conditions at low and intermediate electron-impact energies (from threshold to a few hundred eV) using gas phase molecular targets. Some of the experimental techniques discussed here were developed and applied to atoms but can be applied just as well to molecules with possible minor modifications mainly in the interpretation of the data due to the additional complexity of molecular systems.

It is far beyond the scope of this chapter to make a comprehensive coverage of all the experimental procedures utilized at the present time in laboratories around the world. A brief overview of and references to older and not so widely used techniques will be made. Some of the highly specialized modern techniques (e.g., coincidence measurements, electron scattering in electromagnetic fields)

SANDOR TRAJMAR and DAVID F. REGISTER • Jet Propulsion Laboratory, California Institute of Technology, Pasadena, California 91103.

will be similarly treated. Dissociative excitation, ionization, and attachment is the subject of Chapter 4 and is not discussed here. The emphasis will be on differential (in angle) cross-section measurements using a beam–beam scattering geometry. Such techniques are the ones most widely used, demonstrate most of the basic principles involved in cross-section measurements, and last but not least are the types of experiments with which we are most familiar. Even in this narrow area topics treated in detail in previous publications will be only briefly referred to and details will be given only for recent unpublished methods and considerations which are important for the determination of accurate cross-section data.

2. Definition of Cross Sections

The quantity that characterizes a scattering process is the cross section. It represents the time-independent probability for the occurrence of a particular process. This quantity is well defined in most electron scattering experiments and is the subject of measurements in such studies. One can encounter situations, however, when the definition of scattering cross section is not feasible and only a time-dependent probability can be considered (for example in the case of electron scattering in pulsed high-power laser fields).

The notion of a constant collision cross section used in gas kinetic collisions is not appropriate for electron collision processes. The effective interaction potential between the electron and molecule depends strongly on the electron velocity as well as on the nature of the process and scattering angle being considered. In experiments where only the scattered electrons for a particular excitation process (n) are detected as a function of scattering polar angles $\Omega(\theta, \phi)$, impact energy (E_0), and energy loss (E) we must, in general, define the doubly differential cross section $\partial^2 \sigma_n(E_0, E, \Omega)/\partial E \partial \Omega$.

In conventional electron scattering experiments, several conditions may be applied which reduce the complexity of the defined cross section. Since the cross section away from sharp resonances and thresholds usually does not change appreciably over the energy distribution of the electron beam, it may be well represented by an average doubly differential cross section which is evaluated at the mean energy of the incident energy distribution. (It is not necessary to represent the measured cross section as an integral over the energy distribution of the electron beam). For a discrete process, the doubly differential cross section $\partial^2 \sigma_n(E_0, E, \Omega)/\partial E \partial \Omega$ may be integrated over the energy-loss profile to obtain the differential (in angle) cross section $d\sigma_n/d\Omega$. Finally, since the target molecules are randomly oriented, the measurement can only yield a cross section averaged over molecular orientations which is independent of the azimuthal angle ϕ. Furthermore, the measuring apparatus averages over a finite energy and angular range and the quantity which is determined in most beam–beam experiments is

$$\mathrm{DCS}_n(E_0, \theta) = \overline{\frac{d\sigma(E_0, \Omega)}{d\Omega}} \tag{1}$$

(in area per steradian units). $\mathrm{DCS}_n(E_0, \theta)$ is related to the integral cross section as given below by Eq. (3).

For electron–molecule scattering it is seldom possible to measure a cross section for a well-defined individual process. In most cases one measures the average cross section for a number of experimentally indistinguishable processes. For example cross sections associated with individual isotopes, their hyperfine and magnetic sublevel structure are distinguished only in special scattering experiments. This also applies to electron spin and target orientations. The individual rotational excitations (with the exception of H_2) and sometimes even vibrational excitations (for complex molecules) can not be separated by present-day energy resolutions. For a comparison of experiment with theory, a general averaging procedure which averages over unresolved initial states (i) and sums over unresolved final states (f) has to be carried out:

$$\mathrm{DCS}_n(E_0, \theta) = \sum_i N_i \sum_f \mathrm{DCS}_n^{fi}(E_0, \theta), \tag{2}$$

where N_i is the fractional population in the ith state.

Integration of the DCS over all scattering angles yields the integral $\sigma_n(E_0)$ and momentum-transfer $\sigma_n^M(E_0)$ cross sections:

$$\sigma_n(E_0) = 2\pi \int_0^\pi \mathrm{DCS}_n(E_0, \theta) \sin\theta \, d\theta, \tag{3}$$

$$\sigma_n^M(E_0) = 2\pi \int_0^\pi \mathrm{DCS}_n(E_0, \theta) \left[1 - \frac{k_n}{k_0}\cos\theta\right] \sin\theta \, d\theta, \tag{4}$$

where k_0 and k_n are the magnitudes of the initial and final momenta of the electron, respectively. The total scattering cross section represents the sum of all integral cross sections:

$$\sigma_T(E_0) = \sum_n \sigma_n(E_0). \tag{5}$$

For the special case of coincidence experiments, where the scattered electrons are detected in coincidence with other secondary particles (electrons, photons, ions, neutral fragments), one has to define cross sections differential also with respect to the energy and angular distribution of these secondary species.

All of these cross sections may be further defined in terms of scattering amplitudes, phase shifts, or *S*-, *T*-, or *K*-matrix elements. As these quantities are the usual results of theoretical work and are only rarely amenable to direct experimental measurements, discussion of these topics is left to the introductory and theoretical chapters of this volume.

3. Measurement of Total, Integral, and Momentum-Transfer Cross Sections

3.1. Total Cross Sections

3.1.1. Transmission Method

Total electron scattering cross sections have been measured since the 1920s using the Ramsauer technique.(1) An improvement in the original apparatus was made by Golden and Bandel(2) and detailed descriptions of these methods have appeared previously.(3-6) At the present time, attenuation measurements using straight-through collision chambers are more frequently used.(7,8)

Figure 1 shows a schematic diagram of the experimental arrangement. The discussion of the method given here follows closely that of Ref. 6. An electron beam with energy distribution $f(E_0)$ $[\int f(E_0)dE_0 = 1]$ and current $I(E_0)$ $[=I_T f(E_0)]$ enters the scattering chamber through aperture (or slit) S1. Part of the beam reaches the detector (I^D) and part is captured by the chamber (I^{CH}):

$$I_T = \int I(E_0)dE_0 = I^D + I^{CH} = \text{const.} \tag{6}$$

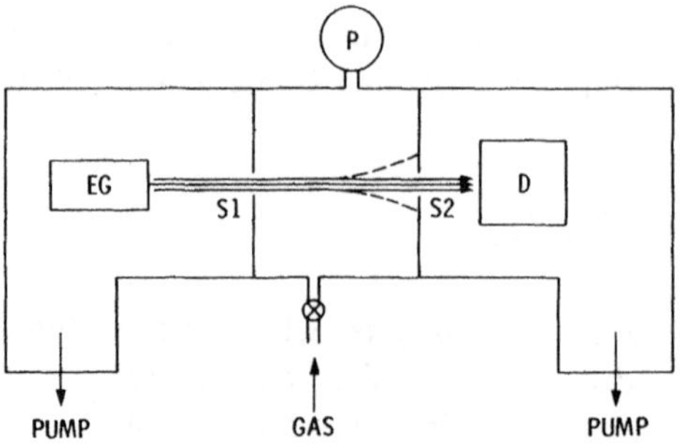

FIGURE 1
Schematic diagram of the transmission method. EG, P, D, S1, and S2 denote the electron gun, pressure gauge, electron detector, and entrance and exit slits, respectively.

At each energy within the distribution $f(E_0)$ we may define the two-dimensional electron-beam shape factor as $T(S, E_0)$ and $\int T(S, E_0) dS = 1$, where dS represents a two-dimensional area element in the surface perpendicular to the electron-beam axis. The current reaching the detector when the impact energy is E_0 in the absence of gas is

$$I_0^D(E_0) = I(E_0) \int_{S2} T(S, E_0) dS, \tag{7}$$

where the subscript zero refers to the absence of gas and the integral is over opening S2.

When the scattering chamber contains a gas (number density ρ), a fraction of the electrons will be deviated from their original paths and/or lose energy as a consequence of collisions with the gas. The attenuation of the incident current as measured at the detector is given by

$$I^D(E_0) = I(E_0) \exp\left[-\rho \iint T(S, E_0) F(E_0, \Omega; P) \frac{\partial \sigma_T}{\partial \Omega} dV\, d\Omega\right], \tag{8}$$

where $\partial\sigma_T/\partial\Omega$ is the total differential cross section for all scattering processes and $F(E_0, \Omega; P)$ is a function which represents the fraction of electrons either scattered out of or into the beam at the point P by primary collisions. The integral is evaluated over all solid angles (Ω) for each collision point (P) in the chamber and this result must be integrated over the entire electron beam–gas target interaction volume (V). To obtain the total detected current; Eq. (8) must be integrated over the initial energy distribution:

$$I^D = I_T \int f(E_0) \exp\left[-\rho \iint T(S, E_0) F(E_0, \Omega; P) \frac{\partial \sigma_T}{\partial \Omega} dV\, d\Omega\right] dE_0. \tag{9}$$

If the energy distribution of the primary beam is sufficiently narrow so that the cross section and the functions T and F are essentially constant over the energy spread of the incident beam, then the detected signal may be written as

$$I^D = I_T \exp(-\rho l \sigma_{\text{eff}}), \tag{10}$$

where

$$\sigma_{\text{eff}} = \frac{1}{l} \iint T(S, E_0) F(E_0, \Omega; P) \frac{\partial \sigma_T}{\partial \Omega} dV\, d\Omega, \tag{11}$$

and l is the length of the interaction region. Only under ideal conditions can the desired total cross section

$$\sigma_T = \int \frac{\partial \sigma_T}{\partial \Omega} \, d\Omega \tag{12}$$

be directly related to the measured effective cross section.

Under experimental conditions where the detector view cone is sufficiently narrow and the pressure sufficiently low that multiple scattering is not significant, it may be assumed that all collisions result in an electron loss with no electrons being scattered back into the beam. The function $F(E_0, \Omega; P)$ may then be taken as 1 for scattering out of the beam and 0 for scattering into the detector. If it may be further assumed that the beam shape factor $T(S, E_0)$ is constant across the electron-beam cross section and along the length of the interaction volume, then σ_{eff} is equivalent to σ_T and Eq. (10) becomes

$$I^D = I_T \exp(-\rho l \sigma_T). \tag{13}$$

A plot of $\ln(I_T/I^D)$ versus ρ should then yield σ_T.

In practice it may be more convenient to measure the current transmitted with and without the gas in the chamber rather than the current entering the chamber. In this case,

$$I_T = I_0^D + I_0^{CH} = KI_0^D, \tag{14}$$

where I_0^{CH} represents current loss at the exit aperture of the chamber in the absence of any gas. Under these conditions, Eq.(13) takes the form

$$\sigma_T = \frac{1}{\rho l} \left[\ln(I_0^D/I^D) + \ln K \right]$$

and a plot of ρ versus $\ln(I_0^D/I^D)$ should be a straight line whose slope should yield the cross section. This method is frequently used and more details may be found in Refs. 3 and 6.

3.1.2. Recoil Method

Although a recoil measurement is carried out in a beam–beam arrangement, it is in fact also a transmission technique but instead of a cell containing the target gas one crosses an electron beam with a molecular beam and the observation of the scattering event is made on the scattered (recoiled) molecules rather than on the electrons. The method has been extensively used on atomic systems[9] but applies equally well, in principle, to molecular species. One has to deal here with narrow, well-defined molecular beams and efficient neutral-beam detectors. The recoil imparted to the target beam particle in the collision process deflects the molecule out of the original beam and the decreased beam intensity is a measure of the total collision cross section. A rough measure of the change of direction (α) of the scattered molecule is the ratio mv/MV of electronic to atomic momentum: $\alpha(\text{recoil}) \approx 140[E_0/MT]^{1/2}$ degrees, where E_0 is the electron energy in eV, M is the molecular weight, and T is the molecular-beam-particle kinetic temperature in K.

In total cross-section measurements, the attenuation of the dc molecular beam by a perpendicular and modulated (ac) electron beam is observed. The total scattering signal I_S for two rectangular beams is related to the total scattering cross section σ_T as

$$I_S = \frac{I_M I_e \sigma_T}{\bar{V}h} \quad \text{or} \quad \sigma_T = \frac{I_S}{I_M} \frac{h\bar{V}}{I_e}. \tag{15}$$

I_M and I_e are the total molecular-beam and electron-beam fluxes, respectively, h is the common dimension for the rectangular molecular and electron beams, and $\bar{V}$ is the average speed of molecules in the beam source.

For the rigorous description of the signal-to-cross-section relations, the same geometrical and flux distribution questions, etc., have to be considered as for conventional transmission experiments discussed in Section 3.1.1. These questions are discussed in Ref. 9, and Eq. (15) was derived there.

The recoil technique has also been utilized for differential cross-section measurements with velocity selected target beams. The change of spin states in a scattering process can also be studied by this technique. In such cases the target beam has to be velocity and state selected before entering the scattering region. These experiments are not discussed here but the reader is referred to Refs. 9, 10, and 11.

3.1.3. Transmission Method with Time-of-Flight Discrimination

Time-of-flight (TOF) techniques for the determination of energy distribution of particles is frequently used in collision experiments. It can be applied to electrons at low energies.[12] The upper energy limit for the application of this technique depends on the time resolution capabilities of the apparatus. Recently, Kennerly and Bonham[13] applied high time-resolution (of the order of 100 ps) techniques to determine total electron scattering cross sections for a number of molecular (and atomic) species in a transmission-type experiment.

A schematic representation of the apparatus used by Kennerly and

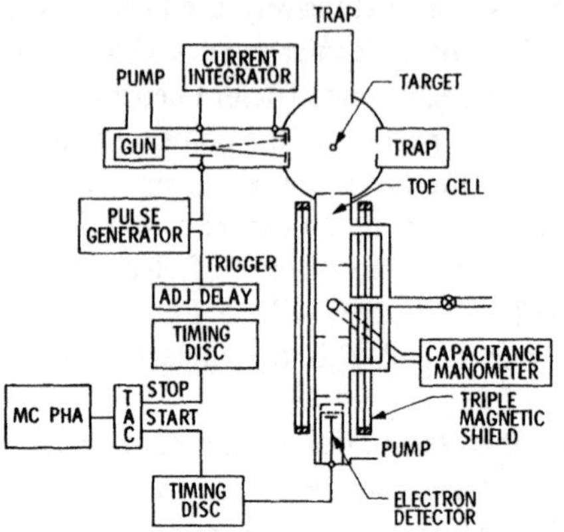

FIGURE 2
Schematic diagram for the time-of-flight apparatus of Kennerly and Bonham.[13]

Bonham is shown in Fig. 2. For total cross-section measurements an electron pulse of 2000 eV energy and with a 100 ps time width is generated. The energy spread of the beam is about 0.5 eV. The electrons are focused on a graphite target. Secondary electrons scattered and ejected by the target in the direction of 90° with respect to the initial beam are transmitted through the cell to the electron detector. This cell serves both as a field-free flight tube and as a gas cell. The electrons enter and exit the cell through circular apertures. When the gas is absent from the cell, the time distribution of the detected electron signal can be converted to the energy distribution of the secondary electrons generated by the graphite target. When gas is present in the cell the attenuation of the secondary electron beam as a function of impact energy can be determined and related to the total scattering cross section as in other transmission experiments. The time-of-flight method, however, avoids some of the problems common to conventional transmission methods.

A timer is started when an electron which has passed through the TOF cell is detected by the channel electron multiplier array. The stop pulse is provided by a delayed trigger pulse from the pulse generator. A time-to-pulse-height converter and multichannel pulse-height analyzer then generates the detected electron count rate as a function of electron energy. This distribution is recorded with and without gas in the cell and properly weighted for data acquisition time and beam current. The difference of these two distributions, together with the measured target density, and path length is used in the Beer–Lambert formula [Eq. (13)] to obtain the total scattering cross section as a function of impact energy.

Since the gas pressure in the TOF cell is low (a few millitorr), the effects of changing gas pressure on the electron gun and detector are negligible if further differential pumping of the detector and graphite target regions is applied. The electron beam current is very low ($\sim 10^{-14}$ A) which corresponds to one or less electron in the flight tube at any time. Space charge and surface charging effects are, therefore, negligible. These precautions and the use of a carefully calibrated capacitance manometer resulted in an overall accuracy in the cross-section measurements of about 3% in the 1–50-eV impact energy range. At lower energies flux problems and at higher energies time resolution problems make the measurements more difficult and less accurate.

3.2. *Integral Cross Sections*

Integral electron-impact excitation cross sections for selected excitation processes or combined ionization and excitation processes for molecules (and even more so for atoms) have been primarily obtained in the past by measuring optical excitation functions (apparent electron-impact excitation cross section as a function of impact energy). The optical excitation function can be corrected to various degrees of approximations to yield electron impact excitation cross sections. In this method the target molecules are excited by a beam of nearly monoenergetic electrons and the photon emission rates by the excited species are quantitatively determined. Details of the experimental procedures, data evaluation, and error sources have been discussed by Heddle[14] and by Heddle and Keesing[15] for atomic gases. The basic principles apply just as well to molecules with the additional complication of vibrational and rotational energy level structures.

A simplified diagram of the pertinent processes is shown in Fig. 3. The volume rate of production of molecules in electronic and vibrational state j by electron impact is given by $\rho_g \rho_e \bar{v} \sigma_{gj}$, where ρ_g and ρ_e are the concentrations of ground-state molecules and electrons, respectively, $\bar{v}$ is the relative velocity, and σ_{gj} is the electron-impact cross section for excitation of state j. Alternatively, one can introduce the electron beam current I and then the production rate is $(I/eS)\rho_g \sigma_{gj}$, where e is the electronic charge and S is the cross-sectional area of the beam. The volume rate of radiative depopulation is $\rho_j \sum_k A_{jk}$, where A_{jk} is the Einstein transition probability for spontaneous transition from state j to k (k may include g) and ρ_j is the concentration of molecules in the excited state j. In an experiment a particular $j \to k$ transition is usually observed. The rate of emission of photons corresponding to this particular $j \to k$ transition per unit length of the electron beam, then, is given as $R_{jk} = \rho_j A_{jk} S$. If all other processes contributing to the population and depopulation of state j can be neglected, then under steady-state conditions we have

$$\rho_g \sigma_{gj} \frac{I}{eS} = \rho_j \sum_k A_{jk} = R_{jk} \frac{\sum_{k'} A_{jk'}}{A_{jk} S}. \tag{16}$$

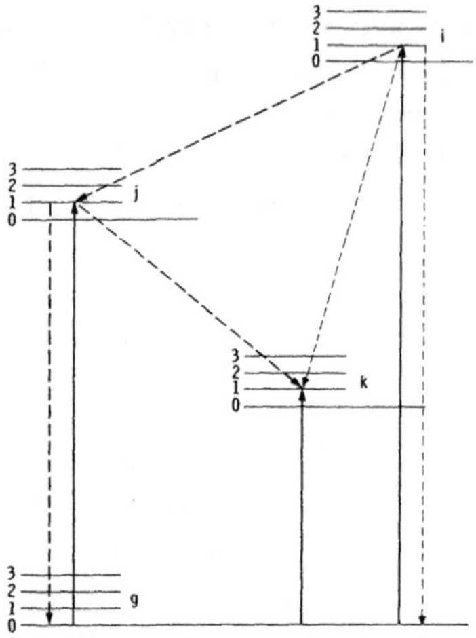

FIGURE 3
A simplified molecular energy-level diagram: g, k, j, and i represent electronic states with vibrational levels 0, 1, 2, 3, etc. Solid lines denote possible electron-impact excitations while dashed lines indicate possible radiative decay processes.

Since rotational structure is not resolved, the usual summing over final and averaging over initial rotational states is implied [Eq. (2)]. To obtain cross sections for the excitations of other vibrational levels of the upper electronic state, the Franck–Condon factors can be utilized.

The experimental determination of $\sigma_{gj}(E_0)$ requires the absolute measurement of the photon emission rate R_{jk}, the knowledge of all spontaneous emission probabilities, and the scattering geometry. A number of experimental problems arise in measuring R_{jk}. In most experiments the photon signal, collected over a small solid angle at 90° with respect to the electron beam, passes through some optical system and monochromator to the photon detector. The overall efficiency of this detector system at the wavelength corresponding to the $(j \rightarrow k)$ photon has to be known. This calibration is usually done by utilizing standard black-body sources which correspond to unpolarized radiation with isotropic spatial distributions. However, the radiation emitted in the actual experiment may be polarized (or equivalently, anisotropically distributed into the 4π solid angle). If the detector system is polarization sensitive and the radiation is polarized, correction for this effect also has to be considered.

Equation (16) requires the total rate of photon emission per unit length of the electron beam $[R_{jk} = \int R_{jk}(\Omega)\, d\Omega]$, but usually the radiation emitted into a small solid angle $\Delta\Omega$ in a direction Ω is measured. One needs to know, therefore, the angular distributions of the photons (or has to determine the polarization of the radiation at the angle of measurement) to obtain R_{jk}. Most frequently the latter procedure is followed. Alternatively, one can carry out the photon measurement at the magic angle (54.7°) with respect to the electron beam which can give information about the total photon flux regardless of the state of polarization of the light.[14]

So far it has been assumed that population and depopulation contributions other than electron-impact excitation of state j and depopulation by spontaneous emission from state j can be neglected. When this assumption is valid, the apparent excitation cross section is equal to the true excitation cross section. This assumption is, however, usually false and a number of other contributions to the population equation may have to be considered. The electron beam excites with some probability all energetically accessible state of the molecule. States which are at a higher energy than the one in consideration (state j) may then decay by radiation to state j. This so-called cascade contribution can be corrected if the spontaneous transition probabilities involved are determined or calculated. The electron beam can also in principle depopulate state j by superelastic collisions or by further excitations, but this is a highly unlikely event. Additional contributions to the population and depopulation of state j may arise from resonant radiation trapping (when state k is the ground state) and from collisional deactivation, respectively. The elimination of both of these effects requires working at low target densities.

Excitation function measurements have recently been carried out on atomic systems with an electron energy resolution of about 0.030 eV[16,17] (as compared to previous results with 0.3 eV resolution). These studies have revealed sharp structures (resonances) in the excitation function. Similar techniques could also be used for investigation of molecular species.

At low impact energies the trapped electron method developed by Schulz[18] can be utilized to obtain excitation functions. Figure 4 illustrates the basic principles of this method. An electron beam with kinetic energy eV enters a chamber containing the target molecules. (V is the potential between the source of the electrons and the entrance slit of the chamber). The electron beam passes along the axis of a cylindrical grid G which is kept at the same potential as the entrance (S1) and exit (S2) slits. The grid is surrounded by a concentric, insulated collector plate (P), which is held positive with respect to G by an amount V_P.

Taking the grid and slit potential as zero, the potentials along the electron path are indicated at the bottom of Fig. 4. The positive trapping potential W, felt by the electrons, is generated by the field V_P penetrating through the grid. This potential well acts as a trap for slow electrons which are produced by inelastic collisions. A magnetic field (B) is applied parallel to the axis of the chamber to align the primary beam. The slow electrons will gradually diffuse in a direction perpendicular to this field to be collected by the detector plate.

The energy of the electrons in the beam within the collision chamber is $e(V + W)$. The electrons which are collected by the plate must have lost an

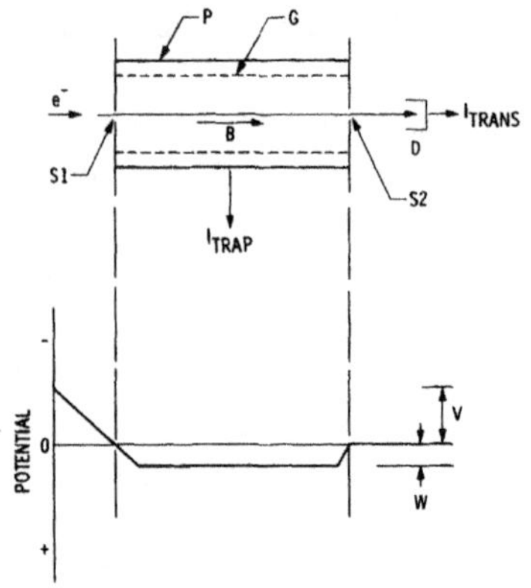

FIGURE 4
Schematic diagram of the trapped electron method (upper figure) and associated potentials. eV is the incident electron energy and W is the trapping field produced by field penetration from the plate (P) through the grid (G). The slits (S1 and S2) and the detector (D) are indicated.

amount of energy between eV and $e(V + W)$. By varying V and measuring the trapped electron current (I_{TRAP}), excitation processes with thresholds between eV and $e(V + W)$ contribute to the signal. This signal can be related to inelastic cross sections, but in practice a number of problems make it difficult to obtain quantitative results.[19] Although a number of improvements have been achieved on the basic trapped electron technique,[20,21] it is mostly used for relative excitation function measurements and for resonance studies.

3.3. Momentum-Transfer Cross Sections

Momentum-transfer cross sections at low electron energies are most commonly determined by the electron swarm technique. (Beam–beam experiments which yield both elastic and inelastic momentum-transfer cross sections are mandatory at higher energies. The beam–beam method will be discussed in detail in Section 4). The question of the steady-state passage of electrons through a gas under the influence of multiple collisions and electric (and magnetic) fields has been considered by Townsend and Baily[22,23] as early as the electron transmission method (single collision conditions) by Ramsauer and co-workers. An excellent and detailed discussion of diffusion and drift of electrons in gases is given by Huxley and Crompton.[24] It is far beyond the scope of the present chapter to discuss the details of these measurements. We briefly summarize the principles.

In the swarm method a stream of electrons enter a diffusion chamber through a small opening from the cathode and proceed through the gas (few-Torr pressure) under the influence of a uniform electric field E. (For simplicity we assume no magnetic field). The stream proceeds to the anode, on its way colliding with the gas and spreading by diffusion as indicated in Fig. 5. The

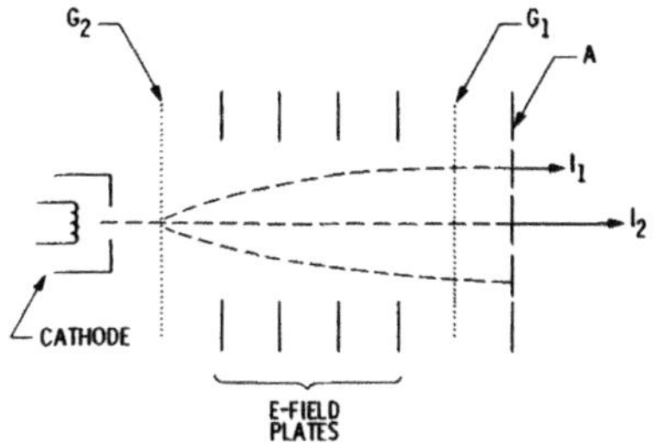

FIGURE 5
Schematic diagram of a swarm experiment. G_2, G_1, and A are the entrance and exit grids and anode, respectively.

behavior of the electrons is governed by the Boltzmann equation. The collisions with the gas manifest themselves by affecting the Boltzmann collision integral and as a consequence the macroscopic transport properties. The distribution of positions and velocities of the electrons at a given instant can be described by the distribution function $F(\mathbf{r}, \mathbf{v}, t)$. The number of electrons within a volume element $d\mathbf{r}$ at position $\mathbf{r}$ which have velocities between $\mathbf{v}$ and $\mathbf{v} + d\mathbf{v}$ is $F(\mathbf{r}, \mathbf{v}, t)$ $d\mathbf{r}\, d\mathbf{v}$ and the number density of electrons is given as

$$n(\mathbf{r}, t) = \int F(\mathbf{r}, \mathbf{v}, t) d\mathbf{v}. \tag{17}$$

The distribution function is found by the solution of the Boltzmann equation

$$\frac{\partial F}{\partial t} + \mathbf{v} \cdot \nabla_{\mathbf{r}} F + \mathbf{a} \cdot \nabla_{\mathbf{v}} F = \left(\frac{\partial F}{\partial t}\right)_{\text{coll}}, \tag{18}$$

where $\mathbf{a}$ is the acceleration produced by an external force and $(\partial F/\partial t)_{\text{coll}}$ is the rate of change of the number of electrons in a unit volume of phase space due to collisions with gas molecules. If only a steady uniform electric field $\mathscr{E}$ is applied, the electron distribution is independent of t and $\mathbf{r}$ and we write $F(\mathbf{r}, \mathbf{v}, t) = f(\mathbf{v})$. In solving Eq. (18) it is a generally accepted practice to expand the distribution function in terms of spherical harmonics and retain only the two terms with $l = 0, 1$ and $m = 0$. That is

$$f(\mathbf{v}) = f_0(v) + f_1(v) \cos\theta. \tag{19}$$

Two equations in $f_0(v)$ and $f_1(v)$ are then derived from Eq. (18). It is $f_0(v)$ that appears in the expressions for the transport coefficients. The Boltzmann equation for a case where both elastic and inelastic collisions occur becomes[24,25]

$$\frac{(e\mathscr{E})^2}{3}\frac{d}{dE_0}\left[\frac{E_0}{\rho\sigma_T^M(E_0)}\frac{df(E_0)}{dE_0}\right]+\frac{2m}{M}\frac{d}{dE_0}\left[E_0^2\,\rho\sigma_T^M(E_0)\,f(E_0)\right]$$

$$+\frac{2mkT}{M}\frac{d}{dE_0}\left[E_0^2\rho\sigma_T^M(E_0)\frac{df(E_0)}{dE_0}\right]+\sum_{jk}(E_0+E_{jk})f(E_0+E_{jk})\rho\sigma_{jk}(E_0+E_{jk})$$

$$-E_0f(E_0)\rho\sum_{jk}\sigma_{jk}(E_0)+\sum_{jk}(E_0-E_{jk})f(E_0-E_{jk})\rho\sigma_{-jk}(E_0-E_{jk})$$

$$-E_0f(E_0)\rho\sum_{jk}\sigma_{-jk}(E_0)\ =0. \tag{20}$$

Here the velocity distribution $f_0(v)$ has been replaced by the energy distribution, the subscript zero has been suppressed, and it has been assumed that the integral inelastic cross sections are small compared to the elastic cross section and that the inelastic scattering is isotropic. The various symbols have the following meaning: $\sigma_T^M(E_0)$ is the energy-dependent effective momentum-transfer cross section, which represents both elastic and inelastic encounters (and is usually assumed to be nearly equal to the elastic momentum-transfer cross section); ρ is the gas number density; m is the electron mass; M is the molecular mass; k is the Boltzmann constant; T is the absolute gas temperature; $\sigma_{jk}(E_0)$ is the integral excitation cross section from state j to state k; $\sigma_{-jk}(E_0)$ is the integral cross section for the $k \to j$ superelastic process; E_{jk} is the difference in internal energy between the jth and kth molecular states.

If $f(E_0)$ were known, Eq. (20) could be solved by varying a set of trial cross sections as a function of E_0 until the equation is satisfied and a self-consistent set of true cross sections is obtained. However, $f(E_0)$ is generally not known and the procedure which is usually followed is to compare the experimentally measured and the calculated drift velocity W and the characteristic energy eD/μ. These quantities are given as

$$W=\mu\mathscr{E}=\left(\frac{2}{m}\right)^{1/2}\frac{e\mathscr{E}}{3\rho}\int_0^\infty\frac{E_0}{\sigma_T^M(E_0)}\frac{df(E_0)}{dE_0}\,dE_0, \tag{21}$$

$$\frac{eD}{\mu}=-\left\{\frac{\displaystyle\int_0^\infty\frac{E_0}{\sigma_T^M(E_0)}\,f(E_0)dE_0}{\displaystyle\int_0^\infty\frac{E_0}{\sigma_T^M(E)}\frac{d}{dE_0}\,f(E_0)dE_0}\right\}. \tag{22}$$

Here μ is the mobility and D is the isotropic (or transverse) diffusion coefficient.

The mobility is usually obtained from the (in principle straightforward) measurement of the time of flight of electrons between two grid electrodes (G1 and G2 in Fig. 5) of the drift tube. The quantity D/μ is usually determined by the Townsend–Huxley method which measures the ratio of currents I_1 and I_2 (in Fig. 5). D/μ is related to this current ratio and to the geometrical parameters of the apparatus by the time-dependent diffusion equation. Detailed considerations for the experimental arrangements concerning the measurements of these quantities are discussed in Ref. 24. The procedure is to calculate $f(E_0)$ from Eq. (20) for a set of trial cross sections. W and D are then calculated from Eqs. (21) and (22) and compared with measured values. The trial cross sections are then varied until agreement for the measured and calculated values of W and D is obtained.

In practice this procedure is applicable only at low electron energies. If only the elastic channel is open, the elastic momentum-transfer cross section can be accurately and reliably determined (more so than by beam experiments). When an additional inelastic process is also possible, the cross sections still can be accurately determined because the inelastic cross section affects only the energy distribution function and $\sigma_T^M(E_0)$ enters only into the transport integrals. However, if more than one inelastic process occurs, the interpretation becomes less clear and ambiguities arise.

Ferrari[26] discussed the generally accepted method of solving the Boltzmann equation by the two-term expression in spherical harmonics and showed that in certain cases the third- and even higher-order terms in the expression (or a different approach) are needed to get acceptable results. The assumption in deriving Eq. (20) was that inelastic (and superelastic) cross sections are small compared to elastic cross sections. This assumption is not valid, e.g., for metal atoms in general and even for N_2.[27] Additional refinements could be introduced and may be necessary when inelastic cross sections with highly anisotropic behavior are important.[28] It is fair to say, however, that at impact energies ranging from about 0.05 to a few eV electron energies, the most accurate (~3%) momentum-transfer cross sections are obtained from swarm experiments.

4. Measurement of Differential Cross Sections

4.1. Instrumentation

The major components of a typical electron-impact spectrometer are the electron gun, target gas source, and detector, all enclosed in a vacuum chamber and shielded from electric and magnetic disturbances. A brief description of these components will be given here. The associated electronics and data handling system will be described in Section 4.4.

The vacuum chamber is usually made of nonmagnetic stainless steel with a number of ports and flanges to accommodate the pumps, valves, electrical, mechanical, and liquid feedthroughs, windows, and auxiliary instrumentation. Vacuum sealing between the chamber and various attachments is achieved by soft metal or elastomer gaskets or O-rings. Inside the vacuum chamber a single- or double-walled μ-metal shield is placed to attenuate the earth's magnetic field (~ 0.3 G). Outside of the chamber a pair (or pairs) of Helmholz coils are located to further decrease the magnetic field in the chamber (which mainly leaks in through the various openings) to the level of a few milligauss or less. The chamber can be pumped down to about 10^{-8} Torr without the introduction of a target gas and the vacuum is maintained at the 10^{-6} to 10^{-4} Torr range when the gas target is introduced. Mercury diffusion pumps have been frequently used to avoid hydrocarbon contaminations, but with the improvements of diffusion pump oils the most widely used pumps today are the oil pumps. Turbo molecular pumps represent cleaner systems than diffusion pumps but their price is still quite high.

4.1.1. Electron Guns and Detectors

The electron gun and detector systems all utilize some kind of electron optics and usually an energy selector element. The most frequently used lenses are electrostatic tube or aperture lenses. The design and the characteristics of electron lenses have been described in a number of publications.[29–31] The general approach to designing such lens systems is to solve the Laplace equation for a given geometry and applied potentials to obtain the fields within the lenses. The electron trajectories in these fields are then calculated and the lens properties (object and image distances, focal lengths, etc.) are obtained usually assuming first-order, paraxial (Gaussian) optics. In a multicomponent lens system each lens is represented by a matrix and the overall characteristics of the system are obtained by the multiplication of appropriate matrices.[32,33]

In principle any system, whose transmission depends on the electron energy, can be used as a monochromator. The demand for better energy resolution has led to the design of a large variety of energy selectors (e.g., barrier-type filters, Wien filter, trochoidal monochromator, Mollenstadt analyzer, parallel plate, cylindrical, and hemispherical analyzers, etc.). At the present time the most widely used analyzers are the 127° cylindrical[34–38] and the 180° hemispherical analyzers.[39–41] Brewer *et al.*[42] described very recently a coaxial cone electrostatic analyzer which is capable of collecting data simultaneously at several angles. We will briefly discuss here only the hemispherical analyzer. An excellent review of the many different energy analyzer designs currently in use has been published by Sevier,[43] Steckelmacher[43] and Read *et al.*[44]

Let us consider two concentric hemispheres as shown in Fig. 6. The radii and voltages applied to the inner and outer spheres are R_1, R_2 and V_1, V_2, respectively. R_0 is the center radius with associated potential of V_0 which is also the potential in the field matching Herzog elements H1 and H2.

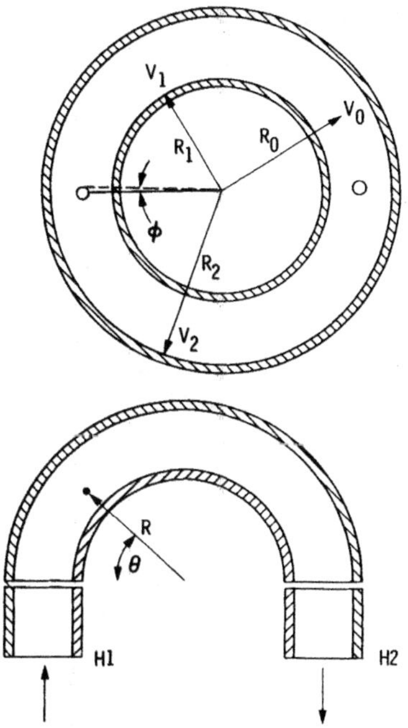

FIGURE 6
Coordinate system for the hemispherical analyzer. R_1, V_1, R_2, V_2 and R_0, V_0 are the radii and voltages for the inner and outer spheres and the sphere centerline. H1 and H2 are the Herzog elements. Upper figure shows the entrance–exit plane and the lower figure is a cut in the $\phi = 0$ plane.

The potential between two hemispheres can be obtained from the solution of the Laplace equation

$$\nabla^2 V(R, \theta, \phi) = 0. \tag{23}$$

With the appropriate boundary conditions (geometry and corresponding potentials) we obtain

$$V(R, \theta, \phi) = V(R) = \frac{R_1 R_2 \Delta V}{R_2 - R_1}\left(\frac{1}{R} - \frac{1}{R_0}\right) + V_0, \tag{24}$$

where $\Delta V = V_1 - V_2$, $R_1 \leqslant R \leqslant R_2$, $0 \leqslant \theta \leqslant \pi$, and $-\pi/2 \leqslant \phi \leqslant \pi/2$. The small field distortion near the entrance and exit planes has been disregarded. The time-independent equation for the path of the electrons between the hemispheres with the proper boundary conditions yields the following trajectory equation in the $\phi = 0$ plane:

$$R(\theta) = \frac{2R_i^2 X' \cos^2\alpha}{\cos\theta(2R_i X' \cos^2\alpha - R_0) - 2R_i X' \cos^2\alpha \tan\alpha \sin\theta + R_0}, \tag{25}$$

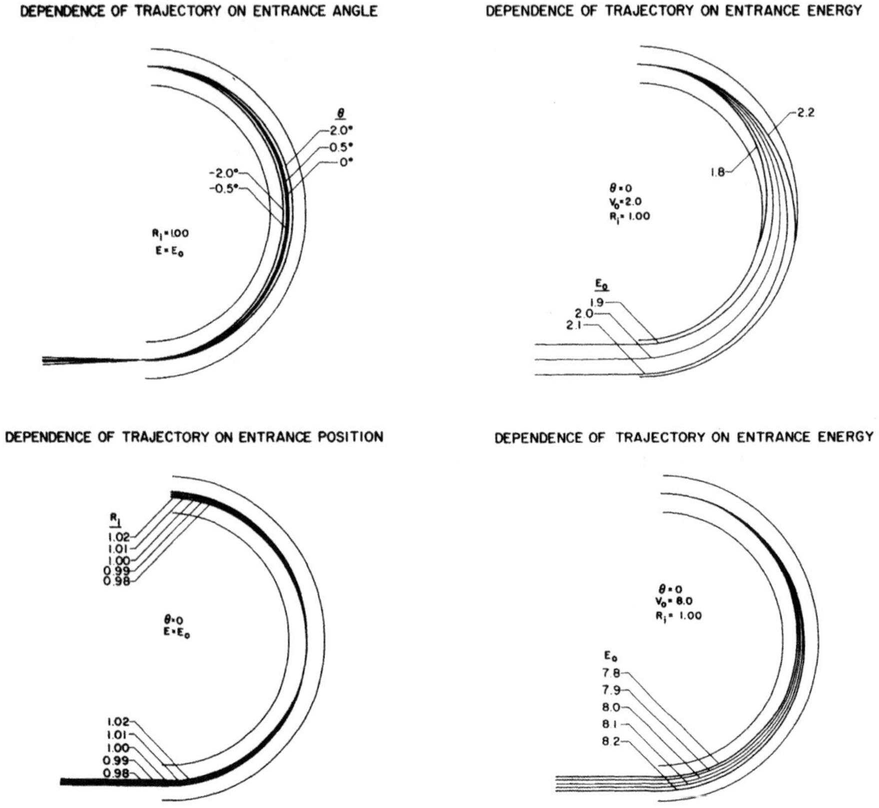

FIGURE 7
Electron trajectories in the $\phi = 0$ plane for the hemispherical analyzer as a function of entrance angle (α), position (R_i) in cm, and energy (E_0) in eV. V_0 is the potential in eV at the center of the gap.[45]

where $X' = (E/2E_0 + R_0/R_i - 1)$ and R_i is the position of entry of the electron into the sphere gap with an angle α with respect to the beam axis. For the details of deriving Eq. (25) see Refs. 39 and 45. Typical trajectories obtained from Eq. (25) are shown in Fig. 7. The system refocuses the entering beam at $\theta = 180°$ to second order in α with a magnification of -1. Images corresponding to different energies are dispersed at the exit plane and the lower the kinetic energy (eV_0) in the sphere the higher the dispersion. The conditions on ΔV for circular orbit for electrons of kinetic energy eV_0 entering at R_0 with $\alpha = 0$ is

$$\Delta V = V_0\left(\frac{R_2}{R_1} - \frac{R_1}{R_2}\right) \tag{26}$$

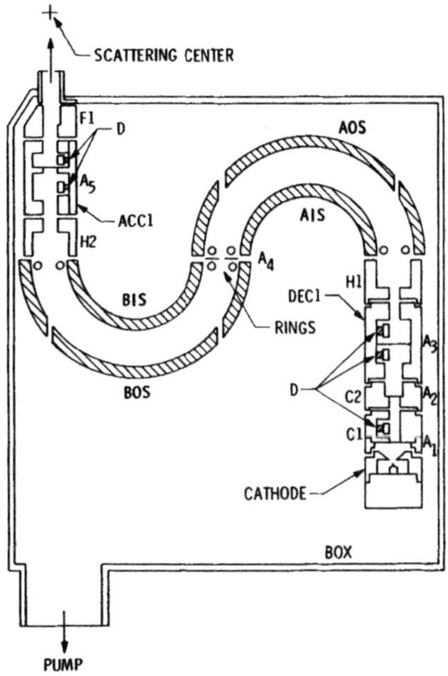

FIGURE 8
Electron gun with tandem hemispherical energy selectors. A and D refer to apertures and deflectors, respectively. The lens elements C1, C2, DEC1, ACC1, F1 and the inner (AIS, BIS) and outer (AOS, BOS) spheres are indicated.[47]

The placement of an appropriate aperture (or usually a virtual aperture) into the exit plane will achieve the energy selection. Energy widths (full width at half maximum, FWHM) of the order of 10 meV have been achieved with this type of selector with beam currents of the order of 10^{-9} A.

Some of the problems associated with the design of high-resolution hemispherical analyzers have been discussed by Brunt *et al.*[41] Various broadening effects (Doppler, stray fields) have been considered by Read,[46] and the relationship between dispersion of the energy selector and analyzer and the width (FWHM) of the observed spectral features has been derived by Hall *et al.*[38]

The electron source is most commonly a thermionic emitter, although photoionization sources are sometimes utilized. Figure 8 shows a gun system used in our laboratory.[47] The electrons are extracted and formed into a well-collimated beam by a multistage lens system, energy selected, and then reaccelerated to the required impact energy E_0 with another composite lens system utilizing a zoom lens. Double tandem hemispheres are used partly for geometrical reasons and partly to obtain better beam quality. A real aperture is placed between the analyzers to eliminate the unwanted part of the beam after the first dispersion. Two concentric rings in the entrance and exit plane of the spheres are utilized to empirically improve the field conditions at the entrance and exit area for producing better line shape.[41] The gun is enclosed in a differentially pumped box to reduce the effect of background gases and to minimize the stray electron background.

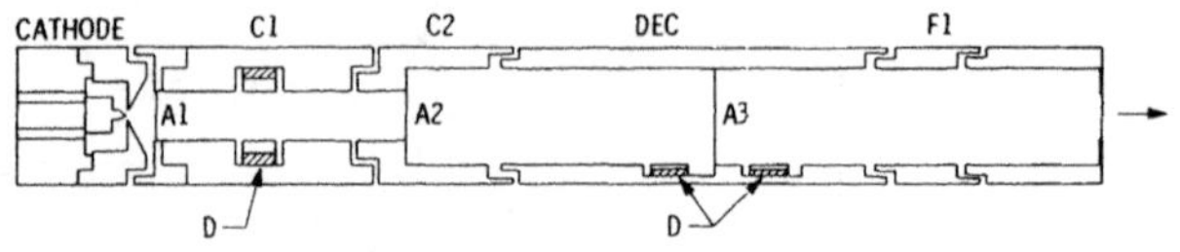

FIGURE 9
Electron gun without energy selection. Notation is the same as for Fig. 8.[47]

Sometimes higher beam currents (with larger FWHM) are desired. A gun without energy selection but capable of producing well-collimated beams of the order of 10^{-6} A in the 3 to 300 eV impact energy range is shown in Fig. 9.

In an attempt to achieve an energy resolution of the order of 1 meV for resonance studies, Brunt and Gallagher[48] used a photoionization electron source together with a supersonic atomic beam. This method is suitable for studying strong resonances but is not practical for extensive measurements because of the low beam intensity ($\sim 10^{-12}$ A).

Figure 10 shows a detector from our laboratory.[47] It consists of electrostatic lens trains and an energy analyzer similar to that utilized in the gun systems. The scattered current is collected from the conical volume defined by two circular apertures at the entrance of the detector (A_1 and A_2). The composite lens focusing the signal into the energy analyzer is somewhat more complicated than the other lenses in the spectrometer. This is due to the special requirement that electrons of very different residual energy must be focused at the same position with the same image size in order to minimize the change of

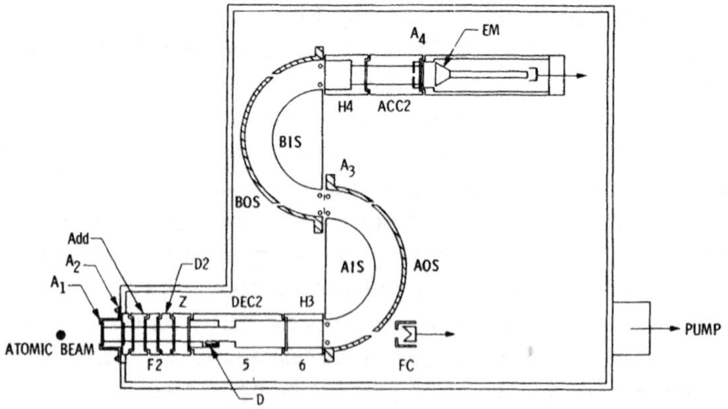

FIGURE 10
Electron detector with tandem hemispherical analyzers. EM, FC, A, and D refer to the electron multiplier, Faraday cup, apertures, and deflectors: Add, F2, Z, DEC2, ACC2, and refer to lens elements, and the notation for the spheres is the same as for Fig. 8.[47]

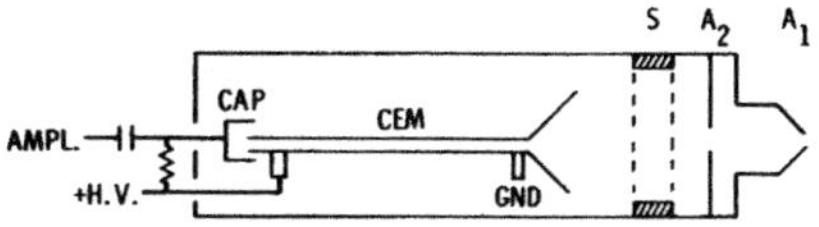

FIGURE 11
Electron detector without energy analyzer. A_1 and A_2 are the view cone defining apertures and S is the screen structure for biasing against inelastic electrons. The channeltron electron multiplier (CEM), signal-collecting cap (CAP), and the pulse coupling circuitry and connection to amplifier (AMPL), high-voltage supply (H.V.), and ground (GND) are indicated.[53]

resolution as a function of residual energy. The special feature of this composite lens is the adder lens which brings back the inelastically scattered electrons to their initial kinetic energy. Adding back the energy affects the properties of the first two lenses, which must then be compensated by sweeping the voltages on elements F2 and Z. For more details see Refs. 31 and 47 and Section 4.4.2. After the energy analyzer a simple lens transfers the electrons to the cone of the electron multiplier and the signal detection occurs in a pulse-counting mode.

When resolution is not the primary concern a detector with a simple retarding potential can be used to increase the detector signal. Such a detector is shown in Fig. 11.

The electron-impact spectrometer can be operated in a number of ways. The most commonly used mode is the energy-loss mode. In this mode the impact energy (E_0) and scattering angle (θ) are kept constant and the scattered signal intensity is recorded as a function of the energy (ΔE) lost by the electron. The resulting energy-loss spectrum is the basic data from which cross sections for various processes can be determined. The location of the features represents the energy-level scheme of the target and the intensities are related to the corresponding cross sections. A typical energy-loss spectrum generated in this mode with the double hemispherical systems (Figs. 8 and 10) is shown in Fig. 12. The operating conditions were $E_0 = 15$ eV, 10^{-9}-A incident beam, 23-meV overall resolution (FWHM) corresponding to 16-meV energy width in the incoming beam. The count rate in the $a\,^1\Pi_g$ $(v = 3)$ feature was 250 Hz. This condition corresponds to a good compromise in resolution, signal level, and line shape.

Energy-loss spectra can also be obtained in the so-called "constant residual energy mode." Here the detector is set to detect electrons only with a specific residual energy (E_R), that is, $E_R = E_0 - \Delta E$ is kept constant together with the scattering angle and the scattered signal is recorded as a function of ΔE (by sweeping both ΔE and E_0 simultaneously). With this method each energy-loss feature is recorded at the same impact energy above its own threshold. The method is frequently used for near-threshold measurements to avoid the effect of electron energy dependence of the detector. See for example Fig. 13.[49]

A third, commonly used method is the impact-energy mode. Here the

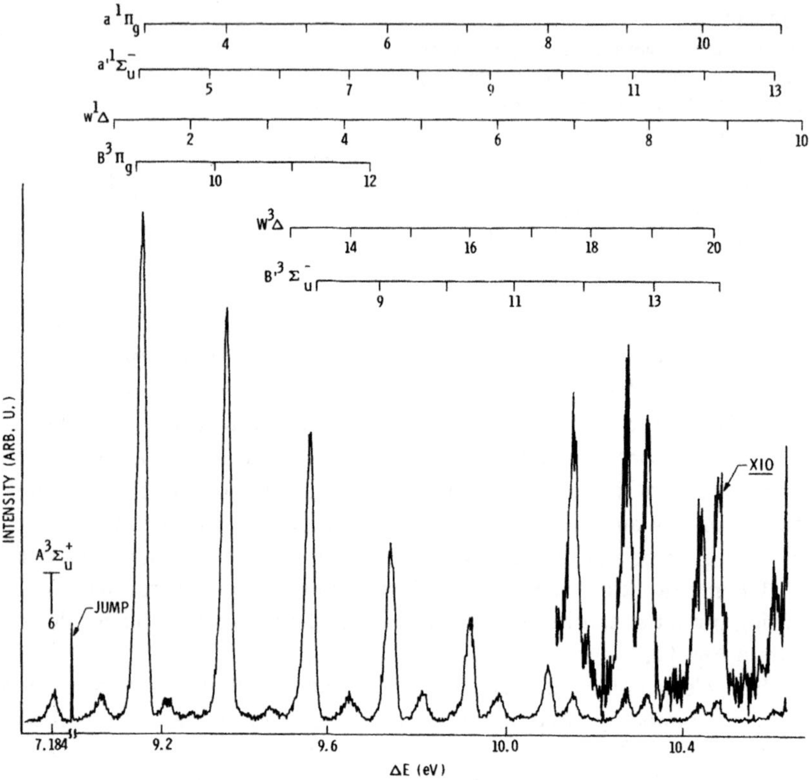

FIGURE 12
Energy-loss spectrum of N_2 at an impact energy of 15 eV, 20° scattering angle, and resolution of 0.023 eV. The electronic states and their vibrational structures and the jump in the energy-loss sweep are indicated. (D.F. Register and S. Trajmar, unpublished.)

energy loss (detection channel) and scattering angle are kept constant and the impact energy is swept. The resulting intensity distribution represents the energy dependence of a particular cross section and can reveal structures (resonances or cusps) in this channel. An example for this operation is shown in Fig. 14.[50]

4.1.2. Targets

Static gas targets have been widely used in early electron-impact cross-section measurements. The relatively large target size assures good signal levels and the uniform target density simplifies the cross section to scattering intensity relation. There are, however, serious problems encountered when differential cross sections are desired. The problems are associated with the change of

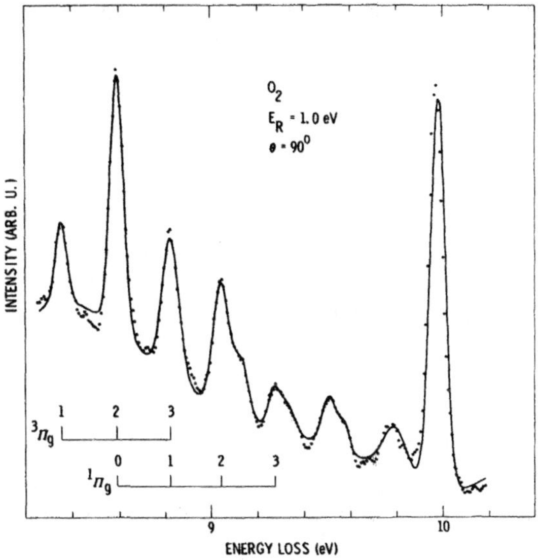

FIGURE 13
Energy-loss spectrum of O_2 obtained in the constant residual energy mode.(49)

scattering geometry with angle. This question is dealt with in more detail in Section 4.2.

At present most of the electron scattering measurements are carried out in a crossed-beam arrangement. A large variety of beam sources are being used ranging from simple orifices to supersonic nozzles. In addition special sources for generating beams of materials with very low vapor pressures at room temperature and of metastable and free radical species have also been used. The ideal target

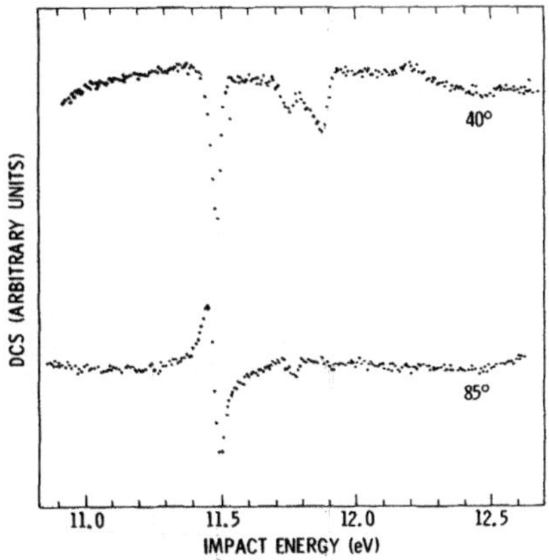

FIGURE 14
N_2 resonance obtained in the impact-energy mode for the elastic channel at 40° and 85° scattering angle.(50)

beam is of small size, with uniform and high density. With this type of beam the scattering geometry is independent of angle and the measured scattering intensities are directly proportional to the cross sections. In actual practice, however, one has to make a compromise among various requirements, and an ideal target beam is seldom obtained.

Brinkmann and Trajmar[51] investigated the spatial distribution of beam density from a variety of commonly used sources on the basis of the flow equations derived by Olander and Kruger.[52] The beam density distribution emerging from a cylindrical tube (under flow conditions where hydrodynamic effects can be neglected) is given as

$$\rho(\mathbf{r}) = \frac{a^2 \rho_s}{4R^2} j(\vartheta) \tag{27}$$

with R the distance from the tube axis, and with

$$j(\vartheta) = \xi_0 \cos\vartheta + \left[\frac{2}{\delta'\sqrt{\pi}} \xi_0 \cos\vartheta \exp(\delta'^2)\right] \times \left\{\frac{R(p)}{2}\left[\operatorname{erf}\left(\frac{\xi_1 \delta'}{\xi_0}\right) - \operatorname{erf}(\delta') + \frac{2\delta'\xi_1}{\xi_0\sqrt{\pi}}\left(\frac{1}{\xi_1} - 1\right)\exp\left(-\frac{\delta'^2 \xi_1^2}{\xi_0^2}\right)\right] + S(p)\right\} \tag{28}$$

for $\tan\vartheta \leqslant \gamma$ and

$$j(\vartheta) = \xi_0 \cos\vartheta + \left[\frac{2}{\delta'\sqrt{\pi}} \xi_0 \cos\vartheta \exp(\delta'^2)\right] S(1) \tag{29}$$

for $\tan\vartheta \geqslant \gamma$, where ϑ is the polar angle with respect to the beam axis. In these equations

$$S(p) = \int_0^p (1 - Z^2)^{1/2} \left(\operatorname{erf}\left\{\delta'\left[1 + \frac{\gamma Z}{\tan\vartheta}\left(\frac{\xi_1}{\xi_0} - 1\right)\right]\right\} - \operatorname{erf}(\delta')\right) dZ,$$

$$\delta' = \frac{\delta}{(\cos\vartheta)^{1/2}},$$

$$\delta = \left[(2\gamma \mathrm{Kn}_D)^{-1} \frac{\xi_0^2}{\xi_1 - \xi_0}\right]^{1/2} = \left[(2\mathrm{Kn}_L)^{-1} \frac{\xi_0^2}{\xi_1 - \xi_0}\right]^{1/2},$$

$$\gamma = D/L, \tag{30}$$

$$\mathrm{Kn}_D = \lambda_s/D, \qquad \mathrm{Kn}_L = \lambda_s/L,$$

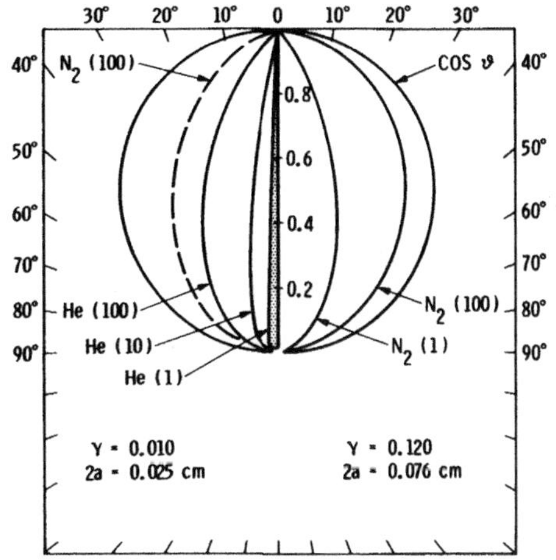

FIGURE 15
Relative density distributions for two tubes ($\gamma = 0.01$ to the left and $\gamma = 0.12$ to the right). The gas back pressure (in Torr) is indicated in parentheses. $2a$ is the tube diameter. For comparison a $\cos \vartheta$ distribution is also shown.[51]

$$\lambda_s = \frac{1}{\sqrt{2}\,\rho_s \bar{\sigma}},$$

$$R(p) = \cos^{-1} p - p(1 - p^2)^{1/2},$$

$$p = (\tan \vartheta)/\gamma,$$

and the symbols have the following meanings: γ is the aspect ratio (ratio of tube diameter to length), D is the tube inner diameter ($= 2a$), L is the tube length, Kn_D is the Knudsen number based upon tube diameter, Kn_L is the Knudsen number based upon tube length, λ_s is the mean free path of the molecules in the source reservoir, ρ_s is the molecular density in the source reservoir, and $\bar{\sigma}$ is the gas kinetic collision cross section. The symbols ξ_1 and ξ_0 denote the tube wall collision rates at the entrance and exit of the tube, respectively, relative to the corresponding rate in the source reservoir. For a thin-walled orifice $j(\vartheta) = \cos \vartheta$, while for tubes with collision-free flow $j(\vartheta) = 1$. Relative density distributions for two types of tubes and an orifice are shown in Fig. 15. The $\gamma = 0.01$ type tube is generally used when collimation is of some concern, while the $\gamma = 0.12$ aspect ratio tube is typical of high-temperature crucible sources used in our laboratory. With increasing pressure the beam divergence increases, but even at 100-Torr pressure behind the tube, it is far away from the $\cos \vartheta$ distribution which is characteristic of aperture sources. The N_2 beam is more diverged than the He beam at the same pressure because of the larger kinetic collision cross section.

TABLE 1
Characteristics of Capillary Arrays Discussed in This Chapter

	Type A	Type B	Type C
d (cm)	0.00508	0.00508	0.00508
L (cm)	0.508	0.508	0.508
γ	0.01	0.01	0.01
D^a (cm)	0.102	0.153	0.076
N^b (approx)	300	675	225

[a] D, array diameter.
[b] N, number of capillary tubes.

Brinkmann and Trajmar[51] derived expressions and calculated density distributions for capillary arrays (assuming no interaction between beams emerging from individual capillaries). Examples of density distributions for a typical capillary array (type A in Table 1) are shown in Fig. 16. Actual number density distributions for the same capillary array are shown in Fig. 17. With further skimmer or aperture collimation the beam distribution can be truncated at the desired angle as indicated for $\vartheta = 10^\circ$ in Fig. 17. A scattering geometry utilizing a capillary array source with further collimation is shown in Fig. 18. The beam chopper is needed to determine the background scattering. This arrangement was used recently by Register *et al.* for accurate differential cross-section measurements[53] and was shown by Brinkmann and Trajmar to require no effective path length correction (see Section 4.2).

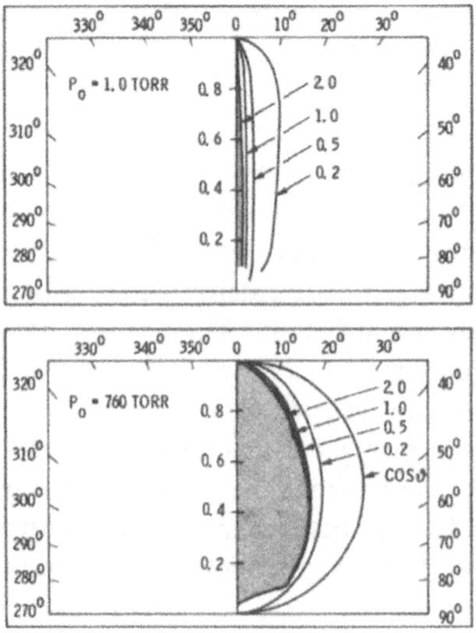

FIGURE 16
Relative density distributions for type A capillary array (Table 1) at back pressures of 1 and 760 Torr. The different curves are for various distances in cm from the center of the array. For comparison, the cos ϑ distribution is also shown at the higher pressure.[51]

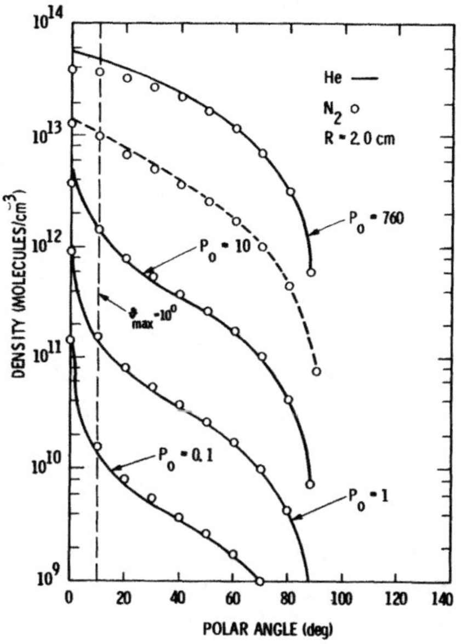

FIGURE 17
Number density distributions for capillary array type A (Table 1) at a radius (R) of 2 cm (measured from the center of the capillary array). The capillary array back pressure P_0 is given in Torr. The vertical dashed line refers to the collimation of the beam at 10°.(51)

Supersonic nozzle beams are used when very high densities with small beam sizes or target species with small internal and kinetic temperatures are required.(54) High-temperature crucible sources have been extensively used to generate beams of metal atoms and highly polar molecular salts. Heating is conveniently done by a resistance heater up to about 800°C.(47) Above this temperature electron beam bombardment(55) or induction heating is needed.(56)

4.1.3. Special Considerations for Near-Threshold and Very Low Energy Measurements

In the past the trapped electron method(18) and SF_6 scavenging techniques(57) have been primarily used for investigating near-threshold-energy

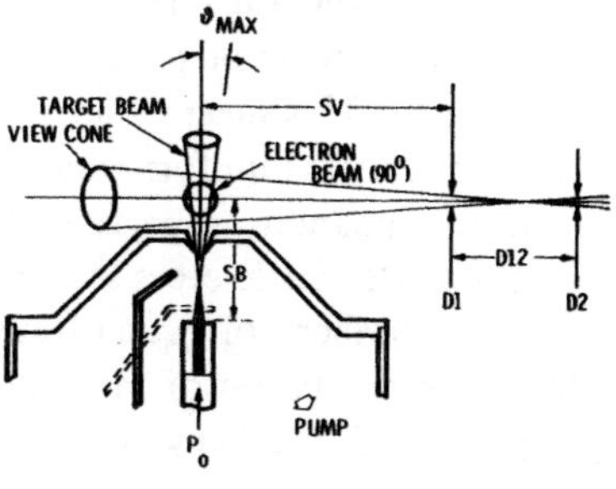

FIGURE 18
A collimated capillary array source used for elastic angular distribution measurements for He.(53) D1 and D2 are the diameters of the detector view cone defining apertures and D12 is their separation.

phenomena. More recently electron beam techniques with high energy resolution have been adapted to very low-energy measurements.[58-66] The handling and detection of electrons with energies below a few tenths of an eV require special techniques. The control of the electron beam becomes difficult because of the divergence due to space charge effects. Small electric or magnetic fields or inhomogeneous surface potentials can cause serious problems. It is mandatory to minimize the surfaces that can be seen by the electron and to keep them clean by baking. The transmission properties of the analyzer and detection system undergo large variations with energy because of the extremely large relative change in electron energy. The efficiency of the system therefore has to be carefully studied in order to perform quantitative measurements. One also faces a serious background problem due to stray electrons which have been thermalized by wall collisions.

Cvejanović and Read[58] developed a technique which gives good energy resolution and is able to detect very low-energy electrons. The idea which they introduced is to allow a small field to penetrate into the collision region. This field produces a saddle point in the potential distribution which enhances the extraction efficiency for low-energy electrons. Although this method generates an excellent signal for near-zero-energy electrons it is not well suited for quantitative work. For further details of this penetrating-field technique see Chapter 5, Section 3.4.

Tam and Wong[64] avoided the intrinsic limitation of the electrostatic system by applying an axial magnetic field to collimate the electron beam. The energy selector part of their apparatus is identical to the trochoidal monochromator of Stamatovic and Schulz.[65] In the detector part of their apparatus they incorporated an electron energy filter in front of the trochoidal analyzer to reduce the background due to elastic scattering. This apparatus operates well for studying zero-angle scattering at very low energies and moderate resolution (~50 to 100 meV) but suffers from several difficulties for quantitative cross-section measurements (e.g., solid angle of collection is energy dependent).

Pichou *et al.*[59] and Joyez[60] utilized a 127° cylindrical energy analyzer and selector with fine-wire grid surfaces and simple aperture optics[38] to study near-threshold excitation of He. They achieved the calibration of the energy sensitivity of the apparatus by noting that at near-threshold energies the excess energy in the ionization process is shared randomly by the two electrons scattered isotropically. This means that the energy distribution of the electrons (primary plus secondary) is uniform and any deviation from uniformity can be attributed to the transmission property of the analyzer. They normalized the relative cross sections to He elastic cross sections.

A number of quantitative near-threshold measurements related to vibrational excitation of molecules have been carried out by the Kaiserslautern group.[61,62,66,67] They used 127° analyzers and aperture optics. The instrumental energy dependence was calibrated against the known behavior of the He elastic differential cross sections. They determined the absolute scale by normalizing to known total scattering cross sections.

Quantitative cross-section measurements at very low impact energies and near-threshold regions require extremely careful and tedious work and are faced with large uncertainties. Perhaps the most practical way to handle this situation is to establish a standard set of cross sections which could be used to calibrate the energy and angular behavior of the apparatus and/or utilize this standard in a relative flow technique to establish the absolute values.

4.1.4. Coincidence Measurements

Coincidence techniques in electron-impact studies have recently seen increased applications for cross-section measurements. The general scope of these experiments may be roughly divided into two areas: electron–ion (or ejected electron) phenomena and electron–photon coincidence measurements. In the former experiments, molecular ions or fragments or ejected electrons from autoionization processes are mass, time-of-flight, or energy analyzed in coincidence with a selected energy-loss primary electron. In the latter case, photons arising from the decay of a particular excited molecular level are detected in delayed coincidence with the selected energy-loss electron which created the excited molecular state. Although the experimental apparatus is quite different for the two types of measurements, the peripheral instrumentation is similar.

Electron–ion (ejected electron) coincidence measurements have recently been reviewed by Brion and Hamnett,[68] and an extensive literature may be found in this reference. Electron–electron coincidence measurements in molecules[69] use two conventional electron spectrometers to energy analyze and detect both the primary energy-loss electron and the ionized (ejected) electron arising from direct or autoionization processes. Recently Wilden *et al.*[70] have also used time-of-flight measurements to study autoionization processes in $(e, 2e)$ experiments on N_2. The TOF technique converts an energy to a time spectrum and avoids many of the instrumental response problems inherent in studying the large energy-loss region. Detection of ions and molecular fragments[71] in coincidence with electrons allows the determination of oscillator strengths (see Sections 3.2, 4.3.4, and Section 5) for ionization and photodissociation of molecules. Mass analysis of the ionic fragments may be performed using mass analyzers or time-of-flight cells. A more detailed discussion of these techniques will be found in Ref. 68.

Electron–photon coincidence measurements have seen extensive applications in atomic collision phenomena.[72–74] These experiments attempt to study the fundamental relationships among the magnetic sublevel populations in electron-impact excitation of a particular atomic energy level. The basic theory of angular correlations[75,76] has recently been extended to molecules.[77] For the molecular case, however, the experimental situation is clouded by the necessity of averaging over rotational and sometimes even vibrational levels. Preliminary work on H_2 electron–photon coincidence measurements has been reported by Malcolm and McConkey[78] and has recently been reviewed by McConkey.[73]

The basic configuration for an electron–photon coincidence measurement in molecules is shown in Fig. 19. In this experiment, uv photons were polarization analyzed in coincidence with the scattered electron. Alternatively, the angular distribution of photons could be detected in coincidence with the electrons to obtain equivalent information.[75] For ejected-electron or ionic-fragment coincidence studies, the photon detector is replaced by an electron monochromator, drift tube, or ion detector (with or without mass analysis). These detectors are typically located in the scattering plane.

In addition to the normal host of precautions inherent in any electron-impact measurement, coincidence studies also must address the problems of timing and pulse shaping of the detected signal, and, quite often, discouraging signal-to-noise (S/N) conditions. Electron–photon coincidence measurements must also take account of the problems associated with radiation trapping.[79,80] The prime sources of timing problems in these experiments are: "jitter" or statistically random time spreading of the coincidence signal, "walk" which results

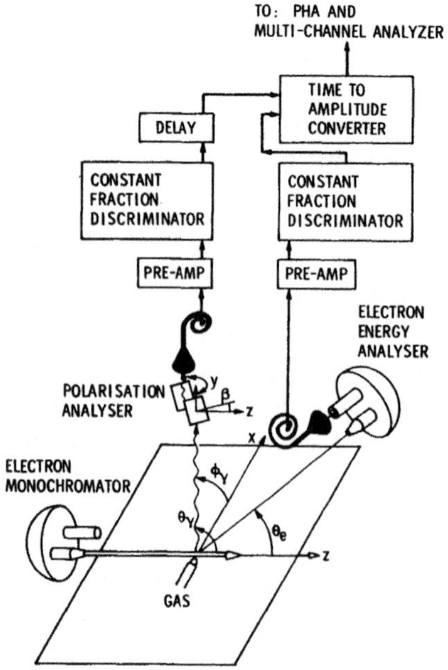

FIGURE 19
Schematic diagram of an electron–photon coincidence measurement using polarization analysis of the emitted photon.[73]

from systematic timing delay errors due to the wide dynamic range of the detected signal amplitudes, and delay and timing window conditions which must be considered in order to obtain acceptable S/N levels.

The primary source of statistically random time spreading is in the path length variations in the electron monochromator (see Fig. 7 and Section 4.1.1). As the maximum spread in path length about the centerline of a well-designed hemispherical analyzer is of the order of 5% to 10%, the energy of the monochromator will primarily determine the timing width of detected electrons. For the double hemispherical analyzer of Fig. 10 with a 2 eV center potential, the transit time is in excess of 200 ns. This leads to an estimated "jitter" of between 10 and 20 ns. Raising the sphere center voltage to 20 V results in a reduction of the timing spread to around 3 to 7 ns, which is acceptable for most electron–photon coincidence measurements. Clearly, small hemispheres (or cylinders) operated at higher energies will reduce the instrumental time spread. However, this trend must be balanced by the energy resolution requirements necessary for the experiment. Similar arguments may be advanced for drift tubes to require accurate path length determinations. Some considerations specific to time-of-flight cells are discussed in Refs. 13 and 70.

Time walk problems arise since low-amplitude pulses from the detector do not pass through conventional amplifier–discriminator systems in exactly the same timing sequence as high-amplitude pulses. Since many detectors have dynamic ranges of the order of 100:1 (maximum signal/minimum signal), this effect may be a serious systematic problem. Channeltrons operated in a gain saturated pulse-counting mode exhibit excellent dynamic range compression, which greatly reduces this effect.[81] However, many experiments call for the use of photomultipliers which do not exhibit as favorable dynamic range characteristics. In order to compensate for this type of timing error, constant fraction discriminators (CFD) are almost universally used for coincidence measurements. These devices sense the height of each input pulse and trigger when the input signal reaches a specified fraction of that pulse height. Timing walk for fast pulses (~ 10 ns) is thus reduced to subnanosecond levels. Since CFDs are borrowed from nuclear and high-energy physics technology, their lower discriminator levels are typically 20–50 mV. For use with channeltrons, this requires the use of a fast linear amplifier or emitter follower circuit[81] (see also Section 4.4.2) to interface the channeltron with the CFD.

In electron–photon and electron–ion coincidence measurements (and in $(e, 2e)$ experiments with drift tubes), long delays of up to 1 μs may be encountered between the arrival of the electron and photon/ion. Although a passive delay line is usually desirable, in case of delay times in excess of a few 100 ns they are impractical. In order to retain good timing information, active delays should use high-speed circuitry such as Schottky transistor-transistor logic (TTL) or emitter-coupled logic (ECL). Monostable multivibrators suitable for delay timing which exhibit excellent temperature and time stability are available in both product lines. Because CFD speeds and logic levels are usually directly compatible with ECL devices, these may be preferred for this application.

Clearly, high-quality components and careful attention to impedance matching and signal levels is essential for the peripheral electronics associated with coincidence measurements. In many cases, the required accumulation time is long (1 to 100 h) and the system stability and reliability are crucial.

4.2. Relation between Cross Section and Scattering Signal Intensity in Beam–Beam Experiments

A typical geometry for an electron beam–molecular beam experiment is shown in Figs. 20 and 21. The target neutral beam is generated by various means as described in Section 4.1.4 and crossed with an electron beam (usually at 90°). The electron beam emerging from the gun can be considered parallel or more generally slightly diverging. The electron flux is highest along the axis of the beam and falls with an exponential-type dependence away from the axis. The intersection of these two beams within the view cone of the detector (defined by two circular apertures) represents the scattering volume from which scattered electron signal can reach the detector. We are, therefore, dealing with an extended source in most practical cases.

A general relation between the scattering intensity and the corresponding cross section is given by the following equation:

$$I(E_0, E, \Omega) = \int_{\mathbf{r}} \int_{E'_0} \int_{E'} \int_{\Omega'(\mathbf{r})} \rho(\mathbf{r}) f(E'_0, \mathbf{r}) F[E'_0, E', \mathbf{r}] \times \left[\frac{\partial^2 \sigma(E'_0, E', \Omega')}{\partial E' \partial \Omega'}\right] d\Omega' \, dE' \, dE'_0 \, d\mathbf{r}, \tag{31}$$

where $I(E_0, E, \Omega)$ is the scattering signal (counts/s) measured by the instrument at a nominal impact energy E_0, energy loss E, and scattering polar angles $\Omega(\theta, \phi)$;

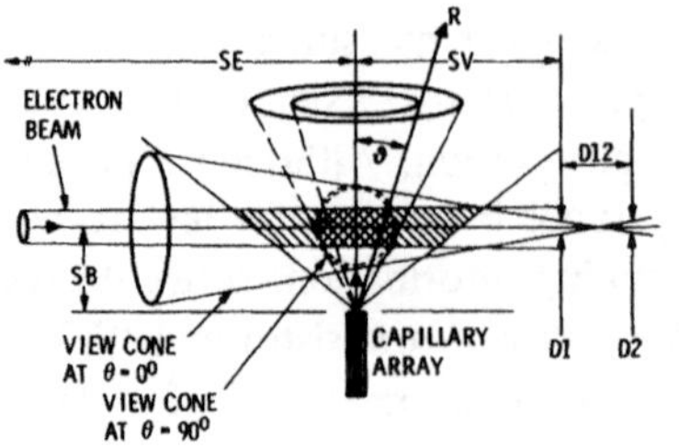

FIGURE 20
A typical scattering geometry (vertical cut). The intersection of the molecular beam, electron beam, and view cone is shown. Two equidensity contours are indicated for the molecular beam.

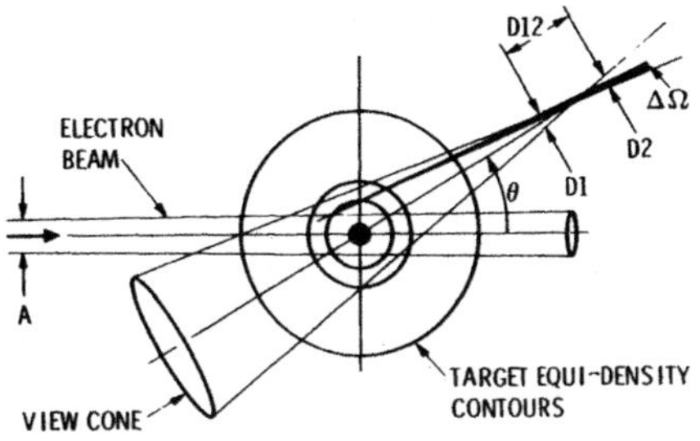

FIGURE 21
Scattering geometry showing the scattering plane (horizontal cut). The detector view cone axis is at the scattering angle θ. The solid angle of detection ($\Delta\Omega$) for a point in the scattering region is shown.

$\rho(\mathbf{r})$ is the spatial density distribution of the target atoms (molecules/cm^3); $f(E_0', \mathbf{r})dE_0'd\mathbf{r}$ is the distribution (both energy and spatial) of the electron flux (number of electrons per second per cm^2 with energy between E_0' and $E_0' + dE_0'$ at points between $\mathbf{r}$ and $\mathbf{r} + d\mathbf{r}$); $F[E_R', \mathbf{r}]dE_R'd\mathbf{r}$ is the response of the detector for detecting electrons which have residual energy between E_R' and $E_R' + dE_R'$ and which were scattered by molecules located at coordinates between $\mathbf{r}$ and $\mathbf{r} + d\mathbf{r}$; $\partial^2\sigma(E_0', E', \Omega')/\partial E'\partial\Omega'$ is the doubly differential cross section for a process with energy loss E at impact E_0 and scattering angle Ω'; $d\Omega'(\mathbf{r}) = \sin[\theta'(\mathbf{r})]\, d\theta'(\mathbf{r})d\phi'(\mathbf{r})$; $\mathbf{r}$ is the position of the individual scattering point which contributes to the observed signal and θ' and ϕ' are the polar angles of scattering from these points. The extraction of the desired cross section from Eq. (31) is complicated. It requires detailed knowledge of the target density and electron beam flux distributions and the overall instrument function. The situation becomes much more manageable in either of two limiting cases: if the cross section is slowly varying in energy and angle with respect to the system response (continuum processes such as ionization) or if the system response is slowly varying with respect to the natural linewidth of the transition under study (discrete processes). In the first case, a cross section evaluated at the nominal scattering angle, energy loss, and impact energy may be removed from the integral. Equation (31) then becomes

$$I(E_0, E, \Omega) = \overline{\frac{\partial^2\sigma}{\partial E\, \partial\Omega}}\, \tilde{V}_{\text{eff}}. \tag{32}$$

Here all the geometrical and instrumental functions are included in $\tilde{V}_{\text{eff}}$, which is obtained by the appropriate integrations. For discrete processes the cross section can be integrated over the energy-loss profile of the process to obtain

$$I_n(E_0, \Omega) = \overline{\text{DCS}_n}(E_0, \Omega)\, \tilde{V}_{\text{eff}}. \tag{33}$$

A number of simplifying assumptions can be made concerning $\tilde{V}_{\text{eff}}$ which are valid to a very good approximation in most scattering experiments:

$$
\begin{aligned}
&\text{(a)} \quad f(E'_0, \mathbf{r}) = f(E'_0) f(\mathbf{r}), \\
&\text{(b)} \quad F[E'_R, \mathbf{r}] = F(E'_R), \\
&\text{(c)} \quad \text{DCS is independent of } \phi \text{ (averaged over } \phi).
\end{aligned}
\tag{34}
$$

Assumption (b) states that the detector efficiency is independent of scattering position and depends only on the residual energy of the electron to be detected. With these assumptions

$$\tilde{V}_{\text{eff}}(E_R, \theta) = C(E_R) V_{\text{eff}}(\theta) \tag{35}$$

with

$$V_{\text{eff}}(\theta) = \int_{\mathbf{r}} \int_{\Omega'(\mathbf{r})} \rho(\mathbf{r}) f(\mathbf{r})\, d\Omega'(\mathbf{r})\, d\mathbf{r}$$

and

$$C(E_R) = \int_{E'} \int_{E'_0} f(E'_0) F(E'_R)\, dE'_0\, dE'_R.$$

The values of the effective scattering volume (or "path length") $V_{\text{eff}}(\theta)$ can be calculated at each impact energy at various scattering angles. Therefore, if $C(E_R)$ is determined, the absolute values of the differential cross section can, in principle, be obtained. This question will be further discussed in Section 4.3. It is, however, an easier and more reliable procedure to treat $C(E_R)$ as an unknown constant at a given impact energy and energy loss.

Brinkmann and Trajmar[51] calculated the values of $V_{\text{eff}}(\theta)$ for a number of frequently used scattering geometries and beam sources and expressed the results in terms of "effective path length correction factors" defined as

$$F_{\text{eff}}(\theta) = \frac{V_{\text{eff}}(90^\circ)}{V_{\text{eff}}(\theta)} . \tag{36}$$

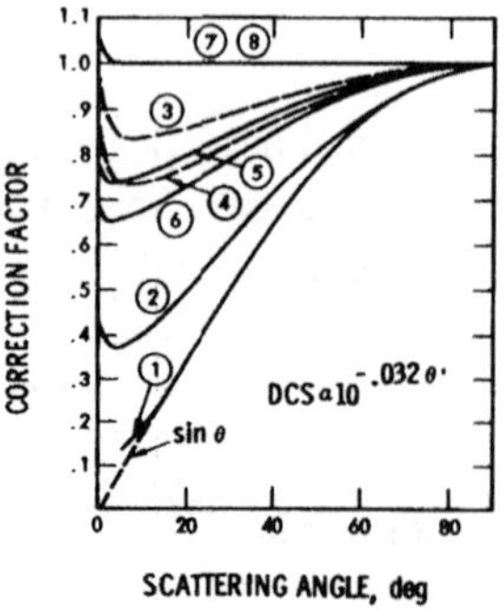

FIGURE 22
Effective path length correction factors for various source and scattering geometries. See Table 2 for explanation of the curves. A moderately forward peaking cross section (DCS $\propto 10^{-0.032\theta'}$) was assumed. The effect of gas kinetic cross section is shown by curves 3 and 4 and the influence of capillary array back pressure is demonstrated by curves 5 and 6.

TABLE 2
Summary of Model Scattering Geometries for the Effective Path Length Correction Curves of Fig. 22[a]

No.	Target type	SE (cm)	SB (cm)	SV (cm)	D1 (cm)	D2 (cm)	D12 (cm)
1	Static: ρ = const; σI	50	–	1.27	0.076	0.076	1.27
2	Orifice: $\rho \propto \cos \vartheta$; σI	50	0.50	1.27	0.152	0.100	1.27
3	Tube: $\gamma = 0.120$; σI; 1 Torr	50	0.25	1.27	0.152	0.100	1.27
4	Tube: $\gamma = 0.120$; σIII; 1 Torr	50	0.25	1.27	0.152	0.100	1.27
5	CA: $\gamma = 0.010$; σI; 0.1 Torr	50	0.25	3.81	0.092	0.050	2.54
6	CA: $\gamma = 0.010$; σI; 760 Torr	50	0.25	3.81	0.092	0.050	2.54
7	CCA: $\gamma = 0.010$; σI; 760 Torr; $\vartheta = \pm 12°$	50	0.70	3.81	0.092	0.050	2.54
8	Jet: ρ = const; $\vartheta = 0°$	50	0.70	3.81	0.092	0.050	2.54

[a]The characteristic parameters for the various sources are indicated. ρ, γ, and ϑ stand for the target density, tube aspect ratio, and target beam divergence angle, respectively. The target gas kinetic cross sections σI and σIII correspond to 2.2 and 100 Å^2, respectively, and the tube back pressures are given in Torr. SE is the distance of the apex of the electron beam cone from the scattering center. SB, SV, D1, D2, and D12 are defined in Figs. 20 and 21. CA and CCA refer to capillary array and collimated capillary array (type A in Table 1), respectively. Numbering (column 1) corresponds to Fig. 22.

With this factor an experimentally measured scattering intensity distribution can be converted to relative differential cross sections at the particular energy in question. Normalization of the relative values to the absolute scale can then be carried out by some of the methods discussed in Section 4.3. Typical effective path length correction factors are shown in Fig. 22 for a cylindrically symmetric view cone, electron beam, and target beam. (See Table 2). It can be seen that the commonly used $\sin\theta$ correction is good down to low scattering angles in the case of static targets and commonly used electron beams. Appreciable corrections are needed for target beams generated by orifices and single tubes and even for capillary arrays. However, with proper collimation, beam targets can be produced for which no effective path length correction is needed at scattering angles above a few degrees. For more details the reader is referred to Ref. 51.

4.3. Normalization Procedures

4.3.1. General Remarks

As described in Section 4.2 the data acquired in an electron–molecule scattering experiment are typically the number of electron counts accumulated in a particular energy-loss channel integrated over the accumulation time of the experiment. In order to relate this number to the cross section, either a direct measurement of all the experimental parameters must be performed or a particular energy-loss count rate must be normalized to a well-known cross section. Any calibration technique may be used on any spectral feature corresponding to a particular electron-impact process, but the most common feature used for normalization is the elastic (or zero energy-loss) peak.

Of the various methods for normalizing differential cross sections at low to intermediate impact energies, the most successful have been the direct normalization approach as used by Bromberg,[82] Chamberlain *et al.*,[83] and Williams and co-workers,[84] the use of known cross sections (e.g., He elastic differential cross sections) for correcting the energy- and angular-dependent instrument response of the apparatus,[62,85] and the relative flow technique as used in our[86] and other laboratories. These procedures will be discussed in detail in the following subsections. Other normalization methods are briefly discussed in Section 4.3.4.

Once a particular transition in the spectrum has been placed on the absolute scale, the entire spectrum is, in principle, calibrated. However, several precautions must be observed in assuming that spectral intensity ratios represent cross-section ratios. In particular, the angular resolution and the focal properties of the electron detector as a function of impact energy and energy loss must be carefully examined.

In the case of electron–rare gas atom and electron–metal atom scattering, deep minima in the elastic differential cross sections have been observed. The depths of the observed minima are a sensitive function of the angular resolution and require detector view cones of the order of 1° for proper investigation. Although such sharp minima have not yet been observed in electron–molecule scattering, this experience should caution against overly coarse angular resolution. This is especially true for experiments where very strongly forward-peaking cross sections can be expected such as in molecules having permanent or large induced dipole moments.

The second problem which requires precaution arises in situations where a large energy-loss region is scanned or where an elastic feature is observed followed by a "jump" to a high energy-loss region (see Section 4.4.2 and Fig. 12). In either case, where a large voltage excursion in the detector is encountered (~10 V) it is usually necessary to ramp one or several lens elements to compensate for the changing focal properties of the lenses during the scan. Recently, zoom lenses for electron detectors have become increasingly popular. While such multielement systems have desirable focal and aberration properties they must be properly compensated to ensure uniform transmission efficiency. For the double hemispherical detector system of Fig. 10, lenses F2, D2, and Z must be properly ramped to ensure good operation over an extended energy-loss region. A typical operating ramp for the F2 element in the energy-loss mode is given in Fig. 23.[87] The generation of such a ramp is discussed in Section 4.4.2.

A final consideration in bringing inelastic intensities to the absolute scale is that of energy resolution. Using presently available instrumentation, an overall system resolution of the order of 15 to 20 meV with 1 to 5 nA of primary

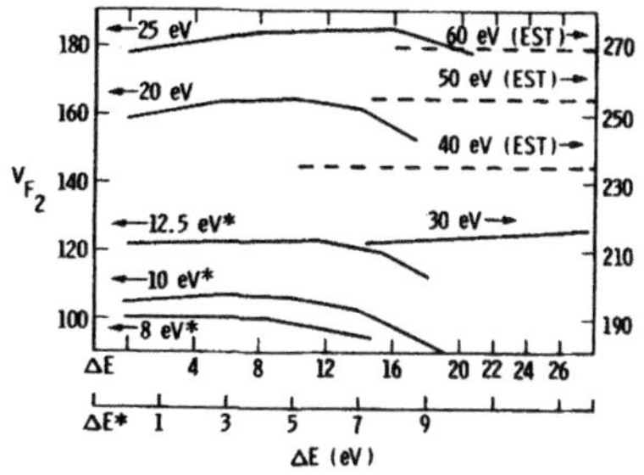

FIGURE 23
Calculated ramp voltages V_{F_2} for the F2 element (Fig. 10) as a function of energy loss for the indicated impact energies.[87]

electron beam current is considered good. However, even under these conditions, individual rotational transitions in most molecules are impossible to resolve. Although the available instrumental resolution is sufficient for most vibrational spacings, the presence of overlapping vibrational bands significantly complicates the observed energy-loss spectra.

To address this problem, several authors have reported "unfolding" techniques for defining the vibrational cross sections. Cartwright *et al.*[88] have used Gaussian beam profiles and the Franck–Condon factors to obtain cross sections for unresolved transitions in N_2. Register *et al.*[89] have applied a similar technique using the experimentally determined elastic scattering line shape in electron–CO_2 vibrational excitation. In neither case has any attempt been made to resolve rotational structures. Wong and Dubé[90] have studied rotational excitation of N_2 and analyzed the vibrational transition line shape to obtain information on rotational excitation. Tanaka *et al.*[91] studied the change of the shape of the elastic scattering feature due to unresolved rotational excitation and utilized the method described by Shimamura[92] (see Chapter 2, Section 3.3) to obtain information on rotational excitation in N_2, O_2, and CH_4. Although these techniques are promising, they do rely on some assumptions (which are well justified for present-day beam experiments on light molecules) and require some improvements in experimental energy resolution.

4.3.2. *Direct Normalization*

In Section 4.2, a relation was developed between the detected signal intensity and the underlying cross section [Eqs. (33) and (35)]. One method for obtaining the cross section from these relations is the direct measurement of all the relevant experimental parameters: instrumental response function, target beam density distribution, the electron beam energy and flux distribution and the detector acceptance angle. Such measurements have been performed at high energies[82,83] and have recently been extended to lower energies.[84] A brief discussion of the relevant experimental considerations will be presented. More detailed analysis may be found in Refs. 82–84.

In order to calibrate the detector efficiency, the instrumental response function must be determined over a wide range of operating currents and energy resolutions. A simplifying assumption is that the measured efficiency is independent of scattering angle and/or detector angular resolution. For this to be reasonable, care must be taken to reduce stray electric and magnetic fields in the scattering chamber and the molecular target should be small compared with the detector view cone so that the electron trajectories will be nearly paraxial. Under these conditions, a measure of the detector efficiency for transmission of the direct electron beam should give a reasonable estimate of the efficiency for the transmission of scattered electrons at any angle. Such detector efficiency measurements have been carried out by Williams and Willis[(84)] and by Preston *et al.*[(93)] for 127° cylindrical analyzers. In these measurements, the current entering the detector system is monitored at a convenient input lens (a deflection plate) and again at the nose cone of the channeltron. The required detector response function may then be determined as a function of energy and resolution and has been found to vary between 9% (0.015 eV resolution) and 97% (0.100 eV resolution) for a typical low-energy measurement (8.7 eV).[(84)]

To evaluate the overlap integral, it is usually necessary to determine both the atomic beam density distribution and the electron beam flux distribution. This problem can be greatly simplified by constructing electron and atomic beams of rectangular cross sections such that the density and flux distribution are constant for the two beams throughout the interaction volume.[(6)] Provided that these conditions can be attained, the measurement of the incident electron beam current in a carefully designed Faraday cup then provides a good determination of $f(\mathbf{r})$. (See, however, the precautions outlined by Kuyatt[(4)] on Faraday cup design).

The determination of the molecular beam density is not nearly as straightforward. The conditions encountered in typical scattering experiments (10^{-3} Torr equivalent target pressures) are not easily amenable to a direct measurement of the target density. For atomic species a convenient, indirect way to determine an average target density is by the utilization of a phase-shift analysis to obtain an absolute cross section. Once the absolute cross section at a particular energy has been determined, the electron flux and detector efficiency may be used to derive the target beam density. Using this technique and phase shift analysis at resonance energies, Williams and co-workers have reported cross sections for atomic hydrogen[(84, 94)] and argon.[(95)]

For molecular systems, well-defined elastic resonances which may be phase-shift analyzed are not generally available. In this case, an atomic resonance may be analyzed and pressure ratios for the molecular and atomic species used to obtain the molecular beam density. Since this is not a direct measurement on the species of interest, some care must be taken to ensure that the atomic and molecular species are being studied under identical flow conditions. This problem is discussed in more detail in the next subsection.

For low-energy and threshold measurements, especially when corrosive gases are under study, it is often necessary to monitor the instrument response function throughout the experiment. This approach has been successfully applied in several measurements (Refs. 62 and 85 and references therein) with e^-–He elastic scattering being used as the standard. In these experiments, the absolute target density was determined and the target beam was periodically replaced by He to monitor drift in the instrument response function due to changes in contact potentials and surface conditions.

4.3.3. Relative Flow Technique

The rigorous determination of all the relevant experimental parameters required by a direct normalization method may be relaxed by use of the relative flow technique (Fig. 24). In this approach an unknown molecular elastic differential cross section may be determined by comparing the electron scattering intensities against that of a known atomic or molecular species.[86] For such measurements the absolute electron and molecular beam densities or fluxes need not be known. The desired elastic cross section may be determined by knowing the relative target densities for the two gases (unknown and standard) provided that the electron beam density and flux distribution, the detector efficiency and

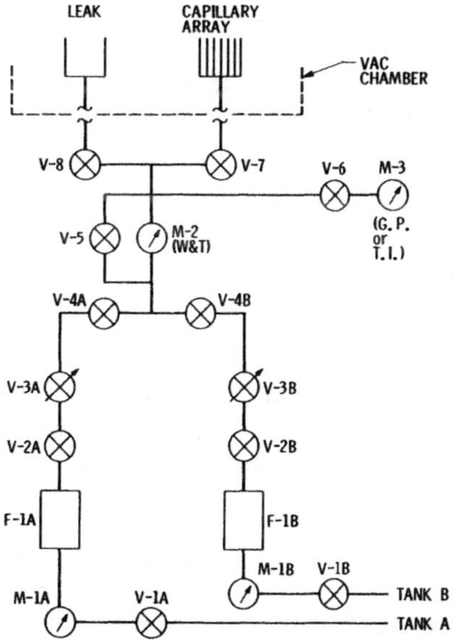

FIGURE 24
Schematic diagram of a relative flow technique. V, M, and F refer to valves (shutoff –⊗, needle –⊗̸), pressure meters, and flow meters, respectively.[89]

the target beam flux distribution are *the same for both gases* during the measurement. Under these conditions, the effective path length corrections for the two gases will cancel to a very good approximation. The fact that the angular dependence of the elastic cross section is in general different for the gas under investigation and for the standard will introduce a small difference into the corresponding effective path length correction and the cancellation will not be perfect. However, this effect needs to be considered only in extreme cases.

In terms of the cross-section definitions [Eqs. (33) and (35)], the scattered electron intensity ratio for two gases in a relative flow measurement may be written as

$$\frac{I_x}{I_{\mathrm{STD}}} = \frac{\mathrm{DCS}_x V_{\mathrm{eff}_x}}{\mathrm{DCS}_{\mathrm{STD}} V_{\mathrm{eff}_{\mathrm{STD}}}} = \frac{\mathrm{DCS}_x \int\int \rho(\mathbf{r})_x f(\mathbf{r})_x \, d\mathbf{r}\, d\Omega'}{\mathrm{DCS}_{\mathrm{STD}} \int\int \rho(\mathbf{r})_{\mathrm{STD}} f(\mathbf{r})_{\mathrm{STD}} \, d\mathbf{r}\, d\Omega'}, \tag{37}$$

where we have assumed that the detector response function $C(E_R)$ is the same for both gases. To further simplify Eq. (37), the electron flux and the relative target density distributions must be the same for both gases. The first condition may be readily achieved by differential pumping of the electron optics and operating at low background pressures in the vacuum chamber ($\sim 10^{-6}$ Torr). The second condition requires further elaboration.

For a single capillary tube, Olander and Kruger[52] have shown that the target angular flux distribution may be written as

$$J(\vartheta) = \frac{\nu_s}{\pi} (\pi a^2)\, j(\vartheta), \tag{38}$$

where ν_s is the rate at which gas molecules enter the tube, a is the tube radius, and $j(\vartheta)$ is a normalized distribution function which depends on the flow conditions in the capillary. It has been assumed that the flow is cylindrically symmetric. The target density is related to $J(\vartheta)$ as

$$\rho(\mathbf{r}) = \frac{J(\vartheta)}{R^2 \bar{v}}, \tag{39}$$

where R is the distance from the tube exit to the target point and $\bar{v}$ is the average thermal velocity of molecules in the beam $[\bar{v} = (3kT/M)^{1/2}]$. Clearly the integral of the flux distribution must equal the gas flow rate

$$2\pi \int_{-\pi/2}^{\pi/2} d\vartheta\, J(\vartheta) = \dot{N} = K\nu_s(\pi a^2), \tag{40}$$

where $\dot{N}$ is the total flow rate and K is the Clausing factor (transmission probability) for the tube. Equations (38) and (40) lead to the relation

$$K = 2 \int_{-\pi/2}^{\pi/2} j(\vartheta)\, d\vartheta. \tag{41}$$

Using Eq. (40) to define ν_s, the target density may then be written as

$$\rho(\mathbf{r}) = \frac{\dot{N} j(\vartheta)}{\pi R^2 \bar{v} K} . \tag{42}$$

Olander and Kruger[52] have shown that for long capillary tubes, intermolecular collisions affect the angular distribution $[j(\vartheta)]$ but do not significantly affect the total flow provided that

$$\gamma < \mathrm{Kn}_L < 10, \tag{43}$$

where γ is the tube aspect ratio and Kn_L is the length Knudsen number (Section 4.2). Under these conditions, the angular flux distribution may be accurately determined and is not a very sensitive function of Kn_L. Since the Clausing factor is determined from $j(\vartheta)$ [Eq. (41)], the angular distribution of target densities for two gases will be the same provided the Knudsen numbers for the two gases are similar.

The scattered electron intensity ratio [Eq. (37)] under these conditions may then be given to a very good approximation as

$$\frac{I_x}{I_{\mathrm{STD}}} = \frac{\mathrm{DCS}_x \dot{N}_x \bar{v}_{\mathrm{STD}}}{\mathrm{DCS}_{\mathrm{STD}} \dot{N}_{\mathrm{STD}} \bar{v}_x} = \frac{\mathrm{DCS}_x}{\mathrm{DCS}_{\mathrm{STD}}} \frac{\dot{N}_x}{\dot{N}_{\mathrm{STD}}} \left(\frac{M_x}{M_{\mathrm{STD}}}\right)^{1/2}, \tag{44}$$

where the average thermal velocity has been replaced by the target mass under the assumption that the two gases are at the same temperatures.

Although this derivation is rigorous for a single capillary tube, Brinkmann and Trajmar[51] have generalized the same arguments to capillary arrays with the restriction that no significant interaction occurs among the beams originating from the different capillary tubes. This is an acceptable assumption for back pressures up to the order of a few Torr, which is well suited for the present applications.

Equation (44) is correct for most flow conditions encountered in electron collision studies (supersonic and high-pressure flows being two obvious exceptions). Quite often it is experimentally more convenient to represent the cross-section ratio in terms of pressure ratio. In this case, further restrictions must be placed on the experiment to require molecular (collisionless) flow in the capillary array. Under conditions of molecular flow the pressure in the gas reservoir may be directly related to the flow rate[86]

$$P_0 \propto \sqrt{M}\dot{N}, \tag{45}$$

which leads to a simpler relative flow expression

$$\frac{\mathrm{DCS}_x}{\mathrm{DCS}_{\mathrm{STD}}} = \frac{I_x P_{0,\,\mathrm{STD}}}{I_{\mathrm{STD}} P}. \tag{46}$$

Either Eq. (44) or (46) is appropriate for conditions of molecular flow in the capillary array. However, in practice, it is difficult to operate under molecular flow conditions.

For a typical capillary structure from Table 1 ($\gamma = 0.01$), the minimum back pressure required for molecular flow may be calculated assuming that the condition

$$\mathrm{Kn}_L \geqslant 1 \tag{47}$$

defines collisionless flow (Olander and Kruger impose the more conservative requirement $\mathrm{Kn}_L \geqslant 10$). Condition (47) requires a mean free path of ~5 mm for the arrays of Table 1 which leads to a number density in the gas reservoir of about 3×10^{15} atoms/cm^3. This density corresponds to a pressure of approximately 0.1 Torr. Since molecular gas kinetic collision cross sections are typically an order of magnitude larger than He, operating pressures which ensure the same flow conditions for molecules are expected to be less than 0.01 Torr (0.001 Torr for $\mathrm{Kn}_L \simeq 10$). At these pressures, data acquisition will be significantly affected by low counting rates and long accumulation times. Therefore, it is usually far more convenient to operate under the flow conditions defined by Eq. (43), in which case Eq. (44) *must* be used for the relative flow measurement.

Since the gas kinetic cross sections for a large number of gases of experimental interest are not well known and since the Knudsen numbers are sensitive

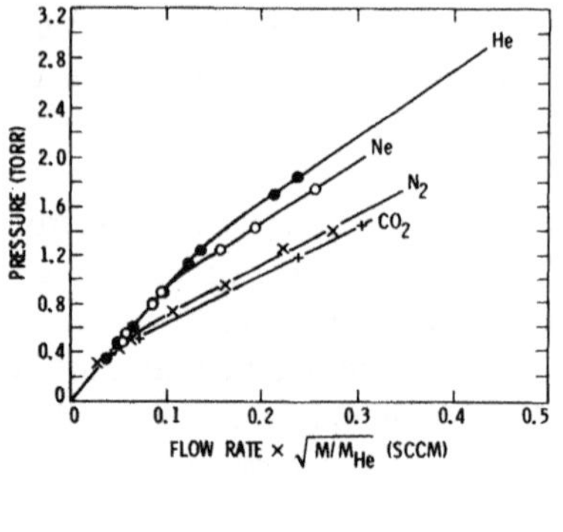

FIGURE 25
Flow rate versus pressure for a number of target gases. The flow rates have been normalized to He by using the square root of the mass ratios.

TABLE 3
Typical Recommended Operating Conditions for Various Gases with Type B Capillary Array[a]

Gas	σ (Å^2)[b]	Pressure (Torr) ($Kn_D \simeq 3$)	Flow rate[c]
He	~ 3.8	3.7	$\simeq$0.6
Ne	~ 5.3	2.7	0.20
Ar	~10.5	1.5	0.08
N_2	~11.2	1.4	0.10
H_2	~ 6	2.4	0.48
CO_2	16.9	1.2	0.08
H_2O	16.6	1.2	0.12

[a] See Table 1 for the characteristics of this array.
[b] Ref. 97.
[c] Flow rate is given in standard cm^3/s.

functions of the particular apparatus, it is best to perform pressure vs. flow rate calibrations to define proper operating conditions. Figure 25 shows typical pressure versus flow rate behaviors for a number of gases.[96] For the case of He this figure indicates that although collisionless flow in the capillary array is not obtained above 0.1 Torr, this does not significantly affect the pressure vs. flow dependence until $Kn_L \simeq 0.1$ ($P \sim 1$ Torr). However, as has been demonstrated by Brinkmann and Trajmar[51] (cf. Fig. 17), the target angular distribution undergoes significant changes in the 0.1 to 1 Torr pressure region. For N_2 and CO_2, the pressure vs. flow dependence changes near 0.2 Torr, which is consistent with the conclusion that collisionless flow for molecules will not be obtained at pressures above about 0.01 Torr.

Since pressure measurement and regulation are difficult in this range, it is better to operate at higher pressure where collisions affect the target distribution but where the gas Knudsen numbers may be more readily adjusted to generate the same flux distribution for the two gases. A summary of suggested operating conditions for a number of gases is given in Table 3 (assuming relative flow calibrations to be made against He and that capillary arrays of the types listed in Table 1 are utilized).

4.3.4. *Other Methods*

Relative DCS may be obtained by applying the appropriate path length and instrumental response functions to the measured scattering intensities. By extrapolating these data to 0° and 180° scattering angles the relative integral and momentum-transfer cross sections may be obtained. Therefore, the normalization of the experimental cross sections (in arbitrary units) to the absolute scale can be achieved by normalizing to any well-known differential, integral, or momentum-transfer cross section which overlaps with the available data.

Optical excitation functions have been measured for several molecules. These apparent excitation cross sections can be corrected for cascade effects to a certain degree of accuracy to represent integral excitation cross sections and could then serve for approximate normalization of the relative data. In each case, however, a careful assessment of the possible error limits has to be made.

High-energy and low-angle electron-impact measurements have been used to generate optical f values by extrapolation of the generalized oscillator strength $f^G(K)$ to zero momentum transfer (see Section 5). One can, however, reverse the procedure and utilize available optical f-values to normalize relative generalized oscillator strengths as measured in electron-impact experiments. The main difficulty with this procedure is the requirement of extrapolating the generalized oscillator strength (obtained from the relative DCS) to zero momentum transfer. This is usually performed by plotting the relative $f^G(K)$ against K^2 and observing the small K^2 tendency in the curve. For an inelastic process the limit of zero momentum transfer is unphysical since even zero degree scattering corresponds to a finite K value. The other problem is that the most important points in this extrapolation are the low-angle DCS points, which are experimentally the most uncertain. This uncertainty in the low-angle data arises mainly from the difficulties of accurate effective path length corrections at low scattering angles. In addition, the $f^G(K)$ curve sometimes changes character at very low K values,[98] and if this change is not recognized an extrapolation to the wrong direction may take place.

Total scattering cross sections with high accuracy (few %) are becoming available for a large number of molecules. One can utilize such data for normalization by adding up the relative integral cross section values for all open channels. At low energies where only a few channels are open this procedure should be quite reliable. At high energies, however, a very large spectral range has to be covered and the error in the experimental sum of the integral cross section could be substantial.

In principle theoretical calculations could also serve as the basis for normalizing relative cross-section data. In general, however, the theoretical calculations have not been accurate enough for this purpose. Lately for simple species (H, He) highly accurate theoretical results have been obtained which certainly satisfy the criteria for a standard. Similar development for simple molecular species can be expected in the future. One always has to carefully consider, however, the limitations of the theoretical model. It may well describe a process at a particular energy or angular range but not in other ranges. It may be reliable for certain types of transitions but not for others. At the present time, theoretical calculations are the most useful where it is difficult to obtain quantitative experimental data (corrosive materials, free radical or excited species, etc.).

4.4. Experimental Techniques and Procedures

4.4.1. Selection of Components and Geometry

Ideally, beam–beam electron–molecule scattering experiments should be performed with well-collimated beams using electron sources of millivolt resolution and microampere current capability. Clearly such conditions are presently unattainable and some compromises must be made to perform useful measurements. The determining factor in most cases will be the count rate required to obtain statistically meaningful results within a reasonable data accumulation time. Table 4 summarises the average range of energy resolution and electron currents available with present hemispherical electron sources and detectors. For the same energy resolution, cylindrical systems may produce larger currents but with a consequent loss in angular resolution.

For most molecular species, the separation between vibrational levels is about 0.1 to 0.2 eV while rotational splittings in a given band are less than a millivolt. Although rotational-structure resolutions cannot be achieved with present systems, vibrational structures in many cases can be separated. The separation of overlapping vibrational bands of different electron transitions in most molecules, however, requires the highest possible resolution. An example of this kind of overlap is shown in Fig. 12 for N_2.

A rough estimate of the count rate which is expected for a particular transition (n) in this type of experiment can be made from

$$I_n \sim I_e \left(\frac{d\sigma}{d\Omega}\right)_n \rho l \Delta\Omega C(E_R), \tag{48}$$

where I_e is the incident electron current, $(d\sigma/d\Omega)_n$ is the differential cross section, ρ is the target density, l is a characteristic length for the interaction

TABLE 4
Typical Currents and Energy Resolutions for a Conventionally Used Electron-Impact Spectrometer with Hemispherical Energy Selector and Analyzer

Incident electron current (nA)	System energy resolution (meV)
500–1000	250–450
50–500	50–250
1–50	15–50

region, $\Delta\Omega$ is the detector view-cone solid angle, and $C(E_R)$ is the transmission efficiency. Using an effusive molecular source ($\rho \sim 5 \times 10^{13}$ molecules/cm^3, $l \simeq 3$ mm) in a high-resolution system [1 nA, 15 meV, $C(E_R) \sim 10\%$, $\Delta\Omega \sim 10^{-2}$ sr], and assuming that all electrons in the beam cross a uniform density target, one obtains a count rate of the order of 10^4 electrons/s for a cross section of 10^{-16} cm^2/sr. For collimated beams the count rate might be an order of magnitude less because of the smaller path length and the larger distance between source and scattering center. Inelastic cross sections for vibrational excitation are typically in the range of 10^{-18} to 10^{-21} cm^2/sr. Count rates for these transitions might be between 100 to 0.1 Hz for effusive sources and between 10 and 0.01 Hz for collimated beams.

From these estimates it is clear that performing meaningful inelastic measurements in collimated beams will be difficult. Conversely, performing elastic cross-section measurements in collimated beams is possible with high signal-to-noise ratio and may be performed under conditions which discriminate against low-lying vibrational levels. This analysis thus suggests that most experimental conditions can be met by a single electron-source-and-detector system capable of moderate to high energy resolution (50 to 15 meV with 50 to 1 nA current capability).

Most molecular target requirements can be met by an effusive source (capillary array) and a comparatively simple gas system for relative flow measurements. Such a system is diagramed in Fig. 24. In this case both a source and a side leak are available for both target and background measurements. Two separate flow meters are provided to avoid gas contamination problems. The cross calibration of the flowmeters is discussed in Ref. 89. The effusive source is most commonly (see Table 1) a capillary array. The array is press fitted into a 1/8 in. o.d. stainless steel tubing to reduce the surface area near the scattering center.

Although relative flow measurements in this system can produce reliable results, for heavy molecules it may be difficult to adjust the gas Knudsen number to match that of the calibration standard. In this case small and large angle geometry corrections may arise due to changes in the wings of the target density distribution for the two gases. This may necessitate the use of a collimated beam for the cross-section calibration (see Sections 4.1.2 and 4.3 and Fig. 18).

Under certain special conditions, high-intensity ($\sim 1\,\mu$A) electron sources may be needed. This type of current can be obtained if the energy resolution requirements can be lowered for the experiment.

4.4.2. Peripheral Electronics

The requirements of each installation and the design of specific electron sources and detectors preclude a detailed characterization of the electronics for a

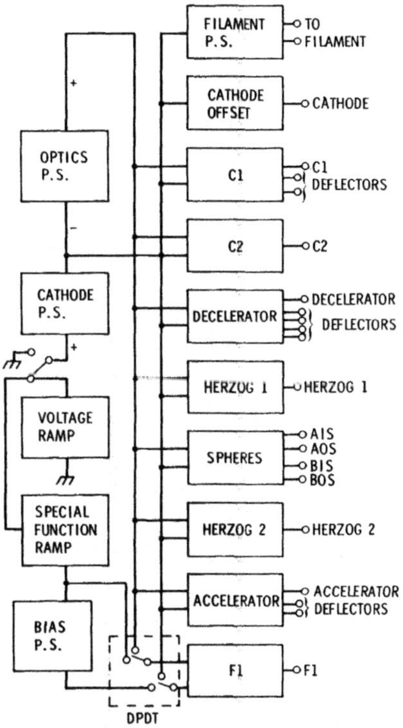

FIGURE 26
Block diagram for the control circuit for the double hemispherical gun of Fig. 8. The lens element voltages are derived from resistive dividers (Fig. 28) indicated by boxes around the element name. The circuit diagram for the voltage ramp and special ramp (F1) are shown in Figs. 29 and 30.

general scattering experiment. However, a number of advances in the understanding of system problems have been made in the past few years. A few of these topics, as they apply to electron scattering, will be discussed in detail.

Figures 26 and 27 present block diagrams of the control circuitries in use in our laboratory[47] for a double hemispherical electron gun and detector, respectively. To achieve energy resolution of 20 meV or better several requirements in this system must be met. In practice it has been found that a single power supply may be used to derive most of the required lens voltages by means of resistive dividers. (A divider circuit for the decelerator is shown in Fig. 28). This approach allows a maximum investment in a single high-quality power supply rather than the less reliable (and more expensive) use of individual power supplies for each function. Minimizing the number of power supplies in the circuitry also reduces the probability of ground loops, which are frequently the greatest cause of noise in these experiments. Ground loops may also be minimized by use of heavy bus wire to connect the various system components to a single grounding point.

To minimize cross talk in the cabling each signal line should be individually shielded. The entire cable bundle should also be enclosed in a grounded jacket to shield against local environmental noise. Control cables inside the

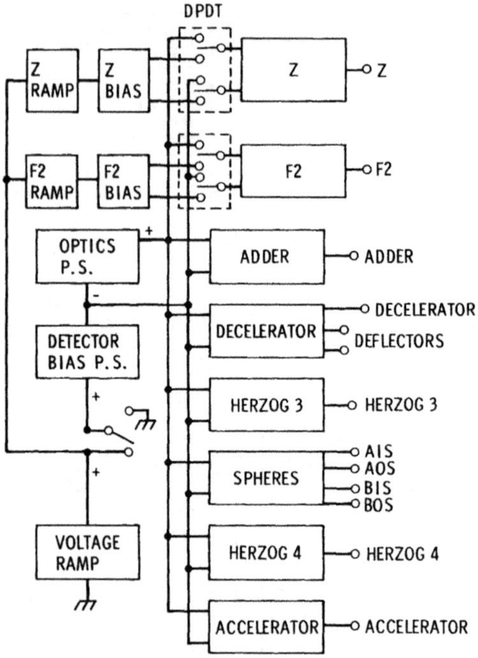

FIGURE 27
Block diagram for the control circuit for the double hemispherical detector of Fig. 10. The circuit diagram for the voltage ramp and special function ramp (F2 and Z) are shown in Figs. 29 and 30.

vacuum system should also be shielded to reduce the possibility of stray electric fields in the interaction region. In some experiments, the scattering center is enclosed in a wire mesh to protect against this problem.[41]

For the double sphere gun control circuit of Fig. 26 a voltage ramp is provided which sweeps the impact energy and provides special focusing control for element F1. This ramp is derived from the x-axis sweep of the multichannel scaler (MCS) data acquisition unit. Since the x-axis ramp from the MCS typically has a limited voltage excursion, it is unlikely to be directly suitable for experimental purposes. Figure 29 shows a schematic diagram for a simple and reliable dc voltage amplifier which converts a 2-V input ramp to a 0.5–1-V output. The

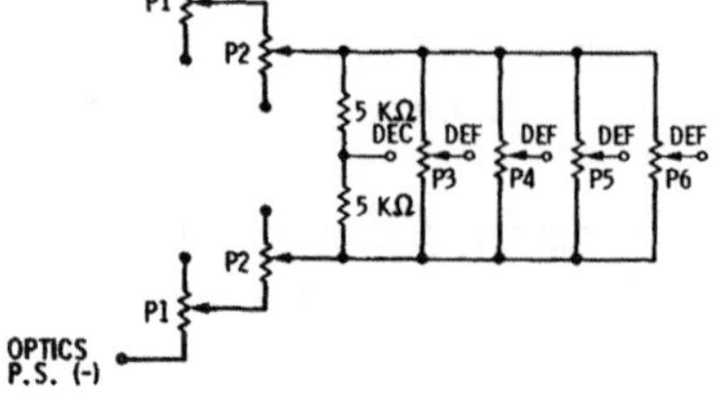

FIGURE 28
Resistive divider for the decelerator element of Fig. 26.[47]

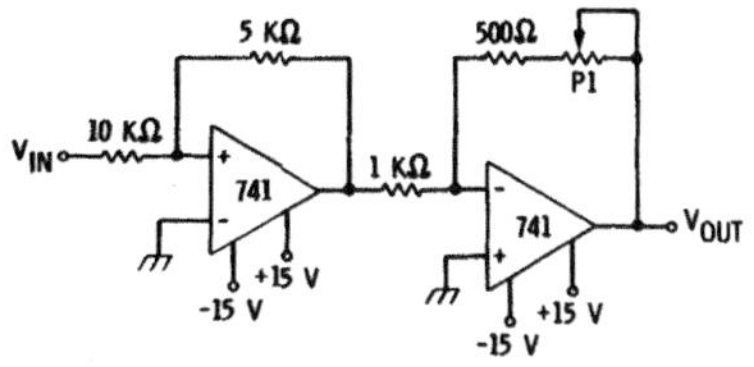

FIGURE 29
Schematic diagram of the ramp amplifier circuit of Figs. 26 and 27.[47]

use of two inverting amplifiers simplifies the design and provides additional buffering to prevent short circuits in the output line from being reflected into the MCS.

For some lens elements, it may be required to change the slope of the ramp voltage as a function of impact energy or energy loss (for example, the F2 ramp of Fig. 27.[87]). This can be most easily accomplished by providing two variable slopes and a break point which may be adjusted (as a function of the basic input ramp). A circuit which performs this function is shown in Fig. 30. The primary ramp is generated from S_1, IC1, and IC3. IC1, IC2, and IC3 are operational amplifiers (741 or equivalent). The secondary ramp (and break) is generated from S_2, IC2, and IC3. Since the outputs of IC1 and IC2 are of opposite polarity and are summed at the IC3 input, the slope after the break point is always less than before.

In many measurements of energy-loss spectra, it is desirable to study features which may be widely separated from one another. It may be useful, for instance, to measure both the elastic scattering intensity and electronic transitions which may be 8 or 9 V removed from the elastic peak. Rather than waste

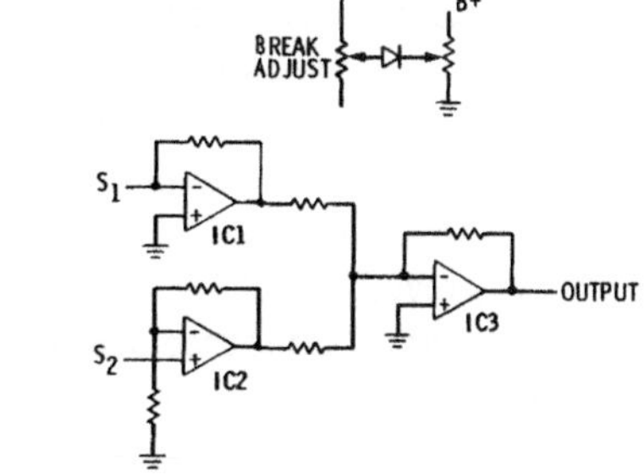

FIGURE 30
A dual-slope ramp amplifier as used for the F1, Z, and F2 ramps.

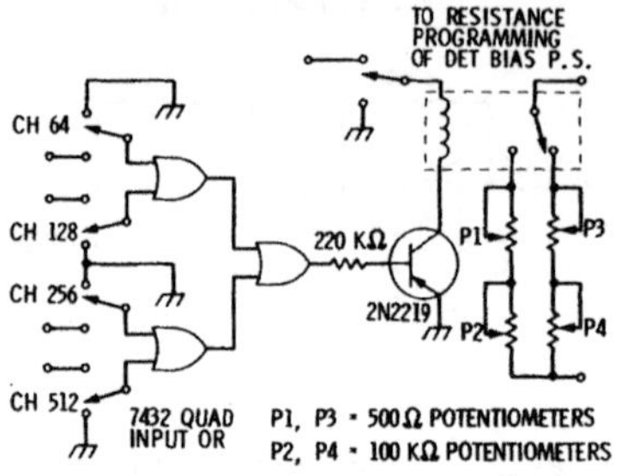

FIGURE 31
Schematic diagram of the jump circuit for use with the detector of Fig. 27 in the energy-loss mode.[47]

machine memory (and time) on the intermediate energy-loss region, it has been found useful to use resistively programmable bias power supplies and a relay circuit which is activated by sensing the channel location of the MCS unit. By providing a switch-selectable choice of jump positions as shown in Fig. 31 a large number of experimental options may be easily programmed. A "jump" spectrum for N_2 is shown in Fig. 12.

In addition to ground loops and environmental noise, a major source of background counts arises from the high-voltage circuitry associated with the electron multiplier. With the recent development of low-current solid-state high-voltage channeltron power supplies many of the corona and arcing problems of tube-type equipment have been eliminated. A secondary source of noise in these circuits results from the resistor–capacitor networks and high-voltage feedthroughs. A potential problem in this area is that the channeltron output pulses are usually carried out of the vacuum system on one of the high-voltage lines. Noise at this feedthrough (caused by micro-arcs due to surface contamination) is introduced on the signal line and contributes to the background at 0.5 to a few Hz rate. Recently it has been found that suitable high-voltage capacitors (disk ceramic construction) may be used inside the vacuum system with no apparent contamination problems. This reduces the micro-arc noise problem to a single high-voltage feedthrough which may be filtered by an *RC* network again placed inside the vacuum system.

To further reduce the probability of noise pickup in scattering experiments proper impedance matching and amplifier–discriminator circuitry is essential. As most electron signals are derived from low-impedance sources and most environmental noise may be considered as deriving from high-impedance sources, the use of 50 Ω cable, terminations, and amplifier circuitry is to be recommended. The 50 Ω impedance standard has been adopted by fast pulse nuclear instrumentation, and a wide variety of pulse shaping and amplifying circuitry is readily available from this source.

For channeltrons used in electron detection, the use of 50 Ω cable preserves the fast rise time of the detected signal. For a typical channeltron gain ($\sim 10^8$) and pulse width ($< 10^{-8}$ s) a 50 Ω terminated system can be expected to produce 30–50 mV pulses at the amplifier input. Since channeltrons operated in the pulse counting mode have a very narrow pulse-height distribution function,[81] setting the discriminator level to 10% of the expected average pulse height will ensure that all true signal pulses will be properly counted. The use of 50 Ω systems thus also leads to rather moderate requirements for the amplifier–discriminators which may be used for electron detection (1 to 10 mV discriminators with 100 to 1000 gain amplifiers).

A recent development in amplifier circuitry is the use of single-element amplifier–discriminator chips which may be directly connected to the channeltron inside the vacuum system. These devices have been developed for space-flight applications and are available from a number of sources (EMI, AMP-TEK, etc.). Precautions must be exercised in output-to-input coupling and proper ground points to prevent oscillation. Another successful approach is the use of an emitter–follower circuit (also inside the vacuum system) to impedance match the channeltron to the external amplifier.[81] In this application, channeltron pulses of several hundred millivolts into a 50 Ω load can be expected. This greatly reduces the threshold and gain requirements for the external pulse-shaping network and provides a signal level directly compatible with most fast constant fraction discriminators.

4.4.3. Data Acquisition and Analysis

A block diagram for a typical data acquisition system used in our and other laboratories is shown in Fig. 32. In principle the use of individual MCS units may be eliminated by multiplexing a central processing unit for time-sharing operation of the individual spectrometers. In practice, however, the speeds required for multiple instrument operation demand machine language instruction sets. Additionally, the programming and display functions are confined to the physical location of the central processing unit (CPU). This may raise difficulties in operating a spectrometer which may be located from several feet to several miles distant from the data logging site.

It has been found more convenient to resort to user-oriented programming languages (FORTRAN or BASIC) in a system where the CPU is dedicated to data logging and analysis and the MCS units are locally controlled (at the spectrometer). In this manner, each operator has the convenience of real-time display

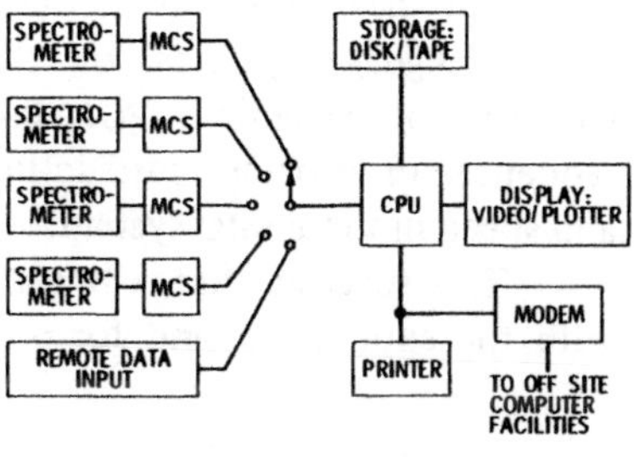

FIGURE 32
Schematic representation for a computer-based data acquisition system.

and operation of the data acquisition sequence. However, the system still retains the capability of accessing the computer for data logging of the results or, when necessary, performing real-time data analysis for monitoring and adjusting the experimental runs.

As most multichannel scaling units suitable for electron-impact spectroscopy are microprocessor based, data communication between the remote site and the CPU is most readily accomplished using standard data communications interfaces. Serial data transfer is accomplished via the standard EIA–RS–232–C interface.[99] In this case the ASCII character set corresponding to a particular channel count in the MCS is transferred serially using a standard 10- or 11-bit character code. This transfer is made on a standard 12- to 25-wire interface whose voltage levels are specified by the RS–232–C designation. Transfer rates of from 100 to 9600 baud (bits/s) are available. At the highest baud rate, 1024 MCS channels may be transferred in a little under 10 s.

Parallel data communication between the MCS and the CPU may also be accomplished via an industry standard General Purpose Interface Bus (GPIB). However, for transferring large data sets (such as the 1024 channels of data typically used in MCS data acquisition), parallel data transfer does not significantly reduce the transfer time. Since the cabling required for this type of transfer is less convenient than the RS–232–C serial data transfer, the latter communication mode is usually preferred.

Central processing units suitable for this type of work may be either a microprocessor-based system or a minicomputer. The difference is primarily one of speed and computational power. The microprocessor systems are usually self-contained units consisting of a processor, tape drive, video display, and RS–232 and GPIB interfaces in a single unit. These systems are usually programmed in BASIC and include a very convenient graphics instruction set for data display. The minicomputer systems (such as the PDP 11 series) are of a modular design in which the CPU (and memory), the data transfer interface, the data display terminal, and the data logging device (tape or disk) are independent units which must be properly cabled together. Most of these devices are programmed in FORTRAN but may, as an option, include a BASIC interpreter.

For systems in which a great deal of analysis of large data sets is anticipated, FORTRAN language systems may have a distinct advantage. Since FORTRAN is a compiled language the execution time for an instruction set is orders of magnitude faster than for an interpreted language such as BASIC. BASIC language systems, on the other hand, have an interactive capability which is not available in FORTRAN. Thus, changing an array size, experimental sequence, or program instruction set may be accomplished with greater ease and speed in the BASIC system.

This suggests that a BASIC instruction set should be used for interacting with the experiment and for preliminary data analysis, display, and storage. A

FORTRAN program may then be needed for detailed data analysis. This may be accomplished via the FORTRAN computer–BASIC interpreter capabilities of the minicomputer systems or by use of a dedicated microprocessor system which can access a large computing facility. The latter approach has been successfully used in our laboratory for several years and provides a minimum amount of interfacing between the CPU, video display, the plotter, and tape drive as these are all integral parts of the system. In addition, the 64K-byte capability of most microprocessor-based systems allows for all but the most complex data analysis routines. In the rare case when this capability is required, connection via a standard MODEM to a large computer system is available. Both data and programs may be communicated over the interface.

4.5. Discussion of Experimental Uncertainties

Perhaps as significant as any reported cross section is the accuracy which may be associated with that cross section. Clearly a result which claims 50% error will not be considered as significant as the same result which claims a 5% precision. It is imperative, therefore, that error estimates associated with experimental results be as circumspect and conservative as possible. In this subsection we will attempt to point out some of the types of experimental uncertainties associated with cross-section measurements, but due to the broad nature of the subject material of this chapter, it is not possible to perform a detailed error analysis which may be useful for any particular application.

In any measurement it must be assumed that there is an underlying absolute parameter (in this case the cross section) which is obscured by the apparatus and personnel attempting to measure this parameter. In this sense, those experiments are usually most accurate which require the minimum number of operations and operators to deduce the desired value. This does not imply that the best experiments are technically unsophisticated. On the contrary, the experiments which are the simplest to interpret are often the most elegant. For instance, crossed-beam differential cross-section measurements are usually performed under conditions where significant path length and background corrections need to be applied (see Sections 4.1.2, 4.2, and 4.3.3). Measurements performed with chopped collimated beams will certainly provide more precise angular distributions, yet the technology (and cost) are considerably greater.

Experimental uncertainties may be classified as arising from either systematic or random sources. Systematic errors arise due to known or unknown corrections which, if not applied, give a systematic bias to the results. Random errors are uncertainties which are due to the statistical nature of any measurement and which may be reduced by repeating the experiment a large number of times. To cite the example mentioned previously, an effective path length correction may be considered a systematic error. If this geometry correction is not included in the analysis, the "correct" angular distribution will *never* be obtained from such measurements. The count rate in such an experiment represents a random error whose statistical uncertainty is $1/\sqrt{N}$, where N is the accumulated number of counts. This error may be reduced by increasing either the count rate or the accumulation time. The background count rate, on the other hand, represents a systematic error with an associated statistical uncertainty.

In many experiments, most systematic errors may be either avoided or corrected. Temperature dependence of pressure readings, temporal drift in electron optics, timing calibrations in coincidence measurements, geometry and background corrections in crossed-beam experiments, and path length corrections in time-of-flight measurements are all examples of correctable systematic errors. As most of these corrections are also subject to statistical uncertainties, it is clearly desirable to reduce the magnitude of the systematic effects within the constraint of retaining sufficient count rate to perform meaningful measurements. Once systematic corrections have been applied, one is still left with a lingering suspicion of unknown systematic errors. This question can only be answered by comparing the results against independent measurements of the same quantities which claim comparable accuracy.

Once the measurement errors have been reduced to statistical uncertainties, standard techniques for data analysis as described in most elementary statistics texts[100,101] may be employed. A conventional analysis defines the mean value for the measurement as

$$\bar{X} = \frac{1}{N} \sum_{i=1}^{N} X_i, \tag{49}$$

where $\bar{X}$ is the mean value, X_i is the value of X in the ith measurement, and N is the total number of observations. Equation (49) assumes that each of the observations X_i is equally weighted. If this is not the case (e.g., different total counts for a spectral feature due to different accumulation times), then the X_i must be given a relative weight d_i and

$$\bar{X} = \frac{1}{N} \sum_{i=1}^{N} d_i X_i, \tag{50}$$

where

$$N = \sum_{i=1}^{N} d_i \tag{51}$$

The probable error in the measurement may then be determined from the standard deviation

$$\delta = \left[\sum_{i=1}^{N}\left(\frac{(X_i-\bar{X})^2}{N}\right)\right]^{1/2}. \qquad (52)$$

Alternatively, the errors in a measurement may be calculated as the root mean square (r.m.s.) value of all the contributing errors. For instance, if a single measurement is made in an electron–molecule scattering experiment, the observed count rate may be a function of the incident electron current, the pressure in the scattering chamber, the flow rate of the molecular gas, the detection efficiency, the noise level in the external circuitry and several other parameters. Each of these parameters has an associated error y_i (usually expressed as a percent) which, in the absence of systematic effects, contributes to the total measurement error as

$$\bar{y} = \left(\sum_i y_i^2\right)^{1/2}. \qquad (53)$$

If, however, the data must be normalized to some external reference which is independent of the experiment under study, the errors associated with this normalization must also be considered. In a relative flow measurement, for instance, the error associated with the data ($\bar{y}$ or δ) must be combined with the estimated errors in the calibration standard to arrive at a total probable error: $[(\bar{y})^2$ (or δ^2) + (error in standard)$^2]^{1/2}$.

In certain special cases, a set of data points $\bar{X}_j$ with an associated error δ_j for each point may represent a process whose behavior as a function of the experimental variables is presumed to be well known. Under these conditions it is further possible to define the χ^2 value for this distribution as

$$\chi^2 = \frac{1}{M-P}\sum_{j=1}^{M}\left(\frac{f_j-\bar{X}_j}{\delta_j}\right)^2, \qquad (54)$$

where M is the number of data points, f_j is the function which is presumed to represent the jth data point ($\bar{X}_j$) and P is the number of independent variables available in f_j. Such a situation arises, for instance, in the elastic angular distributions for low-energy electron–atom scattering which may be represented by a phase shift expression for f_j. The phase shifts are the independent variables and they may be determined by either a variational analysis(53) or a grid search technique.(102) In either case, the criterion is the minimization of χ^2. If the errors in the data are purely statistical and if the functional form of f_j truly represents $\bar{X}_j$, then $\chi^2_{\min}$ is expected to be unity. A careful analysis of the dependence of χ^2 on the variables can then lead to an estimate of the reliability of the data set.[102,103]

5. Electron Collision Measurements at High Energies: Generation of Optical Data

The experimental techniques for cross-section measurements at high impact energies ($E_0 > 100$ eV) are similar to those described here for lower impact energies. The control and focusing of the electron beam is, however, more easily achieved at higher energies because the influence of space charge, electric and magnetic fields, and surface effects are greatly diminished. The scattering geometry and experimental parameters can be accurately determined and absolute cross sections can be obtained directly from the scattered signal intensities. One may encounter some problems associated with secondary electrons generated by high-energy electron–surface collisions, and this has to be kept in mind.

Geiger and co-workers (see, for example, Ref. 104) generated energy-loss spectra of molecules with about 10-meV energy resolution using electrons of around 25-keV energy and Wien filters[105,106] as energy selectors and analyzers. Their spectra represent relative optical excitation cross sections. No attempt was made by them to extract absolute cross-section data from their measurements. Bromberg made accurate elastic differential cross-section measurements at a few hundred eV kinetic energies and low scattering angles utilizing a static gas target, unselected electron beam, and an electrostatic analyzer.[82] He experimentally determined all the parameters relating the measured signal to the cross section and produced absolute cross sections with a few % accuracy. Chamberlain *et al.*[83] followed the same philosophy as Bromberg but also used a hemispherical energy selector in the gun. They generated absolute elastic differential cross sections at 5° in the 50–400 eV impact energy range for He.

A large number of quantitative fast-electron-impact experiments have been carried out with the aim of obtaining optical oscillator strengths. The pioneering work for developing the technique and instrumentation for this type of work was carried out by Lassettre and co-workers.[107,108] Very recently Brion and co-workers and van der Wiel and co-workers have applied elegant coincidence techniques to obtain quantitative photoionization,[69,109] photofragmentation,[71] and photofluorescence[110] cross sections by high-energy electron-impact methods. Normalization of the experimental data to the absolute scale was achieved in some of these experiments by the application of the Thomas–Kuhn–Reiche sum rule.[111] A comprehensive review of these techniques and results is given by Brion and Hamnett.[68]

In generating optical data from high-energy electron-impact measurements, the fact is utilized that these electrons behave like pseudophotons and that the generalized oscillator strength obtained from electron-impact cross-section data converges to the dipole oscillator strength (or, essentially, photoabsorption cross section) in the zero momentum-transfer limit. For detailed discussion of the underlying principles the reader is referred to Refs. 112–114 and Chapter 5, Section 3 of this volume. Here we only briefly state the basic philosophy.

Bethe[112,113] defined the generalized oscillator strength for electron-impact processes as (Chapter 1, Section 6.5.2 and Chapter 5, Section 3)

$$f_n^G(K) = \frac{\Delta E}{2}\left(\frac{k_0}{k_n}\right) K^2 \mathrm{DCS}_n^B(K), \tag{55}$$

where $\mathrm{DCS}_n^B(K)$ is the differential cross section for process n in the Born approximation as a function of momentum transfer K. ΔE is the energy loss and k_0 and k_n are the magnitudes of the electron momentum before and after scattering, respectively. It is easily shown that in the zero momentum-transfer limit $f_n^G(K)$ becomes equal to the optical f value, f_n^O. Lassettre *et al.*[114] extended the use of Eq. (55) to cases where the Born approximation does not hold and defined the apparent generalized oscillator strength, $f_n^{AG}(K)$ by the same relation. They were able to show that $f_n^{AG}(K)$ also converges to the optical f value in the zero momentum-transfer limit.

In practice one either makes several measurements at various angles and energies, plots the $f_n^{AG}(K)$ values as a function of K^2, and extrapolates the resulting curve to zero momentum transfer, or carries out the experiment at very high impact energy and low (preferably zero) scattering angle and assumes that the results correspond to the optical limit. See Chapter 5, Section 3.1 for further details and examples of application.

In recent years, high-energy electrons have also been utilized to investigate inner-shell excitation processes. From the point of view of resolution the electron-impact technique becomes very competitive with or even superior to optical techniques at the large energy losses corresponding to inner-shell excitation (see Chapter 5, Section 3.2, Fig. 32). In addition, it is a much more convenient and inexpensive approach than the use of available uv and x-ray sources or synchrotrons. Some optical absorption data have been obtained by the electron-impact method, but most of the investigations have dealt with spectroscopy and resonances of inner-shell excitation. The reader is referred to Ref. 115 and to Chapter 5, Section 3.2 for more details.

6. Electron Scattering in Electromagnetic Fields

The availability of high-intensity narrow frequency-band lasers has opened up new possibilities for electron scattering studies. These studies can yield information on the interaction of molecules, electrons, and photons which could not be obtained previously and which are of practical importance in the field of laser–plasma interactions. The general treatment of the problem of electron scattering in electromagnetic fields and the interpretation of experimental observations are quite complex. To the usual problem of treating the interaction of an electron with an N-electron molecule additional difficulties arise when the laser field is introduced: (1) the energy states of the molecule may be modified (dressed states); (2) multiphoton excitation and ionization of the molecule may take place; (3) multiphoton free–free electron transitions (bremsstrahlung, inverse bremsstrahlung) can occur; and (4) the close dynamics of the scattering process may be affected by n-photon transitions in the $(N+1)$-electron molecular ion. Both the theoretical and experimental efforts in this area have been limited mainly to atoms and to special situations where one particular interaction dominates and the others can be neglected or treated as perturbations. The experimental investigations carried out so far fall into two categories: (1) electron scattering by laser-excited species, and (2) free–free radiative transitions.

In the first category low-power tunable dye lasers are utilized to excite the target species to specific states and electron scattering by these excited species is studied. In this case the resonant excitation of the target is the dominating process and the interaction of the free electron with the field and the distortion of the energy level scheme of the target can be neglected. Total cross-section measurements on laser-excited Na have been reported by Bhaskar *et al.*[(116)] and Jaduszliver *et al.*[(117)] and differential cross sections for laser-excited Ba by Register *et al.*,[(118)] but the technique has not been applied to molecular species mainly due to the difficulty of producing tunable lasers in the uv or ir regions. When the laser is operated in the single-frequency mode, individual isotopes and hyperfine levels can be pumped. With the selection of proper polarization of the laser light, the population of the magnetic sublevels can also be controlled. Thus a polarized target (aligned and/or oriented) can be prepared for electron scattering measurements. Since the magnetic sublevels can be coherently excited, not only cross sections for individual magnetic sublevel excitation but also scattering amplitudes can be obtained from measuring superelastic scattering intensities. The excitation with polarized light and subsequent measurement of the superelastic electron scattering intensities is equivalent to the time-inverse electron–photon coincidence measurement which has been discussed for atomic species in general by Hertel and Stoll[(119)] and applied to H_2 recently by Malcolm and McConkey.[(78)]

The second category of experiments has been concerned with free–free radiative transitions in strong electromagnetic fields. Here the photon energy is chosen to be far away from resonance (even for multiphoton processes) and the dominant interaction is between the free electron and the photon field. The present situation in this particular area has been summarized by Gavrila and van der Wiel.[120] Elastic scattering by Ar and H_2 in a high-power pulsed CO_2 laser field has been studied by Weingartshofer *et al.*[121] In addition to the regular elastic scattering feature, features appear on both sides of the elastic peak. These additional features are separated by an energy corresponding to a laser photon and represent induced and inverse bremsstrahlung. The overall integrated intensity of these features is equal to the elastic scattering intensity without the laser field (assuming identical scattering conditions) and therefore cross sections for the individual free–free electron transition can be obtained through the calibration of the field-free elastic scattering.[122] Interesting situations concerning resonances can develop in connection with free–free radiative transitions. Resonances can occur whenever the kinetic energy of the electron or the combined energy of the electron and photon is appropriate. Such situations have been studied by Langhans[123] and Langendam and van der Wiel.[124]

7. Closing Remarks

Based on the trends observed in the recent literature for electron–molecule scattering, several encouraging remarks may be made. With the increased understanding of beam–beam geometry and stability problems, the development of more sophisticated electron optics and control circuitry, and the increasingly accurate monitoring of important experimental parameters (flow, pressure, etc.), significant improvements in the reliability of integral and differential cross sections for electron scattering from ground and excited molecular states can be expected. For the particular case of H_2, sufficient beam intensity and resolution are available for a systematic study of vibrational and rotational excitation in this molecule, and work in this direction should be encouraged. For heavier molecules, vibrational structure may now be resolved (or unfolded), but order-of-magnitude improvements in electron energy resolution must be forthcoming before significant progress on rotational excitations can be made.

Increased interest in the more fundamental measurements of scattering amplitudes for excitation processes can be anticipated. The proliferation of coincidence techniques for electron–atom scattering logically suggests that electron–photon studies on molecules will receive increasing attention. Although preliminary results for H_2 indicate rather pessimistic count rates,[73] these are pioneering experiments and evolutionary improvements may be expected. The alternative time-inverse approach (laser excitation of target and electron scattering by the excited species) awaits further improvements in harmonic generation techniques before it may be applied to molecular problems. However, when appropriate uv sources are available, the very high energy resolution of such lasers will allow experiments to be designed which will not only study coherence effects but will also measure excited-state cross sections for individual vibrational and rotational levels.

Inner-shell excitation measurements for molecules may also be expected to see evolutionary improvements. In addition to optical excitation functions, improvements in experimental design may eventually lead to cross-section determinations for optically forbidden inner-shell processes. These same improvements may further enhance the reliability and availability of generalized oscillator strengths for a large number of processes for an increasingly larger set of molecules.

From an instrumental point of view, a number of continuing investigations may lead to interesting innovations. Multichannel arrays have begun to appear in electron scattering experiments.[13,125] Recently, work has been initiated using multichannel electron multiplier arrays and charge-coupled devices for scattered electron energy analysis.[126] Reticons developed for mass spectroscopy purposes may also be used for this application,[127] and these techniques represent a possible immense improvement in accumulation time.

Spin-polarized electron sources for studying spin-dependent interactions have recently been developed by Pierce *et al.*[128] Currents on the order of 10 μA with 0.13-eV energy resolution and about 50% spin polarization have been obtained. Celotta and co-workers have described the technology and possible applications of these devices.[129,130] Although the technological investment in such a source is substantial, the experimental possibilities are extremely interesting.

Finally, the electron-impact instruments described in this chapter may readily be adapted for studying both negative and positive ions. In fact, a great deal of research has been carried out using this type of instrumentation on dissociative attachment processes[131,132] where negative-ion fragments have been detected and their energy and angular distributions determined in the same manner as described here for electrons.

Acknowledgments

The authors wish to express their gratitude to their colleagues and collaborators, many of whom have allowed their work to be quoted in this text. In particular, we would like to thank D.F. Brewer, W.R. Newell, A.C.H. Smith, R.T. Brinkmann, C.E. Brion, A. Hamnett, D.C. Cartwright, R.I. Hall, A. Chutjian, J. Comer, F.H. Read, S.W. Jensen, R.E. Kennerly, R.A. Bonham, S.L. Lin, R.E. Robson, E.A. Mason, Jr., I.C. Malcolm, J.W. McConkey, D. Rassi, K. Rohr, and S.K. Srivastava. We also wish to thank the American Institute of Physics and the Institute of Physics for granting permission to use Figs. 2, and 14 and 19, respectively.

References

1. C. Ramsauer, Über den Wirkungsquerschnitt der Gasmoleküle gegenüber longsamen Electronen. I. Fortsetzung, *Ann. Phys. (Leipzig)* **66**, 546–558 (1921).

2. D.E. Golden and H.W. Bandel, Absolute total electron–helium scattering cross sections for low electron energies, *Phys. Rev.* **138**, A14–21 (1965).
3. H.S.W. Massey, E.H.S. Burhop, and H.B. Gilbody, *Electronic and Ionic Impact Phenomena,* 2nd ed., Vols. 1 and 2, Clarenden Press, Oxford (1969).
4. C.E. Kuyatt, Electron–atom interactions, in *Methods of Experimental Physics* (L. Marton, ed.), Vol. 7A, pp. 1–43, Academic Press, New York (1968).
5. D.E. Golden, N.F. Lane, A. Temkin, and E. Gerjuoy, Low-energy electron–molecule scattering experiments and the theory of rotational excitation, *Rev. Mod. Phys.* **43**, 642–678 (1971).
6. B. Bederson and L.J. Kieffer, Total electron–atom collision cross sections at low energies—A critical review, *Rev. Mod. Phys.* **43**, 601–641 (1971).
7. H.J. Blaauw, F.J. de Heer, R.W. Wagenaar, and D. Barends, Total cross sections for electron scattering from N_2 and He, *J. Phys. B* **10**, L299–303 (1977).
8. R.W. Wagenaar, Total cross sections for electron–atom scattering and their use in the field of dispersion relations. FOM Institute, Amsterdam, Report FOMNR.43.948 (1978).
9. B. Bederson, Crossed beam electron–neutral experiments, in *Methods of Experimental Physics* (L. Marton, ed.), Vol. 7A, pp. 67–95, Academic Press, New York (1968).
10. K. Rubin, B. Bederson, M. Goldstein, and R.E. Collins, Electron alkali-metal inelastic recoil experiments and spin analysis: Experimental method and the small-angle behavior of the $4\ ^2S_{1/2} \rightarrow 4\ ^2P_{1/2,3/2}$ excitation in potassium, *Phys. Rev.* **182**, 201–214 (1969).
11. R.E. Collins, B. Bederson, and M. Goldstein, Differential spin exchange and the elastic scattering of low-energy electrons by potassium, *Phys. Rev. A* **3**, 1976–1987 (1971).
12. J.E. Land and W.R. Raith, High-resolution measurement of resonances in e–O_2 scattering by electron time-of-flight spectroscopy, *Phys. Rev. A* **9**, 1592–1602 (1974).
13. R.E. Kennerly and R.A. Bonham, Electron–helium absolute total scattering cross-sections from 0.5 to 50 eV, *Phys. Rev. A* **17**, 1844–1854 (1978).
14. D.W.O. Heddle, The measurement of optical excitation functions, in *Methods of Experimental Physics* (L. Marton, ed.), Vol. 7A, pp. 43–67, Academic Press, New York (1968).
15. D.W.O. Heddle and R.G.W. Keesing, Measurement of electron excitation functions, in *Advances in Atomic and Molecular Physics* (D.R. Bates and I. Estermann, eds.), Vol. 4, pp. 267–297, Academic Press, New York (1968).
16. O.B. Shpenik, V.V. Sovter, A.N. Zavilopulo, I.P. Zapesochny, and E.E. Kontrosh, Investigation of resonances in the total cross sections for excitation of mercury atoms by slow monoenergetic electrons, *Zh. Eksp. Teor. Fiz.* **69**, 48–58 (1975).
17. O.B. Shpenik, Resonances observed on the optical excitation functions, in *Electronic and Atomic Collisions* (G. Watel, ed.), pp. 99–112. North-Holland, Amsterdam (1978).
18. G.J. Schulz, Measurement of atomic and molecular excitation by a trapped electron method, *Phys. Rev.* **112**, 150–154 (1958).
19. H.S.W. Massey, E.H.S. Burhop, and H.B. Gilbody, *Electronic and Ionic Impact Phenomena*, 2nd ed., Vol. I, pp. 284–288, Clarendon Press, Oxford (1969).
20. D.A. Golden and A. Zecca, An energy modulated high energy resolution electron spectrometer, *Rev. Sci. Instrum.* **42**, 210–216 (1971).
21. F.W.E. Knopp, H.H. Brongersma, and A.J.H. Boerboom, A new method to study molecules by electron impact spectroscopy, *Chem. Phys. Lett.* **5**, 450–452 (1970).
22. J.S. Townsend and V.A. Baily, The motion of electrons in gases, *Phil. Mag.* **42**, 873–891 (1921).
23. J.S. Townsend and V.A. Baily, Motion of electrons in helium, *Phil. Mag.* **46**, 657–664 (1923).

24. L.G.H. Huxley and R.W. Crompton, *The Diffusion and Drift of Electrons in Gases,* John Wiley and Sons, New York (1974).
25. L.S. Frost and A.V. Phelps, Rotational excitation and momentum transfer cross sections for electrons in H_2 and N_2 from transport coefficients, *Phys. Rev.* **127**, 1621–1633 (1962).
26. L. Ferrari, Inelastic collisions and velocity distribution function of electrons in a gas in an electric field, *Physica* **85C**, 161–179 (1977).
27. D.C. Cartwright, Rate coefficients and inelastic momentum-transfer cross sections for electronic excitation of N_2 by electrons, *J. Appl. Phys.* **49**, 3855–3862 (1978).
28. S.L. Lin, R.E. Robson, and E.A. Mason, Jr., Moment theory of electron drift and diffusion in neutral gases in an electrostatic field, *J. Chem. Phys.* **71**, 3483–3498 (1979).
29. D.W.O. Heddle, Tables of focal properties of three-element electrostatic cylinder lenses, University of Colorado, JILA Report No. 104 (October 1970).
30. E. Harting and F.H. Read, *Electrostatic Lenses,* Elsevier, Amsterdam (1976).
31. A. Chutjian, Geometries and focal properties of two electron-lens systems useful in low-energy electron or ion scattering, *Rev. Sci. Instrum.* **50**, 347–355 (1979).
32. C.E. Kuyatt, Lectures on electron optics, NBS (1970), unpublished.
33. D. DiChio, S.V. Natali, C.E. Kuyatt, and A. Galejs, Use of matrices to represent electron lenses. Matrices for the two-tube electrostatic lens, *Rev. Sci. Instrum.* **45**, 566–569 (1974).
34. A.L. Hughes and V. Rojansky, On the analysis of electronic velocities by electrostatic means, *Phys. Rev.* **34**, 284–290 (1929).
35. A.L. Hughes and J.H. McMillen, Refocusing of electron paths in a radial electrostatic field, *Phys. Rev.* **34**, 291–295 (1929).
36. P. Marmet and L. Kerwin, An improved electrostatic electron selector, *Can. J. Phys.* **38**, 787–796 (1960).
37. H. Ehrhardt, L. Langhans, and F. Linder, Die differentiellen Anregungsfunktionen der $n = 2$ Zustände von Helium, *Z. Phys.* **214**, 179–194 (1968).
38. R.I. Hall, G. Joycz, J. Mazeau, J. Reinhardt, and C. Schermann, Electron impact differential and integral cross sections for excitation of the $n = 2$ states of helium at 29.2, 39.2, and 48.2 eV, *J. Phys. (Paris)* **34**, 827–843 (1973).
39. E.M. Purcell, The focusing of charged particles by a spherical condenser, *Phys. Rev.* **54**, 818–826 (1938).
40. C.E. Kuyatt and J.A. Simpson, Electron monochromator design, *Rev. Sci. Instrum.* **38**, 103–111 (1967).
41. J.N.H. Brunt, F.H. Read, and G.C. King, The realization of high-energy resolution using the hemispherical electrostatic energy selector in electron-impact spectrometry, *J. Phys. E* **10**, 134–139 (1977).
42. D.F. Brewer, W.R. Newell, and A.C.H. Smith, A co-axial cone electrostatic velocity analyzer I. Analysis of electron optical properties, *J. Phys. E* **13**, 114–122 (1980); II. Construction and use of high-resolution electron scattering experiments, *J. Phys.* **13**, 123–127 (1980).
43. K.D. Sevier, *Low-Energy Electron Spectrometry*, Wiley-Interscience. New York (1972); W. Steckelmacher, Energy analyzers for charged particle beams, *J. Phys. E* **6**, 1061–1071 (1973).
44. F.H. Read, J. Comer, R.E. Imhof, J.N.H. Brunt, and E. Harting, The optimization of electrostatic energy selection systems for low energy electrons, *J. Electron Spectrosc. Relat. Phenom.* **4**, 293–312 (1974).
45. S. Trajmar, J.K. Rice, and A. Kupperman, A low-energy, high-resolution electron-impact spectrometer, Caltech, JPL Technical Memorandum 33–373 (March, 1968).

46. F.H. Read, Doppler and other broadening effects in electron scattering experiments, *J. Phys. B* **8**, 1034–1040 (1975).
47. S.W. Jensen, Differential cross section measurements for electron impact on barium in ground and laser excited states, Ph.D. thesis, University of California, Riverside, California (1978).
48. R.J. Brunt and A.C. Gallagher, Electron scattering with 1 meV resolution, in *Electronic and Atomic Collisions* (G. Watel, ed.), pp. 129–142, North-Holland, Amsterdam (1978).
49. S. Trajmar, D.C. Cartwright, and R.I. Hall, Electron-impact excitation of the Rydberg states in O_2 in the 7–10 eV energy-loss region, *J. Chem. Phys.* **65**, 5275–5279 (1976).
50. J. Comer and F.H. Read, Electron impact studies of a resonant state of N_2^-, *J. Phys. B* **4**, 1055–1062 (1971).
51. R. Brinkmann and S. Trajmar, Effective path length corrections in beam–beam scattering experiments, *J. Phys. E* **14**, 245–255 (1981).
52. D.R. Olander and V. Kruger, Molecular beam sources fabricated from multichannel arrays III. The exit density problem, *J. Appl. Phys.* **41**, 2769–2776 (1970).
53. D.F. Register, S. Trajmar, and S.K. Srivastava, Absolute elastic differential electron scattering cross sections for He: A proposed calibration standard from 5 to 200 eV, *Phys. Rev. A* **21**, 1134–1151 (1980).
54. J.B. Anderson, Molecular beams for nozzle sources, in *Molecular Beams and Low-Density Gasdynamics* (P.P. Wegener, ed.), pp. 1–91, Marcel Dekker, New York (1974).
55. S. Trajmar, W. Williams, and S.K. Srivastava, Electron-impact cross sections for Cu atoms, *J. Phys. B* **10**, 3323–3333 (1977).
56. D. Bulgin, J. Dyke, F. Goodfellow, N. Jonathan, E. Lee, and A. Morris, A high-temperature furnace for use in photoelectron spectroscopy, *J. Electron Spectrosc. Relat. Phenom.* **12**, 67–76 (1977).
57. R.K. Curran, Formation of SF_6^- by inelastically scattered electrons, *J. Chem. Phys.* **38**, 780–781 (1963).
58. S. Cvejanović and F.H. Read, A new technique for threshold excitation spectroscopy, *J. Phys. B* **7**, 1180–1193 (1974).
59. F. Pichou, A. Huetz, G. Joyez, M. Landau, and J. Mazeau, Electron impact excitation of helium: Absolute differential cross sections of the $n = 2$ and 3 ^3S states from threshold to 3.6 eV above, *J. Phys. B* **9**, 933–944 (1976).
60. G. Joyez, Anisotropic threshold excitation of *S* states of helium by electron impact, in *Electronic and Atomic Collisions, Abstracts of Contributed Papers, XIth ICPEAC* (K. Takayanagi and N. Oda. eds.), pp. 180–181, North-Holland, Amsterdam (1979).
61. G. Seng and F. Linder, Vibrational excitation of polar molecules by electron impact. II. Direct and resonant excitation in H_2O, *J. Phys. B* **9**, 2539–2551 (1976).
62. K. Rohr, Interaction mechanisms and cross sections for the scattering of low-energy electrons from HBr, *J. Phys. B* **11**, 1849–1860 (1978).
63. K. Rohr, Threshold resonances in vibrational excitation of molecules by electron impact, in *Symposium on Electron–Molecule Collisions, Invited Papers* (I. Shimamura and M. Matsuzawa, eds.), pp. 67–76, University of Tokyo, Tokyo (1979).
64. W.C. Tam and S.F. Wong, A magnetically collimated electron-impact spectrometer, *Rev. Sci. Instrum.* **50**, 302–307 (1979).
65. A. Stamatovic and G.J. Schulz, Characteristics of the trochoidal electron monochromator, *Rev. Sci. Instrum.* **41**, 423–427 (1970).
66. K. Rohr, Ph.D. Thesis, University of Kaiserslautern, Kaiserslautern, Federal Republic of Germany (1975).
67. K. Rohr and F. Linder, Vibrational excitation of polar molecules by electron impact. I. Threshold resonances in HF and HCl, *J. Phys. B* **9**, 2521–2537 (1976).

68. C.E. Brion and A. Hamnett, Continuum optical oscillator strength measurements by electron spectroscopy in the gas phase, in *The Excited State in Chemical Physics*, Part 2, *Advances in Chemical Physics* (J.W. McGowan, ed.), Vol. 45, p. 1, John Wiley and Sons, New York (1981).
69. M.J. van der Wiel and C.E. Brion, "Photoelectron" spectroscopy by electron-impact coincidence measurements of scattered and ejected electrons in CO, *J. Electron Spectrosc. Relat. Phenom*. **1**, 309–319 (1972).
70. D.G. Wilden, P.J. Hicks, and J. Comer, Autoionization of N_2 studied using an electron time-of-flight coincidence spectrometer, *J. Phys. B* 9, 1959–1974 (1976).
71. C. Backx, G.R. Wight, and M.J. van der Wiel, Oscillator strengths (10–70 eV) for absorption, ionization and dissociation in H_2, HD, and D_2, obtained by an electron–ion coincidence method, *J. Phys. B* 9, 315–331 (1976).
72. K. Blum and H. Kleinpoppen, Electron–photon angular correlation in atomic physics, *Phys. Rep.* **52**, 204–261 (1979).
73. J.W. McConkey, Electron–photon coincidence measurements as a probe of inelastic scattering of electrons from atoms and molecules, in *Symposium on Electron–Molecule Collisions, Invited Papers* (I. Shimamura and M. Matsuzawa, eds.), pp. 163–178, University of Tokyo, Tokyo (1979).
74. H. Kleinpoppen and J.F. Williams, eds., *Coherence and Correlation in Atomic Collisions*, Plenum Press, New York (1979).
75. J. Macek and D.J. Jaecks, Theory of atomic photon–particle coincidence measurements, *Phys. Rev. A* **4**, 2288–2300 (1971).
76. U. Fano and J. Macek, Impact excitation and polarization of the emitted light, *Rev. Mod. Phys.* **45**, 553–573 (1973).
77. K. Blum and J. Jakubowicz, Electron–photon angular correlation in electron–molecule scattering, *J. Phys. B* **11**, 909–925 (1978).
78. I.C. Malcolm and J.W. McConkey, Electron–polarized-photon coincidence study of H_2 ($^1\Sigma_g^+ - {}^1\Pi_u$) excitation, *J. Phys. B* **12**, L67–70 (1979).
79. M. Eminyan, K.B. MacAdam, J. Slevin, and H. Kleinpoppen, Electron–photon angular correlations in electron–helium collisions: Measurements of complex excitation amplitudes, atomic orientations and alignment, *J. Phys. B* 7, 1519–1542 (1974).
80. M.T. Hollywood, A. Crowe, and J.F. Williams, Coherent excitation of 2 ^1P state of He for large momentum-transfer electron scattering, *J. Phys. B* **12**, 819–834 (1979).
81. K.C. Schmidt, The channeltron electron multipliers, Model CEM4010, Technical Applications Note 9803, Bendix Electro-Optics Division, Ann Arbor, Michigan 48107 (25 April 1969).
82. J.P. Bromberg, Absolute differential cross sections of electrons elastically scattered by the rare gases. I. Small angle scattering between 200 and 700 eV, *J. Chem. Phys.* **61**, 963–969 (1974).
83. G.E. Chamberlain, S.R. Mielczarek, and C.E. Kuyatt, Absolute measurement of differential cross sections for electron scattering in helium, *Phys. Rev* A **2**, 1905–1922 (1970).
84. J.F. Williams and B.A. Willis, Electron scattering from atomic hydrogen I. Differential cross sections for excitation for the $n = 2$ states, *J. Phys. B* 8, 1641–1669 (1975).
85. G. Joyez, A. Huetz, I. Pichou, and J. Mazeau, New measurements of differential and integral cross sections for electron impact excitation of the $n = 2$ states of helium, in *Electron and Photon Interactions with Atoms* (H. Kleinpoppen and M.R.C. McDowell, eds.), pp. 349–364, Plenum Press, New York (1976).
86. S.K. Srivastava, A. Chutjian, and S. Trajmar, Absolute elastic differential electron scattering cross sections in the intermediate energy region, I. H_2, *J. Chem. Phys.* **63**, 2659–2665 (1975).
87. A. Chutjian, private communication.

88. D.C. Cartwright, A. Chutjian, S. Trajmar, and W. Williams, Electron-impact excitation of the electronic states of N_2. I. Differential cross sections at incident energies from 10 to 50 eV, *Phys. Rev. A* **16**, 1013–1040 (1977).
89. D.F. Register, H. Nishimura, and S. Trajmar, Elastic scattering and vibrational excitation of CO_2 by 4, 10, 20 and 50 eV electrons, *J. Phys. B* **13**, 1651–1662 (1980).
90. S.F. Wong and L. Dubé, Rotational excitation of N_2 by electron impact: 1–4 eV, *Phys. Rev. A* **17**, 570–576 (1978).
91. H. Tanaka, Pure rotational excitation of N_2, O_2, and CH_4 by electron impact, in: *Symposium on Electron–Molecule Collisions, Invited Papers* (I. Shimamura and M. Matsuzawa, eds.), pp. 31–35, University of Tokyo, Tokyo (1979); H. Tanaka, L. Boesten, and I. Shimamura, Experimental determination of state-to-state rotational transition cross sections: $e + N_2$, *VII Intern. Conf. Atom. Phys., Book of Abstracts*, pp. 43–44 (1980).
92. I. Shimamura, Systematics in the cross sections for excitation of molecules by electrons, in *Symposium on Electron–Molecule Collisions, Invited Papers* (I. Shimamura and M. Matsuzawa, eds.), pp. 13–30, University of Tokyo, Tokyo (1979); State-to-state rotational transition cross sections from unresolved energy loss spectra, *Chem. Phys. Lett.* **73**, 328–333 (1980).
93. J.A. Preston, M.A. Hender, and J.W. McConkey, A versatile electron spectrometer, *J. Phys. E* **6**, 661–666 (1973).
94. J.F. Williams, Electron scattering from hydrogen atoms II. Elastic scattering at low energies from 0.5 to 8.7 eV, *J. Phys. B* **8**, 1683–1692 (1975).
95. J.F. Williams and B.A. Willis, The scattering of electrons from inert gases I. Absolute differential elastic cross sections for argon atoms, *J. Phys. B* **8**, 1670–1682 (1975).
96. D.F. Register, Flow rate considerations for cross-section calibration (to be published).
97. Landolt-Börnstein, Zahlenwerte und Funktionen aus Physik, Chemie, Astronomie, Geophysik, Technik, 6th Ed., Vol. 1, pp. 325 and 373, Springer-Verlag, Berlin (1950).
98. A. Skerbele and E.N. Lassettre, New selection rule illustrated by collision cross-section studies on the $B\ ^1\Sigma^+ \leftarrow X\ ^1\Sigma^+$ and the $C\ ^1\Sigma^+ \leftarrow X\ ^1\Sigma^+$ transitions in carbon monoxide at kinetic energies of 300, 400 and 500 eV, *J. Chem. Phys.* **55**, 424–434 (1971).
99. EIA Standard, RS-232-C, EIA Engineering Dept., 2001 Eye Street, N.W., Washington, D.C. 20006 (1969).
100. P.R. Bevington, *Data Reduction and Error Analysis for the Physical Science*, McGraw-Hill, New York (1969).
101. H.L. Alder and E.B. Roessler, *Introduction to Probability and Statistics*, 4th ed., W.H. Freemand and Co., San Francisco (1968).
102. D. Andrick and A. Bitsch, Experimental investigation and phase shift analysis of low-energy electron–helium scattering, *J. Phys. B* **8**, 393–410 (1975).
103. N.C. Steph, L. McDonald, and D.E. Golden, Analysis of low-energy scattering cross sections I. Electron–helium elastic scattering, *J. Phys. B* **12**, 1507–1527 (1979).
104. J. Geiger, Intensity perturbations due to configuration interaction observed in the electron energy loss spectrum of N_2, *J. Chem. Phys.* **50**, 7–11 (1969).
105. H. Boersch, J. Geiger, and W. Stickel, Das Auflösungsvermögen des elektrostatisch-magnetischen Energieanalysator für schnelle Electronen, *Z. Phys.* **180**, 415–424 (1964).
106. R.A. Bonham and M. Fink, *High-Energy Electron Scattering*, pp. 239–248, Von Nostrand Reinhold, New York (1974).
107. E.N. Lassettre, A.S. Berman, S.M. Silverman, and M.E. Krasnow, Electron spectrometer for the study of inelastic electronic collision cross sections, *J. Chem. Phys.* **40**, 1232–1242 (1964).
108. A.M. Skerbele and E.N. Lassettre, Higher-resolution study of the electron-impact spectrum of helium, *J. Chem. Phys.* **40**, 1271–1275 (1964).

109. A. Hamnett, W. Stoll, G. Branton, C.E. Brion, and M.J. van der Wiel, A kinematic and experimental investigation of the (*e*, 2*e*) simulation of photoelectron spectroscopy, *J. Phys. B* 9, 945–957 (1976).
110. C. Backx, M. Klewer, and M.J. van der Wiel, Photofluorescence by electron impact: determination of the relative oscillator strength for formation of the $B\ ^2\Sigma$ state of CO^+, *Chem. Phys. Lett.* 20, 100–103 (1973).
111. K.H. Tan, C.E. Brion, Ph. E. van der Leeuw, and M.J. van der Wiel, Absolute oscillator strengths (10–60 eV) for the photoabsorption, photoionization and fragmentation of H_2O, *Chem. Phys.* 29, 229–309 (1978).
112. H.A. Bethe, Zur Theorie des Durchgangs schneller Korpuskularstrahlen durch Materie, *Ann. Phys. (Leipzig)* 5, 325–400 (1930).
113. M. Inokuti, Inelastic collisions of fast charged particles with atoms and molecules–The Bethe theory revisited, *Rev. Mod. Phys.* 43, 297–347 (1971).
114. E.N. Lassettre, A. Skerbele, and M.A. Dillon, Generalized oscillator strength for $1\ ^1S \rightarrow 2\ ^1P$ transition of helium. Theory of limiting oscillator strengths, *J. Chem. Phys.* 50, 1829–1839 (1969).
115. F.H. Read and G.C. King, Inner-shell excited states of molecules, *Symposium on Electron–Molecule Collisions, Invited Papers* (I. Shimamura and M. Matsuzawa, eds.), pp. 155–162, University of Tokyo, Tokyo (1979).
116. N.D. Bhaskar, B. Jaduszliver, and B. Bederson, Scattering of low-energy electrons by excited sodium atoms using a photon and electron atomic beam recoil technique, *Phys. Rev. Lett.* 38, 14–17 (1977).
117. B. Jaduszliver, R. Dang, P. Weiss, and B. Bederson, Total cross sections for the scattering of low-energy electrons by excited sodium atoms, in *Coherence and Correlation in Atomic Collisions* (H. Kleinpoppen and J.F. Williams, eds.), pp. 613–623, Plenum Press, New York (1980).
118. D.F. Register, S. Trajmar, S.W. Jensen, and R.T. Poe, Electron scattering by laser excited barium atoms, *Phys. Rev. Lett.* 41, 749–752 (1978).
119. I.V. Hertel and W. Stoll, Collision experiments with laser excited atoms in crossed beams, in *Advances in Atomic and Molecular Physics* (D.R. Bates and B. Bederson, eds.), pp. 113–228, Academic Press, New York (1977).
120. M. Gavrila and M.J. van der Wiel, Free–free radiative transitions of electron–atom systems, *Comments At. Mol. Phys.* 8, 1–20 (1978).
121. A. Weingartshofer, E.M. Clarke, J.K. Holmes, and C. Jung, Experiments on multiphoton free–free transitions, *Phys. Rev. A* 19, 2371–2376 (1979).
122. D. Andrick and L. Langhans, Measurement of the free–free cross section of e^-–Ar scattering, *J. Phys. B* 11, 2355–2360 (1978).
123. L. Langhans, Resonance structures in the free–free cross sections of e^-–Ar scattering, *J. Phys. B* 11, 2361–2366 (1978).
124. P.J.K. Langendam and M.J. van der Wiel. Fine structure of the neon 18.5–18.7 eV resonances by means of resonant free–free radiative absorptions, *J. Phys. B* 11, 3603–3613 (1978).
125. R.W. Wijnaendts van Resandt and J. Los, Position sensitive detection in atomic physics. in *Electronic and Atomic Collisions* (N. Oda and K. Takayanagi, eds.), pp. 831–842, North-Holland, Amsterdam (1980).
126. S.T. Hood and E. Weigold, A new electron coincidence spectrometer with position sensitive detectors. *J. Electron Spectrosc. Relat. Phenom.* 15, 237–240 (1979); F.H. Read and J. Comer, private communication.
127. C.E. Giffin, Jet Propulsion Laboratory, private communication.
128. D.T. Pierce, R.J. Celotta, G.C. Wang, W.N. Unertl, A. Galejs, C.E. Kuyatt, and S.R. Mielczarek, GaAs spin polarized electron source, *Rev. Sci. Instrum.* 51, 478–499 (1980).

129. G.-C. Wang, B.I. Dunlap, R.J. Celotta, and D.T. Pierce, Symmetry in low-energy polarized electron diffraction, *Phys. Rev. Lett.* **42**, 1349–1352 (1979).

130. R.J. Celotta, D.T. Pierce, G.-C. Wang, S.D. Bader, and G.P. Felcher, Surface magnetization of ferromagnetics Ni(110): A polarized low-energy electron diffraction experiment, *Phys. Rev. Lett.* **43**, 728–731 (1979).

131. S. Trajmar and R.I. Hall, Dissociative electron attachment in H_2O and D_2O: Energy and angular distribution of H^- and D^- fragments, *J. Phys. B* **7**, L458–461 (1974).

132. R.I. Hall, Recent experiments concerning dissociative resonant states in electron–molecule collisions, in *Electronic and Atomic Collisions* (G. Watel, ed.), pp. 25–41, North-Holland, Amsterdam (1978).

7

COMPUTATIONAL METHODS FOR LOW-ENERGY ELECTRON–MOLECULE COLLISIONS

B.D. BUCKLEY, P.G. BURKE, AND C.J. NOBLE

1. Introduction

In the last few years considerable progress has been made in *ab initio* calculations of low-energy electron–molecule collision cross sections. This is partly due to the increasing need for these cross sections in many applications, partly due to the development of new methods and associated computer codes, and finally but not least due to the increasing availability of powerful computer facilities which have enabled the relevant equations to be solved accurately. Thus, in the last few years, accurate static exchange calculations have been carried out for a variety of diatomic molecules, and the first results for electronic excitation based on the close-coupling expansion have been reported.

The primary purpose of this review is to emphasize the various computational methods which have been adopted for performing electron–molecule cross-section calculations. In Section 2 we briefly describe the derivation of the radial equations for both the body-fixed and the space-fixed frames, and we consider the inclusion of vibrational motion. The major computational effort to date has been on the numerical solution of the radial equations within the fixed-nuclei approximation, and thus our discussion of the main theoretical and computational approaches to electron–molecule collisions in Section 3 falls entirely within this approximation. In Section 4 we discuss the attempts which are being made to relax the restriction on the nuclear motion in the fixed-nuclei approximation to enable vibrational excitation and dissociative attachment cross sections to be calculated. Finally we mention in Section 5 those calculations which have so far been reported for electronic excitation.

B.D. BUCKLEY and C.J. NOBLE • Science and Engineering Research Council, Daresbury Laboratory, England. **P.G. BURKE** • Queen's University, Belfast BT7 1NN, Northern Ireland. B.D. Buckley's current address is: Shell Internationale Petroleum, Maatschappij B.V., The Hague, The Netherlands.

2. Derivation of the Radial Equations

In this section we very briefly review the derivation of the radial equations satisfied by the scattered electron for future reference. The nonrelativistic Hamiltonian for an electron incident upon an N-electron diatomic molecule may be written in atomic units as

$$H = -\tfrac{1}{2}\nabla^2_{N+1} + V_{N+1} + H_N + H_{\text{vib}} + H_{\text{rot}} \equiv H_{\text{el}} + H_{\text{nuc}}. \tag{1}$$

The first two terms represent the kinetic energy of the incident electron and its interaction with the target electrons and nuclei, H_N is the electronic Hamiltonian of the molecule, and the vibrational and rotational Hamiltonians are

$$H_{\text{nuc}} \equiv H_{\text{vib}} + H_{\text{rot}} = (-1/2M)\nabla^2_R, \tag{2}$$

where M is the reduced mass of the nuclei.

The total collision wave function for the system is the solution of the Schrödinger equation

$$(H-E)\Psi = 0 \tag{3}$$

with the appropriate collision boundary conditions. It may be expanded in the form

$$\Psi = \mathscr{A} \sum_{\alpha} (\Phi_i \chi_v \rho_j)_\alpha F_\alpha(N+1) + \sum_k d_k \phi_k(1, \ldots, N+1), \tag{4}$$

where Φ_i, χ_v, and ρ_j are the electronic, vibrational, and rotational eigenstates of the target which are assumed closely coupled, whilst the second expansion in Eq. (4) involves known L^2-integrable electron-correlation functions ϕ_k. These are included to compensate for the unavoidable truncation of the first expansion and to include functions which would otherwise be omitted due to the usual orthogonality of the bound and continuum orbitals in Eq. (4). Of course these correlation functions also depend on the nuclear coordinates. The first expansion is frequently augmented by one or more pseudostates to represent the polarization of the target electrons by the incident electron. The wave function is antisymmetrized by $\mathscr{A}$, with respect to exchange of the space and spin coordinates of any pair of electrons.

The solution of the Schrödinger equation (3) using the full Hamiltonian given by Eq. (1) and the collision wave function (4) is neither computationally feasible nor necessary to determine collision cross sections for low-energy electron–molecule scattering. In an analogous fashion to the Born–Oppenheimer separation of the electronic, vibrational, and rotational eigenfunctions for low-lying bound molecular states, it is also possible in the collision process to separate these components of the solution, if not in all, at least in certain distinct regions of the coordinate space of the scattering electron.

In the following only a brief outline of the considerations which justify this simplification is possible and for further details the recent general review articles by Lane and by Burke[1] and a review on highly polar molecules by Norcross and Collins[2] should be consulted. See also other chapters in this book, namely, Chapter 1, Section 6.8, Chapter 2, Sections 3.2, 4.1, and 4.2, and Chapter 3, Sections 4 and 5 and Appendix C.

2.1. Electronic and Rotational Degrees of Freedom

An adiabatic approximation in which the nuclei are held fixed during the collision, corresponding to the neglect of the term H_{nuc} in Eq. (1), may be justified in two circumstances. First, one can argue from the physics that the collision time for an electron of a few eV is, except near narrow resonances, extremely short compared with the rotational and vibrational periods of the molecule (these are typically of the order of 10^{-12} s and 10^{-14} s, respectively) and the scattering electron effectively sees a stationary target. Alternatively, when the incident electron approaches the molecule its angular-momentum vector $\mathbf{l}$ couples (just as the target electrons do) to the internuclear axis $\mathbf{R}$, and $\Lambda = \mathbf{l} \cdot \mathbf{R}$ becomes a well-defined quantum number for the system. If the long-range forces are not too important the range over which Λ remains a collision constant includes the asymptotic region where a scattering matrix can be defined. Under either of these conditions, the collision wave function (4) reduces in the molecular coordinate frame to

$$\Psi \to \Psi_{l'}^{\Lambda}(1, \ldots, N+1) = \mathscr{A}\Phi_0(1, \ldots, N) \sum_{l} r_{N+1}^{-1} f_{ll'}^{\Lambda}(r_{N+1}) Y_{l\Lambda}(\hat{\mathbf{r}}_{N+1}), \quad (5)$$

where to simplify the notation we assume that the molecule is constrained to be in its ground electronic state. This equation when inserted into the Kohn variational principle leads to a set of radial integrodifferential equations of the form

$$[d^2/dr^2 - l(l+1)/r^2 + k^2] f_{ll'}^{\Lambda}(r) = 2 \sum_{l''} [V_{ll''}^{\Lambda} + W_{ll''}^{\Lambda}] f_{l''l'}^{\Lambda}(r), \qquad (6)$$

where the direct potential $V_{ll''}^{\Lambda}$ is defined by

$$V_{ll'}^{\Lambda} = \langle \Phi_0 Y_{l\Lambda} | V_{N+1} | \Phi_0 Y_{l'\Lambda} \rangle, \qquad (7)$$

and where $W_{ll''}^{\Lambda}$ is the nonlocal exchange potential which is discussed in detail in the next section. The solutions of Eq. (6) satisfy the boundary conditions

$$f_{ll'}^{\Lambda}(r) \underset{r\to 0}{\sim} r^{l+1} \tag{8a}$$

and

$$f_{ll'}^{\Lambda}(r) \underset{r\to \infty}{\sim} k^{-1/2}(\sin\theta_l\,\delta_{ll'} + \cos\theta_l\, K_{ll'}^{\Lambda}), \tag{8b}$$

where $K_{ll'}^{\Lambda}$ is the reactance matrix from which the cross sections can be easily calculated. Also

$$\theta_l = kr - \tfrac{1}{2}l\pi - \eta\log(2kr) + \sigma_l, \tag{9a}$$

$$\eta = -(Z-N)/k, \tag{9b}$$

$$\sigma_l = \arg\Gamma(l+1+i\eta), \tag{9c}$$

and Z is the sum of the nuclear charges.

A breakdown of the adiabatic approximation can occur when the long-range forces are important as in polar molecules. In such cases any attempt to define a scattering matrix in the molecular frame leads to a divergent total cross section. Nonetheless, the angular momentum **l** still couples to **R** over a large section of the collision region, though it no longer does so in the asymptotic scattering region. To obtain a scattering cross section, it is therefore necessary to transform the radial wave function $f_{ll'}^{\Lambda}$ at some suitable value of the radius r to the laboratory frame, before continuing the integration to obtain the correct asymptotic solution. This procedure is the essence of the frame-transformation theory proposed by Chang and Fano.(3) See Chapter 2, Sections 4.1 and 4.2 for further details.

A set of equations for the laboratory-frame radial wave function similar in form to Eq. (6) may be derived by noting that, provided vibrations are neglected, the target is equivalent to a rigid rotor whose eigenfunctions ρ_j are, in the case of linear molecules in sigma states, simply the spherical harmonics $Y_{jm_j}(\hat{\mathbf{R}})$ with eigenvalues $Bj(j+1)$. Here m_j is the projection of **j** on an axis fixed in space (laboratory frame) and B is the rotational constant for the molecule. If the orbital angular momentum of the incident electron is **l**, then the total angular momentum $\mathbf{J} = \mathbf{j} + \mathbf{l}$ is a collision constant. The eigenfunctions of **J**, therefore, form a suitable basis for the expansion of the collision wave function Ψ. Again constraining the target to its ground electronic state, Ψ may be expanded as

$$\Psi \to \Psi_{j'l'}^{JM}(1,\ldots,N+1) = \mathscr{A}\Phi_0(1,\ldots,N)\sum_{jl} r_{N+1}^{-1} f_{jlj'l'}^{J}(r_{N+1})\,\mathscr{Y}_{jl}^{JM_J}(\hat{\mathbf{R}},\hat{\mathbf{r}}_{N+1}), \tag{10a}$$

where

$$\mathscr{Y}_{jl}^{JM_J}(\hat{\mathbf{R}},\hat{\mathbf{r}}) = \sum_{m_j} Y_{jm_j}(\hat{\mathbf{R}}) Y_{l,M_J-m_j}(\hat{\mathbf{r}}) \cdot (j,m_j,l,M_J-m_j|jlJM_J). \tag{10b}$$

Here $(j,m_j,l,M_J-m_j|jlJM_J)$ is a Clebsch–Gordan coefficient. These equations together with those satisfied by the radial wave function $f^J_{jlj'l'}$ were first formulated by Arthurs and Dalgarno.[4]

2.2. The Vibrational Motion

The adiabatic approximation may be applied to the vibrational motion of the nuclei provided the collision time is much shorter than the vibrational period, so that the scattering electron sees only a molecule with some fixed internuclear separation R. In this case the vibrational excitation cross section can be calculated from the K matrix with matrix elements

$$K^{\Lambda}_{lvl'v'} = \langle v|K^{\Lambda}_{ll'}(R)|v'\rangle, \tag{11}$$

where $K^{\Lambda}_{ll'}(R)$ is calculated by solving Eq. (6) for a range of values of R.

Unfortunately, this argument often fails in low-energy electron–molecule scattering since narrow resonances occur frequently and it is in the neighborhood of these resonances that vibrational excitation is most important. In resonance regions the collision time increases dramatically and the vibrational motion of the nuclei can be perturbed significantly by the temporary disturbance of the electronic distribution due to the presence of the additional electron.

The Schrödinger equation including this vibrational term in the Hamiltonian is

$$(-\tfrac{1}{2}\nabla^2_{N+1} + V_{N+1} + H_N + H_{\text{vib}} - E)\Psi = 0, \tag{12}$$

where the collision wave function is given by

$$\Psi^{\Lambda}_{l'v'}(1,\ldots,N+1;R) = \mathscr{A}\,\Phi_0(1,\ldots,N) \times \sum_{lv} r^{-1}_{N+1} f^{\Lambda}_{lvl'v'}(r_{N+1}) Y_{l\Lambda}(\hat{\mathbf{r}}_{N+1}) \chi_v(R), \tag{13}$$

and where once again for simplicity we have assumed the molecule is constrained in its ground electronic state. Substitution of Eq. (13) into the Kohn variational principle leads to the following set of coupled integrodifferential equations for the radial function $f^{\Lambda}_{lvl'v'}(r)$:

$$
\begin{aligned}
&[d^2/dr^2 - l(l+1)/r^2 + k_v^2]\, f^{\Lambda}_{lvl'v'}(r) \\
&\qquad = 2 \sum_{l''v''} [V^{\Lambda}_{lvl''v''} + W^{\Lambda}_{lvl''v''}]\, f^{\Lambda}_{l''v''l'v'}(r),
\end{aligned}
\tag{14}
$$

where V and W are the direct and exchange potentials. The K martix in this case is obtained by imposing boundary conditions analogous to Eqs. (8). Equations (14) form the basis of the hybrid theory of Chandra and Temkin.[5] As in the case of the equations obtained by Arthurs and Dalgarno the solution of Eqs. (14) is computationally nontrivial due to the large number of channels which can couple. However, we will consider in Section 4 methods which enable vibrational excitation cross sections to be obtained from adiabatic solutions even in narrow resonance regions.

3. Numerical Solution of the Fixed-Nuclei Equations

We consider in this section the solution of the coupled integrodifferential equations which arise when the nuclei are held fixed during the collision, and a coordinate frame is chosen which is rigidly attached to the molecule. We have seen in the previous section that this approximation leads to equations which are appropriate when the time of collision is short compared with typical rotation and vibration times. It can also be used to describe the collision in an internal region when this condition is not satisfied.

We will mainly be concerned with the solution of the equations in an internal region $r \leqslant a$ where a is chosen so that for larger values of r the exchange potential is negligible and the direct potential assumes a simple analytic form consisting of sums of inverse powers of r. In the asymptotic region $r > a$ the coupled integrodifferential equations reduce to coupled differential equations and we will assume that a complete set of linearly independent solutions can always be determined which can be used to match onto the solution in the internal region. In fact the numerical solution of the coupled equations in the asymptotic region does raise serious problems which have been considered by a number of authors. However, we do not have space to consider them here, but instead refer readers to a review by Burke and Seaton[6] and to more recent discussions by Gailitis,[7] by Le Dourneuf and Vo Ky Lan,[8] and by Crees.[9]

Another topic considered is the inclusion of charge polarization effects. These require special treatment since a significant part of the distortion of the target induced by the incident electron comes from high-Rydberg and continuum states which are not well represented in an expansion of eigenstates of the form of Eq. (4). This is particularly true in the usual situation where only a few of the lowest target states are included. A further complication results from the long-range nature of the polarization. For example, in the case of elastic scattering of an electron by a diatomic molecule, the polarizing interaction may be represented asymptotically by the adiabatic potential

$$V_{\text{pol}}^{\text{asym}}(\mathbf{r}) \underset{r \to \infty}{\sim} -\frac{\alpha_0}{2r^4} - \frac{\alpha_2}{2r^4} P_2(\cos\theta), \tag{15}$$

where α_0 and α_2 are the spherically symmetric and anisotropic polarizabilities, respectively, and $P_2(\cos\theta)$ is the second-order Legendre polynomial. The angle between the position vector of the electron and the internuclear axis (which is taken as the z direction) is denoted by θ.

3.1. Single-Center Expansion Methods

In this subsection we describe the solution of the equations which arise when the scattered electron and the target orbitals are expanded about the center of gravity of the molecule. In this case the radial equations describing the collision of an electron by a diatomic molecule constrained to its ground electronic state are Eqs. (6) which must be solved subject to the boundary conditions (8).

The direct potential $V_{ll'}^{\Lambda}$ defined by Eq. (7) can be written as

$$V_{ll'}^{\Lambda}(r) = \sum_{\lambda=0}^{\infty} \left(\frac{2l'+1}{2l+1}\right)^{1/2} (l'0\lambda0|l'\lambda l0)(l'\Lambda\lambda0|l'\lambda l\Lambda)\, V_\lambda(r), \tag{16}$$

where the $V_\lambda(r)$ are obtained by carrying out a single-center expansion of the static potential of the molecule

$$\begin{aligned} \langle\Phi_0|V_{N+1}|\Phi_0\rangle &= \sum_{\lambda=0}^{\infty} V_\lambda(r_{N+1}) P_\lambda(\cos\theta_{N+1}) \\ &= -\frac{Z-N}{r_{N+1}} - \frac{\mu}{r_{N+1}^2} P_1(\cos\theta_{N+1}) - \frac{Q}{r_{N+1}^3} P_2(\cos\theta_{N+1}) \\ &\quad + \cdots \quad (r_{N+1} \geqslant a). \end{aligned} \tag{17}$$

In this equation μ and Q are the dipole and quadrupole moments of the molecule.

The exchange potential is very much more complex and it is not appropriate to write it down explicitly in this review. Indeed in practical calculations the tabulation of the exchange terms can only be conveniently carried out by computer. However, when the scattered-electron orbitals are assumed to be orthogonal to the target orbitals it has the general form (omitting the second index on the function f for notational simplicity)

$$W_{ll'}^{\Lambda} f_{l'}^{\Lambda}(r) = \sum_{\alpha l'' l''' \lambda} C(l,l',l'',l''',\alpha,\lambda) y_\lambda(u_{l''}^{\alpha} f_{l'}^{\Lambda}; r) u_{l'''}^{\alpha}(r), \tag{18}$$

where the $u_l^\alpha(r)$ are the radial functions arising from the single-center expansion of the αth occupied target orbital

$$\phi_\alpha(\mathbf{r}) = \sum_l r^{-1} u_l^\alpha(r) Y_{lm_\alpha}(\hat{\mathbf{r}}). \tag{19}$$

The y_λ are exchange functions defined by

$$y_\lambda(uv; r) = r^{-\lambda-1} \int_0^r r'^\lambda u(r') v(r') dr' + r^\lambda \int_r^\infty r'^{-\lambda-1} u(r') v(r') dr', \tag{20}$$

and the C's are numerical coefficients.

The main problems in the numerical solution of Eqs. (6) are first to retain enough terms in the single-center expansions in Eqs. (5), (17), and (19) to give convergence, and second to accurately represent the integral terms occurring in the exchange operator in Eq. (18). In the following paragraphs we review how these problems have been treated.

3.1.1. Noniterative Differential Equations Method

An approach suggested by Percival and used by Marriott[10] for electron–atom scattering was extended to electron–molecule scattering by Burke and Sinfailam.[11] This used the fact that the exchange terms defined by Eq. (20) each satisfy the differential equations

$$\frac{d^2}{dr^2}(r y_\lambda) = \frac{\lambda(\lambda+1)}{r^2}(r y_\lambda) - (2\lambda+1)\frac{u(r)v(r)}{r} \tag{21}$$

subject to the boundary conditions

$$r y_\lambda(r) \underset{r\to 0}{\sim} r^{\lambda+1} \tag{22a}$$

and

$$r y_\lambda(r) \underset{r\to\infty}{\sim} r^{-\lambda}. \tag{22b}$$

It follows that the original coupled integrodifferential equations (6) reduce to a set of coupled differential equations where one additional equation (21) is obtained for each exchange term in the expansion (18). These equations can then be solved by standard methods involving numerical integration outwards from the origin, inwards from the asymptotic region, and matching at an intermediate value of r.[12] The basic difficulty with this approach is that the number of exchange terms and thus the number of additional coupled differential equations is very large for all but the simplest molecules. Consequently it has proved very difficult to retain enough terms in the single-center expansions to obtain convergence. Nevertheless some almost converged results for N_2 have been obtained[13] using this approach. Recently Raseev *et al.*[14] have pointed out that the number of additional differential equations can be reduced by combining exchange terms with the same $u(r)$, $v(r)$, and λ in Eq. (21). This gives rise to a considerable saving in computer time but it is still doubtful whether the method is competitive with some of those described below.

3.1.2. Integral Equations Method

This approach was introduced by Sams and Kouri[15] and applied to electron–atom scattering by Henry *et al.*[16] Its application in electron–molecule scattering was reviewed by Morrison.[17] The coupled integrodifferential equations are converted to coupled integral equations which can be written in matrix form as

$$\mathbf{f}(r) = \mathbf{G}^1(r) - \int_0^r \mathbf{G}^{21}(r,r')\mathbf{U}(r')\mathbf{f}(r')dr' - \int_r^\infty \mathbf{G}^{12}(r,r')\mathbf{U}(r')\mathbf{f}(r')\,dr', \quad (23)$$

where $\mathbf{U}$ is the interaction potential which includes both the direct and exchange terms, and where

$$\mathbf{G}^{ij}(r,r') = \mathbf{G}^i(r)\mathbf{G}^j(r') \quad (24)$$

with

$$\begin{aligned} [\mathbf{G}^1(r)]_{ll'} &= \hat{j}_l(kr)\delta_{ll'} \underset{r\to\infty}{\sim} \sin(kr - \tfrac{1}{2}l\pi)\delta_{ll'}, \\ [\mathbf{G}^2(r)]_{ll'} &= k^{-1}\hat{n}_l(kr)\delta_{ll'} \underset{r\to\infty}{\sim} k^{-1}\cos(kr - \tfrac{1}{2}l\pi)\delta_{ll'}, \end{aligned} \quad (25)$$

$\hat{j}_l(kr)$ and $\hat{n}_l(kr)$ being Ricatti–Bessel and Neumann functions, respectively. The integral equations (23) can now be rewritten as

$$\mathbf{f}(r) = \mathbf{G}^1(r)[\mathbf{1} - \int_0^\infty \mathbf{G}^2(r')\mathbf{U}(r')\mathbf{f}(r')dr'] + \int_0^r [\mathbf{G}^{12}(r,r') - \mathbf{G}^{21}(r,r')]\,\mathbf{U}(r')\mathbf{f}(r')\,dr'. \tag{26}$$

Then introducing a homogeneous solution $\mathbf{u}(r)$ by

$$\mathbf{f}(r) = \mathbf{u}(r)[\mathbf{I}^2(\infty)]^{-1}, \tag{27}$$

we see that this solution satisfies the equation

$$\mathbf{u}(r) = \mathbf{G}^1(r)\mathbf{I}^2(r) - \mathbf{G}^2(r)\mathbf{I}^1(r), \tag{28}$$

where the **I** matrices are defined by

$$\mathbf{I}^i(r) = \mathbf{1}\delta_{i2} + \int_0^r \mathbf{G}^i(r')\mathbf{U}(r')\mathbf{u}(r')dr'. \tag{29}$$

In the absence of the exchange potential it is clear that, after the integrals in Eqs. (28) and (29) are represented by quadratures, the solution $\mathbf{u}(r)$ and consequently $\mathbf{f}(r)$ can be determined by a noniterative outward integration procedure. An important aspect of the method is that in order to retain the linear independence of the solutions a stabilization procedure must be adopted. This involves applying a unitary transformation to the solution matrix after every few integration steps. An advantage of the method is that the high angular momentum channels, which are needed to represent the rapidly varying wave function near the nuclei, can be successively dropped as the integration proceeds outward.

The exchange potential defined by Eq. (18) can be included noniteratively in this approach using a method developed by Henry *et al.*[16] for electron–atom scattering. This involves determining auxiliary functions for each exchange term followed by the solution of a set of linear simultaneous equations.

The integral equations method has been widely used to treat electron scattering by diatomic molecules where the exchange interaction is treated by an equivalent local potential such as the free-electron-gas model exchange suggested by Hara.[18] The noniterative treatment of exchange by the method of Henry *et al.*[16] again causes difficulties for complex molecules owing to the large number of exchange terms and consequently auxiliary functions required.

3.1.3. Iterative Method

Recently an iterative method has been developed by Collins *et al.*[19] to treat the exchange terms. In this approach Eq. (6) is rewritten in matrix notation as

$$\left[\frac{d^2}{dr^2} - \frac{\mathbf{l}(\mathbf{l}+1)}{r^2} + \mathbf{k}^2\right] \mathbf{f}(r) = 2\mathbf{V}\mathbf{f} + 2\mathbf{K}, \tag{30}$$

where $\mathbf{l}$ and $\mathbf{k}$ are diagonal matrices and $\mathbf{V}$ and $\mathbf{K}$ are square matrices with

$$\mathbf{K}(r) = \int_0^\infty \mathbf{W}(r,r')\mathbf{f}(r')dr'. \tag{31}$$

The iterative scheme is then based on the following equations:

$$\left[\frac{d^2}{dr^2} - \frac{\mathbf{l}(\mathbf{l}+1)}{r^2} + \mathbf{k}^2\right] \mathbf{f}_m = 2\mathbf{V}\mathbf{f}_m + 2[\mathbf{K}_{m-1}\mathbf{f}_{m-1}^{-1}(r)]\,\mathbf{f}_m(r), \tag{32}$$

TABLE 1
Static Exchange Eigenphase Sums (in Radians) for e–H_2 Scattering Calculated by Collins, Robb, and Morrison[19] Using an Iterative Single-Center Integral Equations Technique[a]

k^2 (Ryd)	$^2\Sigma_g$	$^2\Sigma_u$	$^2\Pi_g$	$^2\Pi_u$	$^2\Delta_g$
0.01	2.9303	0.0134	0.0012	−0.0025	−0.0020
	(56.468)	(0.207)	(0.005)	(0.032)	(0.012)
0.04	2.7246	0.0493	0.0038	0.0045	−0.0034
	(53.034)	(0.676)	(0.006)	(0.010)	(0.013)
0.09	2.5267	0.1233	0.0070	0.0262	−0.0043
	(47.896)	(1.940)	(0.007)	(0.140)	(0.012)
0.16	2.3399	0.2459	0.0110	0.0642	−0.0044
	(41.823)	(4.387)	(0.010)	(0.536)	(0.009)
0.25	2.1679	0.4084	0.0169	0.1155	−0.0031
	(35.483)	(7.583)	(0.015)	(1.164)	(0.006)
0.36	2.0114	0.5797	0.0252	0.1729	0.0001
	(29.467)	(10.10)	(0.026)	(1.843)	(0.003)
0.49	1.8729	0.7269	0.0363	0.2289	0.0054
	(24.053)	(10.95)	(0.044)	(2.370)	(0.001)
0.64	1.7509	0.8361	0.0501	0.2779	0.0128
	(19.427)	(10.45)	(0.068)	(2.645)	(0.003)
1.00	1.5509	0.9591	0.0840	0.3481	0.0324
	(12.481)	(8.092)	(0.131)	(2.554)	(0.002)

[a] Integrated cross sections in atomic units are shown in parentheses. Results are converged to better than 1% and are for the target wave function of Fraga and Ransil.[21]

where m is the number of iterations and where the last term in Eq. (32) is omitted in the zeroth iteration. Clearly, when convergence is obtained the solution of Eq. (32) satisfies Eq. (30). At each step in the iteration Eq. (32) involves a local effective exchange potential which can be included with the direct potential and the equation solved either by direct numerical integration or by using the integral equations method described above. In practice the method converges well for the static exchange solutions of H_2, N_2, LiH, H_2^+, and CH^+[19,20] giving better than 1% accuracy after three iterations. In Tables 1 to 4 the results obtained for H_2, N_2, LiH, and H_2^+ targets are reproduced. The method has the important advantage that inclusion of exchange does not increase the size of the solution matrix over the static case. Unlike the noniterative differential equations method the computation time depends linearly rather than cubically on the number of exchange terms.

A number of other iterative approaches have been used in electron–atom scattering (see review by Burke and Seaton[6]) and in nuclear scattering (see review by Rhoades-Brown *et al.*[24]). However, these have not yet been tested in the electron–molecule collision case so we will not review them here. It is clear from the work done so far that iterative approaches can be rapidly convergent when only the ground electronic state is included in Eq. (4). However,

TABLE 2

Static Exchange Eigenphase Sums (in Radians) for e–N_2 Scattering Calculated by Collins, Robb, and Morrison[19] Using the Target Wave Function of Cade, Sales, and Wahl[22]a

k^2 (Ryd)	$^2\Sigma_g$	$^2\Sigma_u$	$^2\Pi_g$	$^2\Pi_u$	$^2\Delta_g$	$^2\Delta_u$
0.001	3.057	–				
	(89.35)					
0.01	2.874	3.119				
	(87.35)	(10.76)				
0.05	2.544	3.038				
	(77.20)	(2.48)				
0.1	2.311	2.937	0.011	−0.075	0.013	−0.004
	(66.67)	(4.81)	(0.93)	(1.20)	(0.08)	
0.2	2.011	2.749	0.202	−0.189	0.030	−0.004
	(50.60)	(8.69)	(55.77)	(3.82)	(0.18)	
0.3	1.808	2.584	1.461	−0.290	0.057	−0.003
		(11.23)	(82.99)		(0.36)	
0.4	1.654	2.440	2.316	−0.379	0.091	
	(31.12)	(12.71)	(33.29)	(8.00)	(0.63)	
0.5	1.524	2.316	2.454	−0.455	0.132	0.005
	(25.13)	(13.42)	(19.69)	(19.43)	(1.00)	(0.01)
0.75	1.263	2.073	2.484	−0.601	0.247	0.026
	(15.87)	(13.24)	(12.14)	(11.34)	(2.14)	(0.05)
1.0	1.043	1.908	2.453	−0.694	0.361	0.057
	(11.04)	(11.90)	(9.94)	(12.05)	(3.24)	(0.12)

[a] Cross sections in atomic units are shown in parentheses.

TABLE 3
Static Exchange Eigenphase Sums (in Radians) for e–LiH Scattering Calculated by Collins, Robb, and Morrison[19] Using the Target Wave Function of Cade and Huo[23]

k^2 (Ryd)	$^2\Sigma$	$^2\Pi$
0.0025	6.2014	0.4295
0.01	5.1071	0.4538
0.04	4.1238	0.5318
0.1	3.4000	0.6739
0.16	3.0040	0.7807
0.25	2.6045	0.8961
0.49	2.0193	1.0895
0.81	1.6568	1.2403
1.0	1.5620	1.3004

further studies are required when coupled electronic states are included, since in the case of electron–atom scattering they tend to be slowly convergent or even diverge in the neighborhood of narrow Feshbach resonances that then occur.

3.1.4. Linear Algebraic Equations Method

New techniques for solving the integral form of the single-center scattering equations given by Eq. (23) have recently been developed by Collins and Schneider.[25] The method is similar in spirit to the approach used by Seaton[26] to solve the integrodifferential equations for electron–atom scattering. However, the approach is based on the integral equation formalism described in Section 3.1.2 rather than on the direct representation of the integrodifferential equation (6) as in the work of Seaton.

Equation (23) may be recast in a more convenient form by introducing an alternative notation for the Green's function. Let

$$\mathbf{G}^0(r,r') = \begin{cases} \mathbf{G}^{12}(r,r'), & r < r' \\ \mathbf{G}^{21}(r,r'), & r > r'. \end{cases} \tag{33}$$

Then, if we specifically take account of the nonlocal nature of the potential $\mathbf{U}$ which results from the inclusion of exchange terms, Eq. (23) takes the form

$$\mathbf{f}(r) = \mathbf{G}^1(r) - \int_0^\infty \mathbf{G}^0(r,r') \int_0^\infty \mathbf{U}(r',r'')\mathbf{f}(r'')dr''\,dr'. \tag{34}$$

The integrals in Eq. (34) may be replaced by quadratures so that

TABLE 4

Static Exchange Eigenphase Sums (in Radians) for $e + H_2^+$ Scattering Calculated at Various Internuclear Separations r by Robb and Collins[20] [a]

		δ_{sum}(rad)							
$r(a_0)$	k^2 (Ryd)	$^1\Sigma_g$	$^1\Sigma_u$	$^1\Pi_g$	$^1\Pi_u$	$^3\Sigma_g$	$^3\Sigma_u$	$^3\Pi_g$	$^3\Pi_u$
1.4	0.01	−0.1631	0.1404	0.0235	−0.2479	0.3997	1.1116	0.04813	0.2100
	0.09	−0.1757	0.1530	0.0243	−0.2444	0.3949	1.0842	0.0562	0.2128
	0.25	−0.1938	0.1804	0.0248	−0.2386	0.3852	1.0392	0.0703	0.2128
	0.49	−0.2071	0.2231	0.0254	−0.2282	0.3712	0.9881	0.0887	0.2079
	0.81	−0.2098	0.2758	0.0272	−0.2121	0.3546	0.9381	0.1088	0.1986
2.0	0.01	−0.3569	0.3490	0.0450	−0.3488	0.2368	1.4313	0.0917	0.1537
	0.09	−0.3682	0.3712	0.0467	−0.3397	0.2407	1.3828	0.1058	0.1563
	0.25	−0.3807	0.4134	0.0482	−0.3257	0.2464	1.3028	0.1210	0.1537
	0.49	−0.3806	0.4663	0.0507	−0.3055	0.2517	1.2125	0.1565	0.1451
	0.81	−0.3619	0.5150	0.0558	−0.2791	0.2557	1.1250	0.1835	0.1332
2.6	0.01	−0.5071	0.5144	0.0681	−0.4430	0.1409	1.5425	0.1465	0.0762
	0.09	−0.5149	0.5380	0.0701	−0.4272	0.1599	1.4832	0.1667	0.0806
	0.25	−0.5172	0.5758	0.0720	−0.4036	0.1908	1.3848	0.1979	0.0791
	0.49	−0.4969	0.6112	0.0757	−0.3728	0.2240	1.2757	0.2308	0.0722
	0.81	−0.4479	0.6295	0.0832	−0.3364	0.2524	1.1661	0.2587	0.0636

[a] Results are converged to better than 3%.

$$\mathbf{f}(r) = \mathbf{G}^1(r) - \sum_{\alpha} \omega_\alpha \mathbf{G}^0(r,r_\alpha) \sum_{\beta} \omega_\beta \mathbf{U}(r_\alpha, r_\beta)\mathbf{f}(r_\beta), \tag{35}$$

where r_α, ω_α are the quadrature nodes and weights, respectively. By evaluating Eq. (35) on a mesh of values for the radial variable r a set of linear-algebraic equations is obtained which may be written as

$$\sum_{\beta} [\delta_{\beta\gamma} + \sum_{\alpha} \mathbf{G}^0(r_\gamma, r_\alpha)\mathbf{U}(r_\alpha, r_\beta)\omega_\alpha\omega_\beta]\,\mathbf{f}(r_\beta) = \mathbf{G}^1(r_\gamma). \tag{36}$$

By introducing a supermatrix notation in which channel and quadrature-point labels are combined it is easily seen that Eq. (36) may be expressed in the compact form

$$\mathbf{Mf} = \mathbf{g}, \tag{37}$$

where

$$(\mathbf{M})_{\gamma\beta} = (\mathbf{1})_{\gamma\beta} + (\mathbf{G}^0\mathbf{WUW})_{\gamma\beta}, \tag{38}$$

and $\mathbf{W}$ denotes a diagonal matrix of quadrature weights.

Equation (37) may be solved directly for the continuum wave function $\mathbf{f}$ using standard linear algebraic methods. The advantages of this particular technique are that the boundary conditions are incorporated within the Green's function and also the method is stable so that individually chosen high-order quadratures may be employed in each scattering channel. The limitations on the method are effectively imposed by the maximum order of the matrix $\mathbf{M}$ for which computations remain feasible. Even with the use of fast vector-processor computers and matrix partitioning techniques[27] this constraint is rather severe, so a number of extensions of the basic procedure have been devised.

First, it was noted that the linear-algebraic approach is most effective when the potential is strong and nonlocal but offers no particular advantage when the potential is local and of a simple multipolar form. This immediately suggests the adoption of the R-matrix philosophy (for details refer to Section 3.8 of this chapter and Section 4.3 of Chapter 2) and the application of the linear algebraic method to the internal region of configuration space alone. Suitable boundary conditions must then be imposed by the introduction of a Bloch operator;[28] however, while this leads to modifications in the form of the propagator matrices $\mathbf{G}$ and $\mathbf{g}$, the form of Eqs. (37) is preserved. The solution yields the R matrix on the boundary of the inner region. A number of numerical techniques exist which may then be used to propagate the R matrix into the asymptotic region to obtain the K matrix.

The linear algebraic equations method has recently been combined successfully with a separable approximation of the exchange potential which is discussed in Section 3.3 where some recent results will be presented.

3.1.5. Variational Method

We conclude this section by mentioning that the Kohn and inverse Kohn variational methods, which have been developed particularly by Nesbet(29) and by Rountree and Parnell(30) to describe electron–atom scattering, have been extended to treat electron–molecule scattering in the static exchange approximation by Collins and Robb.(31) The trial wave function for the scattered electron was taken as a composite of free-particle, continuum static, and bound orbitals all expanded about the center of gravity. The results, which were obtained for a number of molecules including H_2, N_2, CO, and LiH, show that the method provides an efficient mechanism for producing a set of reasonably accurate cross sections and eigenphases over a wide energy range.

3.2. Expansion in Prolate Spheroidal Coordinates

A basic difficulty with the single-center expansion method is its very slow convergence for all molecules except the lightest. For example in the calculations on e–N_2 scattering in the $^2\Pi_g$ state by Collins *et al.*(19) up to $l = 14$ had to be retained in the expansion of the wave function of the scattered electron in Eq. (19) and up to $\lambda = 28$ had to be retained in the expansion of the static potential in Eq. (16).

An alternative approach which partly avoids these difficulties in the case of diatomic molecules is to use two-center prolate spheroidal coordinates (ξ, η, ϕ) defined by

$$\xi = \frac{r_A + r_B}{R}, \qquad \eta = \frac{r_A - r_B}{R}, \tag{39}$$

where r_A and r_B are the distances of the electron from the respective nuclei and, R is the internuclear separation. The static potential when expanded in these coordinates may be written in the form

$$\langle\Phi_0|V_{N+1}|\Phi_0\rangle = \sum_{\gamma=0}^{\infty} V_\gamma(\xi)P_\gamma(\eta), \tag{40}$$

which converges much faster than expansion (17) with terms up to $\gamma = 6$ giving converged results for N_2 for all ξ. The wave function of the scattered electron can be written as

$$F(\xi,\eta,\phi) = \sum_{\gamma,m} f_{\gamma m}(\xi)^{im\phi}, \tag{41}$$

where $S_{\gamma m}(\kappa,\eta)$ is a spheroidal angular function[32] and $\kappa = \frac{1}{2}kR$. Substitution of this expansion into the Schrödinger equation leads to a set of coupled integro-differential equations for the functions $f_{\gamma m}(\xi)$.

These coordinates have been used for e–H_2 scattering by Nagahara,[33] by Hara,[18] and by Bell,[34] for e–N_2 scattering by Crees and Moores,[35] and for e–H_2^+ by Takagi *et al.*[36] In the work on N_2 the non-exchange equations were solved using the linear algebraic equations method described in Section 3.1.4. However, the differential form of the method developed by Seaton and collaborators[26,37,38] rather than the integral form was used. In this method these equations are transformed into a set of linear simultaneous algebraic equations, where the unknowns are the values of the radial functions $f_{\gamma m}(\xi)$ at a set of mesh points where the interaction is appreciable. In the work of Bell on H_2, exchange was included by a method analogous to that discussed in Section 3.1.1 by including additional differential equations. An adiabatic polarization potential (see Section 3.5) was also included and cross sections calculated up to 60 eV.

See Chapter 1, Section 5.4 for further details of the prolate spheroidal coordinate approach without inclusion of the exchange potential.

3.3. Separable Exchange Potentials

It has recently been pointed out that the large computational effort which is necessary to evaluate the exchange terms is a result of the nonseparable character of the exchange interaction rather than of its nonlocality. The problem may therefore be greatly simplified by introducing an orthonormal basis and constructing a separable approximation to the exchange kernels $W_{ll''}^{\Lambda}$ of Eq. (6). The exchange kernels in the separable approximation may be expressed in terms of a basis set $\{\chi_\alpha\}$ in the form

$$K(\mathbf{r},\mathbf{r}') = \sum_{\alpha} \chi_\alpha(\mathbf{r})\gamma_\alpha\chi_\alpha(\mathbf{r}'). \tag{42}$$

In practice it is found[25] that a 10-term separable representation of the form given in Eq. (42) is sufficient when computing e–LiH static-exchange eigenphase sums to obtain an overall accuracy of 10%. Further, for computations involving a number of different energy points, improvements in computation speed by factors of between 3 and 10 may be obtained over the times required to evaluate the exchange kernels by direct numerical quadrature.

Accurate static-exchange results have been computed using the linear-algebraic method within a single-center expansion framework for H_2, N_2, LiH,

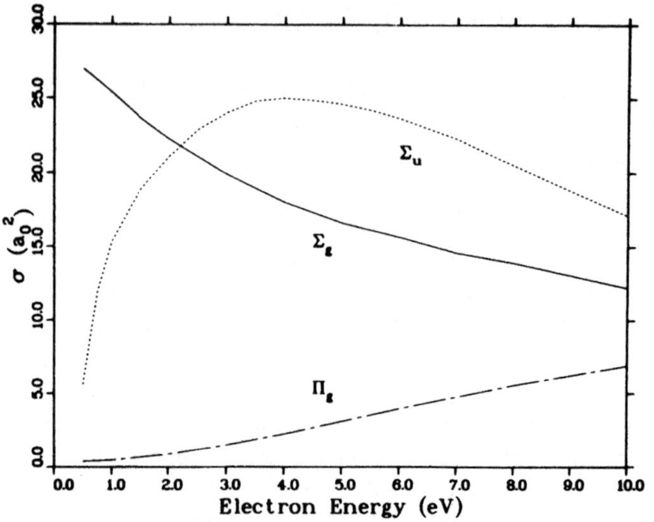

FIGURE 1
Integrated cross sections as a function of energy for Σ_g, Σ_u, and Π_g symmetries for e–CO_2 collisions (from Collins and Schneider[25]).

and CO_2 targets.[25] The cross sections which are obtained for e–CO_2 elastic scattering are illustrated in Fig. 1. The effectiveness of the method has therefore been clearly demonstrated and its application to more complex situations is now of great interest.

The notion that the exchange potential is best treated by a separable representation has also formed the basis of recent work by Rescigno and Orel.[39] When using the T-matrix approximation described in Section 3.9 it is found that it is essential to correct for the truncation implicit in the use of a finite-rank separable-potential representation particularly at low energies and for polar molecules. Rescigno and Orel attribute this problem to the attempt to represent long-range forces in terms of short-range basis functions. They have therefore constructed a model based on the two-potential formalism[40] in which long-range forces are treated exactly by numerical techniques and the exchange interactions, which does not extend appreciably beyond the charge distribution of the target, is treated using an analytic separable approximation.

If a separable approximation is introduced for the exchange kernels of the form given in Eq. (42) using the orthonormal Gaussian basis $\{\chi_\alpha\}$, then the Volterra integral equations (28) may be rewritten as

$$\begin{aligned}\mathbf{f}(r) = {} & \mathbf{G}^1(r) + \int_0^r \tilde{\mathbf{G}}(r,r')\mathbf{V}(r')\mathbf{f}(r')dr' \\ & + \sum_\alpha \gamma_\alpha \int_0^r \tilde{\mathbf{G}}(r,r')\boldsymbol{\chi}_\alpha(r')dr' \int_0^\infty \boldsymbol{\chi}_\alpha(r'')\mathbf{f}(r'')\,dr'',\end{aligned} \tag{43}$$

where $\tilde{\mathbf{G}}$ is the regular Green's function[40] matrix. It is straightforward to show that equations

$$\mathbf{g}^0(r) = \mathbf{G}^1(r) + \int_0^r \tilde{\mathbf{G}}(r,r')\mathbf{V}(r')\mathbf{g}^0(r')dr' \tag{44a}$$

and

$$\mathbf{g}^\alpha(r) = \gamma_\alpha \int_0^r \tilde{\mathbf{G}}(r,r')\boldsymbol{\chi}_\alpha(r')dr' + \int_0^r \tilde{\mathbf{G}}(r,r')\mathbf{V}(r')\mathbf{g}^\alpha(r')dr' \tag{44b}$$

define homogeneous and inhomogeneous terms, respectively, from which the solution $\mathbf{f}(r)$ may be constructed using

$$\mathbf{f}(r) = \mathbf{g}^0(r) + \sum_\alpha \mathbf{C}^\alpha \mathbf{g}^\alpha(r). \tag{45}$$

The coefficients $\mathbf{C}^\alpha$ are determined from a set of linear algebraic equations of dimension equal to the rank of the representation (42). Equations (44) and (45) may therefore be solved by standard noniterative techniques such as the method of Sams and Kouri.[15]

This separable-potential method has been used to compute static exchange results for e–H_2 and e–LiH scattering and is found to accurately reproduce the eigenphase sums using only small basis sets.

The methods described in this section, like some of those described in Section 3.1, share with the R-matrix method the concept of selecting the most appropriate solution technique according to the nature of the interaction, recognizing that the character of the interaction varies qualitatively in different regions of configuration space. It seems likely that exploiting this simplifying feature of the physics of electron–molecule collisions will provide a very significant advantage in future work.

3.4. Model Exchange Potentials

It is apparent from the previous subsections that the inclusion of exchange complicates the solution of the scattering equations and therefore many attempts have been made to find model local potentials to replace the exact nonlocal energy-dependent exchange term.

One of the most important conditions imposed on the total wave function by exchange is the Pauli exclusion principle, and for closed-shell molecules this involves requiring the scattered electron wave function to be orthogonal to all the bound molecular electrons. Thus Burke and Chandra[41] in their pseudopotential method replace $W^\Lambda_{ll'}$ in Eq. (6) by a Lagrange orthogonality condition, thereby reducing the coupled integrodifferential equations to a set of coupled inhomogeneous differential equations

$$\left[\frac{d^2}{dr^2} - \frac{l(l+1)}{r^2} + k^2\right] f_{ll'}^{\Lambda}(r) = 2\sum_{l''} V_{ll''}^{\Lambda} f_{l''l'}^{\Lambda}(r) + \sum_{\alpha=1}^{m} \lambda_{\alpha l'} u_l^{\alpha}(r), \tag{46}$$

where the summation over α includes all bound molecular orbitals with the same symmetry as the continuum orbital, λ is a Lagrange multiplier, and u_l^{α} is defined in Eq. (19).

The pseudopotential method has been applied with some success to many electron–molecule systems[41–43] but suffers from the restriction that it does not make any allowance for exchange effects when no bound orbitals of the same symmetry as the scattering electron exist. Thus for example in electron–nitrogen scattering in the resonant $^2\Pi_g$ channel the pseudopotential method takes no account of exchange. This is a serious defect since it is known that exchange plays an important role in determining the correct position of the $^2\Pi_g$ resonance.

An earlier study of a model exchange potential in electron–molecule collisions was that of Lane and Geltman,[44] who augmented their static electron–molecular hydrogen potential by an exponential short-range attractive term, the parameters of which were adjusted to optimize agreement with measured low-energy elastic cross sections. This model exchange potential is obviously a very crude one and not very suitable for inclusion in *ab initio* type calculations.

A number of different model potentials have been proposed in electron–atom scattering[45] and electron–molecule scattering[46] in the intermediate energy range. For lower-energy electrons, however, the first suitable model for *ab initio* approaches was proposed by Hara.[18] Hara treated the molecular electrons as a free electron gas[47] and the incident electron in the Born approximation, and derived the model exchange potential

$$V_{\mathrm{FEG}}(\mathbf{R},\mathbf{r}) = -\frac{2}{\pi} k_F F(\eta), \tag{47}$$

where

$$F(\eta) = \frac{1}{2} + \frac{1-\eta^2}{4\eta} \ln\left|\frac{1+\eta}{1-\eta}\right|, \tag{48a}$$

$$\eta = \kappa/k_F, \tag{48b}$$

and

$$k_F = [3\pi^2 \rho(\mathbf{R},\mathbf{r})]^{1/3}, \qquad \kappa^2 = k^2 + 2I_P + k_F^2. \tag{48c}$$

I_P is the molecular ionization potential and $\rho(\mathbf{R},\mathbf{r})$ the charge distribution at $\mathbf{r}$, whilst k_F is the Fermi momentum of the free electron gas. For further details see Chapter 1, Section 3.2.

Calculations on low-energy electron scattering within the static-exchange approximation have been performed for H_2,[18,48] N_2,[48] and CO_2,[49] using the Hara free-electron-gas (HFEG) potential for exchange and good agreement with exact static-exchange calculations has been obtained. In a similar application of the HFEG model to the highly polar molecule LiH by Collins and Norcross,[50] spurious resonance structures appeared in their initial calculations. On further investigation they discovered[51] that this resonance arose from the non-orthogonality of the scattered electron's wave function to the bound molecular orbitals. On the introduction of an orthogonality constraint in the same manner as Burke and Chandra[41] [see Eq. (46)] the spurious resonance disappeared and excellent agreement was found with the exact static-exchange results (see Fig. 2).

The role of the orthogonalization procedure has recently been examined by Morrison and Collins[52] in a study of the validity of the HFEG potential. The results obtained using the pseudopotential method, the HFEG model, and an orthogonalized HFEG model (OHFEG) were systematically compared to exact static-exchange results for scattering from H_2, N_2, LiH, CO, LiF, and HC1 targets. It was found that the pseudopotential method provides a poor representation of the exact results and fails to predict resonances, for example in Π-symmetry e–CO scattering, even in situations where bound orbitals with the same symmetry as the scattering electron exist. The HFEG model is more

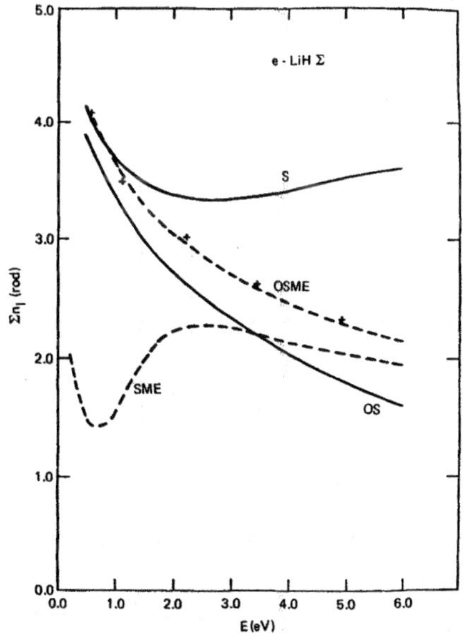

FIGURE 2
The Σ eigenphase sum for e–LiH scattering. Curves S, static; OS, orthogonalized static; SME, static with model exchange; OSME, orthogonalized static with model exchange; crosses, exact static-exchange calculations (from Collins *et al.*[51]).

TABLE 5
Resonance Energies (E_r) and Widths (Γ) for the Π_u Shape Resonance in e–CO_2 Scattering Computed in the Static Exchange Approximation (ESE) are Compared with Results Obtained Using the Pseudopotential Method (OS), the Hara Free-Electron-Gas Equivalent Exchange Potential (HFEG), and an Orthogonalized HFEG Model (OHFEG)[a]

Method	E_r (eV)	Γ (eV)	Method	E_r (eV)	Γ (eV)
ESE	5.4	0.68	HFEG	7.2	0.9
OHFEG	6.7	0.7	OS	~16.0	~5.0

[a] Taken from Morrison and Collins.[53]

successful in producing resonant features in qualitative agreement with exact static exchange. However, in every case studied, it was found that enforcing the correct nodal structure by orthogonalization strengthens the HFEG potential and improves the agreement. In all but a few special cases the OHFEG model is successful in reproducing exact static exchange results. As an example we reproduce in Table 5 parameters obtained for the Π_u shape resonance in e–CO_2 scattering using the different models.[53] Other studies of equivalent local exchange potentials have included investigations of the effect of adjusting the molecular ionization potential I_p in Eq. (48c) for e–H_2[54] and e–N_2[55] scattering.

In conclusion, it appears that ad hoc adjustments are unnecessary and that in most cases accurate results can be obtained for elastic scattering at low energies using a model exchange potential particularly if an orthogonality constraint is simultaneously applied. Those results which have so far been obtained in applications of equivalent local exchange potentials to scattering by nonlinear polyatomic molecules suggest that this is also true in the more general situation. For example the pseudopotential method and an empirical polarization potential have been used successfully by Gianturco and Thompson[56] to treat e–CH_4, e–H_2O, and e–H_2S scattering. Similarly the HFEG model has been used by Salvini and Thompson[57] for e–CH_4 and e–H_2O scattering. The principal outstanding question is therefore whether model exchange potentials can be constructed to give accurate results for low-energy electronic excitation. In the corresponding electron–atom collision case this problem has proven intractable and so far efforts to devise a potential for electronic excitation comparable in accuracy to the HFEG model for elastic scattering have been unsuccessful.[58]

3.5. Polarization Potentials

In many low-energy electron–molecule scattering calculations polarization is included by assuming that the asymptotic adiabatic potential (15) is valid for all values of r greater than the charge cloud of the molecule and is modified at shorter distances by an empirical cutoff factor which is frequently taken to be of the form

$$C(r) = 1 - \exp\left[-(r/r_c)^6\right]. \tag{49}$$

This implies that polarization is effectively negligible for $r < r_c$. The cutoff parameter r_c is usually determined by fitting some "landmark" feature in the collision cross section such as a resonance, for example, the 2.3-eV $^2\Pi_g$ resonance in the elastic scattering by molecule nitrogen.[13,41,59] This approach may be seriously called into question since a meaningful physical interpretation is difficult and frequently the value of r_c will reflect an attempt to compensate for important nonadiabatic or dynamical effects which have been neglected and which will be different in different symmetry states of the scattered electron. Occasionally the value will be masking other deficiencies such as in an equivalent exchange potential. Perhaps the most serious criticism is, however, the fact that it is found that the collision cross section at very low energies and in the vicinity of resonant features can depend sensitively on the value of r_c.

Ab initio adiabatic polarization potentials have been calculated using the finite-field method for e–N_2 and e–CO_2 systems by Morrison and Hay[60] and for e–N_2 by Eades *et al.*;[61] see Chapter 1, p. 19, Fig. 6. These calculations yield rather different potentials from the empirical form described previously. For example, a nonzero effective polarization potential at the origin is obtained. Nonadiabatic effects are important and must be taken into account.

Nonadiabatic effects were taken into account in elastic electron–atom scattering by Temkin,[62] who replaced the ground-state wave function by a wave function which depended on the coordinate of the scattered electron. In electron–molecule scattering this corresponds to replacing Φ_0 in Eq. (5) by $\Phi_0 + \Phi^{\mathrm{POL}}$, where Φ^{POL} depends on $\mathbf{r}_{N+1}$. This polarized-orbital approach gives rise to a direct potential behaving asymptotically as in Eq. (15) as well as an additional exchange contribution. It has been used by Temkin *et al.*[63] for H_2^+ and by Lane and Henry[64,65] for e–H_2 scattering. Nonadiabatic effects have also been considered in a study of a semiclassical exchange potential by Gibson and Morrison.[54] In more recent work[66] these same authors have obtained a polarization potential for low-energy e–H_2 scattering by variationally minimizing the second-order Hartree–Fock energy of the perturbed molecule taking partial account of nonadiabatic effects by (i) switching off the projectile–target interaction whenever the incident electron is within the charge distribution of the bound electrons; (ii) dropping monopole terms. The procedure for weakening the potential at small values of r is therefore somewhat arbitrary although no free parameters are involved.

Other approaches for including polarization effects include adding pseudostates in expansion (4) which are specifically adapted to represent those high-lying bound states and continuum states which give rise to the correct asymptotic polarizabilities defined by Eq. (15). Alternatively the target state orbitals themselves may be modified making some allowances for polarization. Finally an optical potential can be introduced which in general corresponds to the self-energy of the one-particle Green's function.[67] However, it is probably true to say that the accurate representation of the polarization potential in low-energy electron–molecule scattering remains a most difficult outstanding theoretical and computational problem. We will therefore reserve further discussion of these approaches to later sections when we will consider them in the context of their application to specific computational methods and applications.

See also Chapter 1, Section 3.1 that discusses the polarization potential.

3.6. Continuum Multiple-Scattering Method

The continuum multiple-scattering method (CMSM) was developed primarily by Dill and Dehmer[68] and recently comprehensively reviewed by these authors (Dehmer and Dill[69]). This approach has also been independently considered by Ziesche and John.[70] The basic idea, which was first used in bound-state studies (e.g., Johnson[71]) involves the partition of the molecular field into three regions as illustrated in Fig. 3. In region I the electron–molecule potential is represented by direct and local exchange terms; in region II the potential is approximated by a constant; while in region III a polarization potential is added to the direct and local exchange potentials. Several forms have been explored for the local exchange potential but that due to Hara[18] seems most effective. Usually the potentials in regions I and III are spherically symmetric although this limitaion has been relaxed by Siegel *et al.*[72] without undue additional effort.

The form assumed for the wave function representing the scattered electron in the three regions is as follows. In region I surrounding each nucleus

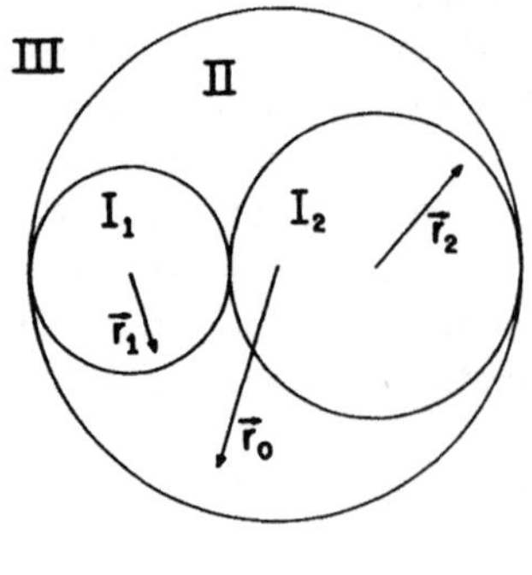

FIGURE 3
Partitioning of the molecular field in the continuum multiple-scattering method for a heteronuclear diatomic molecule (from Dehmer and Dill[69]).

$$\Psi_{\mathrm{I}_i} = \sum_{lm} a_{lm}^{\mathrm{I}_i} f_l^{\mathrm{I}_i}(kr_i) Y_{lm}(\hat{\mathbf{r}}_i). \tag{50}$$

In the region between the inner spheres and the outer sphere

$$\Psi_{\mathrm{II}} = \sum_{lm} a_{lm}^{\mathrm{II}} j_l(kr_0) Y_{lm}(\hat{\mathbf{r}}_0) + \sum_{i} \sum_{lm} b^{\mathrm{II}_i}_{lm} n_l(kr_i) Y_{lm}(\hat{\mathbf{r}}_i). \tag{51}$$

Finally in the outer region

$$\Psi_{\mathrm{III}} = \sum_{lm} [a_{lm}^{\mathrm{III}} f_l^{\mathrm{III}}(kr_0) + b_{lm}^{\mathrm{III}} g_l^{\mathrm{III}}(kr_0)]\, Y_{lm}(\hat{\mathbf{r}}_0). \tag{52}$$

In these equations f_l and g_l are the regular and irregular solutions in the appropriate regions, and j_l and n_l are the regular and irregular spherical Bessel functions. The coefficients a_{lm} and b_{lm} are obtained by solving a set of simultaneous equations obtained by requiring that the wave function and its derivatives match on all the spherical boundaries. The phase shifts and cross sections are then obtained by matching Ψ_{III} to the asymptotic form defined by Eq. (8b). If vibrational excitation cross sections are required the calculation is repeated for a number of different internuclear separations and cross sections determined using the adiabatic-nuclei approximation defined by Eq. (11).

A study of the errors which are associated with certain of the approximations which are introduced to make computations with the model tractable has been carried out by Rumble and Truhlar.[73] They estimate errors of about 30% in the total integral cross section for e–N_2 elastic scattering and therefore confirm the utility of the method particularly as a convenient survey tool.

We show in Fig. 4 the vibrational excitation cross sections calculated using the CMSM for e–N_2 scattering. For the vibrationally elastic cross section the contributions from the various symmetry states are also displayed. The well-known low-energy $^2\Pi_g$ resonance is clearly seen in all the cross sections, although the structure is not given by the adiabatic-nuclei approximation used in the CMSM. What is new is the appearance at higher energies of a broad $^2\Sigma_u$ resonance centered at about 26 eV which stands out most prominently in the vibrational excitation cross sections. This feature had been observed previously in experiments by Pavlovic *et al.*[74] but the CMSM calculation shows conclusively that it is a shape resonance with mainly f-wave symmetry. Similar experimental and CMSM results on e–CO_2 scattering are shown in Chapter 3, p. 247, Fig. 20. The importance of shape resonances in enhancing the vibrational excitation cross section at intermediate energies for many molecules as well as their related role in photoionization of molecules has been emphasized particularly by Dehmer and Dill[69] and by Dill *et al.*[75]

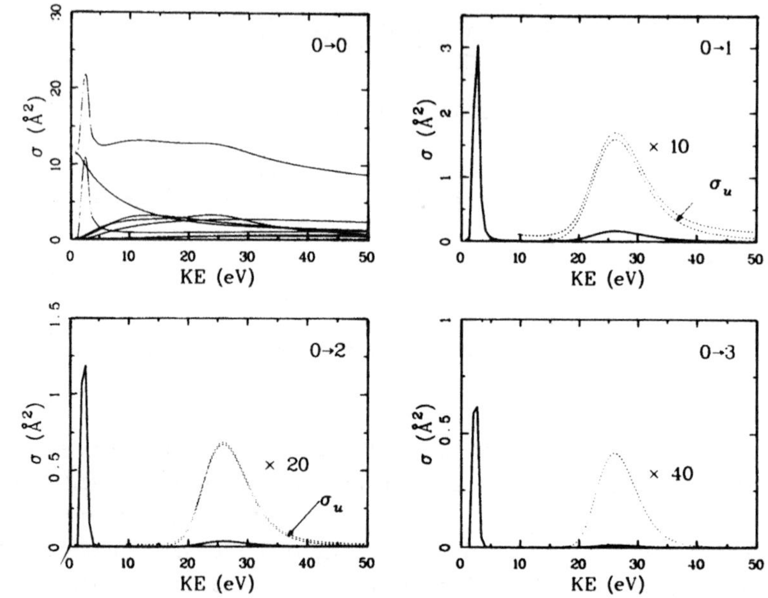

FIGURE 4
Vibrational excitation cross sections for e–N_2 scattering calculated using the continuum multiple-scattering method (from Dehmer and Dill[69]).

3.7. L^2 Methods

We have seen that a basic difficulty with the single-center expansion approach is its slow convergence for all molecules except the lightest. In molecular structure calculations this problem is of course overcome by using a multi-center expansion. It is thus of interest to see if bound-state techniques and if possible computer codes can be used in collision calculations. This is the motivation behind the development of L^2 methods described in this and the following two sections.

The first discussions on the use of L^2 methods in electron scattering calculations were given by Temkin,[76] Hazi and Taylor,[77] and Krauss and Mies.[78] They noted that if the collision wave function is expanded as follows

$$\Psi_k = \sum_j \chi_j(1, 2, \ldots, N+1)a_{kj}. \tag{53}$$

where the χ_j are L^2 discrete basis functions (i.e., they are square integrable) and where the a_{kj} are determined by diagonalizing the electronic part of the Hamiltonian (using a standard molecular structure code)

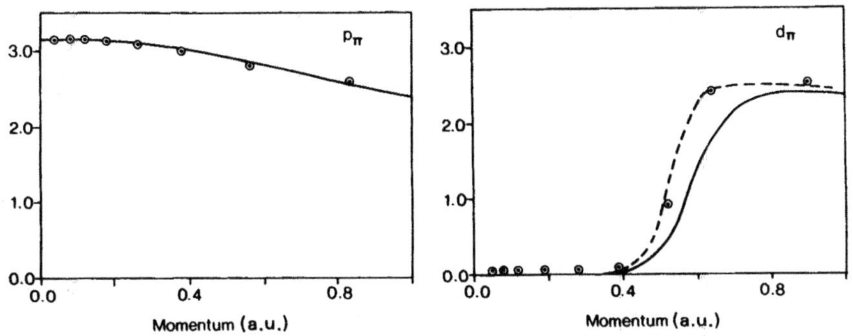

FIGURE 5
The $p\pi$ and $d\pi$ phase shifts for e–N_2 scattering. Solid line, static exchange calculations by Buckley and Burke[13]; circles, L^2 calculations by McCurdy *et al.*[79]; dashed line, calculations by Buckley and Burke including exchange with all target orbitals.

$$\langle \Psi_k | H_{\text{el}} | \Psi_j \rangle = E_k \delta_{kj}, \tag{54}$$

then the Ψ_k provide an accurate representation of the wave function at the energy E_k in the region where the χ_j are nonzero. Since in a typical case most of the E_k lie in the continuous spectrum of H_{el}, this enables the collision wave function to be determined at these energies in an "internal region." An approximate phase shift can then be determined by projecting Ψ_k onto the appropriate channel function in Eq. (5) yielding a radial function

$$r_{N+1}^{-1} R_{kl}^{\Lambda}(r_{N+1}) = \langle \Phi_0 Y_{l\Lambda} | \Psi_k \rangle, \tag{55}$$

where the integral is taken over all electronic coordinates except the radial coordinate of the scattered electron. The phase shift in this channel is then determined by fitting $R_{kl}^{\Lambda}(r)$ to the appropriate combination of spherical Bessel functions near the boundary of the "internal region."

This approach has been used by McCurdy *et al.*[79] to obtain phase shifts for e–N_2 scattering and we present their results for the $p\pi$ and $d\pi$ phase shifts in Fig. 5. The results are seen to be in good agreement with the static exchange calculations of Buckley and Burke, who solved Eq. (6) using the noniterative differential equation method. The main differences with the earlier calculations of Buckley and Burke[13] are due to the neglect of exchange with the $1\sigma_g$ and $1\sigma_u$ orbitals which gives rise to errors in the resonance region (as pointed out by Collins *et al.*[19]).

The results of McCurdy *et al.* exemplify the difficulties of this simple L^2 approach. First, the method produces the phase shift only at a discrete set of energies in the continuum. To obtain the phase at other energies it is necessary either to interpolate or to vary some parameter in the basis functions to move the energies E_k defined by Eq. (54). Further, the definition of the "internal region" is ill-defined leading to difficulties in the extraction of the phase shift. Finally the low-l spoiling approximation, which assumes that only low l values contribute to the cross section and are weakly coupled, is by no means applicable. It is certainly of doubtful validity when there are strong long-range potentials coupling the channels as is the case for polar molecules. Consequently in the last few years interest has turned to two other approaches, the R-matrix and the T-matrix methods, which, while still making extensive use of standard molecular structure codes, avoid these other difficulties.

We conclude this section by mentioning that L^2 methods have recently been extensively and successfully applied to the photoionization of atoms and molecules (e.g., Reinhardt,[80] McCurdy and Rescigno,[81] and Padial *et al.*[82]) and to the calculation of resonance positions and widths (e.g., Hazi[83] and McCurdy and Rescigno[84]).

3.8. R-Matrix Method

This method was first introduced by Wigner[85,86] and Wigner and Eisenbud[87] in a study of nuclear reactions, and extended to electron–atom scattering by Burke and Hibbert[88] and Burke *et al.*[89] Recently it has been developed further and applied extensively to electron–molecule scattering by Schneider,[90,91] Schneider and Hay,[92,93] Morrison and Schneider,[94] Schneider,[95] Burke *et al.*,[96] Buckley *et al.*,[97] and Noble *et al.*[98]

The method automatically takes advantage of the division of configuration space into an internal and an external region suggested by the frame-transformation theory of Chang and Fano.[3] This division of space is illustrated in Fig. 6. In the internal region, where exchange and correlation effects are important, the wave function is expanded using the fixed-nuclei approximation in terms of a multicenter discrete basis in the molecular frame of reference. In

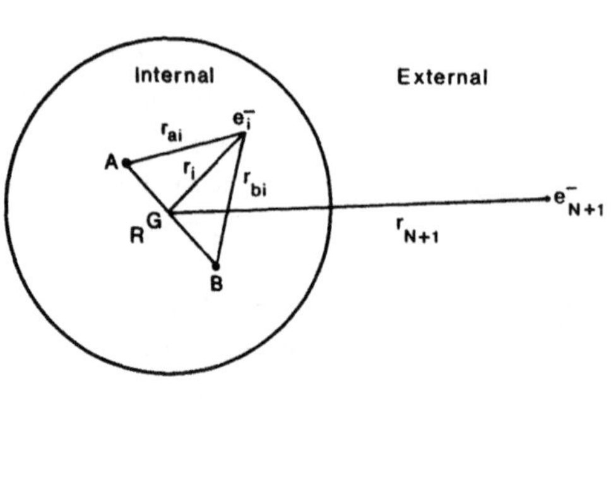

FIGURE 6
Division of configuration space in the R-matrix method.

the external region, where only the long-range components of the direct potential remain, a single-center expansion using only a few partial waves is appropriate and either the molecular frame or the laboratory frame is used depending on the relative importance of H_{rot}. The link between these two regions is provided by the R matrix $\mathscr{R}$ defined by

$$\mathbf{f}(a) = \mathscr{R}\left(r\frac{d\mathbf{f}}{dr} - \mathbf{bf}\right)_{r=a}, \tag{56}$$

where $\mathbf{f}$ is a complete set of solution vectors and $\mathbf{b}$ is an arbitrary constant matrix. The basic problem of the R-matrix method is to calculate this matrix from an expansion of the wave function in the internal region. Section 4.3 of Chapter 2 gives a general account of the R-matrix theory. In the following discussion we describe the R-matrix approach of Burke *et al.*[96] which uses a multicenter Slater-type orbital (STO) basis. The work of Schneider and collaborators uses an approach based on prolate spheroidal coordinates which, as we have already pointed out, has some advantages for diatomic molecules, but which cannot be so easily extended to polyatomic molecules. We expand the wave function describing the electron–molecule system in an internal region bounded by a sphere of radius $r = a$ in the form

$$\Psi_k = \mathscr{A} \sum_{ij} \Phi_i(1, \ldots, N, \sigma_{N+1})\psi_{ij}(\mathbf{r}_{N+1})c_{ijk} + \sum_i \phi_i(1, 2, \ldots, N+1)c_{ik}, \tag{57}$$

which replaces Eq. (5) and generalizes it to the situation where there is more than one electronic state. The channel functions in Eq. (57) do not involve the angular coordinates of the scattered electron but are eigenstates of $\mathbf{S}^2$ and S_Z. The molecular orbitals used in the definition of the target eigenstates and pseudostates and in the correlation functions are expanded in terms of STOs centered on A and B which have effectively vanished by the boundary of the internal region. On the other hand the continuum molecular orbitals ψ_{ij} are expanded in terms of the same set of STOs on A and B but also include STOs centered on the center of gravity G of the molecule which are of longer range and may be nonzero on the boundary. We will see that it is these STOs on G which provide the connection with the single-center expansion used in the external region.

The coefficients c_{ijk} and c_{ik} are determined by diagonalizing the operator $H_{\text{el}} + L_b$

$$\langle\Psi_k|(H_{\text{el}} + L_b)|\Psi_{k'}\rangle = E_k\delta_{kk'}, \tag{58}$$

where the integrals in this equation are carried out over the internal region and L_b is a surface projection operator introduced by Bloch[28] which ensures that $H_{el} + L_b$ is Hermitian. It is defined by

$$L_b = \frac{1}{2} \sum_{il} |\Phi_{il}\rangle \delta(r-a) \left(\frac{d}{dr} - \frac{b_{il}-1}{r} \right) \langle \Phi_{il}|, \tag{59}$$

where the channel functions Φ_{il} are appropriate to the external region where a single-center expansion is used. We can now write the Schrödinger equation (3) as

$$(H_{el} + L_b - E)\Psi = L_b\Psi, \tag{60}$$

which has the formal solution

$$\Psi = (H_{el} + L_b - E)^{-1} L_b \Psi. \tag{61}$$

The inverse operator in this equation can be expanded in terms of the eigenfunctions defined by Eq. (58) giving

$$|\Psi\rangle = \sum_k \frac{|\Psi_k\rangle\langle\Psi_k|L_b|\Psi\rangle}{E_k - E}. \tag{62}$$

We now project this equation onto the channel function Φ_{il} and evaluate it on the boundary $r = a$. If we define the radial functions

$$r_{N+1}^{-1} f_{il}(r_{N+1}) = \langle \Phi_{il} | \Psi \rangle, \tag{63}$$

where the integration in this equation is carried out over the coordinates of all $(N+1)$ electrons except the radial coordinate of the scattered electron, then we obtain immediately Eq. (56) where the R matrix elements are given by

$$\mathscr{R}_{il,i'l'}(E) = \sum_k \frac{\gamma_{ilk}\gamma_{i'l'k}}{E_k - E}. \tag{64}$$

The reduced width amplitudes γ_{ilk} are defined in terms of the value of the continuum orbitals on the boundary. Remembering that we have used a three-center expansion of these orbitals which can be written as

$$\psi_{ij} = \psi_{ij}(\mathbf{r}_A) + \psi_{ij}(\mathbf{r}_B) + \sum_l r^{-1} v_{ijl}(r) Y_{lm}(\hat{\mathbf{r}}), \tag{65}$$

where only the last term centered on G contributes on the boundary, then

$$\gamma_{ilk} = \left(\frac{1}{2a}\right)^{1/2} \sum_j v_{ijl}(a) c_{ijk}^{\Lambda S M_S}. \tag{66}$$

An important point to remember is that the number of l values that need to be retained on G is appropriate to the external region, and not to represent the nuclear singularities which are now represented by the orbitals centered on the nuclei. It has been found in practice that for homonuclear diatomic molecules such as N_2 at most two l values and perhaps 6–8 basis orbitals on G for each value of l are required to give accurate results for energies up to 10 eV.[97]

The most time-consuming part of this calculation is setting up the diagonalizing $H_{el} + L_b$ which must be carried out once for each set of symmetries $\Lambda S M_S$. This can be done by relatively small modifications of standard multicenter integral codes by subtracting the tail from those integrals involving STOs on G which extend beyond the boundary, and by calculating and adding on the Bloch surface terms. These modifications have been made to the ALCHEMY code by Kendrick and Buckley[99] and Noble.[100] Once the R matrix is determined, then the coupled differential equations need to be solved in the external region for each energy where the cross section is required.

Results obtained for e–H_2 and e–N_2 scattering in the static exchange approximation by Morrison and Schneider[94] and by Buckley *et al.*[97] are in good agreement.

TABLE 6
Resonance Parameters for the $^2\Pi_g$ Shape Resonance in e–N_2 Scattering Computed in the Static Exchange Approximation[a]

	Target wave function	E_r (eV)	Γ (eV)
CRM	Cade *et al.*	3.75	1.16
	Nesbet	3.77	1.20
	Ransil	4.49	1.45
Schneider (1981)	Nesbet (c)	3.70	1.16
	Nesbet (d)	3.62	1.11
R-matrix	Cade *et al.*	3.81	1.25
	Nesbet	3.68	1.11
	Ransil	4.59	1.61

[a] The results labeled CRM are those obtained by Collins, Robb, and Morrison[19] using the target wave functions of Cade *et al.*,[22] Nesbet,[101] and Ransil.[102] The results of Schneider[103] are calculated using the *R*-matrix method (c) and a complex version of the *R* matrix method (d). The remaining *R*-matrix results were obtained by Noble, Burke, and Salvini.[98]

In recent work[98] these results have been extended to include fully converged eigenphase sums for the scattering in the Σ_g, Σ_u, Π_g, and Π_u states of both H_2 and N_2 targets. At the lower energies these static exchange results agree in detail with the values reported by Collins, Robb, and Morrison.[19] Parameters of the shape resonance in ${}^2\Pi_g$ N_2 scattering are also successfully predicted as illustrated in Table 6. However, at higher scattering energies the eigenphase sums exhibit an oscillatory behavior which clearly signals a lack of completeness in the continuum basis. This phenomenon is also observed in calculations for LiH, H_2^+, and CO targets. Attempts to improve the representation of the continuum indicate that a successful strategy is to select STO basis sets by taking a constant ratio between successive zeta values and to then choose each principal quantum number so that the STO maxima are evenly spaced across the R-matrix internal region. However, the completeness of the representation is fundamentally limited by the accuracy with which it is feasible to compute multicenter STO integrals since this in turn determines the degree of linear dependence which is tolerable within the basis.

A possible means of extending the range of scattering energies for which accurate results may be obtained is to allow basis functions located on the scattering center G to assume an arbitrary radial dependence, defined numerically, thus circumventing the problems associated with diffuse STOs. It has been found, for example, in the case of electron–atom scattering that the use of numerical basis orbitals allows accurate results to be computed over an energy range of several hundred electron volts. Several alternative schemes for constructing the numerical continuum orbitals $v_{ijl}(r)$ are currently under investigation. In one of these, the basis functions are determined by solving a model electron–molecule collision problem numerically, where orthogonality to the bound orbitals is imposed by the method of Lagrange undetermined multipliers. This approach has been successfully applied to e–N_2 scattering at intermediate energies.[104] An alternative procedure suggested by Nesbet[105] is to allow the numerical functions generated by the solution of the coupled equations in the asymptotic region to extend into the region $r \leqslant a$. This, it is hoped, will mean that fewer functions will be required in this region.

Schneider[95] has reported R-matrix calculations on H_2 in which the polarizability was represented by pseudostates. The method is based on the work of Damburg and Karule,[106] who first pointed out that it was possible to define a pseudostate which could be included in the close-coupling expansion in the same way as a molecular eigenstate but which allowed for the full polarizability of the target states (for details of the formalism refer to the articles by Burke and Mitchell[107] and by Coulter[108]).

The spherical components of the static dipole polarizability of the molecular system are given by

$$\alpha_\mu = 2\sum_{k\neq 0} \int dk \frac{|\langle \Phi_0 | M_1^\mu | \Phi_k \rangle|^2}{E_k - E_0}, \tag{67}$$

where M_1^μ is the usual dipole operator. Damburg and Karule[106] replace the infinite summation and the integration over k in Eq. (67) by a single term for each value of μ by writing

$$\alpha_\mu = \frac{2|\langle \Phi_0 | M_1^\mu | \tilde{\Phi}_\mu \rangle|^2}{E_\mu - E_0}. \tag{68}$$

Polarized pseudostates $\tilde{\Phi}_\mu$ are therefore seen to be solutions of the equation for the target

$$(H_{\text{el}} - E_0)\tilde{\Phi}'_\mu = M_1^\mu \Phi_0, \tag{69}$$

where $\tilde{\Phi}'_\mu = N\tilde{\Phi}_\mu$ and the factor N is included to ensure that the $\tilde{\Phi}_\mu$ are correctly normalized. Once the pseudostates have been computed they may be included in the close-coupling calculation in the same way as real target eigenstates. Preliminary work on the application of this approach to molecular systems has also been reported by Kendrick.[109] An alternative way of including the effect of polarizability has been explored by Schneider and Hay[92, 93] for F_2. They used the F_2^- self-consistent field (SCF) orbitals rather than F_2 orbitals to represent the target, so allowing for part of the relaxation occurring during the collision. A similar approach has been used by Schneider *et al.*[110] for vibrational excitation in e–N_2 scattering and these results will be discussed in Section 4. One difficulty with this approach is that the use of the same relaxed orbitals for each collision symmetry can be unrealistic.

3.9. T-Matrix Method

Another L^2 method, this time based on the expansion of the T matrix in terms of a discrete basis, has been proposed by Rescigno *et al.*[111–113]

The original method starts from the Lippmann–Schwinger integral equation for the transition matrix

$$T = U + UG_0^+T, \tag{70}$$

where U is twice the nonlocal interaction potential between the electron and the target molecule and G_0^+ is the free-particle Green's function with outgoing-wave boundary conditions. We now expand U in a finite basis set of square-integrable functions ϕ_α, taken to be Gaussian-type orbitals (GTOs), giving

$$U^t(\mathbf{r},\mathbf{r}') = \sum_{\alpha\beta} |\phi_\alpha(\mathbf{r})\rangle\langle\phi_\alpha(\mathbf{r})|U(\mathbf{r},\mathbf{r}')|\phi_\beta(\mathbf{r}')\rangle\langle\phi_\beta(\mathbf{r}')|. \tag{71}$$

Inserting this truncated potential U^t into Eq. (70) gives a matrix equation for the T matrix which can be solved yielding

$$\mathbf{T}^t = (1 - \mathbf{U}^t\mathbf{G}_0^+)^{-1}\,\mathbf{U}^t, \tag{72}$$

where the matrix $\mathbf{G}_0^+$ is defined by

$$\langle\phi_\alpha|G_0^+(E)|\phi_\beta\rangle = \lim_{\epsilon\to 0+} \int d^3\mathbf{k}\, \frac{\langle\phi_\alpha|\mathbf{k}\rangle\langle\mathbf{k}|\phi_\beta\rangle}{q^2 - k^2 + i\epsilon}, \tag{73}$$

where $E = \frac{1}{2}q^2$ and $|k\rangle$ and $|k'\rangle$ are the plane-wave states

$$|k\rangle = (2\pi)^{-3/2}\, e^{i\mathbf{k}\cdot\mathbf{r}}. \tag{74}$$

The scattering amplitude for a transition from state $\mathbf{k}_i$ to $\mathbf{k}_f$ is then given by

$$f(\mathbf{k}_i \to \mathbf{k}_f) = -2\pi^2 \sum_{\alpha\beta} \langle\mathbf{k}_f|\alpha\rangle\langle\alpha|T^t(E)|\beta\rangle\langle\beta|\mathbf{k}_i\rangle, \tag{75}$$

where the matrix elements involving the plane-wave states can be evaluated analytically for GTO basis functions (Ostlund[114]).

This theory has been used to obtain static exchange results for H_2, N_2, CO, and F_2 by Rescigno *et al.*,[111–113] Fliflet *et al.*,[115] and Rescigno.[116] In

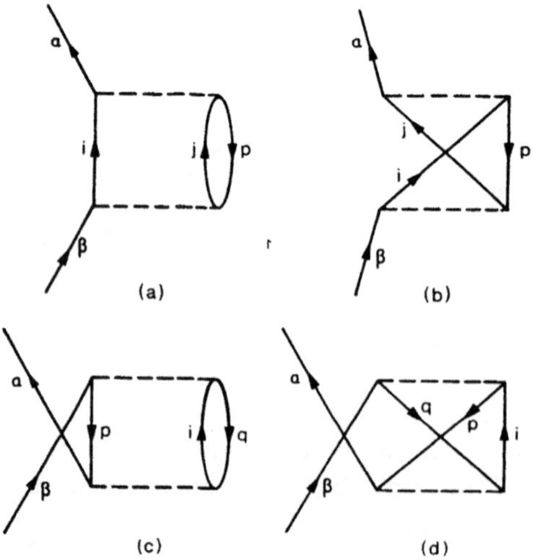

FIGURE 7
Second-order optical potential diagrams. Diagrams (b) and (d) are exchanges of (a) and (c), respectively (taken from Klonover and Kaldor[118]).

addition the polarizability was successfully included for H_2 by Klonover and Kaldor[117–120] using an L^2 basis and second-order perturbation theory. The resulting effective polarization potential V_{pol} is expressed diagrammatically in Fig. 7, and involves the summation over all occupied orbitals with indices (p, q) and over all unoccupied orbitals (i, j) for the given basis.

However, these T-matrix calculations have shown that in general the errors due to the truncation of the potential can only be made small, particularly at energies below 5 eV where the long-range potentials are important, by the use of large basis sets, and this can lead to problems of linear dependence. Although the inclusion of a variational correction using the Kohn method (Fliflet *et al.*[121–123]) helps to overcome this difficulty, it has proved better to use a modification of the method based on a variational principle from the outset. The method has been used in nuclear physics (Lovelace[124]) and involves writing the potential in the separable form

$$U^S = \sum_{\alpha\beta} U|\alpha\rangle(U^{-1})_{\alpha\beta}\langle\beta|U. \tag{76}$$

Substituting into the Lippmann–Schwinger equation then gives

$$\langle \mathbf{k}_f|T^S|\mathbf{k}_i\rangle = \sum_{\alpha\beta} \langle \mathbf{k}_f|U|\alpha\rangle\,[(U - UG_0U)^{-1}]_{\alpha\beta}\langle\beta|U|\mathbf{k}_i\rangle, \tag{77}$$

which can be shown to satisfy the Schwinger variational principle (Adhikari and Sloan[125]). An important feature of this new approach is that the expansion functions ϕ_α only appear in conjunction with the potential, which allows the use of trial functions which do not have the correct asymptotic form. This is of course in contrast to the Kohn variational method. The method involves two new types of matrix element $\langle \mathbf{k}|U|\alpha\rangle$ and $\langle\alpha|UG_0U|\beta\rangle$. The first can be evaluated in closed form when GTOs are used and the second can be approximated by closure. In practice, single-center expansions are used for evaluating the matrix elements. Results involving only three s-type basis functions, one p-type basis function, and one d-type basis function have been obtained by Watson *et al.*[126] for H_2 which are in good agreement with other methods.

As an example of the application of this method we show in Fig. 8 the angular distribution of $v = 0 \rightarrow 1$ vibrational excitation cross sections for e–H_2 scattering calculated by Klonover and Kaldor[119] compared with the measurements of Linder and Schmidt.[127] (For the integral cross sections see Chapter 3, p. 230, Fig. 10.) The calculations were carried out with and without the inclusion of the second-order polarization, and the vibrational excitation cross sections were obtained using the adiabatic theory described in Section 2.2. The calculation used 30 Cartesian Gaussian-type orbitals to describe the occupied H_2 orbitals and the excited orbitals which account for the polarization effects,

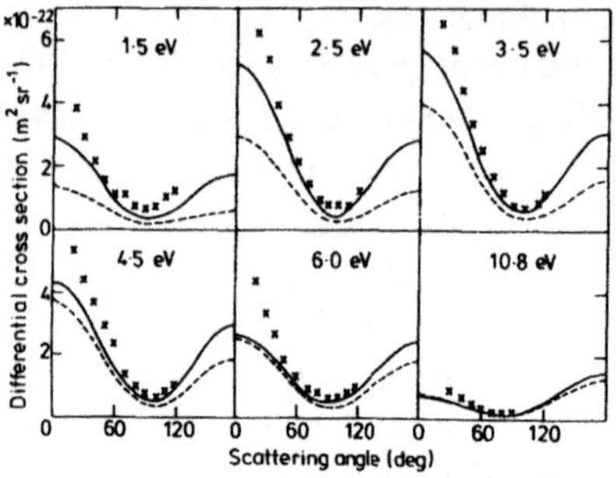

FIGURE 8
Differential $v = 0 \rightarrow 1$ vibrational excitation cross section for e–H_2 scattering. Solid line, T-matrix calculations including polarization; dotted line, T-matrix calculations omitting polarization; both by Klonover and Kaldor.[119] Crosses, measurements of Linder and Schmidt.[127] (From Klonover and Kaldor.[119])

and 61 GTOs to span the scattered electron orbital, some of the latter orbitals being centered on the center of gravity. The calculations were based on Eq. (72) for the T matrix. The results show clearly the importance of polarization effects at low energies in both the forward and backward directions.

In a development of this work Berman and Kaldor[128] have suggested a method for including second-order polarization within the Schwinger variational method without recourse to perturbation theory. A second-order optical potential given by the expression[67]

$$V_{\text{pol}} = \langle p | V^{(2)} R^{(3)}(\omega) V^{(2)} | q \rangle \tag{78}$$

is evaluated using a pole approximation for the response function $R^{(3)}(\omega)$. The method has been applied to predict the $^2\Pi_g$ resonance in electron–nitrogen molecule scattering by Berman *et al.*[128]

Despite the improved accuracy which results from the use of the Schwinger variational principle higher-order corrections to the scattering wave function are still required. The quality of the results which are obtained, however, suggests that the separable potential U^S is a reasonable approximation to the exact potential U and that an iterative procedure based on the difference potential $\Delta U = U - U^S$ might be suitable for computing these corrections.

The result

$$U | \psi^{(+)} \rangle = T | \phi \rangle, \tag{79}$$

where ϕ and $\psi^{(+)}$ are the free-particle and scattering wave functions, respectively, implies that the scattering wave functions may be expressed in terms of the T matrix by the relation

$$\psi^{(+)}(\mathbf{r}) = \phi(\mathbf{r}) + \langle \mathbf{r} | G_0^+(E) T | \phi \rangle. \tag{80}$$

The scattering wave function corresponding to the T matrix of Eq. (77) may therefore be computed by straightforward quadrature using Eq. (80). Lucchese

TABLE 7

Convergence of K-Matrix Elements Obtained by Lucchese, Watson, and McKoy[129] Using the Iterative Schwinger Variational Principle Beginning from Plane Waves for $^2\Sigma_g$ Symmetry in Static Exchange e–H_2 Scattering with $k = 0.5$ a.u.[a]

		$K_{ll'}^{S_n}$				
(l, l')	$n =$	0	1	2	3	4
(0, 0)		0	−2.931	−1.701	−1.552	−1.548
(0, 2)		0	0.128 (−1)	0.133 (−1)	0.134 (−1)	0.134 (−1)
(2, 2)		0	0.163 (−1)	0.163 (−1)	0.163 (−1)	0.163 (−1)

[a] Powers of 10 are indicated in parentheses.

et al.[129] have devised an iterative scheme based on the use of Eqs. (77) and (80). At the nth iteration the T matrix T_n^S is computed using Eq. (77) and an augmented set of expansion functions; the set comprises the initial set $\{\phi_\alpha\}$ and a set of continuum wave functions computed using Eq. (80) with the value of the T matrix obtained on the previous iteration T_{n-1}^S. It may be shown that, provided the wave functions converge, i.e.,

$$\psi_n^{(+)}(\mathbf{r}) = \psi_{n-1}^{(+)}(\mathbf{r}), \tag{81}$$

and provided the relation

$$\langle \psi_n^{(-)} | U - UG_0^+(E)U | \chi \rangle = \langle \phi | U | \chi \rangle \tag{82}$$

is satisfied for the augmented set of expansion functions $\{\chi\} \equiv \{\phi_\alpha, \psi_n^{(+)}\}$, then the wave functions $\psi_n^{(+)}$ satisfy the Lippmann–Schwinger equation for the exact potential U.

As shown in Table 7 the method converges after three or four iterations and yields accurate static exchange results which are independent of the initial choice of expansion functions. It has been applied to scattering by H_2,[129] H_2^+,[130] and LiH.[131] The principal difficulty with this approach is the time-consuming nature of the computations required in order to evaluate the corrections which is mainly a consequence of the complications introduced by exchange. One method for circumventing this problem which is currently under investigation has already been described in Section 3.3. A second possible approach is to improve the effectiveness of the variational principle itself.

A number of new variational functionals have been devised by Takatsuka and McKoy,[132] which retain the fractional form and many of the desirable properties of the Schwinger principle but which also incorporate some of the advantages of the Kohn method. Most importantly the functionals are numerically tractable, generalizable to multichannel situations, and produce symmetric anomaly-free K matrices. This research has also elucidated the relation between the Schwinger and Kohn methods showing explicitly that the former is of higher rank[133] and therefore may be expected to give superior results for the same trial wave function as has been found in practice.

One new functional, which is closely related to the Schwinger functional, has been used to compute static exchange results for electron scattering by H_2, LiH, H_2^+, and CO^+ targets.[134] This C-functional may be written as

$$\tilde{C} = \frac{\langle C|UG_0U|\phi\rangle\langle\phi|UG_0U|C\rangle}{\langle C|U - UG_0U|C\rangle}, \tag{83}$$

where ϕ is the regular solution of the free-particle Schrödinger equation and $C = \psi - \phi$. The K matrix is obtained from the stationary value X of the functional by adding first- and second-Born terms according to the relation

$$-\tfrac{1}{2}K_{ij} = \langle\phi_i|U|\phi_j\rangle + \langle\phi_i|UG_0U|\phi_j\rangle + X_{ij}. \tag{84}$$

The use of functional (83) does not involve significantly greater computational effort than the usual Schwinger method but the results are substantially more accurate. This means that in many cases a single iteration will be sufficient to obtain convergence in the iterative version, a fact which is particularly important for strongly polar molecules such as LiH, which, as has already been seen, are difficult to treat using the Schwinger principle. The development of methods for explicitly including polarization effects is reported to be in progress. The theory has already been extended to include the description of inelastic scattering.[135]

To conclude, we have seen that there now exist a number of highly efficient techniques for accurately solving the fixed nuclei equations. Of these, the results which have been obtained using the L^2 R-matrix and T-matrix variational methods indicate that these two approaches both promise to be particularly powerful ways of calculating electron–molecule collision cross sections.

4. Vibrational Excitation and Dissociative Attachment

In this section we briefly describe calculations carried out within the framework of the adiabatic nuclei approximation and the hybrid theory introduced in Section 2.2. We then discuss the resonance theory method and the R-matrix method in some detail.

4.1. Adiabatic Nuclei Approximation

The application of the adiabatic nuclei approximation to the vibrational motion is based on the use of the Eq. (11) for the K matrix for the transition $v \to v'$. As discussed in Section 2.2, this approximation is valid in situations where the collision time is much shorter than the time of the vibrational motion. In the simplest application the fixed nuclei K matrix is chosen to correspond to the solution of the fixed-nuclei equations over a range of internuclear separations, and thus any of the methods described in Section 3 can be easily extended to obtain vibrational excitation cross sections. Examples include T-matrix results by Levin *et al.*,[136] close-coupling calculations using model exchange and polarization potentials by Rumble *et al.*[137] and by Onda and Truhlar.[138]

The range of validity of Eq. (11) can be extended by using the frame-transformation approach suggested by Chang and Fano.[3] The adiabatic nuclei approximation, which neglects H_{vib}, is used in an internal region $r < a$ (say). The fixed nuclei equations are solved in this region to yield the R matrix on the boundary of the region as a function of the internuclear separation R. This quantity is then transformed using

$$\mathscr{R}_{i'l'v',ilv} = \langle \chi_{v'} | \mathscr{R}_{i'l',il}(R) | \chi_v \rangle, \tag{85}$$

and the coupled equations (14), which include H_{vib}, are solved in the external region using $\mathscr{R}_{i'l'v',ilv}$ to define the boundary condition at $r = a$. However, if this is going to be a practical method of calculation it is necessary that the internal region, where H_{vib} is neglected, also includes the region where the nonlocal terms in the Hamiltonian arising from exchange are nonzero. If this is not the case then nonlocal terms will couple the internal and external regions making the frame transformation difficult to apply. It appears that in many cases of interest this coupling will be important.

Another interesting way of extending the adiabatic theory has been proposed by Nesbet.[139] In this approach, called the "energy-modified adiabatic approximation (EMA)," the S-matrix elements connecting the vibrational states are defined by

$$S_{i'l'v',ilv} = \langle \chi_{v'} | S_{i'l',il}(E - H_i; R) | \chi_v \rangle, \tag{86}$$

where $S_{i'l',il}(E - H_i; R)$ is the S matrix calculated in the fixed-nuclei approximation at the internuclear separation R at an energy defined by the operator $H_i = E_i(R) + H_{\text{vib}}$. This has the effect of including the internal energy of the target and has been used successfully to describe the observed structure in the vibrational excitation in e–N_2 in terms of a parametrized $^2\Pi_g$ resonance potential energy curve.

A detailed assessment of the performance of EMA theory for near-threshold rotational excitation of H_2 molecules has recently been given by Feldt *et al.*[140] as part of a study of the breakdown of the adiabatic nuclei approximation; see Chapter 2, p. 113, Fig. 6. In this work the laboratory-frame coupled channel equations are solved using model exchange and polarization potentials of the type described in Section 3 to yield results against which the approximations of EMA theory may be compared. As expected the integrated cross sections show a progressive breakdown of the adiabatic nuclei approximation as the scattering energy decreases. The EMA results are substantially better but fail to correct deficiencies in the high partial waves so that structure in the differential cross sections is lost in both approximations. The study is currently being extended to include the vibrational motion.[141]

4.2. Hybrid Theory

This is based on the solution of Eq. (14) and has been considered by Chandra and Temkin,[5] Temkin,[142] and by Choi and Poe.[143] In the work of Chandra and Temkin on e–N_2 scattering, the equations were made numerically tractable by replacing the exchange potential by an orthogonalization procedure defined by Eq. (46). They also included a parametrized polarization potential defined by Eqs. (15) and (49) where the cutoff parameter was adjusted to make the calculated position of the $^2\Pi_g$ resonance agree with experiment. With these approximations, and a careful consideration of the variation of the polarizability with internuclear separation (Temkin[144] and Kendrick[145]), recent results are in reasonable accord with experiment as discussed in the following subsection in connection with Fig. 16 of Chapter 3, p. 241.

An interesting new approach to the solution of the equations of electron–molecule scattering theory has recently been advanced by Sullivan and Temkin.[146] The partial differential equations describing the scattering are discretized and solved directly by a noniterative numerical block technique thus avoiding the need to perform any partial-wave expansion. The method has been applied to e–N_2 scattering to obtain $^2\Sigma_g$ eigenphase sums in the static approximation and also to the hybrid theory equations to obtain vibrational excitation cross sections using model exchange and polarization potentials.[147] These results are very promising and demonstrate that the technique can be effective. However, the general application of the method to problems demanding a high spatial resolution may be limited by the size of computation which is feasible.

4.3. Resonance Theory

It was pointed out in Section 2.2 that the adiabatic approximation breaks down in the neighborhood of narrow resonances, and it is just in these energy regions where experiment (e.g., Schulz[148, 149]) shows that the vibrational excitation cross sections are large. It is thus of interest to see if methods can be developed which take explicit account of these resonances from the outset. This is the basis of the methods developed by Herzenberg and collaborators over the last two decades (e.g., Herzenberg and Mandl,[150] Bardsley and Mandl,[151] and Herzenberg[152]) and described in detail in Chapter 3.

As an example of a resonance theory which has been particularly successful in describing vibrational excitation and dissociative attachment we review Fig. 12b of Chapter 3, p. 236, showing the potential energy curves and nuclear wave functions which arise in the "boomerang model" introduced by Herzenberg[153] and Birtwistle and Herzenberg.[154] The incident electron is captured into the resonance state as illustrated in the figure. The nuclei then move apart and are reflected once from the outer turning point of the resonance state before the electron is emitted leaving the molecule in an excited vibrational state. This model is based on the assumption that the dependence of the width $\Gamma(R)$ on the internuclear separation R is such that there can only be a single outgoing and reflected wave as is the case for e–N_2 scattering.

As in the R-matrix method the formulation of the resonance theory starts from a division of configuration space into two regions where the interaction between the scattered electron and the molecule is assumed to be small in the external region (Bardsley *et al.*[155]). For each value of the internuclear coordinate R we define a complete set of resonant states in the internal region which satisfy the equation

$$[H_{\mathrm{el}} - W_n(R)]\,\Psi_n = 0, \tag{87}$$

The resonant states Ψ_n satisfy the Siegert[156] outgoing-wave boundary conditions in the external region

$$\Psi_n \underset{r\to\infty}{\sim} \Phi_i e^{ik_{ni}r}, \tag{88}$$

where Φ_i are the target electronic states, the wave numbers k_{ni} are defined as

$$k_{ni}^2 = 2(W_n - E_i), \tag{89}$$

and for notational simplicity here and below we have omitted the angular and spin channels, but these would have to be included in a detailed theory.

If the total energy W_n is below the first electronic threshold, then all the channels are closed, the wave numbers k_{ni} are purely imaginary, and the wave function defined by Eq. (88) decays exponentially asymptotically. This corresponds to a bound negative-ion state of the electron–molecule system. On the other hand if the energy is such that some of the channels are open then the Siegert prescription makes the eigenvalues complex. This corresponds to a resonance state of the electron–molecule system. We will write in this case

$$W_n(R) = \text{Re } W_n(R) - \tfrac{1}{2}i\Gamma_n(R), \tag{90}$$

where the resonance widths $\Gamma_n(R)$ are real and greater than zero.

In accordance with the Born–Oppenheimer separation of the electronic and nuclear motion we now expand the total electron–molecule wave function in the internal region in terms of this complete set of resonance states, giving

$$\Psi = \sum_n \Psi_n(1, 2, \ldots, N+1)\xi_{niv}(R, \mathbf{k}_{iv}) + \Phi_{\text{inc}}\chi_{iv}(R), \tag{91}$$

where the last term in this equation corresponds to a plane wave incident on the initial state of the target. To obtain equations for the wave functions $\xi_n(R, \mathbf{k})$ describing the nuclear motion we substitute Eq. (91) into Eq. (12), multiply by the functions Ψ_n^*, and integrate over all the electronic coordinates within the internal region. If we now assume that Ψ_n varies slowly with R in comparison to ξ_{niv} so that the nuclear kinetic energy operator acts only on ξ_{niv}, we obtain with the aid of Eq. (87) in the case of zero angular momentum

$$\left[-\frac{1}{2M}\frac{d^2}{dR^2} + W_n(R) - E\right]\xi_{niv}(R, \mathbf{k}_{iv}) = -\zeta_{niv}(R, \mathbf{k}_{iv})\chi_{iv}(R), \tag{92}$$

where $\chi_{iv}(R)$ is the initial vibrational state of the molecule, and $\zeta_{niv}(R, \mathbf{k}_{iv})$ is the so-called "entry amplitude" into the resonance state described by Ψ_n. This is defined by

$$\zeta_{niv}(R, \mathbf{k}_{iv}) = \langle\Psi_n|V|\Phi_i e^{i\mathbf{k}_{iv}\cdot\mathbf{r}}\rangle, \tag{93}$$

where the integral is taken over the internal region and V is the potential interaction between the electron and the molecule.

The amplitude for vibrational excitation is now obtained by calculating the decay of the resonance states into the appropriate final state. If we introduce the Green's function $G_n(R,R')$ of the operator on the left-hand side of Eq. (92) then we obtain

$$\xi_{niv}(R, \mathbf{k}_{iv}) = -\int_0^\infty dR' G_n(R, R')\zeta_{niv}(R', \mathbf{k}_{iv})\chi_{iv}(R'). \tag{94}$$

The amplitude for vibrational excitation is then

$$f_{i'v',iv} = -\sum_n \int_0^\infty dR \int_0^\infty dR'\ \chi^*_{i'v'}(R)\zeta^*_{ni'v'}(R,\mathbf{k}_{i'v'})\,G_n(R,R') \times \zeta_{niv}(R',\mathbf{k}_{iv})\chi_{iv}(R'), \qquad (95)$$

where the appropriate sum over channels is assumed.

Dubé and Herzenberg[157] have applied this theory to calculate the vibrational excitation cross section in e–N_2 scattering. Figure 16a in Chapter 3, p. 241, compares their results for the $v = 1 \rightarrow 2$ transition at 90° with measurements made by Wong.[158] Also shown in Fig. 16b in Chapter 3 are the hybrid theory calculations of Chandra and Temkin[142] compared with the same measurements. This "boomerang model" theory of Dubé and Herzenberg assumes that there is one resonance state of N_2^- with $^2\Pi_g$ symmetry contributing to the cross section and the real and imaginary parts of its potential energy curve are written in terms of six adjustable parameters. In practice the permissible values of all but one of these parameters are determined almost completely by an *ab initio* L^2 calculation by Krauss and Mies,[78] and the remaining parameter is a "correlation energy" which Krauss and Mies adjusted to get the resonance appearing at the right energy. Having parametrized the N_2^- state the theory is then completely determined. Clearly both the "boomerang model" and the hybrid theory reproduce the observed oscillations in the cross section, which arise from the relaxation of the nuclear wave function during the collision. These oscillations would not be obtained in the adiabatic nuclei theory based on Eq. (11), which instead would have given one broad peak in the cross section (see in this connection Fig. 4).

The validity of the physical assumptions implicit in the use of the boomerang model has been studied by Hazi *et al.*[159] by comparing vibrational excitation cross sections for N_2 computed with *ab initio* fixed-nuclei resonance parameters with the results of single-center R-matrix calculations (these will be described in Section 4.4). They conclude that the fundamental assumption of the model, namely, that of a local energy-independent resonance width, is valid in the cases considered, and that realistic results are possible using resonance parameters computed using the Stieltjes-moment technique. It is found that, while the detailed structure in the cross section is sensitive to the parameter values, the cross-section magnitude depends mainly upon the resonance width. The importance of the role played by short-range polarization effects in the target molecule on the vibrational excitation process is also confirmed.

The $^2\Pi_g$ resonance in e–N_2 scattering is discussed also in the next subsection of this chapter and in Section 5 of Chapter 3. The boomerang model has been applied also to core-excited shape resonances; see Chapter 5, Section 2.7.

Dissociative attachment can be described by a small extension of this theory. The electron is captured by the molecule from the same initial vibrational state into an intermediate resonance state. The nuclei then move apart along the resonance potential energy curve. If they separate to the point where this curve becomes real before the electron autoionizes then dissociative attachment occurs. Otherwise we have vibrational excitation as described above. The cross section for dissociative attachment is obtained by calculating the flux of negative ions at large values of the internuclear coordinate giving

$$\sigma_{\mathrm{DA}} = \frac{2\pi^2}{k_{iv}} \frac{K}{\mu} \lim_{R \to \infty} |a(R)|^2, \tag{96}$$

where k_{iv} and K are the wave numbers describing the motion of the incident electron and outgoing ion in the center-of-mass system and $a(R)$ is the amplitude of the outgoing-wave solution of Eq. (92).

This theory of dissociative attachment has had many successful applications. For example in a study of dissociative attachment in e–H_2 scattering Wadehra and Bardsley[160] have successfully explained the effect of vibrational and rotational excitation of the target molecule on the dissociative attachment cross section observed by Allan and Wong[161] (Chapter 4, Section 5.1). As for vibrational excitation the input to this theory is the real and imaginary parts of the resonance potential energy curve. It has also been shown[162] that the two lowest resonant states are both necessary to explain the measured vibrational excitation cross sections for collisions of 10-eV electrons with H_2 and D_2 molecules.

Recently there have been attempts to duplicate some of the very impressive achievements of the empirical models which were described above using a more fundamental approach. Distorted-wave calculations for the dissociation of F_2 molecules have been reported using the T-matrix method.[163] In addition the first completely *ab initio* dissociative attachment cross sections have been computed by Hazi *et al.*[164] also for an F_2 target (Chapter 4, Section 5.4). In this calculation resonance parameters are extracted from large-scale configuration-interaction (CI) calculations using Stieltjes-moment theory[83] in exactly the same way as used to obtain vibrational excitation cross sections. However, the wave equation for the dissociating nuclei is solved by invoking a closure approximation to evaluate the nonlocal potential term but otherwise avoiding the approximations of the boomerang model. An example of the results which are obtained is given in Fig. 9. As expected, the detailed agreement with the measured values of Chantry which is obtained using the empirical

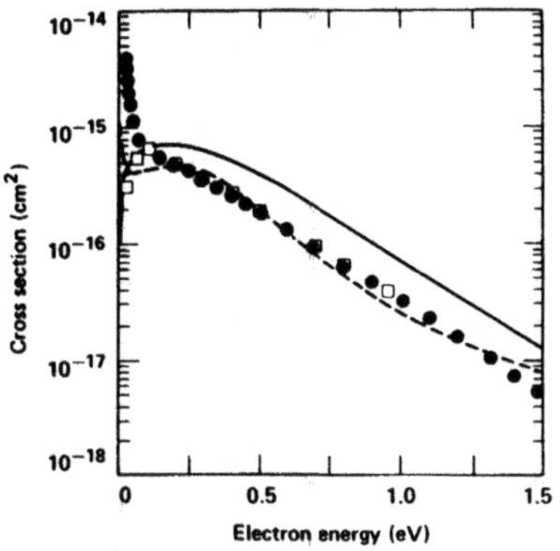

FIGURE 9
Dissociative attachment cross sections for the ground vibrational state of F_2. Solid line, *ab initio* adiabatic-nuclei resonance theory results of Hazi *et al.*;[164] dashed line, semi-empirical calculations of Hall;[165] squares, semiempirical calculations of Bardsley *et al.*; solid circles, experimental data of Chantry (From Hazi *et al.*[164])

approach is not matched by the *ab initio* values which lie a factor of 1.5 to 2 above experiment over the energy range 0.15 to 1.5 eV. Nevertheless they do exhibit the correct energy dependence and represent a very significant theoretical advance.

4.4. *R-Matrix Method*

We have seen earlier that the R-matrix method has proved to be a convenient technique for the *ab initio* calculation of electron–molecule collision cross sections for fixed internuclear separation. It is therefore of interest to see if this approach can be extended to provide an *ab initio* treatment of vibrational excitation and dissociative attachment. This problem has recently been discussed by Schneider *et al.*[110]

The approach of Schneider *et al.*[110] is to adopt an expansion for the electron–molecule wave function which is analogous to the expansion defined by Eq. (91) in the resonance theory case except that the basis states are now R-matrix states. Thus we write

$$\Psi = \sum_n \Psi_n(1, 2, \ldots, N+1)\xi_n(R), \tag{97}$$

where the R-matrix states Ψ_n are defined by diagonalizing $H_{\text{el}} + L_b$ for each fixed value of R as in Eq. (58). In order to obtain an equation for the wave functions $\xi_n(R)$ describing the nuclear motion we project the Schrödinger equation

$$\left(-\frac{1}{2M}\frac{d^2}{dR^2} + H_{\text{el}} + L_b - E\right)\Psi = L_b\Psi \tag{98}$$

onto the R-matrix states. If as before we assume in accordance with the Born–Oppenheimer approximation that the nuclear kinetic-energy operator acts only on the $\xi_n(R)$ we obtain the equation

$$\left[-\frac{1}{2M}\frac{d^2}{dR^2}+E_n(R)-E\right]\xi_n(R) = \langle\Psi_n|L_b|\Psi\rangle. \tag{99}$$

In comparing this equation with Eq. (92) obtained in the resonance-theory case we note that the potential on the left-hand side is now real since the R-matrix eigenvalues defined by Eq. (58) are real and the surface term on the right-hand side replaces the "entry amplitude."

In order to obtain the R-matrix coupling the electronic and vibrational channels we now project the total wave function defined by Eq. (97) onto the appropriate channel functions and evaluate this quantity on the boundary. We obtain

$$\langle\chi_{i'v'}\Phi_{i'l'}|\Psi\rangle_{r=a} = \sum_n \langle\chi_{i'v'}\Phi_{i'l'}|\Psi_n(1,2,\ldots,N+1)\xi_n(R)\rangle_{r=a}. \tag{100}$$

If we now substitute for $\xi_n(R)$ from Eq. (99) by introducing the Green's function $G_n(R,R')$ of the operator $-(1/2M)(d^2/dR^2)+E_n(R)-E$ we obtain

$$\langle\chi_{i'v'}\Phi_{i'l'}|\Psi\rangle_{r=a} = \sum_{ivl}\mathscr{R}_{i'v'l',ivl}\left[\left(\frac{d}{dr}-\frac{b-1}{a}\right)\langle\chi_{iv}\Phi_{il}|\Psi\rangle\right]_{r=a}, \tag{101}$$

where the generalized R-matrix is defined by

$$\mathscr{R}_{i'v'l',ivl} = \sum_n \langle\chi_{i'v'}(R')\gamma_{i'l'n}(R')G_n(R',R)\gamma_{iln}(R)\chi_{iv}(R)\rangle, \tag{102}$$

and where the reduced width amplitudes γ_{iln} are defined by Eq. (66).

If dissociative attachment is possible then the Bloch operator L_b in Eqs. (98) and (99) must be augmented by an additional term allowing for dissociation into $(A + B^-)$-type channels. Schneider *et al.*[110] show that this leads to additional terms in the generalized R matrix coupling the electronic and vibrational channels with the dissociating channels.

We can relate Eq. (102) with the result obtained using the frame-transformation theory of Chang and Fano.[3] If we neglect the kinetic energy operator $-(1/2M)(d^2/dR^2)$ in the internal region then the Green's function $G_n(R',R)$ can be written

$$G_n(R',R) = \frac{\delta(R'-R)}{E_n(R)-E} \tag{103}$$

and then Eq. (102) becomes

$$\mathscr{R}_{i'v'l',ivl} = \int_0^\infty \chi_{i'v'}^*(R) \sum_n \frac{\gamma_{i'l'n}(R)\gamma_{iln}(R)}{E_n(R)-E} \chi_{iv}(R)\, dR, \tag{104}$$

which reduces to Eq. (85) when we use Eq. (64). It is easy to see that, while Eq. (103) is a good approximation to $G_n(R', R)$ for the higher R-matrix levels where $E_n(R)-E$ is large and positive, it cannot be a good approximation when $E_n(R) \cong E$ which occurs when the incident energy is close to the resonance R-matrix eigenenergy. However, since almost all the computational effort is expended in setting up and diagonalizing $H_{\text{el}} + L_b$ for the required set of R values, and very little in determining the Green's function $G_n(R', R)$, it is almost as easy and very much more accurate to use Eq. (102) rather than Eq. (104) to determine the R matrix.

Once the generalized R matrix has been determined, the final step in the calculation is to solve the collision problem in the external region. The appropriate equations in this region are the hybrid equations defined by Eq. (14) without of course the exchange potential, which is zero in this region. These equations are solved subject to the usual asymptotic boundary conditions where the R matrix provides the starting condition at $r=a$. In this way the K matrix and consequently the cross sections are determined.

This theory has been used to calculate the vibrational excitation cross section in e–N_2 scattering by Schneider *et al.*[110] They represented the target by orbitals obtained by an SCF calculation for N_2^- in its $^2\Pi_g$ state. In this way they included part of the target polarizability. They then set up the Hamiltonian matrix for the scattering electron interacting with this distorted target in the static exchange approximation. The Green's functions $G_n(R', R)$ were determined by diagonalizing $-(1/2M)(d^2/dR^2) + E_n(R) - E$ in a basis set of functions $\theta_{nq}(R)$ for each n. Finally the solution of the equations in the external region was obtained assuming that the electron–molecule interaction was zero in this region.

We show in Fig. 10 the potential energy curves and the resonance width calculated by Schneider *et al.*[110] We see that the real part of the resonance energy obtained by fitting the eigenphase shifts from the R-matrix calculation to the usual resonance formula is close to the second R-matrix eigenenergy at small internuclear distances and is close to the first eigenenergy at large internuclear distances. This shows that there is a close but not a one-to-one relationship between the $W_n(R)$ in Eq. (92) and the $E_n(R)$ in Eq. (99). The resonance widths obtained from these calculations are seen to be in reasonable agreement with earlier work.

It is of interest to note in passing that a method of extracting resonance parameters directly from the R-matrix method, thus bypassing the need to

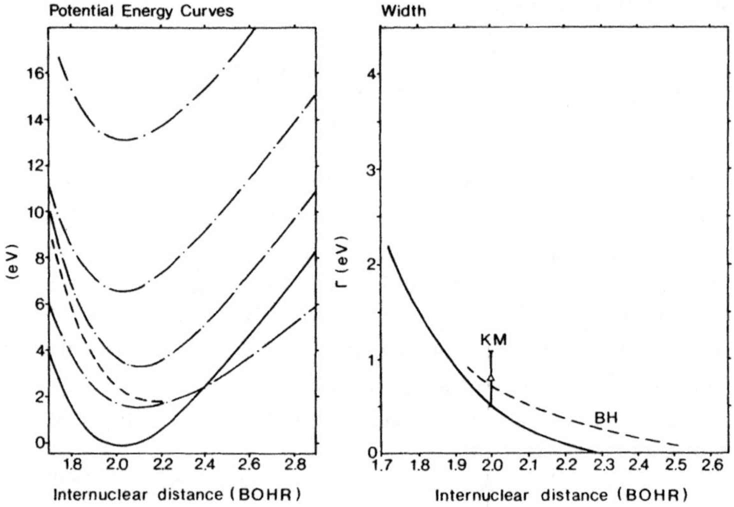

FIGURE 10
The potential energy curves and the resonance width for the $^2\Pi_g$ state in e–N_2 scattering. The potential energy curves: solid line, $^1\Sigma_g$ ground state of N_2; dash-dot lines, R-matrix eigenenergies; dashed line, real part of resonance energy from the R-matrix calculation. The widths: solid line, R-matrix calculation; dotted line, calculated by Birtwistle and Herzenberg;(154) triangle, calculated by Krauss and Mies(78) (from Schneider *et al.*(110)).

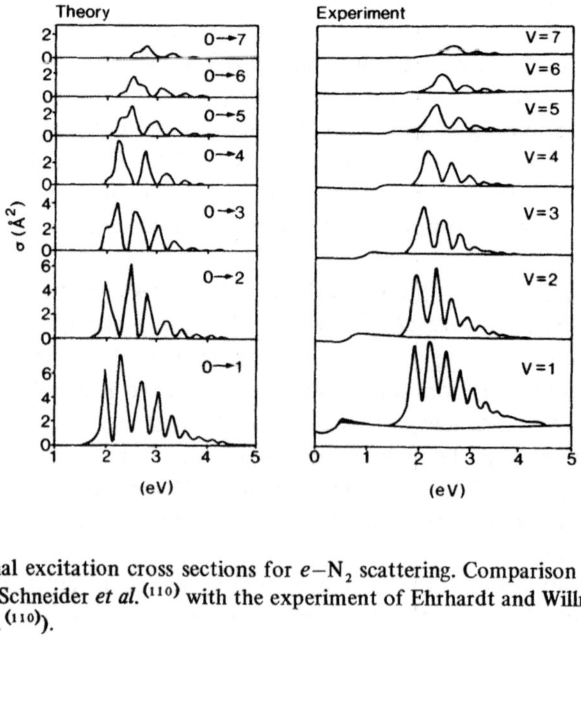

FIGURE 11
Total vibrational excitation cross sections for e–N_2 scattering. Comparison of the R-matrix calculation of Schneider *et al.*(110) with the experiment of Ehrhardt and Willmann(167) (from Schneider *et al.*(110)).

fit the eigenphase shifts, has been described recently by Schneider.[103] Complex boundary conditions are imposed as in the reaction theory of Kapur and Peierls.[166] However, since only the Bloch term[28] becomes complex the construction of the Hamiltonian is little more difficult than in a standard R-matrix calculation. The method is found to converge rapidly with increasing basis sets.

The results obtained by the R-matrix method for the vibrational excitation cross section in e–N_2 scattering are shown in Fig. 11. It is seen that they are in excellent agreement with experiment. This was the first completely *ab initio* calculation for vibrational excitation for this system and indicates that the R-matrix method can be expected to have wide applicability.

In comparing the resonance theory approach with the R-matrix approach we have seen that, while they are both basically *ab initio* approaches, the former lends itself more naturally to a way of fitting and interpreting the data, while the latter can more easily be used to calculate the cross sections from first principles. In this respect they are complementary and both will undoubtedly have a wide role to play in the future.

5. Electronic Excitation

A number of plane-wave-type calculations have been carried out for the electronic excitation of molecules, but these have had little success in explaining the observed low-energy excitation cross sections and will not be considered further. (See Chapter 1, Section 6.5.) Distorted-wave methods might be expected to be most useful in the intermediate energy region above the first ionization threshold. However, since these methods are relatively simple to apply, results have been computed down to low energies and will be described below. Nonetheless there obviously exists a very great need at present for the development of accurate and fast *ab initio* computer codes for the calculation of low-energy electronic excitation cross sections. In principle all of the methods described previously which are based on the close-coupling expansion of the total wave function (4) might be extended to study electronic excitation. In practice, however, the magnitude of the computational task involved has meant that so far only a few of the simplest systems have been considered. Results for the excitation of H_2^+, H_2, and N_2 targets will be described below.

5.1. Excitation of H_2^+

Collins and Schneider[168] have calculated cross sections for both singlet and triplet excitations of the $1\sigma_u$ state of H_2^+ from the ground $1\sigma_g$ state, by solving the two-state and four-state close-coupling equations using the linear algebraic method[25] (see Section 3.1.3). Similar calculations have been carried out by Tennyson *et al.*[169] using the *R*-matrix method. In both cases, positions and widths of the lowest resonances were obtained.

5.2. Excitation of H_2

Two-state close-coupling calculations have been carried out for the electronic excitation of H_2 by Chung and Lin[170] using the noniterative integral equations method of Smith and Henry[16] (described in Section 3.1.2) to solve the radial equations which are a straightforward generalization of Eqs. (6). They computed integrated cross sections for the excitation of the $b\,^3\Sigma_u^+$, $a\,^3\Sigma_g^+$, $e\,^3\Sigma_u^+$, and $c\,^3\Pi_u$ states between threshold and an energy of 40 eV including partial waves $l, l' \leqslant 5$ and symmetries $\Lambda = 0, 1, 2, 3$. Cross sections were also computed for the $B\,^1\Sigma_u^+$ state at energies up to 100 eV using the Born approximation to correct for the contributions from higher partial waves. Each of the target electronic states was obtained from an SCF calculation.

There have been two distorted-wave (DW) calculations for H_2 excitation using a *T*-matrix formalism developed by Rescigno *et al.*[171] In the first of these, Rescigno *et al.*[172] computed differential and integrated cross sections for the two lowest triplet states, $b\,^3\Sigma_u^+$ and $a\,^3\Sigma_g^+$, for incident electron energies up to 20 eV. The electronic target wave functions were determined using the random-phase approximation (RPA)[173] so that both incoming and outgoing distorted waves are computed in the field of the target ground state. More recently Fliflet and McKoy[174] have computed cross sections for the excitation of the $b\,^3\Sigma_u^+$ and $B\,^1\Sigma_u^+$ states using in this case electronic wave functions computed using the Tamm–Dancoff approximation (TDA).[40] The slow convergence of the partial-wave expansion for the spin-allowed excitation process was treated first by introducing an expansion in terms of the angular momentum transferred during the collision as discussed by Fano and Dill[175] and second by computing higher contributions by means of the Born approximation.

The integrated DW–TDA cross section[174] for the excitation of the $B\,^1\Sigma_u^+$ state of H_2 is in reasonable accord with the two-state close-coupling results of Chung and Lin,[170] but both are a factor of 2 larger than the absolute experimental observations of Srivastava and Jensen[176] above 25 eV electron-impact energy; see Chapter 1, p. 64, Fig. 10. Surprisingly there are no signs of improved agreement between theory and experiment even at the highest experimental point, which is at an energy of 60 eV. An examination of the differential cross sections suggests that the DW *T*-matrix approach fails primarily to represent correctly the intermediate- and long-range part of the electron–H_2 interaction, since the largest discrepancy with experiment occurs at small scattering angles in what is a strongly forward-peaked cross section.

There is good agreement between DW–RPA and DW–TDA results for the excitation of the dissociative $b\,^3\Sigma_u^+$ state. In this case the experimental results are of limited accuracy and do not provide a conclusive test of the theory. However, the DW–TDA integrated cross section near threshold does appear to be consistent with the dissociation measurements of Corrigan.[177] The discrepancy with the results of Chung and Lin,[170] which is as much as a factor of 2 at an energy of 20 eV, may be partially accounted for by differences in the target wave functions used in the two calculations.

5.3. Excitation of N_2

In recent work Holley *et al.*[178] have reported two-state cross sections for the excitation of the $a\,^1\Pi_g$ state of N_2 for incident electron energies between 12.5 and 50 eV. The computational techniques used in the calculation were identical to those employed by Chung and Lin[170] for H_2 except that in order to minimize the numerical work, exchange was included by an iterative method wherever possible. Again the slow convergence of the partial cross sections is accounted for by augmenting the close-coupling results with Born-approximation estimates for $l > 7$. The computed cross sections are in quite good agreement with the measurements of Cartwright *et al.*[179] at energies above 30 eV although perhaps tending to overestimate the measured values. At lower energies theory considerably underestimates the measured cross section. This latter result is not, however, unexpected since at these low energies polarization effects, which were not included in the theoretical model, will be important and also the final state will be closely coupled to many other states of the target.

The DW–TDA method has been used by Fliflet *et al.*[180] to compute excitation cross sections for the $B\,^3\Pi_g$, $C\,^3\Pi_u$, and $E\,^3\Sigma_g^+$ states at various electron-impact energies between threshold and 50 eV. The excitation mechanism in these cases is the short-range exchange interaction so that only low partial waves contribute significantly. The distorted-wave results exhibit the large back-angle scattering expected in this situation but otherwise agree only qualitatively with the angular distributions measured by Cartwright *et al.*[179] For energies greater than 10 eV above threshold the cross-section magnitudes are correctly predicted to within a factor of 2 or 3. Also, Lee and McKoy[181] have used the DW–TDA method to obtain differential and integral cross sections for several states of N_2 from threshold to 60 eV.

Lee and McKoy[182] have employed the Schwinger variational principle to obtain continuum wave functions of the negative-ion states and have used these in distorted-wave calculations of the electronic excitation of H_2, CO, and N_2 targets. The calculations are similar to those of Fliflet and McKoy[174] using a single-channel Tamm–Dancoff approximation to obtain the necessary target electronic wave functions.

The results described in this section clearly demonstrate that, although electronic excitation has been extensively studied experimentally, it remains very much a subject for the future as far as low-energy calculations are concerned.

6. Conclusions

We have seen that there has been an enormous advance in the last few years in the accurate calculation of low-energy electron scattering by molecules within the static exchange approximation. Many of these calculations have been carried out using single-center expansion methods; however, an increasing number are now being made using methods based on bound-state multicenter codes. A basic question is how far single-center approximations can be taken before their fundamental simplicity, which makes them relatively easy to program, is overtaken by their inevitable slow convergence for complex molecules. This question of course will become even more acute in the case of electronic excitation. There is also considerable uncertainty in the best way for including electronic polarization. Several methods have been proposed, but the inclusion of all the nonadiabatic and exchange polarization effects is still proving elusive. By way of contrast, the situation with respect to vibrational excitation is now quite encouraging, with several methods capable of giving accurate results at low energy.

Turning to electronic excitation it is apparent that the situation is still at an early stage of development. So far results have only been obtained with four or less states in the close-coupling approximation or using distorted-wave or Born approximations. Experience in electron–atom scattering, however, indicates that accurate results at low energies often require very many electronic states to be coupled. Indeed, in quite simple atoms as many as 10 to 20 electronic states sometimes have to be included, and there is no reason to expect that the electron–molecule collision problem will be any easier. The question of which computational method will be most effective in this situation is still open, but methods which minimize the number of terms used to represent the continuum orbitals will clearly be at an advantage. In conclusion, while considerable strides have been made in recent years, there is still much work to be done before reasonably complete calculations of low-energy electron–molecule collisions can be carried out.

References

1. N.F. Lane, The theory of electron–molecule collisions, *Rev. Mod. Phys.* **52**, 29–120 (1980); P.G. Burke, Theory of electron–molecule collisions, in *Quantum Dynamics of Molecules* (R.G. Wolley, ed.), pp. 483–547, Plenum Press, New York (1980).
2. D.W. Norcross and L.A. Collins, Recent developments in the theory of electron scattering by highly polar molecules. *Adv. At. Mol. Phys.* **18**, 341–397 (1982).
3. E.S. Chang and U. Fano, Theory of electron–molecule collisions by frame transformations, *Phys. Rev. A* **6**, 173–185 (1972).

4. A.M. Arthurs and A. Dalgarno, The theory of scattering by a rigid rotator, *Proc. R. Soc. London A* **256,** 540–551 (1960).
5. N. Chandra and A. Temkin, Hybrid theory and calculation of e–N_2 scattering, *Phys. Rev. A* **13**, 188–203 (1976); N. Chandra and A. Temkin, Hybrid theory calculation of simultaneous vibration–rotation excitation in e–N_2 scattering, *Phys. Rev. A* **14**, 507–511 (1976).
6. P.G. Burke and M.J. Seaton, Numerical solutions of the integro-differential equations of electron–atom collision theory, *Methods Comput. Phys.* **10**, 1–80 (1971).
7. M. Gailitis, New forms of asymptotic expansions for wave functions of charged-particle scattering, *J. Phys. B* **9**, 843–854 (1976).
8. M. Le Dourneuf and Vo Ky Lan. The variable-phase method in multichannel electron–atom or electron–ion scattering, *J. Phys. B* **10**, L35–L42 (1977).
9. M.A. Crees, ASYPCK, a program for calculating asymptotic solutions of the coupled equations of electron collision theory, *Comput. Phys. Commun.* **19**, 103–137 (1980); M.A. Crees, ASYPCK2, an extended version of ASYPCK, *ibid.* **23**, 181–198 (1981).
10. R. Marriott, Calculation of the 1*s*–2*s* electron excitation cross section of hydrogen, *Proc. Phys. Soc.* **72**, 121–129 (1958).
11. P.G. Burke and A.L. Sinfailam, Electron–molecule interactions. II. Scattering by closed-shell diatomic molecules, *J. Phys. B* **3,** 641–659 (1970); A.L. Sinfailam, Electron scattering by closed-shell diatomic molecules, *Comput. Phys. Commun.* **1**, 445–452 (1970).
12. K. Smith, R.J.W. Henry, and P.G. Burke, Scattering of electrons by atomic systems with configurations $2p^q$ and $3p^q$, *Phys. Rev.* **147**, 21–28 (1966).
13. B.D. Buckley and P.G. Burke, The scattering of low-energy electrons by diatomic molecules, *J. Phys. B* **10,** 725–739 (1977).
14. G. Raseev, A. Giusti-Suzor, and H. Lefebvre-Brion, Continuum electronic wave functions in the static-exchange approximation for resonant processes in diatomic molecules. I. Widths for electron capture in the CH^+ closed-shell molecular ion, *J. Phys. B* **11**, 2735–2747 (1978); G. Raseev, Third version of a program for calculating the static interaction potential between an electron and a diatomic molecule, *Comput. Phys. Commun.* **20,** 267–274 (1980); G. Raseev, Electron scattering by closed or open shell diatomic molecules, *Comput. Phys. Commun.* **20**, 275–289 (1980).
15. W.N. Sams and D.J. Kouri, Noniterative solutions of integral equations for scattering. I. Single channels, *J. Chem. Phys.* **51**, 4809–4814 (1969); W.N. Sams and D.J. Kouri, Noniterative solutions of integral equations for scattering. II. Coupled channels, *ibid.* **51,** 4815–4819 (1969); W.N. Sams and D.J. Kouri, Noniterative solutions of integral equations for scattering. III. Coupled open and closed channels and eigenvalue problems, *ibid.* **52**, 4144–4150 (1970); W.N. Sams and D.J. Kouri, Noniterative solutions of integral equations for scattering. IV. Preliminary calculations for coupled open channels and coupled eigenvalue problems, *ibid.* **53,** 496–501 (1970).
16. E.R. Smith and R.J.W. Henry, Noniterative integral-equation approach to scattering problems, *Phys. Rev. A* **7**, 1585–1590 (1973); E.R. Smith and R.J.W. Henry, Application of the noniterative integral-equation method to electron–hydrogen scattering, *ibid.* **8**, 572–575 (1973); R.J.W. Henry, S.P. Rountree, and E.R. Smith, A general program to calculate atomic continuum processes using the NIEM method, *Comput. Phys. Commun.* **23**, 233–273 (1981).
17. M.A. Morrison, The coupled-channels integral-equations method in the theory of low-energy electron–molecule scattering, in *Electron–Molecule and Photon–Molecule Collisions* (T.N. Rescigno, V. McKoy, and B.I. Schneider, eds.), pp. 15–51, Plenum Press, New York (1979).

18. S. Hara, The scattering of slow electrons by hydrogen molecules, *J. Phys. Soc. Jpn.* **22**, 710–718 (1967); S. Hara, A two-center approach in low-energy electron–H_2 scattering, *J. Phys. Soc. Jpn.* **27**, 1009–1019 (1969).
19. L.A. Collins, W.D. Robb, and M.A. Morrison, Low-energy electron scattering by H_2 and N_2: An iterative static-exchange calculation, *J. Phys. B* **11**, L777–L781 (1978); L.A. Collins, W.D. Robb, and M.A. Morrison, Electron scattering by diatomic molecules: Iterative static-exchange techniques, *Phys. Rev. A* **21**, 488–495 (1980).
20. W.D. Robb and L.A. Collins, Iterative static-exchange techniques applied to low-energy collisions of electrons with molecular ions: H_2^+, CH^+, and N_2^+, *Phys. Rev. A* **22**, 2474–2483 (1980).
21. S. Fraga and B. Ransil, Studies in molecular structure. VI. Potential curve for the interaction of two hydrogen atoms in the LCAO MO SCF approximation, *J. Chem. Phys.* **35**, 1967–1977 (1961).
22. P.E. Cade, K.D. Sales, and A.C. Wahl, Electronic structure of diatomic molecules. III. A. Hartree–Fock wave functions and energy quantities for N_2 ($X\,^1\Sigma_g^+$) and N_2^+ ($X\,^2\Sigma_g^+$, $A\,^2\Pi_u$, $B\,^2\Sigma_u^+$) molecular ions, *J. Chem. Phys.* **44**, 1973–2003 (1966).
23. P.E. Cade and W.M. Huo, Electronic structure of diatomic molecules. VI. A. Hartree–Fock wave functions and energy quantities for the ground states of the first-row hydrides, AH, *J. Chem. Phys.* **47**, 614–648 (1967).
24. M. Rhoades-Brown, M.H. Macfarlane, and S.C. Pieper, Techniques for heavy-ion coupled-channels calculations. I. Long-range Coulomb coupling, *Phys. Rev. C* **21**, 2417–2426 (1980); M. Rhoades-Brown, M.H. Macfarlane, and S.C. Pieper, Techniques for heavy-ion coupled-channels calculations. II. Iterative solution of the coupled radial equations, *ibid.* **21**, 2436–2446 (1980).
25. B.I. Schneider and L.A. Collins, A linear algebraic approach to electron–molecule collisions, *J. Phys. B* **14**, L101–L106 (1981); B.I. Schneider and L.A. Collins, Linear-algebraic approach to electron–molecule collisions: Separable exchange approximations, *Phys. Rev. A* **24**, 1264–1266 (1981); L.A. Collins and B.I. Schneider, Linear-algebraic approach to electron–molecule collisions: General formulation, *ibid.* **24**, 2387–2401 (1981).
26. M.J. Seaton, Computer programs for the calculation of electron–atom collision cross sections. II. A numerical method for solving the coupled integro-differential equations, *J. Phys. B* **7**, 1817–1840 (1974).
27. H. Feshbach, Unified theory of nuclear reactions, *Ann. Phys. (N.Y.)* **5**, 357–390 (1958).
28. C. Bloch, Une formulation unifiée de la théorie des réactions nucléaires, *Nucl. Phys.* **4**, 503–528 (1957).
29. R.K. Nesbet, Anomaly-free variational method for inelastic scattering, *Phys. Rev.* **179**, 60–70 (1969); R.K. Nesbet, *Variational Methods in Electron–Atom Scattering Theory*, Plenum Press, New York (1980).
30. S.P. Rountree and G. Parnell, Modified variational method for scattering, *Phys. Rev. Lett.* **39**, 853–856 (1977).
31. L.A. Collins and W.D. Robb, Multichannel variational calculations of low-energy electron–molecule collisions: Static exchange approximation, *J. Phys. B* **13**, 1637–1649 (1980).
32. C. Flammer, *Spheroidal Wave Functions*, Stanford University Press, Stanford (1957).
33. S. Nagahara, The scattering of slow electrons by the diatomic molecules. I. General theory, *J. Phys. Soc. Jpn.* **8**, 165–168 (1953); S. Nagahara, The scattering of slow electrons by the diatomic molecules. II. Elastic scattering by the hydrogen molecules, *J. Phys. Soc. Jpn.* **9**, 52–55 (1954).
34. K.L. Bell, Scattering of electrons with energies below 60 eV by hydrogen molecules, *J. Phys. B* **14**, 2895–2900 (1981).

35. M.A. Crees and D.L. Moores, The use of two-centre expansions in the scattering of electrons by N_2 molecules, *J. Phys. B* **8**, L195–L199 (1975); M.A. Crees and D.L. Moores, Eigenphase sums for the scattering of electrons by N_2 molecules, *J. Phys. B* **10**, L225–L227 (1977).
36. H. Takagi and H. Nakamura, Studies on continuum processes for the e–H_2^+ system, in *Symposium on Electron–Molecule Collisions – Invited Papers* (I. Shimamura and M. Matsuzawa, eds.), pp. 61–65, University of Tokyo (1979); H. Takagi and H. Nakamura, Elastic scattering of electrons from H_2^+: Phaseshifts, quantum defects and two-electron excited states, *J. Phys. B* **13**, 2619–2632 (1980); H. Takagi and H. Nakamura, Two-electron excited states and adiabatic quantum defects of H_2: Analysis of elastic scattering of electrons from H_2^+, *Phys. Rev. A* **27**, 691–708 (1983).
37. M.J. Seaton and P.M.H. Wilson, A frozen cores approximation for atomic structure calculations, *J. Phys. B* **5**, L1–L3 (1972).
38. M.A. Crees, M.J. Seaton, and P.M.H. Wilson, IMPACT, a program for the solution of the coupled integro-differential equations of electron–atom collision theory, *Comput. Phys. Commun.* **15**, 23–83 (1978).
39. T.N. Rescigno and A.E. Orel, Generalization of the separable approximation to scattering in the presence of long-range forces, *Phys. Rev. A* **23**, 1134–1137 (1981); T.N. Rescigno and A.E. Orel, Separable approximation for exchange interactions in electron–molecule scattering, *Phys. Rev. A* **24**, 1267–1271 (1981).
40. J.R. Taylor, *Scattering Theory*, John Wiley & Sons, New York (1972).
41. P.G. Burke and N. Chandra, Electron–molecule interactions. III. A pseudo-potential method for e^-–N_2 scattering, *J. Phys. B* **5**, 1696–1711 (1972).
42. P.G. Burke, N. Chandra, and F.A. Gianturco, Electron–molecule interactions. IV. Scattering by polyatomic molecules, *J. Phys. B* **5**, 2212–2224 (1972).
43. F.A. Gianturco and D.G. Thompson, The Ramsauer–Townsend effect in methane, *J. Phys. B* **9**, L383–L385 (1976).
44. N.F. Lane and S. Geltman, Rotational excitation of diatomic molecules by slow electrons: Application to H_2, *Phys. Rev.* **160**, 53–67 (1967).
45. J.B. Furness and I.E. McCarthy, Semiphenomenological optical model for electron scattering on atoms, *J. Phys. B* **6**, 2280–2291 (1973); R. Vanderpoorten, A local optical potential study of elastic electron–atom scattering, *ibid.* **8**, 926–939 (1975); R. Vanderpoorten, A local optical potential study of elastic e^-–Li scattering, *ibid.* **9**, L535–L538 (1976).
46. D.G. Truhlar and M.A. Brandt, Close-coupling calculations of differential cross sections for elastic scattering and rotational excitation of hydrogen molecules by electrons at 10 and 40 eV, *J. Chem. Phys.* **65**, 3092–3101 (1976); D.G. Truhlar, R.E. Poling, and M.A. Brandt, Infinite-order sudden approximation for rotational excitation of hydrogen molecules by electrons in the energy range 10–40 eV, *ibid.* **64**, 826–829 (1976).
47. J.C. Slater, *Quantum Theory of Atomic Structure*, Vols. I and II, McGraw-Hill, New York (1960).
48. M.A. Morrison and L.A. Collins, Exchange in low-energy electron–molecule scattering: Free-electron-gas model exchange potentials and applications to e–H_2 and e–N_2 collisions, *Phys. Rev. A* **17**, 918–938 (1978).
49. M.A. Morrison, N.F. Lane, and L.A. Collins, Low-energy electron–molecule scattering: Application of coupled-channel theory to e–CO_2 collisions, *Phys. Rev. A* **15**, 2186–2201 (1977).
50. L.A. Collins and D.W. Norcross, Electron collisions with highly polar molecules: Comparison of model, static, and static-exchange calculations for alkali-metal halides, *Phys. Rev. A* **18**, 467–498 (1978).

51. L.A. Collins, W.D. Robb, and D.W. Norcross, Low-energy collisions of electrons with highly polar molecules: Orthogonalization and model exchange potentials, *Phys. Rev. A* **20**, 1838–1840 (1979).
52. M.A. Morrison and L.A. Collins, Exchange in low-energy electron–molecule scattering: Orthogonalization and free-electron-gas approximations for collisions with polar and nonpolar molecules, *Phys. Rev. A* **23**, 127–138 (1981).
53. L.A. Collins and M.A. Morrison, Exchange in low-energy e–CO_2 collisions, *Phys. Rev. A* **25**, 1764–1767 (1982).
54. T.L. Gibson and M.A. Morrison, Comparison of model exchange potentials for low-energy e–H_2 collisions with and without polarisation effects, *J. Phys. B* **14**, 727–739 (1981).
55. J.R. Rumble, Jr. and D.G. Truhlar, Comparison of local exchange potentials for electron–N_2 scattering, *J. Chem. Phys.* **72,** 5223–5227 (1980); A. Jain and D.G. Thompson, Rotational excitation of CH_4 and H_2O by slow electron impact, *J. Phys. B.* **16,** 3077–3098 (1983).
56. F.A. Gianturco and D.G. Thompson, The scattering of slow electrons by polyatomic molecules. A model study for CH_4, H_2O and H_2S, *J. Phys. B* **13**, 613–625 (1980).
57. S. Salvini and D.G. Thompson, Exchange approximations in the scattering of slow electrons by polyatomic molecules, *J. Phys. B* **14**, 3797–3803 (1981).
58. M.E. Stewart and D.H. Madison, Use of local exchange potentials in inelastic scattering, *Phys. Lett.* **87A**, 165–168 (1982).
59. L.A. Collins, R.J.W. Henry, and D.W. Norcross, Electron collisions with polar molecules: Exchange and polarisation in elastic scattering by HCl, *J. Phys. B* **13**, 2299–2307 (1980).
60. M.A. Morrison and P.J. Hay, *Ab initio* adiabatic polarization potentials for low-energy electron–molecule and positron–molecule collisions: The e–N_2 and e–CO_2 systems, *Phys. Rev. A* **20**, 740–748 (1979).
61. R.A. Eades, D.G. Truhlar, and D.A. Dixon, *Ab initio* self-consistent-field polarizabilities and electron–molecule polarization potentials. III. N_2, *Phys. Rev. A* **20**, 867–878 (1979).
62. A. Temkin, Polarization and exchange effects in the scattering of electrons from atoms with application to oxygen, *Phys. Rev.* **107**, 1004–1012 (1957); A. Temkin, A note on the scattering of electrons from atomic hydrogen, *ibid.* **116**, 358–363 (1959).
63. A. Temkin and K.V. Vasavada, Scattering of electrons from H_2^+: The method of polarized single-center orbitals, *Phys. Rev.* **160,** 109–117 (1967); A. Temkin, K.V. Vasavada, E.S. Chang, and A. Silver, Scattering of electrons from H_2^+. II. *Phys. Rev.* **186,** 57–66 (1969).
64. N.F. Lane and R.J.W. Henry, Polarization potential in low-energy electron–H_2 scattering, *Phys. Rev.* **173**, 183–190 (1968).
65. R.J.W. Henry and N.F. Lane, Polarization and exchange effects in low-energy electron–H_2 scattering, *Phys. Rev.* **183**, 221–231 (1969).
66. T.L. Gibson and M.A. Morrison, A parameter-free nonadiabatic polarization potential for low-energy e–H_2 collisions, *J. Phys. B* **15**, L221–L228 (1982).
67. J.S. Bell and E.J. Squires, A formal optical model, *Phys. Rev. Lett.* **3**, 96–97 (1959).
68. D. Dill and J.L. Dehmer, Electron–molecule scattering and molecular photoionization using the multiple-scattering method, *J. Chem. Phys.* **61**, 692–699 (1974).
69. J.L. Dehmer and D. Dill, The continuum multiple-scattering approach to electron–molecule scattering and molecular photoionization, in: *Electron–Molecule and Photon–Molecule Collisions* (T.N. Rescigno, V. McKoy, and B.I. Schneider, eds.), pp. 225–265, Plenum Press, New York (1979).
70. P. Ziesche and W. John, Multiple scattering on stressed clusters of muffin-tin poten-

tials. Application to elastic electron scattering by molecules, *J. Phys. B* **9**, 333–344 (1976).

71. K.H. Johnson, Scattered-waves theory of the chemical bond, *Advances in Quantum Chemistry* (P.O. Löwdin, ed.), pp. 143–185, Academic Press, New York (1973).
72. J. Siegel, D. Dill, and J.L. Dehmer, Continuum and bound molecular electronic wave functions for generalized multiple-scattering potentials, *J. Chem. Phys.* **64**, 3204–3206 (1976).
73. J.R. Rumble, Jr., and D.G. Truhlar, Investigation of the assumptions of the multiple-scattering method for electron–molecule scattering cross sections, *J. Chem. Phys.* **72**, 3206–3210 (1980).
74. Z. Pavlovic, M.J.W. Boness, A. Herzenberg, and G.J. Schulz, Vibrational excitation in N_2 by electron impact in the 15–35 eV region, *Phys. Rev. A* **6**, 676–685 (1972).
75. D. Dill, J. Welch, J.L. Dehmer, and J. Siegel, Shape-resonance-enhanced vibrational excitation at intermediate energies (10–40 eV) in electron–molecule scattering, *Phys. Rev. Lett.* **43**, 1236–1239 (1979).
76. A. Temkin, The art and science of calculating autoionization, in *Autoionization* (A. Temkin, ed.), pp. 55–76, Mono Book Corp., Baltimore (1966).
77. A.U. Hazi and H.S. Taylor, Stabilization method of calculating resonance energies: Model problem, *Phys. Rev. A* **1**, 1109–1120 (1970).
78. M. Krauss and F.H. Mies, Molecular-orbital calculation of the shape resonance in N_2, *Phys. Rev. A* **1**, 1592–1598 (1970).
79. C.W. McCurdy, Jr., T.N. Rescigno, and V. McKoy, A simple method for evaluating low-energy electron–molecule scattering cross sections using discrete basis functions, *J. Phys. B* **9**, 691–698 (1976).
80. W.P. Reinhardt, L^2 discretization of atomic and molecular electronic continua: Moment, quadrature and *J*-matrix techniques, *Comput. Phys. Commun.* **17**, 1–21 (1979).
81. C.W. McCurdy, Jr., and T.N. Rescigno, Complex-basis-function calculations of resolvent matrix elements: Molecular photoionization, *Phys. Rev. A* **21**, 1499–1505 (1980).
82. N. Padial, G. Csanak, B.V. McKoy, and P.W. Langhoff, Photoexcitation and ionization in carbon dioxide: Theoretical studies in the separated-channel static-exchange approximation, *Phys. Rev. A* **23**, 218–235 (1981).
83. A.U. Hazi, Stieltjes-moment-theory technique for calculating resonance widths, in *Electron–Molecule and Photon–Molecule Collisions* (T.N. Rescigno, V. McKoy, and B.I. Schneider, eds.), pp. 281–298, Plenum Press, New York (1979).
84. C.W. McCurdy and T.N. Rescigno, Basis-set calculation of Siegert eigenvalues: Partial resonance widths, *Phys. Rev. A* **20**, 2346–2351 (1979).
85. E.P. Wigner, Resonance reactions and anomalous scattering, *Phys. Rev.* **70**, 15–33 (1946).
86. E.P. Wigner, Resonance reactions, *Phys. Rev.* **70**, 606–618 (1946).
87. E.P. Wigner and L. Eisenbud, Higher angular momenta and long-range interaction in resonance reactions, *Phys. Rev.* **72**, 29–41 (1947).
88. P.G. Burke and A. Hibbert, *R*-matrix theory of electron–atom collisions, in *Proceedings of the VIth International Conference on the Physics of Electronic and Atomic Collisions*, Massachusetts Institute of Technology, pp. 367–369 (1969).
89. P.G. Burke, A. Hibbert, and W.D. Robb, Electron scattering by complex atoms, *J. Phys. B* **4**, 153–161 (1971); P.G. Burke and W.D. Robb, The *R*-matrix theory of atomic processes, *Adv. At. Mol. Phys.* **11**, 143–214 (1975).
90. B. Schneider, *R*-matrix theory for electron–atom and electron–molecule collisions using analytic basis set expansions, *Chem. Phys. Lett.* **31**, 237–241 (1975).
91. B.I. Schneider, *R*-matrix theory for electron–molecule collisions using analytic basis

set expansions. II. Electron–H_2 scattering in the static-exchange model, *Phys. Rev. A* **11**, 1957–1962 (1975).

92. B.I. Schneider and P.J. Hay, Elastic scattering of electrons from F_2, *J. Phys. B* **9**, L165–L167 (1976).
93. B.I. Schneider and P.J. Hay, Elastic scattering from F_2: An *R*-matrix calculation, *Phys. Rev. A* **13**, 2049–2056 (1976).
94. M.A. Morrison and B.I. Schneider, Electron–molecule scattering theory: An *R*-matrix study of low-energy elastic e–N_2 collisions in the static-exchange approximation, *Phys. Rev. A* **16**, 1003–1012 (1977).
95. B.I. Schneider, The *R*-matrix method: Applications to electron–molecule collisions, in *Electronic and Atomic Collisions* (G. Watel, ed.), pp. 257–269, North-Holland, Amsterdam (1978).
96. P.G. Burke, I. Mackey, and I. Shimamura, *R*-matrix theory of electron–molecule scattering, *J. Phys. B* **10**, 2497–2512 (1977).
97. B.D. Buckley, P.G. Burke, and Vo Ky Lan, *R*-matrix calculations for electron–molecule scattering, *Comput. Phys. Commun.* **17**, 175–179 (1979).
98. C.J. Noble, P.G. Burke, and S. Salvini, Low-energy scattering of electrons by diatomic molecules, in *Proceedings of the XIIth International Conference on the Physics of Electronic and Atomic Collisions,* Gatlinburg, Tennessee, pp. 322–323 (1981) *J. Phys. B* **15**, 3779–3786 (1982).
99. J. Kendrick and B.D. Buckley, An adaption of the ALCHEMY atomic and molecular integrals packages for *R*-matrix electron–molecular collisions, Daresbury Laboratory Internal Report, DL/SCI/TM 22T (1980).
100. C.J. Noble, The ALCHEMY linear molecule integral generator, Daresbury Laboratory Technical Memorandom DL/SCI/TM 33T (1982).
101. R.K. Nesbet, Electronic structure of N_2, CO, and BF, *J. Chem. Phys.* **40**, 3619–3633 (1964).
102. B.J. Ransil, Studies in molecular structure. II. LCAO-MO-SCF wave functions for selected first-row diatomic molecules, *Rev. Mod. Phys.* **32**, 245–254 (1960).
103. B.I. Schneider, Direct calculation of resonance energies and widths using an *R*-matrix approach, *Phys. Rev. A* **24**, 1–3 (1981).
104. P.G. Burke, C.J. Noble, and S. Salvini, Electron scattering by nitrogen molecules at intermediate energies, *J. Phys. B,* L113–L120 (1983).
105. R.K. Nesbet, private communication.
106. R. Damburg and E. Karule, A modification of the close-coupling approximation for e–H_2 scattering allowing for the long-range interactions, *Proc. Phys. Soc.* **90**, 637–640 (1967).
107. P.G. Burke and J.F.B. Mitchell, Polarized pseudo-states for low-energy electron–atom scattering, *J. Phys. B* **7**, 665–673 (1974).
108. P.W. Coulter, Formalism for determining pseudostates in modified close-coupling calculations, *Phys. Rev. A* **22**, 372–378 (1980).
109. J. Kendrick, Calculation of pseudo-molecular orbitals for electron–molecule collisions, Daresbury Laboratory Internal Report, DL/SCI/TM 18T (1979).
110. B.I. Schneider, M. Le Dourneuf, and P.G. Burke, Theory of vibrational excitation and dissociative attachment: An *R*-matrix approach, *J. Phys B* **12**, L365–L369 (1979); B.I. Schneider, M. Le Dourneuf, and Vo Ky Lan, Resonant vibrational excitation of N_2 by low-energy electrons: An *ab initio R*-matrix calculation, *Phys. Rev. Lett.* **43**, 1926–1929 (1979).
111. T.N. Rescigno, C.W. McCurdy, and V. McKoy, Discrete basis set approach to non-spherical scattering, *Chem. Phys. Lett.* **27**, 401–404 (1974).

112. T.N. Rescigno, C.W. McCurdy, Jr., and V. McKoy, Discrete-basis-set approach to nonspherical scattering. II. *Phys. Rev. A* **10**, 2240–2245 (1974).
113. T.N. Rescigno, C.W. McCurdy, Jr., and V. McKoy, Low-energy e^-–H_2 elastic cross sections using discrete basis functions, *Phys. Rev. A* **11**, 825–829 (1975).
114. N.S. Ostlund, Polyatomic scattering integrals with Gaussian orbitals, *Chem. Phys. Lett.* **34**, 419–422 (1975).
115. A.W. Fliflet, D.A. Levin, M. Ma, and V. McKoy, Discrete-basis-set calculation for e–N_2 scattering cross sections in the static-exchange approximation, *Phys. Rev. A* **17**, 160–169 (1978).
116. T.N. Rescigno, private communication (1979).
117. A. Klonover and U. Kaldor, *Ab initio* calculations of electron–molecule-scattering cross sections including polarization, *Chem. Phys. Lett.* **51**, 321–324 (1977).
118. A. Klonover and U. Kaldor, *Ab initio* electron–molecule scattering theory including polarisation: Elastic scattering and rotational excitation of H_2, *J. Phys. B* **11**, 1623–1632 (1978).
119. A. Klonover and U. Kaldor, *Ab initio* electron–molecule scattering theory including polarisation: Vibrational and vibrational–rotational excitation of H_2, *J. Phys. B* **12**, 323–330 (1979).
120. A. Klonover and U. Kaldor, Zero-point vibration in low-energy electron–molecule scattering theory, *J. Phys. B* **12**, L61–L65 (1979).
121. A.W. Fliflet and V. McKoy, Variationally corrected discrete-basis-set calculation for electron–molecule scattering in the static-exchange approximation, *Phys. Rev. A* **18**, 1048–1054 (1978).
122. A.W. Fliflet and V. McKoy, Discrete-basis-set method for electron–molecule continuum wave functions, *Phys. Rev. A* **18**, 2107–2114 (1978).
123. D.A. Levin, A.W. Fliflet, and V. McKoy, Low-energy e^-–CO scattering in the static-exchange approximation, *Phys. Rev. A* **21**, 1202–1209 (1980).
124. C. Lovelace, Practical theory of three-particle states. I. Nonrelativistic, *Phys. Rev.* **135**, B1225–B1249 (1964).
125. S.K. Adhikari and I.H. Sloan, Separable expansion of the t matrix with analytic form factors, *Phys. Rev. C* **11**, 1133–1140 (1975).
126. D.K. Watson, R.R. Lucchese, V. McKoy, and T.N. Rescigno, Schwinger variational principle for electron–molecule scattering: Application to electron–hydrogen scattering, *Phys. Rev. A* **21**, 738–744 (1980).
127. F. Linder and H. Schmidt, Rotational and vibrational excitation of H_2 by slow electron impact, *Z. Naturforsch. A* **26a**, 1603–1617 (1971).
128. M. Berman and U. Kaldor, The Schwinger variational method in electron–atom and electron–molecule scattering theory with polarisation, *J. Phys. B* **14**, 3993–4005 (1981); M. Berman, O. Walter, and L.S. Cederbaum, Electron–molecule scattering in the optical potential approach: surpassing second order, *Phys. Rev. Lett.* **50**, 1979–1982 (1983).
129. R.R. Lucchese, D.K. Watson, and V. McKoy, Iterative approach to the Schwinger variational principle for electron–molecule collisions, *Phys. Rev. A* **22**, 421–426 (1980).
130. R.R. Lucchese and V. McKoy, Iterative approach to the Schwinger variational principle applied to electron–molecular-ion collisions, *Phys. Rev. A* **24**, 770–776 (1981).
131. D.K. Watson, T.N. Rescigno, and B.V. McKoy, Schwinger variational calculations for electron scattering by polar molecules, *J. Phys. B* **14**, 1875–1882 (1981).
132. K. Takatsuka and V. McKoy, Variational scattering theory using a functional of fractional form. I. General theory, *Phys. Rev. A* **23**, 2352–2358 (1981); K. Taka-

tsuka and V. McKoy, Variational scattering theory using a functional of fractional form. II. An L^2 approach, *ibid.* **23**, 2359–2364 (1981); K. Takatsuka and V. McKoy, Schwinger variational principle for multichannel scattering, *Phys. Rev. Lett.* **45**, 1396–1399 (1980).

133. K. Takatsuka, R.R. Lucchese, and V. McKoy, Relationship between the Schwinger and Kohn-type variational principles in scattering theory, *Phys. Rev. A* **24**, 1812–1816 (1981).
134. M.-T. Lee, K. Takatsuka, and V. McKoy, Application of a new variational functional for electron–molecule collisions: An extension of the Schwinger variational principle, *J. Phys. B* **14**, 4115–4123 (1981); K. Takatsuka, M.-T. Lee, and V. McKoy, The application of a fractional $\tilde{C}$-functional for electron–molecule scattering, in *Proceedings of the XIIth International Conference on the Physics of Electronic and Atomic Collisions,* pp. 303–304, Gatlinburg, Tennessee (1981).
135. K. Takatsuka and V. McKoy, Extension of the Schwinger variational principle beyond the static-exchange approximation, *Phys. Rev. A* **24**, 2473–2480 (1981).
136. D.A. Levin, A.W. Fliflet, and V. McKoy, Low-energy rotational and vibrational-rotational excitation cross sections for H_2 by electron impact, *Phys. Rev. A* **20**, 491–498 (1979).
137. J.R. Rumble, D.G. Truhlar, and M.A. Morrison, Calculation of vibrational excitation of N_2 by electron impact at 5–50 eV using extended-basis-set Hartree–Fock wave functions, *J. Phys. B* **14**, L301–L305 (1981).
138. K. Onda and D.G. Truhlar, State-to-state cross sections for elastic and inelastic electron scattering by N_2 at 20–35 eV, including resonant enhancement of vibrational excitation, *J. Chem. Phys.* **72**, 5249–5262 (1980).
139. R.K. Nesbet, Energy-modified adiabatic approximation for scattering theory, *Phys. Rev. A* **19**, 551–556 (1979).
140. A.N. Feldt and M.A. Morrison, Breakdown of the adiabatic-nuclear-rotation approximation for near-threshold e–H_2 collisions, *J. Phys. B* **15**, 301–308 (1982).
141. A.N. Feldt, T.L. Gibson, and M.A. Morrison, Theoretical investigations of low-energy e–H_2 scattering, in *Proceedings of the XIIth International Conference on the Physics of Electronic and Atomic Collisions,* pp. 324–325, Gatlinburg, Tennessee (1981).
142. A. Temkin, Theory and calculation of electron–molecule scattering, in *Electron and Atomic Collisions* (N. Oda and K. Takayanagi, eds.), pp. 95–107, North-Holland, Amsterdam (1980).
143. B.H. Choi and R.T. Poe, Vibrational and rotational transitions in low-energy electron–diatomic-molecule collisions. I. Close-coupling theory in the moving body-fixed frame, *Phys. Rev. A* **16**, 1821–1830 (1977); B.H. Choi and R.T. Poe, Vibrational and rotational transitions in low-energy electron–diatomic-molecule collisions. II. Hybrid theory and close-coupling theory: An l_z-conserving close-coupling approximation, *ibid.* **16**, 1831–1840 (1977).
144. A. Temkin, Internuclear dependence of the polarizability of N_2, *Phys. Rev. A* **17**, 1232–1235 (1978).
145. J. Kendrick, *Ab initio* calculations of the derivative of the static polarisability of N_2 and CO, *J. Phys. B* **11**, L601–L603 (1978).
146. E.C. Sullivan and A. Temkin, A non-iterative method for solving PDEs arising in electron scattering, *Comput. Phys. Commun.* **25**, 97–110 (1982).
147. K. Onda and A. Temkin, P.D.E. approach to electron–molecule scattering, in *Proceedings of the XIIth International Conference on the Physics of Electronic and Atomic Collisions,* pp. 293–294, Gatlinburg, Tennessee (1981).
148. G.J. Schulz, Vibrational excitation of nitrogen by electron impact, *Phys. Rev.* **125**,

229–232 (1962); G.J. Schulz, Vibrational excitation of N_2, CO, and H_2 by electron impact, *Phys. Rev.* **135**, A988–A994 (1964).
149. G.J. Schulz, Resonances in electron impact on diatomic molecules, *Rev. Mod. Phys.* **45**, 423–486 (1973).
150. A. Herzenberg and F. Mandl, Vibrational excitation of molecules by resonance scattering of electrons, *Proc. R. Soc. London A* **270**, 48–71 (1962).
151. J.N. Bardsley and F. Mandl, Resonant scattering of electrons by molecules, *Rep. Prog. Phys.* **31**, 471–531 (1968).
152. A. Herzenberg, Strong excitation of nuclear motions in electron–molecule collisions, in *Electronic and Atomic Collisions* (G. Watel, ed.), pp. 1–22, North-Holland, Amsterdam (1978).
153. A. Herzenberg, Oscillatory energy dependence of resonant electron–molecule scattering, *J. Phys. B* **1**, 548–558 (1968).
154. D.T. Birtwistle and A. Herzenberg, Vibrational excitation of N_2 by resonance scattering of electrons, *J. Phys. B* **4**, 53–70 (1971).
155. J.N. Bardsley, A. Herzenberg, and F. Mandl, Vibrational excitation and dissociative attachment in the scattering of electrons by hydrogen molecules, *Proc. Phys. Soc.* **89**, 321–340 (1966).
156. A.J.F. Siegert, On the derivation of the dispersion formula for nuclear reactions, *Phys. Rev.* **56**, 750–752 (1939).
157. L. Dubé and A. Herzenberg, Absolute cross sections from the "boomerang model" for resonant electron–molecule scattering, *Phys. Rev. A* **20**, 194–213 (1979).
158. S.F. Wong, unpublished results reproduced by A. Herzenberg in *Electronic and Atomic Collisions* (G. Watel, ed.), pp. 1–22, North-Holland, Amsterdam (1978).
159. A.U. Hazi, T.N. Rescigno, and M. Kurilla, Cross sections for resonant vibrational excitation of N_2 by electron impact, *Phys. Rev. A* **23**, 1089–1099 (1981).
160. J.M. Wadehra and J.N. Bardsley, Vibrational- and rotational-state dependence of dissociative attachment in e–H_2 collisions, *Phys. Rev. Lett.* **41**, 1795–1798 (1978).
161. M. Allan and S.F. Wong, Effect of vibrational and rotational excitation on dissociative attachment in hydrogen, *Phys. Rev. Lett.* **41**, 1791–1794 (1978).
162. J.N. Bardsley and J.M. Wadehra, Dissociative attachment and vibrational excitation in low-energy collisions of electrons with H_2 and D_2, *Phys. Rev. A* **20**, 1398–1405 (1979).
163. A.W. Fliflet, V. McKoy, and T.N. Rescigno, Dissociation of F_2 by electron impact excitation of the lowest $^3\Pi_u$ electronic state, *Phys. Rev. A* **21**, 788–792 (1980).
164. A.U. Hazi, A.E. Orel, and T.N. Rescigno, *Ab initio* study of dissociative attachment of low-energy electrons to F_2, *Phys. Rev. Lett.* **46**, 918–922 (1981).
165. R.J. Hall, Dissociative attachment and vibrational excitation of F_2 by slow electrons, *J. Chem. Phys.* **68**, 1803–1807 (1978).
166. P.L. Kapur and R.E. Peierls, The dispersion formulae for nuclear reactions, *Proc. R. Soc. London A* **166**, 277–295 (1938).
167. H. Ehrhardt and K. Willmann, Angular dependence of resonance scattering of low-energy electrons by N_2, *Z. Phys.* **204**, 462–473 (1967).
168. L.A. Collins and B.I. Schneider, Linear algebraic approach to electronic excitation of atoms and molecules by electron impact, *Phys. Rev. A* **27**, 101–111 (1983); B.I. Schneider and L.A. Collins, $^1\Sigma_g$ resonances in low-energy e–H_2^+ collisions, *Phys. Rev. A* **28**, 166–168 (1983).
169. J. Tennyson, C.J. Noble, and S. Salvini, Low-energy e–H_2^+ collisions using the R-matrix method, *J. Phys. B* **17**, 905–912 (1984).
170. S. Chung and C.C. Lin, Application of the close-coupling method to excitation of

electronic states and dissociation of H_2 by electron impact, *Phys. Rev. A* **17**, 1874–1891 (1978).

171. T.N. Rescigno, C.W. McCurdy, Jr., and V. McKoy, A relationship between the many-body theory of inelastic scattering and the distorted wave approximation, *J. Phys. B* **7**, 2396–2402 (1974).
172. T.N. Rescigno, C.W. McCurdy, Jr., and V. McKoy, Low-energy electron-impact excitation of the hydrogen molecule, *Phys. Rev. A* **13**, 216–223 (1976).
173. C.W. McCurdy, Jr., T.N. Rescigno, D.L. Yeager, and V. McKoy, The equations of motion method: An approach to the dynamical properties of atoms and molecules, in *Methods of Electronic Structure Theory 3* (H.F. Schaefer III, ed.), pp. 339–386, Plenum Press, New York (1977).
174. A.W. Fliflet and V. McKoy, Distorted-wave-approximation cross sections for excitation of the $b\ ^3\Sigma_u^+$ and $B\ ^1\Sigma_u^+$ states of H_2 by low-energy-electron impact, *Phys. Rev. A* **21**, 1863–1875 (1980).
175. U. Fano and D. Dill, Angular momentum transfer in the theory of angular distributions, *Phys. Rev. A* **6**, 185–192 (1972).
176. S.K. Srivastava and S. Jensen, Experimental differential and integral electron impact cross sections for the $B\ ^1\Sigma_u^+$ state of H_2 in the intermediate-energy region, *J. Phys. B* **10**, 3341–3346 (1977).
177. S.J.B. Corrigan, Dissociation of molecular hydrogen by electron impact, *J. Chem. Phys.* **43**, 4381–4386 (1965).
178. T.K. Holley, S. Chung, C.C. Lin, and E.T.P. Lee, Electron-impact excitation cross section of the $a\ ^1\Pi_g$ state of the N_2 molecule by the close-coupling method, *Phys. Rev. A* **24**, 2946–2952 (1981).
179. D.C. Cartwright, A. Chutjian, S. Trajmar, and W. Williams, Electron impact excitation of the electronic states of N_2. I. Differential cross sections at incident energies from 10 to 50 eV, *Phys. Rev. A* **16**, 1013–1040 (1977); D.C. Cartwright, S. Trajmar, A. Chutjian, and W. Williams, Electron impact excitation of the electronic states of N_2. II. Integral cross sections at incident energies from 10 to 50 eV, *Phys. Rev. A* **16**, 1041–1051 (1977).
180. A.W. Fliflet, V. McKoy, and T.N. Rescigno, Cross sections for excitation of the $B\ ^3\Pi_g$, $C\ ^3\Pi_u$ and $E\ ^3\Sigma_g^+$ states of N_2 by low-energy electron impact in the distorted-wave approximation, *J. Phys. B* **12**, 3281–3293 (1979).
181. M.-T Lee and V. McKoy, Cross sections for electron-impact excitation of the electronic states of N_2, *Phys. Rev. A* **28**, 697–705 (1983).
182. M.-T. Lee and V. McKoy, Distorted-wave-approximation cross sections for excitation of several electronic states of H_2 and N_2 by low energy-electron impact, in *Proceedings of the XIIth International Conference on the Physics of Electronic and Atomic Collisions*, pp. 330–331, Gatlinburg, Tennessee (1981).

INDEX

GPSR Compliance

The European Union's (EU) General Product Safety Regulation (GPSR) is a set of rules that requires consumer products to be safe and our obligations to ensure this.

If you have any concerns about our products, you can contact us on ProductSafety@springernature.com

In case Publisher is established outside the EU, the EU authorized representative is:

Springer Nature Customer Service Center GmbH
Europaplatz 3
69115 Heidelberg, Germany

Batch number: 10755248

Printed by Printforce, the Netherlands